111° 112° 113° 5°

CW00631988

Java Sea

Rembang • Lasem
• Pati Pegunungan North Kapur Tuban Paciran Arosbaya PULAU MADURA Sumenep 6°
Blora S. Bengawan Bangkalan Pamekasan
Cepu Babat Gresik Ferry Bull Races
• Purwodadi Bojonegoro EAST ■ ✈ Surabaya Madura Strait
Ngawi Mojokerto Sidoarjo Situbondo Baluran National Park
Sragen Kertosono Jombang Bangil Teak Plantations
Madiun Nganjuk Arjuna Lalijiwo Reserve Pasuruan Probolinggo Bondowoso Kawah Ijen BALI 7°
•akarta Kediri Pare Purwodadi Botanical Gardens Gn. Argopuro 3088m Yang Plateau Reserve Ketapang
(Solo) Wonogiri Malang Bromo Jember Banyuwangi
n Ponorogo Gn. Butak 2868m Tengger-Semeru National Park Lumajang Rogojampi Genteng DENPASAR
•Wuryantoro Wlingi Kepanjen Tempeh JAVA Benculuk Bali Strait Kuta NUSA PENIDA
W. Gajahmungkur Trenggalek Ngunut Blitar Sempu Island Nature Reserve Meru Betiri Reserve Grajagan
•A • Pacitan Nusa Barung Nature Reserve P. BARUNG Sea Turtles Sukamade Surfing Blambangan Nature Reserve
Stalactite Caves

Indian Ocean 8°

Bali Sea

Kubutambahan
Singaraja Tejakula
Lovina
•erbuk Seririt Gn. Batur 1717m
385m BULELENG D. Tamblingan D. Buyan BANGLI Hot Spring Bali Aga Village
Taman Nasional Bali Barat Gn. Lesong 1860m D. Bratan Penelokan D. Batur
(West Bali National Park) Gn. Sangiyang 2093m Bedugul Gn. Agung 3014m
EMBRANA Gn. Pohen 2063m Pura Besakih (Mother Temple) KARANGASEM
Jembrana Gn. Batukau 2278m Besakih Gn. Seraya 1175m
Negara TABANAN Kukup GIANYAR Rendang Amlapura
Bali Aga Village
Bali Strait Antosari Ubud Goa Gajah KLUNGKUNG Candidasa
Tabanan Mengwi Gianyar Klungkung
Kapal Blahbatuh Kerta Gosa Lombok Strait

Bali

↑
N

| km | 5 | 10 | 15 |
| miles | 5 | 10 | |

Pura Tanah Lot BADUNG Batubulan
■ Denpasar Sanur Bandung Strait NUSA
Legian Teluk Kuta PENIDA
Kuta Tuban P. SERANGAN
Teluk Jimbaran
Jimbaran
Pura Luhur Uluwatu Nusa Dua
Uluwatu

THE ECOLOGY OF JAVA AND BALI

THE ECOLOGY OF INDONESIA SERIES

VOLUME II

THE ECOLOGY OF INDONESIA SERIES

Volume II: The Ecology of Java and Bali

Other titles in the Series

Volume I: The Ecology of Sumatra
Volume III: The Ecology of Kalimantan
Volume IV: The Ecology of Sulawesi
Volume V: The Ecology of Maluku and Nusa Tenggara
Volume VI: The Ecology of Irian Jaya
Volume VII: The Ecology of the Indonesian Seas

Produced by
Environmental Management Development in
Indonesia Project, a cooperative project of the
Indonesian Ministry of the Environment
and
Dalhousie University, Halifax, Nova Scotia
under the sponsorship of the
Canadian International Development Agency

The Ecology of Java and Bali

Tony Whitten
Roehayat Emon Soeriaatmadja
Suraya A. Afiff

OXFORD UNIVERSITY PRESS
1997

Copyright © 1996 Dalhousie University
All maps in this edition copyright © 1996 Periplus Editions
All rights reserved

Published by Periplus Editions (HK) Ltd.

ISBN 962-593-072-8

Publisher: Eric Oey
Typesetting and graphics: JWD Communications Ltd.
Copy editing: Sean Johannesen

Distributors:
Australia:
University of New South Wales Press Ltd
Sydney NSW 2052

Indonesia:
C.V. Java Books
Jalan Kelapa Gading Kirana, Blok A14 No. 17
Jakarta 14240

Japan:
Tuttle Shokai Ltd
21-13, Seki 1-Chome, Tama-ku,
Kawasaki, Kanagawa 214

Singapore and Malaysia:
Berkeley Books Private Ltd.
5 Little Road #08-01
Singapore 536983

United States:
Charles E. Tuttle Co., Inc.,
RRI Box 231-5, North Clarendon,
VT 05759-9700

Published in the UK, continental Europe, the Middle East,
India and Africa exclusively by Oxford University Press

Printed in the Republic of Singapore

Table of Contents

EMDI

The Environmental Management Development Project (EMDI) was designed to upgrade environmental management capabilities through institutional strengthening and human resource development. A joint project of the Ministry of State for Environment (LH), Jakarta, and the School for Resource and Environmental Studies, Dalhousie University, Halifax, Nova Scotia, EMDI supported LH's mandate to provide guidance and leadership to Indonesian agencies and organizations responsible for implementing environmental management and sustainable development. Linkages between Indonesian and Canadian organizations and individuals in the area of environmental management are also fostered.

EMDI received generous funding from the Canadian International Development Agency (CIDA). CIDA provided Cdn$2.5 million to EMDI-1 (1983-86), Cdn$7.7 million to EMDI-2 (1986-89), and contributed Cdn$37.3 million to EMDI-3 (1989-95). Significant contributions, direct and in kind, were made by LH and Dalhousie University.

EMDI-3 emphasized spatial planning and regional environmental management, environmental impact assessment, environmental standards, hazardous and toxic substance management, marine and coastal environmental management, environmental information systems, and environmental law. The opportunity for further studies was offered through fellowships and internships for qualified individuals. The books in the Ecology of Indonesia series form a major part of the publications programme. Linkages with NGOs and the private sector were encouraged.

EMDI supported the University Consortium on the Environment comprising Gadjah Mada University, University of Indonesia, Bandung Institute of Technology, the University of Waterloo, and York University. Included in EMDI activities at Dalhousie University were research fellowships and exchanges for senior professionals in Indonesia and Canada, and assistance for Dalhousie graduate students undertaking thesis research in Indonesia.

For further information about the EMDI project, please contact:

Director
School for Resource and Environmental Studies
Dalhousie University
1312 Robie Street
Halifax, Nova Scotia
Canada B3H 3E2
Tel. 1-902-494-3632
Fax. 1-902-494-3728

Foreword

It is always gratifying when something for which one has long waited finally materializes. So it is that over five years after its inception by the EMDI project, *The Ecology of Java and Bali* is now available to Indonesian and English readers.

One could argue that this is the most important of the seven volumes in the *Ecology of Indonesia* series, not just because Java and Bali are home to more people than the rest of the country combined, but because the ecological issues on Java and Bali – such as the consumption and growth of the population, determining the carrying capacity of the land, securing sufficient and healthy food and water, controlling the use of dangerous chemicals, developing a sustainable tourism industry, stemming the loss of biological diversity, providing just and reasonable access to and control of resources, reviewing the nature of current economic paradigms, and dealing with the general ignorance of ecology – are more acute than anywhere else in Indonesia, and because these issues impinge on the lives of the most senior decision makers. Advances in tackling the issues in Java and Bali can and must be translated into appropriate action elsewhere in the archipelago.

The authors must be congratulated on bringing together such varied and detailed information, but I must also pay tribute to the many individuals from all sections of society – researchers, librarians, students, government employees, and villagers who have contributed to the undoubted success and value of the book.

Finally, my hope and expectation is that this book is not just for reading but for implementing. If Java and Bali are truly to have a sustainable future, then the lessons from this book must be heeded and acted upon.

Jakarta, February 1996
State Minister for Environment

Sarwono Kusumaatmadja

Jakarta 10110, Indonesia
Tel. 62-21-3807566
Fax. 62-21-351515

Important Notes

This is a large book and only the most diligent or enthusiastic people with time on their hands are likely to read it from cover to cover. We expect most people to use it primarily as a reference book, and the comprehensive index and cross-referencing will facilitate this. There are five parts:

- Ecological Concerns which includes a background to the writing of the book, and a simple statistical and graphical description of Java and Bali. The second chapter discusses the twelve most critical ecological issues currently facing these islands (and indeed all of Indonesia);
- Ecological Components which comprises an inventory and description of the past and present physical, biological, and human components;
- Ecosystems which covers the major habitats, and the influence of people on them;
- Conservation which examines human attitudes to nature, the loss of species, and the means by which it is hoped the important elements will be conserved; and
- Finding a path for the future which looks at the need to encourage a stronger personal environmental ethic, and at other changes that will need to be made if we are not to miss the path of sustainability and further endanger the integrity of the environment for future generations.

Notes on names

Few Indonesian vernacular names are used in the English version of this book, but as many as possible are used in the Indonesian version. English vernacular names are used where there is little or no room for confusion. It is often the case that the vernacular names are not sufficiently specific for scientific use. Efforts have been made to ensure that the scientific names used are accurate and up-to-date. Some readers will find that these names are markedly different from those taught or used in older books. This is because discoveries of new species, or new insights into the relationships between species, must be accommodated by the system of naming governed by internationally agreed rules.

After the first mention of a plant species in each chapter there follows a four- or five-letter abbreviation in parentheses, such as (Legu.), (Podos.), and (Saxi.). This indicates the plant family to which the species belongs, and a key to these can be found in Appendix 1.

Notes on copyright

Strenuous efforts have been made to seek the permission of copyright holders for the reproduction of material used in this book. We apologize if our attempts have been imperfect. A few publishers have insisted that 'with permission of the publishers' appear with a particular figure or table and we have obliged, but this does not indicate that permission has not been given for the others.

Acknowledgements

It is hackneyed to write that a book such as this would never have been written without the selfless and unstinting help of a large number of people; nonetheless it is true, and we wish to thank them all. If any are omitted, we apologize.

Most importantly we must look to the early EMDI management under the guiding hand of Prof. Dr. Emil Salim, former Minister of State for Population and Environment, Republic of Indonesia. He, together with Prof. Arthur Hanson (now of the Winnipeg-based International Institute for Sustainable Development), supported the concept of regional ecology books back in 1982, and were wonderful apologists in meetings in Jakarta and Ottawa. Similar support for this volume on Java and Bali book has been given by the present Minister of State for Environment, Prof. Dr. Sarwono Kusumaatmadja and various of his staff, and by EMDI managers (Heather Johannesen, Diane Blachford, Dr. Shirley Conover, Gerry Glazier, George Greene, Pauline Lawrence, Lynne Norrena, Barbara Patton).

We have been supported excellently by Siti Juliani (Lia) (for the first part of the book preparation), Ir. Agus Widyantoro, and especially Ir. Sri Nurani Kartikasari (Ani), who have delved in dusty places for information, have encouraged people to talk about their feelings on ecological conditions, have arranged meetings, have covered for our lapses of memory, and have made a multitude of tasks easier. Back in 1988 Ani covered the length and breadth of Java and Bali collecting written material to compile the initial bibliography (Kartikasari and Whitten 1989). Ani's irritatingly precise and tireless editing and translation skills have been used to produce a fine translation. In addition, Jane Whitten has edited the English text and nursed the large and complex bibliography into its final form. Dewi R. Purbawati has been our faithful secretary through the preparation of the book and special thanks are due to her.

The typesetting, graphics, and copyediting were done by Marylouise Wiack and Sean Johannesen of JWD Communications Ltd., Halifax, Nova Scotia, and great thanks are due to them for all their efforts in making the manuscript into a book.

We have been able to contract out research projects to a number of individuals who have produced interesting data and interpretations. That is, Dr. Soetikno Wirjoatmodjo, Drs. Boeadi, Reni K.S, Dr. Dedi Darnaedi, Dr. Arie Budiman (all of Puslitbang Biologi LIPI, Bogor), Dr. Jatna Supriatna and Drs. Martarinza (Universitas Indonesia, Jakarta), Yus Rusila Noor and teams of the Asian Wetland Bureau, Dr. Maurice Kottelat (Cornol, Switzerland), James Haile and Dr. Nigel Stork (Natural History Museum, London), and members of the Biological Sciences Club (Universitas

Nasional, Jakarta).

The general contents and structure of the book owe much to a workshop held in 1991 at the Centre for Environmental Education at Seloliman, East Java, at which we were joined by Prof. Hermien Hadiati Koeswadji (PSLK Universitas Airlangga, Surabaya), Mochtar Lubis (Yayasan Obor Indonesia, Jakarta), Drh. Suryo W. Prawiroatmodjo (PPLH, Seloliman), Dr. Soedarmanto (PSLK Universitas Brawijaya, Malang), Ir. Danu Widjaja (Bappeda Jawa Timur, Surabaya), and Dr. Arie Budiman and Dr. Dedi Darnaedi (both of Puslitbang Biologi LIPI, Bogor). At the end of 1993 we subjected the text to a peer review meeting attended by: Ir. Harry Suryadi (*Kompas*, Jakarta), Prof. Dr. Ida Nyoman Oka (formerly of Department of Agriculture, Jakarta), Johan Iskandar (PPSDAL, Unpad, Bandung), Dr. Meutia Farida Swasono (FISIP UI, Jakarta), Dr. Nono Makarim (Jakarta), Ir. Yusak J. Pamei (PT. Bra Saktirealindo, Bogor), and Prof Dr. Siti Sundari (Unair, Surabaya).

Sections of the book have been read, corrected and criticized by: S. van Balen (Birdlife International, Bogor), Drs. Boeadi (Puslitbang Biologi LIPI, Bogor), Dr. L.A. (Sampurno) Bruijnzeel (Free University, Amsterdam), Veronica Brzeski (Dalhousie University, Halifax), Dr. C. Burrett (Sultan Qaboos University, Oman), Dr. Shirley Conover (Dalhousie University, Halifax), Dr. Rokhmin Dahuri (IPB, Bogor), Dr. Julian Evans (Forestry Authority, Wrecclesham), Dr. Friedhelm Göltenboth (UKSW, Salatiga), Dr. Atmadja Hardjamulia (formerly of Freshwater Fisheries Research Institute, Bogor), Simon Hedges (Southampton University), Dr. Robert Hefner (Boston University), Derek Holmes (Indonesian Ornithological Society, Jakarta), Drs. Moh. Indrawan (Aberdeen University), Dr. Kuswata Kartawinata (MacArthur Foundation, Chicago), Ir. Mahsyur (BBAT, Sukabumi), Victor Mason (Bali Bird Club, Ubud), Ir. Wardono Saleh (Manggala Wanabhakti, Jakarta), Prof. Emil Salim (Jakarta), Drs. Effendy Sumardja (PHPA, Jakarta), Martin Tyson (Southampton University), Stephen Walker (Natural Resources Institute, Chatham), and Elizabeth Widjaja (Herbarium, Bogor).

And then there are enormous numbers of other people around the world who have provided help in a wide variety of ways. Despite all this help we take full responsibility for the contents of the book and any or all criticism must be laid at our feet.

In Australia: Dr. B.A. Barlow (CSIRO, Canberra), David Bishop (Victor Emmanuel, Kincumber), Dr. G. Cassis (Australian Museum, Sydney), Dr. Peter Daniels (CSIRO, Burrimah), Dr. Colin Groves (ANU, Canberra), Dr. Darrell Kitchener (Western Australian Museum, Perth), Dr. T.R. New (La Trobe University, Bundoora), Dr. D.E. Symon (The Botanic Gardens of Adelaide and State Herbarium, Adelaide), Dr. Adrian Vickers (Wollongong University), Dr. Alice Wells (NT Museum of Arts and Science, Darwin).

In Canada: Anne Wilson (Toronto), Dr. Christopher Darling (Royal

Ontario Museum, Canada), Dr. Mark E. Taylor (Geomatics International, Ontario), Kevin Boehmer (Ontario), Brian Yates (Halifax N.S.). Elsewhere in Europe: Dr. F. Cassola (Rome), Louis Deharveng (Toulouse), Malcolm Hadley (Unesco, Paris), Prof. Ernst Heiss (Tiroler Landesmuseum, Innsbruck), Dr. R.D. Hoogland (Lab. de Phanerogamie, Paris), Dr. I. Ineich (National Natural History Museum, Paris), Dr. P. Lehmusluoto (Helsinki University), Dr. Hans Malicky (Lunz am See), Dr. Harald Riedl (Natural History Museum, Vienna), Dr. Michael Riffel (Zoologische Gesellschaft für Arten-und Populationsschutz, München), Dr. P. Zwick (Max Planck Institute, Schlitz).

In Indonesia: Anak Agung Gde Agung (Gianyar Palace, Gianyar), Ir. Moh. Amir (Puslitbang Biologi LIPI, Bogor), Dr. Russ Betts (WWF, Jakarta), Dr. Jim Davie (World Bank/PHPA, Bogor), Bunjamin Dharma (Jakarta), Dr. Sean Foley (World Bank, Jakarta), Wim Giessen (AWB, Bogor), Ir. Sarkan Tena Ari Hadi (Sub-Balai KSDA, Jember), Ir Hardjito Haknojosoebroto (Kanwil Kehutanan, Bandung), J.R.E. Harger (Unesco, Jakarta), Cindy Jardine, (EMDI, Jakarta), Hira Jhamtani (Konphalindo, Jakarta), Ir Rachmat Kosasih (BKSDA, Malang), the late Dr. A.J.G.H. Kostermans (Puslitbang Biologi-LIPI, Bogor), Dr. Rusdian Lubis (Bapedal, Jakarta), Dr. N.K. Mardani (Kanwil Pariwisata, Denpasar), Martoto (L.P. Batu, Nusa Kambangan), Victor Mason (Bali Bird Club, Ubud), Ani Pakpahan (IPB, Bogor), A. Hadi Pramono (Universitas Indonesia, Jakarta), Dr. Mien Rifai (Puslitbang Biologi-LIPI, Bogor), Ir Ristiyanti (Puslitbang Biologi LIPI, Bogor), Dr. Fred Rumawas (IPB, Bogor), Asep Setiadi (Perhutani, Probolinggo), Dr. Frances Seymour (Ford Foundation, Jakarta), Marcel Silvius (AWB, Bogor), Siswoyo (BKSDA III, Bogor), Prof. Dr. Otto Soemarwoto (Institute of Ecology, Pajajaran University, Bandung), Ir. Isra'jadi Soerjokoesoemo (Kanwil Kehutanan, Surabaya), M. Noor Soetawijaya (Taman Nasional Bali Barat, Gilimanuk), Subari (Resort KSDA, Baderan), Sunarto (Sub-Seksi KSDA, Cilacap), Dr. Jatna Supriatna (Universitas Indonesia, Jakarta), T. Suryadi (P.T. Nirmala Agung, Sukabumi), Ir. Puja Utama (SB KSDA, Jember), Caroline van der Sluys (EMDI, Jakarta), Ir. Tri Wibowo (PHPA, Jakarta), Bambang Yuwono (PT. Vivaria, Jakarta).

In Japan: R. Yoshii (Kyoto).

In the Netherlands: Dr. Ing. C. van Achterberg (National Museum of Natural History, Leiden), Dr. F.A. Adema (Rijksherbarium/Hortus Botanicus, Leiden University), Dr. Pieter Baas (Rijksherbarium/Hortus Botanicus, Leiden University), Dr. M.M.J. van Balgooy (Rijksherbarium/Hortus Botanicus, Leiden University), Christa Deelesman-Reinhold, Dr. René Dekker (National Museum of Natural History, Leiden), Prof. L.B. Holthuis (National Museum of Natural History, Leiden), Dr. Rienk de Jong (National Museum of Natural History, Leiden), Dr. C. Kalkman, Dr. Jan Krikken (National Museum of Natural History, Leiden), C.W.J. Lut (Rijksherbarium/Hortus Botanicus, Leiden University), Dr. Hans Nooteboom (Rijksherbarium/Hortus Botanicus, Leiden University), Dr. Fred Smiet (Foreign Ministry, The Hague), Dr. W.B. Snellen (International Institute for Land Reclamation and Improvement Wageningen), Dr. Jan van Tol (National Museum of Natural History, Leiden), Dr. J.F. Veldkamp (Rijksherbarium/Hortus Botanicus, Leiden University), Dr. Jaap Vermeulen (Rijksherbarium/Hortus Botanicus, Leiden University), Dr. John de Vos (National Museum of Natural History, Leiden), Dr. D.J. van Weers (Instituut voor Taxonomische Zoologie, Amsterdam), Dr. Willem J.J.O. de Wilde (Rijksherbarium/Hortus Botanicus, Leiden University).

In Singapore: Dr. John N. Miksic (National University of Singapore), Dr. D. Murphy (National University of Singapore), Dr. Peter K.L. Ng (National University of Singapore).

In the United Kingdom: Robaire Beckwith (Cambridge University), Dr. P.E. Bragg (Nottingham), Dr. Paul D. Brock (Slough), Josephine M. Camus (Natural History Museum, London), Dr. Barry Clarke (Natural History Museum, London), Jim Comber (Southampton), Anne Disney (Cambridge), Dr. John and Soejatmi Dransfield (Royal Botanic Gardens, Kew), John Edwards Hill (Edenbridge, Kent), Dr. Mark Hobart (School of Oriental and African Studies, London), Dr. Jeremy Holloway (International Institute of Entomology, London), Michael Long (Brunel Technical College, Bristol), C.J. McCarthy (Natural History Museum, London), Lewis and Sue McKenzie (Cambridge), Dr. Nick Polunin (Newcastle University), Mark and Simon Rowland (Cambridge), Libby Saxton (Oxford University), Rosemary Smith (Royal Botanic Gardens, Edinburgh), Stephen Walker (Natural Resources Institute, Chatham), Dr. David Wall (Natural Resources Institute, Chatham), Sue Wells (Newcastle University), Dr. Tim Whitmore (Cambridge University), Ruth and Peter Whitten (Cambridge).

In the USA: Dr. John Clark (University of Miami), Dr. Kathy MacKinnon (World Bank, Washington D.C.), Dr. Willem Meijer (University of Kentucky), Dr. David J. de Laubenfels (Syracuse University, New York), Dr. Larry Hamilton (Vermont).

Last but foremost, we acknowledge the sustaining and guiding hand of God in the preparation this book, and the feeling that this is somehow part of His agenda.

We note that a Balinese temple is never regarded as finished because the people feel that the reaching of a goal is rather an anticlimax. The same must be true for this book; what you hold is simply our latest draft when the final deadline fell. Parts of this book will be like the carvings which, exposed to the elements, weather rapidly, with algae covering increasingly imprecise detail. These should be replaced when funds and craftsmen are available but, even if they are not, it is to be hoped that the basic structure will still be both recognizable and useful. Such are our hopes for this book.

Part A

Ecological Concerns
and Principles

The economic developments that have led to progress in human economic well-being in Java and Bali are well known and have been much discussed, and the pattern of progress has become a typological model for the outer islands of Indonesia. However, many of the ecological, social and physical impacts resulting from past and present economic developments are now the cause of great concern, to such an extent that it is now questioned whether the available natural resources are able to sustain future economic growth. Java and Bali are able, of course, to import goods such as basic natural resources, energy and other materials (with inevitable impacts at their origins), but the inexorable decline in overall environmental quality on Java and Bali is still a serious ecological problem, and the issue of sustainable development needs to be addressed seriously very soon.

Sustainable development needs ecological guidelines, but this demands that people, all people, should absorb and apply the constraints set by the ecological systems in which we live. Although much is being said about ecology and environment in Indonesia at the present time, very little is actually being done because economists still tend to ignore the environment, or to include it occasionally as an elective subject. If they continue to do so, the economic development of the nation will certainly suffer a serious blow, for sustainability will be threatened by environmental degradation and resource depletion, not to mention international pressures and boycotts (Soemarwoto 1991b, 1992).

Conventional economic development does not yet have any way of measuring the ecological consequences of development activities, and most economists operate without a specific theory of ecological value, relegating ecological impacts to the category of 'externality'. Therefore exhaustion of non-renewable resources, irreversible damage to renewable resources, pollution, and general degradation of the environment, all of which are ecological concerns, are evaluated arbitrarily in the economic sense so that significant actions to prevent and/or mitigate further ecological deterioration are not accountable. Many attempts are being made to address these issues, but there is a singular lack of easily accessible ecological information to support these activities. We hope that this book will

1

make a contribution here.

In this first part we introduce the rationale behind the book, the geographical setting of Java and Bali, and then discuss what we regard as the twelve most important ecological issues.

Chapter One

Introduction

Of Java in 1861 A.R. Wallace made the following judgement:

> Taking it as a whole, and surveying it from every point of view, Java is probably the very finest and most interesting tropical island in the world. Scattered through the country, especially in the eastern part of it, are found buried in lofty forests, temples, tombs and statues of great beauty and grandeur; and the remains of extensive cities, where the tigers, the rhinoceros, and the wild bull now roam undisturbed (Wallace 1869).

Had Wallace spent longer than a few days on Bali and travelled further than just Buleleng on the north coast, he would doubtless have included Bali in his praise as did many later writers. Despite the countless and varied changes that have been wrought in the Java and Bali landscapes, many areas of both islands are staggeringly beautiful. Java has probably always borne the mark of people more clearly and more deeply than anywhere else in the archipelago, not least because of its fertile volcanic soils. While Java is unique, many of the problems that were first found and experienced there, are becoming common and spreading elsewhere in Indonesia: effects of population growth, inappropriately managed material wealth, land degradation, loss of agricultural land, crises in water availability and quality, fragmentation and loss of natural habitats, and species extinction. We therefore see Java as an excellent and crucial test case. The challenge is to deal with the current problems on Java, the island where power is centred, so that decisions and actions can then be adapted and adopted for other areas of the archipelago.

Java is indeed the economic, social, political and cultural hub of Indonesia. In terms of development, Bali is unique in being neither a typical 'outer' island, with relatively low population density providing resources for foreign exchange, nor a typical 'inner' island, with highly organized agriculture, industrial centres, and urban support. In Indonesia, where the dichotomy of 'inner-outer' has developed into 'centre-periphery', Bali is rather peripheral of the centre. It has many of the problems of Java (high population density, intensive cultivation), but not the advantages of centrality and accessibility (McTaggart 1984). Even so, it receives more attention than might be supposed from its position, because of its importance in earning foreign exchange though tourism, and because many of those with decision-making authority in the 'centre' are familiar with the island from regular holidays there, and may also have business interests.

In a world often dominated by economists, it is good to know that environmental concerns can enter their equations. Professor Emil Salim, a professional economist and Indonesia's first environment minister, was appointed for the first of three five-year terms in 1978, six years after he represented Indonesia at the 1972 Stockholm Environment Conference. From self-confessed ignorance concerning ecology and the environment he became one of the world's most respected environmentalists. This shows that if the appropriate information is provided, then great and important changes can result. He once said, "When it comes to the implementation of development plans, projects, and programmes, mistakes often occur because of sheer ignorance and not ill-will". No one would deny that much of that sheer ignorance relates to ecology, and it is with the aim of making important ecological information about Java and Bali available, that this book is written.

There are now a fair number of immediately relevant ecology books available in Indonesian (e.g., Anwar 1984; Heddy et al. 1986; Nontji 1987; MacKinnon 1986; Whitten et al. 1987a; Resosoedarmo et al. 1989; Soemarwoto 1983; Program PHT 1991; Deshmukh 1992; Irwan 1992; Prawiroatmodjo 1992; MacKinnon et al. 1996a), and English (Whitmore 1984; MacKinnon 1992 ; Whitten et al. 1987b,c; Whitten and Whitten 1992; MacKinnon et al. 1996b), but none is wholly devoted to the ecology and sustainable development of Java and Bali. It is our hope that this book may be as influential in Java and Bali, even all Indonesia, as *Silent Spring* (Carson 1962, 1990) was in the USA and the rest of the western world. We hope that it will embolden decision makers to tackle the difficult issues posed by putting theories on sustainable development into practice. Each member of the authorship team has brought his or her own biases, views, and cultural background to this book, and we hope we have integrated the Eastern approach of seeking harmony, with the Western insistence on realism and objectivity (Lubis 1990).

For the purposes of this book we understand 'ecology' to mean the interrelationships between living organisms, including people, and the complex of biotic, climatic, edaphic, social, political, and other conditions, termed the 'environment', which comprise the immediate habitat of those organisms (Lincoln et al. 1982). Ecology has great potential relevance in deciding how the environment is managed, but those with ultimate power in directing development must understand its principles, its warnings, its benefits and constraints.

Ecology has developed two strands. The scientific strand develops theories which help to explain relationships between organisms and their environment. While we are unlikely ever to be able to explain everything, good science is essential if government policies are to be based on sustainable principles. But data and graphs are not enough because ecologists find themselves facing disturbing and even unsavoury conclusions, and while their arguments for ecological reason are now being heard, they are not always heeded. This leads to the second strand of ecology, the emotional, which is becoming increasingly conspicuous as interest and frustrations grow.

BASIC GEOGRAPHY AND STATISTICS

Java is an island of about 130,000 km^2 (different sources have different precise figures), slightly larger than New York State, with a length similar to that of Italy or Finland, and rather longer than the United Kingdom (table 1.1, fig. 1.1). Just north of East Java, of which it forms part politically, lies the island of Madura (5,620 km^2). Unless otherwise stated, the word 'Java' in the rest of the book includes Java, Madura, and the few other offshore islands. In 1995 Java had about 114 million inhabitants living at an average density of 862 people/km^2, ranging from nearly 40,000 in some parts of Jakarta to virtually zero in some of the remaining wild areas (table 1.2). The extremely high population density on Java is largely a result of historical influences, and the very fertile soils which lend themselves to terracing for irrigated rice.

Bali is a much smaller island of 5,560 km^2, with a population nearing three million living at an average density of 520 people/km^2, although most of these live in the fertile southern quarter of the island.

Table 1.1. The areas of Java and Bali in relation to some other islands, countries, states, and provinces.

Island/Country/ State/Province	Area (km^2)	Island/Country/ State/Province	Area (km^2)
Singapore Island	570	Netherlands	40,844
Oahu	1,580	Costa Rica	51,010
Mauritius	2,304	Nova Scotia	55,490
Rhode Island	3,140	Sri Lanka	64,850
Bali	**5,561**	**Java** (excl. Madura)	**126,566**
Madura	**5,621**	New York State	127,190
Buru	9,320	Peninsular Malaysia	132,030
Seram	17,440	Nepal	140,797
Wales	20,521	Sulawesi	182,870
Rwanda	26,338	Great Britain mainland	218,024
Timor	28,000	New Zealand	269,057
Switzerland	40,808	Japan	365,360

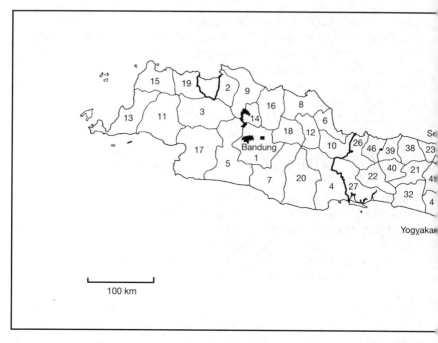

Figure 1.1. Provinces, Provincial Capitals and kabupatens on Java and Bali.
West Java: 1 - Bandung, 2 - Bekasi, 3 - Bogor, 4 - Ciamis, 5 - Cianjur, 6 - Cirebon, 7 - Garut, 8 - Indramayu, 9 - Karawang, 10 - Kuningan, 11 - Lebak, 12 - Majalengka, 13 - Pandeglang, 14 - Purwakarta, 15 - Serang, 16 - Subang, 17 - Sukabumi, 18 - Sumedang, 19 - Tangerang, 20 - Tasikmalaya; **Central Java**: 21 - Banjarnegara, 22 - Banyumas, 23 - Batang, 24 - Blora, 25 - Boyoali, 26 - Brebes, 27 - Cilacap, 28 - Demak, 29 - Grobogan, 30 - Jepara, 31 - Karanganyar, 32 - Kebumen, 33 - Kendal, 34 - Klaten, 35 - Kudus, 36 - Magelang, 37 - Pati, 38 - Pekalongan, 39 - Pemalang, 40 - Purbolinggo, 41 - Purworejo, 42 - Rembang, 43 - Semarang, 44 - Sragen, 45 - Sukoharjo, 46 - Tegal, 47 - Temanggung, 48 - Wonogiri, 49 - Wonosobo; Yogyakata: 50 - Bantul, 51 - Gunung Kidul, 52 - Kulon Progo, 53 - Sleman; **East Java**: 54 - Bangkalan, 55 - Banyuwangi, 56 - Blitar, 57 - Bojonegoro, 58 - Bondowoso, 59 - Gresik, 60 - Jember, 61 - Jombang, 62 - Kediri, 63 - Lamongan, 64 - Lumajang, 65 - Madiun, 66 - Magetan, 67 - Malang, 68 - Mojokerto, 69 - Nganjuk, 70 - Ngawi, 71 - Pacitan, 72 - Pamekasan, 73 - Pasuruan, 74 - Ponorogo, 75 - Probolinggo, 76 - Sampang, 77 - Sidoarjo, 78 - Situbondo, 79 - Sumenep, 80 - Trenggalek, 81 - Tuban, 82 - Tulungagung; **Bali**: 83 - Badung, 84 - Bangli, 85 - Buleleng, 86 - Gianyar, 87 - Jembrana, 88 - Karangasem, 89 - Klungkung, 90 - Tabanan.
After RePPProT 1990

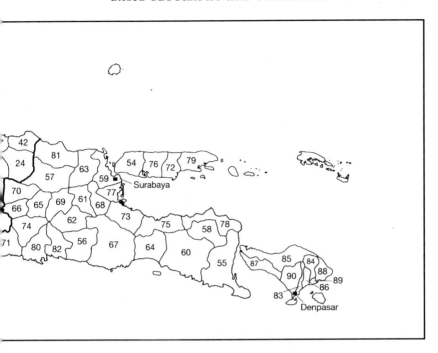

Table 1.2. Essential statistics of the six major administrative areas of Java and Bali. Na = not available.

	DKI Jakarta	West Java	Central Java	DI Yogyakarta	East Java	Bali
Area (km²)	661	46,300	34,206	3,169	47,922	5,561
% of country's area	0.03	2.41	1.78	0.17	2.50	0.29
Population 1995 (millions)	9.16	38.34	29.69	2.9	33.88	2.9
% of country's population	4.60	19.73	15.90	1.62	18.13	1.54
Population density (per km²)	13,858	851	868	920	707	522
Annual population growth 1980-1990	2.41	2.57	1.18	0.58	1.08	1.18
Natural forest area (km²)	0	4,997	2,031	13	5,409	1,009
Natural forest area (%)	0	10.8	5.9	0.4	11.3	18.1
Protection forests (km²)	na	1878	407	na	3,050	963
Asphalt road (all types) length (above) and length/km² (below)	na	11,340 0.25	12,468 0.36	1,766 0.56	13,990 0.29	2,855 0.51
Consumer price index 1990 (1988/1989 = 100)	112	111	112	111	114	120
Inflation rate 1995	9.54	6.4	8.45	9.64	8.69	5.77

Since 1961, Java has been unique among the islands of Indonesia in that the management of all forests outside nature reserves, game reserves, recreation forests, and national parks is entrusted to Perhutani, a State-owned enterprise with the dual aims of benefitting social welfare and making a profit, and able to plough back profits into its own development and projects. It thus manages about 5,300 km^2 of protection forests (steep lands and mangroves protected for their hydrological functions), as well as about 19,000 km^2 of plantations (p. 589). Perhutani land totals nearly 2.5 million ha or 19% of Java, distributed more or less evenly among the major provinces. When considering these figures it is important to realize that forest land (land designated to be under the authority of Perhutani or other parts of the Department of Forestry) and forested land (land covered with trees) are not necessarily the same thing, and the boundaries between forested land and cultivated land can move up the mountain slopes even though the boundaries of the formal forest land remains stationary.

There follow a number of maps taken from the Overview Atlas of the Regional Physical Planning Programme for Transmigration (RePPProT 1990) which illustrate the overall patterns of land use on Java and Bali (fig. 1.2-1.7) These are preceded by a table showing areas of land use types for each of the regencies and metropolitan areas.

Table 1.3. Summary of the area (km²) of major land use types in the kabupatens of Java and Bali. It is important to note that the scale of mapping which generates these figures could not recognize small areas such as for the first three categories in Jakarta where the totals are small but more than the zero listed.

	Forest	Scrub	Fields/ gardens	Perm. upland cult.	Sawah	Tree crops and estates	Water	Settle-ments	Total km²	% of total Java & Bali area
West Java										
Bandung	678	388	181	464	626	300	48	494	3,305	2.39
Bekasi	50	15	0	87	879	57	50	259	1,475	1.07
Bogor	389	539	176	504	261	537	0	862	3,297	2.39
Ciamis	113	487	336	274	337	1,192	0	132	2,871	2.08
Cianjur	521	1,171	142	832	436	428	39	108	3,682	2.66
Cirebon	17	42	0	47	502	231	50	157	1,067	0.77
Garut	760	661	0	734	160	507	0	187	3,034	2.20
Indramayu	0	59	3	22	1,344	251	147	206	2,063	1.49
Karawang	78	116	0	96	1,196	108	140	202	1,942	1.41
Kuningan	123	184	11	343	227	178	0	133	1,219	0.88
Lebak	287	766	460	883	183	542	0	36	3,176	2.30
Majalengka	57	221	21	289	362	250	0	102	1,318	0.95
Pandeglang	565	730	219	494	472	380	0	30	2,896	2.10
Purwakarta	109	99	3	214	87	140	82	140	884	0.64
Serang	74	153	62	513	667	148	55	247	1,945	1.41
Subang	137	166	23	277	876	355	111	230	2,180	1.58
Sukabumi	716	1,301	221	908	318	620	0	113	4,212	3.05
Sumedang	135	363	69	401	146	379	0	107	1,603	1.16
Tangerang	0	14	0	138	595	42	58	467	1,329	0.96
Tasikmalaya	188	970	41	193	284	1,143	0	119	2,965	2.15
Total	**4,997**	**8,445**	**1,968**	**7,713**	**9,958**	**7,788**	**780**	**4,331**	**46,463**	**33.62**
Jakarta Total	**0**	**0**	**0**	**37**	**90**	**18**	**16**	**96**	**658**	**0.48**
Central Java										
B. Negara	80	222	0	318	174	171	0	97	1,082	0.78
Banyumas	164	218	0	216	334	187	7	286	1,466	1.06
Batang	72	52	0	255	108	129	1	132	767	0.55
Blora	8	0	0	345	510	860	0	191	1,914	1.38
Bayolali	19	84	0	354	285	87	0	207	1,084	0.78
Brebes	217	147	51	152	436	483	89	162	1,742	1.26
Cilacap	391	261	0	264	630	384	5	333	2,278	1.65
Demak	0	0	0	9	477	26	51	247	810	0.59
Grobongn	29	22	62	186	743	484	0	413	1,945	1.41
Jepara	96	22	11	149	197	202	14	305	1,014	0.73
Kranyar	22	77	0	189	262	49	0	246	847	0.61
Kebumen	10	196	0	250	414	218	0	261	1,477	1.07
Kendal	114	70	0	117	252	341	31	182	1,111	0.80
Klaten	2	16	0	65	298	11	0	222	623	0.45

Table continues.

Table 1.3. *(Continued.)* Summary of the area (km^2) of major land use types in the kabupatens of Java and Bali. It is important to note that the scale of mapping which generates these figures could not recognize small areas such as for the first three categories in Jakarta where the totals are small but more than the zero listed.

	Forest	Scrub	Fields/ gardens	Perm. upland cult.	Sawah	Tree crops and estates	Water	Settle- ments	Total km^2	% of total Java & Bali area
Kudus	19	8	34	32	203	15	0	99	455	0.33
Magelang	39	95	0	312	297	100	0	218	1,081	0.78
Pati	29	31	38	251	594	197	61	332	1,553	1.12
Pekalongan	295	75	0	108	267	113	3	152	1,013	0.73
Pemalang	41	84	0	138	279	315	13	262	1,135	0.82
Purbolinggo	56	148	0	140	205	137	0	146	837	0.61
Purworejo	38	259	0	62	335	156	0	202	1,136	0.82
Rembang	19	62	4	360	256	279	7	60	1,047	0.76
Semarang	21	134	14	338	326	291	40	286	1,485	1.07
Sragen	0	30	9	225	411	40	40	228	990	0.72
Skoharjo	0	18	0	56	249	38	0	186	584	0.42
Tegal	100	18	0	69	399	236	7	202	1,043	0.75
Temanggung	88	213	0	249	228	63	0	122	966	0.70
Wonogiri	5	305	141	593	260	147	58	331	1,920	1.39
Wonosobo	57	242	0	263	68	107	5	56	832	0.60
Total	**2,031**	**3,109**	**364**	**6,065**	**9,497**	**5,866**	**432**	**6,166**	**34,237**	**24.77**
Yogyakarta										
Bantul	0	31	0	63	222	78	0	71	473	0.34
Gunungkidul	0	333	189	288	240	152	0	142	1,383	1.00
Kaliprogo	0	86	0	80	98	152	5	134	595	0.43
Sleman	13	5	0	18	277	26	0	201	592	0.43
Total	**13**	**455**	**189**	**449**	**837**	**408**	**5**	**548**	**3,043**	**2.20**
East Java										
Bangkalan	51	8	0	782	258	46	49	289	1,489	1.08
Banyuwangi	1,252	97	0	66	840	976	9	312	3,620	2.62
Blitar	105	269	12	246	379	358	33	355	1,767	1.28
Bjnegoro	0	0	0	157	760	943	20	377	2,264	1.64
Bondowoso	324	240	0	316	259	195	0	197	1,591	1.15
Gresik	0	8	0	153	327	197	319	145	1,251	0.91
Jember	1,104	73	0	203	969	655	0	399	3,410	2.47
Jombang	51	15	0	107	299	523	4	103	1,113	0.81
Kediri	79	84	0	128	689	195	0	375	1,617	1.17
Lamongan	0	17	0	74	902	364	38	365	1,769	1.28
Lumajang	330	107	0	345	424	339	0	190	1,788	1.29
Madiun	22	5	0	153	246	454	1	218	1,136	0.82
Magetan	32	44	0	140	261	20	0	216	740	0.54
Malang	786	259	3	858	549	824	36	421	3,839	2.78

Table continues.

Table 1.3. *(Continued.)* Summary of the area (km²) of major land use types in the kabupatens of Java and Bali. It is important to note that the scale of mapping which generates these figures could not recognize small areas such as for the first three categories in Jakarta where the totals are small but more than the zero listed.

	Forest	Scrub	Fields/ gardens	Perm. upland cult.	Sawah	Tree crops and estates	Water	Settle- ments	Total km²	% of total Java & Bali area
Mojokerto	168	29	0	58	275	370	0	58	983	0.71
Nganjuk	31	36	0	115	372	445	0	231	1,279	0.93
Ngawi	8	9	0	110	443	466	10	324	1,398	1.01
Pacitan	0	613	135	444	30	112	0	101	1,449	1.05
Pamekasan	0	48	0	318	210	30	28	162	805	0.58
Pasuruan	85	176	85	352	260	401	52	68	1,527	1.10
Ponorogo	105	221	0	96	480	223	2	217	1,482	1.07
Probolinggo	285	164	11	509	216	318	12	169	1,776	1.29
Sampang	0	33	17	621	137	36	79	294	1,224	0.89
Sidoarjo	0	0	0	0	260	263	209	14	754	0.55
Situbondo	432	299	19	239	180	129	0	181	1,610	1.16
Sumenep	66	143	81	990	166	301	56	266	2,124	1.54
Trenggalek	2	418	0	239	153	195	0	87	1,209	0.87
Tuban	10	56	0	512	590	539	6	173	1,945	1.41
Tulungagung	81	187	5	135	297	229	4	206	1,211	0.88
Total	**5,409**	**3,658**	**368**	**8,466**	**11,231**	**10,146**	**967**	**6,513**	**48,170**	**34.85**
JAVA TOTAL	**12,450**	**15,667**	**2,889**	**22,730**	**31,613**	**24,226**	**2,200**	**17,654**	**132,571**	**95.92**
Bali										
Badung	15	47	0	117	114	28	1	66	388	0.28
Bangli	29	68	0	267	43	37	15	14	532	0.38
Buleleng	420	196	0	269	186	242	14	26	1,368	0.99
Gianyar	0	0	0	142	165	4	0	67	380	0.27
Jembrana	373	14	0	23	121	342	1	10	886	0.64
Karangasem	80	200	0	258	125	103	0	12	857	0.62
Klungkung	0	35	54	48	59	47	0	17	318	0.23
Tabanan	92	73	0	185	269	169	4	108	904	0.65
BALI TOTAL	**1,009**	**633**	**54**	**1,309**	**1,082**	**972**	**35**	**320**	**5,633**	**4.08**
JAVA AND BALI TOTAL	**13,459**	**16,300**	**2,943**	**24,039**	**32,695**	**25,198**	**2,235**	**17,974**	**138,204**	**100.00**

Source: RePPProT (1989) in which more detailed land use divisions by kabupaten can be found.

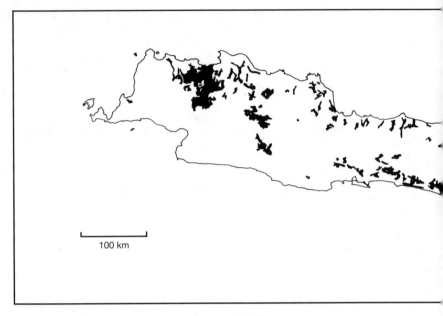

Figure 1.2. Settlement areas on Java and Bali.
After RePPProT 1990

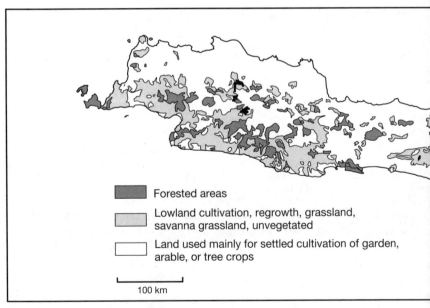

Figure 1.3. Generalized land cover of Java and Bali.
After RePPProT 1990

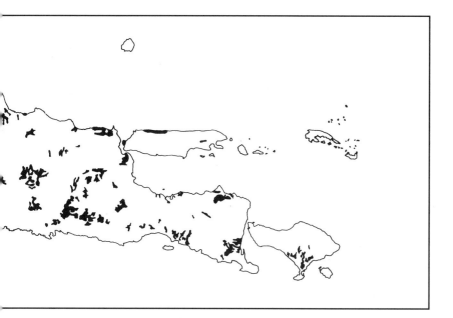

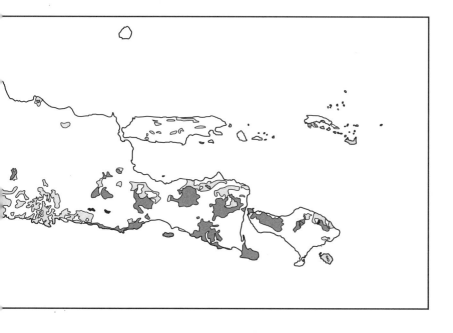

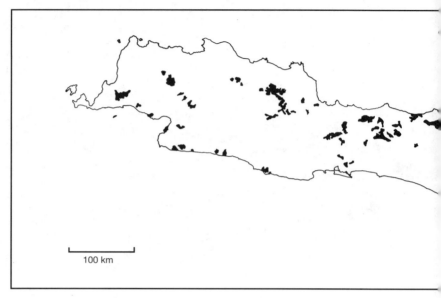

Figure 1.4. Reforestation and forest plantation (mainly teak) areas in Java and Bali.
After RePPProT 1990

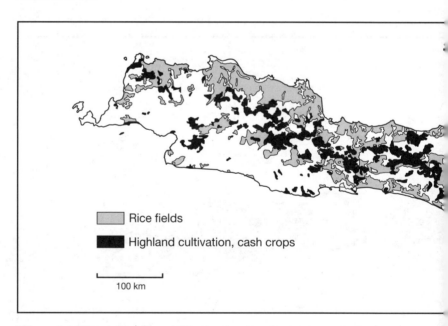

Figure 1.5. Wet rice areas and highland cultivation in Java and Bali.
After RePPProT 1990

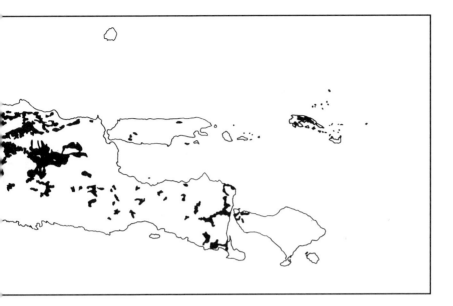

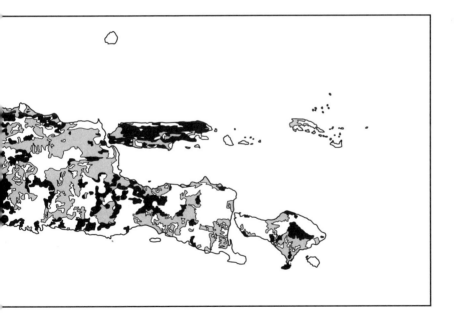

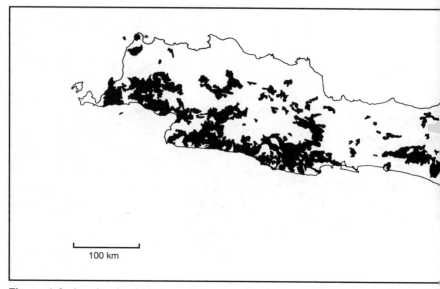

Figure 1.6. Lowland cultivation, shifting cultivation, regrowth, and grassland on Java and Bali.
After RePPProT 1990

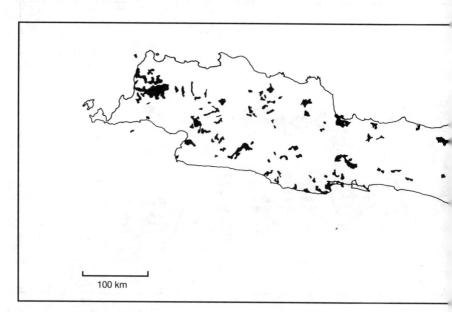

Figure 1.7. Tree crops/estates (non timber) on Java and Bali.
After RePPProT 1990

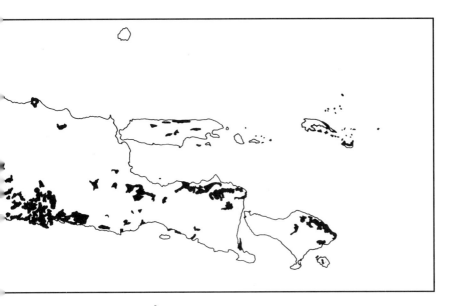

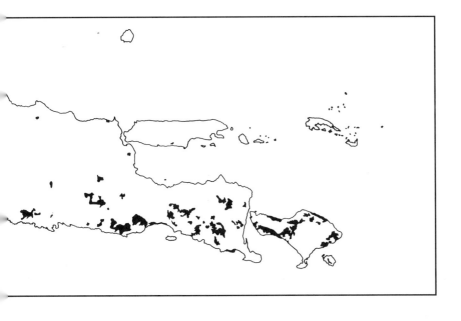

Chapter Two

Ecological Issues

Ecology is such a broad science that there is no widely accepted definition. Examples, however, are "the study of the Earth's life-support systems" (Odum 1989), "the scientific study of the interactions that determine the distribution and abundance of organisms" (Krebs 1985), "the study of the interrelationships between living organisms and their environment" (Lincoln et al. 1982), and it is generally agreed to cover individuals, populations, and communities (Begon et al. 1990). Some sociologists use the term ecology to mean the way people live (e.g., Suzuki and Ohtsuka 1987), but their literature does not always use the term as they deal with an extremely adaptable species subject to government rules and regulations, capable of rational thought and of developing codes of behaviour. Our emphasis is on the natural or semi-natural environment but, as is apparent in this chapter, we examine human roles and impact wherever possible and relevant.

IGNORANCE OF ECOLOGY

The subject matter of ecology is apparent to everybody, and most people have at some time pondered or at least observed nature, but in spite of this the failure to comprehend the principles and consequences of ecology has been a major factor in causing the environmental problems, some might say crises, affecting us. Although this situation is changing, there is not yet enough time or money spent on the laborious and time-consuming collection, identification, analysis, and interpretation of relevant ecological field data, or on training skilled practitioners in order to answer all the questions which arise when development decisions need to be made. Greater ecological understanding often requires many hours of patient study, and may not require high technology equipment, and has led to the view that this is a soft science of only peripheral importance. Now that the issue of sustainability of our use of natural resources is a major subject on political agendas the world over, it is clear that our understanding of the interactions between organisms (including ourselves) and the environment must be improved. Indeed, it is bizarre that we now know more about some of our neighbouring planets, black holes, and distant galaxies, than we do about many of the ecological communities of our own planet (Lawton 1989).

An example: tractors have replaced buffaloes in part of the rice plains of North Java. This makes some sense because it saves significant human labour costs. It does mean, however, that the free fertilizer of buffalo dung and urine is no longer available, and more has to be spent on inorganic fertilizers. Herding buffaloes when not working is generally done by young children in exchange for pocket money, and if buffaloes are not used then this source of income is lost. There are other problems. Tractors disrupt the soil structure, and this has adverse affects on water retention and yield, whereas the trampling by buffaloes damages weeds, improves water retention, and produces ideal conditions for planting out the rice seedlings. The wallows used by buffaloes are important as refuges for aquatic animals during the periods when the rice fields themselves are dry. Some of these animals will play roles as predators on pests, and the greater the distance from a refuge to a rice field, the smaller the likelihood of it being recolonized when flooded. The wallows are also used by rat snakes, otters, and monitor lizards. The rat snake, as its name suggests, preys on rats and mice, while the lizard and otter prey on the freshwater crabs which weaken and destroy the sawah bunds.

This is not an ecology textbook, and readers should seek up-to-date publications to establish a firm grounding in ecological science (e.g., Deshmukh 1986, 1992; Odum 1989; Begon et al. 1990; Ricklefs 1990; Real and Brown 1991; Chapman and Reiss 1992; Beeby 1993; Colinvaux 1993; Krebs 1994). For the sake of those who are too busy, too far from a good bookshop or library, or unable for other reasons to find and read such books, a very brief overview of some of the most important concepts is given below.

Ecosystems

The 'ecosystem' is probably the most important concept in ecology (Cherrett 1989). The word is generally agreed to constitute all the interacting parts of the physical, biological, and human environments, but it has no fixed size - a pond and the entire biosphere of Earth are both ecosystems - and no time limit, for the worst ecological problems are those that extend beyond a single generation. All ecosystems, natural and artificial, can be said to have functions: of energy transfer, maintenance of biogeochemical cycles, protection of biological diversity (p. 61), purification, regulation, and information, but their scale and relative importance will vary among the types of ecosystem.

The study of ecosystems can be said to encompass the circulation, transformation, and accumulation of energy and matter through the medium of living things and their activities (Evans 1956). As these flows are measured, so it becomes possible not just to describe a state, but to predict ecological responses to some change or disturbance (Waring 1989), and this is what an ecologist in an environmental assessment team attempts to do. Given the paucity of our knowledge, however, this is a difficult exercise full of unknown quantities.

The development and changes in plant communities after an area has been disturbed is known as succession. The composition of the animal communities generally changes also and is determined to a large extent by the vegetation. The speed, direction, and composition of the succession is determined by which species are present or are blown in immediately after the disturbance, climatic events, and so forth. Certain species are, of course, better able to cope with and compete in particular environmental conditions than others, and so a general pattern of species and structure is found that is recognizable between locations. What is not possible, is to predict the precise mix and relative abundance of species which will eventually become established.

The life strategies of organisms occurring in the successional stages are markedly different. The early stages are dominated by species, both animal and plant, with short life spans, broad food and habitat preferences, and high potential for population increase. Many agricultural pests and weeds are of this type. As succession proceeds, so an ecosystem is understood to mature (Odum 1969). For example, a lake will, over time, develop into a swamp and eventually into dry land, and grassland will develop into forest. Mature ecosystems tend to have more species than immature ones, and the niches or 'roles' of species (p. 692) tend to be more narrow so that the use of time, space, and resources are divided up to diminish competition. Immature ecosystems tend to have a low biomass but high productivity. This relationship reverses with maturity, and in a mature forest the productivity more or less equals respiration because energy is being used to maintain the system rather than to produce new material. It is thus clear that agriculture and plantation forestry artificially create and maintain immature ecosystems in order to exploit the high net productivity in terms of accumulating biomass or a crop (Dover and Talbot 1987).

Interactions

The interactions between organisms are typically expressed as food webs (collections of food chains), and examples of these can be found elsewhere (pgs. 450, 560, 579). While useful at a certain level, these have their limitations. First, they reflect only trophic relationships (what eats what) and generally miss other forms of interaction such as competitive relationships, differences among a group of related species, mutualisms, relative abundances, intensity and frequency, and so forth (Lawton 1989).

Even with these failings the webs serve to show, as with management structures of government departments or companies, that the loss of one species (or office) forces a juggling in the rest of the web. The loss may be critical, or it may even go unnoticed. At the present time far more thought goes into the possible effects of losing offices as a result of cutting costs, than of losing species.

It should be noted here that the conventional wisdom of complex ecosystems being inherently stable (MacArthur 1955; Elton 1958) stands challenged, although the relationship between complexity and stability is not simple. To begin with, stability covers at least three distinct ecological properties. Stability may be viewed as:

- **resilience** or the ability of an ecosystem to restore its original condition after some disturbance (Dover and Talbot 1987);
- **persistence** or the ability to remain much the same without disturbance (Margalef 1969); or
- **resistance** or the capacity to withstand disturbance (Holling 1973).

It does seem, however, that environments with relatively unchanging climates, such as the aseasonal lowlands of much of West Java (p. 122), support complex ecosystems with low resilience, while more variable environments allow only the relatively resilient ecosystems to persist. The former ecosystems, being more susceptible to outside interferences, are in greater need of protection. Relatively unchanging environments will tend to have species with high competitive abilities, high inherent survivorship, low reproductive output, rather stable populations, and low resilience. Relatively changeable environments will tend to have species with low stability but high resilience.

Natural Healing

Many natural ecosystems have great potential for healing themselves after being disturbed. Logging, widespread tree falls caused by high winds, and bombing of coral reefs, all have serious effects, but if subsequent disturbance is avoided, or at least moderated, then these natural ecosystems will rebuild themselves. Java and Bali have enormous areas in need of restoration in order to increase their usefulness. This process can be accelerated by people through:

- **restoration**, a tactic to return degraded land to its original condition or some close approximation; and
- **rehabilitation**, a management strategy to arrest the degradation of a landscape and to make it useful.

Restoration can be achieved using either *passive* or *active* means. A passive approach abandons disturbed areas in conditions designed to be habitable to the succession of organisms that were originally present or others that approximate the original inhabitants, creating opportunities for natural regeneration. Active restoration aims at managing an area to achieve an accurate re-creation of the site condition existing prior to the disturbance with due consideration being given to all environmental components whether valued or not (Brinck et al. 1988; Schrenkenberg et al. 1990). Active restoration efforts may appear to be akin to gardening, but restoration and gardening have different objectives. Gardeners try to improve on nature (breeding bigger, more colourful flowers and fruit) whereas restoration can strive to recreate an entire ecosystem, including those elements many would find unpleasant such as floods or biting insects. Restoration also differs from gardening in that it requires long-term commitment: it can take just days to destroy a patch of forest or coral reef, but decades to bring it back.

Rehabilitation and restoration should be viewed as integral parts of the development process. The restored landscapes do not have to become hands-off exhibits with oppressive protection, but rather places that can be widely and fully exploited on a sustainable basis. Success depends on:

- the choice of an appropriate site;
- agreement on what is to be considered the 'original' habitat;
- the social desire to commit mind, heart, and wallet;
- an explicit and public agreement of management goals;
- adequate time;
- the focus of resources on priority sites;
- collaboration of economic and social forces; and
- active and professional decisions on strategy and tactics from the outset (Janzen 1988).

Success can be judged to have been reached when:

- the new system is self-perpetuating;
- the new biological community resists invasion by other species;
- the new system is as productive as the original;
- nutrient cycling is efficient; and
- all key animal and plant species are present (Janzen 1988).

Cycles

Perhaps the best known of the ecological cycles is the hydrological cycle which traces the path of water molecules from clouds and back again via rainfall and evaporation from lakes, soil, the sea, and other surfaces, and the transpiration from plants. Also well known is the carbon cycle. In this, carbon molecules in the form of carbon dioxide are incorporated by plants into complex organic molecules such as sugar, fat, protein, or cellulose using energy from the sun via the process of photosynthesis. This forms the gross primary productivity, part of which is consumed in the process of respiration. What remains is the net primary productivity. The organic molecules are consumed, defaecated, assimilated, consumed, defaecated, assimilated, and so

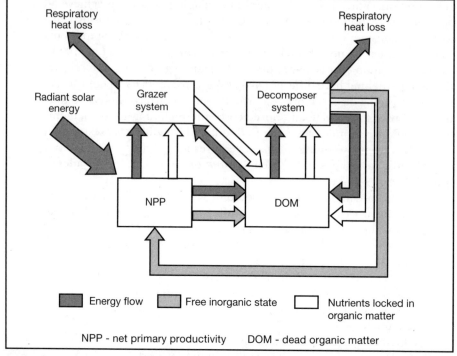

Figure 2.1. Relationship between energy flow and nutrient cycling. Nutrients locked in organic matter are distinguished from the free inorganic state.
After Begon et al., 1990

forth until they are used to provide energy. The carbon molecule is then released into the atmosphere again as carbon dioxide as a product of the respiration of a consumer organism such as an animal or fungus, or of a plant before it is consumed (Begon et al. 1990). The carbon cycle comprises two major components: the above-ground biomass and the soil organic matter. In undisturbed ecosystems the quantities and proportions of the two components remain fairly constant, with organic matter produced by the vegetation eventually being returned to the soil. Natural events such as fire, tree falls, and landslips cause localized changes, but forest clearing or annual cropping causes major changes.

Other nutrients such as nitrogen and phosphorus can also be shown to cycle: they become available to plants as simple inorganic molecules in the atmosphere or dissolved in water, become incorporated into complex organic chemicals, and available again when the chemicals are metabolized, either within the organism or as a product of the decomposer system (fig. 2.1). Forest plants allow very few nutrients to leak out of the ecosystem. This is achieved in part by having the roots very close to the soil surface,

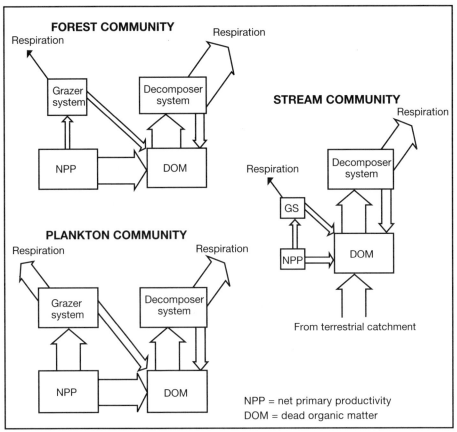

Figure 2.2. General patterns of energy flow in three generalized ecosystems. Relative sizes of arrows and boxes are proportional to the relative sizes of flows and compartments within each type of ecosystem, and they cannot be compared among the ecosystems.

After Begon et al., 1990

and also by having mycorrhizae, intimate associations of fungal hyphae and the plant roots which transfers nutrients from fallen leaves and other organic material in the soil into the plant. Nutrients intercepted by branches and boughs also remain in the system through their utilization by epiphytic plants such as certain orchids and ferns.

The flow of energy through ecosystems is basically very similar to the carbon cycle except that the heat energy which is produced by the breakdown of organic molecules is dissipated and eventually lost to the atmosphere rather than recycled, its loss balanced by the energy arriving on the Earth from the sun. Energy flow is similar among ecosystems, but there are differences in the relative sizes of the components (fig. 2.2).

As indicated above, agricultural and plantation ecosystems are immature, and a cost of maintaining this state is that the nutrient cycling is open, with nutrients 'leaking' out through harvesting and soil erosion, rather than closed, or relatively so, as it is in mature ecosystems.

Scope and Value of Ecology

Natural and artificial ecosystems are diverse and complex, and as one investigates them in closer and closer detail, many of the terms and concepts used in their study, such as 'ecosystem', 'food web', 'community', and 'niche', seem to have less meaning, and they seem to escape meaningful measurement. There are, unfortunately, few universal principles and fewer unchallenged theories. But the value of ecology is that it can explain and even predict although its scope does not include prescription (Clark et al. 1979). Prescription is the responsibility of politicians and we hope that this book will show that ignoring ecology will, like it or not, result in ecological responses which can have sociological and economic impacts, some positive, some negative. Decision making with an ecological perspective will not eliminate unexpected effects, but it will reduce environmental degradation.

POPULATION GROWTH AND PATTERNS OF CONSUMPTION

The current world population of nearly 6 billion is growing at an annual rate of 1.7%. A gradual slowing of this rate would mean that by 2025 there would be 8.5 billion people, and growth could level off by about 2150 with the total at 11.6 billion, but some would claim this view is overly optimistic (Daily and Ehrlich 1992). The global population currently consumes, co-opts or eliminates some 40% of the net terrestrial primary productivity (the basic energy supply of all terrestrial animals) (Vitousek et al. 1986), and it is not known how much more can be exploited before the ecosystems upon which humans depend cease to function.

The population of Indonesia is roughly 200 million, making it the world's fourth largest population after China, India, and the United States. This book does not labour the issue of population, because it is accepted at all levels of government and society that achieving a stable or even declining population of adequately fed, healthy, employed people is an overriding priority. There has been some progress in this direction, but no one is complacent. We write this book very aware of continuing population growth and of the many awesome problems it brings, confident that every possible humane measure will be taken to control it. Some background follows.

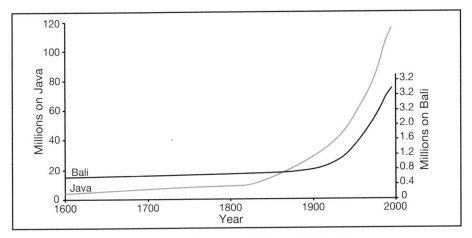

Figure 2.3. Human population of Java and Bali from 1600 until present.
After data in Foley 1987; Hugo et al., 1987

Population Growth

Nearly 200 years ago a Dutch colonial official described Java as 'over-crowded', and the population along its northeast coast as far too large for the area under cultivation in a given village or district (van der Kroef 1956). Not long after, the General Commissioner of the Dutch colonial administration wrote that he foresaw that one day the population of Java would crowd the whole island as it did even then in the fertile and irrigated river plains. Millions of tenants, he continued, would have to live on fractions of a hectare growing nothing but rice, with incomes no more than those of a poor field labourer, barely covering their day-to-day needs (du Bus de Giesignies 1827). The population at that time was about six million (one-nineteenth the present total) living on an island that was predominantly covered with forest—much of it of a type we shall never know because it grew on the extensive fertile soils on the flood plains and adjacent areas which have now been converted to rice fields (Donner 1987). Indeed, Raffles (1817) estimated that only one-eighth (12%) of Java was cultivated. This had increased to about 18% by 1870, to 50% by 1920 (Booth 1988), and to 64% at present (p. 328).

Even though the population of Java was dense locally, it did not increase dramatically during the nineteenth century, not least because of crop failures, famines, floods, volcanic eruptions, localized fights, and military engagements. The large population could find employment (albeit meagre) within the production of export crops forced by the Dutch, and this forced labour and separation of husbands from their families acted to reduce fertility. The rate of population increase accelerated at the end of the nineteenth century (fig. 2.3). Many explanations have been put forward

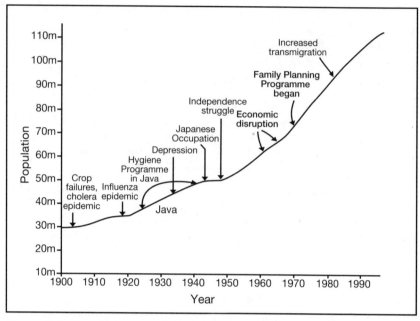

Figure 2.4. Population growth in Java since 1900.

Adapted from Hugo et al., 1987. With permission of the publishers.

to explain this (Hugo et al. 1987), but sustained growth was probably due to the rapid increase in cultivation of the dry uplands, the growing commercial activities in the towns, the building and improvement of irrigation systems, the intensification of *sawah* cultivation by more workers (the agricultural involution of Geertz (1963)), and an ever diversifying economy (McDonald 1980). During this century the rate of growth has continued to increase, but by no means smoothly (fig. 2.4), due to economic depression, strife, and disease (Hugo et al. 1987).

Java and Bali now have populations of about 115 and 3 million respectively. They are islands of population contrast, however, because there are still stretches of land with no human inhabitants, yet parts of Jakarta, currently the world's eighth largest city, has a population density of 14,000 per km² (figs. 2.5-6), 17 times the island's average. In 1980 the populations of Java and Bali were increasing at 17 million and 38,000 per year, but the rates of increase have dropped in ten years from 2.0% and 1.5% to 1.8% and 1.2% respectively. These figures represent impressive reductions resulting from the intense and well-supported family planning programme, but even a rate of population increase of 1% means a doubling of population in 70 years. The population increase on Java and Bali is not entirely due to biological growth (reproduction), however, since Java and Bali

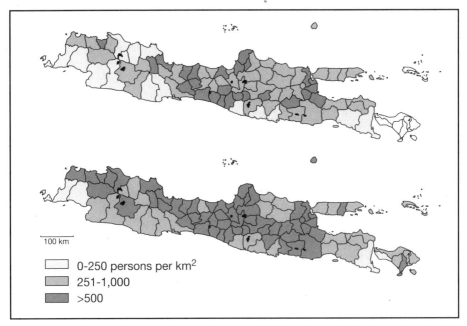

Figure 2.5. Population densities on Java and Bali in about 1930 (above) and 1990 (below). Densities for Bali from 1930 not available.

Adapted from KNAG 1938, RePPProT 1990

Figure 2.6. A view of Tanah Abang market, Jakarta, just before the Idul Fitri holiday of 1993.

With permission of Kompas

attract large numbers of people from elsewhere in the archipelago (Hardjono and Hill 1989). About 60% of the interprovincial migration in the outer islands is to Java, and 78% of the interprovincial migration within Java is to the Jakarta-Bandung corridor (World Bank 1990a) because of the greater opportunities perceived as available there. In general, though, people appear to be very sedentary and attached to their homes: in 1985 over 80% of people in West Java had never lived outside the village of their birth (Haskoning 1988a, b). Indeed, this is a major concern in Bali where the indigenous population feel their strong cultural identity is threatened by 'human dilution' from outside, mainly Java.

With the fall in population growth rates it is predicted that there will be just over 120 million people on Java by 2000, and this actually appears quite favourable compared with early predictions of 160 million for the same date (MacLeish 1971). The first effects of family planning, the decreasing rate of growth, is thus being felt, but it will be some years before the rate of growth of the potential labour force will be so affected. As a result, population will clearly remain a consideration if not a constraint for many decades to come for all policy decisions with impacts on the environment.

As a result of the government's attention over several decades to developing agriculture and increasing rural incomes, the percentage of the population in cities is lower than it might otherwise have been. Even so the annual growth of urban populations has been more rapid than overall population growth – some 5% during the 1980s – reflecting rapid industrialization and increasing employment opportunities in the urban centres. If extrapolation can be made from the experience of other countries, it is likely that this trend will continue for at least the next few decades. The urban population of Java was some 15 million or 20% in 1970, is 38 million or 36% at present, and could exceed 85 million or 60% by 2020. The

Table 2.1. Population (in thousands) of Java and its provinces and their percentage growth from 1930 to 1990.

	1930	1961	1971	1980	1990	1930-1961	1961-1971	1971-1980	1981-1990
Java	41,718	62,993	76,103	91,282	107,517	1.3	1.9	2.0	1.8
Jakarta	811	2,907	4,576	6,506	8,223	4.2	4.6	4.0	2.6
West Java	10,586	17,615	21,633	27,490	35,379	1.7	2.1	1.7	2.9
Central Java	13,706	18,407	21,877	25,365	28,517	1.0	1.7	1.7	1.2
Yogyakarta	1,559	2,241	2,490	2,745	2,913	1.2	1.1	1.1	0.6
East Java	15,056	21,823	25,527	29,175	32,488	1.2	1.6	1.5	1.1

Sources: Repetto (1986, 1989), BPS (1991)

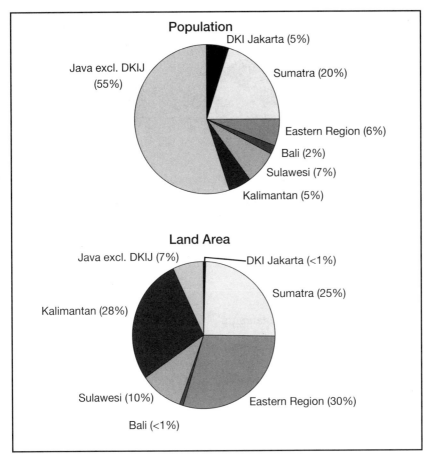

Figure 2.7. Relative contribution to Indonesian population and land area of seven major regions.
After RePPProT 1990

positive side of these figures is that Java's rural population should start to experience absolute decline by 2000, and that most of those moving to the cities will be from the poor and landless, which will help relieve pressures on ecologically sensitive areas such as the steeper uplands and lands bordering conservation areas. The negative side, of course, is that the urban areas will have to find accommodation, employment, and space for up to 1.5 million additional residents each year.

It is well known that the distribution of the population between the different regions of Indonesia is not related to land area, with Java and Bali having a disproportionately large population (fig. 2.7). In the past it was possible for ousted family members to find less-intensively cultivated areas

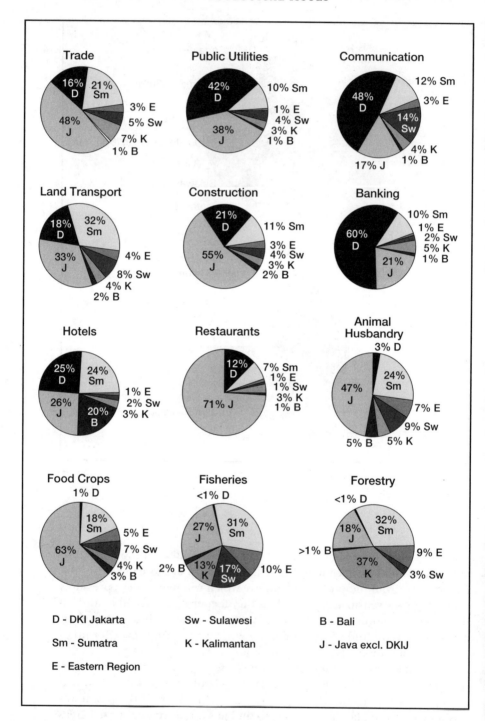

Figure 2.8. Distribution of the gross domestic product among the major Indonesian regions.

After RePPProT 1990

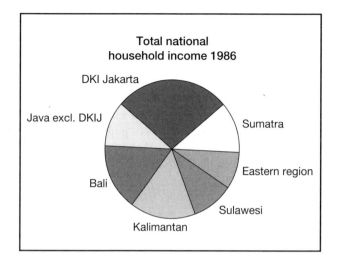

Total national
household income 1986

DKI Jakarta

Java excl. DKIJ

Sumatra

Eastern region

Bali

Sulawesi

Kalimantan

Figure 2.9. Total national household income among the major regions.
After RePPProT 1990

and open sawah, but intense population pressure has foreclosed that option. Sawah is still being opened under government programmes in Java, but this generally benefits people already living in the area rather than attracting and settling people from other areas. Many of the poor have left Java under the transmigration programme, but this has had little detectable impact on Java or Bali (in contrast to the receiving areas) except in very local cases, such as the Gunung Kidul area near Yogyakarta. This is not surprising when one considers that even during the period when transmigration was most active (Repelita III), the numbers leaving were equivalent to only one quarter of the annual recruitment rate of people joining the work force (Whitten et al. 1987d; World Bank 1988).

Java and Bali are by far the richest islands in the archipelago in terms of their contribution to national sectoral gross domestic product (fig. 2.8). This in part relates simply to their very large population, but it also reflects the manner in which Java and Bali 'parasitize' the rest of the country - Sumatran coal is brought to Java to be converted to electricity to feed Java's industries, Kalimantan timber is brought to Java for construction and other purposes, and so forth.

The share of total national household income is extremely high in Jakarta (fig. 2.9) indicating that, on average, the inhabitants are relatively well off. Such statements, however, mask the enormous gulf between the rich and the poor of this city and the highly skewed nature of income distribution. The number of people in Java and Bali living in a state of poverty has recently been a matter of active debate and conflicting data. No matter what the definitions are, it is clear that too many people are living at too low a standard of living, and that sustainability will be unattainable if the situation persists. Capital ventures clearly seek cheap labour to maximize profits and competitive advantage, and the temptations to thus foster poverty must be resisted.

Population Pressure

As population increases so pressure increases on natural and social resources. Some of the more obvious consequences of population pressure on natural resources are:

- the long-term decline in forested area, particularly on higher, drier land (McTaggart 1983);
- major coral reef destruction and coastal erosion as a result of coral mining;
- concentrations of human refuse blocking watercourses;
- overloading the natural oxidizing capacity of inland river systems with village effluent, resulting in increases in skin and intestinal diseases; and
- loss of even 'common' animal species mainly resulting from conversion of complex habitats to simple agricultural land, the use of airguns, and excessive use of agrochemicals (Foley 1987).

The effects of population pressure on social structures were examined in a major survey of inland and coastal villages in East Java (Collier 1978b). This found that when heirs to a small area of land realize that it is too small to be divided usefully between them, they either sell the land and divide the money, or one of them buys out the others who may migrate to urban centres in search of employment. Thus as the minimum tenable area of holding is reached so the number of landless people increases, and this puts strains on the village institutions which were designed to maintain the welfare of all the inhabitants. These institutions were founded at a time when land ownership was more or less evenly distributed and when disadvantaged residents could easily move away and open new land. When the concentration of land into fewer hands coincides with the emergence of new economic opportunities (improved rice varieties, intensification of brackish water ponds, and, in the uplands, the growing of high value vegetables), then some of the major landowners will use their assets and local power to introduce these innovations and make good profits. Conservative or less able landowners will observe the success and risk following the same course. As these changes proceed a number of identifiable groups are formed in the village:

- modern, innovative, landed farmers with adequate assets but few cares for the traditional village way of life;
- traditional, landed farmers with adequate assets but reluctant to destroy the traditional village way of life;
- farmers owning small parcels of land, dependent on the traditional

village institutions, who have too few assets to innovate even if they wanted to;

- landless farmers who labour for, benefit from, and are indebted to the innovative farmers, with the result that welfare institutions are further weakened;

- and a larger group of tenant farmers whose incomes and opportunities are steadily declining and for whom the traditional village institutions are unable to provide the necessary welfare.

As these groups move apart so the attitudes towards the social and political structure of the village changes. Those with substantial assets see that their traditional role of supporting the poor is impossible because of the ever-increasing number of poor. They also see and seize new opportunities for furthering their own and their family's economic welfare. The members of the last three groups described above begin to recognize that their plight has a connection with the advancement of a small group. At this point, the poor either move to the cities in the hope of a better life, or social tensions in the village increase (Surjohudoyo 1973; Collier 1978a, b; Collier and Soentoro 1978).

Thus, for many people even the progress in agricultural technology, family planning programmes, and rural industrialization cannot overcome the negative impacts of population pressure, and there always seems to be a segment of the community that is marginalized, landless, unemployed, with poor housing and subject to extreme poverty. Even involving as many family members in as many occupations as possible does not guarantee bringing them out of the abyss (Marzali 1992).

An interesting case is that of Gunung Kidul Mt. Sewu, an area quite unlike the rest of Java in its extensive karst or limestone topography (p. 539), underground water resources, characteristic soils, vegetation, and human living standards. The only natural resource available for exploitation in the area at the end of last century was stands of planted teak trees. As the human population increased the teak was felled and coffee was planted as a faster income generator. The coffee then suffered from a blight, and by the 1940s and '50s Gunung Kidul had become desperately poor with few available options (Khan 1963). At this time a *Bupati*, a former head of the Yogyakarta forestry office, started to plant trees including teak in an early agroforestry exercise (now known as Wanagama). This, together with an improved road system, the introduction of elephant grass *Pennisetum purpureum* for livestock fodder, and a major migration from the area under the transmigration programme, made the situation of the remaining farmers less desperate, allowing them to take a longer term view of development. Gunung Kidul today is surprisingly green, especially in the wet season, and valuable young teak trees are very common. Indeed, the last century of agricultural development in Java has shown that small farmers are willing and able to innovate when such changes are in their economic interests, and do so even when official extension or other leading is absent (Booth 1988; Nibbering 1991).

Consumption Patterns

Even if the goal of stabilizing Indonesia's population is achieved by 2030, there will still be very serious problems because population growth is only one side of the problem, the other side being consumption. Thus, if the population total is stable but standards of living increase, then there has been no gain in terms of use of resources. If those involved with reducing population growth feel that their work is fraught, they should be consoled by the knowledge that at least their overall aim has more or less blanket global support. This is not the case when discussing changes in consumption or standards of living.

In global terms the one billion people in the developed world have a living standard some 15 times higher than the four billion people in the Third World. The effects of changing the degree of consumption are shown in a simplistic manner in table 2.2, but these reveal just what a serious position we are in even if we accept equitability so that the average living standards of the poor should be quadrupled, and our own living standards reduced to about a quarter of their present levels. It is easy, of course, to set this scenario outside our own experiences and responsibilities, but anyone reading this book would probably represent the 'rich' 20%. In Jakarta the differences between lowest and highest salaries (and hence living standards) are at least 200-fold. Clearly, failure to increase anybody's standard of living would be an injustice to the 'have nots'.

Table 2.2. Hypothetical situations to show the need for the rich of the world to reduce their living standards.

	Rich	Poor	Total consumption units
Billions of people (a)	1	4	
Consumption units per billion (b)	15	1	
Present total consumption units (a x b)	15	4	19
Total consumption units if minimum acceptable consumption is 4 units per billion	15	16	31
But in the time taken to achieve this, the total population of poor may double to 8 billion, causing overall consumption to increase	15	32	47
If the rich reduce their consumption to 4 units per billion too	4	32	36

Source: based on Holway (1992)

Unfortunately the majority of the remainder of the population do not regard themselves in the 'have' group, but rather as in the 'have-some-but-could-do-with-more' group. This is the section of the community that drives relentless economic growth.

This growth and consumerism are stimulated by businesses which stand to gain financially by seeding our material desires and drawing us into debt. At some point, however, the finite world in which we live will force us to distinguish 'need' from 'want', and to take a negative view of needless consumption. Changing consumption patterns requires conviction and will, and while much is written about the 'poverty trap', the 'consumption trap' is also difficult to escape from.

Global initiatives are afoot to try to affect the impacts of our consumption. Industries are being encouraged to manufacture more with less, to use recycled and recyclable materials, and to produce less waste (particularly hazardous and toxic wastes), and pricing policies are being devised which will place a levy on the less environmentally friendly goods. The cost of such levies will be internalized in the setting of prices so that an equivalent product which is relatively environmentally-friendly will be relatively cheaper. The newest initiative concerns ecolabelling which will provide consumers with some readily-understood indication of how damaging or friendly a particular product is to the environment. Labels produced by the manufacturers themselves are starting to appear, but some of these can be misleading and their use requires regulation by some independent authority. Such approaches are an advance, but they still advocate short-term material gain within a system which stops when growth slumps (p. 77). Sharing and equitability are rare traits but at least reducing consumption among the 'affluent' (a less problematic word than 'rich') must become widely accepted as a moral imperative. To encourage and effect this change is a real challenge to a democratic government.

CARRYING CAPACITY AND LAND CAPABILITY

Carrying capacity is the maximum size of a population that the resources of a given area can just sustain ('carry'), and at this density the population may cease to grow. More likely, however, is that the population will, for a period, overshoot the carrying capacity of a limiting resource causing a decline in its size due to increased mortality, reduced fecundity and emigration (where this is possible). If this occurs, then it is likely that the excess population will damage the environment and in some way reduce its carrying capacity, a form of negative feedback loop, with the result that the population will fall to a new, lower level until the environment is restored (assuming restoration is even possible). The need to use space within its

environmental limits is at the core of the Act No. 24 1992 concerning
Spatial Use Management. Unfortunately, as at mid-1994, none of the necessary government regulations for the Act had been issued.

Quality of Life

Carrying capacity is routinely calculated for livestock on pasture, but it is a
more problematic concept to apply to humans. Resources are often
brought in from outside peoples' immediate environment, and the subjective issue of quality of life must also be addressed. For example, an
area could support a certain maximum number of people eking out a
wretched existence in squalor and despair, fomenting unrest and very
sensitive to any decrease in the resource base, but an optimum carrying
capacity that still allows a reasonable quality of life would be preferred. To
decide the meaning of 'optimum', a series of value judgements must be
made.

The analogy of the 'Boiled Frog Syndrome' is relevant here. This relates
to the fact that it would be possible, if cruel, to place a frog in a saucepan
of water and to put it on a stove to heat up. As long as the heat is increased
gradually the frog will not try to escape, and will eventually allow itself to be
boiled. Conversely, if a frog is thrown into a saucepan of very hot water, it
will try to jump out immediately. Humans react in a similar way to environmental degradation - at a slow rate we suffer it and adapt. If, however,
we are thrown into a dramatically worse situation we react and complain.
Thus, someone who has had a life of comfort and ease might declare that
life in a small village on marginal land is intolerable, while the inhabitants
have a relatively high level of satisfaction.

Assessing carrying capacity is also difficult because the environmental
impacts of individuals differ. Thus the carrying capacity of a population
depends on the lifestyle and the resource base used to maintain it. Indeed,
this 'intensity' of use has been incorporated into a definition of carrying
capacity by Catton (1987): the volume and intensity of use that can be sustained without degrading the environment's future sustainability for that
use. The intensity of use and the number of people are, of course, inversely
related such that as intensity increases so the number of people that can
use those resources decreases. This is not hard to grasp, but is very hard to
translate into acceptable political moves, since it requires an imposed
reduction in the intensity of resource use, at least for some people, and
those people are likely to have major political influence. Lifestyle is relevant
here (p. 848), since many of the problems of carrying capacity would be
greatly reduced if the population comprised highly cooperative, anti-mate-
rialistic, ecologically-sensitive vegetarians. Some hold that technological
innovations will lower per capita impacts so that lifestyle need not change,
but this view probably represents vain hope more than a likely scenario

(Daily and Ehrlich 1992).

Interestingly, carrying capacity is set in part by social standing and wealth. For example, in the exclusive Jakarta suburb of Bukit Golf, the human density is about $270/km^2$, among the lowest in the Metropolitan Area, but the residents and owners (and the guards patrolling the estate) would not permit any increase in population. That is, they regard the carrying capacity of the estate to have been reached already. In contrast, the population density in Kelurahan Jelambar (to the west of Jl. Latumenten in northern Jakarta) is about $28,000/km^2$, but official records indicate a steady annual increase of 1% in the number of people living in Jelambar.

The human reaction to approaching the carrying capacity is to redefine where the barrier should be. This has been done and will continue to be done, but it is clear that there must be an ultimate limit to all exploitative uses of resources, particularly if sustainable development (pgs. 81, 846) is to be genuinely adopted. The decision must be made whether we crash into this limit and experience life beyond the carrying capacity, or whether our way of life is managed (by politicians or the general populace) to keep below it. Eventually the carrying capacity will be set, not by one of our definitions, but by some limiting factor. In many parts of Java and Bali the limiting factor against increasing carrying capacity is the supply of water.

Land Capability

As stated above, it is very difficult to quantify carrying capacity for humans, but an attempt has been made to assess just one aspect of it by examining land capability or suitability for agriculture (RePPProT 1989). The assessment was objective in the use of land systems and land suitability, but subjective in the ranking of the land systems, and there may be disagreement as to the precise capability or susceptibility to hazards of different land systems. The RePPProT analysis scored land capability from one to ten with the highest score given to land suitable for wetland arable agriculture, and the lowest to land suitable only for rattan cultivation. Using this information, the mean land capability for each of the kabupatens was calculated (table 2.3).

The carrying capacity of a kabupaten should be related to its land capability, and plotting capability against the 1985 population does indeed show this relationship (fig. 2.10). There is, however, considerable scatter around the calculated regression line which indicates the relative degrees to which the kabupatens are under- or over-populated with respect to land capability (table 2.4). On the basis of the analysis it appears that Yogyakarta is least, and Bali is most, able to support its population from local agriculture. The most critical kabupatens on Java and Bali, in terms of support from local agriculture, are Gunung Kidul/Wonosari, Klungkung (most critical), Pamekasan, Kebumen, Pacitan, Kuningan, Tasikmalaya, Garut,

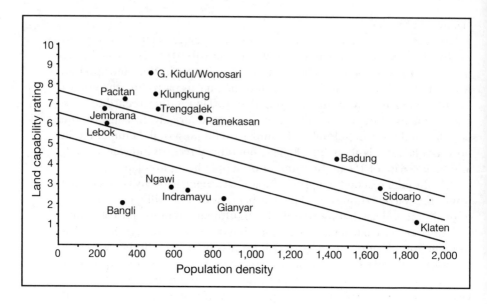

Figure 2.10. Plot of land capability score against population density for kabupaten in Java and Bali showing extreme kabupatens.

After RePPProT 1989

Table 2.3. The two best (maximum = 1) and two worst (minimum = 10) mean scores for land capability, as measured by suitability for agriculture, among the kabupaten of each province of Java and Bali.

Land capability score		Land capability score	
WEST JAVA		**YOGYAKARTA** (continued)	
Tangerang	2.6	Kulon Progo/Wates	5.4
Cirebon	2.8	G. Kidul/Wonosari	8.5
Tasikmalaya	6.6	**EAST JAVA**	
Garut	6.7	Kediri	2.5
CENTRAL JAVA		Sidoarjo	2.8
Klaten	1.2	Sumenep	6.7
Demak	2.4	Pacitan	7.3
Banjarnegara	6.2	**BALI**	
Rembang	6.3	Bangli	2.1
YOGYAKARTA		Gianyar	2.3
Sleman	2.1	Jembrana	6.8
Bantul	3.6	Klungkung	7.5

Source: RePPProT (1989)

Table 2.4. The two most overpopulated (positive) and underpopulated (negative) kabupaten relative to land capability and the one most balanced (closest to zero) in each province of Java and Bali.

Degree of overpopulation		Degree of overpopulation	
WEST JAVA		**YOGYAKARTA** *(continued)*	
Kuningan	+1.38	Kulon Progo/Wates	-0.49
Tasikmalaya	+1.38	Sleman	-0.96
Cirebon	+0.02	**EAST JAVA**	
Karawang	-1.41	Pamekasan	+1.53
Indramayu	-1.75	Pacitan	+1.43
CENTRAL JAVA		Blitar	-0.05
Kebumen	+1.43	Kediri	-1.63
Banyumas	+1.36	Ngawi	-1.88
Semarang (Ungaran)	-0.05	**BALI**	
Sragen	-1.38	Klungkung	+1.90
Demak	-1.61	Badung	+1.31
YOGYAKARTA		Buleleng	-0.10
G. Kidul/Wonosari	+2.89	Gianyar	-1.78
Bantul	+0.64	Bangli	-3.06

Source: RePPProT (1989)

Banyumas, Banjarnegara, Sumenep, and Pemalang. Clearly this analysis needs to be refined to take account of local industrial, service, off-farm, or seasonal migratory employment (RePPProT 1989).

Such scientific approaches are all very well but, as was shown at the start of this section, the issue of carrying capacity is primarily a social one. The science can provide numbers and graphs, but the crux is rather what kinds of living standards people are prepared to accept, and how much are they prepared to change.

LAND USE EFFICIENCY AND FOOD SUFFICIENCY

Much of Java and Bali is blessed with wonderfully productive agricultural land, and the rice and other crops amply feed both their own populations and also those of other parts of Indonesia. Although Java is best known for its rice, as the population has increased, so more people have had to turn to dryland farming, and 70%-80% of all maize, soybean, groundnut, and tuber crops grown in Indonesia are produced there (Donner 1987). Very few areas outside Java and Bali have comparable existing or potential productivity, and with the country's growing population one would expect such a national asset to be guarded jealously.

In fact, many prime agricultural areas are being lost (fig. 2.11). Some

Figure 2.11.
Displaced farmers.

With permission of Kompas

parts of Java appear to have become vast building sites, and farmers are selling their soil to construction companies - voluntarily or because they have no choice. Some are changing from farmers to brick makers, or selling the holes that used to be productive farmland as uncontrolled landfill sites with unknown and unmonitored effects on water supply and future soil productivity. Indeed, about 0.3 million ha of the 1.2 million ha provided with technical irrigation on Java between 1969 and 1985 is no longer used for farming (Hardjono 1991a), and as industrial and urban developments continue apace in the fertile northern plains, so this loss is sure to increase. The waste of government money, much of it loan funds, does not bear thinking about. In response to this problem, a Presidential Instruction was issued in early 1994 stating that no more irrigated fields in the northern Java 'rice belt' should be converted to other uses. Not many weeks later it was announced that a further 600 ha of irrigated rice fields were to be taken for the latest extension to Jakarta's Soekarno-Hatta International Airport. To be fair, the irrigation in this area does not depend on high-investment infrastructure, but its loss flies in the face of the spirit of the Instruction.

Figure 2.12.
Popularizing sport. . .
And bringing sport to
the people.

With permission of Kompas

One of the newest and most contentious land-use issues is that of the establishment of golf courses (fig. 2.12). Not that there is anything inherently or ecologically wrong with golf or golfers, but it transpires that productive agricultural land (and land of conservation value) has been converted into recreation centres for the very rich. The nearly 100 courses on Java and Bali also provide 'cheap' golfing for expatriates and tourists. For example, there are 12 million golfers in Japan competing to use the 1,700 courses. As a result the membership fees are extremely high, and it is cheaper for Japanese people to fly to Java and Bali for the weekend than to play golf at home. Local communities generally benefit very little, and the number of local people employed is less than the number employed by agriculture. The courses also use considerable volumes of water, often in areas where water supply is an issue itself (p. 52), in order to keep the imported grasses green and healthy, and this same close-cropped grass receives heavy doses of fertilizer and pesticides which, it is claimed, leak into surrounding areas (Pearce 1993a). Indonesia is not the only country experiencing these problems because large numbers of golf courses are also being built in India, Malaysia, the Philippines, and Thailand. Golf courses are probably the most environmentally rapacious and socially divisive form of property development and tourist infrastructure.

The food self-sufficiency programme, initiated in 1968 was originally interpreted in a narrow sense as meaning a rice development programme, with the result that new reservoirs inundated large areas of productive land to serve irrigation schemes watering virtually a single variety of rice protected, temporarily, from pests by chemical pesticides. The ratio of consumption of rice to other staples climbed from 0.4 in the late 1960s to over 0.8 by the early 1980s (BPS 1985). Preoccupation with rice production has served to marginalize certain upland areas (Nibbering 1991), and this monophagy (Soemarwoto 1991b; Soemarwoto and Conway 1991) carries with it perpetual potential calamity: a climate extreme or a new resistant pest could cause enormous rice shortages. For example, the drought brought about by the El Niño climatic event in 1982/1983 necessitated larger imports of rice. If the same or some other calamity affected major rice areas, there could be serious socioeconomic and even political consequences (Soemarwoto 1991b). Failure to feed the people will as surely destroy the rural ecosystem as inappropriate agricultural models. There must be change and development but making the development sustainable is imperative.

Irrigated Rice

Enormous investment has been made in irrigation systems throughout Java to support and intensify rice production as part of the thrust to reach self-sufficiency. The short growing season varieties, the increased cropping intensity, upgrading or rehabilitation of certain areas, and improved water supplies have all contributed to the success of the self-sufficiency programme. The achievements are undeniable and laudatory, but the signs are that the battle to maintain this hard-won position may have to be fought hard, and there are those who claim the battle is already lost. The problems are that the demand for rice is increasing at a rate at least as fast as population growth, while the area of sawah (often of the very best quality) is decreasing by at least 4,000 ha each year as this land is used for other purposes. There are only about 100,000 ha (0.8% of Java's surface) of potential new sawah in Java, but this is not of the highest quality (RePPProT 1989). If self-sufficiency is to be sustained on Java and Bali, it will only come from improved operation and management, and from improved cooperation among water users. In fact, the single most important step along the path of increasing rice production on Java and maintaining self-sufficiency is the relatively easy and cheap provision of basic irrigation infrastructure to every possible area of rain-fed sawah in Java. In 1988 this totalled nearly 900,000 ha, only about 15% of which produced two crops each year. Most of this is in Central Java; there are only about 1,000 ha of rain-fed sawah in Bali. The other options available are limited but include the commitment of time and money to improving the operation and maintenance of existing irrigation schemes that are functioning below their full potential (RePPProT 1989). Previous improvement programmes have had limited effectiveness because more attention has been given to buildings than to reducing field-level wastage of water, or to improving control and management of water distribution. Those with authority over water resources must be highly motivated to optimize water supply to the farmer's fields. Another means of increasing rice production is to direct research and extension to reducing the 6% annual waste that occurs between harvest and consumption because of moulds and rats.

Lessons can perhaps be learned from elsewhere in Asia. In the developed countries of Asia, Japan and South Korea, rice yields are over 6 tons/ha (compared with the Indonesian average of 4.3), the people are abandoning rice as their staple food, and farmers' children are abandoning the muddy fields for the cities. Increasing affluence tends to turn people towards the consumption of more potatoes, noodles, bread, meat, and vegetables. All the Asian countries now produce surplus rice in good years - although buyers and 'borrowers' can be hard to find. The logical strategy to be followed is to grow more rice on less land at lower cost produced by fewer farmers earning more. In 1991 President Soeharto exhorted people to eat less rice and turn to other crops, and there are moves to encourage decentralized agricultural planning, and changes in land ownership. The once unthinkable strategy of planting on a rice/non-rice/non-rice system is now practised, and from a farmer's point of view this is acceptable since the planting of rice does not give particularly great returns, indeed barely different from the net returns on sweet potatoes or peanuts. If such a strategy succeeded, then water demand would be reduced such that the need for dams which are expensive in terms of social, environmental, as well as financial costs, would decline (Soemarwoto 1987b) and virtually all Java's river basins would have surplus water (Soemarwoto 1991b).

In this context the apparent 'abuse' of first-class farming land by developers seems less of an issue, but it is not certain to us that the pattern of development described is necessarily desirable or worth following given that the migration of people to the cities is encouraged by patterns of resource use that are grounded in old-style economics (p. 77).

Quality of Land

The production of food has much to do with the quality of land. It is salutary to consider that 15% (1.9 million ha) of Java is regarded as 'critical' or subject to serious erosion, and that this area is inhabited by some 12 million people. Soil erosion on Java has long been recognized as a problem. Even in 1866 soil erosion on Java was described as "a danger that creeps ever closer" (Holle in Donner 1987). In the 1870s all Javanese and European farmers were required to protect the soil on sloping fields with the result that large areas were terraced. In subsequent years, however, the area under cultivation increased so rapidly, and onto more and more marginal land by poorer and poorer farmers, that it was impossible to control land management.

The loss of soil on Java is enormous, averaging 6-12 tons/ha/year on volcanic soils and 20-60 on limestone soils, but on agricultural land the loss across Java averages 123 tons/ha/year (World Bank 1990a). While high losses are to be expected because of the quantity and intensity of the rainfall, and the large amounts of volcanic material raining down after eruptions, the situation is seriously aggravated by inappropriate management of the land (p. 134). The soil erosion on Java carries with it an annual cost of about US$400 million, 80% accounted for by reduced agricultural productivity. The remaining 20% represents the off-site costs of the siltation of reservoirs, ports and rivers, flooding and disruption in water supplies. This smaller portion has, however, received nearly two-thirds of the money spent on upland programmes, such as for afforestation, check dams, and gulley plugs. Clearly, the most attention needs to be paid to the on-site practices where good terraces can reduce soil loss ten-fold (World Bank 1990a). However, no matter how willing the farmers are, they realize that building terraces takes time, and time is money, so unless the conservation measures provide a net gain, they will not be adopted (Barbier 1988). The challenge, then, is to develop efficient and cost-effective upland models appropriate to the many and varied conditions, and one significant candidate is the planting of contour grasses that build up their own terraces (p. 144).

Major aid-supported watershed projects, such as those in the Citanduy and Solo basins, have reinstated bench terracing for upland soils and have reforested the steepest slopes. Socioeconomic conditions are sometimes so severe or complex, however, that the preferred environmental solution of removing people from the steepest land is not possible to execute. This is seen in the Puncak and Tengger regions of West and East Java respectively where crops such as cabbage, potato, carrot, and apples are grown on upslope raised beds, guaranteed to maximize erosion (fig. 2.13). In remote areas of Tengger the ingenuity and doggedness of the people who carry market goods to collecting points across the harsh Bromo 'sand sea' confound a city-based planner's predictions. Solutions depend on the degree to which the authorities are prepared to insist. These farmers have not traditionally cultivated the steepest slopes. Rather they are reacting to policies pressing for agricultural produce for export markets in situations that favour investors rather than local people. Everyone agrees that Indonesia needs to reduce its reliance on oil and gas exports, but finding crops that do not cause major environmental impacts is extremely difficult. Vegetable growing has the advantage that it is a labour-intensive form of agri-

Figure 2.13. Steep cultivated land on the northern Tengger slopes.
By A.J. Whitten

culture, and processing industries can further increase the labour demand. It will, however, remain unsustainable, unless (Hardjono 1991b):

- the land is terraced;
- the steepest land is taken out of cultivation and reforested;
- the reforestation programme is expanded and strengthened;
- the local people are involved in the care of these trees through incentive programmes;
- the removal of timber from the remaining forest is prohibited; and
- the quantities of pesticides used are reduced.

SECURING WATER QUANTITY AND QUALITY

Water is a vital resource, regarded by Muslims as the origin and posterity of all life and as a spiritual blessing by Balinese Hindus. Freshwater accounts

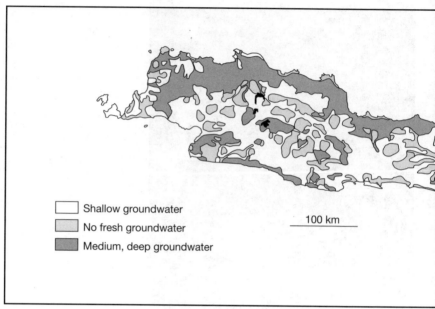

Figure 2.14. Groundwater potential of Java and Bali.
After RePPProT 1990

for only 2.6% of all water on Earth, and almost all of it is held in ice caps, glaciers, and as groundwater. Only 0.007% is in lakes, 0.005% in soil moisture, and 0.0001% in rivers, and some 110,000 km^3 of water is cycled through evaporation and precipitation each year. Water joins the atmosphere by evaporation from water bodies or by evapotranspiration from plants. Under appropriate conditions it falls as rain, and either runs directly into rivers and out to sea, recharges groundwater reservoirs, or is taken up by plant roots.

All aspects of water - quantity, quality, allocation, and storage - have been of increasing concern over the last couple of decades. Like other major issues, they are not discrete problems but are both cause and effect of other decisions and actions. Water is the ultimate cause of many of the environmental problems experienced by urban residents. The supply of water from the upstream areas is not generally a problem, but its storage and distribution to the urban and burgeoning suburban areas certainly are. New reservoirs (causing the loss of more productive land) or long-distance water transmission are possible, but without some firm decisions on limiting growth and/or consumption these will only be short-term solutions. The importance of managing water resources is highlighted in considering the competition for water between irrigated land, domestic urban users, and industry (Hadiwinoto and Clarke 1990).

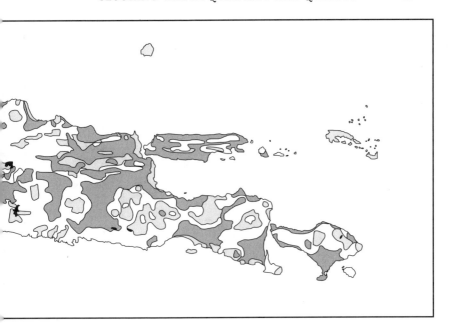

Water Resources

Because Java is long and narrow, most of the rivers are short, less than 50 km long. Thus, despite the adequate rainfall over most of the islands, the combination of small catchments and inappropriate vegetation cover have led to water shortages, particularly in relatively dry years (p. 119) (Trihadiningrum 1991). In addition, most if not all urban rivers are polluted, no matter what measure or standard one chooses. This complicates the work of the mains water company, endangers health, reduces the options on uses of the water (e.g., recreation, food fishing), damages the life of the coastal zone, and causes direct economic losses in areas under aquaculture.

Under natural conditions only a small proportion of the rain that falls will flow directly into rivers, the majority infiltrating into the ground (fig. 2.14). As more land is built upon, particularly in the important recharge areas in the hills, so more rain falls on impervious man-made surfaces, and is channelled along drains to flow directly into rivers (Hadiwinoto and Clarke 1990). During periods of heavy rain this causes floods, particularly in urban areas, and these are in stark contrast to the low flows experienced during the dry season. The fact that less water flows through the ground means that the rate at which the groundwater is replenished is reduced, and this is aggravated by the extraction of water from ever deeper wells as major users compete with each other for the diminishing resource. Reports suggest that the groundwater surface in the Jakarta area is falling

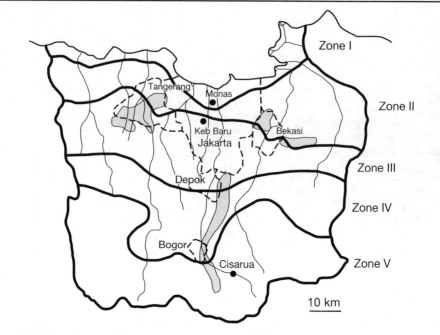

ZONE I: Avoid urban development
Low lying coastal strip.
Flat, thus bad drainage.
Subject to flooding.
Agriculture suited to fish ponds.
Groundwater saline and undrinkable.
Poor soil-bearing capacity for building.
This area encroaching on Zone II as saline
intrusion increases in urban areas.

**ZONE II: Agriculture intensification - limited
urban development**
Low lying plains.
Flat, thus bad drainage.
Subject to flooding.
Excellent for rice growing especially if irrigated.
Groundwater fresh but easily polluted.
Poor soil-bearing capacity.

**ZONE III: Major urban development -
agricultural intensification**
Higher lands rising from the coastal plains.
Reasonable gradient thus good natural
drainage.

Groundwater fresh and leaching soils limit
pollution.
Poorer agriculture.
Reasonable soil-bearing capacity.

**ZONE IV: Limited urban development -
agriculture intensification**
Steeper sloping zone.
Good natural drainage.
No flooding.
Limited groundwater and no deep aquifers.
Reasonable agriculture due to more rainfall.
Reasonable soil bearing capacity.

**ZONE V: Upland forest plantations - irriga-
tion and conservation, avoid agricultural
intensification**
Steep mountainous zone.
Rapid runoff but limited by vegetation.
Natural forest areas.
Agriculture limited to complex terrace con-
struction.
Subject to rapid erosion if forests are
cleared.

Figure 2.15. The five advizable development zones of the 'Jabotabek' area (Jakarta, Bogor, Tangerang, Bekasi) showing the approximate positions of the national monument (Monas), the residential area of Kebayoran Baru, and Cisarua, the main industrial areas (shaded), and municipality boundaries.

After World Bank 1990a

at some 0.5 m each month. Floods are worsened by the large quantities of urban waste that finds its way into the rivers obstructing their flow. An example of these problems is found in the Brantas watershed where the water is carefully controlled, with the high wet season flows directed mainly through the Porong delta to avoid flooding in Surabaya, and in the dry season, the Porong is deliberately blocked in order to ensure that enough flows through Surabaya to flush the open sewer that is the R. Surabaya. In 1987 the rubbish and other waste, as well as deposited sediment, was so bad in the R. Surabaya that five million m^3 of water had to be released over seven hours (200 m^3 per second) from the Sutami Dam in Karangkates south of Malang in order to flush the river clean and deepen the channel. This has been repeated subsequently, but at best it represents only a partial and temporary solution.

Groundwater is also subject to pollution, with seepage from waste disposal sites and intrusions of salt water from the sea reducing its quality. Groundwater regimes below the northern alluvial plains are complex, reflecting geology, the seaward advance of the coast, and the excessive exploitation of the resource by the population and industries of Jakarta and other major cities. Since water is such a critical resource, its quality and quantity should have a major influence on development activities (fig. 2.15). The situation in Jakarta is acute and serves to illustrate what could happen elsewhere. The recharge rate of the Jakarta aquifer is well below the extraction rates, and so seawater has been intruding rapidly from the coast, such that concern is being expressed about the corrosive effects of this saline water on the reinforced concrete foundations of the tall business buildings along Jakarta's 'money mile' of Jalan Thamrin and Jalan Sudirman. In addition, many urban dwellers without access to piped water must buy drinking water in cans at high rates (fig. 2.16). This situation refers to the 'unconfined' groundwater that lies 20-60 m below most of Jakarta. Lower, some 100-150 m below the streets, is a confined freshwater aquifer which gradually rises towards the surface in the hills, being about 50 m deep around Depok and Cisalak.

There are numerous reservoirs on Java, but the rationale of building the many so-called multipurpose reservoirs - serving irrigation, flood control, hydropower generation, and water supply for industrial and domestic users - appears to have been to promote national stability by achieving self-sufficiency in rice, rather than being based on an economic cost-benefit analysis. In fact, it is said that there is no truly multipurpose reservoir on Java (World Bank 1990a). Economics apart, the potential of the reservoirs will not be wholly realized. One of the main reasons is the accumulation of eroded soil from the upper catchment area. In addition to filling the reservoirs, sediment settles in the rivers which require further embankment (which increases the threat of floods) or dredging (which is expensive).

Figure 2.16.
Water, Waaaater.
With permission of Kompas

When floods flow over raised banks, the flood waters are unable to drain away because they are lower than the drainage channels or rivers. The manifold problems with dams and reservoirs are such that any further development is hard to justify. The World Bank (1990a) believes that dam building in certain areas of Java may be justified, but all the costs and benefits would need to be calculated with great care.

Water use on Bali appears to be extremely efficient at least in part because the *subak* system (p. 567) ensures that each adult male has a role and some responsibility for the operation of the system on which the communities depend.

Water Use

The allocation of water resources is becoming an increasingly critical issue given that use already exceeds supply in some areas (table 2.5), with industries in Surabaya having been known to close for lack of adequate water. In

Java, irrigated agriculture uses nearly 77% of the river water in a once-in-five-year dry year, whereas municipal uses account for only 2% of this. Much of the surface water available for domestic and industrial use is of low quality, but there is intense competition to gain access to it. The relative safety and convenience of the groundwater resources and the ease of access has long made these the preferred source for municipal use, such that wells supply about 60% of the water for rural domestic use, and most of the water for city dwellers, particularly where surface waters are grossly polluted.

The integration of water management activities has long been an issue, with the most frequent recommendation being that each river basin should have a central authority to integrate and coordinate the many competing uses of water resources, and that each of the major basins or groups of small basins (e.g., in northern West Java) should have its own Water Resources Planning Authority to convene and supervise consumer groups (farmers, PLN, industries, and drinking water agencies) to ensure they use water effectively and neither pollute, nor waste, the limited supplies (RePP-ProT 1989; World Bank 1990a). The Government Regulation No. 5 of 1990 provided the framework for this, and the water affairs of the Brantas River are now the responsibility of a state-owned company. Many of Java and Bali's problems of water supply and use can in fact be met by improving the efficiency of water use through a better system of pricing and metering, both for surface and for groundwater. Unfortunately even for this there is tremendous institutional confusion and overlap in responsibilities and authority, and the position is unlikely to improve unless this problem is met head on.

Table 2.5. Annual water supply and use on Java.

	billion m³
Total average rainfall	352
Total average surface flow	175
Total average usable (divertable) surface flow	126
Once-in-five-year total surface flow	78
Volume impounded by dams	7
Total average surface water used for agriculture	59.4
Total average surface water used in urban and rural areas	1.3
Total groundwater extraction	9
Total projected average surface water used for agriculture in 2010	63.7
Total projected average surface water used in urban and rural areas in 2010	3.3
Total projected groundwater extraction in 2010	?10

Source: World Bank (1990a)

Water Quality

The generally low water quality of the rivers, particularly during the dry season, has impacts on ecology, human health, and industrial growth. The most pervasive pollutant is faecal coliform bacteria from human excreta (fig. 2.17), and concentrations are commonly three orders of magnitude greater than the recommended maximum. Organic pollution is indicated by very high COD and BOD levels, particularly near major urban and industrial areas (fig. 2.18). The cost to the economy of this pollution in terms of sickness, reduced industrial output, fisheries losses, and other factors has not been calculated, but it could probably be measured in billions of dollars. Anti-pollution legislation is finally being applied, but the massive problems of solid waste and sewerage hang over Jakarta like a descending sword ready to cut its lifelines.

The lowering of water quality from industrial effluent is a much more localized but common problem along the northern alluvial plains, but because of the toxic nature of some of the pollutants, it can be extremely important. Major strides have been made in the control of industrial water pollution since the start of this decade, primarily through the Prokasih initiative. Prokasih was designed to be a 'crash' campaign to last from 1990 to 1994 with the dual objectives of pollution abatement, particularly where the pollution is caused by industrial processes, and institution building, concentrating on specific rivers, most of them in Java. The abatement was to be achieved by identifying pollution sources, meeting the parties responsible, determining acceptable standards, setting agreed targets and deadlines, monitoring, evaluating, imposing fines and sanctions, and enhancing the legal apparatus. Progress has been slow but steady, and there have been some well-publicized cases of reluctant companies being pressured to follow the letter and the spirit of the programme. Hazardous wastes were chosen as the Prokasih focus because their point sources could be identified, the perpetrators of the pollution were able to pay for abatement, and toxic pollutants have dangerous health effects. The latest move is to adopt a rating system for industries which will take into account the levels of treatment, recycling, and waste reduction. It must be stressed, however, that it is domestic sewage which is the dominant source of water pollution in both rural and urban areas accounting, for example, for 96.5% of the organic load in the Ciliwung River (Bansgrove 1991), and the source of many diseases. Tackling this problem is probably more difficult even than preventing industrial pollution.

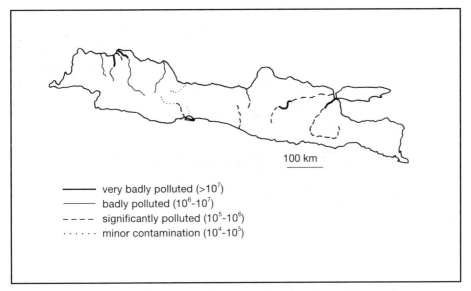

Figure 2.17. Average dry season coliform numbers in selected Javan rivers.
After World Bank 1990a

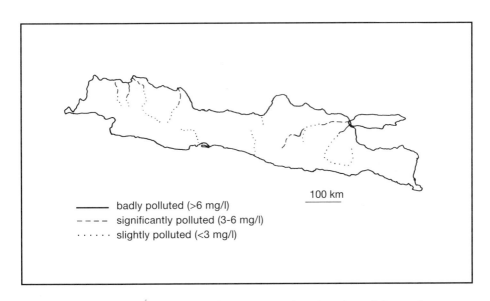

Figure 2.18. Average dry season BOD concentrations in selected Javan rivers.
After World Bank 1990a

Figure 2.19. Industrial and domestic wastes introduce large numbers and large quantities of chemicals into the environment.
With permission of Kompas

UBIQUITOUS CHEMICALS

Virtually all human activities contribute in some way to the increasing numbers and quantities of chemicals entering the environment (fig. 2.19). There are so many types of pollutants that it is impossible to detail all their environmental effects, particularly as some of the chemicals react with others to produce compounds which are more toxic than the originals. Chemicals with proven dangerous effects, termed hazardous wastes, are indiscriminately dumped into rivers, roadside ditches, and left in rotting containers on public dumps, all of which threaten human and environmental health. The necessary infrastructure, regulations, human resources, enforcement, and information are gradually being established, particularly in the worst-affected industrial areas, spearheaded by the Board for Environmental Impact Control. Encouragement, in part through economic incentives, is given for the adoption of cleaner technologies with fewer and less wastes, and for the recycling of wastes into useful or at least environmentally benign materials.

Of less immediate threat, but of largely unknown long-term effect, are the thousands of toxic manufactured and natural chemicals required in industrial processes. These are generally used in small amounts, but there is a disturbing lack of information concerning the life and environmental effects of most of them. The information that is available is often anecdotal, statistically insignificant, or of limited application to 'real-world' conditions. Some toxic chemicals are well known to accumulate in body tissue with the result that even small quantities in the environment can be amplified along a food chain. Unfortunately, awareness concerning the existence and dangers of toxic chemicals is not widespread, and it is difficult to convince decision makers of the need to act when direct and dramatic links with human deaths cannot be made. Pollution effects tend to be slow to appear and may be hidden by causing a predisposition to a condition or disease which would be recognized as an ultimate cause of death. Other chemicals or lower doses can cause sub-lethal but debilitating effects. Deeper understanding of the ecological role and fate of these chemicals is an urgent priority, but useful books for an initial understanding are Harte (1991), Soetikno (1992), and Ekha (1991).

Pesticides are a major group of toxic chemicals, and concern over their effects is twofold: first, the development of resistance, particularly where the insecticide has been sprayed at low dosages, and when a vector becomes resistant even before spraying is directed against it such as the early spraying of rice with organochlorine pesticides such as DDT against rice pests which produced resistance in malarial mosquitoes (Oomen et al. 1990). Although DDT is the best known of the chemicals causing resistance in pests, there is no biological reason why insects should not acquire resistance against any chemical. The long-term effects of organochlorine pesticides have been known for over 30 years (Carson 1962, 1990; Abdullah 1991). While they remain in use, even in restricted and regulated quantities, there is an acute need for widespread monitoring of their residues in the environment, in foodstuffs, and in organisms, including humans. Unfortunately, studies on residue concentrations from Indonesia are scattered, often not comparable, and rarely repeated to determine trends. The second concern is that there are health risks for the farmers who apply the chemicals to their crops when they do not take all the recommended but somewhat onerous precautions.

One of the most frightening compounds, especially in busy urban areas, is lead. It enters the environment from the exhaust fumes of engines running on petrol to which it is added to improve performance. Lead is not essential to engine operation, however, and a slightly higher-priced refining process produces a petrol which does not need lead. The reason for concern is that lead is a cumulative poison, and breathing lead-tainted air can result in lead being absorbed and retained in the body. Lead from engine exhausts is found not just in the air we breathe but also on vegetables growing along highways and roads (Djuangsih and Soemarwoto 1985). It is obvious that city residents will be closer to the threshold at which poisoning effects appear than those in rural areas, and children have a threshold only about one-half that of adults. Although lead poisoning can kill, in milder but less identifiable cases, it leads to mental retardation. Even levels below the poison threshold seem to cause subtle changes in learning ability, loss of appetite, and lack of interest in play. The connection between lead and children's health was recognized over twenty years ago, and in 1974 the USA was the first country to introduce lead-free petrol. Most modern cars can be converted very easily to run on lead-free petrol, and one wonders how much the potential of urban youth will have been cramped before lead-free petrol eventually starts to be introduced in Indonesia in mid-1995. Lead can also found in very high concentrations in certain paints, particularly primers. This can cause serious problems when children chew painted wooden toys and other articles. Paints produced by the leading companies in Indonesia no longer contain lead as a drying agent or as a pigment, but the use of lead in cheaper paints still appears to be widespread. Despite the global awareness of the problem there is no Indonesian regulation which bans the use of lead.

TOURISM

Tourism is a major industry and a significant source of foreign exchange earnings - some 14% of total non-oil/gas earnings. Tourism falls into a number of categories, such as cultural, beach, and nature, or combinations of these, each with markedly different effects. The effects are not just one-way, however, because the total environmental condition of an area plays an important role in attracting tourists, as well as deterring them (Soemarwoto 1990a). Despite the fact that Java has an enormous number of cultural, physical, and biological attractions, mass foreign tourism is attracted to relatively few of them, largely because there has been inadequate investment in the necessary infrastructure. There is, however, a high volume of domestic tourism on Java which, because of increasing wealth, is of significant value to the economy. Tourism on Bali, on the other hand, caters mainly for foreign tourists, and it is by far the most popular tourist destination in Indonesia. The scale and pace of tourism there has raised many concerns (Guppy 1990; Manuaba et al. 1990), and it has been subject to a major review and analysis (Hassall and Associates 1992).

Bali

In February 1597 a Dutch ship moored off Kuta, and three men were sent ashore to scout for the supplies needed for their return trip to the Netherlands. Earlier, in Banten, they had been warned that they should expect a hostile reception, and so they were surprised by the very friendly welcome and the generous hospitality they received. The people and environment of Bali seem to have made such an intense impression on two of the crew that they stayed behind when their ship left two weeks later, and one of these spent the rest of his life there (Covarrubias 1937; Boon 1977; Francillon 1979; Agung 1991).

Since then Bali has characterized 'paradise' for many Westerners seeking traditional societies, rich culture, sumptuous ceremonies, living and vibrant arts, a beautiful setting, sun, and sea. Organized tourism began in 1908 and increased slowly as illustrated books were published (e.g., Nieuwenkamp 1910; Krause and With 1922), shipping lines included Bali in their routes, and artists, film stars, and writers set up house, some for many years, fuelling the idea of Bali's magic. Several hundred foreign tourists visited Bali annually in the late 1920s, several thousand annually during the 1930s, and nearly 30,000 annually in the late 1960s. The first Indonesian Five-Year Plan paved the way for a comprehensive study of tourism development on Bali (Francillon 1974, 1990), and it was proposed that a new tourism focus at Nusa Dua, on the dry limestone *Bukit* in the south be constructed. During the 1980s arrivals on Bali directly from overseas increased at an average of 13%/year, compared with 4.6% for world tourism, and in the same period there were large increases in the numbers of Europeans, who displaced Australians as the major group of visitors, and ASEAN members who represent a small but rapidly-growing part of the market. In 1990 about 1.9 million visitors arrived in Bali, only about 60% of them foreigners, indicating that it is now by no means foreigners alone who find Bali a desirable tourist destination.

Sixty years ago people were writing that the manifold beauties of Bali were poised for extinction under the influence of tourism (Covarrubias 1937). That is still being written. While there has been enormous change in infrastructure and accessibility, Bali's culture has proved to be remarkably robust, and the interest of visitors has provided an economic outlet for the renowned artistic abilities of the people, and fuelled the sense of pride of the Balinese. In many ways tourism is already part of the Balinese culture and economy, and its influence cannot be stopped. Even if it were, there would still be the effects of mass media, homogenization of culture (p. 342), national education, growing mobility, and centrally-planned development to contend with (Francillon 1979; Picard 1986, 1990; Wall 1995). Indeed, the worst impacts are being seen not as cultural pollution caused by the tourists themselves, but social desecration and destruction as powerful conglomerates ride rough-shod over provincial plans and local aspirations. Tourist-related demand for land has been judged the most important factor affecting agricultural and land-use planning on Bali, and the loss of rice fields amounts to at least 1,000 ha/year thereby reducing the island's ability to be self-sufficient in rice. With a human population growth of 1.4%, likely to decline to 1.1% in 20 years, the total population will be 3.8 million by 2010 - 1.4 times the population in 1990. If built-up areas and settlements increase by the same proportion, then an area equivalent to one quarter of current irrigated rice fields will be lost. Already in some areas the views of carefully-tended rice terraces are obstructed by ribbon developments of small enterprises selling items to tourists attracted to the views, and it has been suggested that no freehold land purchases be allowed for 50 years to ensure continued equity to rural Balinese landowners (Hassall and Associates 1992). Another possibility is to allow the displaced farmers and other villagers to become shareholders.

Most tourists visiting Bali spend at least some of their time on the beaches. While coral and sand mining have had important impacts, the tourist industry has had dramatic effects on the beaches. Developments of hotels and other facilities have ignored the fact that beaches and the sand ridges behind them are dynamic systems. If they are built upon, then the natural processes of sand deposition and removal along the beach are disturbed and the width of the beach decreases. This provokes landowners to build unsightly groynes into the sea to retain sand for their waterfront, thereby aggravating the problem for their neighbours who are forced to build groynes of their own. This process has been understood for some years, and a provincial regulation prohibiting developments within 100 m of the shore was issued in 1989. Despite the obvious and proven results of not heeding such conditions, such as the disfigurement of parts of Candi Dasa, Nusa Dua, and Sanur resorts, the regulation is not always enforced. The best solution is to remove the groynes and offending developments and allow the beach to reach a new natural equilibrium.

Eco-tourism is being developed on Bali for visitors who wish to appreciate some of the wild landscapes and the wildlife. Snorkelling and scuba diving are experiencing major growth, partly as a result of publicity around the world and beautifully-illustrated guides (Muller 1992) (p. 345). Whitewater rafting through the steep canyons of southern Bali is one of the most popular activities, but walks in the forest appear to have less appeal. Wildlife such as the holy eels at Toya Bubuh, monkeys at various temples, and the bats at Goa Lawah temple and Alas Kedaton attract some interest, but not a significant proportion of the market. At least this way the attractive features are less threatened. The most exciting wildlife species on Bali is without doubt the stunning Bali starling (p. 227), but it is so rare that only a very few specialist groups are allowed to see it.

There must be a limit to tourism on Bali, and if it is not set by the planners, then it will be set by some critical limiting resource such as water, space on the roads, tolerance of the people, or by whether it is perceived by tourists as maintaining its advertized attractiveness. As major developments continue to be approved, one wonders whether Bali could develop along the lines of Oahu in Hawai'i, where the original inhabitants are granted 'reservations' in which to live, and many of the profits flow to outside the island, particularly to Japan.

LOSS OF BIOLOGICAL DIVERSITY

Biological diversity is used in different ways to mean all or any one of:
- ecosystem variety;
- the variety of species;
- the genetic variation amongst members of a species or a population; and
- human cultural variety as manifested in language, beliefs, land use, arts, food preferences, and so forth.

These four meanings are closely interrelated since, if ecosystem variety is reduced, then the number of species is likely to be reduced, and species' populations decline leading to genetic erosion. Different groups of humans selectively increase and decrease diversity to different degrees. For this section, primary attention is given to species diversity. Genetic erosion and preservation of genetic variation is discussed elsewhere (pgs. 295, 704).

It should be noted that diversity is not necessarily the same as abundance or richness. Thus a community comprising ten species from the same family is less diverse than one comprising ten species each from a diffent family. Also, a landscape is more diverse if it contains swamp, lakes, mountains, and savanna, than if it contains a number of types of lowland forest.

Something between 1.4 -1.8 million valid species of animals and plants have been described (Stork 1994), but this is far from being the total number of species actually living. Reasoned estimates for a global total vary from three to 30 million, depending on the numbers not yet known to science, with 50%-90% of them confined to tropical forests. The numbers game requires a great deal of intelligent guesswork, approximation, and extrapolation, but even the more conservative figures reveal an enormous biological wealth as yet undescribed, let alone assessed in utilitarian terms, or understood in terms of ecology, behaviour, or abundance (May 1992). The birds and mammals are known quite well – although the rats, bats, and shrews probably still hold some surprises – but the less 'attractive' a group is, the fewer collections have been made, the fewer specialists there are likely to be, and the less well-known they are (p. 204).

The loss of biological diversity is a global phenomenon, and the Global Biodiversity Strategy lists six fundamental causes:
- the unsustainable high rate of human population growth and natural resource consumption;
- the steadily narrowing spectrum of traded products from agriculture, forestry, and fisheries;

- economic systems and policies that fail to value the environment and its resources;
- inequity in the ownership, management, and flow of benefits from both the use and conservation of biological resources;
- deficiencies in knowledge and its application; and
- legal and institutional systems which promote unsustainable exploitation (WRI/IUCN/UNEP 1992).

Although extinction is an entirely natural process (p. 699), the rate of extinctions is probably higher now than it has ever been. At the global level 83 species of mammal, 113 species of birds, and 384 species of plants are known to have become extinct since 1600 (Reid and Miller 1989). Although the rate is very high, available data from around the world do not indicate that there is a massive 'spasm' of daily species extinctions occurring at present, nor that it is necessarily very imminent (Whitmore and Sayer 1992). It should be stressed, however, that the base data are often inadequate, and there are few informed people who would be bold enough to say that the huge losses of tropical forest and other natural ecosystems will not inevitably lead to numerous extinctions in the long term. Indeed, most people are aware of species which have been lost from former haunts, and this is simply the thin edge of the wedge of extinction. Most of these species destined for extinction will be lost step-by-step by virtue of severely-reduced habitat and small populations. Some of these species may currently have one or more breeding populations, but they are little more than 'living dead' (p. 701) with no prospects of long-term survival.

There are a variety of ways of measuring diversity, and new approaches have been devised which incorporate taxonomic and functional ecological measures and which may be useful in environmental assessments (Cousins 1991). To some extent these exercises are merely theoretical, and spending time discussing them is akin to fiddling while Rome burns. Enough is already known about biological diversity to know the urgent cases for conservation (Ehrlich 1993), even if they have yet to be understood at all levels of society and government.

Reasons for Alarm

Biological diversity is declining: this fact is incontestable, but why should decision makers be concerned? A decision maker might acknowledge that Java and Bali have already lost species primarily as a result of habitat loss, but she or he observes that life continues. Take Central Java as an example. It is Indonesia's least-forested province, supports an average human density of $833/km^2$, is one of Indonesia's largest per capita producers of rice, and has some 60 universities. It is unlikely that any of these facts would change markedly if the remaining pockets of natural forest on the mainland of the province, such as on Mt. Slamet, were felled to make way for sustainable agriculture or forestry. A cynic might ask what the point of biological diversity was when its great loss in Central Java seems to have had such little effect. The point is, of course, that Central Java depends for many of its needs on biological diversity from elsewhere, importing natural resources from outside the province, much of which is consumed at an unsustainable rate. In addition, recent research related to the national encouragement of integrated pest management techniques indicates that the greater the ecosystem variety around rice growing areas, the greater and more varied are the populations or predators and parasitoids that help control rice pests. If Central Java is to have truly sustainable agriculture, then it should look to increasing ecosystem diversity.

Alarmist early literature on conservation made one believe that if any species went extinct, the human race would follow through as yet unknown ecological linkages. This is patently untrue since extinctions have always been a fact of life. More accurate is the analogy of a man popping rivets out of an aircraft wing just before it was due to take off. An alarmed passenger expressed his concern but was told that there was no need to worry because the man had popped many other rivets from the wing before, and nothing serious had yet happened (Ehrlich and Ehrlich 1982). This illustrates that not all species are equal, and the loss of some is of more significance than of others (Lawton and Brown 1993). Many ecosystem processes appear to have considerable redundancy built in to them, but species that appear to be 'redundant' at one level of analysis may take on important roles in succession, or be important in fluctuating environments. The truth is we have little idea of the impact of any given extinction, but we do know that we cannot recreate a species which has been forced over the brink, and that the extinction of certain 'keystone' species (Bond 1993; Paine 1995) will lead to a cascade of other extinctions by virtue of their position and importance in a food web. Examples of such 'keystone' species would include pollinators of economic crops, species which provide food during seasonal periods of food scarcity, predators on pest species, and organisms playing roles in decomposition. The insects are a significant group in this context and yet much overlooked. They are food for all sorts of carnivores which are generally very particular about which insect species is eaten; they are killers of seeds and thus have a major influence on vegetation composition; they are the most important pollinators, and certain species are often totally mutually dependent on a particular plant species; the fruits arising from the pollinators are major sources of food for a wide range of larger animals; they harvest large quantities of biomass as leaves or decaying matter, and without them decomposition processes would be quite different (Janzen 1987b).

Other species have a utilitarian value to humans, and the natural ecosystems of Indonesia are enormous, untapped reservoirs of plants for timber, cellulose, fibres, fruits, tubers, green manure, drugs and other chemicals

(Heyne 1913, 1987; Burkill 1966; Sastrapradja and Kartawinata 1975; Whitmore 1976; Sastrapradja 1977, 1989; Kartawinata 1990). Many species may not be of immediate significant use, but one of the greatest values of biological diversity is probably the opportunities it provides for humanity to adapt to local and wider changes, for example by allowing crops to be tailored to new climatic conditions.

Genetic riches are not just present in terms of species, but also as varieties. For example, a staggering 234 varieties of bananas have been recognized in the Yogyakarta region alone (Subroto 1992). Such genetic diversity of agricultural crops faces two major problems: first, it is hard for national bodies to control, and second, it is not immediately convertible into cash. It is difficult for commercial bodies to find and exploit this diversity when it is maintained on small and scattered farms, but that does not necessarily argue the case for expensive gene banks, because the farmers are the ones who know the varieties best. In any case, it would be an enormous task to collate, catalogue, and disperse all the information and to maintain a significant quantity of seeds. Thus, what is required is policies that promote informal innovation rather than wholesale adoption of a few varieties. In this way the farmers make the most of what they have in great abundance: local knowledge and local genetic resources (Reid 1992).

Decision makers generally look to balance sheets for guidance, and much has been made of the economic benefits of biological diversity, and of the economic incentives and disincentives that can be devised for looking after it (e.g., McNeely 1988, 1992). The premise is that decision makers will only be prepared to stem the loss of biological diversity if it can be shown to work in economic terms. This has led to initiatives and plans for biological diversity prospecting, that is, the exploitation of biological diversity for commercially valuable genetic and biochemical resources (Reid et al. 1993). This is very necessary and laudable, but the expectations by some of the scale of profits and immediate financial benefits which can be realized by such exercises are probably unrealistic. The real returns may not even offset the cost of proper management of conservation areas. In addition, it is not yet clear how local people with indigenous knowledge will be guaranteed their share of the benefits of any success.

Fortunately, people value the natural world for reasons other than purely economic — recreational, religious, nationalistic, psychological, and ethical (Orr 1991)— and to suggest that Indonesia and other tropical countries should seek economic reasons for conservation is perhaps patronizing, given that this has not been the major reasoning elsewhere (Robinson 1990). Ethical values of biological diversity are founded on the notion that life forms other than our own have intrinsic value and deserve protection from the destruction wrought by humans. This protection is due simply because longstanding existence in Nature carries with it the unimpeachable right to continued existence until natural extinction processes become dominant.

It is relevant here to mention the results of an international survey of scientists, conservationists, teachers, and others with an interest in tropical rain forests conducted a few years ago (Murray 1990). Each respondent was asked to review a list of arguments used in various publications as reasons to support the conservation of rain forests. The list covered a variety of arguments related to aesthetic values, biological uniqueness, economics, education and research, moral responsibility, and the regulation of the physical environment. Respondents were to indicate the importance of each argument according to both their own personal beliefs and to their perceived usefulness in convincing others. In brief, it appeared that arguments concerned with economics and the uses of rain forests were rated higher when used to persuade others than when defining personal beliefs. Personal beliefs tended to be based on biological uniqueness, moral responsibility, and regulation of the physical environment. Arguments believed to be useful in convincing others were those emphasizing economic benefits and regulating the physical environment. The lesson to be learned is that while it might be felt that the stress on economics is simply realistic when talking with decision makers, it may be that the best argument will be the one that is truly believed and about which the apologist can be genuinely enthusiastic.

Microorganisms and Invertebrates

Microorganisms and invertebrates are among the most numerous and most diverse of living groups (table 2.6). Almost all biological processes are directly or indirectly governed by microorganisms, but this is generally overlooked, not least because microbiology is a relatively new science, and to identify and evaluate the organisms requires laboratory equipment. Yet microorganisms and invertebrates have successfully colonized every possible habitat. Their presence and activity is essential to the health and functioning of ecosystems and the whole environment through, for example, mineralization and recycling of important elements, photosynthesis, proper conditioning and fertility of the soil, pest and vector control, and detoxification of pollutants (Altieri 1991; Olembo 1991; Waage 1991). They are also important in the direct service of people in the synthesis of foods (such as tempe, tahu, and beer), antibiotics and other medicines, pest control, and chemicals. Indeed, through the burgeoning science of genetic engineering, they are potentially of major importance in three major environmental fields: energy production, food production, and waste-free technologies (Beringer et al. 1991; Nisbet and Fox 1991; Olembo 1991).

Given the extreme importance of these small organisms, it is worrying

Table 2.6. Described and estimated total numbers of species of major types of organisms.

	Described species	Estimated species
Macroorganisms		
Mammals	4,000	4,000
Birds	9,000	9,100
Reptiles and amphibians	9,000	9,500
Ferns	10,000	>10,000
Mosses and liverworts	17,000	>17,000
Fishes	19,000	21,000
Monocot plants	50,000	>50,000
Dicot plants	170,000	>170,000
Microorganisms and invertebrates		
Bacteria	5,000	30,000
Algae	40,000	60,000
Protozoa, etc.	30,000	100,000
Viruses	5,000	130,000
Nematodes	15,000	500,000
Fungi	69,000	1,500,000
Insects	800,000	2,000,000 - 80,000,000 or 5,000,000 - 10,000,000

Source: Hawksworth and Mound (1991)

that so little work is being conducted on making an inventory of the 'species', let alone assessing their potential benefits. Furthermore, the biological diversity of these organisms must surely be diminishing because as soil is used more intensively, so the number of species decreases (Stewart 1991). Since the upper layers of soil contain very high numbers and diversity of microorganisms and invertebrates, so soil erosion is not just about physical loss, but also about the loss of the very organisms which cause the production of fertile soil by forming stable aggregates, mixing in the organic surface matter, and making burrows which aid porosity and infiltration, thereby reducing runoff (Lal 1991; Lee 1991).

CONSERVATION AREAS AND LOCAL PEOPLE

Conservation areas are the most important element in conserving biological diversity. The natural ecosystems in conservation areas on Java and Bali are more strictly protected than anywhere else in Indonesia, and each area has fixed boundaries, many of them marked in the field. Large-scale conversion of forest land to arable uses is a relatively small problem (except for some lowland forest fragments in West Java (e.g., p. 793) requiring small boundary corrections to the conservation areas. This is the good news. The bad news is that not all the necessary conservation areas have been gazetted, and almost all forests, regardless of their status, are used for a variety of subsistence and other needs such as the gathering of timber, food, fodder, fuelwood, fruits, and traditional medicine. There is nothing inherently wrong with these activities, but their scale causes attrition and degradation of the natural forests (Smiet 1990b). In time, the natural forests become so degraded that the areas become dominated by grasses or shrubs, and neither conservation nor the people have won. Similar problems exist with other natural ecosystems.

Reasons for the Problems

This exploitation of forests is primarily by people from low socio-economic groups who have few if any alternative livelihoods, and to whom the legality of an activity is not the over-riding consideration (Nibbering 1989; Smiet 1990b). The socio-economic status of the people taking forest products such as bamboo, black leaf fibres of the wild sugar palm *Arenga pinnata*, yams, and rattan, from the north of Meru Betiri was studied and it was found that according to the criteria of Sajogyo (1978) they were very poor, and they worked hard and long in the forest in order to get what little money they could earn. They knew little of any benefits forest can provide beyond the products they were extracting, and the majority were uneducated. Because of this they were also very hard to reach in terms of the communication of ideas, restrictions, and extension (Syafi'i and Kasim 1981). The authors of the report suggested that in order to limit the damage to the National Park, the existing regulations had to be made known, even to these groups of people, and enforced. This alone would not be adequate or reasonable, and the key must be rather to reduce demand for the bamboo used in vast quantities for drying racks by a local tobacco company, and to find alternative sources of income for the collectors. Even then there would be demand for bamboo, and it was suggested that some of the unmanaged, deteriorating teak forest in the buffer zone be converted to bamboo. Bamboo could also be planted in critical land outside the reserve under the supervision of Perhutani (Syafi'i and Kasim 1981). In the decade since the report was issued there has been a reduction, albeit not total, in the use of bamboo by the tobacco company, but alternative ways for the people to generate income have not been introduced, and recommendations to Perhutani to plant four adjacent areas with economic species of bamboo are still being considered (T. Wibowo pers. comm.).

Conservation problems are worsened where the local communities have no role in the planning and management of natural common resources. The resources in question generally have no clear owner (other than the government) which leads to a lack of concern over their future. Where local communities are given responsibility to protect a resource, they may impose sanctions such as temporary ostracism on members of a village body who have contravened local regulations on the use of that resource. This was the case on Bali during the Dutch colonial period and was considered to have been very effective (A.A.G. Agung pers. comm.). These same resources have changed into perceived open access resources, such as 'protection forests' which are progressively degraded. With forest resources outside conservation forests severely limited, it is no wonder that people look within the boundaries to the relative riches of the conservation areas.

Conventional approaches to the management of conservation areas appear to be unable to balance the needs to conserve biological diversity and the needs of local people. As a result, new initiatives, termed integrated conservation development plans (ICDPs), have arisen to integrate conservation and rural development projects. This approach is discussed elsewhere (p. 760), and it is shown that, like so many elements in the complex equation for sustainable development, ICDPs depend as much on political motivation and commitment as on funding.

Adat Law - Conservation Friend or Foe

It is generally recognized that enforcement of government regulations relating to conservation areas is weak. The necessary enforcement is characteristic of centralized state control and sits uncomfortably with the traditional exercise of community authority exemplified in *musyawarah*. In addition, rural people do not readily accept restrictions to their daily life, particularly with regard to harvesting products from 'free land' even if the state regards such areas as its own. It is commonly the case that when someone is apprehended for some form of trespass on this land, the defence presented is based on adat rights. These rights are part of a complex system of customary rules of behaviour that evolved in and were accepted by local communities to regulate daily life in such a way that the community survived and harmonious interactions with the environment ensued. It is generally true that in former times rural people did live in harmony with their environment, but if people living adjacent to conservation areas today were allowed to exercise the same rights as then, disaster would ensue from the conservation point of view. The wildlands are now much smaller in area and the number of people very much greater (Rijksen 1984).

Some politicians and others will attempt to corner an advocate of protective conservation by asking "What is more important, nature or people?". The answer must be 'people', and such a simplistic answer must not be taken as a justification for allowing the destruction of forests, coral reefs, or wetlands. When those in authority do this, they are demonstrating that they have misunderstood the position and function of adat, the concept of the state, the function of government, and not least the value of their task (Rijksen 1984).

Appeals are often made to adat rights when in fact an individual or group of individuals are temporarily and conveniently adopting a selected part of an otherwise abandoned lifestyle in order to seek financial gain. This is not a clash between tradition and the modern state, but rather between individual or small group interest against the interests of the state community which may be difficult to quantify. Conflict in these situations can be avoided if 'multiple-use' or utilization is allowed 'within reason' even when that reason is hard to define let alone enforce. It is sometimes hoped that people will harvest at a rate that will serve the community's long-term interest if harvestable resources are given a value. The only relevant example of this is that of the beleaguered Inner Baduy people of West Java, amongst whom lifestyle and population density are strictly controlled with sanctions imposed on offenders (p. 340). In almost all cases today, however, utilization by village poor or urban elite alike is so intense that it leads to destruction and is in direct conflict with the concept of conservation which requires the protection of natural resources to ensure the ecosystem is kept in good working order.

The case of the Inner Baduy is consistent with the view of Rijksen (1984) who, after reading many ethnographies of large numbers of peoples, concluded that their traditional culture-religion-adat codes led their adherents not to take more from nature than was necessary for their daily subsistence. The most important factor in limiting environmental degradation was the low density of people. This was achieved through wars, minor conflicts, and infanticide which are not appropriate today, although the family planning programme may be seen as the modern equivalent! Flouting the adat codes by taking more than was necessary was not regarded as immodest or illegal, but rather as a sin that would be punished by a Divine Force. Whether or not this happened, a violator might be so ostracized and feel so isolated and unsupported that she or he would fall ill from psychological causes. When pagan behaviour codes were abolished or deserted, the dominant exogenous religions did not fill the void, but the 'rights' remained more or less unscathed. Unfortunately, the rights without the prohibitions do not result in a balance. The State now acts as a super-clan, and many adat regulations have been translated into laws, but religious awe has not been translated into obedience of the law or respect for the law enforcers. As we all lurch towards a free, materialistic, global, and homogeneous culture, it may be interesting to note that the rich and

diverse cultures of the past flourished within the adherence to institutions, regulations, and restrictions.

Thus conservation is a group concept and does not depend on the
interests of individuals within that group or, indeed, on whether every
individual assents to the idea. Managers of conservation areas should therefore reject the notion that enforcing the protection of a natural area is
adopting alien or modern concepts and realize that even the strictest
forms of protection are entirely legitimate with closer ties to adat than the
assertion of private or individual rights (Rijksen 1984).

ACCESS TO AND CONTROL OF RESOURCES

While the pressure of people on natural resources is very important, the
increasing pressure of people on people may be even more important
when it deprives the poor of access to resources (Schiller 1980). In recent
years there has been much concern at all levels of society over the questions
of who should have access to and control over natural resources. In one
sense the answer is simply 'the state', the decision expressed by the 1945
Constitution which was reiterated in the 1960 Agrarian Act, and has been
upheld ever since. This law asserts that the state has the sole authority to
regulate and implement the allocation, use, supply, and care of all
resources, that the national interest is placed above that of the individual,
and that all rights to land imply a social function (Hardjono 1991a).

Despite 30 years of this decision being enforced, the traditional systems
of resource allocation which allow free access to resources such as riverine
gold, firewood, timber, fish, and other game have not been forgotten and
occasionally surface, causing acrimonious conflict with government
departments. The issues fall into two categories. The first is caused by the
tradition that new or unused land belongs to all (or none) until someone
cultivates it. At this point the person concerned can assume ownership and
pass the land down as an inheritance. The other type of conflict issue is due
to the tradition that forests are public property to which all have equal
rights (Hardjono 1991a). Conflict over access to forest land and products
by local people has a long history, but exploded into open resistance
movements and widespread non-cooperation in the turbulent years of
1942-1966. Those in authority were losing their source of power, profit, and
legality as the Dutch-imposed infrastructure of forest exploitation broke
down. The only way this could be recovered was by insisting on the rein-
statement of the philosophy of total state control of forests (Peluso 1990,
1992, 1993). Thus, the New Order government has changed neither the
underlying structure nor the ideology of the colonial Dutch forest management for Java. The priority of modern forestry on Java has been to

produce export material and luxury domestic goods, with social projects being more or less incidental, at least until recently.

The Dutch did require the local people to police the use of natural resources to some extent however, since some of the responsibility for controlling the protection of steep forest lands on Bali was devolved to the local institutions. Punishments for transgressions could involve not just a fine or imprisonment but also, and to some extent far worse, temporary ostracism from the *banjar* or community (A.A.G. Agung pers. comm.).

Analysis has shown that the tensions, sometimes violent, between forest communities and Perhutani have been due to long-standing disputes over the tenure of forest land and its trees, a history of corruption and theft, and the failure of central policies to respond to the diverse socioeconomic and ecological conditions across Java. Awareness of these allowed the social forestry initiative to begin in 1986 (p. 595), initially as pilot projects, but then as an island-wide programme. The initiative aimed to resolve disputes, improve villager-staff relations, and increase the appreciation of, and responsiveness to, the necessary uses of forest by local communities (Peluso et al. 1990). Perhutani is continuing to examine itself, exploring ways in which preventive, punitive, and repressive approaches can be discarded and replaced with cooperative and mutually beneficial ones (Peluso 1990, 1992a). For example, there is a gradual intensification of the use of plantation forest lands in order to benefit people living in the area. Programmes include the intercropping of elephant grass *Pennisetum purpureum* (Gram.) among the young trees to improve the forage used in cattle raising, planting *Calliandra calothrysus* (Legu.) trees for fuelwood, and encouraging the keeping of bees for honey, and silkworms for silk (p. 621).

A question that begs asking in this context is whether resources would be any better managed under local or traditional control than under government control. This is a complex issue, but it can be understood intuitively that if a resource is perceived by local people as having long-term material or spiritual value, then those same people should evolve some means of control over its exploitation. The answer seems to be that traditional controls, where they exist in their entirety, can be valid means of resource management, but the pressures to seek short-term economic and financial gains even at local level are so acute that local control is not necessarily going to lead to better management than government controls. This is dealt with in part elsewhere in relation to forested conservation areas (p. 68), but the coastal situation should also be mentioned because marine resources are not covered by the Agrarian Act. Whereas in parts of the Pacific there is a wealth of traditional and 'modern' controls over the exploitation of marine and coastal resources, there are few equivalents in Java and Bali. Today, at least along the north coast of Java, there appear to be no traditional controls on the exploitation of coastal or marine resources, and there is no memory of times when such controls were exercised

(Sya'rani and Willoughby 1985). Means of avoiding resource conflicts between coastal communities have evolved, but the situation is not so peaceful between communities and commercial interests whose philosophy generally disavows attention to local economies, traditions, or sensibilities. Commercial operators generally have few if any long-term interests in a particular resource because company profits could be from fish ponds today, logging tomorrow, and real estate with golf courses next week.

IMPLICATIONS OF FUTURE CLIMATE CHANGE

As shown elsewhere (p. 117) the world's climate has changed dramatically in the distant past, and we are currently enjoying relatively warm conditions. The mechanisms that cause the cycles of climate change are poorly understood, although it is clear that numerous variables, such as changes in solar radiation, interact in complex ways. One of the most important and best known factors is the 'greenhouse effect' in which short-wave radiation in the form of light passes through our atmosphere, but some of the heat or long-wave radiation this forms is prevented from radiating back into space by natural trace gases which are transparent to light but not to heat. These gases are responsible for making the Earth habitable, because without them the surface would be some 30°C cooler (IPCC 1990). The most abundant of the gases is carbon dioxide (CO_2), but others include water vapour (H_2O), methane (CH_4), ozone (O_3), and nitrous oxide (N_2O). Not all these gases make the same contribution (a molecule of relatively rare CH_4 having an effect 21 times that of CO_2), but the dominant effect of CO_2, the knowledge of its emissions and concentrations over time, make it the gas on which almost all predictions are made. Most CO_2 emissions are produced by industrial nations, with the USA being both the largest absolute producer and the highest per capita producer. As other countries adopt a policy of industrialization, so their contributions will increase, although those, like Indonesia, able to adopt the relatively clean technologies associated with using natural gas will contribute less. Massive tree planting campaigns could eventually sequester carbon in significant amounts as an effort to avert the worst effects of some of the projected scenarios of the greenhouse effect (Myers and Goreau 1991).

The concentration of gases in the atmosphere varies over time, and CO_2 concentrations have varied from 180-280 ppm (parts per million) over the last 160,000 years, according to analyses of small bubbles frozen into layers of Antarctic ice, and these correlate with known changes in global temperatures over that period. In the recent past the concentration of CO_2 was about 275 ppm in 1790, 315 ppm in 1952, and 350 ppm in 1990. The start of the major increase in the concentration of this gas coincided with the beginning of the Industrial Revolution which relied on the burning of carbon-rich fuels such as coal to generate energy. Between 1890 and 1990 the average world temperature has risen by 0.4°-0.7°C (Maull 1992). While this is within the range of natural variation, it is often interpreted as a consequence of human activities. In fact, unequivocal evidence of a greenhouse effect enhanced by the impacts of human activity is not expected before the end of the century (IPCC 1990). Meanwhile, estimates of future CO_2 emissions and global temperatures have been made, and the consensus seems to be that by 2050 the CO_2 concentration will have risen to 550 ppm, and by 2090 to 700 ppm by which time the average global temperature will be 2°-6°C higher than now (Beckerman 1992). This would represent a rate of change far in excess of anything the Earth has ever experienced. Of course, governments have been alerted to this, and there is some hope that emissions of CO_2 will decrease, but it may be that the die has already been cast for an unavoidable global temperature increase of 1°-2°C (Douthwaite 1992). This temperature rise would be most marked in the arctic and temperate regions rather than the tropics, leading to reduced temperature gradients across the globe. This would result in altered patterns and intensities of the major pressure belts and therefore the strengths and directions of winds, and the patterns of rainfall.

There are, of course, enormous uncertainties associated with these figures, and they could equally be overestimates or underestimates. For example:

- only about half of the carbon produced by human activities since the start of the Industrial Revolution can be accounted for. There is thus a 'sink' for carbon somewhere, and many believe that this is the carbonate deposits fixed by plankton in the upper levels of the oceans which then accumulate as sediment on the sea bed. It may be that this system is unable to absorb any additional load of carbon, in which case future emissions would cause a proportionately greater rise in levels of atmospheric CO_2. Alternatively the system could be disrupted when large quantities of freshwater from melting glaciers float on top of the heavier sea water. In addition the plankton may be seriously damaged by the increasing levels of ultraviolet radiation arising from the erosion of the ozone layer by CFCs (chloro-fluoro-carbons) from discarded and leaking air conditioners and refrigerators;

- increasing temperatures in large areas of permafrost, such as Siberia, could cause the ice within the soil to melt with the consequent release of huge amounts of methane as the organic matter decays. Methane is a much more potent 'greenhouse' gas than CO_2; and

- the structure of the ice fields on the Antarctic mainland is such that certain patterns of thawing could cause them to break up and release huge quantities of water into the oceans, raising sea levels way beyond current predictions.

In the meantime, the figure of a 3°C rise in global temperature by 2090 is used to assess impacts, and five aspects are considered below.

Impacts on Sea Level

The sea level rise over the last century has been about 15 cm, and the consensus figure for the increase over the next century is about 65 cm, although many specialists, recognizing the uncertainties involved, work with the figure of 1 m (IPCC 1990; Parry et al. 1992). Large areas of northern coastal Java and around Segara Anakan lie below 1 m in elevation and so would become inundated, but a much more extensive area of flat lowlands would be affected by increasingly saline conditions. Most of the areas to be affected are on the north coast of Java where the country's main centres of commerce and government are located. Java and Bali are fortunate in having some hills and mountains; some nations such as the Maldives, Tuvalu, and Kiribati, not to mention the enormous and densely populated floodplains of south Asia will all but disappear.

Impacts on Population

Rising sea levels will undoubtedly affect the lives of millions of people and have indirect effects on almost everyone. Some populations could be protected by building sea walls, but these would need to be maintained for generations. The alternative is that people would have to relocate, causing inevitable friction in areas which are already 'full' by most criteria.

Impacts on Agriculture

Patterns of land use, cropping seasons, crop growth rates, and water availability, would obviously be affected by rises in sea level, rises in temperature, and changed rainfall patterns, but precisely how, and whether the overall impact would be positive or negative is difficult to judge (Rosenweig and Hillel 1993). For example, it is estimated that agriculture in the USA would not be affected in economic terms by the impending global climate change, since the predicted $10 billion loss caused by the above factors would be balanced by an equal gain due to the same factors, equivalent to only plus or minus 0.2% of the total economy (Beckerman 1992). This balance is possible in the USA because of its great area and range of habitats from hot deserts to permafrost. It may be that in Java and Bali there would also be some positive impacts, but precisely what they would be is hard to judge since rice yields would probably decrease because of the increase in temperature, and also because large areas of the fertile and productive coastal plains in areas such as Karawang and Subang would become inundated with saline water (Parry et al. 1992). In any case, the worst effects would be felt by people living from subsistence agriculture in habitats already marginal by reason of low or uncertain rainfall. The increasing unpredictability of weather that is likely to occur is one of the aspects with which peasant farmers are least able to cope.

Changes in crops will be essential, some of which will be achieved by swapping species and varieties between areas as climate changes. Genetic improvement will also be critical, but this will depend on how well this aspect of biological diversity is conserved. The greatest genetic reserves are in the non-industrial countries, but these are the countries which are losing them fastest and least able to invest in them.

Impacts on the Conservation of Biological Diversity

At the most basic level warming will have effects on the physiology, feeding habits, reproduction, and longevity of many organisms, any or all of which may threaten the survival of populations and species. When the effects of climate change are truly felt, small conservation areas distant from each other will be of little use. Species will need to be able to move through and across altitude and habitat zones to find suitable conditions, and where this and emigration are not possible, they will die out. Because of this, a major goal now is to establish systems of conservation areas which are as large, as representative, and as interconnected as possible. This is the least that can be done before a period of likely ecological chaos begins (Schneider 1993). Even so, it is probable that many species with narrow niches will become extinct. The problems will not be restricted to the land, and a taste of the possible marine impacts has been the widespread 'bleaching' of corals following sea warming as part of the 1982-83 El Niño effect (p. 364).

Impacts on Global Order

Not all countries will be affected equally by the predicted future climate changes. Rather unfairly, the wealthy, industrialized countries with no dependence on subsistence agriculture and with well-established economies and infrastructures, will be able to adapt with relative ease. Even though some countries, notably Bangladesh, will be very seriously affected by sea level rise, it makes more economic sense to pay compensation, even to the extent of building sea walls around the low-lying parts of the country, than to invest in avoiding climate change through expensive adjustments to existing economic activities and patterns of economic growth (Beckerman 1992). It has been estimated that a 60% overall reduction in CO_2 emission is required if CO_2 concentrations are to be kept at current levels (IPCC 1990). Recommendations following from this are given on the following page.

Climate Change in the Context of Gaia

Gaia was the name of the ancient Greek goddess 'Mother Earth,' and the hypothesis that bears her name states that "the biosphere is a self-regulating entity with the capacity to keep our planet healthy by controlling the chemical and physical environment" (Lovelock 1979, 1988). The Earth is thus seen as a super-ecosystem with multitudinous interactions and feedbacks which together moderate extremes of temperature and keep the chemical composition of the atmosphere and the oceans relatively constant. Living organisms are said to play a role in controlling the global environment, a role begun soon after the first anaerobic organisms (those not requiring oxygen) first appeared more than three billion years ago; at this time their respiration changed the Earth's atmosphere making it suitable for aerobic (oxygen-using) organisms. Those who oppose this view opine that purely geological and other physical processes produced conditions that happened to be favourable for life.

Indirect evidence for the Gaia hypothesis can be found if one compares the composition of the atmosphere of Mars, Venus (our closest planetary neighbours), and Earth both as it is now and as it would be without life (table 2.7). The influence of life processes, particularly on the buildup of oxygen, reduction of carbon dioxide, and the accumulation of nitrogen, is quite startling. Indeed, it is difficult to theorize how those changes could have come about without life. If organisms did not produce ammonia (NH_3), then the entire surface of the earth would be acutely acidic such that few organisms could survive. Nitrogen-fixing and denitrifying organisms play important roles in moving ammonia between the physical and biological worlds.

The Gaia system is hugely complex and we understand only a small part of it, and the regulatory role of the oceans is little understood. Considering all the unpredictable and catastrophic physical events, such as volcanic eruptions and comets crashing into the Earth, the Earth's biological controlling systems are clearly very resilient (assuming the Gaia hypothesis to be correct). The danger now is that as people directly and indirectly alter ecosystems and the composition of the atmosphere, and as species extinctions proceed, this resilience cannot necessarily be expected to continue (p. 22).

Table 2.7. Atmospheric conditions on Mars, Venus, Earth without life, and Earth as it is now.

	Mars	Venus	Earth without life	Earth as it is
Carbon dioxide (%)	95	98	98	0.03
Nitrogen (%)	2.7	1.9	1.9	79
Oxygen (%)	0.13	+	+	21
Mean temperature (°C)	-53	477	290±50	13

Source: Lovelock (1979). With permission of Oxford University Press

While there are many uncertainties regarding global warming and its effects, there are a number of actions which would offer some insurance. The least controversial of these (termed 'no-regret options') include:

- improve the efficiency with which energy is used;
- encourage all sectors to change to less carbon-rich sources of energy such as gas;
- encourage all sectors to change to appropriate non-carbon sources of energy such as solar, wind, geothermal, and microhydro – nuclear power generation is still a controversial issue;
- identify and remove market imperfections which allow environmental costs to be externalized, or which discourage the introduction of new technologies;
- continue efforts to reduce the production of greenhouse gases;
- invest strongly in environmental education as a means to extend awareness and understanding of the climate change issue and its implications at all levels of society.

INADEQUATE ECONOMIC PARADIGMS

While moral and spiritual values (p. 833) are indeed unscientific and generally subjective, there is increasing and, to some, overwhelming evidence that current, logical, 'scientific' economic paradigms are not serving the best interests of the majority of the world's people or the environment because their perspectives are too narrow (Røpke 1994). In the last 40 or so years the world's population has doubled, and gross world product and fossil fuel consumption has quadrupled. Further economic growth is likely to be 'uneconomic', and when growth in the classic sense is no longer possible, then politicians will be faced with appalling decisions. The view has been expressed that there will be a need for highly authoritarian governments to oversee the transition to declining economies (Heilbronner 1980), but realistic alternatives which deal with the real and pending issues do exist (Daly and Cobb 1990; Goodland et al. 1991). They require, however, that economists face up to some hard truths.

The economic system into which we are born and within which we live is based on principles that began to be laid down some 300 years ago when no thought was given to the exhaustion of natural resources, to the damage done to human health by industrial wastes, to the loss of biological diversity, or to undesirable social, environmental, and international consequences of gross indebtedness and 'free' trade (George 1992; Daly and Goodland 1994; Ekins et al. 1994; Røpke 1994). Standard economic theory, through industrialization, has increased the goods and services available to huge numbers of people, although most benefits have accrued to the North. However, the competition and materialism it has spawned have bred aggressiveness, insatiable desires for more goods, and unrealistic expectations of freedom (Henderson 1978). For several decades now it has been clear that taking increasing quantities of materials from the environment to feed the 'growth monster' has produced exhaustion of resources and widespread pollution. Economists have exhibited a surety, some might say arrogance, that the prosperity generated by the growth will be able to take care of the problems, which they anyway consider to be exaggerated, by sharing the benefits of growth and providing the capital to pay for technological fixes.

Standard economics is therefore in the paradoxical position of being regarded as both enormously successful and being severely criticized. The success is due in part to humans being seen to, and required to, seek personal gain. Such behaviour is deemed by economists to be rational and capable of quantification, whereas action directed towards the well-being of others or the environment is neither. Indeed, any restraint of self-interest is regarded as confusing, unnecessary, and harmful to growth.

Despite this, the voices uttering disquiet with the system are becoming louder, more articulate, and more authoritative, if not yet dominant over the babel of those hanging on to the system that has seemingly served so well in the past: we do not deny that huge numbers of poor farmers in Java and Bali have improved their lot enormously over the last few decades (Collier et al. 1993; World Bank 1993). However, it is reported that even more than ten years ago there was a feeling among two-thirds of American economics professors at 50 universities, that economics had lost its moorings (Maital 1982). A new economic system must abandon the key unit of humans-in-isolation, if not all its axioms, and consider instead persons-in-community. The axioms need to be corrected and expanded, and the overriding concept of the market has to be subordinated when circumstances are not served by its promotion and support.

The paradigm or economic model proposed by Daly and Cobb (1990) in their book, *For the Common Good: Redirecting the Community toward Community, the Environment and a Sustainable Future*, envisions aiming for an optimal 'unit' size in which communities can provide their members with a good and sustainable life and in which the person-in-community is the basic unit. The senior author was a Senior Economist at the World Bank, and the junior author a professor of philosophy; they are not ill-informed green radicals. They believe that the current economic system is leading humans to a dead end and increasing poverty even in countries of the North attest to this (although poverty is not prevalent among most of the politicians and economists). The discussion is not about whether the form of the economy will change, for it must do so in the next few decades, but how to change it and whether to wait until virtually all the existing options are closed to us. The concern is not just focused on the environment because all related issues are involved too: poverty alleviation, trade policies, measurement of economic progress, and changing attitudes. One of the keys is finding a better way to measure the economy than gross domestic product (GDP) and gross national product (GNP).

The conventional view, encouraged by some development banks and national governments, is that the health of a country's economy is best measured by the value of goods and services produced within the country (GDP), and the national income (GNP). This view is ostensibly reasonable, but many or all environmentally-damaging activities increase both measures: for example, a foreign-owned polluting industry would increase GNP by paying its taxes to the government of the host country, and inefficient disposal of household waste creates jobs among those who sort the rubbish to be recycled, and increases GDP. GDP is also flawed as a measurement of income because it does not differentiate windfall profits (from the sale of stocks and shares) from true income. In addition, it takes no account of the depletion of natural resources or of pollution. GDP may thus be envisaged as a speedometer, but not as a fuel gauge. This means it fails to give any information about efficiency or sustainability (Tinbergen and Hueting 1992).

Keeping track of national accounts, the balance sheet of incomes and expenditure, has been conducted in more or less the same way by almost all countries for the last 25 years. Unfortunately, again, these measures do not serve environmental concerns well because they omit the depletion and degradation of natural capital (remaining exploitable resources) (Cleveland 1993). If an economy is to be sustainable, then the capital stock per capita of a growing population must stay the same, either by increasing the total capital stock, or by processing the natural capital more efficiently. Some economists suggest that so long as loss of natural capital is substitutable by other capital such as buildings, machines, or software, then there is no cause for worry. But this is really a fudge of economists uncommitted to a truly sustainable future. To achieve genuine sustainability the stocks of natural capital would have to be kept at some predetermined level (Costanza 1992). These sorts of considerations have not yet been incorporated into the mainstream national accounting systems.

To be fair, quantifying and monitoring useful measures of resource depletion is fraught with difficulties, but it is not impossible. One of the first attempts to conquer these was made in Indonesia a few years ago. It found that the GNP grew at an average of 7.1%/year, but this was nearly halved when natural resource depletion was incorporated. This would have been reduced still further had resource depletion throughout the country been incorporated (Repetto et al. 1989). Indeed, environmental damage in terms of depreciation of natural capital in Indonesia appears to be the highest in the world at 17% of GDP/year, contributing to serious unsustainability (Pearce and Atkinson 1992). The study also found that far too little of the profits made from the exploitation of natural resources were being invested in productive assets such as education and health to guarantee future income for the people.

The conventional measures of a nation's economic well being do not in any way reflect its ecological well-being, but part of the reason that the measures have remained as indicators of economic health for the last 50 years is because more meaningful measures have been hard to find. After much pressure the UN Statistical Office, which instituted the standard accounting rules, has formulated modifications to the established GNP measurement which incorporate life expectancy, educational attainment, modified GDP, and deprivation, but it has yet to incorporate these into the standard accounting procedures, possibly because to do so would show the true economies of many countries in rather a bad light, but possibly also because of the difficulties of collecting and organizing the appropriate data (Friend 1993). So business-as-usual makes assessments of sustainability distorted. An alternative index, the Index of Sustainable Economic Welfare (Daly and Cobb 1990) has been proposed, however, which takes into account parameters such as personal consumption, income distribution equality, expenditures on services, loss of natural resources, pollution, capital growth, and long-term environmental damage.

Daly and Cobb explain their own understanding and commitment as rational and religious, and observe that others who share this passion and commitment are deeply religious in varied ways. They conclude that the real possibility for change depends on an awakening of the religious depths in a world whose secularity has gone stale. They believe that the changes so desperately needed will occur within a religious context and with religious passions stirred. The issues cannot be decided by reason, and the necessary overcoming of idolatry in the form of overconfidence in money, the GNP, the market, and so forth, will be a religious task.

Other economists feel that the current paradigms can serve the environment correctly, and list the many policy options which are available, such as imposing taxes on goods produced with polluting technology, deposit-refund schemes, tradable pollution permits, tradable resource rights, and liability laws. These actions will alter prices, and so the market, with government direction, will have to place a value on the environment in order for it to benefit. Since governments are run by politicians whose professional lives depend on not offending the general populace or major business interests, the orchestration of appropriate relative prices is critical. Meanwhile, there is frequently a degree of cynicism that greets politicians' 'green' rhetoric, but this is only matched by the politicians' own private cynicism that if they do actually deliver the somewhat bitter pills of genuine sustainability, then the consumer will cause a rumpus and refuse to pay the costs. Thus getting the price right is vital and is taxing the minds of many economists around the world.

Sustainable Development

The term 'sustainable development' became popular, almost to the point of overuse, with the publication of the Brundtland Report, *Towards a Common Future*, by the World Commission on Environment and Development (WCED 1987). The Commission defined sustainable development as "meeting the needs of the present without compromising the ability of future generations to meet their own needs." The work of the Commission was to make world leaders recognize the legitimacy of the 'sustainability' concept without delving into its detailed workings, and in this they have been remarkably successful. It should be borne in mind, however, that the Brundtland Report defines a concept, not an operation, and poses many problems, such as what is a 'need'? Does everyone in Java and Bali 'need' a car? Who is to judge sustainability and by what criteria? How much or how little do we have to leave for future generations?

There have been many attempts to describe how sustainable development would operate (e.g., Dixon and Fallon 1989; Pezzey 1989; Pearce and Atkinson 1993), and it is clear that those who applaud the general concept are polarized in their understanding of it. The dominant view of economists, perhaps because it demands the fewest changes, is that of 'sustainable growth' which is very different from sustainable development. The oxymoron of sustainable growth, allows the consumption of natural resources beyond their regenerative capacity, so long as the declining value of a nation's resources appear on its national income accounts, and the proportion of its resulting revenues (representing the depreciation or depletion of the resources) is invested in other forms of human or human-made capital to maintain a comparable stream of benefits in the future. This assumes that such capital can substitute for natural resources to sustain overall growth, and that technological advances will allow an ever-expanding production despite the declining natural resource base (Pearce and Warford 1993).

Sustainable development is rather a process in which qualitative development is maintained and prolonged, while quantitative growth in the scale of the economy becomes increasingly constrained by the capacity of the ecosystem to perform, in the long term, two essential functions: to regenerate the raw material inputs, and to absorb the waste outputs of the human economy (Daly 1989). Any system which insists on growth within a finite system (such as the Earth) will one day cease to grow and is unsustainable from the moment it begins to operate. Thus, in this view, sustainable growth is self-contradictory. To achieve sustainable development the world has to learn a maintenance culture in which enough is best and more is not better (Adams 1990; Jacobs 1992; Costanza and Daly 1992; Daly 1992b; Goodland 1995).

The technical and economic problems involved in achieving sustainability are not that difficult. The real challenge is overcoming our addiction to growth (Brown et al., 1992). We dare not admit that for the elimination of poverty the distribution of the material benefits of growth through the population must be controlled so that the benefits of development do not fall disproportionately into the laps of the wealthier sections. Poverty alleviation cannot simply be achieved by faster growth. There must be limits to per capita resource use, limits to population growth, and limits to inequality. If this is not accepted by decision makers, then fewer people will be able to live in conditions of material sufficiency, and many of those existing below material sufficiency will suffer premature death (Daly 1992b Goodland 1992; Goodland and Daly 1994).

It would be difficult today to find any politician who did not espouse 'sustainable development', and there is no lack of written and verbal government commitments to sustainable development and conservation. So, there are good policies and sound commitment, but this book demonstrates that even in Java and Bali, many of the policies and regulations are not being implemented or enforced, and economic development which is genuinely sustainable is far off. To bring it closer is not just a precondition for person-focused development but also for life itself. In this period before the wholesale and genuine adoption of sustainable development, it is necessary to care for the integrity of ecosystems by preventing the conversion or modification of the remaining natural or other complex ecosystems wherever possible, halting the degradation of fertile land, and ensuring that the quantity and quality of effluents discharged into ecosystems do not exceed those systems' assimilative capacities. It is also necessary to care for biological diversity to ensure that the links in ecological processes are not broken. To do this, the largest possible areas of representative natural or near-natural ecosystems need to be conserved, with or without forms of sustainable utilization. Aspects of environmental management required to achieve these aims are the prevention of pollution, the reclamation and rehabilitation of disturbed natural ecosystems, intensification increase in the production capacity of agricultural and forestry ecosystems, and an improvement in public education (Djajadiningrat and Amir 1992) as well as stabilizing the human population total, and controlling the overconsumption of the rich (p. 36).

The rising middle classes of the urban areas are both a boon and a bane to the cause of genuine sustainability. They are a boon because they are more interested than others in ecology and environmental problems, relatively well informed, more vocal about their opinions, and are major forces for change. They are also a bane in that they are trapped in their consumption patterns which worsen environmental pollution and degradation (p. 36). Meanwhile, their ability and willingness to pay for, for example, holiday villas in the Puncak area or green fees on the burgeoning golf courses are the root causes of many social and ecological conflicts.

Since 1990 the Bali Provincial Government has been engaged in studies and analyses, supported by Canadian universities, to determine how sustainable development should proceed (Martopo and Mitchell 1995). Bali is an excellent area to choose:

- it is an island within defined boundaries;
- it is a single political unit;
- it is relatively small;
- it is culturally distinct;
- it has strong human resources;
- its people are known for their independent thought;
- it is self-sufficient in food;
- it has no heavy industries;
- its major industry provides considerable foreign exchange; and
- there are existing models of sustainable development already working on the island.

It is probably true to say that there is nowhere else in the world to which the above would all apply. Given all these factors in its favour, then, everyone concerned with sustainable development should watch closely how policies, actions, and measures develop on Bali. The frightening possibility is, however, that there will not be sufficient political will from the provincial and the central governments to see through the necessary changes. If that is the case, what hope is there for the rest of us?

Part B

Ecological Components

Ecology concerns the physical environment, the organisms found in it, and the hand of people; those three components of the system are considered here. Ecology has great relevance in deciding how natural resources and the environment can be managed sustainably, but until the existence, diversity, function, use, and behaviour of these components is fully realized, then those decisions will be found lacking.

We all suffer from cultural amnesia and rarely view our present condition in an historical context. To counter this, the past as well as present conditions are described wherever possible. This enables the present to be seen in its historical context and helps to develop an understanding that the present is merely a snapshot in time whose conditions have been determined by past events and processes. The present is the starting point for the future, and past changes may teach valuable lessons about wise environmental management.

Chapter Three

Physical Conditions

GEOLOGY

Formation

About 250 million years ago most of the Earth's continental crust was gathered into a single land mass known as Pangea. Then, rather more than 200 million years ago, during the Triassic Period (table 3.1), Pangea rifted into two super-continents called Laurasia, comprising present-day North America, Europe, and some of central and east Asia, and Gondwanaland, comprising present-day South America, Africa, India, Australia, and the remainder of Asia. Later, fragments of continental plates fractured from these super-continents, drifted away, and collided with one another.

There have been a number of interpretations of the geological evidence to determine the continental origins of the different islands of Indonesia (e.g., Audley-Charles 1981, 1987; Audley-Charles et al. 1981), and the latest scheme (Burrett et al. 1991) is shown in figure 3.1. This is based on the distribution of collages of crustal fragment known as 'terranes', that have dissimilar geological histories. It is not possible to plot the positions of Java and Bali with any reliability on these maps, first because there has been so little dating of their rocks, and second because Java probably did not exist in any form before the Miocene, and Bali probably did not emerge above the sea until even later, about three million years ago (C. Burrett pers. comm.).

Some 250 km to the south of Java and Bali is the very deep Java Trench (fig. 3.2). South of this is that part of the Earth's crust known as the Indo-Australian Plate, which is formed in the ocean depths south of India and Australia. The northward movement of this plate continues even now at a rate of about 6 cm/year (Hamilton 1979) (fig. 3.3). It pushes against the Sunda Plate on which Southeast Asia sits, and over millions of years the forces produced by these movements have folded the layers of older sediment into mountain ranges. The Indo-Australian Plate dips beneath the Sunda Plate along the Java Trench, and earthquakes are caused by occasional rapid slippage as the plates rub past each other, and the heat-gen-

Table 3.1. Geological timetable of the last 365 million years with major geological and biological events.

Period	Epoch	Million years before present	Climate-Habitat-Fauna trends
Quaternary	Recent	0.01-present	Generally benign climate: no ice ages. Rise of humans: domestication of plants and animals. Faunal diversity reduced: body size of many forms reduced.
	Pleistocene	1.6-0.01	Great climatic fluctuations: four major ice ages in Eurasia causing considerable changes in sea level. Arrival of 'Java Man' *Homo erectus* and 'modern man' *Homo sapiens*; first manufacture of human stone tools and first control of fire. Extinction of many genera of large mammals, some possibly due to human hunting.
	Pliocene	1.6-5	Bali born as submarine volcanoes, possibly emerging three million years ago. Periodic catastrophic eruptions would have killed all living organisms present. Cooler, drier climate, with savanna conditions widespread during much of the time. Mangrove forests were present in the centre of what is now C. Java. Few fossils, but many modern genera and all modern families had probably appeared.
Tertiary	Miocene	5-26	'Java' present as volcanic islands, eventually coalescing to form a single island. Falls of burning ash would have eliminated plants and animals from all or part of the area. Relatively warm and wet. Continuous forest belt from Spain to China, with extensive grasslands further north; mammals at height of evolution, with great diversification; many modern families and subfamilies, including Hominidae (represented by human-like apes), first appeared.
	Oligocene	26-38	Warm climate. Archaic mammals became extinct; forerunners of most modern families and genera appeared, including a diversification of herbivores and the first monkeys and apes.
	Eocene	38-54	Climate warm and wet: southern continents separated from northern. Most modern orders established: placental mammals diversified and specialized. India in first contact with Asia.
	Paleocene	54-65	Mild climate. Diversification of archaic mammals; no modern families, but some modern orders had appeared (rodents, insectivores, primates, etc.).
Cretaceous		65-140	Angiosperms became dominant. Dinosaurs reached peak, became extinct; toothed birds became extinct, while first modern birds appeared. Archaic mammals common.
Jurassic		140-200	First birds (toothed); dinosaurs larger and specialized; archaic insectivorous marsupial mammals appeared.
Triassic		200-245	Pangea rifts to form Laurasia and Gondwanaland again. First dinosaurs, turtles, and egg-laying mammals; extinction of primitive amphibians; conifers dominant plants.
Permian		245-290	Pieces of Gondwanaland start to rift and float northwards. Extinction of many kinds of marine animals including trilobites.
Carboniferous		290-365	Pangea formed by the fusion of Laurasia and Gondwanaland. Extensive coal-forming forests with first coniferous trees. Sharks and amphibians abundant, first reptiles.

Source: Adapted from Ricklefs (1990)

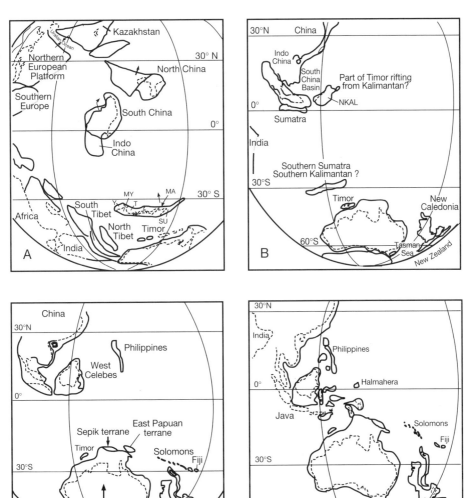

Figure 3.1. Drifting of Indonesian Islands to their present positions from the late Permian to present. A - Late Permian, 255 million years ago; B - Paleocene, 60 million years ago; C - Oligocene, 30 million years ago; D - late Miocene, 10 million years ago. J - Java, MA - Malay Peninsula, MY - Myanmar, NKAL - North Kalimantan, SU - Sumatra, T - Thailand, Y - West Yunnan.

After Burrett et al., 1991; C. Burrett pers. comm.

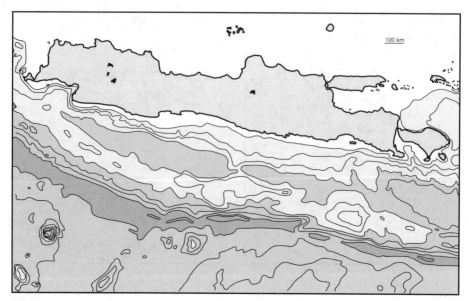

Figure 3.2. Depth contours in the sea around Java and Bali to show the Java Trench. Contours are shown at 200 and 1,000 m and, thereafter, at every 1,000 m. Note that the sea north of Java is only about 50 m deep.

After Hamilton 1979

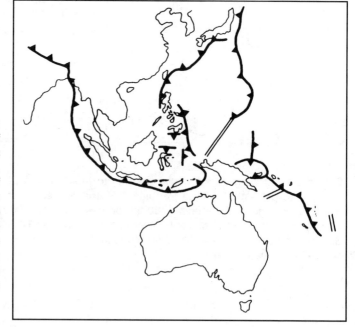

Figure 3.3. Position and movement of tectonic plates around Indonesia. double line: zone of spreading from which plates are moving apart; line with triangles: zone of underthrusting (subduction), with triangles on the upper plate.

After Hamilton 1979

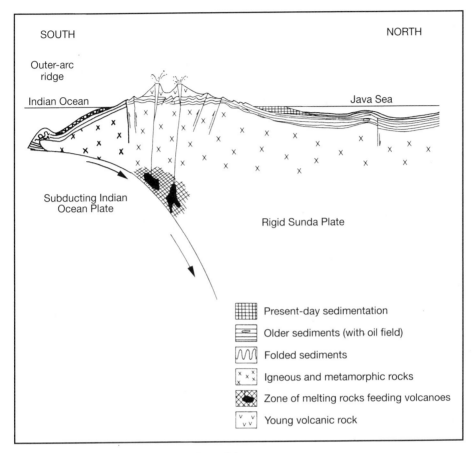

Figure 3.4. Schematic cross section of Java.

After RePPProT 1989

erating friction of the plates' movements form pockets of molten rock under high pressure. These pockets can leak to the surface where they form volcanoes (fig. 3.4).

The 1:100,000 geological maps produced by the Geology Research and Development Centre, Bandung, cover almost all of Java and Bali (except southeast Java and the Kangean Islands), but these are too detailed for smaller-scale reproduction in a book such as this, for which a simplified map suffices (fig. 3.5). Clearly, although volcanic rocks dominate, there are also extensive sedimentary areas. North and south of the main range recent sediments from the erosion of the modern mountains are deposited over older sediments which have been uplifted due to the massive engagement below of molten rock. Not all the sedimentary rocks are erosion products, however, because there are large areas of uplifted limestone

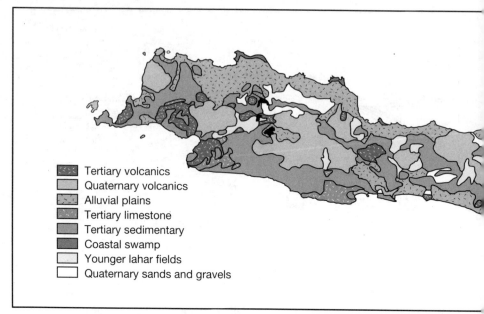

Figure 3.5. Geological map of Java and Bali.

After RePPProT 1990

from periods when reef-building organisms such as corals could thrive. The entire alluvial plain of North Java has been formed in the last 8,000 years during which time sea level has fallen some 5-6 m. This plain has been formed partly by alluvial fans of volcanic debris and partly as a raised post-Pliocene peneplain. These processes have continued until the present. For example, about 2.5 km of alluvial fan accreted around the mouth of the Cidurian River (that flows to the north through Jasinga) between 1927 and 1945 (Verstappen 1954), and the growth in delta areas along the north coast of Java has been dramatic.

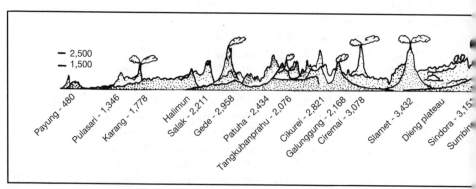

Figure 3.6. The distribution and relative heights of Java's volcanoes.

After Koorders 1911

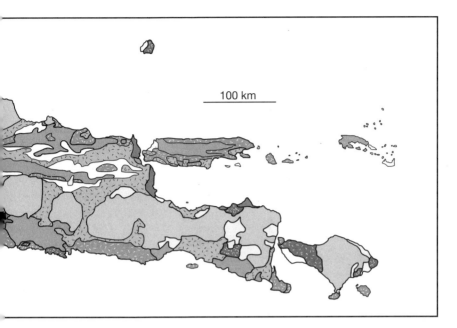

Volcanoes and Earthquakes

Indonesia has some 155 centres of active volcanism, and Java and Bali are the most volcanically active islands in the world with 20 of their volcanoes having been active in historical times. In addition, 13 older volcanoes have active solfatara (vents emitting hydrogen sulphide and other sulphurous gases from which pure sulphur may condense) and fumaroles (vents emitting high pressure steam and sometimes other gases) (fig. 3.6). Some peaks such as Semeru, Merapi, Agung, and Ciremai have classic shapes, and some, such as Tengger and Batur, have dramatic calderas where the peak has been blown off leaving a plain or lake with smaller peaks within. The Batur complex has been described as one of the world's

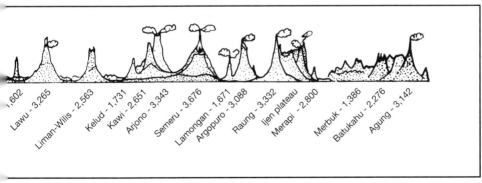

Table 3.2. Volcanoes with eruptions recorded since about 1600 (A) and volcanoes with solfatara and fumarole fields (B). VEI - Volcanic explosivity index (see text and table 3.3).

Island	Volcano	Type	Number of active periods and year of most recent activity	VEI of most recent activity (VEI of largest eruption with year)
Java	Pulasari	B	1942	
	Karang	B	-	
	Kiaraberes-Gagak	A	6/1936	1
	Salak	B	6/1938	2 (3-1699)
	Gede	A	21/1956	2 (3-1748)
	Tangkuban Perahu	A	9/1969	2
	Patuha (Kawah Putih)	B	1941	
	Wayang Windu	B	-	
	Guntur	A	25/1847	2 (3-1690, 1843)
	Papandayan	A	2/1923	2 (4-1772)
	Galunggung	A	3/1982	? (5-1822)
	Telaga Bodas	B	1925	
	Ciremai	A	6/1973	2 (3 - 1698)
	Slamet	A	36/1973	2
	Dieng Volcanoes:			
	Pakuwodjo	A	2/1847	2
	Other	B	3/1979	2 (3+ -1375)
	Butak Petarangan (Timbang)	A	3/1939	1
	Sindoro	A	7/1971	2
	Sumbing	B	1/1730	1
	Merbabu	B	4/1797	2 (3+ -1586)
	Merapi	A	59/1992	2 (4 -1006)
	Ungaran	B	-	
	Lawu	B	1908 ?	
	Kelud	A	35/1990	4
	Arjuno-Welirang	B	1/1952	0
	Bromo	A	51/1980	1 (3-1915)
	Semeru	A	63/1994	3
	Lamongan	A	44/1953	2
	Yang Argapura	B	1597	
	Raung	A	47/1977	2 (3-1586, 1638, 1730, 1817)
	Kawah Ijen	A	5/1952	2
Bali	Batur	A	21/1974	2 (3-1968)
	Agung	A	4/1963	4

Sources: van Bemmelen (1970), Kusumadinata (1979), Simkin et al. (1981)

Table 3.3. Criteria for assessing volcanic explosivity index.

Description	Non-explosive 0	Small 1	Moderate 2	Moderate large 3	Large 4	Very large 5
Volume of ejecta (m³)	<104	104-106	106-107	107-108	108-109	109-1,010
Height of cloud column (km)	<0.1	0.1-1	1-5	3-15	10-25	>25

Source: Simkin et al. (1981)

Figure 3.7. Ash falls during an eruption of Mt. Kelud in 1990.
With permission of Kompas

largest and finest calderas (van Bemmelen 1970). The centres of active volcanism are listed in table 3.2 (see also fig. 11.3), together with an indication of their activity and the volcanic explosivity index. This index is determined by the occurrence of a number of qualitative and quantitative indicators (volume of volcanic products, height of cloud, descriptive terms used by observers, etc.); the criteria are shown in table 3.3. West Java also has remains of old volcanoes such as the huge Bajah of which Mt. Halimun formed an outer, northern wall. Among the oldest volcanoes on Java are those represented by the dramatic 'plugs' seen near Plered on the southwest shores of the Jatiluhur reservoir (Blanke 1939; Witkamp 1940).

Volcanoes have played a crucial role in the geological and human history of Java and Bali. Their impact has been largely positive because they create land through lava flows, ash deposits (fig. 3.7), and mud flows (lahars). Natural erosion carries volcanic material as alluvium to the plains forming thick layers of fertile sediment (fig. 3.8). This fertility varies across Java and Bali, however, being higher in Central and East Java and Bali where the volcanoes produce basaltic lavas, and lower in West Java where the volcanoes produce more silica-rich andesitic lavas. The benefits are not limited to the immediate vicinity of the active volcano, however, because

Figure 3.8. Planting rice in the fertile ash ejected from Mt. Galunggung.
With permission of Antara

fine ash is transported great distances from the erupting crater, providing a top dressing of soil-enriching material over wide areas.

Volcanoes have been agents of major landscape changes, even in the quite recent past. For example, the Citarum River used to flow through the hills at Padalarang rather than along its current course through the western gorge near Rajamandala. The change began perhaps 2,000-6,000 years ago when mud, ash, and rocks from the erupting Tangkuban Perahu blocked the Padalarang valley. With the only outlet blocked, the Bandung basin began to fill, and it eventually became a lake more than 700 km^2 in extent (for comparison, Lake Toba in North Sumatra, southeast Asia's largest lake, is 1,146 km^2). Eventually the rising water found its new course and the lake level fell. The newly fertile and dry Bandung plain did not appear immediately because it took time for the new channel to erode down. About 150 years ago the lake was still quite large, and wild animals including banteng, rhinos, and tigers were abundant (de Wilde 1830), and even now the southern part of the plain is prone to flooding (Bennett and Bennett 1980; Darsoprajitno 1990).

It has been suggested that the cataclysmic eruption of Tangkuban Perahu that spread thick layers of white ash over a wide area may have been the origin of the name Sunda for the local people and the region (van Bemmelen 1970). The reason is that 'Sunda' is related to the Sanskrit "*cuddha*" which means 'pure' or 'white', and the only place in Java given the name Sunda is just north of Bandung. It thus seems that the massive eruption, the creation of the lake, and its drainage have all been witnessed by succeeding generations of people.

Volcanic eruptions have also had political repercussions. A manuscript called *Maha Pralaya Kerajaan Mataram* recounts how mudflows from Mt. Merapi buried the centre of the first Mataram Kingdom in 1006 (Darsoprajitno 1990), and the shift of the main government seat in the 13th century from Central to East Java could well have been caused by eruptions of Mt. Merapi (van Setten van der Meer 1979).

Volcanoes can, of course, be catastrophically hazardous for people living around them, but when compared with the deaths caused by floods, earthquakes, wars, civil disturbance, disease, or even traffic accidents, they are not among the major causes of human loss of life. Among the most destructive eruptions were those of Mt. Agung in 1963 which claimed 1,148 people (p. 685) and 18,000 cattle (Suryo 1965 in Kusumadinata 1979). In 1930 Mt. Merapi claimed 1,369 lives when flows of hot, dry particulate material moved along the ground up to 10 km from the peak. Another hazard is the invisible emission of gases (such as carbon monoxide, hydrogen sulphide, and sulphur dioxide) particularly on still, cloudy days. Such gases, released by an eruption in Dieng (C. Java) in 1979, claimed the lives of 149 people fleeing the eruption. Fiery volcanic blocks are also a danger. These shot out of Mt. Agung in 1963 and burned houses up to 7 km away from the crater (Situmorang and Sudradjat 1987).

Lahars (an Indonesian term adopted into international usage) are also very damaging. These are mudflows of volcanic debris mixed with water that can be caused either by eruptions occurring through a crater lake (primary) or by heavy rains disturbing unconsolidated volcanic material (secondary). The worst known primary hot lahar in Indonesia was from the squat Mt. Kelud in 1919 in which 5,110 people were killed and 135 km² of cultivated land was buried (fig. 3.9). This disaster stimulated a series of engineering projects to reduce the threats from this volcano, and horizontal tunnels and siphons connecting the crater lake with the outer slopes were completed in 1926 to prevent the lake from becoming too full again (van Bemmelen 1949). After the 1982 eruption of Mt. Galunggung a secondary *lahar* flowed for over 5 km destroying houses and cultivated land. The absence of loss of life on this occasion pays tribute to the early warning systems in operation (Katili and Sudradjat 1984). In early 1994 hot lahars streaming off the southern slopes of Mt. Semeru claimed several lives, and the view of the steaming river winding to the sea became quite a local attraction.

Submarine volcanoes displace water during eruptions, and this causes

Figure 3.9. Destruction of a village by a lahar on the slopes of Mt. Kelud in 1919.
Photographer unknown, with permission of the Rijksherbarium/Hortus Botanicus, Leiden University

'tidal' waves or tsunamis. The worst the world has known were created by the Krakatau explosion in 1883. The waves reaching Anyer on the west-facing coast of Java reached 36 m in height and resulted, all told, in the loss of 36,417 lives (Kusumadinata 1979). One large coral boulder caught in the wave was found 300 m inland at Tanjung Seraga (Joongjai and Rosengren 1983). It should be noted that Krakatau (a group of four islands), while often considered part of Java, actually falls within Lampung province, Sumatra, and its interesting ecological features are covered in Whitten et al. (1987).

Mt. Galunggung between Garut and Tasikmalaya was very active in the middle half of 1982 with 36 major and 300 minor eruptions. The ash and lava ejected destroyed crops, livestock, and pond fish but caused no loss of human life. Ash was blown 20 km into the stratosphere, and by September a layer about three centimetres thick had fallen 15 km from the crater, representing an estimated 200 million m³ of debris (Bakornas PBA 1982). The ash cloud in the atmosphere caused the breakdown of all four engines of a jumbo jet which was flying through it, but despite a totally scoured windscreen it landed safely at Jakarta airport. A second jumbo had three of its engines damaged and that too landed safely (Tootell 1985).

It is possible to drive up to the craters of a few volcanoes although it should be remembered that they are potentially very dangerous. For example, just over 200 years ago, Mt. Papandayan did not have its famous triple crater. Then, on the night of 11 August 1772 about 4 km^3 of mountain top blew sideways across Garut, a town surrounded by a ring of 14 volcanoes, killing over 3,000 people (Bennett and Bennett 1980). This volcano now has a road leading almost to the summit.

All the active volcanoes are continuously monitored for changes in activity and systems set up for warning people who might be in danger. The areas around all volcanoes on Java and Bali are designated according to their risk (Kusumadinata 1979) although the existing land use does not always reflect this (RePPProT 1989). Part of the reason for this is the pressure on scarce land (p. 34), part the belief that another eruption is unlikely to happen and, around Mt. Merapi at least, because of very deep traditional beliefs (Triyoga 1991). Indeed, volcanoes have had (and some still have) intense local religious significance. Gases shrieking out of small fissures make noises which could easily be perceived as the sounds of discontented spirits suffering unutterable torments (d'Almeida 1864). As a result they gain a spiritual significance, and ancient statues have been found near the craters of Mt. Salak and Mt. Malabar where they were supposed to represent petrified gamelan musicians (Bennett and Bennett 1980).

Climbing volcanoes (and other mountains) is increasingly popular and, if the necessary precautions are taken regarding clothing, food, and reporting to officials before and after the climb, can be very rewarding (Petheram and Lowry 1983; Perhutani KPH Bantim 1986; Maria 1988).

There have been about a dozen destructive earthquakes in Java and Bali this century, the worst being in Bali in 1917 when about 1,500 people were killed. Other destructive earthquakes occurred in 1924 in Wonosobo when 727 people were killed, 1976 in West Bali when 559 people were killed and over 850 seriously injured, and 1943 in Yogyakarta when 213 people lost their lives (Soetoardjo et al. 1985). The most recent serious earthquake was in 1994 when over 200 people were killed by tidal waves caused by two earthquakes off the south-east shore of Java in Alas Purwo National Park. The loss through earthquakes of an average of 34 lives each year does, however, pale into insignificance against the approximately 350 and 5,750 lives lost in traffic accidents in Jakarta and all of Java respectively each year.

Mineral Wealth

Neither Java nor Bali have particularly significant mineral wealth (fig.

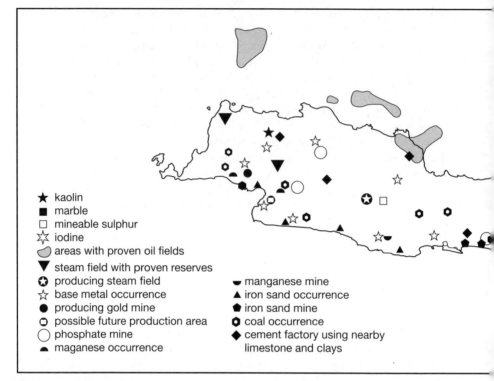

★ kaolin
■ marble
□ mineable sulphur
☆ iodine
◗ areas with proven oil fields
▼ steam field with proven reserves
✪ producing steam field
☆ base metal occurrence
● producing gold mine
○ possible future production area
◯ phosphate mine
▲ maganese occurrence

◥ manganese mine
▲ iron sand occurrence
⬟ iron sand mine
◙ coal occurrence
◆ cement factory using nearby
   limestone and clays

Figure 3.10. Mineral and energy resources of Java and Bali.
After RePPProT 1990

3.10). Where folding and faulting of sedimentary rocks have provided satisfactory conditions, pockets of oil and gas are found. These are present on both the north and south of the mountain chain running across Java, but only those to the north are currently exploited. There are onshore wells near Cepu, Surabaya, and Indramayu as well as offshore wells north of Jakarta and Madura. The Western Sunda and Laut Jabar oil fields together hold an estimated 16.3% of future national capacity, whereas the Laut Jatim and Jatim fields contain only about 0.1% each (RePPProT 1989).

The young volcanoes of Java and Bali also have potential as sources of geothermal energy relatively close to the surface which heats up ground water under pressure. Dry steam from wells sunk southeast of Bandung powers the Kamojang electricity station which currently delivers 140 MW. A further 55 MW is planned to come into operation by 1995-96. Even so, this will represent only 10% of the potential in the province. A small station on the Dieng plateau delivers 2 MW from a major source of wet steam, but this will soon be expanded to 55 MW (Tjinda 1993). This source of energy is set

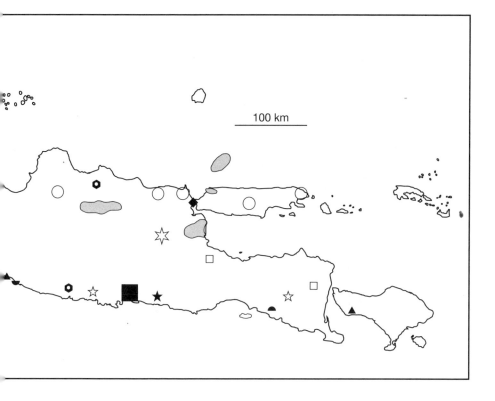

100 km

to undergo expansion in the future, and detailed feasibility studies are being made elsewhere.

Java has only three active metal mines: Cikotok and Mt. Pongkor (gold) (p. 788), and Cilacap (iron sands). These are not of particular national significance, but are nonetheless of considerable local importance. Gold is also sought by individual prospectors whose activities can be dangerous and disproportionately damaging to the environment. Limestone is quarried for cement and lime, marble for decorative building stone, and clay is dug in large quantities for bricks and tiles. In addition, alluvial gravels and sands are dug from and around rivers and these are of major importance in the construction industry (RePPProT 1989) (figs. 3.11-14).

Figure 3.11. Iron sand mine on the beach east of Cilacap.
By A.J. Whitten

Figure 3.12. Panning for gold in a river near Blitar.
With permission of Kompas

Figure 3.13. Quarrying of limestone for lime near Karangasem, Lamongan.

With permission of Kompas

Figure 3.14. Digging of river sediments near Serpong for the construction industry.
With permission of Kompas

PHYSIOGRAPHIC REGIONS

The physical structure of Java and Bali is complex, and 128 land systems were recognized in the appropriate volume of the detailed Regional Physical Planning Project for Transmigration study (RePPProT 1989). Land systems recognize combinations of rock types, hydro-climatology, landforms, soils, and organisms, and the links between them. A land system is not unique to one locality but rather recurs wherever the particular combination of characteristics is found. Similar land systems can be grouped in different ways to produce 'physiographic types' and 'physiographic regions'.

Each land system is found in only one physiographic type, and the nine types are mapped in figure 3.15. The grouping of land systems (or similar units) into general types was also used in detailed surveys of the Cimanuk River basin (Dent et al. 1977), in the lower Citanduy River basin (Verstappen 1975), the entire Citanduy basin (RMI/PRC 1986), and Ujung Kulon National Park (Hommel 1987) (p. 769).

Java and Bali can also be divided into four major physiographic regions among which certain land systems may be shared. These are similar to the three physiographic zones used by Pannekoek (1949) to describe the geomorphology of Java, except the two northern RePPProT regions are considered as one. The boundaries between the RePPProT regions (fig. 3.16) are not defined precisely, but can be divided into subregions with unique characteristics. Parts of the four regions can be found in each of the three major provinces of Java, but only two are found in Bali.

A - Northern Alluvial Plains: A1 - Serang-Jakarta-Cirebon alluvial plains, A2 - Brebes-Pekalongan alluvial plains, A3 - Semarang-Rembang alluvial plains, A4 - Solo alluvial plain, A5 - Brantas alluvial plain;

B - Northern Foothills and Plains: B1 - Ujung Kulon-Banten rolling plains, B2 - Cipamingkis-Cimanuk foothills, B3 - Losari-Pemali embayment, B4 - Cacaban-Layangan embayment, B5 - Kali Bodri embayment, B6 - Kendeng foothills, B7 - Blora-Drajat karstic hills and plains, B8 - Madura karstic hills and terraces;

C - Central Volcanic Mountains: C1 - Karang volcanic complex, C2 - Cangkrang volcanic complex, C3 - Halimun-Gede volcanic complex, C4 - North Bandung volcanic complex, C5 - South Bandung volcanic complex, C6 - Pembarisan volcanic complex, C7 - Slamet volvanic complex, C8 - Ragajembangan mountain complex, C9 - Dieng volcanic complex, C10 - Merapi volcanic complex, C11 - Lawu volcanic complex, C12 - Liman volcanic complex, C13 - Arjuno volcanic complex, C14 - Tengger volcanic complex, C15 - Yang volcanic complex, C16 - Ijen volcanic complex, C17 - Bali volcanic complex, C18 - Muria volcanic complex;

D - Southern Dissected Plateaux and Plains: D1 - Jampang-Pangandaran dissected plateaus, D2 - Citanduy-Serayu-Progo ridges and plains, D3 - Sewu-Lengkong karstic plateaus, D4 - Betiri hills, D5 - Negara plains, terraces, and hills, D6 - Blambangan-Nusa Dua-Nusa Penida karstic terraces.

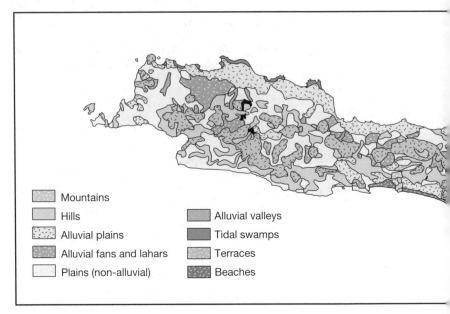

Figure 3.15. Physiographic types on Java and Bali.

After RePPProT 1990

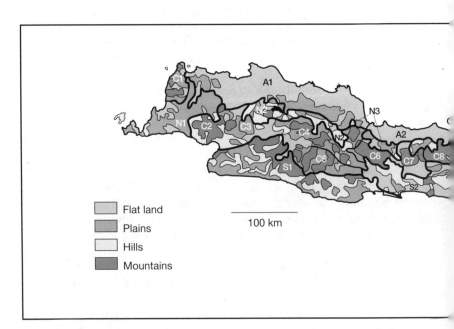

Figure 3.16. Physiographic regions of Java and Bali with their subdivisions.

From RePPProT 1990

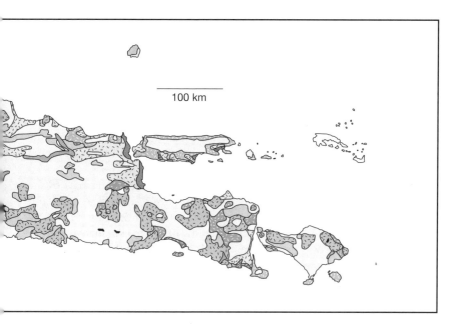

100 km

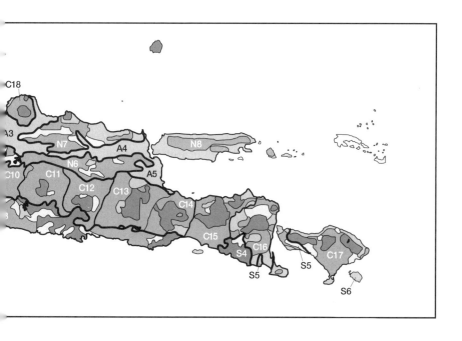

The four physiographic regions of Java and Bali are described in detail in RePPProT (1989) and summarised in RePPProT (1990) as follows:

1. The Northern Alluvial Plains. This region, like the three other regions in Java, is influenced strongly by the east-west structural trend of the island, and is long in an east-west direction and narrow. It occupies 21,219 km^2 and for descriptive convenience includes the Thousand Islands. As the name implies, it is dominated by alluvial plains, mainly derived from sediments laid down by many rivers draining the interior volcanic mountains. The rate of sedimentation is high and the plains are advancing seawards rapidly.

The mostly fine-grained sediments which form the greater part of the plains do not vary widely from west to east and are of mixed riverine and estuarine origin with a high proportion of volcanic minerals. The landforms are also very uniform and consist mainly of flat backplains crossed by slightly incised, narrow, meandering floodplains. Along the coast at intervals are fringing estuarine and deltaic plains and at the inland margin there are commonly gently sloping and slightly dissected alluvial fans which overlap onto the riverine backplains. Some of these fans are extensive and comprise a number of coalescent fans.

Five sub-regions can be recognised. From west to east these are the Serang-Jakarta-Cirebon Alluvial Plains, the Brebes-Pekalongan Alluvial Plains, the Semarang-Rembang Alluvial Plains, the Solo Alluvial Plains, and the Brantas Alluvial Plains.

The region has a markedly seasonal climate, at its wettest from December to April, but with some rain throughout the year and no months of absolute drought. It includes some of the driest areas of Java.

2. The Northern Foothills and Plains. This region has a total area of 22,226 km^2 and extends from Panaitan Island and the peninsula of Ujung Kulon in the far west of Java to Madura and the nearby Kangean Islands in the east. For descriptive convenience the outlying islands of Karimunjawa, Bawean, and Kangean are included in the region. Its southern boundary with the Central Volcanic Mountains (the following region) coincides basically with the change from dissected hills and plains over sedimentary rocks to the radial valley and spur patterns of volcanic cones.

The rocks are almost entirely sedimentary in origin and are derived from both volcanic and marine sources. They are young but are increasingly folded and faulted the nearer they are to the Central Volcanic Mountains. There is a clear tendency for volcanic materials to be dominant in the west and for marine sediments to become increasingly common in the east. A substantial proportion of the rocks are soft or weakly compacted marls and claystones, especially in the east. In the west there are superficial young deposits of volcanic ash.

The landforms of the region range from undulating to rolling and hilly plains. Dissection of the landscape by dense stream networks has resulted in low ridge systems with short but usually steep side slopes which are commonly aligned east-west, parallel to the main structural trend of the rocks. Soils are predominantly deep, well drained and medium to fine-textured. They vary from being slightly to strongly weathered, depending largely on slope stability, and from being low to rich in certain mineral nutrients, depending on the soil parent materials. The richer soils are invariably found over marine sediments, but so are the most shallow and rocky soils of the region. In the drier east there are considerable areas of soils which shrink and swell in response to seasonal climatic changes. These are difficult to use for agriculture and pose problems for engineers.

The region can be divided conveniently into eight sub-regions - the Ujung Kulon-Banten Rolling Plains, Cipamingkis-Cimanuk Foothills, Losari-Pemali Embayment, Cacaban-Layangen Embayment, Kali Bodri Embayment, Kendeng Foothills, Blora Drajat Karstic Hills and Plains, and the Madura Karstic Hills and Terraces.

3. The Central Volcanic Mountains. The total area of this, the largest of the four physiographic regions is approximately 60,139 km^2. It comprises essentially clusters of young volcanic mountains which emerge at more or less regular intervals down the entire length of Java and Bali. The great majority of the region is continuous, although very narrow at one point.

Eighteen subregions are identified on the basis of separate volcanic clusters. Most of these are surrounded by foothill fans and volcanic alluvial plains, which in many cases join adjacent volcanic centres. Starting from the west there are the Volcanic Complexes of Karang, Cangkrang, Halimun-Gede, North Bandung, and South Bandung. These are followed in the centre by the Pembarisan Mountain Complex, Slamet Volcanic Complex, and Ragajembangan Mountain Complex. Then in the east there are the Volcanic Complexes of Dieng, Merapi, Lawu, Liman, Arjuno, Bromo, Yang, Bali, and Muria. Most of their boundaries with the adjacent regions are clear.

On the whole, the western complexes tend to be more heterogeneous, containing inclusions of older volcanic, sedimentary, or even metamorphic rocks, while in the east the materials tend to be more homogeneous. Two of the central areas in fact contain little or no young volcanic material, and it is debatable whether they should be treated as subregions as is done here or whether they are sufficiently distinct to be classed as mini-regions.

The landforms are predominantly conical strato volcanoes and calderas with classical radial drainage patterns. This symmetry is disturbed wherever older volcanic or sedimentary strata occur with less orderly drainage and ridge patterns. The majority of the volcanoes have very steep but lightly dissected upper slopes, less steep but deeply dissected middle slopes, and lightly dissected lower slopes which are made up of coalescent fans, lahars, and alluvial plains.

4. The Southern Dissected Plateaux and Plains. This region of 30,620 km^2

is unequal in its distribution on the southern side of Java. In the west and centre it comprises a broad, continuous band, but the eastern part is thinner, more attenuated, and divided into separate components.

The basic lithology of the region is one of young, mixed volcanic and calcareous marine sediments lying on the southern flank of the series of young volcanic piles aligned along the centre of the islands. There are common intrusions of older volcanics, and one subregion is dominated by such material. Most areas have been uplifted and tilted to the south; folding is not an important feature.

The landforms are varied but all the non-alluvial plains and hilly areas are moderately to strongly dissected by a fine pattern of streams. There are several broad plateaux of tuffaceous sediments comprising undulating to rolling summit areas; in other places they have been tilted south and dissected to leave flat-topped ridges, widely separated by deep valleys. Karst scenery is very distinctive where well developed over pure limestones, but impure tuffaceous limestones and marls are more common, and karstic landscapes with some karst features mixed with non-karst topography are more usual. The limestones of the small eastern subregions are impure and exhibit the usual features of raised reefs and atolls. Large alluvial plains of riverine and estuarine origin are a characteristic of one subregion only. They are flat except for lightly incised floodplains and coastal beaches.

Six sub-regions are distinguished: the Tampang-Pangandaran Dissected Plateaux, Citanduy-Serayu-Progo Ridges and Plains, Sewu-Lengkong Karstic Plateaux, Betiri Hills, Negara Plains, Terraces, and Hills, and the Blambangan-Nusa Dua-Nusa Penida Karstic Terraces.

NATURAL HAZARD VULNERABILITY

As already indicated, Java and Bali are subject to various forms of natural hazard, and the vulnerability of the different areas was assessed in detail by the RePPProT project (1989). Short-term hazards, often referred to as natural disasters, include volcanic eruptions, earthquakes, landslides, and floods. Erosion is a long-term hazard (fig. 3.17) which may ultimately have greater overall impacts than the more dramatic hazards listed above, albeit with fewer deaths. Earthquakes generally leave no lasting marks, whereas eruptions and floods can actually benefit the local economy by enriching the soil. The timing of natural disasters cannot be accurately predicted, but certain areas are clearly more at risk than others, particularly where human management of the land is inappropriate.

The volcanoes of Java and Bali are well documented, and the surrounding areas are divided into three categories of risk - a forbidden zone, a primary danger zone, and a secondary danger zone. These assessments of risk assume that future eruptions will occur vertically through existing craters and will not be massive (that is, not of the scale of Toba, Tambora,

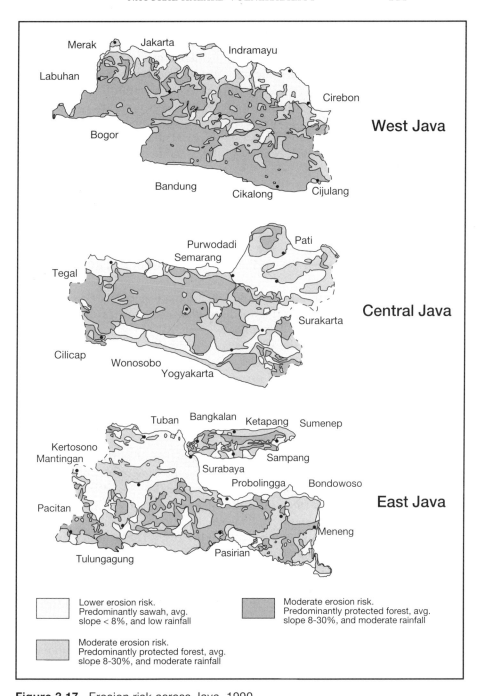

Figure 3.17. Erosion risk across Java. 1990.

Simplified after World Bank

or Krakatau) (Kusumadinata 1979).

Earthquake centres are concentrated in the offshore area between the south coast of Java and Bali and the Java Trench. The size of earthquakes are measured in two ways: the intensity is measured using the subjective Modified Mercalli Index (MMI) which is based on visible damage, and the magnitude by the logarithmic Richter Scale which relates to the actual energy released in the bedrock. The worst earthquakes in Java and Bali over the last century rate about 9 on the MMI, and 8 on the Richter Scale (Soetoardjo et al. 1985). Neither of the scales necessarily correlates with loss of life.

Landslides are one of the most dangerous results of earthquakes, even quite small ones, especially in areas of erosion risk, where heavy rain falls on already saturated soil, and where land use is inappropriate. For example, the conversion of steep *talun-kebun* gardens (p. 617) to rain-fed sawah caused a landslide that buried and killed over 50 people in late 1992. Especial danger is faced where such hazardous areas lie near geological faults.

Flooding is a natural phenomenon and many people are very familiar with it, particularly along the northern plains of Java (figs. 3.18-19). The land systems most susceptible to floods cover 14% of Java and Bali although nearly one quarter of the islands is liable to flooding. Most flooding is shallow, of short duration, and agriculturally beneficial, even if it is also unpleasant and damaging to property. However, severe storms on already saturated, deforested catchments can cause floods that lead to serious losses of life, property, crops, and livestock. Although the events may be rare, every area liable to flooding is statistically likely to experience severe flooding at some time.

Management of a river course sometimes involves the raising of the banks for flood 'protection' in order to compensate for the deposition of sediment in its lower reaches. When the river does flood, unfortunately, these banks prevent the water returning to the river as it falls, such that flood waters remain on the land far longer than under natural conditions, unless they can be pumped back up to the river. The planning and management of floodplain areas needs to integrate engineering skills with an appreciation of natural process to a greater extent than is evident at the moment. There is a school of thought which suggests that allowing rivers to flood into natural or artificial floodplains is more cost-effective in the long term than conventional methods such as straightening or deepening channels, or raising their banks.

Erosion is a natural process which has formed fertile valleys and plains, and the resulting deposition has extended the northern shore of Java. The rate of erosion has increased dramatically over the last 100-150 years, however, as landless farmers have cut down forests and attempted agriculture on marginal, unterraced soils. Much has been done to improve the

Figure 3.18a.
Flooding in
Jakarta in
1993.

With permission of
Kompas

Figure 3.18b. Novel road transport is required when the flood waters rise.
With permission of Kompas

situation, but the intensive growing of vegetables on extremely steep volcanic slopes of, for example, the Tengger and Puncak regions is tantamount to 'agricultural mining' and is laying the land open to serious erosion and landslides. In the RePPProT study six of the most eroded land
systems (all in limestone areas) are not regarded as current hazard areas.
In general, land systems on weakly hardened (indurated) marls are
regarded as a greater potential hazard than land systems on volcanic

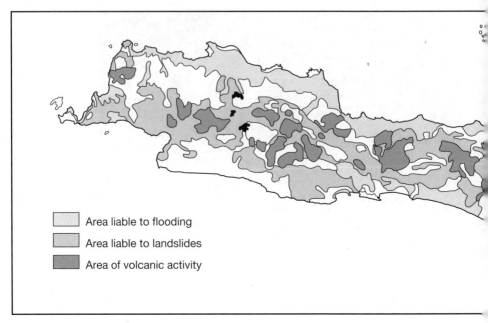

Figure 3.19. Areas of volcanic, landslide, and flooding hazard on Java and Bali.
After RePPProT 1990

materials. Nine land systems, covering 1% of Java and Bali, have such a high hazard potential that it was felt they should never be used for agriculture even with terraces. Some 23 other systems covering 29% of the islands present a severe or very severe erosion hazard and should only be used with permanent vegetation cover. It is some of these that are currently being used to grow vegetables in the Tengger and Puncak regions. These various hazards are indicated on figure 3.20, although much greater detail is given on the 1:250,000 maps of RePPProT (1989).

Among the unfathomable potential future hazards is the effect of increased ultraviolet radiation (UV) on the genetics of plants and animals. Penetration of the atmosphere by shortwave UV of a narrow waveband is very sensitive to the concentration of ozone, and this is why the amount of UV radiation reaching the earth's surface increases with altitude, by 14%-18% for every 1,000 m. It has now been shown that a family of manufactured compounds known as CFCs released into the atmosphere from refrigerators and air conditioners react in the atmosphere over a long period to reduce ozone concentrations. Since the natural concentration of ozone over the equatorial regions is lower than elsewhere, further reductions as a result of CFCs entering the atmosphere mean that Java and Bali would be likely to receive high levels of damaging UV radiation which could cause skin carcinomas in humans, as well as genetic changes,

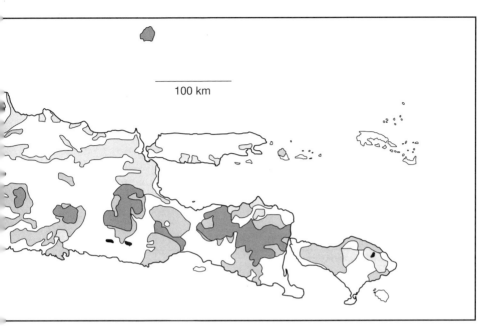

altered reproductive biology and chemistry in other organisms, all of which could amount to major ecosystem changes in terms of competitive balance and relative abundance of species (Caldwell et al., 1989).

SOILS

The various soil classifications and the different intensities of surveys in different parts of the island lead to considerable confusion. The soil classification used here is based on the Soil Taxonomy of United States Department of Agriculture (USDA 1985) which, while not entirely suited to the Indonesian situation, enjoys wide international usage (albeit with almost unpronounceable names) and is suitable for small-scale differentiation. Thus distribution of soil types is complex, but it reveals two basic divisions based mainly on the climatic differences between the moist and seasonally-dry areas, although altitude, surface rocks, and topography are also influential (table 3.4).

Ustropepts, calciostolls, paleustults, paleustalts, and haplustalfs are found only in the seasonal areas, as are eutropepts characterized by their high base saturation. In the humid areas (the highlands and most of western Java and Bali), soil weathering is essentially continuous because of the absence of long dry periods. As a consequence, dissolved minerals gradually leach away in mature soils to produce acid, kaolinitic, aluminium-rich clays that are unable to hold dissolvable minerals because of the rain. Fertility is only maintained by the presence of humus and litter on the soil surface.

Most of the soils in the hilly parts of the region are relatively immature, however, and because they retain moisture, mineral nutrients, organic matter, and phosphorus they are potentially very productive. Exceptions are soils over limestones and marls. Both the mature and immature soils are subject to severe erosion when left uncovered, and maintaining cover and

Table 3.4. Soils of Java and Bali using USDA soil taxonomy 1985.

Soil type	Characteristics
Hydraquents	Undeveloped, permanently saturated soft muddy soils
Troporthents	Undeveloped rocky soils of hot climates
Ustorthents	Undeveloped rocky soils subject to seasonal moisture stress
Tropopsamments	Undeveloped sandy soils of hot climates
Ustipsamments	Undeveloped sandy soils subject to seasonal moisture stress
Tropaquepts	Slightly weathered, permanently saturated soils of hot climates
Dystrandepts	Slightly weathered volcanic ash soils with low base saturation and thick, black topsoil
Eutrandepts	Slightly weathered volcanic ash soils with high base saturation and thick, black topsoil
Vitrandepts	Slightly weathered volcanic ash soils with dominantly coarse textures and thick, black topsoil
Dystropepts	Slightly weathered soils of hot climates with low base saturation
Eutropepts	Slightly weathered soils of hot climates with high base saturation
Ustropepts	Slightly weathered soils subject to seasonal moisture stress or with soft powdery lime concentrations, and with high base saturation
Calciustolls	Moderately weathered, weakly acid to neutral soils subject to seasonal moisture stress
Rendolls	Moderately weathered, shallow soils on calcareous parent materials
Tropudalfs	Well weathered soils of hot climates with finer-textured subsoil and high base saturation
Paleustalfs	Well-weathered soils with thick, uniform finer-textured subsoil and high base saturation subject to seasonal moisture stress
Haplustalfs	Well-weathered soils with finer-textured subsoil and high base saturation subject to seasonal moisture stress
Paleudults	Strongly weathered acid soils with thick, uniform, finer-textured subsoil and low base saturation
Tropudults	Strongly weathered acid soils of hot climates with finer-textured subsoil and low base saturation
Palenstults	Strongly weathered acid soils subject to seasonal moisture stress with thick, uniform, finer-textured subsoil and low base saturation
Haplustults	Strongly weathered acid soils subject to seasonal moisture stress with low base saturation

Source: RePPProT (1990)

an organic topsoil are essential elements of soil conservation programmes.

In the seasonally-dry areas, evaporation from the soil during the dry periods pulls the dissolved minerals up through the soil to the surface by capillary action where they eventually crystallize out of solution. Thus the loss of nutrients is less than in the humid zone. Therefore, in the seasonally dry zone, especially over calcareous rocks, fertile clays form in the mature soils but unfortunately they swell, become impervious, intractable, and sticky when the rain comes, and in the dry periods they tend to shrink, crack, and become very hard. Immature soils on the young volcanoes in this zone are less problematic.

Top dressings of ash from volcanic eruptions periodically fall on the soils of Java and Bali, and if these are not too thick, they weather rapidly to provide a free application of useful minerals. The alluvial soils of Java and Bali develop more or less independently of climate because they occur where the water table is always high. They are generally fine-textured and poorly drained.

CLIMATE

Palaeoclimate

Climate has changed throughout the ages. In Cretaceous times, 100 million years ago, tropical surface waters may have been up to 5°C warmer than they are today, and further back in the Carboniferous, somewhat cooler than today. One consequence of the higher Cretaceous temperatures was that the reefs were not made of coral (which are intolerant of high salinity and temperature) but of clams. Indeed, corals could not live closer than about 3,000 km from the equator at that time, whereas today that is the area to which they are more or less confined (Hecht 1990).

Throughout the last million years or so there have been considerable fluctuations in sea level throughout the world as water has alternately been incorporated in and released from the polar ice caps, although on average sea level has been lower than it is at present. Evidence from various sites in southeast Asia suggests that the most recent sea level maximum was about 4,500-6,000 years ago when it reached five metres higher than present, and temperatures were 1°-2°C warmer, but sea levels of 25-50 m above present levels may have occurred during the last two million years (Tjia 1980, 1984; de Klerck 1983). The higher Pleistocene sea levels are evidenced by beach ridges, abandoned marine cliffs, and coral well above sea level, although some may have been lifted by tectonic movements (Joongjai and Rosengren 1983). For the last 4,500 years sea level has been falling at

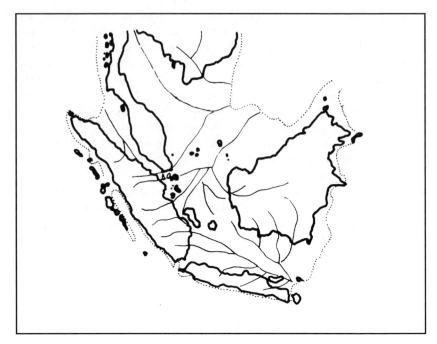

Figure 3.20. The Sunda Shelf during the driest period of the Pleistocene showing the main rivers and present coastlines.
After Whitten et al., 1987c

irregular rates, but there is every indication to suggest that sea level is now rising (p. 74). The non-alluvial coasts of, say, west-facing Java have had roughly the same alignment and character for the last 6,000 years (Bird and Ongkosongo 1980). Since then, in response to the fairly stable (though oscillating) sea level, fringing reefs have become established, rock and earth cliffs and platforms have formed, and beaches have developed. It is important to understand, however, that the present geographic extent of Java and Bali are most atypical of the Pleistocene/Holocene period as a whole. For much of the time sea level has been lower than at present, up to 150 m lower when three times the present land area of the Sunda Shelf (the continental part of the Sunda Plate) was exposed, and depth measurements reveal the submerged courses of large rivers on the sea bed (fig. 3.20). Even when sea levels were only 40 m below present it would have been possible to walk between Java, Borneo, Sumatra, and the Malay Peninsula, although Bali would have been separated from Java by a strait, slightly narrower than that found today. During these periods of lower sea level, there was probably lower rainfall and humidity, greater diurnal and sea-sonal variations in temperature, and the seasonal areas of Java and Bali would have been more extensive than now (Verstappen 1975). During these times the areas of monsoon forest would have been greater, and exchange of monsoon species through a virtual corridor of monsoon forest between Asia and Australia could have been possible.

The last major cooling was in the late Pleistocene 18,000 years ago when the sea temperature was only about 2°C lower than now and the sea level about 100 m below its present level. The atmosphere and climate were drier than now, as indicated by the lapse rate (the rate at which air cools with altitude) being about 0.8°C per 100 m from 17,000-18,000 B.P., com-pared with 0.6°C now. Thus at 1,500 m altitude the average temperature would have been 7°C lower than today (12° against 19°), and at 3,600 m (e.g., Mt. Semeru) about 12°C lower (-6° against 6°) (Flenley 1985).

Present Climatic Patterns

The yearly climatic variations experienced today in Java and Bali are gov-erned by the oscillations of air masses within the inter-tropical conver-gence zone. These oscillations are caused by the passing of the sun between the Tropics of Cancer and Capricorn crossing the equator every six months. Low atmospheric pressure occurs at the moving 'heat equator' (where the sun is overhead) with the result that airstreams from the northern and southern hemispheres converge causing air uplift, cloud formation and heavy rain. The variations in sea temperature through the year are small, but nevertheless important. A warm pool of water, about 200 m deep, oscillates in an irregular manner between Indonesia and the eastern Pacific. The warm phase of this so-called El Niño event results in extended and drier-than-normal dry seasons and droughts. Such events have occurred in 1972, 1976, 1982-83, 1986-87, and 1991, and have resulted in water shortages, forest fires, and reduced rice harvests (fig. 3.21). Severe droughts before 1970 were not necessarily associated with El Niño events, but since then all droughts have been. Enough is now known of the initial stages of the events that, provided appropriate data are collected by each June, the duration of the rest of the dry season and the likely average monthly rainfall in the following few months can be calculated (Wasser and Harger 1992).

Rainfall exhibits marked seasonal differences, though more in some parts of the islands than in others (fig. 3.22), whereas temperature and rel-ative humidity vary more between night and day than between months or years. Wind speed and sunshine show some seasonality in their behaviour. Rainfall is probably the most important ecological variable because it has direct relevance to plant growth in the lowlands. Rainfall records are avail-

Figure 3.21. Praying for rain during an El Niño year.

With permission of Kompas

Table 3.5. Areas of Java and Bali receiving different quantities of rainfall (mm).

	Arid		Dry		Moist		Wet		Very wet		
	<1,000		1,000-1,499		1,500-2,999		3,000-4,999		>5,000		Total
	km²	%	km²	%	km²	%	km²	%	km²	%	km²
Java and Bali	112	<1	8,301	6	88,397	64	40,945	30	449	1	138,204

Source: RePPProT (1990)

Figure 3.22. The dry season in parts of Java and Bali can be severe.

With permission of Kompas

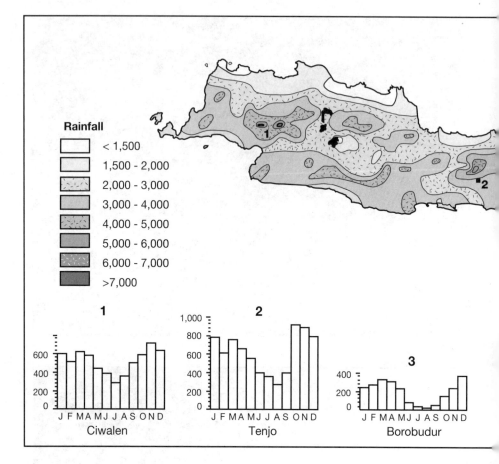

Figure 3.23. Mean annual rainfall in Java and Bali with six examples of distribution patterns.
After RePPProT 1990

able from 3,200 stations in Java, and some of these have been operating more or less continuously for over 100 years. Less than 1,500 mm falls in many parts of the north and northeast coasts (fig. 3.23-24), and it is common for the annual rainfall around Baluran National Park on the northeast tip of Java to receive less than 1,000 mm. At the other extreme, more than 6,000 mm is regularly received on the upper slopes of the central mountains of western Java. Over 90% of Java and Bali receives at least 1,500 mm (table 3.5). Eastern Java is drier than western Java. The wettest places in Java are in the Ragajembangan group of mountains in Central Java, including the mountains of Kendalisodo, Prahu, Sindoro, and Sumbing

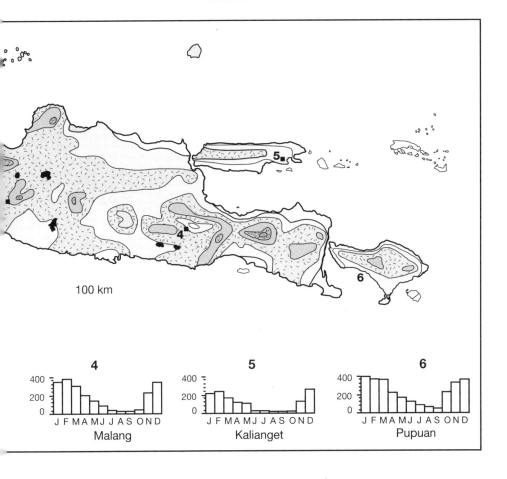

5

4

6

| 4 | 5 | 6 |

400
200
0
J F MA MJ J A S O N D
Malang

400
200
0
J F MA MJ J A S O N D
Kalianget

400
200
0
J F MA MJ J A S O N D
Pupuan

100 km

Figure 3.24. Part of the dry northeastern coast of Bali near Tejakula.

By A.J. Whitten

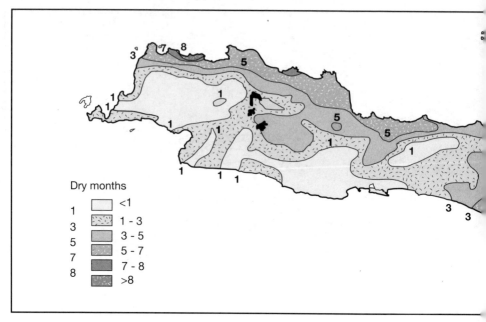

Figure 3.25. Numbers of dry months experienced on Java and Bali.
After RePPProT 1990

where the annual rainfall reaches more than 7,000 mm — just 15 km down the slopes, the average annual rainfall is less than 3,000 mm (RePPProT 1989). The atlas of rainfall maps by Sijatauw (1973) allows the distribution of areas experiencing different numbers of wet and dry months to be calculated. For these purposes, 'wet' months are defined as those with an average monthly rainfall of at least 200 mm, enough to grow a crop of irrigated rice. Conversely 'dry' months are those in which less than 100 mm falls, a rainfall which is inadequate for most upland crops (fig. 3.25).

Temperature can vary on a relatively local scale. For example, highly urbanized areas may be four degrees hotter than neighbouring rural areas, because buildings trap the air and heat. The opposite effect is seen in the forest, where temperatures are typically four degrees cooler than in rural areas, and up to ten degrees cooler than in an urban centre. Thus shading and evaporative cooling can provide a new stimulus for the design or redesign of tropical cities (Tay 1989).The increase in reflective building materials and the loss of vegetation have led to an increase of 1.65° in the average temperature of Jakarta over the last 125 years (Wasser and Harger 1992). As is well known (and positively enjoyed by many city dwellers at

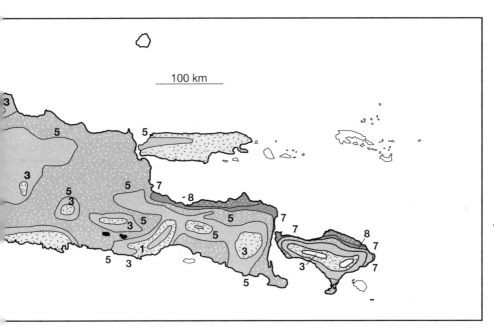

weekends), temperature decreases with altitude by about 0.6°C every 100 m. However, given that over 90% of Java and Bali lies below 500 m, the temperature experienced by most people is a maximum of 31°-33°C and minimum of 22°-24°C. On mountain tops the mean minimum may fall below 10°C, but frosts can occur in certain places above 1,500 m where cold air collects in 'frost pockets' (van Steenis 1972). Indeed, both slope and aspect impose striking variations on the general picture.

The general hot, wet, and humid regime of many parts of the tropics is disturbed in those areas affected by trade or monsoon winds. The dry southeast monsoon winds blowing from Australia in the middle of the year have a major influence in central and east central Java and Bali, and this extends along the north coast as far as Jakarta to the west and eastern Bali to the east. This season can begin in March extending, in recent years, almost to January giving a dry season of between one and nine months. The mountains complicate this picture because high rainfall can be expected on the southeast slopes facing the wind, leaving very dry areas to the north and west. These dry periods affect the flora in many ways, not least through climate-related soil features. For example, in the dry areas

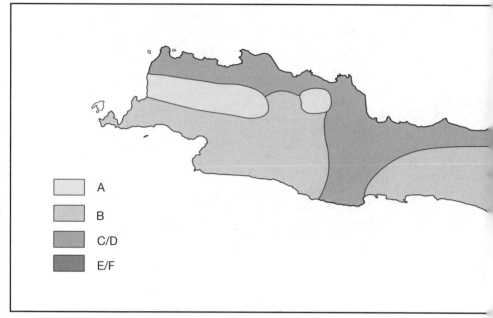

Figure 3.26. The Schmidt-Ferguson climate classification for Java and Bali.
After Schmidt and Ferguson 1951; Whitmore 1984

evaporation exceeds precipitation for much of the year, and so soils do not become leached, remain richer in nutrients, have less humus, and are hence less acid.

Thus climate clearly plays a major role in the distribution of plants, and relatively minor differences may have a notable influence. Various attempts have been made to give meaningful pictures of climate and its ecological effects (van Bemmelen 1916; Boerema 1931 (both in van Steenis and Schippers-Lammerste 1965); Mohr 1933; Schmidt and Ferguson 1951; Oldeman 1977; Fontanel and Chantefort 1978). Clearly, total rainfall alone is inadequate as a descriptor of climate in a generally wet area because evaporation rates and daily or monthly distributions may differ more in some areas than in others. It is thus the distribution of rain through the months and years that really counts, in particular the strength and duration of the annual drought.

The classification with the widest currency is that of Schmidt and Ferguson (1951) based on the ratio (termed Q) of dry months to wet months expressed as a percentage, where the former received less than 60 mm and the latter more than 100 mm (fig. 3.26). They formed six categories

(table 3.6), the distribution of which correlate quite well with vegetation zones (Whitmore 1984). Detailed analyses of climatic preferences and lists of plant species indicating moist climates and extreme drought are given in van Steenis and Schippers-Lammerste (1965). Examination of lists of forest trees from different parts of Java and Bali showed that the distribution of vegetation types was delimited well by the number of dry months and rainfall, and this was used to map the original distribution of the natural

Table 3.6. The six categories of the Schmidt-Ferguson climate classification encountered on Java and Bali.

Climate type	Category	Percentage (Q) of wet months to dry months
Perhumid	A	Q = 0-14%
Slightly seasonal	B	Q = 14-33%
Seasonal	C	Q = 33-60%
Seasonal	D	Q = 60-100%
Strongly seasonal	E	Q = 100-167%
Strongly seasonal	F	Q = 167-300%

Sources: Schmidt and Ferguson (1951), Whitmore (1984)

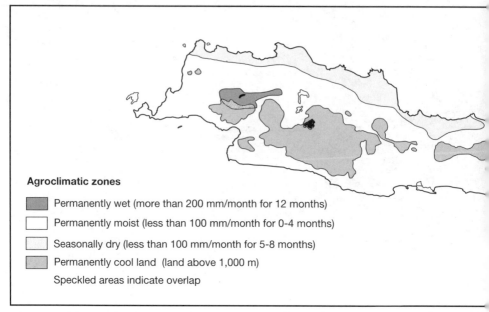

Figure 3.27. Agroclimatic zones.
After RePPProT 1990

vegetation types (p. 190). The latest agroclimatic map is that produced by RePPProT (1990) (fig. 3.27) which divides the islands into three zones of permanently wet, permanently moist, and seasonally dry. The permanently wet zone is restricted to a single block east and west of Bogor, and the seasonally dry zone is found predominantly along the northern coasts and in a block around Wonosari, Madiun, and Jombang.

Climate is, to a certain and an increasing extent, determined by people's activities. For example, the loss of vegetation with consequent increase in reflectivity of the ground surface and the decrease in evapotranspiration lead to hotter and drier climates, and this is most obvious in urban sites. The issue of global warming is addressed elsewhere (p. 72).

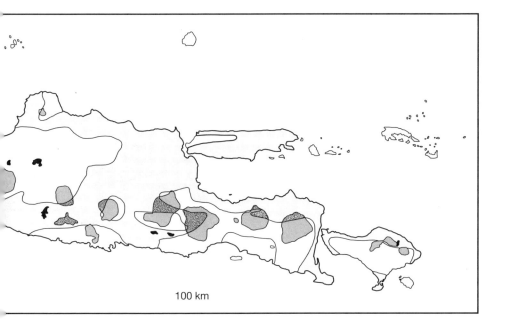

100 km

HYDROLOGY

Water Catchments

Most of the water catchments in Java and Bali are short (30-70 km), narrow, steep, and with an area less than 250 km^2. Java, not surprisingly, has the larger rivers, and 24 catchment areas are greater than 750 km^2, including the 'critical watersheds' in which the government pays special attention to developing and managing the riverbanks and uplands. It is regrettable that political boundaries do not follow the natural boundaries of watersheds because this complicates the control and monitoring of pollution and land management (p. 49). Seven of the catchments exceed 3,000 km^2, and the largest are the Brantas (11,050 km^2) and Bengawan Solo (15,400 km^2) (table 3.7, fig. 3.28) which are described in more detail below (fig. 3.29).

Bengawan Solo River. The waters of the Solo River flow from the slopes of Mt. Lawu and Mt. Merapi in Central Java to a peculiar 'single-finger', human-originated muddy delta north of Surabaya. Last century the river

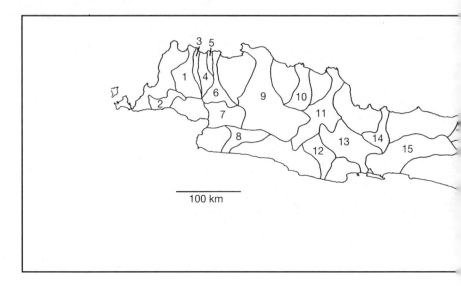

Figure 3.28. The major catchment areas of Java and Bali. 1 - Ciujung, 2 - Cibaliung, 3 - Cidurian, 4 - Cilontar, 5 - Cimauk, 6 - Cisadane, 7 - Cimandiri, 8 - Cidalog, 9 - Citarum, 10 - Ciliwung, 11 - Cimanuk, 12 - Ciwulan, 13 -

Table 3.7. Catchment areas and estimates of runoff for major rivers in Java and Bali. Full data are not available for Bali, and the Ayung data below is for only the upper part above Buangga which records some 60%-70% of the flow (S. Walker pers. comm.).

	Catchment area (km²)	Rain (mm)	Runoff Depth (mm)	Run-off m³/s	Province
Cipaloh, Ciliman	940	3,190	1,790	53	W
Cibaliung, Ciujung	1,980	3,050	1,650	105	W
Cimanceur	760	2,490	1,090	27	W
Cisadane	1,440	3,840	2,440	110	W
Cimandiri Citarik	1,980	3,380	1,980	125	W
Cidalog Cibumi	1,330	3,720	2,320	98	W
Citarum	7,250	2,600	1,200	275	W
Cipunegara	1,330	2,550	1,150	49	W
Cimanuk	3,260	2,530	1,130	115	W
Ciwulan	1,460	3,640	2,240	105	W
Citanduy	3,380	3,280	1,810	195	W+C
Pemali	1,100	2,780	1,380	48	C
Serayu	3,700	3,710	2,310	270	C
Bodri	800	3,160	1,760	45	C
Progo	2,380	2,780	1,380	105	C+Y
Tuntang	2,160	2,710	1,310	90	C
Serang	3,860	2,450	1,050	130	C
Oyo	1,600	2,160	760	39	Y
Juwono	1,380	2,060	660	29	C
Bengawan Solo	15,400	2,100	700	340	C+E
Brantas	11,050	2,220	820	290	E
Bedadung, Petung	1,050	2,420	1,020	34	E
Sampean	1,320	1,940	540	23	E
Mayang	1,330	2,290	890	38	E
Ayung (above Buangga)	179	2,850	1,450	8	B

Source: RePPProT (1989).

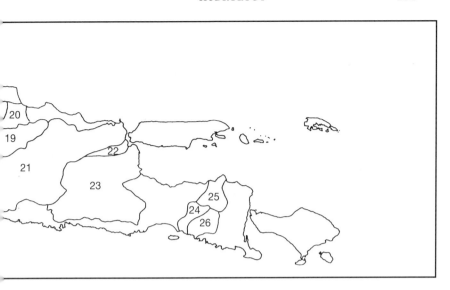

Citanduy, 14 - Cimanuk, 15 - Serayu, 16 - Pemali, 17 - Progo, 18 - Tuntang, 19 -
Serang, 20 - Juwono, 21 Bengawan Solo, 22 - Porong, 23 - Brantas, 24 - Bedudung,
25 - Sampean, 26 - Sanen.

After RePPProT 1989

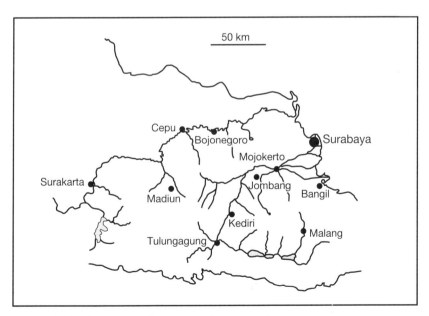

Figure 3.29. Courses of the Solo and Brantas rivers.

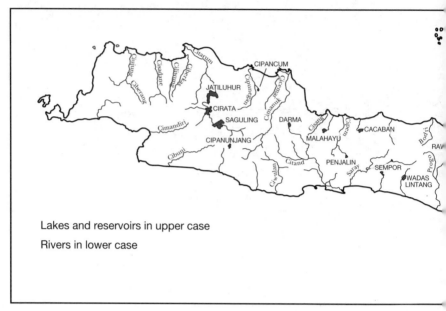

Figure 3.30. Major rivers, lakes, and reservoirs on Java and Bali.
After RePPProT 1990

flowed into the Madura Strait, and the sediment it brought down hampered navigation, so in the 1880s the channel was forced somewhat north but this was ineffective. In the final years of last century, a channel was dug parallel to the eastern coast so that the water flowed north into the Java Sea. Some small alternative channels have formed, but almost all the water flows along the main channel. There are about 11 million cubic metres of sediment deposited by the river each year resulting in an annual extension of about 70 m (Erftemeijer and Djuharsa 1988), although this has slowed as sediment is trapped behind large dams, check dams, and sluices. The part of Java nearest the estuary is very flat, and during the dry season tidal influence can be detected 100 km upstream.

Brantas River. The Brantas is a peculiarly-shaped river draining water from an area of over $11,000$ km^2 from the southern slope of Mt. Kawi-Kelud-Butak, Mt. Wilis, and the northern slopes of Mt. Liman-Limas, Mt. Welirang, and Mt. Anjasmoro. At Mojokerto the river starts to divide between the R. Porong to the south and the R. Surabaya to the north. Between the two is an alluvial area comprising some tens of metres of alluvial deposits sitting on mid-Pleistocene marine deposits. Indeed

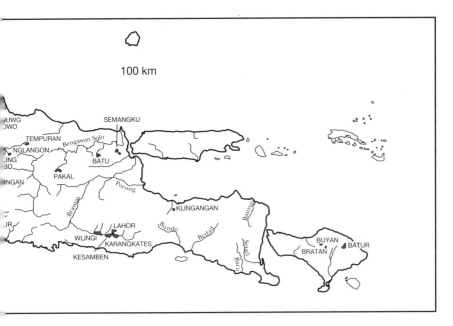

Mojokerto, now 30 km from the sea, was a port in a major estuary and was used by sea-going vessels from at least the tenth century until the end of the fourteenth century. The change was due to the increase in river-borne silt, and since 1880 the delta at the mouth of the Porong has been actively forming (Hoekstra 1987; Erftemeijer and Djuharsa 1988). Eighty years ago the serious condition of the Brantas watershed as a result of deforestation had already been noted (Altona 1913). In the dry season water was not always available for irrigation, and siltation made rivers more susceptible to flooding and obstructed shipping around Surabaya. The situation had been aggravated by the large qualities of ash spewed out from Mt. Kelud near Blitar, and the mud flows or lahars which plagued the country around the volcano. Considerable land management has been and is being effected to control erosion and sedimentation, and an integrated resource and land use plan for the Brantas watershed has recently been completed (Taylor and Soetarto 1993).

Rivers, Lakes, and Reservoirs

In no part of Indonesia is the hydrology better known than on Java, because of the exhaustive and long-term efforts to support the population and the national rice needs by optimal use of land and water resources. As a result of this intensive use there is no free-flowing or 'wild' major river remaining on Java, all of them having been raised at some point along their length for electricity, flood retention, dry season water supplies, irrigation, or all four (fig. 3.30). Most reservoirs are said to be multipurpose, but the World Bank does not regard this to be true of any of them (World Bank 1990a). Java has 36 reservoirs and 13 lakes, and Bali has just one reservoir and four lakes (p. 415).

The irrigation networks on Bali are generally better developed and maintained, more intensively used, and better organized than are the networks on Java. Bali grows a larger percentage of the rice crop on double cropping per year using simple and semi-technical irrigation than any of the Java provinces. The physical characteristics of Bali do not in any case lend themselves to the construction of reservoirs because the valleys are too steep and narrow (Walker et al. 1977), and only one small reservoir has been built, Palasari, in West Bali. Bali's small rivers are also unsuitable for major hydroelectric projects, but micro-hydro schemes of up to 250 KW using the low heads but reliable supplies could make a major contribution to the development of remote villages (RePPProT 1989).

Land Use and Effects

The massive investment in reservoirs, dams, irrigation systems, water supply works, power stations, and flood control structures is jeopardized by the effects of decades of inappropriate land use in the uplands (fig. 3.31). To correct and rehabilitate is increasingly expensive, but it is well worth doing because there are many indications that the productivity of the uplands could be much increased (Tarrant et al. 1989), and it is widely recognized that while the hydrological potential of the mountainous areas is high, the watersheds must be managed carefully to maximise this potential (Martopo 1984). Indonesia must come to grips with these acute problems, and it must succeed.

Many of the efforts expended to rehabilitate Java's critical watersheds have attempted to influence streamflow and erosion not only by mechanical measures such as bench terracing, but also by the management of the vegetation cover. Forest cover is the best form of insurance against soil loss, decreases in water yield, and flooding, and the mossy upper montane forests (p. 521) have a particularly important role in water capture and slope stability (Bruijnzeel and Proctor 1993; Bruijnzeel et al., 1993). Forests do not, however, in any way eliminate the problems as is sometimes intimated, and this is discussed below. Indeed, these topics have been surrounded by misinterpretation, misunderstanding, misinformation, and

Figure 3.31. The critical land monument at Jumantono, near Surakarta, on the western slopes of Mt. Lawu is preserved to show the crippling effects of inappropriate land management.

With permission of Kompas

myth, and it is important that the situation is clarified in the minds of politicians and decision makers (Hamilton and King 1983; Bruijnzeel 1990). There is, of course, no single hydrological model that can be applied to every watershed situation because of the differences in morphology, soils, and geology. Even so, the figures determined for a range of variables such as annual evapotranspiration and rainfall interception in a wide variety of situations are remarkably consistent between sites, thereby allowing guarded extrapolations to be made.

Regreening

The Indonesian Regreening Programme began in the early 1970s under Presidential Instruction and was administered and managed by the Ministry of Forestry. Regreening involves the planting of perennial crops or grass in combination with the construction of artificial structures such as check dams to prevent erosion in areas outside state forest land. The dual aims of the programme were to stabilize erosion on the vast areas of critical lands (table 3.8) and to improve hydrological conditions in the critical watersheds. 'Critical land' is defined by criteria embracing hydrological conditions, as well as areas:

- which are or will be used for water supply;
- subject to flooding or drought, areas subject to inappropriate agricultural activities and/or a high ratio of degraded to protected land;
- where the understanding of soil conservation among the local people is low; and
- areas with a high density of human population.

The tree species used most widely, the native (Sumatran) pine *Pinus merkusii*, encountered problems because it was a forestry species rather than a species giving direct benefits to the army of villagers who were encouraged to give of their time and energy to support the programme. Latterly *Acacia auriculiformis* from eastern Indonesia and Australia was distributed for planting because villagers could eventually exploit it for timber, but neither this nor *P. merkusii* have any significant role in the biological ecosystem since they are hardly visited by birds which could speed up the rehabilitation of the critical lands by dispersing seeds.

Acacia auriculiformis has dramatically changed the landscape of Madura over the last couple of decades, and the people talk of an improved and more equitable rural environment, and many farmers have found that the trees provided a profitable crop. This was the experience of the Mt. Mere Farmers Group, southeast of Bangkalan on Madura, which received a Kalpataru environment award in 1988. Unfortunately many areas of Madura have lost the benefits of the trees because many of them have been felled and often not replanted, and in general the trees are lopped so heavily that they offer very little protection for the soil.

Another tree, lamtoro *Leucaena leucocephala*, was planted in drier areas, but it is inappropriate unless it is regularly and heavily pruned for fodder, and in such areas deciduous species make much more sense if water and soil protection are the main goals. Regreening has had very disappointing

Table 3.8. Areas (ha) of critical lands inside and outside forest land in 1990.

	Inside forest land	Outside forest land	Total
West Java	84,700	488,000	572,000
Central Java	0	316,300	316,300
Yogyakarta	3,600	24,900	28,300
East Java	0	359,500	359,500
Bali	9,400	74,400	83,800

Source: *Directorate-General of Reforestation and Rehabilitation*

results with very low survival and no marked decrease in the sediment loads of rivers. It has been argued that the programme actually lowers the carrying capacity of the land and increases population pressure. It has been proposed that best results would be achieved if the trees were allowed to grow by natural regeneration, and investment diverted to economic development of new income sources for the rural people (Soemarwoto and Soemarwoto 1983; Soemarwoto 1991b).

Regreening, it should be remembered, was deemed to be necessary because inappropriate uses of land led to such degradation that many areas were no longer productive. In most cases the degradation was caused by landless or small-scale farmers who were seeking land and food at the literal and environmental margins of cultivable land. It is clear that unless the symptoms of land degradation are dealt with, then no real advance will be made, and the best way to do this is to create jobs outside the agriculture sector and of a type that will not add to the problems facing land. Unless this enormous problem is solved, many people on Java really have no future to look forward to.

Water Yield

It has long been believed that a forest acts like a sponge, soaking up rain water in its soil, litter, and roots, and yielding it up gradually, prolonging water flow into the dry season. To support this contention there are reports of streams drying up after forest cover is removed. Conflicting with this, however, are equally valid reports of streams drying up following tree planting higher in the catchment area. There are in fact two aspects of water behaviour relevant here: the water yield or total volume of the runoff entering a stream, and the flow regime or seasonal distribution of flow (Bruijnzeel 1989, 1993b). These are affected by a number of factors such that forest clearance and other actions will not necessarily have the same effect in different areas.

Where water is not limiting, areas of forest or mature plantations lose by evapotranspiration approximately 1.5 times as much water annually as areas of short vegetation (fig. 3.32). In addition, the evaporation losses of rain from wet stems, trunks, and leaves in a forest or mature plantation during and after rain may account for 15%-25% of the total rainfall, and this is 2-5 times greater than from short vegetation under similar conditions. These differences become progressively less as the short, young vegetation matures (assuming that rainfall remains the same), but the water yield may stabilize at more than the original where the new vegetation is grassland, annual crops, tea, rubber, or cocoa. Where the conversion of forest is to timber plantations or where natural regrowth is allowed, the eventual resultant yield may be more or less the same as the original. This is because both transpiration and evaporation from tall, and deep-rooted

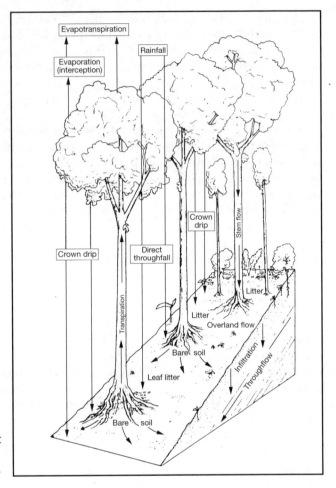

Figure 3.32.
The hillslope forest
hydrological cycle.

After Douglas in Bruijnzeel 1990

forests and plantations is greater than from short and shallow-rooted pioneer plants and most crops, reflecting their relative abilities to intercept
rainfall during rainy periods, and to take up water during the dry season
(Bruijnzeel 1990, 1993b). Thus, one can conclude that the afforestation of
grasslands or areas of degraded cropland with fast-growing (generally
exotic) trees will result in a decrease in water yield (e.g., Waterloo et al.,
1993). Conversely the replacement of forest with grassland will tend to
result in an increased water yield. The total picture is not simple, however.

Major reductions in water yield have often been blamed on 'deforestation'. As shown above, however, loss of forest (at least in experimental
watersheds) generally results in a higher total water yield, but most of this
occurs during the wet season. The solution to this apparent contradiction

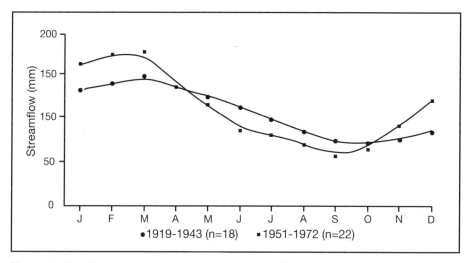

Figure 3.33. Changes in average seasonal distribution of streamflow in the Kali Konto catchment between 1919-1943 and 1951-1972.

After Rijsdijk and Bruijnzeel 1991

lies in considering the net impact of major land use change on evapotranspiration and infiltration. Infiltration of rain into the soil is reduced by:

- compaction caused by the use of heavy equipment in forest clearance;
- prolonged exposure of topsoil to the impact of rain drops;
- overgrazing;
- inappropriate agricultural practices; or by
- an increase in the area occupied by impervious surfaces such as roads, industrial or residential estates and villages.

In such cases the flow of water across the ground and into rivers during heavy rainfall will exceed the gains associated with the lower evapotranspiration of a cleared surface. As a result the wet season flow will increase and the dry season flow will decrease. This has been demonstrated for the Konto River basin, even after taking account of different rainfall patterns (fig. 3.33). Analysis suggests that the cause has not been the increase in scrub or changes in agricultural land use, but rather the expansion of residential areas with their extensive, largely impervious surfaces. Rainfall that is directed into drains clearly does not get the chance to infiltrate into the soil, or to contribute to the groundwater supply or the dry season flow (Rijsdijk and Bruijnzeel 1991). Conversely, the combined effects will indeed show up as increased average flow during the dry seasons where infiltration opportunities are maintained (or at least not too severely impaired) because of:

- well-planned roads;
- hard surfaces which allow water to penetrate to the soil;
- careful forest clearance practice;
- controlled grazing; and
- appropriate soil conservation measures.

The fact that total water yield from degraded crop or grassland is usually reduced considerably following reforestation, indicates that changes due to increased evapotranspiration may easily override those due to increased infiltration, particularly on shallow soils (Bruijnzeel 1989, 1993b). Most of these trends have been identified in small and simple experimental catchments. Although 'real' basins may act somewhat differently, the lesson that can be learned is that the deterioration of regimes of water flow is more to do with inadequate attention to the techniques and effects of land clearance than to the clearing itself.

Concern has also been expressed that exotic trees, particularly *Eucalyptus*, are voracious consumers of water, and various evidence from around the world seems to confirm this. Considering what is known about the biology of the many species and about their ecological effects in different environmental conditions, however, it is clear that no unsupported extrapolation of results from one site to another is justified (Poore and Fries 1985). Work in Australia, the centre of diversity for *Eucalyptus*, found that there are, in fact, two groups, one of species possessing no stomatal control and which therefore act as water pumps, and another in which species do have stomatal control and which transpire no more than other tree species. The first group do have specific uses, such as in poorly drained lowlands where rising water tables threaten to bring salts to the surface and thus ruin agriculture. In such situations, however, the 'mining' of the water is not constant and will last only as long as water can be reached by the roots. In fact, plantations of the second group of *Eucalyptus* appear to use no more water than natural forest (Cahoun et al., 1984; Calder 1991). It is unfortunate that it does not seem to be known to which group the two species planted in Java and Bali (*E. alba* and *E. urophylla*) belong. Until these species are better understood, it would be wise not to plant them, especially on a large scale, without a careful and intelligent assessment of the likely social and economic consequences, and an attempt to balance advantages against disadvantages. This can probably best be done by a sympathetic examination of the ecological circumstances and of the needs of local people. In the case of the ecology, this will be assisted by a knowledge of water balance, nutrient use, and so forth (Poore and Fries 1985).

Seasonal Variations

Where the rainfall is not too heavy, the soils relatively deep and permeable, and the slopes straight or convex, then most of the water flow resulting from a storm enters rivers as throughflow. Where these conditions are not all met, and particularly during heavy rains, the water flowing over the surface becomes the dominant form of runoff, significantly increasing peak volumes. Forest cover tends to reduce the size of the flood peak and to increase the duration of the flood event. This delaying of the runoff also allows more water to infiltrate the soil to recharge the groundwater, prolonging dry season flows but, as described above, a significant proportion of the rainfall is lost from a forest through evapotranspiration. Depending on which of these effects (infiltration or evapotranspiration) is greater, dry season flows will be higher and longer or lower and shorter respectively. When a catchment is cleared, a greater volume of water enters the runoff and this reaches the streamflow rapidly, thereby causing brief but high floods. Less water is able to enter the groundwater, and dry season flows are less. Thus dry season flows may be lower and end sooner due to reduced recharge, or be higher and/or last longer due to reduced evapotranspiration from the short vegetation.

Even when forest clearance is conducted carefully so as to avoid compaction, storms are likely to cause increases in surface runoff (Bruijnzeel 1993b). The relative increase of streamflow after such clearance is greater for small storms, becoming smaller with increasing rainfall. In other words, the heavier the rainfall, the less visible becomes the effect of land use on peak flow (Bruijnzeel 1990).

Erosion

Erosion is the wearing away of the soil surface by some agent, primarily water, but also others of lesser importance. A well-developed litter or understorey layer greatly reduces erosion, and a tight network of tree roots increases slope stability against the forces of rain and overland flow. Both these benefits are lost when a forest is cleared or the soil surface disrupted, although both features are recovered under secondary growth (fig. 3.34). It should be pointed out that high rates of erosion are of serious concern not just because of the loss of the physical medium of soil, but also because of the loss of soil-forming and soil-binding organisms within it.

Java experiences some of the highest rates of erosion in the world (Meijerink 1977). This is generally equated with 'soil' erosion, but a major source of river sediments is recent volcanic products – over 500 million m^3 of ash and dust were deposited on Java between 1900 and 1990, or an average of 5.7 million m^3/year. The main cause of the erosion is inappropriate land management, and much erosion occurs from fields, bare roads, roadside paths, open village areas, landslips, incised riverbanks, and from

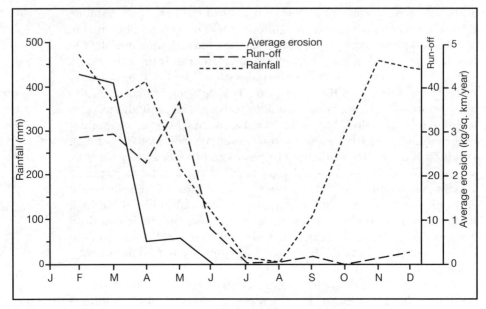

Figure 3.34. Effect on erosion of clearing a plot in February and leaving it undisturbed. Note that the figures for erosion rate and run-off in January are not known but were probably similar to those measured in November and December.

After Coster 1938; Soemarwoto 1991b. With permission of the publishers.

Table 3.8. The relative area and contribution to total erosion of different land use types in the 233 km^2 of the Konto catchment. (*) - natural bank erosion and landslips, (#) - mainly along roads and tracks.

	% of area	% contribution to total erosion
Forest and scrub*	65	9
Rainfed fields	19	42
Irrigated rice	9	assumed negligible
Roads and trails	3	24
Housing	2	18
Lake Selorejo	2	-
Non-forest banks	-	5
Non-forest slides#	-	2

Source: Rijsdijk and Bruijnzeel (1991)

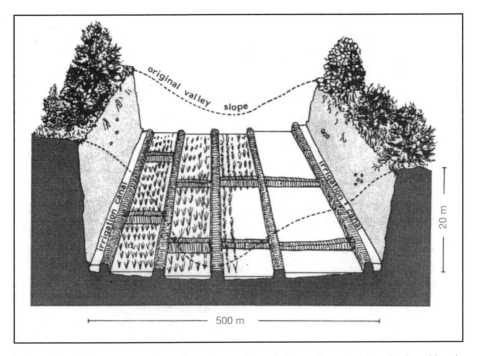

Figure 3.35. Ngagugur, the deliberate erosion of slopes to extend agricultural land.
After Diemont et al., 1991

rice field areas where *'ngagugur'* is practiced (fig. 3.35). This term (in Sundanese) denotes activities where running water in a man-made canal is used to remove and transport any surplus of soil. One use of this is in enlarging rice fields in a valley bottom (Diemont et al., 1991). Although not studied systematically, the impact may be very large given that about one million ha of Javan sawah (23% of the total) is in areas where the dominant slope is in excess of 30%.

The first attempt in the humid tropics to quantify natural and various man-caused erosion for a major river basin was in the upper Konto watershed, East Java (table 3.8). The results suggest that the sediment originating from residential areas, dirt roads, and footpaths are just as important to the overall basin sediment yield as the traditional 'culprits' of dryland agricultural fields. Landslips were relatively unimportant but bank erosion was locally important (Rijsdijk and Bruijnzeel 1991). In summary the findings were that:

- conversion of young volcanic uplands to rainfed agricultural land and settlements in rain-fed areas of East Java do increase erosion after a few years; and

- erosion could be reduced if more care were taken with the disposal of storm runoff and sediment from built up areas and dirt roads.

All the commonly-used soil conservation measures in rainfed fields such as contour cropping, bunding, grass strips, terracing, mulching, and tree planting reduce the quantity of sediment from surface erosion and shallow land slips entering water courses. Evidence from trials in the Solo and Citanduy basins have shown that bench terracing combined with modified cropping patterns can greatly reduce soil erosion rates and be economically viable on slopes of 15%-50%. These are the sort of slopes that have received the most attention, but cropping systems on slopes exceeding 50% have rarely been investigated, despite the fact that these are the most prone to slippage and subject to very high rates of soil erosion. Unfortunately the farmers in these areas have virtually no alternatives (McCauley 1991). Bench terracing is not, however, a panacea. Work in the Gunung Kidul area has found that on steep slopes where the soil is shallow, bench terracing may be deeper than the subsurface flow. As a result water runs out of the riser face and on to the terrace below, with the result that 90% of the rainfall becomes runoff causing serious erosion.

There is clearly a need to strengthen technologies for the management of steep and otherwise unstable slopes, and to develop appropriate strategies of soil conservation, mixed agroforestry, silvopasture, and livestock (Muljadi et al., 1986). This is far more than an academic task because even an excellent farming model will come to nothing if it is not adopted by the farmers. For example, an analysis of farming and the condition of natural resources in the Cimanuk River basin found that farmers are too poor to invest part of their incomes or time in soil conservation measures, with the result that the land is becoming increasingly critical. At the same time a small proportion of the people who are economically much stronger exhibit selfish greed and tacitly encourage the continuing drain of resources and diminishing environmental quality. For this reason it has been recommended that soil conservation requires full attention be given to job creation, improved farm technology, and encouragement to the people to work outside the agricultural sector, as well as to strictly soil conservation techniques (Anwar 1982).

Vetiver

One of the greatest potential future aids in transforming the uplands is vetiver grass *Vetiveria zizanioides*, an introduced but long-resident species which looks disconcertingly like alang-alang, is often confused with citronella grass *Cymbopogon nardus*, and grows literally under the noses of many rural people. It is best known as the source of an aromatic oil which is distilled from the roots, and is used in perfumes, soaps, candles. Java is

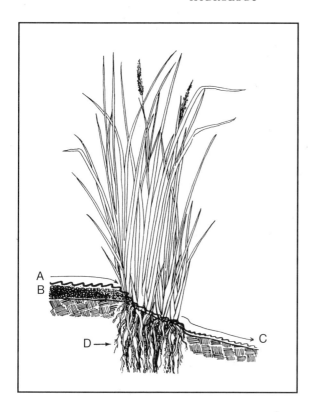

Figure 3.36.
Cross section of a vetiver
hedge to demonstrate its
soil-retaining ability.
After World Bank 1990b

the world's major producer, most of it from a 10,000 ha area near Garut. The use of this plant for erosion control would seem, to some, to be sheer folly since it is commonly regarded as aggravating erosion, and apart from in the above area, its cultivation is generally banned. In actual fact the erosion associated with it is due only to the practice of pulling it up so that the roots can be cut off. This leaves an unprotected trench which can indeed result in erosion, but it is the management, not the grass itself that is at fault (BOSTID 1993).

Experience from different parts of the tropics suggest that vetiver has a major impact on erosion when single plants or culms (stem sections) are planted along the contour (World Bank 1990b). They grow to form an extremely dense 'wall' of grass which prevents the passage of creeping weeds, snakes, and light grass fires. Even if they do burn the growing tip is below soil level, so regrowth is not hindered. But perhaps the most important passage blocked is that of soil and even water washing down a slope (fig. 3.36). So effective is a wall of this grass that it builds terraces across the hillsides and grows up them as they are formed. This retention of soil, and thus water, is of major importance in upland management. The grass walls are also extremely strong and are known to have survived a storm in which 500 mm of rain fell in three hours. The mature leaves are very tough and are unpalatable even to hungry goats and cattle, so a hedge can operate as a stock barrier. The young leaves can be cut as forage for stock and vegetarian fishes such as grass carp. The mature leaves can be chopped and used as a mulch. In areas subject to flooding vetiver can easily survive a whole month of total submergence. Under normal conditions it grows very rapidly, reaching 2 m tall in just a few weeks, and growing roots down to 2 m in just three months. Most of the roots grow more or less straight down so that crops can be grown right up to the grass without deleterious effects of competition. The 'wall' of roots beneath are apparently even a deterrent to burrowing rats, presumably because they dislike the smell of the roots' oil (BOSTID 1993). As if this were not enough, the grass grows in many types of soil and climate and is tolerant of some shade.

Trials using this grass began in Indonesia in 1990, and given its 'wonder' qualities it might have been expected to have been adopted rapidly. This is not the case, although the new World Bank project in the Cimanuk basin has been able to negotiate its use. It seems that the only problem with vetiver is that it is just too simple. It does not need expensive project infrastructure or government intervention or well-trained engineers–just farmers, some extension, and some rootstock. There is also the problem that the bottom could fall out of the market for vetiver oil if the grass were grown widely. Indonesia is being left behind in Asia–for example, the King of Thailand has put all this authority behind a nation-wide programme and has instituted international awards for insights into the best uses for vetiver. When one considers the major problems of erosion and water, it is about time the farmers of Java and Bali were let into the secret of vetiver.

Chapter Four

Flora and Vegetation

PREHISTORIC SPECIES AND VEGETATION

Studies of fossil plants in Java started over 150 years ago, and much of this early information was pulled together by Göppert (1854). The identification of the partial and sterile specimens is very difficult, but he was of the opinion that Java's Tertiary flora was quite similar to the present flora. Work on the structure of wood and the differences between species began to be published in the first quarter of this century, and dozens of species have been identified from the common fossil tree trunks that originate from the Miocene to early Pleistocene. All of them can be referred to species or genera still present in lowland forests (Posthumus 1929; Schweizer 1958; Kramer 1974a, b), although fossil records of two extinct dipterocarp species *Dryobalanops spectabilis* and *D. javanica* are the only evidence of the genus from Java (den Berger 1927; Ashton 1982). Examples of these wood fossils are nowadays sold on roadsides as polished decorations or as interesting house facings.

As has been shown (p. 117), the climate of the region has fluctuated significantly during at least the last two million years between hot, humid conditions during interglacial periods such as the one we are currently in, which began 8,000-4,000 years ago, and cooler, drier, and more seasonal conditions during the glacials (Morley and Flenley 1987). These changes have caused the areas under the various types of forests to expand and contract. The timing of these changes can be determined by examining fossil pollen. The shape and size of pollen grains is characteristic of a species or genus in many types of plants, although others, such as the members of the grass family Graminae, show virtually no differences within the family.

It had once been thought that significant vegetation changes during the late Pliocene, Pleistocene, and Holocene occurred only in the temperate regions, but examination of cores of sediment containing layers of ancient pollen from the bottom of old tropical swamps and lakes have shown this to be in error. The oldest sediment core taken on Java, possibly dating back to the Pliocene, was taken at the hominid fossil site of Sambungmacan in Central Java. The sequence of pollen types within that core suggests that the site was influenced by major volcanic activity and records the recolonization

147

of sparsely vegetated soil leading to a vegetation with affinities to lowland tropical rain forest (Sémah 1984). Pollen analyses from sediments at nearby Sangiran, nowadays situated roughly 70 km from both the north and south coasts of Central Java, indicate that mangrove forest was present in the area in the upper Pliocene (Sémah 1982; van Zeist et al. 1979).

Studies tend to be made of cores from sites above 1,000 m because these show more marked changes as a result of altitudinally determined vegetation types. For example, the cores from Situ Bayongbong, 55 km southwest of Bandung at 1,250 m, covered a period from 16,500 years ago to present. The earliest layer up to about 12,000 years ago showed a pre-dominance of *Dacrycarpus imbricatus* (Podo.), a conifer nowadays found mostly in upper montane forest at around 1,800 m. Pollen of trees such as chestnuts *Castanopsis* (Faga.) and oaks *Quercus* (Faga.) were also found, indicating a proximity to lower montane forest where these trees are abun-dant. The next period, from about 12,000 years to 10,000 years ago, is characterised by a gradual increase in *Engelhardia* (Jugl.) which may pos-sibly indicate some forest disturbance in the vicinity. Between 10,000 and 8,000 years ago the pollen in the core is more typical of relatively rich, and somewhat open, lower montane forest, indicating a general rise in tem-perature. The next zone, between 8,000 and 7,000 years ago, is character-ized by an increase in *Schima* (Thea.) and Melastomataceae pollen, a decline in *Altingia* (Hama.), and an abundance of urticaceous pollen, all of which suggest disturbance, perhaps by people or perhaps as a result of storms or widespread landslips caused by violent tectonic activity. The final zone reflects the gradual change to present vegetation with an increase in pollen from pioneer trees of secondary vegetation such as *Trema* (Euph.) and *Macaranga* (Euph.) (Stuijts 1984, 1993). Cores from two further sites in the vicinity confirm that significant vegetational changes occurred in the last 8,000 years (van Zeist et al. 1979; van Zeist 1984). Similar results have been obtained from sites in the ancient lake basin in which Bandung sits today (Polhaupessy 1980).

A peat core from Rawa Pening/Ambarawa covers the last 4,000 years, and it shows a number of periods when the original swamp forest on the peat was reduced in area. The pollen record indicates that the arrival of people on the plains was probably prior to the tenth century, roughly when the temples of the first Mataram kingdom were being built (p. 318). The last historical period of disturbance was in about the 14th century, and probably represents extensive human colonization of the plain (Sémah et al. 1992).

Results from Java, Sumatra, and New Guinea are all encouragingly consistent. In summary, these show that during the Pleistocene, the land was predominantly covered by forest, either evergreen or deciduous (van Zeist et al. 1979; Maloney 1981; Polhaupessy 1980; Sémah 1982; Whitmore 1984), and that in the late Pleistocene montane vegetation zones were everywhere depressed and compressed. Open and lightly wooded grasslands still occur in Southeast Asia, but the evidence suggests that all these are the direct or indirect effect of human activity, and that grassy savannas did not occur on the Sunda Shelf during the Pleistocene (Whyte 1972, 1974; Whitmore 1984; Flenley 1985). Further support for these findings is found in the composition of the fossil faunas (p. 195). The gradual change to present-day vegetation zones occurred between 14,000-8,600 years ago (and especially between 12,000-9,000 years ago. The fossil pollen suggests that the vegetation found after about 12,400 years ago in the areas sampled was characteristic of climates that were cooler, rather than drier, than today. Some vegetation disturbance, presumably by people, can be detected from about 7,000 years ago, and is widespread from about 4,000 years ago (Stuijts 1984, 1993; Flenley 1985; Stuijts et al. 1988).

PRESENT INDIGENOUS SPECIES

Flora

The total number of plant species recorded on Java, including weeds and cultivated species, is over 6,500, and of this total, over 4,500 are native species (table 4.1). From a botanical point of view, Java is the best known island in Southeast Asia (Ashton 1989; table 4.2), although there are considerable differences between its constituent parts (table 4.3). Although the number of collections has obviously increased greatly since these figures were calculated 20-40 years ago, the comparison between areas is probably still similar. In an incomplete list of alien plants naturalized in Java, Backer (1909) recorded 153 species, but thirty years later he noted 300 (Backer 1936), and a total of 413 had been recognized by 1965 (p. 182) (van Steenis and Schippers-Lammerste 1965). The total has surely increased since then. The vast majority of introduced plants in Java and Bali are native to the Americas, among which are the important food crops cassava *Manihot esculenta* (Euph.), and maize *Zea mays* (Gram.).

Table 4.1. Numbers of native, alien, and cultivated plant species on Java.

	Pteridophytes (ferns and fern-like plants)	Gymnosperms (conifers and related plants)	Monocotyledons (Plants with seeds producing a single leaf)	Dicotyledons (Plants with seeds produc-ing two leaves)	Totals
Native plants	497 (95%)	9 (31%)	1,311 (68%)	2,781 (68%)	4,598
Alien and naturalized plants	7 (1%)	-	58 (3%)	348 (9%)	413
Cultivated introduced plants	15 (4%)	20 (69%)	555 (29%)	933 (23%)	1,523
Totals	519	29	1,924	4,062	6,534

Source: van Steenis and Schippers-Lammerste (1965)

Table 4.2. Relative plant collecting efforts in different parts of Sundaland up to 1972.

	Area (km^2)	Approximate number of herbarium sheets (collections)	Collecting intensity (no./100 km^2)
Java and Madura	132,000	263,000	199
Bali	5,770	3,224	58
Sumatra and adjacent islands	473,630	103,000	22
Borneo			
Whole island	736,695	255,000	35
Indonesian part	539,665	65,000	12
Sarawak and Brunei	117,815	90,000	76
Sabah	79,235	100,000	126
Palawan	11,655	9,000	77

Source: Ashton (1989) after van Steenis-Kruseman (1974)

Table 4.3. Relative plant collecting efforts in different areas of Java and Bali up to 1950.

	Area (km^2)	Herbarium sheets/100 km^2
West Java	39,830	332
Central Java	41,492	92
East Java	44,301	140
Bali	5,770	58
Nusa Kambangan	121	2,083
P. Sempu	7	600
Nusa Barung	56	357
Kangean	667	649
Madura	5,298	70
Bawean	199	400
Karimunjawa	43	16
Thousand Islands	5	40,000

Source: van Steenis-Kruseman (1950)

Endemic Species

It is often stated that Java is poor in endemic species (e.g. Ashton, 1989; van Balgooy 1989), but this should be seen in the context of its neighbours which are extremely rich in endemics. Relatively poor in endemic species Java may be, but these still total at least 325 (see following tables), including 217 of the 731 orchid species. There are, however, no endemic genera, whereas Borneo has 60, Peninsular Malaysia 41, and Sumatra 17. This dramatic difference may be due to the turbulent volcanic history of the island (pgs. 87, 93) which simply has not given time for higher levels of endemism to develop. It had earlier been thought that Java had four endemic genera. Of these four, *Grisseea* (Apoc.) and *Heynella* (Ascl.) will probably be demoted to species rank within a related genus when their families are revised (M. van Balgooy pers. comm.). The monotypic orchid *Silvorchis* is an ephemeral, saprophytic plant which could easily have been overlooked in southern Sumatra and elsewhere, and *Semeiocardium* (Bals.) from Madura has already been synonymized (Grey-Wilson 1989). Only one plant species appears to be endemic to Bali, and that is the white-flowered ground orchid *Calanthe baliensis* (Wood and Comber 1986). The best known endemic species, although most people do not realize that its wild populations are confined to Java and Bali, is the popular orchid *Vanda tricolor*. This has large flowers with a purple lip and red-brown spots on the other petals and sepals. It is collected from forest areas, and since most orchid enthusiasts select their plants on the basis of size or colour, the genetic base of the cultivated populations is becoming smaller.

The known endemic plant species of Java and Bali are shown below in three tables: dicotyledonous families (those with two seed leaves and net-like leaf veins), monocotyledenous families (those with a single seed leaf and parallel leaf veins) excluding orchids (table 4.4-4.6). The orchids are separated because information is presented for them that is not available for other families as a result of the detailed surveys summarized by Comber (1990). For information on families not yet revised or published in *Flora Malesiana* (up to volume 11(2) in series I, and volume 2 in series II) specialists were consulted, and it appears that the following families do not have any endemic species on Java and Bali: Caryophyllaceae (K. Larsen pers. comm.), Cucurbitaceae (W.J.J.O. de Wilde pers. comm.), Cunoniaceae (R.D. Hoogland pers. comm.), Cycadaceae (D.J. de Laubenfels pers. comm.), Daphniphyllaceae (T.-S. Huang pers. comm.), Elaeocarpaceae (M. van Balgooy pers. comm.), Lecythidaceae (K. Kartawinata pers. comm.), Meliaceae (D. Mabberley and C. Pannell pers. comm.), Passifloraceae (W.J.J.O. de Wilde pers. comm.), and, for ferns, Davalliaceae (H. Nooteboom pers. comm.), and Grammitidaceae (B.S. Parris pers. comm.).

Table 4.4. Endemic plant species from certain dicotyledonous families indicating which national botanic gardens, if any, hold living specimens.

	Distribution	Remarks
Acanthaceae		
Blepharis exigua	E	Creeping herb found in very seasonal areas up to 400 m, prefers slightly shaded spots.
Anacardiaceae		
Mangifera lalijiwa	C	Grew wild only between Semarang and Yogyakarta; now widely cultivated in Java and Bali and not known if wild stocks remain.
Actinidiaceae		
Saurauia bogoriensis	W	Known only from a 1975 collection made in the Ciapus Gorge, 650 m, on Mt. Salak, Bogor.
S. bracteosa	W C E B	
S. cauliflora	W	
S. lanceolata	W	
S. microphylla	W C E	
Araliaceae		
Aralia javanica	W C	In mountain forests, 2,000-3,000 m on Mt. Papandayan, Malabar and Dieng (possibly synonymous with A. dasyphyella).
Macropanax concinnus	W C E	Widespread in damp mountain forests down to 400 m, but local and uncommon.
Schefflera reiniano		
Boraginaceae		
Cynoglossum sp.	W	
Burmanniaceae		
Burmannia steenisii	E	Known only from Mt. Lamongan and Pasuruan areas.
Burseraceae		
Canarium kipella	W	Apparently rare, in hill forests on Mt. Salak and near Pelabuhan Ratu. BOGOR
Callitrichaceae		
Callitriche sp.	W	Known only from the patch of gravel in the courtyard of the Cibodas Botanical Gardens' guesthouse.
Celastraceae		
Cassine koordersii	E	Known only from a now deforested and seasonally very dry area in the Puger area in the Lampesan valley and on the Watangan hills. Probably extinct in the wild. BOGOR
Clethraceae		
Clethra javanica	E	Known only from near the lake at Taman Hidup on the Yang Plateau in forest and forest borders from 2,000-2,300 m.
Combretaceae		
Terminalia kangeanensis	E	Known only from beach forest on the islands of Karimunjawa and Kangean.
Compositae		
Anaphalis maxima	W C E	Very local in glades, landslips, and scrub from 2,000-2,800 m.
Vernonia zollingerianoides	E	Known only from Nusa Barung and nearby Puger from near sea level.

W = West Java, C = Central Java, E = East Java, B = Bali

Table continues

Table 4.4. *(Continued.)* Endemic plant species from certain dicotyledonous families indicating which national botanic gardens, if any, hold living specimens.

	Distribution			Remarks
Convolvulaceae				
Erycibe macrophylla	C			Known from a single collection made by Junghuhn on Mt. Ungaran. BOGOR
Cruciferae				
Rorippa backeri				Found in burnt forests on Mts. Merbabu, Wilis, Yang and Ijen from 1,600 - 3,000 m.
Dipterocarpaceae				
Dipterocarpus littoralis	C			Known only from Nusa Kambangan Island, where it is apparently common. BOGOR
Shorea sp.	W			One adult tree known near Leuweung Sancang Reserve but many seedlings observed in 1992.
Vatica bantamensis	W			Known only from Ujung Kulon where it is very rare. BOGOR
Epacridaceae				
Styphelia javanica			E	In sunny, dry, sandy, stony places from Mt. Penangungan to the Yang Plateau, generally 2,100-3,350 m, though 1,650 m on Mt. Penang-gungan.
Ericaceae				
Diplycosia pilosa	W			Found in montane forests, 1,200-2,200 m, locally common, from Banten to Preanger.
Gaultheria solitaria			E	Known from a single collection from 2,900 m on Mt. Arjuno.
Rhododendron album	W	C		In mountain forest 1,200-1,700 m.
R. loerzingii		C		Rare on Mt. Tlerep, Merbabu and Sumbing in scrub or grassland on rather dry but fertile soil, 1,800-2,000.
R. wilhelminae	W			Found once only at 1,350 m near the crater on Mt. Salak; may be a natural hybrid between *R. javanicum* and *R. malayanum* (both found on Mt. Salak).
Fagaceae				
Lithocarpus crassinerivis	W	C		Rare, in primary forests, 200-2,000 m.
L. indutus	W	C		Common in submontane forests up to 1,800 m east to Mt. Slamet.
L. kostermansii	W			Hill forest up to 1,000 m.
L. platycarpus	W	C		Known from lowland forests in South West corner of Java and Nusa Kambangan Island.
Flacourtiaceae				
Casearia flavovirens			E B	From mixed, primary or disturbed forest up to 800 m.
Gesneriaceae				
Cyrtandra elbertii		C		Known only from Mt. Lawu in forest from 1,500-1,700 m.

W = West Java, C = Central Java, E = East Java, B = Bali

Table continues

Table 4.4. *(Continued.)* Endemic plant species from certain dicotyledonous families indicating which national botanic gardens, if any, hold living specimens.

	Distribution	Remarks
Labiatae		
Plectranthus petraeus	E	Known only from Mt. Ijen where locally common between grass, herbs and seedling trees on old lava streams in exposed places between 1,100-1,450 m.
P. steenisii	E	Known from three collections made on stony slopes in Casuarina forest, 2,000-2,650 m on Mt. Arjuno, Tengger, Ijen.
Lauraceae		
Dehaasia acuminata	W C	Found in lowland forests near Pelabuhan Ratu and a former forested area near Pringombo.
D. pugerensis	E	Collected last century in lowland forest near Meru Betiri National Park.
D. chatacea	W	Known from a single collection made before 1850, probably from West Java.
Notaphoebe javanica	W	Known from a single specimen collected in Ujung Kulon National Park (Mt. Payung).
Leguminosae		
Derris danauensis	W	Known only from Ranca Danau Nature Reserve.
Erythrina euodiphylla	C E B	Thorny, short tree of dry forest.
Ormosia sp.	W	Single tree in the remnant lowland forest of Cibodas Botanical Gardens.
Loganiaceae		
Mitrasacme saxatilis	E	Known only from Banyuwangi and Madura in open vegetation, on cliffs and dry limestone hills, locally gregarious, from sea level up to 400 m.
Loranthaceae		
Ameya longipes	B	Known from just three collections, the latest from 1936, from 1,600-1,935 m.
Dendrophthoe praelonga	W C E	Widespread from sea level to 1,400 m on a wide range of host trees.
Lepeostegeres gemmiflorus	W	From 200-1,600 m on a range of hosts.
Mitrasacme bogoriensis	W	From three collections around Bogor in grassy fields and roadsides.
Scurrula didyma	E	Local, from 2,500-2,950 m.
Melastomataceae		
Melastoma zollingeri	E	Known only from the Tengger Mts. in thickets from 1,500-2,000 m.
Myristicaceae		
Myristica teijsmanii	C	From lowland mixed forests up to 700 m from Pacitan to Kawi.
Myrtaceae		
Eugenia ampliflora	W	Known only from Mt. Galunggung in forest from 1,300-1,400 m. If this represented its entire range it may not have survived the 1982 eruption.
E. discophora	C	Known only from Mt. Wilis in forest from 1,300 1,500 m.

W = West Java, C = Central Java, E = East Java, B = Bali

Table continues

Table 4.4. *(Continued.)* Endemic plant species from certain dicotyledonous families indicating which national botanic gardens, if any, hold living specimens.

	Distribution			Remarks
Proteaceae				
Heliciopsis lanceolata	W	C		Known from hill forests, 600-1,200 m. BOGOR
Rosaceae				
Alchemilla villosa	W	C	E	The only member in Malesia of this very widespread genus. Found in grassy places and open casuarina forest from Mt. Papandayan eastwards from 2,100-3,300. Last collected in 1981.
Prunus adenopa	W	C	E	Few collections but apparently found in lowland and coastal forest from Ujung Kulon to Malang up to 500 m. Last collected in 1960.
Rubiaceae				
Lasianthus tomentosum	W			Known only from Mt. Salak at about 1,700 m.
Rutaceae				
Limnocitrus littoralis		C		Believed to be endangered (T. Uji pers. comm.).
Zanthoxylum penjaluensis	W			Known only from very few collections Penjalu area at 720 m in forest.
Solanaceae				
Solanum alpinum			E	Known only from the Tengger Mts. from 2,000-2,900 m, last collected in 1925.
S. anacamptocarpum			E	Known only from the Tengger Mts., last collected 1905.
S. rhinozerothis			E	Known only from the Yang Plateau at about 2,000 m.
S. viscidissimum			E	Known only from the original 1844 collection from the Tengger Mts.
Sterculiaceae				
Heritiera percoriacea				Known only from Ujung Kulon and from Sukawayana, a forest remnant near Pelabuhan Ratu.
Stylidiaceae				
Stylidium inconspicuum	W			Known only from seasonally moist lowland grasslands near Indramayu.
Symplocaceae				
Symplocos costata	W	C		From high mountain forest, 900-2,000 m, scattered. Known as far east as G. Telomojo.
S. junghuhnii	W			From mixed montane forest, c. 1750 m; probably very rare.
Violaceae				
Viola javanica			E	Known only from grasslands on G. Arjuna, Tengger, and Yang, 2,000-3,000.

Sources: Flora Malesiana, van Steenis (1972), Jafarsidik and Soewanda (1985), Frodin (1988), Kostermans and Bompard (1993), also H.P. Nooteboom (1975, pers. comm.) for Symplocaceae, B. Barlow (pers. comm.) for Loranthaceae, R.D. Hoogland (1978) for Actinidiaceae, C. Kalkman (pers. comm.) for Rosaceae, A.J.G.H. Kostermans (pers. comm.) for Lauraceae, H. Riedl (pers. comm.) for Boraginaceae, D. Symon (pers. comm.) for Solanaceae, T. Uji (pers. comm.) for Rutaceae.

Figure 4.1. Javan Lady's Mantle *Alchemilla villosa* (Rosa.), endemic to the mountains of Java, and the only member of this widespread genus found in Malesia. Scale bar indicates 1 cm.

After van Steenis 1972

Figure 4.2. Javan Plum *Prunus adenopa* (Rosa.), a small (<12 m) endemic tree of Java's lowland forests. Scale bar indicates 1 cm.

After Koorders and Valeton 1913

Figure 4.3. Flowering branch of *Saurauia bogoriensis*, a 10 m tree first discovered in 1975 in the Ciapus Gorge on Mt. Salak, Bogor. Scale bar indicates 1 cm.
After Hoogland 1978

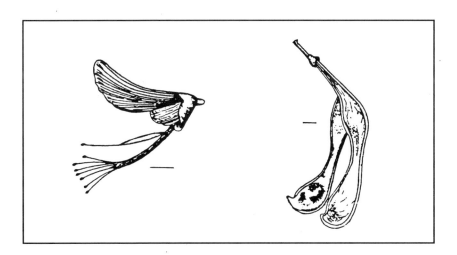

Figure 4.4. Flower and fruit of *Erythrina euodiphylla* (Legu.), endemic to central and East Java and Bali. This is the only green-flowered *Erythrina* on Java and Bali. Scale bars indicate 1 cm.
After Jafarsidik and Soewanda 1985

Table 4.5. Endemic plant species from certain monocotyledenous families indicating which national botanic garden, if any, holds specimens. Many of the ginger species may in fact be found on Borneo, Sumatra, and/or the Malay Peninsula. Most of the types are in poor condition or were destroyed in Berlin during World War II, and there is a need for fresh herbarium material (and for motivated taxonomists).

	Distribution			Remarks
Cyperaceae				
Fimbristylis subdura		C		Collected only twice, both times in teak forest.
Hypolytrum humile	W			In damp lowland and hill forests, 400-1,000 m.
Dioscoreaceae				
Dioscorea blumei	W			Known from a single collection made 150 years ago by Junghuhn on Mt. Salak.
D. madiunensis	W		E	In hills and mountains (Mt. Gede, Ponorogo). Underground parts unknown.
D. platycarpa		C		Known from a single collection made at Besuki.
D. vilis	W	C	E	Rare in mountains, 650-2,400 m; also in Casuarina forest.
Graminae				
Bambusa cornuta			E	Known from 100-600 m in Lamongan and Meru Betiri.
Gigantochloa manggong			E	Known only from Meru Betiri from which it has been much collected to provide raw material for the Jember chopstick industry. PURWODADI
Nastus elegantissimus	W			Found only near the forest edge around Mt. Tilu Nature Reserve, Pangalengan, where it is abundant between 1,500-2,000 m. CIBODAS
Schizostachyum biflorum	W			Known only from Mt. Salak.
Palmae				
Coryphoideae				
Licuala gracilis	W			Known only from Mt. Payung (Ujung Kulon) where it is abundant at the summit, notable for being one of the very few dioecious members of the genus.
Calamoideae				
Calamus adspersus	W		E	Uplands to 1,400 m.
C. asperrimus	W			Uplands to 900 m.
C. burckianus	W			Lowlands to 800 m.
C. heteroideus	W			Uplands to 1,400 m; widespread in remaining forest fragments.
C. melanoloma	W		E	Montane forest to 2,000 m.
C. sp.	W			Massive species with excellent cane found in 1971 in Ujung Kulon National Park.
Ceratolobus glaucescens	W			Lowlands, now only known from Pelabuhan Ratu. BOGOR
Daemonorops rubra	W			Lowlands to 900 m, locally common in remaining forest fragments. BOGOR
Daemonorops sp.	W			Found in 1974 at Cibarengkok, south of Ciwidey in montane forest.

Table continues

Table 4.5. *(Continued.)* Endemic plant species from certain monocotyledenous families indicating which national botanic garden, if any, holds specimens. Many of the ginger species may in fact be found on Borneo, Sumatra, and/or the Malay Peninsula. Most of the types are in poor condition or were destroyed in Berlin during World War II, and there is a need for fresh herbarium material (and for motivated taxonomists).

	Distribution	Remarks
Korthalsia junghuhnii	W	Poorly known taxon, upland forest, 500-900 m. BOGOR
Plectocomia longistigma	E	Lowland forest in southeast Java - e.g., Meru Betiri.
Arecoideae		
Pinanga javana	W C	Scattered in montane forest from Banten to Mt. Slamet, 1,000-1,600 m. BOGOR
P. sp.	B	Massive, single-stemmed species found in montane forest near Bedugul; found in 1973.
Zingiberaceae		
Amomum blumeanum	W ?C E	50-650 m, primary forest.
A. gracile	W C	<100 m, teak forest, perhaps in mountains. BOGOR
A. hochrentineri	W	1,000-1,400 m, primary forest.
A. pseudofoetens	W	Known from about 1,400 m, Gede-Pangrango only, rare.
Etlingera foetens	W	Found in lowland forest below 500 m. BOGOR
E. hemisphaerica	W C	50-700 m, primary forest, locally cultivated.
E. heynianum	W	Collected only once, in a cemetery near Sentiong, Jakarta.
E. parvum	W	Collected only once, 450 m, primary forest.
E. solaris	W C	In Central Java only known from Mt. Merapi, 800-1,650 m, forest.
E. walang	W	Known from lowland forest below 1,200 m, also cultivated. BOGOR
Hedychium roxburghii	W C E	1,000-2,250 m, forest, scrub, forest edges. BOGOR, CIBODAS
Hornstedtia horsfieldii	W	1,000-1,450, forest, local.
H. mollis	W	900-1,000 m, forest, local. BOGOR
H. paludosa	W	1,000-2,500 m, forest, common.
H. rubra	W	450 m, forest. BOGOR
Zingiber gramineum	W C E	Not known east of Rembang-Kediri, 5-300 m, teak forest, bamboo forest. BOGOR
Z. inflexum	W C E	5-1,600 m, primary forest, sometimes secondary. BOGOR
Z. oderiferum	W C	Western half of Java, 600-1,500 m, primary and secondary forest. BOGOR

W = West Java, C = Central Java, E = East Java. B = Bali

Sources: Flora Malesiana, J. Dransfield (pers. comm.) for palms, S. Dransfield and E. Widjaja (pers. comm.) for bamboos, R.M. Smith (pers. comm.) for gingers, Backer and Bakhuizen van der Brink (1963).

Figure 4.5. *Nastus elegantissimus*, an endemic bamboo species from Mt. Tilu Nature Reserve, West Java. Scale bar indicates 1 cm.

After Sastrapradja et al. 1977

Figure 4.6. Endemic bamboo *Bambusa cornuta*. Scale bars indicate 1 cm.

After Koorders 1913

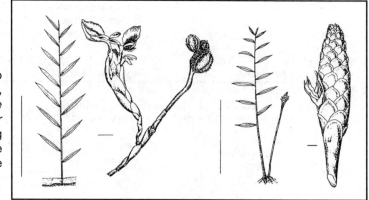

Figure 4.7. Two endemic gingers, *Amomum gracile* (left) and *Zingiber odoriferum*. Long scale bars indicate 50 cm, short scale bars indicate 1 cm.

After Koorders 1913

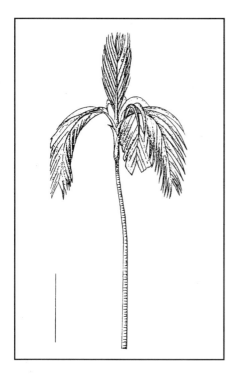

Figure 4.8. Endemic palm tree *Pinanga javana*. Scale bar indicates 1 m.

After Koorders 1913

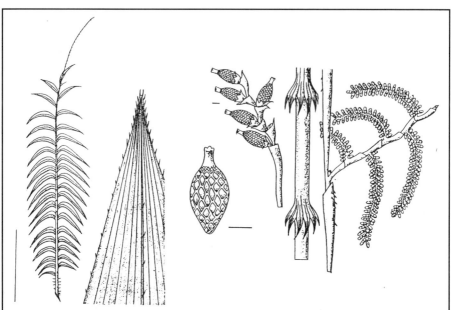

Figure 4.9. Endemic rattan *Calamus melanoloma*. Small scale bars indicate 1 cm, long scale bar indicates 1 m.

After Koorders 1913

Table 4.6. Endemic orchids of Java and Bali. Status categories: OK - of no present concern, I - insufficiently known to form an opinion except that the species is neither widespread nor common, R - rarely encountered despite searches, V - vulnerable, En - apparently endangered, Ex - apparently extinct.

	Distribution			Remarks	Altitude	Range	Status
Agrostophyllum latilobum	W	C			825	1,700	OK
Anoectochilus flavescens	W				1,450	2,000	I
Appendicula congenera	W				400	900	OK
A. imbricata			E		500	1,000	OK
Bogoria raciborskii	W	C	E		50	1,000	OK
Bulbophyllum ardjunense			E	Tiny plant growing only on trunks or branches in quite open situations on slopes facing north, thus experiencing a long dry season.	1,370	2,500	R
B. binnendijkii	W				1,000	1,100	OK
B. capilligerum			E		600	600	I
B. devium			E		1,460	1,650	OK
B. grudense		C			?	?	I
B. hamatipes	W		E		750	1,600	OK
B. hydrophilum	W				1,000	1,200	I
B. igneum	W				1,000	1,800	I
B. inaequale	W		E		1,050	1,050	OK
B. mucronatum	W				?	?	I
B. obscurum	W		E		760	1,220	OK
B. pahudii	W	C			900	1,000	OK
B. papillatum	W				1,000	1,200	I
B. peperomiifolium	W				1,000	1,000	I
B. petiolatum	W				1,000	1,000	OK
B. sarcoscapum	W				?	?	I
B. scotiifolium	W			In foothills and mountain forests.	250	1,070	R
B. semperflorens	W				800	1,000	OK
B. sp.			E		1,800	1,800	I
B. sp.			E		1,450	1,450	I
B. submarmoratum	W	C			900	1,000	I
B. tenellum	W				1,000	1,700	I
B. truncatum		C			2,290	2,300	OK
B. xylocarpi		C			1	1	I
Calanthe abbreviata	W	C	E		1,400	2,100	OK
C. baliensis				B Known only from forests behind Bali's Botanic Garden at Candi Kuning.			I
C. ecallosa	W			Known from two collections, one of them from the Mt. Halimun area.	1,400	1,400	R
Ceratochilus biglandulosus	W		E		1,000	2,000	OK
Ceratostylis anceps	W	C	E		1,000	2,000	OK
C. anjasmorensis			E		1,770	1,890	OK
C. backeri		C			2,470	2,470	OK
C. braccata	W		E		1,000	2,100	OK
C. brevibrachiata			E		1,520	1,520	OK

Table continues

Table 4.6. *(Continued.)* Endemic orchids of Java and Bali. Status categories: OK - of no present concern, I - insufficiently known to form an opinion except that the species is neither widespread nor common, R - rarely encountered despite searches, V - vulnerable, En - apparently endangered, Ex - apparently extinct.

	Distribution			Remarks	Altitude	Range	Status
C. capitata	W	C			1,600	610	OK
C. crassifolia	W				1,500	1,600	OK
C. latifolia	W	C			1,350	2,400	I
C. simplex	W				1,500	2,400	I
Chamaeanthus brachystachys	W	C			1,350	2,400	I
Cheirostylis javanica			E		1,400	1,700	OK
Chiloschista javanica	W	C	E		1,400	1,700	OK
Cleisostoma montanum		C	E		1,200	1,800	OK
Coelogyne simplex	?				?	?	I
C. tumida	W				1,000	1,500	I
Corybas acutus		C		Known only from Dieng Plateau among ericaceous shrubs esp. *Rhododendron javanicum.*	1,900	1,900	R
C. imperatorius		C	E	Especially common in grass lands in East Java. Reforestation on these fire-prone areas will reduce populations.	1,280	1,730	V
C. praetermissus	W			Known only from Mt. Gede-Pangrango at 1,600 m on mossy banks.	1,000	1,600	R
C. umbrosus	W	C	E		1,070	1,590	OK
C. vinosus	W				1,500	1,500	
Cryptostylis filiformis	?				?	?	
C. javanica	W		E		250	1,600	OK
Dendrobium arcuatum			E		1	800	OK
D. atavus	W		E	Nowhere common.	790	900	R
D. capra			E	Known only from E. Java from planted teak forests around base of Mt. Penanggungan and Mt. Lamongan. Possibly found in Lesser Sundas.	100	250	R
D. corrugatilobum	W				1,000	1,000	?R
D. jacobsonii			E	High mountain orchid from eastern half of Java, partic-ularly frequent on Mt. Lawu and Mt. Semeru. Very bright scarlet-red flower.	2,300	3,000	R
D. kuhlii	W				760	1,800	OK
D. paniferum	W		E		760	1,000	OK
D. prianganense	W			Known only from 950 m in a rather dark and wet valley at Cibarengkok south of Bandung.	915	915	R
D. tenellum	W	C	E		1,500	2,600	OK
Dendrochilum abbreviatum	W	C	E		700	2,000	OK
D. aff. cobolbine	W				?	?	I

Table continues

Table 4.6. *(Continued.)* Endemic orchids of Java and Bali. Status categories: OK - of no present concern, I - insufficiently known to form an opinion except that the species is neither widespread nor common, R - rarely encountered despite searches, V - vulnerable, En - apparently endangered, Ex - apparently extinct.

	Distribution			Remarks	Altitude	Range	Status
D. brachyotum	W				1,800	1,800	?R
D. edentulum	W	C			1,300	2,560	I
D. vaginatum	W				?	?	
Didymoplexis cornuta	W			Known only from under bamboo near Bogor.	225	250	?R
D. flexipes	W				1,200	1,200	I
D. minor	W				250	250	?R
D. obreniformis	W				1,000	1,000	?R
D. striata	W			Known only from Mt. Salak.	?	?	?R
Diplocaulobium noesae	W	C		Known only from the forests of the prison island of Nusa Kambangan. Last collected before 1910.	200	550	R
Disperis javanica		C	E		900	1,200	OK
Epigeneium triflorum	W	C	E		1,000	1,800	OK
Eria bogoriense	W	C	E		250	1,000	OK
E. coffeicolor	W			Known only from Mt. Malabar.	2,000	2,300	R
E. junghuhnii	W	C	E		1,600	2,000	OK
E. rhynchostyloides	W		E		760	760	OK
E. valida	?	?	?		?	?	I
E. verruculosa	W		E		1,200	1,500	OK
Eulophia exaltata			E	Known only from among tall grass in E. Java.	15	760	R
E. javanica	W	C			50	1,000	I
Flickingeria dura	W				750	1,000	I
F. integrilabia							I
F. puncticulosa	W		E		600	1,000	OK
Gastrodia abscondita	W			Known only from bamboo forest near Bogor.	250	250	?R
G. callosa	W			Known only from lowland forest.	250	250	?R
G. crispa	W			Fairly common only in the few damp primary forests in W. Java.	820	1,500	R
Goodyera glauca			E		1,400	1,600	OK
Habenaria backeri	W				1,800	2,000	R
H. bantamensis	W		E		60	150	OK
H. curvicalcar	W				200	700	I
H. giriensis			E	Known only from Gresik area near Surabaya just above sea level in shady spots. Virtually all this area now developed.	100	100	Ex?
H. horsfieldiana	?	?	?	Last collected in early 1800s somewhere in Java.	?	?	?R
H. javanica	?	?	?		?	?	I
H. loerzingii		C	E		760	1,070	OK
H. multipartita	W	C	E		1,300	2,500	OK
H. parvipetala	W	C	E		900	1,600	I

Table continues

Table 4.6. *(Continued.)* Endemic orchids of Java and Bali. Status categories: OK - of no present concern, I - insufficiently known to form an opinion except that the species is neither widespread nor common, R - rarely encountered despite searches, V - vulnerable, En - apparently endangered, Ex - apparently extinct.

	Distribution			Remarks	Altitude	Range	Status
H. salaccensis	W			Currently known only from near Mt. Halimun.	1,000	1,150	R
H. tosariensis		C	E		2,000	2,800	OK
H. undulata	W	C	E		760	760	OK
H. zollingeri			E	Last collected last century on Mt. Ijen from 800-1,150 m. May still be in Mt. Raung/Ijen.	800	1,150	R
Hetaeria cristata	W	C	E		1,450	2,100	OK
H. lamellata	W		E		750	1,500	I
H. micrantha	?	?	?	Not collected since last century.	?	?	?R
H. purpurascens	W			Not collected since last century.	?	?	?R
H. velutina		C			1,200	1,200	OK
Hymenorchis javanica	W			Tiny plant known only from 900-1,000 m in Mt. Halimun.	900	1,000	R
Liparis affinis	W				?	?	I
L. bilobulata	W				1,500	1,500	I
L. bleyi		C			1,000	1,000	I
L. clavigera	?	?	?	Not collected since last century.	?	?	?R
L. decurrens	W	C			1,000	1,200	I
L. javanica	W	C	E		1,150	1,500	OK
L. lauterbachii			E	Collected only this century on Mt. Arjuno in forest. No forest remains at that altitude and this species was not found in seven years of searching.	900	900	?Ex
L. odorata	W	C	E		250	1,650	OK
L. prianganensis	W				1,000	1,220	OK
Luisia taurina	W		E		450	650	OK
Malaxis crepidium	W				1,400	2,000	I
M. cuprea	W				750	1,000	I
M. humerata	W			Not collected for many years despite searches.	1,000	1,000	?R
M. junghuhnii	W				1,000	1,700	OK
M. kobi		C	E		900	2,100	OK
M. koordersii		C	E		400	1,220	OK
M. lobatocallosa	W	C			1,700	1,700	I
M. obovata	W				400	1,000	I
M. purpureonervosa			E		800	1,220	OK
M. ridleyi	W	C	E		600	1,500	OK
M. sagittata	W			In dark and humid localities from 1,500 (Puncak) to 2,000 m (Mt. Gede).	1,525	2,000	R
M. slamatensis		C	E		1	1,100	OK
M. soleiformis	W				1,280	1,670	OK
M. tenggerensis			E	Known only from Mt. Tengger.	?	?	R
M. tjiwideiensis	W	C	E		1,500	2,000	I
Malleola forbesii	W			Not collected for many years despite searches.	1,050	1,050	?R

Table continues

Table 4.6. *(Continued.)* Endemic orchids of Java and Bali. Status categories: OK - of no present concern, I - insufficiently known to form an opinion except that the species is neither widespread nor common, R - rarely encountered despite searches, V - vulnerable, En - apparently endangered, Ex - apparently extinct.

	Distribution	Remarks	Altitude	Range	Status
S. sigmoideum	C	Known only from southern slopes of Mt. Slamet.	1,200	1,600	R
Sacroglyphis comberi	W	Known from only one part of Mt. Halimun.	1,000	1,000	R
Schoenorchis juncifolia	W C E		500	2,500	OK
Silvorchis colorata	W	Endemic monotypic genus, collected only once, ninety years ago near Garut.	1,600	1,600	R
Stigmatodactylus javanicus	W E	Minute orchid from Mt. Gede, Mt. Argopuro, and Mt. Anjasmoro just west of Mt. Arjuno. Maybe overlooked.	1,670	1,670	R
Taeniophyllum aurantiacum	W		1,000	1,000	I
T. bakhuizenii	W		1,000	1,000	I
T. biloculare	W		1,000	1,600	I
T. biocellatum	W C E		300	1,520	OK
T. calyptrochilum	W		500	500	?R
T. djampangense	W		500	1,000	I
T. doctersii	W		?	?	I
T. glandulosum	W E		2,400	2,700	OK
T. hirtum	W E		1,370	1,900	OK
T. mamilliferum	W		500	500	?R
T. pantjarense	W		600	870	I
T. reynvaaniae	W		80	80	OK
T. rostellatum	W		1,000	1,000	?R
T. tenerrimun	W		1,550	1,550	R
Tainia elongata	W		550	1,200	OK
Thelasis javanica	W		?	?	?R
Thelymitra javanica	W C E		1,400	3,300	OK
O. salakana	W	Not collected for many years despite searches.	?	?	?R
O. similis	W C		600	2,100	OK
O. subligaculifera	W		1,500	1,500	I
O. tjisokanensis	W		750	750	I
O. valetoniana	C		150	150	I
O. zimmermanniana	? ? ?		?	?	R
Omoea micrantha	W		1,000	1,600	R
Paphiopedilum glaucophyllum	W E	Once very common, but between 1965-1980 extensively collected for sale, so it is now very hard to find.	200	770	En
Pennilabium aurantiacum	W		1,800	1,800	I
Peristylus djampangensis	W E		50	1,200	I
Phalaenopsis javanica	W	Collected to extinction at only known site near Cianjur.	700	1,000	En
Pholidota camelostalix	E		1,680	1,830	OK
Phreatia acuminata	W		1,000	1,500	OK

Table continues

Table 4.6. *(Continued.)* Endemic orchids of Java and Bali. Status categories: OK - of no present concern, I - insufficiently known to form an opinion except that the species is neither widespread nor common, R - rarely encountered despite searches, V - vulnerable, En - apparently endangered, Ex - apparently extinct.

	Distribution	Remarks	Altitude	Range	Status
P. subsaccata	W		1,830	1,830	OK
Plocoglottis latifolia	E	Known only from a single collection made nearly 200 years ago, possibly from an area now deforested.	?	?	?Ex
Pseudovanilla affinis	W E	Originally recorded from West Java, but found once on south coast of East Java.	450	450	En
Pteroceras fraternum	W	Not collected for many years despite searches.	1,000	1,000	?R
P. javanica	E	Collected only once, near Puger, Jember, in 1930 on a *Barringtonia* tree close to the sea.	1	1	En
P. zollingeri	W E		50	610	OK
Saccolabium pusillum	W C	From a few localities in western Java.	915	2,000	R
S. rantii	W	From humid forests.	1,500	1,600	R
Thrixspermum conigerum	W		1,400	1,500	I
T. doctersii	E	Known from a single specimen.	1,400		I
T. javanicum	W C E		1	800	I
T. obtusum	W C E		760	1,650	OK
T. patens	W E		50	1,250	OK
T. purpurascens	W E		800	1,400	OK
T. roseum	W		1,400	1,500	OK
T. squarrosum	W		1,400	1,700	OK
Trichoglottis javanica	W E	From lowland forest in southern half of the island.	?	?	R
T. rigida	W		900	1,000	?R
T. tricostata	W	From moist forest.	700	1,800	R
Vanda tricolor	W C E B	Grows on fairly open tree branches.	700	1,600	OK
Vrydagzynea purpurea	W	Known only from damp forest.	700		?R
V. uncinata	W E	Known only from forest.	300	1,000	R
Zeuxine tjiampeana	W	Known only from a limestone hill near Ciampea, now virtually destroyed for the rock.	350	350	?Ex

Source: Comber (1990, pers. comm.)

Distributions

Java and Neighbouring Regions. Java has about half of the 580 plant genera
which are decidedly Malesian in their distribution, and some families typical of Sundaic rain forest are surprisingly poorly represented, such as
Sapotaceae, Palmae (especially the rattans *Calamus* and *Daemonorops*),
Myristicaceae, Ebenaceae, Annonaceae, Gesneriaceae, and Dipterocarpaceae. In the last-named family there are just 10 named species (two
endemic) on Java, against 267 (155 endemic) on Borneo, and 105 (11
endemic) on Sumatra (Ashton 1982). Fossil evidence, however, suggests
that they were once more abundant and that at least one genus (*Dryobalanops*) is now extinct (den Berger 1927). Sixteen genera which have a distribution centred in Asia are found (within the biogeographical region of
Malesia which stretches from Malaya to New Guinea) only in Java, eight of
them from mountains, six from deciduous forest. Conversely, some widely-
distributed genera without exacting requirements for soil or climate are
absent from Java for no obvious reason, such as *Agathis* (Arau.), *Baeckea*
(Myrt.), *Cladium* (Cype.), *Dacrydium* (Podo.), and *Rhodomyrtus* (Myrt.) (van
Steenis and Schippers-Lammerste 1965).

It is interesting that the mountain flora of Java and Bali has great
affinity with Sumatra (with its volcanoes and high plateaux), and is singularly different from that of Borneo (with its much smaller mountains and
no volcanoes).

By virtue of its position on the shallow Sunda Shelf and its repeated
land connections with Kalimantan and Sumatra (and thence to the Malay
Peninsula), it is not surprising that Java shares many species with those
areas (fig. 4.10). There are, however, some basic differences (van Steenis
1933a; van Steenis and Schippers-Lammerste 1965). For example, there are
111 (predominantly tree) genera and three families that might be considered typically Sundaic which are absent from Java. Some of these are
absent simply because of the lack of specific habitats, such as heath forest
and deep peat forest. It has been suggested that the absence of many
Sundaic genera from Java is due to major volcanic activity in the Tertiary,
aggravated by the destruction of the relatively species-rich lowland forest by
humans (Endert 1935). Weight is added to the second argument by the
occurrence of certain abundant Sundaic genera in Java being in forest remnants such as on Nusa Kambangan and forest pockets at the western
extreme of Java. Ujung Kulon, at the western tip of Java should intuitively
hold more typical Sundaic genera than elsewhere, but apparently this is not
the case (A.J.G.H. Kostermans pers. comm.). However, much of it was
once cultivated, and it experienced a major disturbance in 1883 when
Krakatau erupted and perhaps should not be expected to be 'typical'
Sundaic forest. The degraded forest in Yanlappa Reserve (p. 465) is more
similar to Sundaic forest (K. Kartawinata pers. comm.).

In summary, then, the paucity of the Java flora is due to climatic factors

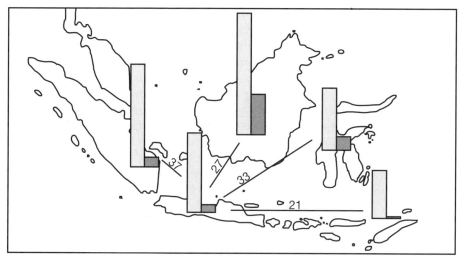

Figure 4.10. Relative richness (shaded), endemism (black), and overlap (figures) of the plant flora of Java and Bali and their neighbouring islands and island groups. 'Overlap' is the percentage of the total species list that is shared between them.
After MacKinnon and MacKinnon 1986

(the highly seasonal climate of the north and east), soil (the absence of certain soil types), and forest destruction.

Changes West to East. If the numbers of species in the three divisions of Java are calculated, there are several clear trends (table 4.7): West Java has the richest native flora (3,882 species), followed by central (2,851) and East Java (2,717). West Java also has the most species not found in the other two parts, followed by East Java. Some 50% of the orchids are known only from West Java. Central Java has very few species not found in west or east

Table 4.7. Analysis of native species of plant and endemic orchids found in, and confined to, west, central, and East Java. Note: west, central, and east designate areas only roughly the same as the administrative provinces of the same names.

	Native species found in			% of native species present confined to		
	W	**C**	**E**	**W**	**C**	**E**
Number of ferns	368	342	315	94	8	10
Number of gymnosperms	6	6	6	2	1	0
Number of monocots	701	656	561	415	35	71
Number of dicots	1,689	1,721	1,454	607	82	210
Total	2,764	2,725	2,336	1,118	126	291
Number of endemic orchids	167	59	90	97	9	25

Source: van Steenis and Schippers-Lammerste (1965), orchid data calculated from Comber (1990, pers. comm.)

Java. The primary reasons for this general picture are the relatively large areas of species-rich everwet forest in West Java, whereas these are restricted to the southeast slopes of the mountains in the rest of the island. The drought-tolerant plants are more common in the northern parts of central and East Java but are also present along the north of West Java from Indramayu, Karawang, to Jakarta, so these are added to its total. The west of Java is also immediately adjacent, and was once connected, to the much more species-rich Sumatran flora, whereas East Java is adjacent to the dry part of Bali as the first in a chain of relatively small, dry islands. It is, of course, true that central Java has very little remaining forest (indeed a lower percentage than any Indonesian province), and the species totals represent the total ever recorded, rather than total now present, but the long history of forest destruction there (p. 328) may well have caused local extinctions before serious plant collecting began.

North and South Coasts. The north and south coasts are strikingly different in many ways. Most of the south coast comprises a continuous sand beach interrupted by rocky headlands and lengths of cliffs and local areas of mangrove; there are also a few dune areas. The predominant current is east to west and sand dykes often form across the mouths of rivers during the dry season. When the rains fall, the flooding rivers burst a path through these to the sea again. Various stretches of this coast, especially in the southeast, experience a dry season, but as a whole it is wetter than the north coast. There are few major centres of human habitation, largely because the sea is rough and the coast offers few safe anchorages. As a result there are some important and sizeable remnant expanses of forest along the south coast.

The north generally has a shallowly sloping muddy coast, parts of which were originally bordered by deep mangrove forests. Wide sandy beaches, dunes, and rocky headlands are rare. Swamps were present behind the mangrove fringe but have now largely been converted to sawah and tambak. Since they lie in the rain shadow of the mountains, the northern coasts experience a marked dry season. Four of the largest towns of Java are on the northern coast (Jakarta, Cirebon, Semarang, and Surabaya) and there is now virtually no primary or old secondary vegetation within 35 km of the coast. As a result of the above differences there are four times as many beach formation plant species confined to the south coast as there are on the north coast (16 species against 4), and only one muddy beach mangrove plant confined to the south against 23 confined to the north (van Steenis and Schippers-Lammerste 1965).

Madura is interesting in that the southern coast facing Java across the strait is akin to the north Java coast, while its north coast has sandy beaches like Java's south coast. Bali's east and west coasts have considerable waves and would have had beach formations along them before they were converted to agriculture, as would the north coast which is quite steep and sandy, lacking any year-round rivers bringing sediment from the hills.

Altitudinal Distribution. The plants of Java and Bali's mountains form an extension of the flora of the Himalayas, and detailed analyses of the species distributions can be found elsewhere (van Steenis 1934a, b, 1936a; Smith 1986). The altitudinal distribution of plants within Java has also been analysed in detail, and the methodology is given in van Steenis and Schippers-Lammerste (1965). In those analyses it was argued that there were a number of disjunctions in the general decrease in species numbers at 1,000 m, 1,500 m, 2,000 m, 2,400-2,500 m, and 3,000 m. Looking at the shape of the graph it may to some extent simply illustrate people's reluctance to record altitude as 900, 1,100, 1,300, 1,400, or 1,900 m which sound more exact than 1,000, 1,500, or 2,000 m. If this is the case, then the curve might more genuinely be much smoother. Similar disjunctions at 1,000 and 2,000 m are apparent among alien weeds and cultivated plants, which adds to the suspicion that the altitudinal data have been rounded off by the plant collectors and other botanists. Altitudinal zones are discussed elsewhere (p. 503).

Data on altitudinal distribution of orchids in Comber (1990) shows that 99% of the 217 endemic orchid species have been recorded (though not exclusively) between 800-1,200 m, and only two of the endemic species are found over 2,800 m. The (past) importance of lowland forest is again demonstrated by endemic orchids, 57% of which have been found from sea level to 400 m.

Some plants occasionally occur well outside their 'normal' altitudinal range where certain environmental conditions are duplicated; such as coastal plants in inland, well-lit, and mineral-rich salt springs, or lowland plants around the steamy hot fumaroles at high altitude (van Steenis 1935a, 1936c). These interesting cases are discussed by van Steenis and Schippers-Lammerste (1965).

Offshore Islands. The flora of the islands off West Java shows no appreciable difference from that of the mainland, although the coastal tree *Serianthes dilmyi* (Legu.) is found both on the islands in Jakarta Bay and in the Sunda Straits but not on the mainland. The Karimunjawa Islands north of Semarang have some species absent from the mainland such as the tree *Ouratea arcta* (Ochn.), and the floral affinity is closer to Bangka and Kalimantan than to Java. Nusa Kambangan is separated from the south coast of Central Java and has kept more of its forest for longer than elsewhere in the province because it is off-limits on account of its prisons. It was explored botanically at the start of the century but further collecting has not been allowed. Some of the species there are endemic and others are very rare. The southern half of the island is still forested (RePPProT 1989), and it has been proposed to gazette this as a reserve to join the existing three small reserves into a single meaningful conservation area (p. 800). Among its more dramatic plants is the giant voodoo lily *Amorphophallus decus-silvae* (Arac.). Its hooded spadix is about 120 cm tall, and this sits upon a stem of the same or greater height.

The most remote of Java's islands is Bawean, halfway between Java and Kalimantan. Bawean has a distinctly Javan flora although, again, some species occur which are not known from the mainland, such as the tall, lowland, very hard tree *Irvingia malayana* (Sima.) known from southeast Asia, Sumatra, and Kalimantan. The seeds are very fatty and can be used to make soap, wax, and candles (Nooteboom 1962; Heyne 1913, 1987). Interestingly some of its non-Java plants (shared with Kangean) have eastern affinities such as *Canarium vulgare* (Burs.) (though it is grown in Java as a shade tree), *C. asperum*, and *Tristiropsis canarioides* (Sapi.) (also cultivated as an ornamental). The Kangean Islands in the east, like Bali and Madura, differ little from East Java. The sedge *Fimbrystylis adenolepis* (Cype.) is known only from mainland Southeast Asia and a moist, grassy field in Kangean (Kern 1974). Madura also has some very rare (but widespread) plants such as *Utricularia baouleensis* (Lent.), which grows up grass stems and is abundant in rice fields (Taylor 1977).

South of East Java is the sizeable, uninhabited limestone island of Nusa Barung, which has some botanical interest (Jacobs 1958) (p. 806). The entire island is a nature reserve (at least in part because it has no permanent fresh water).

Notable Species

Java and Bali's plants have enormously important and varied roles. There is no space to describe all of Java and Bali's plants which are notable in some way for their appearance, usefulness, role in human culture, or important ecosystem function. Instead, just three distinctive plants have been chosen for their especial interest. Strictly useful plants are dealt with in the following section.

Rafflesia. One of Java's most interesting plants is the large *Ràfflesia* (Raff.). There were once thought to be three native species: *R. rochussenii* from lower montane forest in West Java (van Steenis 1941b), *R. patma* in lowland forest along the south coast of West and Central Java, and *R. zollingeriana* from lowland forest in Meru Betiri National Park. In fact, both lowland forms are now regarded as *R. patma* (fig. 4.11) which is found also in

Figure 4.11. *Rafflesia patma* on Nusa Kambangan, 1919.

By M.L.A. Bruggeman, with permission of the Rijksherbarium/Hortus Botanicus, Leiden University

Sumatra (Meijer 1984). *R. rochussenii* has been found only a handful of times, and 73 years elapsed between its first and second records on Mt. Salak after being found by a nature lovers' group from IPB in 1990 (H. Suryadi pers. comm.); this species too is also known from Sumatra. *R. patma* is the more common of the two species, and was discovered for the first time in 1825 by Blume on Nusa Kambangan (Haak 1889). The two valid Javan species can also form hybrids (Jafarsidik and Meijer 1983), and are sought after for medicinal purposes (p. 182).

Rafflesia is remarkable in many ways for the plant has no stem, no leaves, and no proper roots. Instead, the seed grows almost fungus-like 'roots' into the stem of climbing *Tetrastigma* (Vita.) vines. At intervals along the stem, flower buds form and after several months of growth they open into the worlds largest blooms. Carrion flies (Zuhud 1988; Priatna et al. 1989) and beetles are attracted by the flowers' cadaveric stench of rotting snakes (Bänziger 1991), which advertises carbohydrate 'bait'. The insects may also find its appearance appealing since its lurid colour, crater shape, and whitish blotches resemble a festering sore. The stench is a long-range marker and once the fly is inside the flower, a fruitier smell emanates from the depths of some species, although this is not yet confirmed for Javan rafflesias. The flies venture into the depths of the flower, and pollen is smeared onto the backs of the flies' thorax which is then rubbed off onto the stigma of the female flower (Beaman et al. 1988; Bänziger 1991). The pollen is produced in the form of a thick mush which

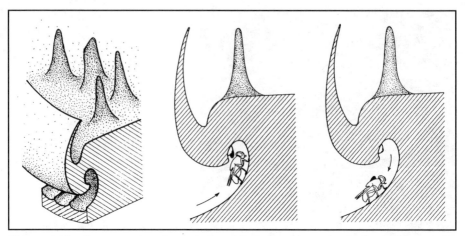

Figure 4.12. Cross section through the disk and column of a *Rafflesia* flower (left), and a calliphorid blowfly venturing up one of the channels about to receive a drop of pollen mush on its back, and then retreating, carrying the mush (middle, right).
After Bänziger 1991

thickens when taken outside the flower, allowing it to be carried around for a long time (fig. 4.12).

When the flower dies, it falls into itself and forms a black, mouldering mass within which the seeds mature. A wide range of animals from elephant to squirrels have been implicated in dispersal of the seeds (Meijer 1984; Emmons et al. 1991). Hooves or claws were suggested as means by which the seeds would get access to the *Tetrastigma* roots, but it has been suggested recently that various soil arthropods and nematodes would be just as effective since they make microscopic wounds through which the tiny seedling roots could inoculate the host. *R. rochussenii* seeds were successfully implanted into a *Tetrastigma* stem in the Bogor Botanical Gardens in 1924, and the plant flowered four and a half years later. Once the buds have started growing, over 18 months pass before they flower, and it is a further six months before ripe seeds are produced.

It had been assumed (Meijer 1984) that male and female rafflesias must bloom at more or less the same time, and in reasonably close proximity, in order for successful pollination to occur. Now that the pollen is seen as durable and that long-living and strong-flying calliphorid flies are recognized to be the principal pollination agents, even single flowers take on a new significance. Even so, for long-term survival rafflesias clearly require (Bänziger 1991):
- a forest rich in *Tetrastigma* vines;
- *Tetrastigma* seed dispersers such as frugivorous bats and other appropriate terrestrial or arboreal mammals;

- populations of large mammals to provide, when dead, brood sites for large populations of the calliphorid flies;
- dispersers of rafflesia seeds such as squirrels and other mammals of the forest floor; and
- healthy soil microfauna to allow infection of the vine root.

Edelweiss. Edelweiss *Anaphalis javanica* (Comp.) is the most famous mountain plant of Java and Bali. It can reach 8 m in height with a stem as thick as a human leg (van Steenis 1972), but such wonderful plants are now very rare, if indeed any remain. Small areas of woods comprising 5 m specimens of its close relative *A. viscida* (with green rather than woolly leaves), however, can still be seen on the Yang plateau. The edelweiss is an early colonizer of new montane volcanic soils and is able to survive in such poor soils because it forms mycorrhiza with certain soil fungi which effectively enlarge the area covered by the roots and increase their efficiency at finding nutrients. Its flowers appear to be singularly popular with insects, and at least 300 species of bugs, thrips, butterflies, flies, wasps, and bees have been seen visiting them. If allowed to branch and become quite sturdy, the edelweiss also provides nest sites for the Sunda whistling thrushes *Myophonus glaucinus* (Docters van Leeuwen 1933).

Pieces of the edelweiss are frequently picked and taken down mountains for spiritual or aesthetic reasons, or simply as a souvenir of a mountain hike: between February and October 1988, 636 stems were recorded as having been taken from Mt. Gede-Pangrango National Park (Aliadi et al. 1990). Within certain bounds and so long as only small pieces are taken, this pressure can be sustained. Unfortunately greed and false expectations have taken their toll of many populations, particularly those close to the paths. The plant was cultivated for some decades on the slopes of Mt. Agung, and research has shown that it can be propagated quite easily from stem cuttings (Aliadi et al. 1990). It may therefore become possible to sell cuttings to visitors to relieve pressure on the wild populations as has been done in the past (fig. 4.13).

Wijaya Kusuma. To the south and east of Nusa Kambangan are some very small islands which were one of the very few sites in Java of the *wijaya kusuma*, sometimes known as the lettuce tree or Moluccan cabbage *Pisonia grandis* (Nyct.) (fig. 4.14). This is a relative of *Bouganvillea* and is naturally more or less confined to guano-rich bird islands, although its continued presence off Nusa Kambangan has not been confirmed for some years. It figures in Javanese shadow plays, and its flowers were not to be gathered or possessed (on pain of death) except for the coronation ceremonies of the Sultan of Surakarta. In order to collect them officially, a legation of 60

Figure 4.13. Cultivation of edelweiss on the slopes of Mt. Agung, in the early part of this century.

By C.N.A. de Voogd, with permission of the Rijksherbarium/Hortus Botanicus, Leiden University

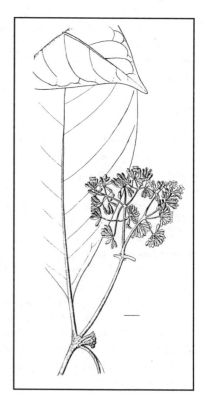

Figure 4.14. Wijaya kusuma *Pisonia grandis.*

After Koorders 1913

people, led by a priest, would cling to the steep rock using a ladder fixed to a boat, and place the flowers (their number proportional to the number of prosperous years in the reign of the new prince) in a golden box wrapped in silk. This in turn was placed in a decorated box on two bars protected on the voyage back by a golden umbrella. On arrival at the palace the flowers were placed in the hands of the Sultan, who took them to his room of sacred objects. It was also believed that the Sultana should eat the flowers when pregnant to ensure a victorious male heir (hence the Sanskrit-derived name meaning 'flower of victory'). Strangely, the flowers are neither fragrant nor showy, and it may be that the plant's apparent rarity and isolation on the top of a wave-swept rock generated its special esteem (Stemmerik 1964; Ekowati 1991). A cultivar of this plant with pale yellow-green chlorotic leaves is cultivated widely, but it hardly ever flowers. *Pisonia grandis* is also recorded in its wild state from Bali, the Karimunjawa Islands, Jakarta Bay, the Padalarang hills near Bandung, and throughout other parts of the Malesian region, but nowhere else does it have an important cultural role.

Major Publications

As stated above, the flora of Java is one of the best known in Southeast Asia, and shares with Sri Lanka and Taiwan the distinction of having a complete local flora (van Balgooy 1989): the three-volume *Flora of Java* (Backer and Bakhuizen van den Brink 1963) which covers all flowering plants but which regrettably has no illustrations or indications of total distribution. An earlier work, the four-volume *Exkursionflora von Java* (Koorders 1911, 1912a, b, 1913) has some illustrations in the first three volumes, but the final volume has excellent line drawings of hundreds of species. Unfortunately, some of Koorders' names are out of date. The trees of Bali are listed and described briefly in the appropriate volume of the *Tree Flora of Indonesia Checklist* (Whitmore et al. 1989). The lists of vernacular names with scientific equivalents available for the trees of Bali (Prawira et al. 1972; Whitmore et al. 1989) and Java (Prawira and Oetja 1976, 1977a, b) are useful, but only if used in conjunction with botanical texts to confirm the identifications so made. The reason for doing this is because many vernacular names are applied to more than one species (Altona 1912). Also out of date, but singularly useful if used with care, is the four-volume *Atlas der Baumarten von Java* (Koorders and Valeton 1913, 1914, 1915, 1918), which has line drawings of nearly 800 tree species. *The Mountain Flora of Java* (van Steenis 1972) is a celebration of the beauties of this group of plants, having 459 species illustrated at life size and in colour. It also has fascinating introductory sections. Similar is the recent book on the *Orchids of Java* (Comber 1990) in which almost all the island's 731 orchid species are illustrated by a colour photograph.

The *Flora untuk sekolah di Indonesia* (van Steenis 1981) covers (in Indonesian) those 500 or so native and introduced species most likely to be encountered, primarily in disturbed habitats. Another useful but rare book is *Ken je die plant* (de Voogd 1950) (in Dutch), which covers 150 of the most commonly encountered herbs and other plants. Its line drawings are printed on different coloured papers corresponding roughly with the colour of the flowers. For ricefield areas the *Weeds of Rice in Indonesia* (Soerjani et al. 1987) has detailed line drawings of 265 species (and their seedlings). Other useful weed books are *Aquatic Weeds of Southeast Asia* (Pancho and Soerjani 1978), *Weeds of Vegetables in the Highlands of Java* (Everaarts 1981), *Geillustreerd handboek der Javaansche theeonkruiden* (on the weeds of tea plantations) (Backer and van Slooten 1924), and *Onkruidflora der Javasche suikerrietgronden* (on the weeds of sugarcane plantations) (Backer 1934, with the final atlas volume published in English in 1973).

In a quite different category, small but lavishly illustrated books on common *Flowers of Bali* (Eiseman and Eiseman 1988a) and *Fruits of Bali* (Eiseman and Eiseman 1988b) are available, although the species covered, and more, are dealt with also in *Plants and Flowers of Singapore* (Polunin 1987), the comprehensive and practical *Gardening in the Tropics* (Holttum and Enoch 1991), and the *Collins Guide to Tropical Plants* (Lötschert and Beese 1981).

In addition to the above is the series of books on selected species from different groups of plants (covering all Indonesia) produced by the Biology Research and Development Centre of the Indonesian Institute of Sciences, such as *Ubi-ubian* (root crops) (LBN-LIPI 1977a), *Jenis rumput dataran rendah* (lowland grasses) (LBN-LIPI 1980), *Rumput pegunungan* (mountain grasses) (LBN-LIPI 1981), *Polong-polongan perdu* (legumes) (LBN-LIPI 1984a), and *Kerabat beringin* (figs) (LBN-LIPI 1984b).

The standard work on ferns, *Varenflora voor Java* (Backer and Posthumus 1939), has an illustration of one species from each genus but is now somewhat out of date in terms of taxonomy. To gain easier access to ferns, interested readers should consult *Ferns of Malaysia in Colour* (Piggott and Piggott 1988).

For lower plants, reference should be made to *Seaweeds of Singapore* (Teo and Wee 1983), and to *A Handbook of Malesian Mosses* (Eddy 1988, 1990), although *Mosses of Singapore and Malaysia* (Johnson 1980) is easier to use for a non-specialist. The liverworts or Hepaticopsida of Java were detailed in a book at the beginning of this century (Schiffner 1900), and some further collecting was conducted in the years prior to the Second World War (Meijer 1953). A major thrust, clarification and popularization of liverworts was made by Meijer (1954a, b, 1958a, b, 1959a, b, 1960).

Information on the useful plants (see below) can be found in the new Indonesian translation of Heyne's exhaustive *Nuttige planten van Nederlandsche Indie* (Heyne 1913, 1987), *A dictionary of the economic products of the*

Malay Peninsula (Burkill 1966) in which considerable information from Java is included, and in the series of publications produced by the Plant Resources of South-East Asia (PROSEA) project based in Bogor. Among the volumes available to date are a basic list of species and commodity groupings (Jansen et al. 1993), pulses (van der Maessen and Somaatmadja 1992), edible fruits and nuts (Verheij and Coronel 1992; Wulijarni-Soetjipto and Siemonsma 1993), and forages (Mannetje and Jones 1992).

In addition to all these are of course the fundamental family revisions of ferns and flowering plants in the volumes of Flora Malesiana, the first volume of which appeared in 1947. The series currently covers 170, or 57%, of the families, encompassing over 5,000, or more than 20%, of the estimated total species (updated from Geesink 1990).

Useful Species

The uses to which the plants of Java are, and have been, put are legion, as can be proved by perusal through the standard dictionaries of useful plants and related texts (Heyne 1913, 1987; Burkill 1966; Wijayakusuma et al. 1992, 1993). The uses include a wide range of medicinal drugs for humans and animals, building materials, household containers, preservatives, fruits, edible roots, pesticides, starch, leafy vegetables, beans, sources of oil, teas, fodder, jewellery, weaponry, gums, fibres, weed control, flavourings, magic, ornaments, poultices, furniture, green manure, soaps and shampoos, paper, cloth, fuel, alcohol, sugar, status, roofing, and fish poison. Java played a pioneering role in the modern science of experimental physiology when, in 1809, Francois Magendie began experiments into the fabled arrow poison made by boiling the bark from one of the Javan species of *Strychnos* trees. Small quantities were jabbed into dogs, which died rapidly (Neill 1973). The chemical responsible was strychnine.

Among the plants Java has given to the world are various cultivated species such as lipstick plants *Aeschynanthus radicans* (Gesn.), *Agalmyla parasitica* (Gesn.), coleus *Plectranthus scutellarioides* (Labi.), moon orchid *Phalaenopsis amabilis* (Reader's Digest 1971), and the Cheribon cultivar of 'noble' sugar cane *Saccharum officinarum* (Gram.). Most of the commercial sugar cane cultivars used around the world can be traced back to this large Javan grass (Purseglove 1972).

Indeed, grasses form the most useful family of plants, being the staple diet of most of the world's population, the main diet for many domestic and wild animals, the source of sugar and fragrant oils, the raw material for alcoholic and other beverages, thatch, packing, fibres, and paper. They can also play important roles in erosion control. On the down side, many species are important agricultural weeds. Bamboos are a tribe of giant grasses, which are used extensively in Java and Bali, primarily as a substitute for wood in building and for scaffolding but also for musical instruments,

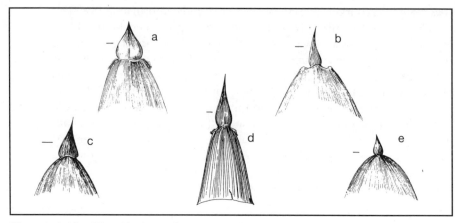

Figure 4.15. Sheath tips of common Javan bamboos. a - *Schizostachyum brachycladum*, b- *Gigantochloa atter*, c - *G. apus*, d - *Bambusa blumeana*, e - *G. verticillata*.

After LBN-LIPI 1977b

food (bamboo shoots), wicker baskets, mats, walls, fishing tackle, weapons, water pipes, and landscaping. They are both gathered from the 'wild' and planted close to habitation, particularly in steep areas unsuitable for other plants. Bamboo groves occupy some 30,000 ha in East Java, and some of the bamboo produced is used as an important raw material for a paper pulp factory. Over 30 species of bamboo are found in Java, although less than half of these are cultivated (Widjaja 1980). Of these, four achieve major importance and can be found throughout. These are *Bambusa blumeana, Gigantochloa atter, G. apus,* and *G. pseudoarundinacea* (fig. 4.15), all of which have probably been introduced to Java in the distant past. Three native bamboos are cultivated locally: *Schizostachyum iraten, S. brachycladum,* and *S. zollingeri* (Hildebrand 1954).

Javanese traditional medicine, jamu, is based mainly on plants and is highly developed both at the village and industrial level (figs. 4.16 and 4.17). The most frequently consumed plants are the gingers (Zing.) such as turmeric *Curcuma domestica* and ginger *Zingiber officinale,* as well as parsleys (Umbe.) such as fennel *Foeniculum vulgare* and *Hydrocotyle asiatica,* and the pepper *Piper retrofractum* (Pipe.). Many jamu recipes contain the products of more than five plants (some up to fourteen) (Barbier and Courvoisier 1980). Jamu are used widely even though modern, scientifically-based medicines are easily obtainable, and most are drunk as general tonics rather than as cures for specific ailments. Some have proven therapeutic value and others may have the desired effect more for psychosomatic reasons. About one third of 583 jamu plants from West Java analysed

Figure 4.16. Traditional jamu seller.
With permission of Kompas

Figure 4.17. View inside a modern jamu factory.
With permission of Kompas

were regarded as poisonous because they contained alkaloids, cyanide, and oxalates. Most of these were members of the Compositae, Leguminosae, and Euphorbiaceae (Ginting et al. 1981). In small quantities these are probably harmless, although some countries ban the import of Indonesian jamu because of alleged carcinogenic effects. Another major concern is that moisture in and around the packets may cause the otherwise harmless spores of the mould *Aspergillus flavus* to germinate, thereby producing aflatoxins - a highly potent liver carcinogen. So, the promotion of packeted jamu, which is given government support (e.g., Dep. Kesehatan 1980), needs to be considered against the generally unknown action of their active constituents in the human body, and their possible long-term effects.

In the Meru Betiri area the most commonly used medicinal plants among a wide range of trees, shrubs, and herbs were *Spigelia anthelnia* (Loga.), *Phyllanthus niruri* (Euph.), and *Alstonia scholaris* (Apoc.), all of which are gathered from within the Park. Inventories of medicinal plants growing in other conservation areas have been conducted in various parts of Java and Bali (Jafarsidik and Sutarto 1980; Jafarsidik and Sutomo 1983; Sidiyasa and Sukawi 1984; Sangat-Roemantyo and Riswan 1990) with some species being more common in some areas than in others, and a number of important medicinal species such as *Rauvolfia serpentina* (Apoc.) (from which reserpine is extracted for the treatment of hypertension and mental disturbances) and *Rafflesia patma* being decidedly rare throughout. Indications are that unbridled exploitation will lead to at least local extinctions of such plants, although this is not the policy of the jamu manufacturers. Indeed, commercial factories for jamu are increasingly sophisticated, employing large numbers of manual labourers, and growing many of the plants themselves. The cultivation and harvesting of medicinal plants would be an important element in the development of a buffer zone for the park, but the initiative, will, and funding have yet to materialize. In any case *Rafflesia* has not yet been cultivated commercially anywhere, and it is clearly going to be many years before village-level techniques are developed. Until then the minimum requirement should be that collection of its flowers must never be made by cutting through the stem of the host (p. 172).

Introduced Species

Few if any natural ecosystems are saturated with species (Cornell and Lawton 1992), and so the potential exists for species to be introduced and to thrive in alien but acceptable conditions. The exact numbers of introduced plants (p. 149) can of course be contested, and arguments can be made that an ancient introduction be considered native, but the general picture would remain the same. Discussions of whether species are native/indigenous or alien/introduced are often subjective, being based on local patriotism, uncritical acceptance of earlier opinions, and so forth. The criteria that should be considered are (Webb 1985):

- fossil evidence;
- historical evidence;
- habitat (species found only in human-modified habitats are generally alien);
- geographical distribution (disjunct distributions sometimes indicate introductions);
- existence as a naturalized (self-sustaining) alien elsewhere;
- genetic diversity (reckoned to be generally lower in introduced species);
- reproductive pattern (plants not reproducing by seed in one area are likely to be alien); and
- an idea of how an alien was introduced.

The complications are illustrated by species that behave as weeds in dry areas but are in fact indigenous plants of the lost deciduous forest. For example, at Baluran National Park there are problems with the invasive native plants, *Veronica cinerea* (Scro.) and *Thespesia lampas* (Malv.).

The origin of many plants, particularly those used by people spiritually, or as food, medicine, or clothes, is lost in time since early waves of migrating people would have doubtless carried plants with them; modern Javanese and Balinese transmigrants nurse saplings under their arms and carry seeds of favoured plant varieties as treasures in their luggage. Also, traders have been dealing in ever more improved or novel plants, and containers and soil are often contaminated with seeds – many weeds have spread outwards from ports. A notable species about which the truth of origin will probably never be known is teak: van Steenis and Schippers-Lammerste (1965) found no sound arguments to assume teak was not native to Java, yet others have used evidence from ancient documents reporting the carrying by Indian Hindus of pots with teak seeds, as evidence that this is an alien species (p. 591).

While introduced animals and plants can cause enormous and expensive problems, it should be remembered that very few of the plant species moved around the world by people actually fall into that category - for example, only just over 20 species of woody plants have become serious tropical weeds. There do not seem to be any general rules to determine which introduced 'weedy' species will become successfully invasive, and even the general view that invasions occur more easily in temperate ecosystems rather than tropical ones does not bear testing (Fryer 1991). And although most weeds are introduced species, this is not always the case.

The Bogor Botanical Gardens has been a major source of exotic plants that have become naturalized and spread. The most famous and expensive 'mistake' was the dumping, early this century, of surplus water hyacinth *Eichhornia crassipes* into the Ciliwung River that runs through the Gardens.

Figure 4.18. Dense growth of the introduced water hyacinth.
With permission of Kompas

This attractive, blue-flowered plant is a native of Brazil and was brought to the Gardens in 1894 through the agency of the wife of the Brazilian ambassador. It was washed out to the plains, was sent to people with ornamental ponds, and was used to provide shade for fish ponds, until it reached its ubiquitous position throughout the archipelago today (fig. 4.18). The same route of the Ciliwung River was taken by the North American water plant *Sagittaria platyphylla*, first found 'wild' in ditches and rice fields near Depok in 1949 but which has since spread throughout Java (van Steenis and Schippers-Lammerste 1965).

Another pestilential weed from South America which causes serious problems in agricultural and plantation areas is the climbing *Mikania micracantha* (Comp.) (fig. 4.19), sometimes known as mile-a-minute for good reason. Specimens were brought to Bogor Botanical Gardens from Paraguay in 1949, and in 1956 it was distributed widely as a cover crop in rubber estates. It now needs no encouragement to distribute itself, and it is detrimental both because it smothers plants, outcompeting them for water, light, and nitrogen, and because its exudates may inhibit tree growth and soil nitrification processes (Kostermans et al. 1987).

A small tree that has experienced a dramatic spread from the Gardens is *Piper aduncum* (Pipe.) (fig. 4.20), native in forest edges and open

Figure 4.19. *Mikania micracantha*, a pestilential introduced weed. Scale bar indicates 1 cm.

After Kostermans et al. 1987

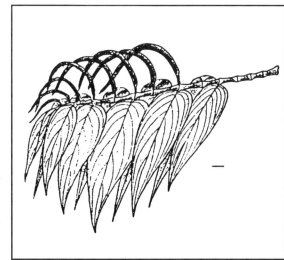

Figure 4.20. *Piper aduncum*, a widespread tree introduced from Central America. Scale bar indicates 1 cm.

After Koorders 1913

sites from Argentina, Mexico, and the West Indies from which it was introduced over 100 years ago. A tree produces at least one ripe infructescence
each night, each with hundreds of seeds, and dog-faced fruit bats *Cynopterus*
spp. bite these off, and spread the seeds far and wide. The tree is now
common in West Java and has reached Sulawesi and even logging roads in
Irian Jaya.

A particularly noticeable introduced tree in and around Bogor,
Cianjur, and Sukabumi is the neotropical pioneer *Cecropia peltata* (Cecr.)
with large leaves in the shape of a hand, the leaf stem joining the leaf in its
centre (like a lotus), and ringed trunk and branches. The first trees were
planted in the Bogor Botanical Gardens in the 1920s and have been spread
by fruit-eating bats and birds (Rumantyo pers. comm.). In its native Central
and South America this tree is well known for the habitual presence of
biting *Azteca* ants which are believed to reduce the incidence of leaf
damage from herbivores, decrease the infestation from climbing plants,
and increase growth rates (Janzen 1973; Schupp 1986). In fact, observation
of the same tree in Peninsular Malaysia where it has 'escaped' from plantations, even without its ants, grew faster, suffered less herbivore damage,
and are generally less laden by vines than the neighbouring pioneer trees
(Putz and Holbrook 1988).

Other centres of exotic plants are around the Botanic Gardens at
Cibodas and Purwodadi, the old private botanic gardens near the latter at
Nongkojajar, agricultural stations, and market garden centres such as at
Lembang (near Bandung), Bogor, Salatiga, and Dieng where various crops
were given trials. In addition, a host of ornamental and other useful plants
were introduced by individuals, commercial estates, and other arms of
government (van Steenis and Schippers-Lammerste 1965). Most were
introduced for specific purposes, such as ground cover or erosion control,
without thought to the possibility that they could spread out of control and
have major negative impacts. Among the plants that 'seemed a good idea
at the time' are *Chromolaena* (formerly *Eupatorium*) *odorata* (Comp.) (fig.
4.21), *E. inulifolium*, *Mimosa invisa* (Legu.), and *Lantana camara* (Verb.).
The fact that stall feeding of livestock is now more popular than grazing
may be due in part to the spread of the unpalatable *Chromolaena*. These 3
m plants have seeds which germinate rapidly and compete successfully
against *Acacia villosa*, *Lantana*, and alang-alang *Imperata cylindrica* (Gram.)
on fallow land, and thereby reduce grazing potential, although in the
long term they improve soil conditions as a result of their copious litter.

An interesting plant discovery was made in 1975 when the grass
Desmostachya bipinnata (fig. 4.22) was found in Bali Barat National Park.
This was the first occasion that this species had been found in the Malesian
plant region, its natural distribution being from north Africa to south
China. It may have been introduced accidentally by tourists who had
passed through India, but another more intriguing possibility exists. This

Figure 4.21.
Chromolaena (formerly
Eupatorium) *odorata*.
Scale bar indicates 1 cm.
After Kostermans et al. 1987

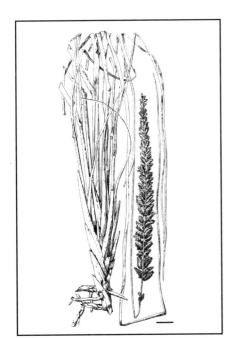

Figure 4.22. *Desmostachys bipinnata*,
One of the nine holy herbs of Hinduism,
introduced at some time in the past to
Bali. Scale bar indicates 1 cm.
After Bor 1970

grass is one of the nine holy herbs of Hinduism, and it may have been introduced to Bali hundreds of years ago, found the climate unsuitable, but settled in its more 'natural' habitat of *Borassus* savanna. It is certainly rare on Bali, and its use as raw material for making shrouds for corpses has been replaced by alang-alang (J.F. Veldkamp pers. comm.).

Weeds

A weed is any plant growing where it is not wanted, but the term is most often applied to introduced plants competing with crops. An enormous amount of information is available on the ecology of most weed species found in Java and Bali, starting with Backer and van Slooten (1924) and Backer (1934), on tea and sugarcane weeds respectively, and Kooper (1927), and finishing for the time being with Soerjani et al. (1987). Strangely, these texts do not seem to be exploited to the extent they could be in the basic modern texts on Indonesian weed science (e.g., Sastroutomo 1990). While there has been a great deal of money spent on the control of weeds using chemicals, it is undesirable for many reasons to control them by blanket applications of herbicides. Far better to have a good understanding of each plant's ecology and physiology and to apply that to develop more subtle means of control.

The fundamental reason why weeds are a problem is that they compete with crops for limited resources, the quantity and availability of which change through the crop cycle. Competition is an extremely complex subject and its study is fraught with problems. Crops may develop some advantages through genetically-controlled features such as height and form of stem, leaves, and roots, or by features such as the relative densities of seeds and seedlings of rice and weeds. At the densities at which rice gives maximum yield, for example, weeds also fare extremely well, and must be crowded out by the rice if it is to survive. Competition occurs both above ground for light, and below ground for water and nutrients. Competitive advantage is sometimes gained by a weed producing growth-inhibiting substances (allelopathy).

Weeds exhibit numerous strategies for ensuring they outcompete other plants. They generally produce large numbers of seeds (50,000 per plant is a typical quantity), but such great quantities are achieved at the expense of size, and such small seeds can, of course, only produce small seedlings with a relatively small competitive advantage. Plants that have effective means of vegetative reproduction tend to have fewer, larger, seeds capable of dispersal over large distances, and with large food reserves to ensure, or at least give maximum possibility of, successful establishment. Examples are alang-alang and purple nutsedge *Cyperus rotundus* (Cype.), for which the main problem is the production of rhizomes and tubers respectively, and relatively few of their seeds are viable. The water hyacinth also produces few seeds, but they remain viable for about 15 years. Most of its growth is from stolons; a single plant can, under favourable conditions, be the source of 3,000 others in just 50 days, and the offspring from a single plant can cover 600 m^2 in one year (Kostermans et al. 1987).

Weeds, almost by definition, have wide distributions, and most of those in Indonesia have been accidentally or deliberately introduced – amongst the most recent being *Sagittaria platyphylla* (Alis.) (first found wild in 1949) and *Porophyllum ruderale* (Comp.) (first found wild in 1945). Some have seeds or other parts which are dispersed in or floating on irrigation water. For the floating weeds such as water fern *Salvinia* (Salv.) and water lettuce *Pistia stratiotes* (Arac.) the whole plant is dispersed by water. Numerous species have seeds dispersed by wind, and many of those have a pappus, or feathery, umbrella-like structure such as those of members of the Compositae and alang-alang. Most of these seeds will fall close to the parent plant, but they have the potential to travel large distances when conditions are right. Some seeds pass through the alimentary tract of grazing cattle unharmed and are dispersed in that manner, while others such as spanish needles *Bidens pilosa* (Comp.) have barbs on the seed cases that get snagged on animal fur. In fact, people are one of the major agents in weed dispersal– rice seed is sometimes contaminated with other seeds (though this is not serious where only the rice panicles are harvested), and seeds mixed with mud are carried on vehicle wheels and implements. Only those weed seeds present in the soil that experience optimum conditions of moisture, warmth, and light, soil oxygen concentration, germinate at any one time. Those remaining (the 'seed bank') maintain their viability until a suitable time arises. This is important to the plant because otherwise local extinctions could occur if all the seeds germinated in a bad year, resulting in no viable seed and no living seeds being left in the soil. Some seeds remain viable for decades, while others may remain viable for only a few years.

Alang-alang is one of the world's most pernicious weeds and justifiably finds a place in the book *The World's Worst Weeds* (Holm et al. 1977), although it does have cultural and economic importance in some areas (p. 676). It is a robust perennial grass with a persistent rhizome below ground. Its favoured habitats are sunny, slightly moist grasslands created by humans on moderately fertile soils, although it is found in a far wider range of habitats than this would suggest. It is regarded as a serious pest in coconut, rubber, oil palm, and citrus plantations, as well as in upland rice and cassava fields. Its furry fruiting heads hold about 3,000 wind-dispersed seeds, but not all these are fertile and most of its reproduction is achieved vegetatively by rapidly-spreading rhizomes and dispersed pieces of rhizome. These rhizomes are highly competitive and can even pierce the roots of other plants, causing them to rot or die, perhaps with some chemical action too. This species is most dominant where burning is common, because the fire's heat kills most plants, but leaves the deep rhizomes of alang-alang untouched. Burning occurs frequently in many areas because the young shoots that grow up quickly after fire are highly palatable to livestock. Indeed, the establishment view against alang-alang is somewhat illogical since the much-vaunted 'superior' grasses such as elephant or Napier grass *Pennisetum purpureum* are no more nutritious (Soewardi and Sastrapradja 1980). Species of grass that are genuinely more nutritious require so much management and capital that villagers are reluctant to adopt them (Dove 1986).

Many minds have worked hard and long to find additional uses for alang-alang, but all ideas, such as to use it in paper making, have come to nought because its yield is not particularly high in economic terms, and the costs of transporting the bulky material (compared with softwood or bamboo) are too high. Its control requires repeated close cutting, heavy grazing, ploughing, heavy shade, or the establishment of creeping plants such as *Pueraria phaseoloides* (Legu.) and butterfly pea *Centrosema pubescens* (Legu.), which can smother the grass (Kostermans et al. 1987; Swarbrick and Mercado 1987).

VEGETATION

The vegetation types on Java and Bali can be classified in a number of different ways. A map of the natural vegetation of Java and Bali based on the agroclimatic map of Oldeman (1977) has been published (MacKinnon et al. 1982; MacKinnon and MacKinnon 1986), but this did not distinguish the forests of the weakly seasonal areas (such as the Ujung Kulon peninsula) and did not bring out the essential difference between western and eastern Java. The vegetation types of Java were mapped by van Steenis and Schippers-Lammerste (1965), but this included all the types existing in the middle of this century including rice fields, estates, and grassland. A new map of natural vegetation types has been prepared based on altitude, dry seasons, rainfall, and available floristic information (fig. 4.23) adapting previous definitions (van Steenis and Schippers-Lammerste 1965; Champion and Seth 1968; Kartawinata 1980; Whitmore 1984). The forest types mapped are as follows:

- evergreen rain forest - forest growing below 1,200 m, where rainfall exceeds 2,000 mm, and where there are less than two dry months, and with typical trees being *Artocarpus elasticus* (Mora.), *Dysoxylon caulostachyum* (Meli.), *Lansium domesticum* (Meli.), and *Planchonia valida* (Lecy.);
- semi-evergreen rain forest - forest growing below 1,200 m, where rainfall exceeds 2,000 mm, where there are two to four dry

months, a few deciduous trees, and where *Garuga floribunda*
(Burs.) and *Kleinhovia hospita* are present;

- moist deciduous forest - forest growing below 1,200 m, where rainfall is 1,500-4,000 mm, where there are four to six dry months,
  where at least half the trees are deciduous, and where *Acacia leucophloea* (Legu.) and *Salmalia malabarica* (Bomb.) are typical trees;
- dry deciduous forest - forest growing below 1,200 m, where rainfall
  is less than 1,500 mm, where there are more than six dry months,
  where almost all the trees are deciduous, and where typical plants
  include the tall palm *Corypha utan*;
- aseasonal montane forest - forest growing above 1,200 m, above
  evergreen or semi-evergreen forest; and
- seasonal montane forest - forest growing above 1,200 m, above
  monsoon forest.

In addition to these are forest types which occupy small areas, unmap-
pable at the scale this book allows, such as:

- mangrove forest on periodically inundated brackish soils;
- swamp forest on soils permanently or seasonally inundated by
  freshwater;
- lower montane forest, upper montane forest, and subalpine vegetation;
- dry evergreen forest; and
- thorn forest where there is an average annual rainfall of about
  1,000 mm and about nine dry months.

There are also lowland forests on limestone which differ in minor
respects, described elsewhere (p. 482).

The forest types are described in detail in the next part of the book.
The vegetation of a wide variety of sites and areas in Java and Bali has been
described by many authors over the last century. These are indicated in
Appendix 2. This may be of singular use in environmental assessment
studies when the 'original' condition needs to be known. Very little of the
natural forest remains (p. 7, fig. 4.24), and its distribution is highly skewed,
with a large proportion of montane forest remaining, but only a tiny percentage of the drier types (table 4.8).

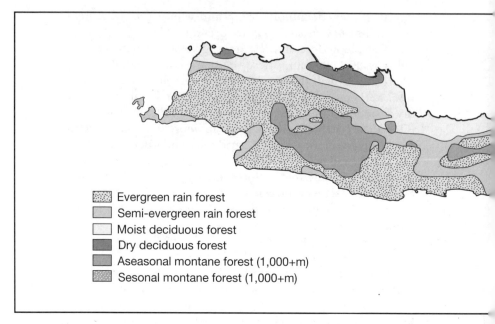

Figure 4.23. Natural distribution of major vegetation types on Java and Bali.
Based on climatic maps in RePPProT 1990; see text for details

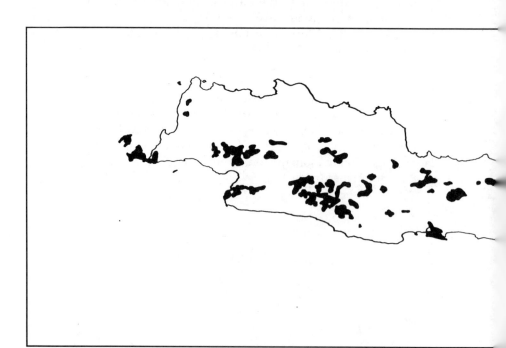

Figure 4.24. Current distribution of forest on Java.
After RePPProT 1990

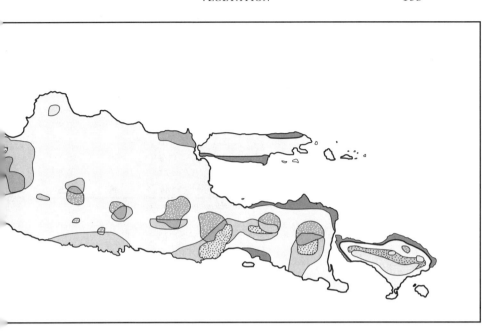

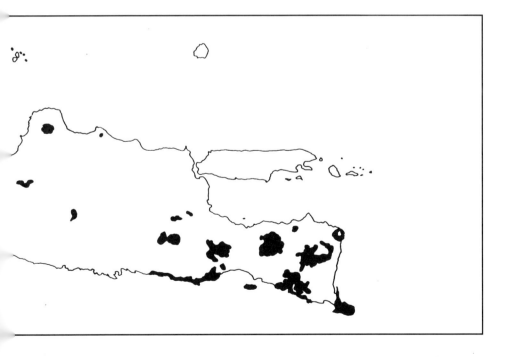

Table 4.8. Original and present areas (km^2) of forest on Java and Bali with their percentage remaining.

	Java			Bali		
	Original area	Present area	Percentage remaining	Original area	Present area	Percentage remaining
Evergreen rain forest	26,949	1,902	7.0	1,082	238	22.0
Semi-evergreen rain forest	22,235	1,764	7.9	805	247	30.7
Moist deciduous forest	61,292	1,436	2.3	2,445	266	10.9
Dry deciduous	4,902	167	3.4	1,018	89	8.7
Seasonal montane forest	3,065	907	29.6	77	0	0.0
Aseasonal montane forest	13,704	4,227	30.8	141	98	69.5

Source: Planimetric analysis of Figs. 4.23 and 4.24 by the World Conservation Monitoring Centre, Cambridge

Chapter Five

Fauna

PREHISTORIC SPECIES

The first reports of large animal fossils on Java date back nearly 150 years (Shutler and Braches 1985), but most of the existing knowledge finds its origins in the collections of Eugene Dubois made at the end of last century, mainly from Central Java (fig. 5.1). The remains, often no more than a piece of tooth or bone, have been interpreted and identified (Hooijer 1947, 1952, 1954, 1955, 1960, 1962a, b, 1964, 1973, 1974a, b, 1982). Despite the scant remains, there are a plethora of opinions held about their significance, origin, relative ages, and relationships but, as Groves (1985) has remarked, as time moves on we actually appear to be sure of less and less about Java's prehistoric fauna.

It must be remembered in any discussion of fossils that only very specific environmental conditions are suitable for the mineralization of bones in order to form fossils, and that the large animals with robust bones tend to be overrepresented. Most bones rot to their constituent elements in moist tropical regions, but in open plains, along lake shores and rivers, and in dry caves preservation may occur. Thus the paucity of Pliocene and Pleistocene fossils from West Java should not in any way be interpreted as evidence of a pauce fauna.

There are many hypothetical routes along which Java's terrestrial palaeofauna could have arrived, but it is most likely that it was across the Sunda Shelf during periods when parts of it were exposed. Dry land is not a precondition, however, because many animals are capable of swimming sufficiently well to cross narrow or not-so-narrow straits (Shutler and Braches 1985). The route by which the fauna arrived would most likely have been down the Malay Peninsula or possibly by an exposed route across what is now the South China Sea.

Mammals

The oldest mammal remains from Java date from the Pliocene, and are of species similar to those found in the Siwalik Hills of India: a primitive elephant-like mastodon *Trilophodon bumiajuensis*, a hippopotamus *Hexaprotodon*,

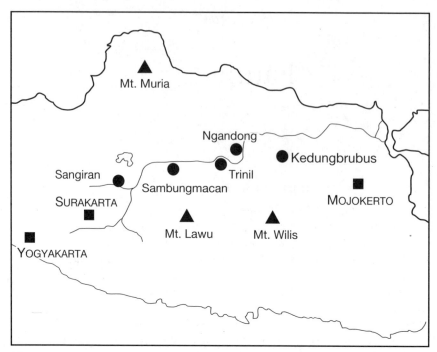

Figure 5.1. Major locations of animal fossils.

and deer. These all date from 1.5-3 million years ago (Groves 1985).

The Pleistocene faunas from perhaps 0.8-1.5 million years ago of Jetis, Trinil, and Ngandong (Leinders et al. 1985), are very similar and related as much to the Siwalik as to the prehistoric Chinese faunas, and some of the differences, particularly between the first two, may simply be due to chance (table 5.1, figs. 5.2-5; Hooijer 1952, 1975; Groves 1985). It is not known how long these species persisted in Java, but the giant pangolin *Manis palaeojavanica*, which grew to 2.5 m long, is known from sediments in Niah Cave, Sarawak, from just 40,000 years ago (Hooijer 1974a). The most famous of the Pleistocene species is the Java Man (Person?) *Homo erectus* which is described in more detail later (p. 309).

Unfortunately the major site of Sangiran has never been the subject of a detailed palaeontological programme, and so the erection of theories regarding the significance of species found at the various sites which are or are not found at Sangiran is premature (Groves 1985). New sites continue

Figure 5.2. The genus of dirk-toothed cats *Megantereon* was found in North America, Africa, Europe, and in Asia as far east as Java.

After Savage and Long 1986, with permission of the Natural History Museum, London

Figure 5.3. The giant tortoise *Geochelone* from the Pliocene of Java, and the more recent but extinct giant pangolin *Manis palaeojavanica* are drawn next to a person 1.65 m tall.

By A.J. Whitten

Figure 5.4.
Two species of stegodonts such as these are known from the Pleistocene of Java.

By A.J. Whitten

Figure 5.5. The extinct buffalo *Bubalus palaeokerabau*.

By A.J. Whitten

Table 5.1. Mammals found in prehistoric deposits in Java. †† = extinct species; † = no longer found on Java but extant elsewhere. Note that Trinil H.K. is a particular layer at the Trinil site and that the 'Trinil' of von Koenigswald (1934, 1935) was a composite fauna from different localities (Sondaar 1984).

		Satir 1.5 mill. years ago part man-grove	CiSaat ±1.2 mill. years ago open wood-land	Trinil H.K. ±1.0 mill. years ago open wood-land	Kedung Brubus ±0.8 mill. years ago open wood-land	Ngan-dong open wood-land	Pun-ung humid forest	Uncer-tain posi-tion
INSECTIVORA								
Moon rat	Echinosorex sp. †	–	–	–	–	–	+	–
PRIMATA								
Silvered leaf monkey	Semnopithecus auratus	–	–	+	–	–	–	–
Javan leaf monkey	Presbytis comata	–	–	–	–	–	–	+
Pig-tailed macaque	Macaca nemestrina †	–	–	–	–	–	–	+
Long-tailed macaque	Macaca fascicularis	–	–	+	–	+	+	–
Siamang	Hylobates syndactylus †	–	–	–	–	–	+	–
Orangutan	Pongo pygmaeus†	–	–	–	–	–	+	–
'Erect people'	Homo erectus ††	–	–	+	+	–	–	–
Modern people	Homo sapiens	–	–	–	–	–	+	–
RODENTIA								
S.E. Asian porcupine	Hystrix brachyura †	–	–	+	–	–	+	–
Giant pocupine	Hystrix gigantea ††	–	–	–	–	–	–	+
Small porcupine	Hystrix vanbreei ††	–	–	–	–	–	–	+
Trinil rat	Rattus trinilensis ††	–	–	+	–	–	–	–
LAGOMORPHA								
Stone hare	Caprolagus lapis ††	–	–	–	–	–	–	+
Hispid hare	C. hispidus †	–	–	–	–	–	–	–
PHOLIDOTA								
Giant pangolin	Manis palaeojavanica ††	–	–	–	+	–	–	–
CARNIVORA								
Sun bear	Helarctos malayanus †	–	–	–	–	–	+	–
Trinil tiger	Panthera trinilensis ††	–	+?	+	+	–	–	–
Tiger	Panthera tigris †	–	–	–	–	+	+	–
Leopard	Panthera pardus	–	–	–	–	–	–	+
Dirk-toothed cat	Megantereon sp. ††	–	–	–	–	–	–	+
Sabre-toothed cat	Hemimachairodus zwierzyckii ††	–	–	–	–	–	–	+
Scimitar-toothed cat	Homotherium ultimum ††	–	–	–	–	–	–	–
Clouded leopard	Neofelis nebulosa	–	–	–	–	–	–	+
Leopard cat	Prionailurus bengalensis	–	–	+	–	–	–	–
Slender-clawed otter	Lutrogale palaeolep-tonyx ††	–	–	–	–	+	–	–
Robust otter	L. robusta ††	–	–	–	–	–	–	+
Short-snouted hyaena	Hyaena brevirostris ††	–	–	–	+	–	–	–
Spotted hyaena	Crocuta crocuta	–	–	–	–	–	–	+
Trinil dog	Mececyon trinilensis ††	–	–	+	–	–	–	–
Merriam's dog	Megacyon merriami ††	–	–	–	–	–	+	–
Wild dog	Cuon sp.	–	–	–	–	–	–	+
PROBOSCIDEA								
Bumiayu mastodont	Tetralophodon bumia-juensis ††	+	–	–	–	–	–	–
Triangular-headed stegodont	Stegodon trigonocephalus ††	–	–	+	+	+	+	–

Table continues

Table 5.1. *(Continued.)* Mammals found in prehistoric deposits in Java. †† = extinct species; † = no longer found on Java but extant elsewhere. Note that Trinil H.K. is a particular layer at the Trinil site and that the 'Trinil' of von Koenigswald (1934, 1935) was a composite fauna from different localities (Sondaar 1984).

		Satir 1.5	CiSaat ±1.2	Trinil H.K. ±1.0	Kedung Brubus ±0.8	Ngan-dong	Pun-ung	Uncer-tain posi-tion
		mill. years ago	mill. years ago	mill. years ago	mill. years ago			
		part man-grove	open wood-land	open wood-land	open wood-land	open wood-land	humid forest	
Large-crested stegodont	*S. hypsilophus* ††	–	–	–	+?	–	–	–
Early elephant	*Elephas hysudrindicus* ††	–	–	–	+	+	–	–
Asian elephant	*E. maximus* †	–	–	–	–	–	+	–
PERISSODACTYLA								
Curve-clawed horse	*Nestoritherium* cf. *sivalense* ††	–	–	–	–	–	–	+
Javan rhinoceros	*Rhinoceros sondaicus*	–	–	+	+	+	+	–
Kendeng rhinoceros	*R. unicornis* †	–	–	–	+	–	–	–
Asian tapir	*Tapirus indicus* †	–	–	–	+	+	+	–
ARTIODACTYLA								
Stremm's pig	*Sus stremmi* ††	–	+	–	–	–	–	–
Short-jawed pig	*S. brachygnathus* ††	–	–	+	+	+	–	–
Large-jawed pig	*S. macrognathus* ††	–	–	–	+	+	–	–
Dwarf pig	*S. sangiranensis*	–	–	–	–	–	–	+
Wild pig	*S. scrofa*	–	–	–	–	–	+	–
Bearded pig	*S. barbatus* †	–	–	–	–	–	+	–
Early hippopotamus	*Hexaprotodon simplex* ††	+	–	–	–	–	–	–
Sivalik hippopotamus	*H. sivalensis* ††	–	+	–	+	+	–	–
Dwarf hippopotamus	*Merycopotamus nanus*	–	–	–	–	–	–	+
Muntjak deer	*Muntiacus muntjak*	–	+	+	+	–	+	–
Unidentified deer		+	+	+	+	+	+	–
Lydekker's deer	*Axis lydekkeri* ††	–	+	+	+	–	–	–
Mountain goat	*Capricornis sumatraensis* †	–	–	–	–	–	+	–
Thin-horned ox	*Epileptobos groeneveldtii* ††	–	–	–	–	+	–	–
Santeng buffalo	*Duboisia santeng* ††	–	–	+	+	–	–	–
Buffalo	*Bubalus palaeokerabau* †	–	–	+	+	+	+	–
Unidentified bovids		–	+	–	–	–	–	–
Early banteng	*Bibos palaesondaicus* ††	–	–	+	+	+	+	–

Sources: Dawson (1971); Hemmer (1971); Schütt (1972; Aimi (1981, 1989); Sondaar (1984, pers. comm.); Leinders et al. (1985); van Weers (1985, 1992); Willemsen (1986); de Vos and Aziz (1989); Hardjasasmita (1987); Aziz (1989); J. de Vos (pers. comm.); C. Groves (pers. comm.)

to be found, but one at Miri, at least, has been submerged beneath the Kedung Ombo reservoir, albeit after an intensive last minute survey.

Prehistoric but 'recent' animal remains have been found in Sampung Cave, near Ponorogo, Central Java, and Wajak, 'Hoekgrot', and Gua Jimbe near Tulungagung, and Blitar in East Java. It was believed that the sites had been destroyed by limestone mining (e.g., van den Brink 1982), but they have recently been rediscovered intact and available for palaeontological excavation (Aziz and de Vos 1989). It is not known for sure how old these deposits are, but they are probably between 3,000-5,000 years old (van den Brink 1982). A full listing of all Sampung animal groups is given in Dammerman (1934), and a full list of Sampung molluscs (which includes land, freshwater, brackish water and marine species) is given by van Bentham Jutting (1932). The mammals are listed below (table 5.2). The remains included bones of modern people, and are particularly interesting because they had been painted with red ochre.

Of these recent remains, the tiger, jungle cat, elephant, tapir, buffalo, and 'Bawean' deer, are now extinct on Java. Interestingly, the 'Bawean' deer which remain on the island of Bawean were described in 1845 as coming from Java, so it is likely that their extinction on the mainland is relatively recent (van den Brink 1982), and that the Bawean animals are a remnant population. The apparent extinction between the Mesolithic (Wajak and Gua Jimbe) and Neolithic (Hoekgrot) would have coincided with the 'agricultural revolution' of the intermediate period (Leinders et al. 1985, Storm 1991). It is also interesting to speculate whether any of Java's buffaloes today are descendants of the native wild species *Bubalus arnee*, or whether that became extinct before the domestic species *Bubalus bubalis* from the mainland was introduced. If it did become extinct, the cause is unknown (p. 298).

Old historic records of Javan rulers riding on elephants postdate contacts from India where these grand animals could have come from. Java did export ivory to China before this, but there is no way of knowing whether this was from Sumatra or from remnant herds in Java. It is intriguing, however, that the old Javanese for elephant is 'liman' which means 'the beast with a hand'. The Sanskrit word has the same meaning. The similarity between the trunk and a hand was noted by Aristotle more than 2,000 years ago (Dammerman 1934). The Sampung finds also indicate that the Javan Rhinoceros was found considerably further east than historical records indicate.

While the palaeofauna of Java is quite diverse, there are none of the horse-like, giraffe-like, or camel-like animals found in east and south Asia. These are animals that today are more typical of extensive open areas, and this too suggests that Java was predominantly covered by forest (Pope 1985). Even so, it has been suggested that the presence of the long-horned *Bubalus palaeokerabau*, with a spread across the horns of 2.25 m, and of

Table 5.2. Mammal species found at Sampung and Wajak. † = species extinct on Java.

		Sampung	Wajak	'Hoekgrot'	Gua Jimbe
DERMOPTERA					
Flying lemur	*Cynocephalus* sp.	–	–	–	+
CHIROPTERA					
Dog-faced fruit bat	*Cynopterus sphinx*	–	–	–	+
Common rousette	*Rousettus amplexi-caudatus*	–	–	–	+?
PRIMATA					
Slow loris	*Nycticebus coucang*	+	–	+	+
Long-tailed macaque	*Macaca fascicularis*	+	–	+	+
Silvered leaf monkey	*Semnopithecus auratus*	+	+	+	+
Human	*Homo sapiens*	+	+	+	+
PHOLIDOTA					
Pangolin	*Manis javanica*	–	+	–	+
LAGOMORPHA		–	–	–	–
?Black-naped hare	*Lepus* cf. *nigricollis*	+	–	–	–
RODENTIA					
Giant flying squirrel	*Petaurista petaurista*	+	–	–	–
Giant bicoloured squirrel	*Ratufa bicolor*	+	–	–	+
Palm squirrel	*Callosciurus notatus*	+	+	–	–
Giant rat	*Leopoldamys sabanus?*	+	–	–	+
Javan flat-nailed rat	*Kadarsanomys sodyi*	+	–	–	+
Field rat	*Rattus tiomanicus*	+	+	–	–
Polynesian rat	*Rattus exulans*	+	–	–	–
Javan porcupine	*Hystrix javanica*	+	+	+	+
CARNIVORA					
Wild dog	*Cuon javanicus*	+	–	–	+
Domestic dog	*Canis familiaris*	–	–	+	–
Small-clawed otter	*Aonyx cinerea*	+	–	–	–
Yellow-throated marten	*Martes flavigula*	–	–	–	+
Common palm civet	*Paradoxurus hermaphroditus*	+	–	+	–
Palm civet	*Arctogalidia* sp.	–	–	–	+
Tiger	*Panthera tigris* †	+	+	–	–
Jungle cat	*Felis chaus* †	+	–	–	–
Forest cat	*Prionailurus bengalensis*	+	–	–	–
PROBOSCIDEA					
Elephant††	*Elephas maximus* †	+	–	+	–
PERISSODACTYLA					
Tapir	*Tapirus indicus* †	–	+	–	+
Javan rhinoceros	*Rhinoceros sondaicus*	+	–	+	+
ARTIODACTYLA					
Wild boar	*Sus scrofa*	+	+	–	+
Warty pig	*S. verrucosus*	–	–	–	+
Java deer	*Rusa timorensis*	–	+	+	+
Bawean deer	*Axis kuhli* †	+	+	–	–
Barking deer	*Muntiacus muntjak*	+	+	+	+
Mouse deer	*Tragulus javanicus*	+	–	+	+
Bovid	??	–	–	+	–
Banteng	*Bos javanicus*	+	?	–	+?
Buffalo	*Bubalus arnee* †	+	+	–	+?

Sources: Dammerman (1934); Dawson (1971); Hemmer (1971); van Weers (1979); van den Brink (1982, 1983); Musser (1982); Burgers (1988); Storm (1990a, b); Boedihartono (1993); J. de Vos (pers. comm.); C. Groves (pers. comm.)

some stegodonts, with tusks that are four metres long, indicates that the forest habitat could not have been entirely dense tropical rain forest but rather open woodlands with grassland and swamps (von Koenigswald 1976). The existence of cooler, wooded habitat is also supported by the fossil occurrence of the jungle cat *Felis chaus*, which today is found in seasonal forests north of peninsular Thailand (Lekagul and McNeely 1977).

Birds

Although less well known than the mammals, the fossil birds from Central Java are very interesting from the point of view of understanding the palaeoenvironment at the time of *Homo erectus*. The remains of seven species have been found; all are water birds and one is extinct, and only one is still found on Java (table 5.3). The presence of most of them in the past suggests that the climate was cooler and that a more 'temperate' community of migratory lowland birds used to visit Java.

Table 5.3. Species of Pleistocene fossil birds found in Central Java. †† = extinct species, † = no longer found on Java but extant elsewhere.

Vulture †	Aegypiinae/Accipitridae	The closest vultures are now found in northern Peninsular Malaysia and Thailand.
Adjutant stork †	*Leptoptilos* cf. *dubius*	*L. dubius* is now confined to northeast India, Burma, Thailand, and Indochina. The remains are of a bird much larger than the increasingly rare lesser adjutant *L. javanicus* found today.
Giant adjutant stork††	*L. titan*	A very large extinct stork.
Black-necked stork †	*Ephippiorhynchus* cf. *asiaticus*	Today this is found in India, Pakistan, and Sri Lanka to Vietnam and the northern Malay Peninsula. A fresh skull of this species was found near Jakarta in 1898.
Red-breasted goose †	*Branta* cf. *ruficollis*	Migratory species breeding nowadays in Siberia and wintering as far south as southeast Iraq.
Australian shelduck †	*Tadorna tadornoides*	Today breeds in southern Australia, reaching the far north of the continent in summer.
Peafowl	*Pavo muticus*	Known today from Indochina, Thailand, and Java (but nowhere else on Sundaland).
Grey crane †	*Grus grus*	Found today from Scandinavia to Africa and Hainan (China).

Sources: Wetmore (1940); Weesie (1982)

Reptiles and Fishes

Fossil remains of giant tortoises *Geochelone atlas* have been found in Pliocene deposits of Java, Sulawesi, and Timor (fig. 5.3) (Sondaar 1981). Living relatives can now be found only on small, remote islands (Galapagos in the Pacific Ocean, and Aldabra in the Indian Ocean), but in the Tertiary and Pleistocene they were present across both the Old and New Worlds. In northern America they suffered attacks by large carnivores, but most probably became extinct by continued and wholesale destruction of their eggs and young. Stretches of water would not have prevented their dispersal because they float quite well and could have survived long periods away from land (Hooijer 1982).

A variety of turtles, catfish, snakeheads, and sharks have been described from Central Java Pleistocene deposits (Dubois 1908; Boeseman 1949; Koumans 1949; Williams 1957).

PRESENT INDIGENOUS SPECIES

Like its flora, Java's fauna is less rich than that on the other Greater Sunda islands, but it does have relatively high levels of endemism. The presentation of totals of species and percentages of endemic species is fraught with difficulties for all groups, but particularly so for the invertebrates, because of inadequate information. For example, 80 species of booklice (psocopterans) are known from the small and presumably depauperate Krakatau Islands as a result of intensive collecting as part of large scientific expeditions, but only 54 species are known from Ujung Kulon and 41 from Carita, both on the adjacent Java mainland (Thornton et al. 1988). In addition, the species that are known tend to be those that are relatively easy to catch because they are in reach of a hand-held net, on the ground surface, or attracted to a light. Critical analysis of the way arthropod faunas are divided up between the main habitats reveals in fact that huge numbers of many species are actually restricted to the canopy and within the soil, and these are habitats which are generally not sampled (Stork 1988).

Lists of endemic species are presented below because, as with the plants, these are the elements of Java and Bali's biological diversity for which the provincial and national authorities in Indonesia have global responsibility. While a few may attract international attention and support, the future of the vast majority of endemic species depends totally on decisions made in Java and Bali.

Tracing apparently endemic species is possible in many cases only with considerable work, examining old species catalogues for species limited to Java and then ploughing through more recent literature to discover range additions, synonymies, and new species. Another problem is that many groups are sorely in need of revision, that is, a critical reappraisal of a group of species in the light of new material or new interpretations. Revisions can reveal when two named 'species' are actually only variations of a single species, or conversely when a single-named species comprises a number of different species. In many cases, as has been found during the preparation of this book, there is literally no one in the world with the interest or expertise to do this work.

For the groups that follow, acknowledged experts have been able to provide lists and even some ecological information. For some relatively large and attractive groups, such as the butterflies, the degree of reliability of the data is probably quite high and comparable to the vertebrates. For most, however, inadequate collecting both in Java and elsewhere results in two problems. First, a species that has been found to date only in Java may be lurking undetected on Sumatra, Borneo, or elsewhere. Second, genuinely endemic species may not yet have been found. It is virtually certain that with larger sample sizes from Java and elsewhere the number of endemics will decline. For the groups that follow, however, no matter how shaky the assumptions for including species in the endemism lists, it is felt that *they should be regarded as endemic until proved otherwise.*

Many of the species are confined to the wetter, western part of the island, but the eastern mountains have appreciable numbers of endemic species. Many species, according to the information available, are restricted to Mt. Gede-Pangrango. Whether this accurately reflects the zoogeography of Java is open to question because so much more collecting has been conducted here than elsewhere, given its proximity to the biological centre of Bogor and to the national capital, its juxtaposition to the Cibodas Botanic Gardens, and the availability of cool air and pleasant accommodation. This might be referred to as the 'Mt. Gede-Pangrango effect'.

MAMMALS

Fauna

This section concerns only the terrestrial species of mammal. Some 18 species of whale and 10 species of dolphins might be found around Java and Bali (Payne et al. 1985), but the only time they impinge on terrestrial systems is when they are washed ashore (fig. 5.6).

The indigenous terrestrial mammal fauna of Java comprises 137 species, including 18 rats and mice, and 68 bats (Sody 1989a Kitchener and Maryanto 1993; R. Melisch pers. comm.). Knowledge of the mammals of Bali lagged behind that of Java, and when Alfred Russel Wallace visited the island in 1856, the only mammal species collected there had been tiger and

a

b

Figure 5.6. Stranded whales: a - a 17 m blue whale, the world's largest living animal, washed ashore near Sampang, Madura, in the first part of this century; b - part of a group of 85 pilot whales washed ashore in Bali in 1985.

a. *photographer unknown, with permission of the Tropen Instituut, Amsterdam, b. with permission of* Kompas

long-tailed macaque (Zollinger 1845a). Given that he spent only a few days there, restricted to the north coast, it was a step of faith that led him to state that "Bali and Lombok differ far more from each other in their birds and quadrupeds (i.e., mammals) than do England and Japan" (Wallace 1869). Even nearly eighty years later Sody (1933) was of the opinion that, as regards mammals, Bali was "one of the most unknown of the larger islands of the entire archipelago". At that point 31 species were known, and by 1993 a further four (all bats) had been added (Kitchener and Foley 1985; Kitchener et al. 1993; D. Kitchener pers. comm.; Boeadi pers. comm.). Sody (1933) was correct in stating that the Bali mammal fauna is really "no more than a poor representation of that of Java".

The mammal fauna of Java and Bali is relatively poor when compared

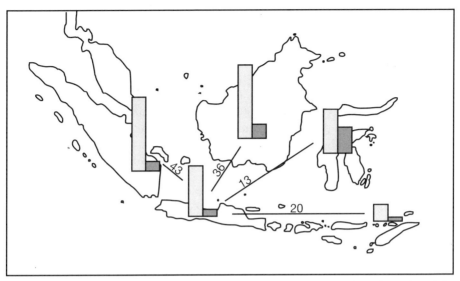

Figure 5.7. Relative richness (light shade), endemism (dark shade), and overlap of the mammal fauna of Java and Bali and their neighbouring islands and island groups. 'Overlap' is the percentage of the total species list that is shared between them.
After MacKinnon and MacKinnon 1986

with its neighbours', but is still appreciable. Its composition is most similar to that of Sumatra (fig. 5.7).

Even now the mammal fauna of Java and Bali is probably not completely known: an entirely new species of fruit bat, *Megaerops kusnoti*, was caught by the doyen of Indonesian mammalogy, Drs. Boeadi, near Lengkong in 1973 (Hill and Boeadi 1978). New bats and a rat were found in 1993, and further records surely await discovery by diligent observers.

Collections of mammals made on the Kangean Islands in 1982 and 1984 resulted in records of 10 bat species not previously known from the islands, one of which was described as a new subspecies restricted to the islands. All except one of the 15 bats known from the islands live in caves, of which there are many in the uplifted limestone deposits that make up these islands. The caves have also served, until recently, as lairs for leopard *Panthera pardus* (de Iongh et al. 1982; van Helvoort 1985b) of which few if any probably now remain. Among the other Kangean mammals, all are species which have been deliberately or accidentally transported in sailing ships: commensal rats, deer, long-tailed macaques, and palm civets. Collections made 80 years before included the large, black fruit bat Pteropus alecto, but it now appears to be absent and it is possible that hunting (by visiting Balinese or local people) may have caused its extinction there (Bergmans and van Bree 1986).

The occurrence of the leopard *Panthera pardus* on Java is intriguing. It is not known in the extensive fossil record of Java except in one deposit of unknown age (p. 199) and has never been recorded on neighbouring Sumatra or Borneo. It was not recorded on Ujung Kulon until 1939, and it seemed to become somewhat more common over the following twenty years, possibly as a response to the decline of the tiger there (Hoogerwerf 1970). It seems plausible that animals were brought to Java from India (where it is native) during the days of close cultural, commercial, religious, and political ties between the two countries. Large animals have special significance in certain Hindu ceremonies (the King of Klungkung requested, and was sent, a Javan rhinoceros by the Dutch authorities in 1839 for one such ceremony), and a festival called *rampok macan* ('large cat plunder') was celebrated on Java during which tigers and leopards were ritually killed. The Kangean Islands were used for hunting by the rulers of Madura and Java, and it is quite possible that the leopard population there derived from animals initially deliberately introduced (Bergmans and van Bree 1986).

Endemic Species

Surprisingly, perhaps, the degree of endemism among Javan mammals is really quite high, with 22 or 16% of Java's mammals (15 or 22% if the bats are excluded) not found beyond its shores (table 5.4). All but two of the species are found in West Java. Of the 18 native species of rats and mice found on Java, seven of them are endemic species and one (*Kadarsanomys*) is an endemic genus (Musser 1981, 1982). This number of endemic rats is surprising—it is four more than the Malay Peninsula, one more than Sumatra, and only one less than endemic-rich Borneo. This situation, and the fact that four species of murids found on Sumatra, Borneo, and Malaya have never been collected on Java, suggests that Java has long been separated from the rest of Sundaland (Musser 1982). But the relatively small number of endemic genera reflect an underlying uniformity among the island faunas of the Sunda region (Groves 1985). Bali has no endemic mammals and they are all found on Java (D. Kitchener pers. comm.).

Although this book is not overly concerned with subspecies, it is noteworthy that many of Java's mammals have subspecies that are known only from the ungazetted Mt. Slamet in Central Java, such as the lesser giant flying squirrel *Petaurista elegans*, the endemic Javan mountain spiny rat *Maxomys bartelsi*, and the endemic Javan leaf monkey or surili *Presbytis comata* (also possibly from Mt. Lawu). There are current suggestions that the last of these be regarded as a distinct species *Presbytis fredericae* (Eudey 1987).

Table 5.4. Endemic mammals of Java and Bali. * = known from fossils only. W = West Java, C = Central Java, E = East Java, B = Bali.

INSECTIVORA
Soricidae

Crocidura orientalis	Javan mountain shrew	W				Known only from Mt. Gede-Pangrango, at 1,800-2,700 m.

CHIROPTERA
Pteropididae

Megaerops kusnotoi	Javan pleated-cheeked fruit bat	W			B	Known only from specimens caught in 1977 in Lengkong, south of Sukabumi, and in Bali in 1993.

Hipposideridae

Hipposideros madura	Madura horseshoe bat			C E		Known from 15 specimens caught 50 years ago in Madura and Landak cave, Karang Sagung, near Semarang.
H. sorenseni	Sorensen's horseshoe bat	W				Known from 7 specimens caught in 1976 in Kramat cave in Pangan daran.
H. sp.	Cidolog horseshoe bat	W				Found in 1992 in Cidolog cave near Sukabumi.

Vespertilionidae

Pipistrellus mordax	Fungus pipistrelle	W				Known only from a handful of specimens: the type described in 1866 and a few specimens reported by Sody in 1937.
Glischropus javanus	Javan thick-thumbed bat	W				Known only from specimens caught on Mt. Pangrango nearly sixty years ago. May be a subspecies of a widespread congener.

Molossidae

Otomops formosus	Javan mastiff bat	W				Known only from around Mt. Gede Pangrango.

PRIMATA
Cercopithidae

Semnopithecus auratus	Javan lutung	W C E B				Lowland and lower montane forests. Introduced to Lombok.
Presbytis comata	Javan leaf monkey or surili	W C				Lowland and lower montane forests.

Hylobatidae

Hylobates moloch	Javan gibbon	W C				Lowland forests (<1,500 m).

CARNIVORA
Mustelidae

Melogale orientalis	Javan ferret-badger	W		E B		Found primarily in mountain areas, although it may use agricultural land.

RODENTIA
Sciuridae

Hylopetes bartelsi	Bartels' flying squirrel	W				Known only from Mt. Gede-Pangrango.

Table continues

Table 5.4. *(Continued.)* Endemic mammals of Java and Bali. * = known from fossils only. W = West Java, C = Central Java, E = East Java, B = Bali.

Muridae			
Kadarsanomys sodyi	Javan flat-nailed rat	W C*	Forests (bamboo) on Mt. Gede-Pangrango at 1,000 m.
Maxomys bartelsi	Javan mountain spiny rat	W C	Found in forests and forest edges on mountains from 1,350-2,600 m where it can be relatively common. When caught it is very docile and can be handled easily.
Mus vulcani	Javan volcano mouse	W	Mountains.
Niviventer lepturus	Woolly mountain rat	W C E	Mountains.
Pithecheir melanurus	Javan monkey rat	W	Known only from Mt. Gede-Pangrango.
Sundamys maxi	Javan giant rat	W	Mountains, 900-1,350 m, largest Javan murid with very long tail.
ARTIODACTYLA			
Suidae			
Sus verrucosus	Javan pig	W C E	Found in lowland areas of the mainland, Madura, and Bawean.
Cervidae			
Axis kuhli	Bawean deer	C	Known only from Bawean, although fossils are known from the mainland.
Cervus timorensis	Javan deer	W C E	Introduced widely in eastern Indonesia.

Sources: Sody (1937a, 1989a); van Peenen et al. (1974); Musser et al. (1979); Musser (1981, 1982); Musser and Newcomb (1983); Groves (1985); Kadar (1985); van Strien (1986); Becking (1989); Schreiber et al. (1989); Boeadi (1990); Corbet and Hill (1992); Kitchener and Maryanto (1993); D. Kitchener (pers. comm.)

Notable Species

Bats. Fruit bats are generally divided into two groups: the nectar-feeding Macroglossinae and the fruit-eating Pteropodininae. The former have long, fuzzy-tipped tongues which are adapted to reaching pollen and nectar sources. They are represented in Java and Bali by the cave fruit bat *Eonycteris spelaea* (famous for its presence at the temple of Goa Lawah, Klungkung), and two much smaller long-tongued fruit bats *Macroglossus* spp., which commonly roost in palm and banana trees. Fruit bats play an exceptionally important ecological role by pollinating and dispersing plants, many of which have great commercial value (Fujita 1991; Fujita and Tuttle 1991). Perhaps the best known example is the cave fruit bat *Eonycteris spelaea* (fig. 5.8) and its central role in pollinating flowers of durian *Durio zibethinus* (Bomb.) (Start and Marshall 1975), the cultivated petai *Parkia speciosa* (Leg.), and wild petai *P. timoriana* (formerly *P. javanica*)

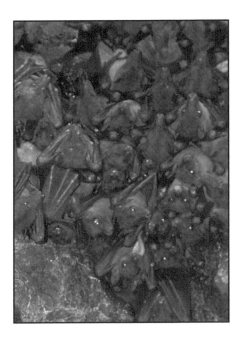

Figure 5.8. Cave fruit bat
Eonycteris spelaea in the colony
at Goa Lawar, Klungkung, Bali.
By A.J. Whitten

(Docters van Leeuwen 1938; Hopkins 1992), but mangrove trees *Rhi-zophora* (Rhiz.) and *Sonneratia* (Sonn.), *Terminalia catappa* (Comb.), kapok *Ceiba pentandra* (Bomb.), and many others owe their existence and dispersal to bats. Bat-pollinated flowers tend to be quite large, of a pale colour, and to hang such that bats are able to land and take off with ease (van der Pijl 1936).

One of the most common and most widely distributed fruit bats in Java and Bali and the whole of tropical Asia is the lesser short-nosed fruit bat *Cynopterus brachyotis*. It is particularly common in urban and rural situations, where it roosts in groups in treetops and especially under the leaves of palms. It is generally regarded as an important seed disperser (as well as a flower pollinator) (Docters van Leeuwen 1935; van der Pijl 1957; Gould 1978), although the seeds of different species experience different fates depending on fruit weight, seed size, and fruit texture (Phua and Corlett 1989). For example, fruits weighing less than 20 g are generally carried off to feeding roosts for peeling and processing, although the smallest seeds (less than about 2 mm in length) are swallowed and are defaecated either while the bat is flying or roosting. Larger seeds are simply dropped to the ground from the feeding roost after processing, which must make them highly susceptible to predation and disease. However, at least a few are dropped accidentally between the fruiting tree and the roost. Thus this common little bat is probably a major disperser of seeds for many tree species (Phua and Corlett 1989).

The larger fruit bats, known as flying foxes, are hunted extensively in Indonesia. Two species are found in Java: the more common species *Pteropus vampyrus* has a grey-black back and is the largest bat in the world. It has famous seasonal roosts such as in Bogor Botanical Gardens, West Java, and Taman Ayun, Bali. The less common species, the island flying fox *P. hypomelanus*, has a brown back, is smaller (though much larger than any other Javan bat) and roosts on islands, visiting the mainland only to feed. Its main roost in on P. Panjang in Cilegon Bay, and they have also been recorded on P. Dua (Boeadi pers. comm.). Moslems are not allowed to eat bat flesh, but vendors in live (or half-dead) bats can be found in some markets to meet the demand from Menadonese and Chinese communities. One Menadonese restaurateur in Jakarta is reported as selling up to 2,000 flying foxes in a good year, purchasing them from the south coast of West Java and from around Surabaya (Fujita and Tuttle 1991). Some flying foxes are taken for medicinal purposes–a few are taken from the roost each day (when present) in the Bogor Botanical Gardens in order to sell the liver, which is thought to be a potent medicine against asthma. Hunters report that finding *Pteropus* roosts is increasingly difficult even during the fruiting peaks when these bats are easiest to find (Fujita and Tuttle 1991).

Many fruit bats of all sizes are shot by owners of *rambutan, langsat,* and rose apple orchard even though the bats tend to take only the ripest fruit, which is too ripe to take to market. Their nocturnal raids can be effectively halted by shining bright lights or by lighting smoky fires (Fujita and Tuttle 1991).

There is no indication of the extent to which forest dynamics have been affected by the decline in fruit bat populations, but their important roles in flower pollination and seed dispersal suggest that major changes may result. Details may never be known because the effects will be so hard to distinguish from those of other environmental changes in Java and Bali.

Insectivorous bats are exploited primarily for their nutrient-rich faeces which fall to the floor of caves. This guano is classified as a class C mineral deposit and licences are granted by the kabupaten to collectors. The value of the resource is reflected in the double row of barbed wire fences across the mouth of the famous cave outside Pelabuhan Ratu. Although many bats roost in caves, renewable, and economic quantities of fresh guano are produced only by the huge roosts of the wrinkle-lipped free-tailed bat *Tadarida plicata* which number hundreds of thousands of individuals (p. 544). The guano industry on Madura relies on removing large quantities of deposits from the many small caves and is not in any way sustainable.

Primates. Of all the 29 species of primate species in Indonesia, none is as threatened as the endemic Javan gibbon *Hylobates moloch* and surili leaf monkey *Presbytis comata*, which have only 4% of their original habitat remaining (MacKinnon 1986b, 1987). They are found only in lowland forests in a number of remaining forest patches in West Java, and on and around Mt. Slamet in Central Java (Ruhiyat 1983; Kappeler 1984a, b; Ghofir 1989; Pasang 1989; Kool 1992b; Putra 1992; Supriatna et al. 1992; Asquith 1993; Mulcahy 1993; Melisch and Dirgayusa in press; Sözer and Nijman in press a; R. Beckwith pers. comm.). Their natural and preferred altitudinal limit is probably about 1,250 m, although they are sometimes found higher than this, particularly where lowland forest has diminished in area. It is most unlikely, however, that populations would be able to sustain themselves in lower montane forest because many of their food species are unavailable (p. 530). The forests of Mt. Halimun National Park support more of these species than any of the other gazetted conservation areas, many of which are too small or too high to be suitable (Kool 1992b). Ujung Kulon National Park, Mt. Simpang Nature Reserve, and Mt. Salak Protection Forest probably still harbour substantial populations, but all are threatened by forest degradation on their perimeters. Even more at risk are the unprotected areas supporting gibbon and surili. Two forests with potentially large gibbon populations surround Mt. Wayang and Mt. Kendang, but neither area is protected (Asquith et al. 1995). Mt. Slamet is the easternmost locality for the Javan gibbon, and Banjarnegara, 50 km further to the east, the furthest east for the surili.

The lutung *Semnopithecus auratus* is also endemic, but is more widespread across Java and Bali, and is found in mangrove, lowland, and montane forests up to 2,200 m (Manullang et al. 1982; Brotoisworo 1983; Weitzel and Groves 1985; Kartikasari 1986; Brotoisworo and Dirgayusa 1990; Kool and Craft 1992; Kool 1992a, 1993; Beckwith in prep.). Three subspecies are recognized‡: the relatively small *S. a. sondaicus* with jet-black but grey-tipped fur from West and Central Java, the larger *S. a, auratus* from East Java (Indonesia's largest leaf monkey), and the relatively small and very dark *S. a. kohlbruggei* from Bali (introduced to Lombok) (Weitzel and Groves 1985; Weitzel et al. 1988), which may now be represented by a mere handful of groups, all confined to the Bali Barat National Park (Wheatley et al. 1993). Some individuals in Central and particularly East Java are a red colour, although the proportion varies among areas. The ancestor of the lutung is thought, according to an old East Javan story, to be a human prince who had once been hit on the head by a rice ladle. He immediately changed into a monkey, and the occasional plaintive calls these monkeys make are believed to be cries of longing that they be taken

---

‡ A further subspecies has been described from Indo-China, and Java's subspecies have been reassigned (Brandon-Jones 1995).

back into human society (d'Almeida 1864).

The other two primate species are the nocturnal slow loris *Nycticebus coucang* and the common long-tailed macaque *Macaca fascicularis*.

Large Cats. The most famous of the carnivores on Java and Bali are the tigers, belonging to two of eight subspecies, that on Java being *P. t. sondaica*, and on Bali *P. t. balica* (table 5.5). Both should now be regarded as extinct, and the stories of their decline is given later (p. 706). Of course, it is impossible to be absolutely sure that extinction has occurred, but even if one is left, the subspecies is extinct; if two are left, they may be the same sex and the subspecies is extinct; and even if there is a viable pair remaining, the pressures on the protected areas and their wildlife are such that, again, the subspecies is effectively extinct, and a captive breeding programme would be a singular waste of financial resources given the other pressing and under-funded conservation priorities in Indonesia. Unfortunately, it is difficult to be optimistic about the future of the last Indonesian tigers in Sumatra. The same pressures that caused the extinction of the other two Indonesian subspecies are acting there: habitat destruction, ecosystem disturbance, and hunting. If the existing regulations are not enforced, then the Indonesian Republic will have watched all of these wonderful beasts in its care pass into oblivion.

The leopard *Panthera pardus* is an extremely adaptable predator feeding on a range of prey from rats and bats to barking deer (Hoogerwerf 1970). There may remain some 350-700 within Java's conservation areas alone, but many of these populations are isolated in remnant patches of forest (Santiapillai and Ramono 1992). Unfortunately the leopard scavenges (feeds on

Table 5.5. Characteristics of Indonesia's tiger subspecies.

	Sumatran *sumatrae*	Javan *sondaicus*	Bali *balica*
Size	as *sondaicus*	as *sumatrae*	smaller than either
Background colour of coat	lightest	darker	darker
Colour of inner side of forelegs	whitish	as background but lighter	as background but lighter
Nasal bones	short and wide	long and narrow	long and narrow
Occipital plane (back of head)	broad	narrow	narrow
Frontal line (forehead)	most flat	more vaulted	most vaulted
Bullae	normal	normal	rather flat

Sources: Sody (1937a); Seidensticker (1987)

animals killed by some other factor), and the greatest threat facing the species is the widespread use of poisons hidden in meat baits set deliberately either for the leopards or against crop-raiding pigs (Santiapillai and Ramono 1992). Leopards are still quite widespread, venturing even into populated rural areas under cover of darkness, but their range is diminishing; in 1929 they were still present in Mt. Gede-Pangrango National Park (Dammerman 1929b), ten may have remained in 1986 (Purbawiyatna 1987), but they are no longer present (Boeadi pers. comm.). They used to be present in Pangandaran Reserve but were extinct there by 1973 (Halder 1975).

Wild Dog. The second largest carnivore on Java is the wild dog *Cuon alpinus*. It used to be widespread, but now is in isolated and uncounted populations in at least Ujung Kulon, Baluran, and Alas Purwo National Parks, although it may also be present in other lowland forest patches. Wild dog ecology and behaviour has been described (Sody 1929; Hoogerwerf 1970; Fox 1984; Indrawan et al. in press). The wild dogs' vernacular name is *ajag* (meaning 'urging on'), and a pack is a formidable predatory unit, capable of exhausting and pulling down a buffalo.

Pigs. The endemic Javan pig *Sus verrucosus* is similar to, and just as variable in size, shape, colour, and ecology as the common wild boar *Sus scrofa* which is found throughout the temperate and tropical regions. It is distinguished from the wild boar, however, by three pairs of warts on the face of the adult male at the corner of the jaws, below the eyes, and on the snout. The Javan pig roams in small groups of four to six, whereas the wild boar can sometimes be seen in aggregations ten times this size. In addition, the adult female Javan pigs are much smaller than the males (less than half of the weight), whereas the female wild boar is 85% of the respective male's weight (Bartels 1942; Groves 1981). A naturally-occurring hybrid has been found, but this may be due to the ever-reducing habitats available to the populations (Blouch and Groves 1990). The Javan pig is not found above 800 m, and its preferred habitat is extensive areas of lowland secondary vegetation, particularly teak plantations in Central Java where there is a mixture of different-aged trees and grasslands with clumps of bush or heavily disturbed forest. It also frequents coastal forests where it is often the only pig species present. In contrast, the wild boar can be found at all altitudes in most habitats from primary forest to agricultural land close to human habitation, where it can cause considerable damage to crops (Zuhud 1983). Many means of controlling them have been tried, but among people whose religious code forbids contact with or consumption of pig, this animal often gets the upper hand. Two control methods which

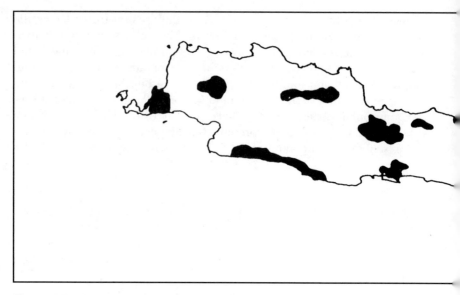

Figure 5.9. Approximate distribution of the Javan pig *Sus verrucosus*.
After Blouch 1988

were said to be effective in the early 19th century were leaving out the solids
remaining after the production of *brem* or rice wine. This intoxicated them,
making them easier to hunt. Alternatively, urine-soaked rags were left on
sticks around plantations or gardens; the pigs found this so repulsive that
they avoided the area (Raffles 1817).

West Java, with its relatively greater area of forest, generally has most of
the major populations of native species, but because its forests are mainly
at high altitude, and because the Javan pig has only a small population in
the lowland forests of Ujung Kulon National Park, the province has a relatively small population of this species. In Ujung Kulon, the Javan pig was
already outnumbered by the wild boar in the 1960s (Hoogerwerf 1970)
since when clearings and young secondary growth have matured, producing habitat more favoured by the latter. The approximate distribution
of the pig is shown in figure 5.9. Typical Javan pig habitat in West Java is
found in the hills north of Bandung where a sparse human population, few
roads that can be used by sport hunters, and dryland crops are found in the
same area. All the pigs here seem to be of the endemic species. But the
greatest concentration of Javan pigs in Java is, perhaps surprisingly, in the
extensive teak plantations between Semarang and Surabaya (Blouch 1988).

The status of the Javan pig recently caused some concern but it does
not seem to be as rare as was feared, and could even be described as
common in some areas, with people its only significant predator. Despite

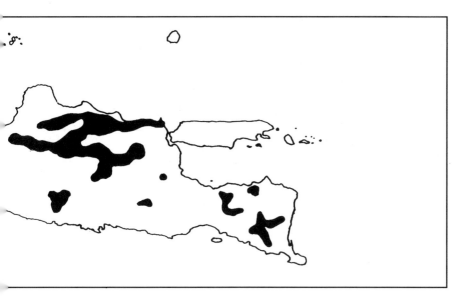

the fact that 90% of Java's population adhere to the tenets of Islam and therefore cannot eat pig meat, pigs are a popular quarry among those who enjoy hunting. They are hunted, using firearms or spears and dogs, for pleasure, and are also snared and illegally poisoned to protect crops, although Javan pigs raid crops less frequently than the wild boars. The meat is then sold in the markets to those whose religion permits them to eat it (Blouch 1988). The conservation problems of the endemic pig are discussed elsewhere (p. 746).

It is possible that the endemic pig would be of interest to pig breeders because of the small size of the sow (making it appropriate for household stock) and its genetic 'distance' from the common pig.

Javan Rhino. The largest animal on Java, weighing up to 2 tons (Lekagul and McNeely 1977), is the Javan rhino *Rhinoceros sondaicus* (fig. 5.10), a pale-skinned and smiling model of which became the mascot for Visit Indonesia Year 1991. It is a somewhat inappropriately named beast since its natural distribution, as far as can be ascertained, once extended to the Brahmaputra River in Bangladesh and to Vietnam and southwest China, and was first described from specimens caught in Sumatra. Its original full distribution will never be known because it has at different times and at different places been confused with the Sumatran rhinoceros *Dicerorhinus*

Figure 5.10. Two Javan rhinoceroses at a wallow in Ujung Kulon National Park.

By A. Hoogerwerf, with permission of the National Museum of Natural History, Leiden

sumatrensis and the Indian rhinoceros *Rhinoceros unicornis*. In the historical past it was known only from southern West Java and from Mt. Slamet in Central Java, although fossil remains have been found north of Yogyakarta (p. 200). When Junghuhn climbed Mt. Pangrango in 1839 (the first recorded ascent by a European), he surprised two Javan rhinos near the top, one immersed in a small stream and the other grazing on the bank (Junghuhn 1854). Indeed, many mountain paths tended to follow rhino tracks, and today's routes up Java's mountains may be the last vestige of the presence of these grand animals. The last twelve Javan rhinos in Sumatra were shot by a Dutch hunter between 1925-1930, and after one was shot near Tasikmalaya (West Java) in 1934, it was believed that the only remaining population was confined to Ujung Kulon National Park from which they were first reported in 1861 (Sody 1941b; Hoogerwerf 1970). In 1989, however, a handful were found surviving in southern Vietnam (Schaller 1990). Details of its decline in Sumatra and Java are given in Sody (1941b).

Javan rhinos are browsers, that is they feed on young leaves, shoots, and twigs growing above ground level. If these are out of reach, a rhino will try to break a stem by pushing against it or crushing the stem with its teeth. Over 150 species of plants have been identified as food (Djaja et al. 1982;

Ammann 1985), and it is possible that all accessible species of a suitable size are eaten. Foods are eaten in a wide variety of vegetation types, although most feeding occurs in unshaded spots such as gaps caused by tree falls or in treeless shrubland. The variety of foods taken may be a response to the need to consume a diet in which toxin intake is limited or mitigated, certain minerals maximized, and the seasonal variations coped with (Ammann 1985). Since almost all the records of food plants are from indirect observations, it is relevant to note that 'typical' rhino damage to saplings can also be produced by banteng and deer (S. Hedges pers. comm.).

The problems of conserving and managing Javan rhinos are discussed elsewhere (p. 750).

Deer. The Bawean deer *Axis kuhli* is one of the rarest deer in the world. It is restricted to the 200 km^2 volcanic island of Bawean (Notowinarno 1988), 200 km north of Surabaya, making it the most restricted deer species in the world. It is quite a small, brown deer reaching only 70 cm at the shoulder, with a short face, white throat, a dark stripe down the middle of its back, and a bushy tail. It was recognised as a distinct species in 1836 when the German explorer Salomon Müller examined a domestic herd kept by the Governor of the Dutch East Indies. It is a close relative of the spotted deer *Axis axis* of the low Himalaya, many of which can be seen in the grounds of the Bogor Presidential Palace (Hasanuddin 1988b), but closer still to the hog deer *Axis porcinus* of the Himalayas, Burma, Thailand, Laos, and south Vietnam. Indeed, it has been suggested that Bawean deer are descended from an early introduced population of hog deer (Groves and Grubb 1987; Corbet and Hill 1992).

Very little of Bawean now has natural forest, nearly a quarter has teak plantations, and much of the remainder is thick secondary forest. Dense, bushy vegetation has about four times as many deer as teak forest with understorey, and 'clean' teak forest supports virtually no deer at all. Grassy glades are favoured when there is young growth after a fire, and primary forest is used for resting, but only where not disturbed by people (Blouch and Atmosoedirdjo 1987). The total population is 200-400 animals and, thanks to the gazetting of a reserve, high quality guards, and Government programmes to raise the awareness of the 75,000 islanders concerning the protected status of the deer, hunting is far less common now than in the past. Uncontrolled fires, and the conversion of the remaining steep forest land to dryland agriculture, however, remain serious threats. It is hoped that those responsible for the production of teak and those charged with protecting wildlife can come to practical management decisions to ensure the survival of this remarkable deer. It has been reported that Bawean deer have been successfully maintained in ranching trials (Anon. 1988), but this has little or no conservation significance.

Figure 5.11. A Java deer stag in Ujung Kulon National Park.

By A. Hoogerwerf, with permission of the National Museum of Natural History, Leiden

   The Java deer *Cervus timorensis* (fig. 5.11) is now widespread throughout the archipelago because people have taken it on their migrations, and it is a major game species in many areas of eastern Indonesia. Its optimum habitat is savanna and savanna woodlands, and thus this species is more abundant in the fire-influenced grass areas of eastern Java than in the rain forest areas in the west. For example, Junghuhn estimated seeing about 50,000 deer on the Yang Plateau (p. 807), and on one day at about 2,300 m in 1844 he saw 75 herds varying in size from 25-1,500 (Junghuhn 1854; Hoogerwerf 1970). This deer has become rather uncommon as the lowlands have been converted to rice and sugarcane, and the hills to coffee and other crops, while the smaller muntjac deer *Muntiacus muntjak* has persisted in many areas where there is some forest cover.

   The third and smallest deer on Java and Bali is the mouse deer *Tragulus javanicus*. It is still numerous and can be seen easily in many areas such as the Tourist Park part of Pangandaran Nature Reserve.

Banteng. The second largest wild animal in Java and Bali is the banteng *Bos javanicus* (fig. 5.12) which stands 1.8 m or more at the shoulder and weighs up to 900 kg (Pfeffer 1965; Hoogerwerf 1970), although average figures for bulls would be 1.6 m and 635 kg, and for females 1.4 m and 400 kg (NRC 1983). It is distributed from northern India and Burma to Borneo and Bali but is missing from Malaya and Sumatra, although it did once occur in Malaya. Banteng are large, handsome cattle with the male and female very

Figure 5.12. A group of banteng in Ujung Kulon National Park.
By A. Hoogerwerf, with permission of the National Museum of Natural History, Leiden

distinct: the male is predominantly black and the female mid-brown, but both have a contrasting white rump patch and stockings. They are the ancestors of the domestic 'Bali' cattle, and in some areas where both are found it can be difficult, even for the cattle owners, to tell young wild and domestic animals apart at a distance. The adult male wild banteng is highly conspicuous, however, by his colour and the high ridge along his back. Some of the 'banteng' in conservation areas, such as Pangandaran, are in fact the product of varying amounts of cross breeding between wild banteng and Bali cattle.

Banteng are found in forested areas of West and East Java, but their 'preferred' habitat is not clear since they are found in grasslands, scrub, and forest. It may be that they favour mosaics of different habitats. They live in loose herds with the cow and calf as the fundamental unit, and generally only one adult male in each herd. The surplus males group together to form bachelor herds. Herding is to their advantage because animals in a group would have been less likely to fall prey to tigers (Hoogerwerf 1938a, b, 1947a).

Banteng do not discriminate much between day and night in terms of their activity, but in areas where there is human disturbance and hunting they may tend to more nocturnal activity. When guns became available, the wild populations suffered severely, and they were exterminated from many areas (Zwerver 1941). As a result, banteng are now regarded as a vulnerable species. They seem able to adapt to living in disturbed forest, not least because many of their food plants colonise open ground, but they are unable to cope with the extremes of habitat change which come with forest clearance. Apart from the obvious threat of habitat destruction, banteng also face the threat of interbreeding with domestic 'Bali cattle' and hence losing their genetic identity, but this has yet to be investigated in detail. Even so, it is known that the beasts at Pangandaran Reserve at the southeast corner of West Java are hybrids, with weakly developed horns and off-white hind parts, and this may also be the case in other reserves.

The problems of banteng conservation and management are discussed elsewhere (p. 747).

Major Publications

Although there is no field guide specifically for Java and Bali, the identification of most of the species can be achieved using the colour field guides for the mammals of the Malay Peninsula (Medway 1983) and of Borneo (Payne et al. 1985). First-rate information about most of the land animals, particularly the larger species, can be found in the book about Ujung Kulon (Hoogerwerf 1970), and a recent, revised checklist of mammals of Java and Bali has been prepared based on a series of early papers by Sody (1989a) which includes local names, rough distributions, and other notes. A more accessible and even more recent checklist has been produced by the Asian Wetlands Bureau (Melisch 1992). In addition, data on the breeding of Java's mammals can be found (Sody 1940), and drawings of mammal tracks are available (Sody 1936, 1937b; Pluygers 1952; Seidensticker and Suryono 1980; van Strien 1983).

BIRDS

Fauna

Raffles (1817) thought that Java was home to about 200 bird species, but the total now recorded is just over 430, about 340 resident and the remainder regularly migrating from Australia or Asia to breed or over-winter. An additional fifty or so species have been recorded as accidentals. Not surprisingly, given its relatively small size, Bali has only 320 species recorded of which 204 are confirmed or presumed residents (Mason and Jarvis 1989; V. Mason pers. comm.), but new records are being made as interest increases. All but six of the Bali resident species are also resident in

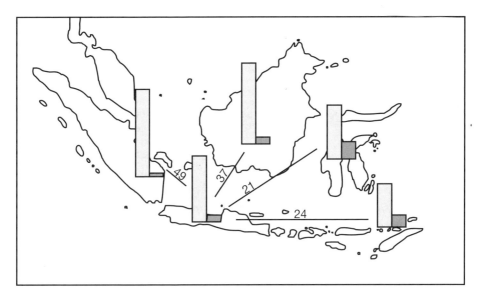

Figure 5.13. Relative richness (light shade), endemism (dark shade), and overlap of the bird fauna of Java and Bali and their neighbouring islands and island groups. 'Overlap' is the percentage of the total species list that is shared between them.
After MacKinnon and MacKinnon 1986

Java, the exceptions being the Bali starling *Leucopsar rothschildi,* rainbow lorikeet *Trichoglossus haematodus,* elegant pitta *Pitta elegans* (Nusa Penida only), red-chested flowerpecker *Dicaeum maugei* (Nusa Penida only), brown honeyeater *Lichmera indistincta,* and black-backed fruit-dove *Ptilinopus cinctus* (S. van Balen pers. comm.).

The bird fauna of Java and Bali is relatively poor compared with its Sundaland neighbours, although some of this is due to their relatively small size. A significant proportion of it is endemic, however, and its composition is most similar to that of Sumatra (fig. 5.13).

Endemic Species

There are 30 endemic species of birds on Java and Bali (9% of the resident species), with one restricted to Bali and 20 to Java. Two of the endemic species, the pygmy tit *Psaltria exilis* and the Bali starling *Leucopsar rothschildi* represent monotypic genera (table 5.6). All but one of the endemic

Table 5.6. Endemic birds of Java and Bali.

FALCONIFORMES			
Accipitridae			
Spizaetus bartelsi	Javan hawk-eagle	W C E	Rare at all altitudes.
GALLIFORMES			
Phasianidae			
Arborophila javanica	Chestnut-bellied partridge	W C E	Not uncommon in montane forest, 1,000-3,000 m.
CHARADRIIFORMES			
Charadriidae			
Charadrius javanicus	Javan plover	W ?C?E	Found in coastal lowlands. Only recently recognized as a distinct species.
Hoplopterus macropterus	Javan lapwing	W C E	Probably extinct.
PSITTACIFORMES			
Psitiidae			
Loriculus pusillus	Yellow-throated hanging-parrot	W C E B	Quite common in forest up to 2,000 m.
CUCULIFORMES			
Cuculidae			
Centropus nigrorufus	Javan coucal	W C E	Rare lowland bird in coastal areas.
STRIGIFORMES			
Strigidae			
Otus angelinae	Javan scops-owl	W	Known only from Mt. Pangrango and Mt. Tangkuban Perahu.
Glaucidium castanopterum	Javan owlet	W C E B	Found wherever there are trees in the lowlands and hills.
APODIFORMES			
Apodidae			
Aerodramus vulcanorum	Volcano swiftlet	W	Rare inhabitant of high volcanoes, e.g., Mt. Gede-Pangrango.
CORACIIFORMES			
Alcedinidae			
Halcyon cyanoventris	Javan kingfisher	W C E B	Not uncommon in fairly open country, up to 1,000 m.
PICIFORMES			
Capitonidae			
Megalaima armillaris	Orange-fronted barbet	W C E B	Probably Java's most common barbet; found in primary forest up to 2,500 m, mainly above 900 m.
M. corvina	Brown-throated barbet	W C	Found only in moist montane forests.
M. javensis	Black-banded barbet	W C E	Found in lowland and hill forests up to 1,500 m.

Table continues

Table 5.6. *(Continued.)* Endemic birds of Java and Bali.

PASSERIFORMES			
Aegithalidae			
Psaltria exilis	Pygmy tit	W C	Java's smallest bird; rare but locally common in forests and forest edges above 1,000 m.
Turdidae			
Cochoa azurea	Javan cochoa	W C	Rare bird of high mountains.
Timaliidae			
Alcippe pyrrhoptera	Javan fulvetta	W C	Locally abundant above 1,000 m.
Crocias albonotatus	Spotted crocias	W	Uncommon, found only on some of the highest mountains.
Garrulax rufifrons	Rufous-fronted laughing thrush	W C	Not uncommon in montane forests.
Macronous flavicollis	Grey-cheeked tit-babbler	W C E	Not uncommon bird of open lowland forest.
Stachyris grammiceps	White-breasted babbler	W C E	In lowland forests; rather rare.
S. melanothorax	Crescent-chested babbler	W C E B	Not uncommon bird in forests 500-1,500 m.
S. thoracica	White-bibbed babbler	W C E	Uncommon bird of submontane and montane forests.
Sylviidae			
Tesia superciliaris	Javan tesia	W C	Uncommon but locally abundant in montane forests, 1,000-3,000 m.
Muscicapidae			
Rhipidura euryura	White-bellied fantail	W C E	Locally common in hill forest.
R. phoenicura	Red-tailed fantail	W C	Locally common in montane forests 1,000-2,500 m.
Sturnidae			
Leucopsar rothschildi	Bali starling	B	Found only at western tip of Bali.
Nectarinidae			
Aethopyga eximia	White-flanked sunbird	W C E	Locally common in montane forests.
A. mystacalis	Violet-tailed sunbird	W C E	Found at forest edges up to 1,200 m.
Zosteropidae			
Lophozosterops javanicus	Grey-throated darkeye	W C E B	Locally common on the highest mountains.
Ploceidae			
Padda oryzivora	Java sparrow	W C E B	Rather rare, lowland areas; much traded and introduced to SE Asia, Australia, Hawaii.

Sources: MacKinnon (1988, 1990); Marle and Voous (1988); Andrew (1992); MacKinnon and Phillipps (1993, in press); S. van Balen (pers. comm.); D. Holmes (pers. comm.).

Figure 5.14. Indonesia's national bird, the Javan hawk-eagle *Spizaetus bartelsi*.

After van Balen 1991a

birds found in Java occur in West Java. It is a measure of the degree to which the Bali bird fauna is relatively unknown, that whereas all but four of Java's endemics were described by 1850, the Bali starling *Leucopsar rothschildi* was discovered only in 1911 and described the following year. The first endemic Javan bird to be described was the conspicuous, and once very common, Java sparrow *Padda oryzivora* which was made known to science in 1758 (Sibley and Monroe 1990). Nineteen of the endemic species are relatively common and do not give cause for concern. However, of the others, the Javan lapwing *Hoplopterus macropterus* appears to be already extinct, and the status of seven others is precarious (MacKinnon 1988, 1990). Most of the endemic species are now confined to montane forests (though some may have been found at lower altitudes when lowland forest was more extensive).

Notable Species

Javan hawk-eagle. The Javan hawk-eagle *Spizaetus bartelsi* (fig. 5.14) is probably one of the rarest birds in the world and possibly one of the most endangered with a maximum population of 50-60 pairs remaining (Meyburg 1986; Thiollay and Meyburg 1988; Meyburg et al. 1989; Sözer and Nijman 1995, in press b). This total population is divided into small populations distributed from Ujung Kulon in the west to Alas Purwo in the east, primarily in montane forest (Sözer and Nijman 1995, in press b). Each bird needs an estimated 20-30 km^2 of forest and forest edge to support it, and so forest loss has a severe impact. Apart from forest loss, these birds were, and possibly still are, having to contend with trapping for sale to status-seeking collectors (Thiollay and Meyburg 1988; van Balen 1991a). In recognition of its threatened status and its resemblance to the mythical *garuda* which appears on Indonesia's coat of arms, it was declared as the national bird in 1993.

Javan scops-owl. The Javan scops-owl *Otus angelinae* is known from just 10 museum specimens (Marshall 1978), one from Mt. Tangkuban Perahu north of Bandung, and the remainder from Mt. Gede-Pangrango. This bird has been observed in the field on only a handful of occasions, and it seems to emphasize its apparent rarity by being singularly quiet, unlike its close relative *O. spilocephalus* from Sumatra and Borneo. It may in fact be relatively common within its apparently limited range but simply difficult to find. Detailed surveys may find it to be widely distributed within forested western Java (Andrew and Milton 1988), although there is no indication of whether it is tolerant of disturbance.

Javan lapwing. Zoologically speaking this is an enigmatic bird as is shown by it having had six generic names (*Chettusia, Hoplopterus, Lobivanellus, Rogibyx, Vanellus,* and *Xiphidiopterus*) and three specific epithets (*tricolor, macropterus,* and *cucullatus*) in various combinations since it was first described by the American scientist Thomas Horsfield in 1821. It is known from 39 skins and mounted specimens in the National Museum of Natural History in Leiden and a handful in Bogor, and their origins are restricted to what are now intensively cultivated agricultural areas, brackish ponds, or industrial estates (figs. 5.15-16).

It is not entirely clear when the bird became extinct, and no author appears to have expressed concern over its status. Koningsberger (1909) regarded it as 'not very common', but Kuroda (1933) gave no indication of scarcity. The reason usually given for the demise of the Javan lapwing is habitat loss due to the spread of sawah and fish ponds in the coastal lowlands, but this is probably not the whole reason. They would have been shot or trapped for food, and their eggs, laid on the ground, would have been quite easy to find even if they were camouflaged.

Bali starling. Bali's only endemic bird is the stunningly beautiful crested Bali starling *Leucopsar rothschildi* (fig. 5.17), which is white with black wing and tail tips, and has blue facial skin. It was discovered only in 1911 (Strese-

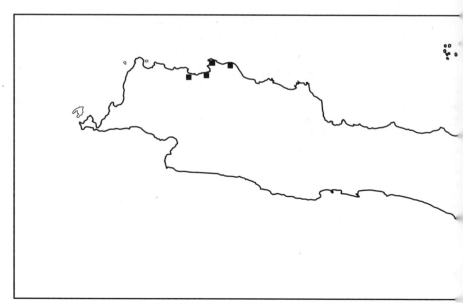

Figure 5.15. Known localities of the Javan lapwing. Localities from the records of the National Museum of Natural History, Leiden.

Figure 5.16. The extensive dune system at Meleman, one of the last locations of the Javan lapwing, as it appeared in 1935.

By J.G. Kooiman, with permission of the Tropen Instituut, Amsterdam

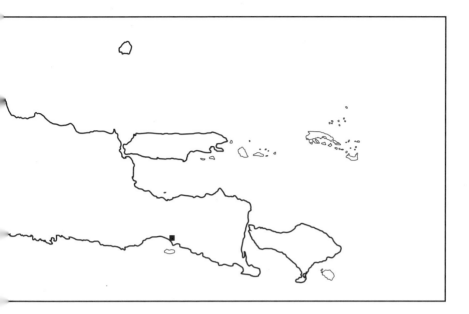

Figure 5.17. Bali starlings *Leucopsar rothschildi.*

With permission of Birdquest, Stonyhurst, Lancs.

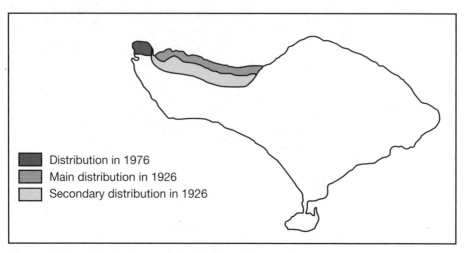

Distribution in 1976
Main distribution in 1926
Secondary distribution in 1926

Figure 5.18. Distributions of the Bali starling *Leucopsar rothschildi* in 1926 and 1976. according to der Paardt (1926) they were most numerous in the area shown with medium shading. It is now confined to the northern part of the Prapat Agung peninsula.

After der Paardt 1926; Sieber 1978

mann 1912), and even at that time its entire distribution was apparently confined to the western tip of the island, east to about Bubunan nearly 60 km away, where the first-known specimen was shot. It is remarkable that there is no record of the Bali starling ever having been seen in Baluran National Park in northeast Java, despite the fact that the vegetation and climate are very similar, the distance across the strait is less than 2 km, and the distance between the two Parks is only 20 km. It is possible that the starling did once live in eastern Java but became extinct long ago as a direct or indirect result of volcanic activity or when the Baluran area was first converted to agricultural land (p. 817).

By 1970 almost all the Bali starlings were confined to what was then the West Bali Nature Reserve (now National Park) in which considerable numbers of large trees were being illegally removed, and its status caused such great concern that it was protected by Indonesian law. Unfortunately, its continued popularity among fanciers of caged birds within and outside Indonesia ensured that trappers were prepared to risk being caught, and the birds' habit of roosting and nesting at known and obvious trees had made them very vulnerable (de Iongh 1983) (fig. 5.18). It was declared a globally endangered species by IUCN in 1977 (IUCN 1977), and in 1979 it was estimated that no more than 200 birds remained (de Iongh ct al. 1982). During the 1980s the population crashed, with a study in 1986 revealing only 54 birds (van Helvoort et al. 1986; Pujiati 1987), and by 1990 only 13-18 individuals remained in the wild. The population increased steadily to 55 birds in late 1991, but decreased again to just 34 by the end of 1993. Meanwhile over 360 captive Bali starlings are now registered in Jakarta alone, and several thousand birds, most the result of captive breeding, are held overseas. Some of these captive-bred birds are being brought into genetically-appropriate groups to produce offspring for release as part of a major conservation effort led by PHPA, Birdlife International, the American Association of Zoological Parks and Aquariums, and the Jersey Wildlife Preservation Trust, to ensure the birds' survival in the wild state. It is unfortunate that few of those people holding birds are willing to give them up for the breeding programme which is being conducted to assist the conservation of the species. The outlook for the wild starlings is not good (Simmonds 1993).

Java sparrow. The strangest of fortunes applies to the beautiful endemic Java sparrow *Padda oryzivora*. Not only has it become rather scarce on Java and Bali because of trapping for the caged bird trade, it has also been introduced throughout Southeast Asia as a result of cage bird escapes, and now faces extinction even in those areas because of further trapping. It even appears that many of the Java sparrows found in Javan bird markets today are imported from introduced populations on other Indonesian islands (van Helvoort 1984). These birds used to be extremely common in Java, and last century there are reports of myriads of them roosting in a single tree and even of them being trained to fly to a hand when called (d'Almeida 1864). They were not uniformly common, however, being less abundant in West Java than the other provinces, although even here they were serious rice pests (Kuroda 1936; Hoogerwerf and Siccama 1938). In certain areas they can still be seen in flocks and were one of the most common birds in the fields around the Gadjah Mungkur reservoir, Central Java, in 1982 (Kuslestari 1983). Large flocks can also be seen in Baluran National Park (M. Indrawan pers. comm.). The decrease in numbers of Java sparrow may in part also be due to the efficiency of modern rice mills that no longer provide easy pickings of grain (van Balen 1989a, c).

Junglefowl. The red junglefowl *Gallus gallus* is one of the few species in Java found from the coasts to over 3,000 m, but it is rather sparsely distributed and mainly restricted to remote areas including the primary forests where it prefers the fringes. It is also found in remote parts of agricultural areas, but it is almost always an extremely shy and wary species (Hoogerwerf 1969). The green junglefowl *Gallus varius*, on the other hand, prefers open places with low ground cover, and in Ujung Kulon, for example, it is never encountered in 'real' forest. In West Java it is found up to 1,500 m, but in East Java it is found even on mountain summits. This difference is possibly because the vegetation in East Java is drier, more often burnt, and has more grass.

The interest in cage birds that has so affected Bali starlings and Java sparrows, also threatens the green junglefowl. The offspring from mating a wild cock with a domestic hen is a sterile bird known as 'bekisar', which is very popular and traditionally used in bird singing contests. Elaborate cross-breeding used to be the monopoly of the Madurese people on Kangean (Vorderman 1893), but now it is commonly practiced in many parts of East Java and Bali (van Balen and Holmes in press). A decree from the State Minister of Population and Environment requiring the parents of contest birds to have been bred in captivity has been issued, but is little enforced. Unfortunately pressure on the wild males has increased since 1990 when the Minister of Home Affairs required each province to have a faunal and floral symbol to promote a sense of pride and awareness of the natural environment. East Java chose the bekisar, and suddenly every hotel and financially able citizen wanted one of these birds in their forecourt or drive. Since these are not used in singing contests, the already weak decree on origin does not apply, with the result that wild male green junglefowl are avidly sought (van Balen and Holmes in press).

Exploitation

Cage birds. The keeping of birds in cages has a history lost in time, and in Javanese tradition it was thought desirable for every man to have five things: a job (*narpadha*), a house (*wisma*), a horse or carriage (*turangga*), a wife (*wanita*), and a bird (*kukila*). The possession of the peaceful dove *Geopelia striata* in particular is believed to bring good luck to the owner (Brotoisworo and Iskandar 1984). The modern inhabitants of Java have been called 'inveterate aviculturalists' (Morrison 1980), and bird cages can be seen outside shops and houses, with many of them, usually containing doves, suspended high above the roofs where their occupants are believed to appreciate the view and breeze. Different regions seem to have different preferences for species and cages, and among the most remarkable cages are the coffin-like pigeon cotes supported on poles around Meleman on the south coast of East Java.

Most people buy birds because of their voice, either their singing such as Asian pied starling *Sturnus contra*, black-naped oriole *Oriolus chinensis*, and white-rumped shama *Copsychus malabaricus*, mimicing such as cockatoos and hill mynah *Gracula religiosa*, or cooing such as zebra doves *Geopelia striata*, and spotted doves *Streptopelia chinensis*. Others buy birds because of their attractive appearance such as Java sparrow *Padda oryzivora*, and munias

Lonchura spp., because of their attractive flight, such as pigeons, or because of their rarity and value as a status symbol, such as eagles, Bali starling, and lineated barbet *Megalaima lineata.* Still others buy birds as toys for children, or for release as part of Chinese and other ceremonies.

About one million birds are sold in the bird markets of Java and Bali each year, and many fruit doves *Ptilinopus,* parrotfinches *Erythrura,* and munias *Lonchura* are exported. Most of the market birds are unprotected, although protected endemic birds such as the orange-fronted barbet *Megalaima armillaris,* Javan kingfisher *Halcyon cyanoventris,* Javan barbet *Megalaima javensis,* and rufous-fronted laughing thrush *Garrulax rufifrons* are also traded. In most cases neither the stall-holder nor the purchaser recognizes their status. Most of the market birds sold belong to a few species and are moderately priced, although unusual or especially attractive birds sell for a premium. Species that are rare and nondescript have no special value; for example, in May 1992 one of the very few Javan scops-owls seen this century was being offered for sale in Jakarta's Jl. Pramuka market, the largest wildlife market in southeast Asia, at a price no different from the other scops-owls. Most of the birds are trapped in the rural and forested parts of Java and Bali, although tens of thousands of babblers, bulbuls, and zebra doves (some common native species, others not) are imported each year from China, Malaysia, and Thailand. It is not known to what extent the trade in wild-caught cage birds adversely affects the wild populations because there are inadequate baseline data, but considered opinion is that the straw-headed bulbul *Pycnonotus zeylanicus* has become very rare largely because of its popularity as a cage bird (Nash 1994).

Shore bird netting. Probably the largest exploitation of wild animals in Java or Bali, in terms of numbers of individuals killed, is the night netting of migratory and resident birds along the north coast of Java, particularly between Indramayu and Cirebon. The migrants are generally birds that breed in eastern Asia and spend the northern winter in Australia and Indonesia. While claimed to be a traditional harvest, investigation has shown that it probably began only in 1946.

In the late 1970s about one million birds were caught annually, but by 1984 a decrease in the catch had caused the labour force to be reduced by half. While a variety of factors are probably involved in the decline, the contribution of the unmanaged and unrestricted trapping is likely to have played a major part. For example, in the early years of this century, millions of shorebirds were killed in North America bringing down the populations of some of them to critical levels. After the signing of the Convention on Migratory Birds in 1917, many species recovered there, but some have not, remaining in danger of extinction (Milton 1984a).

In the North Java harvest, 56 species from 20 families were recorded to be caught, but just five species accounted for two-thirds of the catch (table 5.7). Over 70% of the total catch comprised migrant birds which are caught mainly between November and March. When the migrants fly on, the percentage of resident birds netted increased to over 70% in late May, not just because of the fewer migrants but also because large numbers of immature birds entered the resident populations.

The birds provide a potentially sustainable, though limited and meagre, livelihood for the netters and their families. Daily income per netter in 1984-85 was only Rp1,000-3,660. The carcasses are cheaper than chicken, and thus represent a locally important source of protein. If the harvest is going to be sustained and international ire not heaped on Indonesia for failing to manage those migrant species that are protected in their breeding and wintering areas and by international convention, then a number of steps will be necessary. The objectives would be to eventually license all netters and wholesalers, to monitor the numbers caught, to institute quotas for daily and annual catches, to reduce and then eliminate the harvest of species internationally recognized as being under threat and those species protected under Indonesian law, and to place greater reliance on resident species whose abundance can be encouraged by appropriate local management. A five-year, step-by-step strategy for reaching those objectives has been proposed (Milton 1984a), but to date there has been no agreement on whether or not to begin. Meanwhile, the PHPA, with the Asian Wetland Bureau and the Australasian Wader Studies Group, have begun putting leg bands and flags on hunted species so that objective monitoring can begin (Johnson et al. 1993).

Swiftlets. One of the most remarkable bird-human relationships in Java involves certain species of swiftlets, Apodidae, a family whose members look like swallows but are in fact more closely related to the hummingbirds of America. The family comprises swifts (of 15 cm or more in length) and the smaller swiftlets of which there are five species in Java: the edible-nest swiftlet *Aerodramus fuciphagus*, volcano swiftlet *A. vulcanorum*, black-nest swiftlet *A. maximus*, Himalayan swiftlet *A. brevirostris*, moss-nest swiftlet *A.*

Table 5.7. The five most common birds netted off the coast of northern West Java in order of abundance. [m] = migrant species.

Watercock [m]	*Gallicrex cinerea*
Pintail snipe [m]	*Gallinago stenura*
Yellow bittern [m]	*Ixobrychus sinensis*
Slaty-breasted rail	*Gallirallus striatus*
Common moorhen	*Gallinula chloropus*

Source: Milton (1984a)

salangana, and the commonest and smallest of all, the linchi swiftlet *Collocalia linchi* (Somadikarta 1986; Andrew 1992) which tends to breed in relatively light caves, crevices and buildings. These birds are insectivorous and fly with their mouths open scooping up the aerial plankton of small day-flying insects.

The birds have attracted attention because the edible-nest and black-nest species construct nests of hardened saliva which are harvested and made into a soup. This is particularly favoured by those of Chinese descent and, increasingly, Arabs in Saudi Arabia. The dark red nests command the highest prices, yellow the next, and white the lowest. It is not certain when this harvest was first initiated, but the swiftlet caves on the south coast of Java near Kebumen were apparently first exploited during the eighteenth century. The prising of nests from the high clefts and ledges is not without considerable danger, and the collection of the nests from sea caves is shrouded in traditional beliefs (p. 673).

Edible-nest swiftlets have made some people very wealthy because the dried saliva nests sell for Rp500,000-1 million/kg depending on quality. One kilogram is the equivalent of about 140 nests. These small birds will sometimes decide to occupy an ordinary house, but in certain towns such as Pasuruan and Pemalang people have built special houses for the birds, with a few very small 'windows' and only one door, specifically for the birds to nest in. A single house 7 x 12 m can produce 50 kg of nests every year. A proportion of the nests are collected every three or four months and are sold to dealers before, for the majority, they are exported to Hong Kong. Indonesia accounts for 60% of the Southeast Asian production, and almost all this derives from Java. The market is buoyant and there are handbooks and guides available to encourage people to build or dedicate houses to these birds (Marzuki 1987; Whendrato and Madyana 1989).

Peafowl. The fact that such a large bird as the peafowl *Pavo muticus* has survived until now on Java is remarkable, but hunting has only recently become a problem. In some areas it used to be believed that peafowl groups would follow behind hunting tigers, waiting for it to defaecate and then picking the intestinal worms out of its stools (Cabaton 1911). Feasting on peafowl therefore had no attraction.

Despite being a protected species, the peafowl is now sought in many areas partly for its meat, partly because of its decorative and valuable tail feathers, partly because mounted specimens command a good price, and because it is regarded as a pest in corn growing areas. In a single episode about a decade ago, near the knobbly, ungazetted, but forest-clad Mt. Ringgit (Clason and Clason-Laarman 1932), a farmer killed 'hundreds' using DDT-tainted bait, and the population has never really recovered. One reason for this has been hunting, to provide souvenirs at the market in the nearby resort town of Pasir Putih, situated on the north coast between Probolinggo and Situbondo, which is believed to be one of the largest trading centres for peafowl in Java (Setiadi and Setiawan 1992). Suppliers pay Rp2,500-5,000 for an egg, and nearly Rp100,000 for an adult male. There is also a substantial trade in feathers for house decorations, for earrings, and for the traditional *Singabarong* masks used in *reog ponorogo* performances in Central and East Java. A single mask requires at least 1,000 peacock tail feathers, and given that a peacock has about 100 tail feathers with the desired round 'eye' tips, and that there are 200-300 reog groups, the demand for peacocks is clearly significant (Edwin 1988).

Conservation efforts have been helped by the 1990 Conservation Act, and villagers, especially former trappers, around Mt. Ringgit have been encouraged to help conservation activities in the area. By so doing they have become less inclined to hunt the peafowl (Setiadi and Setiawan 1993). Even so, while consumers are prepared to buy the feathers, there will be a demand that others will seek to satisfy. More attention needs to be given to increasing public awareness of the threats facing the birds, to conducting more field surveys, and to using the chicks confiscated by PHPA for reintroduction to existing and potential conservation areas that have lost their populations, such as Mt. Muriah and Cikepuh (van Balen et al. in press).

Altitudinal Distribution

The altitudinal distribution of Javan birds has been analysed (Hoogerwerf 1948a), and this revealed that 420 species of birds are found between sea level and 800 m, compared with about 300 species between 800 m and 2,000 m. By examining the maximum height at which each species has been found, it seems that there is a major boundary at about 1,300-1,600 m where 105 species reach their highest altitude, but the other apparent 'boundaries' found by Sody (1956) (table 5.8) are subject to the same misinterpretations as those found for plants (p. 171). Even so, the basic message that many bird species are found only in the lowland areas is clear.

Table 5.8. Number of bird species found in altitudinal zones on Java.

Species living from sea level to mountaintops	72
Species living exclusively at sea level	98
Species not found above 800 m	187
Species living between sea level and 1,500 m	233
Species living between 1,000 m and 3,000 m	134

Source: Sody (1956)

Major Publications

The birds of Java and Bali are well served by recent Indonesian and English field guides covering both the entire range of species encountered (MacKinnon 1988, 1990; MacKinnon and Phillipps 1993, in press; see also van Balen 1993b) and selections of the more common or interesting species (Iskandar 1989; Holmes and Nash 1989; Mason and Jarvis 1989). Some of the earlier books and papers (many in Dutch) are still extremely useful, such as the works of Hoogerwerf (1948, 1949a,b,c, 1953a,b,c,d,e,f, 1962a,b, 1965, 1969), Hellebrekers and Hoogerwerf (1967), and Sody (1925, 1956, 1989a). A checklist of all birds occurring on Java and Bali (and the six other faunal regions of Indonesia) has been published (Andrew 1992) by the Indonesian Ornithological Society (P.O. Box 4087, Jakarta 12040) which also produces a regular English-language journal, *Kukila* (the old Javanese word for bird). Bird clubs are strong at Universitas Gadjah Mada in Yogyakarta, Universitas Pajajaran in Bandung, and elsewhere, and there is a small but growing number of professional Indonesian field ornithologists.

With the general increase of interest in the natural environment, the publication of the recent books on the birds of Java and Bali, and the direct and indirect impact of the activities and studies of the Asian Wetland Bureau, Birdlife International (formerly ICBP), Bali Bird Club, and student groups, many papers have been published on new observations or new records for Java (e.g., Compost and Milton 1986; van Balen 1986a, 1991b, 1992b; Allport and Milton 1988; Andrew and Milton 1988; van Balen and Compost 1988; Whitten 1989; Prawiradilaga 1992; Andrews 1993; van Balen and Holmes in press), and Bali (e.g., Klapste 1984; van Helvoort 1985a; van Helvoort and Soetawijaya 1987; Wiegant and van Helvoort 1987; Bishop 1988; Mason 1988, 1990, 1993; Erftemeijer 1989; Meeth and Meeth 1989; Wells 1989; Green 1991; Indrawan 1991; van Balen 1991b; van Balen and Noske 1991; Ash 1993).

Surveys and Studies

Java and Bali are fortunate compared with the rest of Indonesia in the number and quality of bird studies conducted in a wide range of habitats and areas. The sites and the authors are shown in Appendix 2. Most of these studies are surveys; the number of long-term studies of forest bird ecology on Java or Bali is very few, perhaps the most detailed being on the effects of forest fragmentation (van Balen 1987b; in prep.), and the species composition and feeding behaviour of mixed flocks of birds (Erftemeijer 1987). The most detailed coastal study has been of the breeding of little egrets *Egretta garzetta* on P. Rambut (Pakpahan et al. 1991; Pakpahan 1992).

Reptiles and Amphibians

Fauna

Reptiles and amphibians are found in a wide variety of habitats, although most amphibians are tied, at some point in their life history, to water. Some species are common in disturbed habitats and are virtually dependent on people and their activities for their survival. The greatest service provided by reptiles is probably the killing of rats by snakes, but people tend to have such an aversion to them that this role is not as great as it could be. Monitor lizards and crocodiles are also used for their skins. In Asia certain frogs and toads have been used in poisons (Burkill 1966), but without any great refinement. Secretions from glands in frogs skins seem to have great potential if work on Australian and South American species is any indication (Anderson 1993). Medical conditions as varied as schizophrenia and septic wounds can be treated, and one burrowing frog produces a glue that could be used instead of stitches after surgery.

Java and Bali have 87 species of terrestrial and freshwater snakes, 42 lizards, skinks, geckos, and monitors, 8 freshwater turtles, and 36 amphibians (Whitten and McCarthy 1993). It is likely that new species and new records will be made; in 1993 a frog new to Java, *Rana baramica*, was found in fields just north of Bogor by the Asian Wetland Bureau (Budi-Asmoro 1993). That site has now been built on.

More reptile and amphibian species are found in West than in East Java, but some species found in the east are not found in the west such as the Indo-chinese sand snake *Psammophis condanarus*, Russell's viper *Vipera russelli*, and Fruhstorfer's mountain snake *Tetralepis fruhstorferi*. The first two of these seem to be relics of a period when much of Sundaland was subject to a more seasonal climate, because their closest neighbours are now in continental Asia (Hodges 1993). The reptile and amphibian fauna of Java and Bali is relatively poor compared with its neighbours' in terms of absolute numbers, but it has a rather high density of species (table 5.9). Among its neighbours, the composition of the reptile and amphibian fauna of Java and Bali is most similar to that of Sumatra (fig. 5.19).

Table 5.9. Number of snake species, and density of species, for the four major components of Sundaland.

	Number of species	No. species/1,000 km^2
Java	84	0.64
Sumatra	131	0.28
Borneo	133	0.24
Malaya	117	0.88

Sources: de Haas (1950); Haile (1958); Tweedie (1983); Hodges (1993); Whitten and McCarthy (1993)

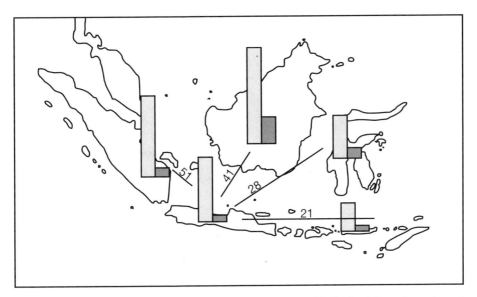

Figure 5.19. Relative richness (light shade), endemism (dark shade), and overlap of the reptile fauna of Java and Bali and their neighbouring islands and island groups. 'Overlap' is the percentage of the total species list that is shared between them.
After MacKinnon and MacKinnon 1986

Endemic Species

From the latest information available, six reptiles (5%) and 11 amphibian (30%) species are endemic to Java and Bali (Whitten and McCarthy 1993) (table 5.10; figs. 5.20-23). One of the endemic frogs and one of the snakes represent endemic genera. One of the endemic snakes, the Javan krait *Bungarus javanicus*, was discovered only in 1931 (Kopstein 1932a).

Notable Species

Venomous snakes. Of 106 species of snakes found on and around Java, it is of interest that only 30% are at all dangerous to humans, and this percentage reduces to just 12% if the 18 species of sea snakes are excluded. The potentially dangerous snakes fall into four groups. The triangular headed pit vipers (Viperidae), species of front-fanged cobras, kraits, coral snakes, and sea snakes (Elapidae), and the non-venomous but constricting reticulated python *Python reticulatus* and rock python *P. molurus* (Boidae).

Table 5.10. Endemic reptiles and amphibians of Java.

Reptiles			
SQUAMATA			
Agamidae			
Harpesaurus tricintus			Known from a single specimen of unknown locality caught in the middle of last century.
Typhlopidae			
Typhlops bisubocularis	W		Known from a single specimen (?) caught late last century.
Colubridae			
Oligodon propinquus			Known from a single (?) specimen caught at an unknown locality in the middle of last century.
Pseudoxenodon inornatus	W		Known from just four specimens from near Mt. Gede, and near Garut.
Tetralepis fruhstorferi		E	An endemic genus known from a handful of museum specimens caught last century. These, and some recent sightings (Hodges 1993) have all been made in the Bromo-Tengger-Semeru National Park.
Elapidae			
Bungarus javanicus	W	E	Known only from the Cirebon-Indramayu area. The records from E. Java have not been con-firmed (Hodges 1993).
Amphibians			
GYMNOPHIONA			
Ichthyophiidae			
Ichthyophis bernisi			Known from a single specimen of unknown origin.
I. hypocyaneus	W		Known from three speciments caught early last century near Serang.
I. javanicus			Known from a single specimen of unknown origin.
ANURA			
Bufonidae			
Bufo chlorogaster			Scarcely known with very few specimens.
Leptophryne cruentata	W		Known only from Mt. Gede, 1,400-2,500 m.
Microhyla achatina	W	E	Sea level to 1,500 m (fig. 5.20).
Rhacophoridae			
Nyctixalus margaritifer	W		An endemic genus, known only from a handful of specimens from Mt. Wilis. Possibly a syn-onym of *Philautus aurifasciatus* (Liem 1971).
Philautus jacobsoni	C		Known from a single (?) specimen from Mt. Ungaran.
P. pallidipes	W		Known only from a single (?) specimen caught near the summit of Mt. Pangrango caught early this century. Possibly a synonym of *Philautus aurifasciatus* (Liem 1971)
P. vitiger	W		Known only from a single (?) specimen caught at 1,200 m
Rhacophorus javanus	W		Known from around Mt. Gede and Mt. Malabar at about 1,500 m.

Sources: de Rooij (1915, 1917); van Kampen (1923); Brongersma (1950); de Haas (1950); Taylor (1960, 1968); Salvador (1975); C. McCarthy (pers. comm.)

Figure 5.20. The relatively widespread endemic frog *Microhyla achatina*. Note that the webbing on the feet should be rather more extensive (B. Clarke pers. comm.). Previously unpublished, this painting was executed in about 1820 for the Natuurkundige Commissie under the patronage of King William I for the series of books edited by C.J. Temminck (1839).

With permission of the National Museum of Natural History, Leiden

Figure 5.21. *Harpesaurus tricinctus*, an endemic horned lizard known from a single specimen caught somewhere in Java 150 years ago.

By A.J. Whitten adapted from de Rooij 1915 and Brygoo 1988

Figure 5.22. *Pseudoxen-odon inornatus*, an endemic snake known from a single specimen caught some-where in Java over 170 years ago.

After de Rooij 1917

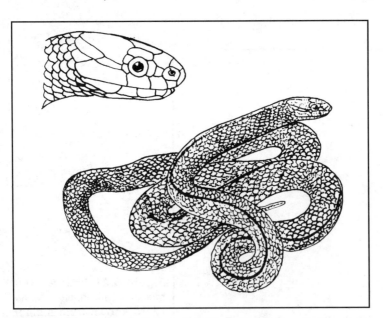

Figure 5.23. *Tetralepis fruhstorferi*, the only species of an endemic genus known from a single specimen caught in the Tengger Mountains a century ago, and recently rediscovered in the Tengger Caldera.

After de Rooij 1917

Unfortunately some of the most dangerous species are quite common in scrubby vegetation close to human habitation, such as the Malayan krait *Bungarus candidus,* banded krait *B. fasciatus,* king cobra or hamadryad *Naja hannah,* common cobra *N. sputatrix,* ground pit viper *Calloselasma rhodostoma,* and the white-lipped pit-viper *Trimeresurus albolabris.* The common reaction of most people on seeing a snake is to kill it if possible, and if certain harmless species are not to be wiped out altogether in Java, it is important that relevant information be given to and absorbed by at least scouts, nature lovers' groups, and other field personnel. Snakes do not attack people unless provoked, and many species play an important role in controlling rats, particularly in rice fields, if given a chance.

The closest contact with snakes for some people is at the few stalls in large cities on Java where snakes, particularly the king cobra, are displayed. Customers select one from a cage, and it is then killed, drained of its blood which is caught in a mug and mixed with its gallbladder, chopped marrow, whisky, and herbs. This raw potion is drunk to increase, it is said, sexual prowess, and to cure asthma, diabetes, skin disease, and kidney stones.

House geckos. House geckos are ubiquitous in Java and Bali everywhere that air conditioning is not used, and they breed throughout the year (Church 1962). All three species are very similar in size, weight, and gonad morphology, but they do have separate habitat preferences–*Cosymbotus platyurus* being found in the relatively dry, bright commercial district, *Hemidactylus frenatus* in relatively damp, well-lit residential districts, and *Gehyra mutilata* in the darker areas of both of the above. It is hard to find any other example among the vertebrates of such closely related species in such similar habitats (Church and Lim 1961).

Common toad. The intrepid common toad *Bufo melanostictus* is found from India and the Himalayas to southern China, south through the Malay Peninsula to Sumatra, Java, and Bali. In 1930 it could be claimed that *B. melanostictus* had never been reported from Bali (Mertens 1930), and in 1923 its range apparently did not even include the eastern tip of Java. A collecting trip in 1958, however, found this successful animal in Banyuwangi on the eastern tip of Java, and in Negara in West Bali, but it was absent from Singaraja in the north and Ubud and Denpasar in the south and Klungkung in the east (Church 1960b). In these areas the smaller *B. biporcatus,* with a pair of horny parietal ridges between the eyes, was present in abundance (Church 1959). *B. melanostictus* is now found throughout Bali.

Thus *B. melanostictus* is a recent immigrant to eastern Java and Bali and joins a short list of invasive amphibians from around the world (cane toads *Bufo marinus* in the Philippines, Bermuda, Hawaii, Solomons, and Australia, and *Kaloula pulchra* in Sulawesi, Kalimantan, and Flores). The relatively large size and aggressiveness of *B. melanostictus*, the resilience or hardiness of the larvae which are almost unique among amphibians in being able to stand brackish water (yet morphologically indistinguishable from those of *B. biporcatus*) (Schijfsma 1932), and its ability to adapt successfully to almost any habitat from sea level to mountains, suggested that it would not remain limited to Negara for long, and would proceed eastwards, displacing the smaller, less active species *B. biporcatus* as it went. Even so, the displacement has not been total because *B. melanostictus* is primarily an urban species being found only rarely in rice fields and streams where populations of *B. biporcatus* continue to reside. Thus, *B. biporcatus* is most common where *B. melanostictus* is absent (Church 1960).

Bufo melanostictus breeds throughout the year with a peak coinciding with the start of the rainy season in November/December. Ovulation follows a pattern related to the lunar cycle, with most females ovulating just before or after the time of the full moon. *B. biporcatus* exhibits a similar pattern, and reproductive cycles of neither species appears to be influenced by heavy rainfalls on particular days (Church 1961). The calling of the males, however, exhibits individual cycles unrelated to any apparent external cues. The female's ovaries can occupy 30% of the gross body weight; a considerable proportion which emphasizes the importance of quantitative reproduction (Church 1960, 1961). The breeding behaviour of the rice field frog *Rana cancrivora* is also influenced by the moon but with ovulation concentrated in the relatively dark period before the full moon (Church 1960).

Marine turtles. All five of the turtles found in Indonesia are believed to nest along Java's south coast but only the green turtle *Chelonia mydas*, leatherback *Dermochelys coriacea*, and hawksbill *Eretmochelys imbricata* have been confirmed lately, and the first of these is by far the most common (fig. 5.24). It seems unlikely that turtles still nest on any of Bali's southern beaches, although they are reported from around Bali Barat National Park in the northwest. The turtles are very particular about the beaches they use, and although their preferences are not entirely understood, nesting beaches tend to be protected from prevailing winds and to have sand moisture with a lower salinity than beaches not used for nesting (Johannes and Rimmer 1984). Hawksbills are said to favour steeper beaches than green turtles, and also to prefer small islands (Nuitja and Uchida 1983). However, many laying sites of hawksbill turtle were found in Bali Barat National Park on the Bali mainland opposite Menjangan Island by Nuitja and Ismail (1984), most of them at the top of the beach above the high tide level with only 50% sand. They nest throughout the year, but most laying occurs during the rainy season from September to December.

Observations of truly undisturbed turtle beaches in remote parts of the globe demonstrate that under natural conditions enormous numbers of females haul themselves up on the beaches to lay their eggs. Thus the situation in Java (and Bali), even at the famous beaches of Cikepuh (Ujung Genteng), Meru Betiri, and Ujung Kulon, is probably the faintest shadow of what once was. Numbers have plummeted because of people taking the eggs, developing the beaches, and hunting the adults, but even well-intentioned interest can be damaging. For example, female turtles can easily be put off coming up the beach by noise, torch light, or being touched by humans. These may upset her breeding cycle because during the nesting season a single female may lay several clutches at intervals of about every two weeks. The reduced population sizes mean that the relatively small numbers of eggs laid are susceptible to predation by wild boar and monitor lizards.

Aspects of turtle conservation are discussed elsewhere (p. 756).

Water monitors. Water monitors *Varanus salvator* are large, shy lizards of riverine habitats. They have been hunted remorselessly for their valuable hide and because they will feed on domestic chickens, but they can still be found throughout Java and Bali, albeit in relatively low numbers. In fact, Java has an annual export quota of about 20,000 monitor skins, but since most of this is given to Jakarta, it reflects numbers of traders rather than origin (Luxmoore 1989).

A year's study of water monitors (Vogel 1979) was made twenty years ago, and it was found that their activities are focussed around rivers, swamps, and the coast, where they feed on large arthropods, small vertebrates, eggs of birds and reptiles, and occasionally carrion. Monitors are strictly diurnal, though they may rest for longer than a single night after a large feed. They have no fidelity to any particular location, spending the night in the closest available shelter. Their main competitor, at least in Ujung Kulon where they were studied, is the wild pig *Sus scrofa,* and the adults are preyed upon by people, leopards, wild dogs, and pythons. They are not territorial, but there is considerable intraspecific competition around a carcass, where only individuals over one metre are generally seen, perhaps because the smaller ones are avoiding being set upon and possibly eaten by their elders. The largest individual observed in the study was a male at 2.1 m, although they can grow to 2.4 m. (This compares with 2.8 m for a large Komodo monitor). Those below 60 cm were hardly ever seen, indicating that they are highly secretive in their habits.

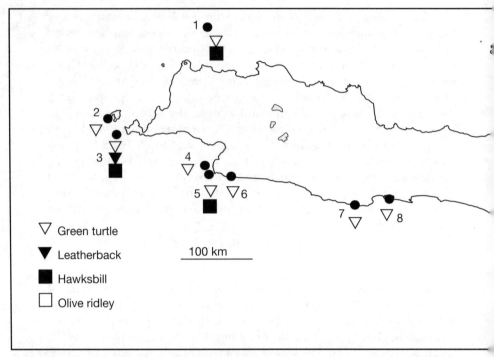

Figure 5.24. Turtle beaches on Java and Bali with indication of species nesting. 1 - Pulau Seribu, 2 - Pulau Panaitan, 3 - Ujung Kulon, 4 - Cibulakan, 5 - Pangumbahan, 6 - Sidang Kerta, 7 - Cokalong, 8 - Nusa Kambangan, 9 - Karimunjawa, 10 - Bawean, 11 - Nusa Barung, 12 - Sukamade, 13 - West Blambangan, 14 - Blam-

Crocodiles. The largest reptile known from Java is the estuarine crocodile *Crocodylus porosus,* but this has been ousted from virtually all its former haunts with the exception of a few rivers in Ujung Kulon National Park, due to persecution, hunting for its valuable skin, and habitat disturbance and destruction. The Thai crocodile *C. siamensis* is known from Java from a single specimen caught in Rawa Danau, Banten (p. 773), where an unidentified crocodile was also observed in late 1992 (M. Silvius pers. comm.).

Exploitation

Indonesian reptiles and amphibians are exported overseas, and while those species in Java and Bali are not particularly popular in the main markets of Japan, North America, and Europe, the main exporting company in Jakarta, PT. Vivaria, has had some 86 Javanese species passing through their hands in recent years. Of these, none was endemic save the Javan

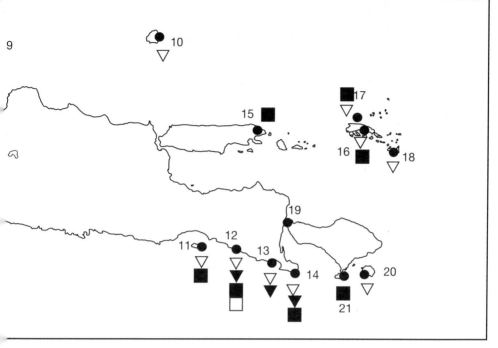

bangan, 15 - Gili Yang, 16 - Sagubing/Saobi, 17 - Araan, 18 - Sepanjang, 19 - Bali Barat, 20 - Nusa Penida, 21 - South Bali.

After Salm and Halim 1984

krait or *Bungarus javanicus* which has been found very few times and only in the Indramayu area. The majority of species traded in large numbers come from rice fields and scrub areas around Bogor, but even with this exploitation the populations remaining of the most popular 'chameleon' *Calotes cristatellus*, the very long-tailed true lizard *Takydromus sexlineatus*, and various frogs are appreciable. The absence of the endemic species from the trade lists is probably because they are largely forest species (perhaps living in the canopy) and thus relatively hard to find, because collectors are reluctant to catch species for which there is no certain market, because the endemic species are not (as far as is known) particularly attractive, and because the species being traded tend to be those that are known to thrive in captivity on a diet of mice or fish (B. Yuwono pers. comm.).

Amphibians appear to be experiencing a global decline due to habitat destruction, introduced predators, pesticides and other pollution, and human consumption (Blaustein and Wake 1990). The last of these has recently been suggested as a major cause of frog decline in Indonesia, particularly on Java, but the vast majority of exports are legs of the introduced and cultivated American bull frog *Rana catesbiana,* which has been developed as an export commodity since 1980 (p. 302) (Susanto 1989).

Altitudinal and Habitat Distribution

The frogs and toads in and around Mt. Gede-Pangrango National Park were studied in the 1960s by Liem (1971). Animals were sought from the peak and *alun-alun* of Mt. Gede down through the forest to the Cibodas Botanic Gardens and into the surrounding agricultural areas (table 5.11). A total of 19 species were encountered, including two endemic species, the red-spotted *Leptophryne cruenata* (known only from this mountain and the only frog or toad in Java to be found above 2,250 m), and *Rhacophorus javanus.* The second species has also been recorded from Mt. Malabar and, 70 years ago, from around Bogor. However, with the disappearance of the remnant forests it had become extinct there by 1963 (Liem 1971). Interestingly, five of the 10 species found in cultivated land were not found in the Botanic Gardens, and none of the remaining five were found in forest. *Rana chalconata* and *Bufo biporcatus* are found in lowland forest in Ujung Kulon (Hoogerwerf 1970; Anon. 1990), however, and it appears that these two species were originally lowland forest dwellers which have adapted to secondary growth and certain agricultural conditions. The forest and non-forest species were also found to occupy different microhabitats, and these are shown in table 5.12.

Throughout the 1930s some 3,500 snakes were caught in and around two tea estates in West Java at 700-900 m altitude (de Haas 1941): Nanjung Jaya near Sumedang, and Banjarwangi southeast of Cikajang near Garut. Interestingly, although the areas were similar in altitude and land use, the climate of Nanjung Jaya is among the driest in West Java, and Banjarwangi has a very high rainfall. The wetter area had more snake species and individuals and by far the most abundant species *Elapoides fuscus,* which was caught just once at the dry Nanjung Jaya site. The most abundant snake at Nanjung Jaya was the keeled slug snake *Pareas carinatus,* a rather surprising discovery since it feeds exclusively on slugs and snails. A total of 31 and 35 species respectively were taken at each locality (24 common to both). One of the most common species was the very poisonous white-lipped pit viper *Trimeresurus albolabris,* but it hardly ever bit the tea workers. Two species of deadly coral snake *Maticora* were also fairly common, but under normal conditions the small mouths of these beautiful but deadly animals prevent them from being any danger to people.

Table 5.11. Altitudinal distribution of frogs and toads in and around Mt. Gede-Pangrango National Park.

	Cultivated land 1,250-1,350 m	Botanical Garden 1,350 m	Primary rain forest		
			1,600-1,700 m	1,800-2,000 m	2,250-2,500 m
Bufo melanostictus	+	+	-	-	-
B. biporcatus	+	-	-	-	-
B. asper	+	-	-	-	-
Leptophryne cruenata	-	-	+	+	+
Megophrys montana	-	-	+	+	_
Leptobrachium hasseltii	-	-	+	+	_
Microhyla achatina	+	+	-	-	-
M. palmipes	-	-	+	-	-
Rana limnocharis	+	-	-	-	-
R. cancrivora	+	-	-	-	-
R. kuhli	-	+	+	-	-
R. microdisca	-	-	+	-	-
R. chalconota	+	+	-	-	-
R. nicobariensis	+	-	-	-	-
Amolops jerboa	-	-	+	+	-
Philautus aurifasciatus	-	+	+	+	-
Rhacophorus javanus	-	+	+	-	-
R. reinwardti	+	+	-	-	-
Polypedates leucomystax	+	+	-	-	-
Total number of species	10	8	9	5	1

Source: Liem (1971)

Table 5.12. The distribution of toads and frogs in different habitats in and around Mt. Gede-Pangrango National Park. f = feeding, b = breeding.

	Forest species	Non-forest species
Terrestrial	*Megophrys montana (f)* *Microhyla palmipes (f)* *Leptobrachium hasselti (f)*	*Bufo melanostictus (f)* *B. biporcatus (f)*
Swift-moving streams	*Amolops jerboa (f, b)* *Leptophryne cruentata (f)*	*Bufo asper (f, b)*
Slow-moving streams	*Leptobrachium hasseltii (b)* *Leptophryne cruenata (f)* *Megophrys montana (b)* *Rana kuhli (f, b)* *R. microdisca (f, b)*	*Bufo biporcatus (f)* *B. melanostictus* *Rana cancrivora (f, b)* *R. chalconota (f)* *R. limnocharis* *R. nicobariensis*
Standing water	*Leptophryne cruenata (b)* *Microhyla palmipes (b)* *Rana kuhli (f, b)* *R. microdisca (f, b)* *Rhacophorus javanus*	*Bufo biporcatus (f)* *B. melanostictus* *Polypedates leucomystax* *Rana cancrivora (f, b)* *R. chalconota (f, b)* *R. limnocharis (f, b)* *R. nicobariensis (f, b)* *Rhacophorus reinwardti (b)*
Arboreal	*Rhacophorus javanus (f, b)*	*Polypedates leucomystax (f, b)* *Rana chalconota (f)* *Rhacophorus reinwardti (f, b)*

Source: Liem (1971)

Major Publications

The reptiles of Java are described in detail in the standard works by de Rooij (1915, 1917) although not all of the taxonomy is still usable. People's fascination with snakes has resulted in a number of books: *Ophidia Javanica* (van Hoesel 1959) in which the 40 most common species are described in Indonesian and English; *De Javaansche gifslangen* (Kopstein 1930a) which gives portraits of the sea snakes and eight other poisonous terrestrial species; and three more recent books on the poisonous snakes in Indonesian: *Ular berbisa Indonesia* (Supriatna 1981), *Ular berbisa dan pembelit raksasa* (Santoso 1985), and *Ular-ular berbisa di Jawa* (Suhono 1986). Poisonous and constricting snakes in Java have been described by Santoso (1985) in terms of species recognition, pattern of teeth marks of a bite, methods of treating different bites, methods of capturing snakes, means of removing constricting coils, and their benefits to man. As he says, most snake stories are exaggerated and not realistic. Other helpful books which do not give any undue weight to the relatively small proportion of dangerous species are *The snakes of Malaya* (Tweedie 1983), and *Fascinating snakes of Southeast Asia* (Lim and Lee 1989). Colour photographs of 11 species of Javan snakes, including the first ever photograph of the endemic Fruhstorfer's mountain snake, were published in Hodges (1993), which also includes a checklist of snakes from East Java. Checklists of the reptile species found on the different Javan islands (Mertens 1957, 1959) are available, but these need updating. A checklist for the reptiles recorded from Java and Bali was published recently (Whitten and McCarthy 1993).

The amphibians have never been dealt with in a popular form, and the only source of reference is the somewhat out-of-date *Amphibia of the Indo-Australian Archipelago* (van Kampen 1923). A checklist of species is given in Whitten and McCarthy (1993). A large variety of tadpoles is described by Schijfsma (1932), and Liem (1971) has interesting ecological notes. However, many of the more common reptiles and amphibians likely to be encountered in Java and Bali are illustrated in colour in a guide written for use in Singapore (Lim and Lim 1992). Colour illustrations of 10 of Java's 33 frogs and toads can be found in a book intended for use in Sabah (Inger and Stuebing 1990), and in which useful information is given on ecology and behaviour. Those 10 species are generally abundant in, or even dependent on, ecosystems disturbed by people.

Neither the reptiles nor amphibians have been studied to any great extent, although various aspects of the natural history of the common toad *Bufo melanostictus* and of house geckos were studied closely in the Bandung area by Church (1959, 1960a, b, c, 1961, 1962, Church and Lim 1961). Frog communities have been studied in the mountains (Liem 1971) and the lowlands (Premo 1985; Atmowidjojo and Boeadi 1986). Some of the early papers on distribution by Kopstein (see Bibliography) have interesting snippets of ecological information, such as the find that eggs of the small skink *Lygosoma sanctum* from the mountains of West and East Java are laid in the nest of *Nasutitermes* termites' nests (Kopstein 1932b).

FRESHWATER FISHES

Fauna

The term 'freshwater fishes' actually covers a wide range of ecological preferences which can be divided into four categories:

- primary division families are those whose members are strictly intolerant of saltwater in the present as well as in the past;
- secondary division families are those whose members now live in freshwater but are able or supposedly able to tolerate saltwater for a short period;
- diadromous species are those which migrate between marine and freshwater at different periods of their life cycle; and
- vicarious species are strictly freshwater representatives of primarily marine groups (Kottelat et al. 1993).

There are 132 freshwater fishes known from Java and Bali. All but a handful of species recorded from Bali are also known from Java. Further collections are quite likely to reveal new distribution information and even new species. For example, a new goby was discovered just below the popular tourist attraction of Gitgit waterfall in Bali in 1993, and a small gold-spotted goby *Pandaka trimaculata,* originally described by Japanese Emperor Akihito in 1975 from specimens caught in Japan and Philippines, was caught during 1992 as a new record for Indonesia in a rather unpleasant small stream flowing across Lovina Beach, northern Bali.

Java has many fewer species than Sumatra or Borneo but the density of species is the same as for Borneo and greater than for Sumatra (table 5.13).

Java's native fishes are both less abundant and less diverse than they were because of the loss of forest, the pollution of water, the digging of sediment, and the interuption of the river by dams which may have both stopped certain migratory fish species from breeding and removed the seasonal changes in water flow which acted as cues for breeding behaviour. Surveys of Bali's rivers and lakes in 1992 found that there was little in the way of an indigenous fish fauna. In some respects the paucity is not surprising:

- Bali is a small island at the edge of the Sunda Shelf, separated from Java by a relatively deep channel (p. 90) that would have provided land/freshwater linkages only during the lowest of sea level falls, and

Table 5.13. Total species number and number of endemic species of fish native to Java, Sumatra, Borneo, and Java, indicating percentage of endemic species.

	Sumatra		Borneo		Java	
	sp/end	%	sp/end	%	sp/end	%
Primary division families						
Osteoglossidae	1/0	0	1/0	0	0	0
Notopteridae	2/0	0	2/0	0	2/0	0
Cyprinidae	107/15	14	138/46	33	44/6	14
Balitoridae	13/6	46	47/35	74	6/1	17
Cobitidae	10/0	0	19/7	37	7/0	0
Gyrinocheilidae	0/0	0	1/1	100	0	0
Bagridae	19/3	16	26/10	38	8/0	0
Siluridae	22/1	5	28/9	32	9/0	0
Schilbidae	2/0	0	2/0	0	0	0
Pangasiidae	6/1	17	8/3	38	3/0	0
Akysidae	5/1	20	5/2	40	3/1	33
Parakysidae	1/0	0	2/1	50	0	0
Sisoridae	3/0	0	4/0	0	2/0	0
Clariidae	5/0	0	6/0	0	4/0	0
Chacidae	1/0	0	1/0	0	0	0
Nandidae	1/0	0	1/0	0	0	0
Pristolepididae	2/0	0	2/0	0	1/0	0
Anabantidae	1/0	0	1/0	0	1/0	0
Belontiidae	14/2	9	27/20	74	4/0	0
Helostomatidae	1/0	0	1/0	0	1/0	0
Osphronemidae	1/0	0	1/0	0	1/0	0
Luciocephlidae	1/0	0	1/0	0	0	0
Channidae	9/1	11	9/0	0	4/0	0
Chaudhuriidae	0	0	1/0	0	0	0
Mastacembelidae	5/0	0	6/1	17	4/0	0
Sub-total	232/30	13	340/135	40	104/9	9
Secondary division families						
Sundasalangidae	0	0	1/1	100	0	0
Adrianichthyidae	0	0	0	0	0	0
Aplocheilidae	1/0	0	1/0	0	1/0	0
Sub-total	1/0	0	2/1	50	1/0	0
Diadromous and vicarious species						
Dasyatidae	1/0	0	2/0	0	0	0
Clupeidae	1/0	0	3/1	33	0	0
Hemiramphidae	2/0	0	7/4	57	2/1	50
Belonidae	1/0	0	1/0	0	0	0
Oryziidae	1/0	0	1/0	0	1/0	0
Telmatherinidae	0	0	0	0	0	0
Chandidae	3/0	0	5/1	20	1/0	0
Teraponidae	0	0	0	0	0	0
Datnioididae	0	0	1/0	0	0	0
Toxotidae	1/0	0	1/0	0	0	0
Rhyacichthyidae	1/0	0	0	0	1/0	0
Eleotrididae	4/0	0	3/0	0	1/0	0
Gobiidae	18/0	0	17/5	29	19/2	11
Tetraodontidae	6/0	0	9/2	22	2/0	0
Sub-total	39/0	0	52/13	25	27/3	11
Total	272/30	11	394/149	38	132/11	8
Land area km^2	469,000		548,000		132,000	
Spp/1,000 km^2	0.58		0.72		1.00	
End. spp/1,000 km^2	0.06		0.27		0.08	

Source: Kottelat et al. (1993)

Figure 5.25. The Rasbora *Rasbora baliensis*, one of two possibly endemic species of freshwater fish from Bali.
By A.J. Whitten

thus few opportunities for primary freshwater species to reach Bali;

- the rivers in the west, northeast of Bali are small and are entirely dry for the majority of the year. Thus, the near-total absence of cyprinids (carps and minnows) may be due to the fact that even during the cool periods with low sea level, none of Bali's permanent rivers were connected to the rivers draining Sundaland (p. 118);

- in parts of the east where there are recent cinder soils, rainwater percolates rapidly, and surface water is present in the headwaters only during the wettest part of the rainy season; and

- none of Bali's rivers have the long stretches of slow-moving water favoured by many cyprinids and catfish. Rather the rivers tend to be rapid with rocky bottoms.

These factors may account for the paucity of primary division species, but not for the paucity of species more tolerant of marine or brackish conditions. The situation on Bali can be compared with the situation on seasonal Sumbawa, 200 km to the east, where surveys were also made in 1992, and where there are no native primary division species but a comparatively rich fauna from the other groups. The conclusion must be that Bali must have had a rather poor indigenous fish fauna, and that human activities have reduced this still further (pgs. 334, 718).

Table 5.14. Endemic fishes of Java and Bali.

CYPRINIFORMES		
Cyprinidae		
Barbodes platysoma	C	Known from a single specimen, caught last century near Surakarta.
'Cirrhina breviceps'	W	Known from a single dried specimen, now lost, collected over 150 years ago in the Banten River. Probably represents an extinct species of *Osteochilus*.
Labeo erythropterus	W	Known only from near Jakarta, Lebak, and Parongkalong; not caught since last century.
Lobocheilos lehat	W	Known only from Parongkalong.
Puntius aphya		Known from a single specimen, caught last century in an unrecorded locality.
P. microps	C	Known only from cave waters in the Mt. Sewu area near Yogyakarta. Some specimens have small eyes, some much reduced eyes, some no eyes, and some eyes on one side only. This species is possibly a synonym of *P. binotatus*. It was, however, given legal protection in 1991, which will complicate clarifying its taxonomic position.
Rasbora aprotaenia	W	Known only from NW Java, e.g., Rawa Danau.
R. baliensis	B	Known only from Lake Bratan and rivers to the south (fig. 5.25). It is possible that *R. 'lateristriata'* in eastern Java is the same species.
Balitoridae		
Nemacheilus chrysolaimos	W	A specimen similar to this was caught recently in a survey conducted for this book in the Ciliwung River.
SILURIFORMES		
Akysidae		
Akysis variegatus	W	Known from a single specimen caught at an unknown location last century.
CYPRINODONTIFORMES		
Hemiramphidae		
Dermogenys pusilla		Possibly also occurs in Sumatra and Malaya.
PERCIFORMES		
Gobiidae		
Lentipes whittenorum	B	Discovered in 1992 just below Gitgit waterfall. May be found on islands in the Lesser Sundas.
Pseudogobiopsis neglectus	W	Known only from a Jakarta canal; last caught 150 years ago. Possibly a synonym of the widespread *Redigobius roemeri*.
Sicyopterus parvei	W	Known only from rivers near Garut and Pelabuan; some specimens were caught in the late 1960s.

Source: Kottelat et al. (1993), Roberts (1993), Watson and Kottelat (1994)

Endemic Species

Eleven (10%) of the 132 species on Java, and two of Bali's species appear to be endemic (table 5.14). Two of the Javan endemic species are said to have been found at Parongkalong (one only from that locality), but this name is not included in the Dutch gazetteers. The closest is Parong Koleng in Cikalong near Cianjur. It might also refer to the village of Parung near Pekalongan.

Notable Species

The largest freshwater fish known from Java is the giant catfish *Bagarius yarrelli* which can grow to over 2 m in length. It has been recorded from West and Central Java, but is now probably no longer present. The world's largest eel *Thyrsoidea macrurus* grows to 2.4 m and has been recorded from estuaries in West Java (Weber and de Beaufort 1916; Kottelat et al. 1993). At the other end of the scale the tiny goby *Pandaka trimaculata* from the north coast of Bali grows to only 13 mm.

A relatively uncommon but magnificent and native cultured fish in Java is the tambra or kancera *Tor douronensis, T. soro,* and *T. tambroides,* all of which have excellent tasting flesh. They are similar in shape to the carp but they have larger scales, a less rounded body, and can grow to a metre in length. *T. soro* at least are kept in ponds in the Tasikmalaya, Garut, and Majalengka areas of West Java, and the Wonosobo, Purwokerto, and Magelang areas of Central Java (Budiman and Djajasasmita 1981; Rachmatika 1990). Although very popular before the introduction of the carp, they are now relatively rare. The main reason for this is that the fry have generally been collected from rivers. River fish have, however, been much abused by overfishing, destructive fishing (explosives, pesticides), and pollution, and it is often not worthwhile for fishermen to seek them. Breeding them in ponds has been attempted but faces a number of cultural problems (Budiman and Djajasasmita 1981), although these could probably be overcome with a modicum of research and development.

Snakeheads are among the largest freshwater predators in Java and Bali, as well as being a favoured food fish. They are able to breathe atmospheric air, and they are reported to slither across fields and roads looking for new habitat, and to bury themselves in mud during droughts. Three of the four species (fig. 5.26) build bubble nests among dense vegetation in a swampy area or slow moving river. The fourth species, *Channa gachua*, found in mountain streams, is a mouth brooder. It is not unknown for fishermen or swimmers who have gone too close to the fry of the large *C. micropeltes* to be attacked, sometimes with fatal results (Kottelat et al. 1993). It is the voracious habits of this species that gives them all a bad name in fisheries circles.

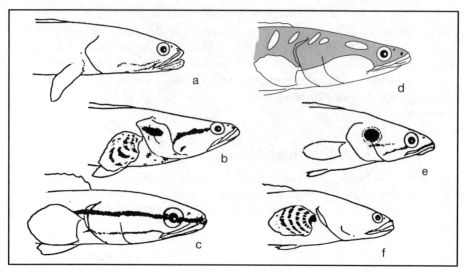

Figure 5.26. Heads of snakehead *Channa* species found in Java and Bali a. - *C. striata,* b. - *C. lucius,* c. - *C. micropeltes* (young), d. - *C. micropeltes* (adult), e. - *C. pleurophthalma,* f - *C. gachua.*
After Ng and Lim 1990

Exploitation

Fish is a very popular food in Java and the per capita consumption of fish is still increasing–from 14.2 kg/year, in about 1987 to 15.0 kg/year in 1991. Much of this comprises dried fish from Kalimantan (MacKinnon et al. 1994), and the vast majority originating in Java is from ponds, partly because these hold favoured species, and partly because all the rivers have experienced considerable overfishing, particularly with poison and electricity (fig. 5.27) (despite having been banned by the Fisheries Act No. 9 of 1985), and many previously fish-supporting habitats are now of very poor quality. Fish is not a particularly favoured food on Bali. Aspects of aquaculture are dealt with elsewhere (p. 631).

Distribution

The different histories of connection and separation from the northwards-flowing Great Sunda River which drained most of Sundaland (p. 118) are reflected in the present fish faunas of the major rivers of Java. Thus those draining north tend to have a richer fauna than those draining south into the Indian Ocean. Another major difference between the fish faunas in the northern and southern rivers is the occurrence of large *Anguilla* eels (fig. 5.28). These fish breed in deep waters of the Indian Ocean from which the young swim to the Sunda region. When they reach water about 200 m

Figure 5.27. Using electricity to catch fishes, an illegal activity, has caused local extinctions of many species.

With permission of Kompas

deep, they metamorphose into elvers which enter rivers, swim up to the headwaters, and grow before swimming back again to breed and die. Given that they appear to frequent headwaters, the effect of irrigation and other weirs will presumably cut the cycle. These eels are present only in those rivers emptying to the south, west, and east of Java (and presumably the southeast and southwest rivers in Bali). As for the north coast they appear to be totally absent from the Cimanuk and Bengawan Solo and every river between (Delsman 1926a, b). They can grow up to 160 cm in length and 40 cm in girth and have given rise to many folk beliefs.

Major Publications

The recent comprehensive book on the freshwater fishes of western Indonesia (Kottelat et al. 1993) provides, for the first time, the means of identifying all the species encountered in the rivers, lakes, and estuaries of Java and Bali. Ecological information is scarce for almost all species except those of proven economic value, and references are given in the above book. In addition, two new colourful guides to the coral fishes have also been published, and some of the species covered are found in estuaries. Both are useful, but that by Myers (1991) is more comprehensive than that by Kuiter (1992).

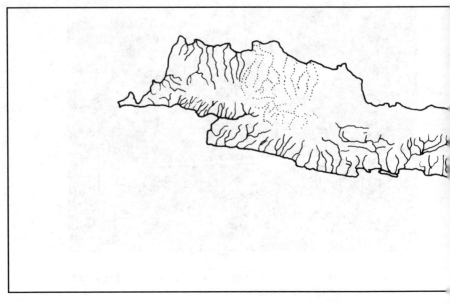

Figure 5.28. Rivers in which *Anguilla* eels have been occasionally (dotted) and consistently reported. Note that eels have never been reported in the

Surveys

Most fish surveys conducted on Java and Bali are of less ecological value than could be the case because their focus and catching methods provide information only on economic species. Thus, for the purposes of this book, extensive fish collections were made all over Bali and in selected major Javan rivers in 1991–1992, the latter by the LIPI Centre of Biology Research and Development, and most with the assistance of M. Kottelat. The results (Kottelat and Whitten in prep.) are summarized elsewhere (p. 718).

BUTTERFLIES

Fauna

There are over 600 species of butterflies found on Java and Bali, nearly 40% of which are endemic at subspecies level. In order to improve the state of knowledge of the butterflies, a joint project between the Zoological

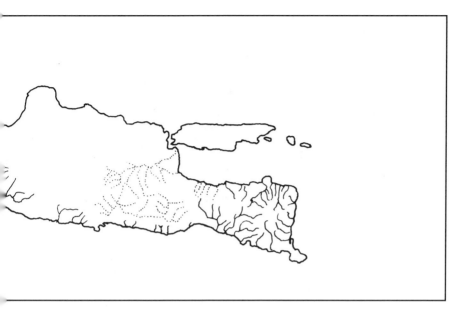

Bengawan Solo and other northern rivers.
After Delsman 1926a, b

Museum of LIPI (the Indonesian Institute of Sciences) and the National Museum of Natural History in Leiden, Holland, is attempting to assess the current distribution of Java's butterflies and status changes over the past few decades by comparing old and recent collections (including those by commercial collectors) and making new collections in selected localities. An initial review of the data reveals that eastern Java has received far less attention than the west (R. de Jong pers. comm.).

Although the data are not yet available for all the butterfly families, the relative similarity between the faunas of Java and neighbouring islands can be calculated for the swallowtail butterflies (Papilionidae) (fig. 5.29) and milkweed butterflies (Danaidae) (fig. 5.30). This shows that the members of these families have great similarity with Sumatra, rather less with Borneo, and little with the islands to the east.

Endemic Species

Preliminary analysis from the Indonesian-Dutch collaboration shows that 46 butterfly species are endemic to Java and Bali, none endemic solely to

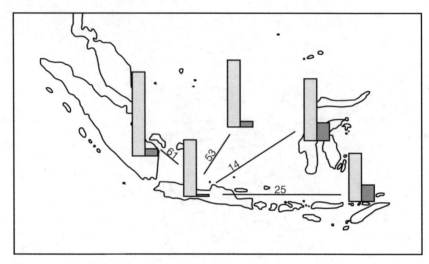

Figure 5.29. Relative richness (light shade), endemism (dark shade), and overlap of swallowtail butterflies of Java and Bali and their neighbouring islands. 'Overlap' is the percentage of the total species list shared between them.

Data from Collins and Morris (1985)

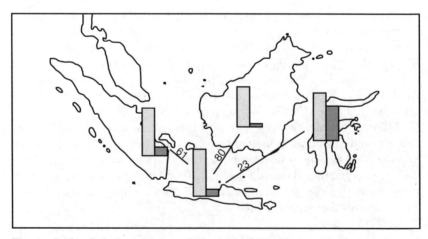

Figure 5.30. Relative richness (light shade), endemism (dark shade), and overlap of milkweed butterflies of Java and Bali and their neighbouring islands. 'Overlap' is the percentage of the total species list shared between them.

Data from Ackery and Vane-Wright (1984)

Bali, and that the percentage of endemic species differs between families (tables 5.14-15).

Notable Species

The most spectacular of the endemic butterflies are the two endemic papilionids or swallowtail butterflies (fig. 5.31) both of which, from single mountains at opposite ends of Java, must be considered rare (M. Amir pers. comm.), although the *IUCN Red Data Book* does not regard *Papilio lampsacus* as particularly threatened (Collins and Morris 1985). Despite their very limited range and the possibility that they are threatened by loss of habitat, neither is among the 25 swallowtails protected by law in Indonesia (PHPA 1990).

Major Publications

The species are almost all dealt with in the series *The Rhopalocera of Java* (Piepers and Snellen 1909, 1910, 1913, 1918) and *Rhopalocera Javanica* (Roepke 1942). An entry into the group is also possible using the more recent and more available *The Butterflies of the Malay Peninsula* (Corbett and Pendlebury 1978), where many of the species can be found. For focussed coverage on the Danaidae, *Milkweed Butterflies* (Ackery and Vane-Wright 1984) should be consulted. In addition, there are the splendid but hugely expensive books on the *Butterflies of Southeast Asia* (Tsukada 1982a, b,

Table 5.14. Numbers and percentage of endemic butterflies on Java and Bali (in parentheses) with the number of those species found in neighbouring parts of the Archipelago. nc = not calculated.

	No. species	No. endemic species	% endemic	No. species also in Bali	No. species also in Sumatra	No. species also in Borneo	No. species also in Nusa Tenggara	No. species also in Sulawesi
Papilionidae	35	2	6	14	32	26	21	8
Pieridae	52	10	19	31	32	25	27	15
Nymphalidae	226	18	14	115				
Danainae	36	1	3	.	30	26	23	13
Satyrinae	44	9	20		29	23	13	7
Amathusiinae	13	1	8		8	3	1	
Libytheinae	3	0	0		2	1	3	2
other subfams	130	7	5		nc	nc	nc	nc
Lycaenidae	179	13	7	30	nc	nc	nc	nc
Hesperiidae	137	3	2	40	nc	nc	nc	nc
Total	629	46	7	230				

Source: R. de Jong (pers. comm.)

Table 5.15. Endemic butterflies of Java and Bali.

	W	C	E	B	
Papilionidae					
Atrophaneura luchti			E		Found only around Mt. Ijen, last collected in 1979, 1980.
Papilio lampsacus	W				Known only from Mt. Gede, last collected in 1979.
Pieridae					
Appias lucasii	W				Rare; last record 1937.
Delias aurantia			E		Found from 800-1,500 m; rare.
D. dorylaea	W				Found from 1,300-1,600 m; uncommon.
D. fruhstorferi			E		Found from 1,500-2,100 m; rare.
Elodina leefmansi			E		Known only from a few males caught in 1934.
Eurema beatrix	W				Found from 700-1,500 m; last record 1939.
E. tilaha	W	C	E	B	
Ixias balice	W	C	E		
I. venilia	W	C	E		Also known from Kangean Is.
Prioneris autothisbe	W				Found from 1,300-2,000 m.
Nymphalidae					
Satyrinae					
Elymnias ceryx	W	C			Found from 1,500-1,800 m; rare.
Lethe manthara	W	C	E	B	Found from 150-1,700 m.
L. samio	W	C	E		Found from 100-200 m; very rare.
Mycalesis moorei	W		E		Found from 500-800 m in forests; no recent information.
M. nala	W		E		Lower elevations, primary forest, no recent information.
M. sudra	W	C	E	B	Often very common, in forest and cultivation.
Yptima eupeithes	W				Found from 0-1,500 m; last record 1937.
Y. gaugamela	W		E		Found from 0-1,350 m.
Y. nigricans	W	C	E	B	Found from 0-1,400 m.
Amathusiinae					
Zeuxidia dohrni	W				Known only from Mt. Gede, no recent information.
Danainae					
Euploea gamelia	W	C	E		Fairly abundant over 1,000 m.
Other subfamilies					
Cynitia iapis	W		E	B	Found up to 1,000 m; not rare.
Cyrestis lutea	W	C	E	B	Found up to 1,400 m.
Hestinalis mimetica	W	C	E		Found from 0-1,200 m; rare.
Kallima paralekta	W	C	E	B	
Neptis nisaea	W	C	E		Found over 1,300 m.
Rohana nakula	W		E	B	Fairly rare.
Tanaecia trigerta	W	C	E	B	Not rare.
Lycaenidae					
Arhopala weelii					No information available.
Dodona chrysapha					No information available.
D. vanleeuweni					No information available.
Nacaduba deliana					No information available.
N. glauconia					No information available.
Oreolyce quadriplaga					No information available.
Prosotas datarica					No information available.
P. norina					No information available.
Ptox catareus					No information available.
Rapala renata					No information available.
R. vajana					No information available.
Rhinelephas cyanicornis					No information available.
Udara ceyx					No information available.
Hesperiidae					
Celaenorrhinus saturatus	W				Found from 0-1,600 m.
C. toxopei	W				Found from 1,500-1,800 m; known only from two females and four males caught between 1885 and 1943.
Halpe zandra					Known only from the holotype, a male with no locality or other details.

Sources: Roepke (1935); D'Abrera (1982-1986; Tsukada (1982-1991); R. de Jong (pers. comm.)

Figure 5.31. Java's two endemic swallowtail butterflies: *Papilio lampsacus* (male) left, and *Atrophaneura luchti* (female) right. Scale bars indicate 1 cm.

1985a, b, 1991) and *Butterflies of the Oriental Region* (D'Abrera 1982, 1985, 1986), but none of these is in any institution library in Indonesia.

MOTHS

Fauna

Moths are broadly and unscientifically divided into two groups, the larger (Macrolepidoptera) and the smaller (Microlepidoptera). The latter are very numerous, sometimes economically important as pests of food and other crops, but often extremely difficult to identify. It is not possible yet to give an accurate total of the moth species in Java and Bali, but it is certainly many hundreds.

Endemic Species

Since no catalogue of even the larger moths of Java and Bali has been compiled, it is not possible as yet to list the endemic species. A few are known for certain, however, and these are shown below (table 5.16).

Notable Species

Many species of moths are key pests of food crops and trees (Dammerman 1929c; Kalshoven 1981; Barlow 1982). Examples are the kenari moth *Cricula trifenestrata* the larvae of which can strip avocado, cocoa, rose apple, mango, and kenari trees of their leaves, the banana skipper *Erionota thrax* the larvae of which roll themselves in strips of banana leaves, the sweet potato leaf roller *Tabidia aculealis*, the large fruit-sucking moth *Ophideres fullonica* which spoils citrus, mangoes, banana, and other fruit, a range of rice borers *Scirpophaga*, and the death's-head hawkmoth *Acherontia lachesis* whose larva can reach 17 cm long and do great damage to many food plants (LBN-LIPI 1980).

Java also has two of the largest moths in the world, the pale green and long-winged moon moth *Argema maenas*, and the patterned, russet-colour atlas moth *Attacus atlas* with four scaleless 'windows' on its wings.

Although not yet observed on Java and Bali, some species of Javan notodontid, geometrid, and pyralid moths are known elsewhere to drink tears from large mammals (including humans), generally when they are sleeping. There is also a noctuid moth *Calyptra minuticornis* which is a skin-piercing, blood-sucking species that feeds on cattle, deer, and other

Table 5.16. A few of the endemic species of larger moths of Java and Bali.

Limacodidae		
Altha purina	E	Known only from the Tengger Mts.
Eupterotidae		
Ganisa javana		No information available.
Lasiocampidae		
Syrastrena tamsi	B	Known only from three males caught at Baturiti.
Lymantridae		
Dura helicta		No information available.
Geometridae		
Spaniocentra bulbosa	E B	In Java known only from the Tengger Mountains, from two specimens caught in 1934.
S. clavata	E	Known from a single male caught near Tretes in 1932.

Source: Barlow (1982)

domestic animals, but not (as far as is known) on humans (Bänziger 1988, 1989a, b, pers. comm.)

Major Publications

Lay people really only stand a chance of identifying the larger species, and the only realistic source of information is *An Introduction to the Moths of Southeast Asia* (Barlow 1982), which has colour photos of almost all the genera if not species. Many of the species commonly encountered are widespread in the Sunda region. Microlepidoptera are dealt with by Diakonoff (1973) and Meyrick (1969). Good illustrations and interesting details of species (both 'micros' and 'macros') considered agricultural pests are provided by Dammerman (1929c) and Kalshoven (1981).

CADDISFLIES

Most of the Java and Bali caddisfly (trichopteran) material in existence was collected during the German Sunda expedition in the late 1920s, and reported upon by Ulmer (1951). Relatively little work has been conducted since then. The world authority on this group for the Sunda region, Dr. Hans Malicky, provided the following information (tables 5.17-18), but it was not thought worthwhile to provide a species-by-species breakdown because of the very limited collections which probably indicate a far greater localization of distribution than is in fact the case (p. 204). In addition, he believes there are many more species to be found.

BEETLES

Fauna

Beetles (Coleoptera) represent the largest animal order with perhaps 250,000 species known worldwide and probably several millions yet to be described. Their abundance and variability is not always appreciated because their larvae and adults are often concealed in timber, bamboo, stems, dung, or soil where some have symbiotic relationships with fungi, termites, or ants, and many are small and inconspicuous. The largest species on Java and Bali is the three-horned dynastine *Chalcosoma atlas* which grows up to 13 cm in total length, and the smallest are among the Corylophidae and Ptiliidae, which are barely 0.5 mm long. Many species are crop pests, and it is the larval stage that is most injurious (Kalshoven 1981).

For the purposes of this book just two groups of beetles were reviewed, the ground beetles (Carabidae) and six families of water beetles. Ground beetles are robust, long-legged, and active beetles with large mandibles that are generally found on or near the ground. They are usually coloured black or brown with little patterning, although some are brightly coloured. Most species are predatory in adult and larval forms, and some species have been used in the biological control of pestilential moths. The best known species are the bombadier beetles (Brachyini) which, when disturbed,

Table 5.17. Caddisfly species known from Java.

| | Total species | Endemic | | | | |
		Total 'endemic'	West only	Central only	East only	More than one	Locality unknown
Rhyacophilidae	3	2	2	-	-	-	-
Hyabiosidae	1	1	-	1	-	-	-
Glossosomatidae	3	1	1	-	-	-	-
Hydroptilidae	11	8	3	3	2	-	-
Philopotamidae	5	2	1	1	-	-	-
Polycentropidae	3	3	-	-	-	2	1
Dipseudopsidae	2	0	-	-	-	-	-
Ecnomidae	3	0	-	-	-	-	-
Psychomydae	5	2	1	-	-	1	-
Hydropsychidae	36	19	10	4	1	1	3
Calamoceratidae	5	0	-	-	-	-	-
Odontoceridae	2	2	2	-	-	-	-
Leptoceridae	10	4	1	-	1	2	-
Goeridae	2	2	1	-	-	1	-
Lepidostomatidae	4	2	1	-	-	1	-
Helicopsychidae	1	0	-	-	-	-	-
Total	96	48	23	9	4	8	4

Source: Ulmer (1951); H. Malicky (pers. comm.)

Table 5.18. Caddisfly species known from Bali.

	Total	'Endemic'
Rhyacophilidae	1	0
Hydroptilidae	2	2
Philoptamidae	2	0
Hydropsychidae	2	0
Leptoceridae	1	2

Source: Ulmer (1951); H. Malicky (pers. comm.)

can direct a jet of boiling hot secretion forward out of the tip of the abdomen. The ground beetles are a widely distributed family, and there are about 25,000 described species worldwide. The tiger beetles Cicindelinae and the Rhysodinae, formerly separate families, are nowadays included in the Carabidae. These have prominent eyes, large, curved mandibles, long legs, and often bright coloured spots and metallic sheens.

Water beetles tend to be found in primary or little-disturbed habitats. Most water beetles live in water, but the members of the Gyrinidae live on the surface of part-shaded streams (p. 434). Dytiscid beetles live in a range of habitats, but members of the other families tend to be restricted to clear running waters. Adults of the Psephinidae are terrestrial. All these beetles scavenge or predate on smaller organisms including fish, and are able to remain submerged because they carry a layer of air under their hardened wing cases.

Endemic Species

It was possible to extract some information about non-cicindeline and non-rhysodine ground beetles found on Java and Bali from an unpublished checklist of ground beetles at the Natural History Museum, London, prepared by P.J. Darlington up to 1970, prior to revising the New Guinea carabid fauna. He died before this could be completed. Additions to this list have been found by searching *Zoological Record* since 1970 (N. Stork and J. Hailes pers. comm.). The analyses revealed that:

- 502 species (in 191 genera) of ground beetles have been recorded from Java and Bali;
- 29 of these are known from both Java and Bali, nine from Bali alone;
- 209 species (42%) appear to be endemic.

The species are not listed because there is virtually no distribution or ecological information for any of the species.

A total of 45 tiger beetle species are known from Java including eight (18%) endemic species (Pearson and Cassola 1992). The information on these species is, again, almost entirely based on single, old collections. No new species have been described since 1900, and 62% had been described by 1829. Their ecology is not known in much detail (Kalshoven 1981), but members of the genus *Neocollyris* (representing seven of the eight endemics) are generally arboreal leaf dwellers in primary and secondary lowland forest, so the massive loss of lowland forest on Java may have hit these species hard (F. Cassola pers. comm.).

Some 128 species of water beetles are known from Java and Bali, and 10 of these are not known from Java. In addition, there may be a further 60 species which await description (M. Balke pers. comm.). The known levels of endemism are 28% for Java and 14% for Bali (table 5.19), although the levels among the families vary from 0%-100%. Definite figures will not be

Table 5.19. Water beetles endemic to Java and Bali.

Gyrinidae		
Orectochilus agnatus		No details available.
O. bataviensis		No details available.
O. bipartitus		No details available.
O. drescheri		No details available.
O. javanus		No details available.
O. lumbaris		No details available.
O. panoembanganus		No details available.
Dytiscidae		
Allodessus thienemanni		No details available.
Allopachria barongia	B	No details available.
Copelatus regimbarti		No details available.
C. uludanuanus	B	No details available.
?Hydroglyphus aberrans		No details available.
H. erranus		No details available.
?Hydrovatus pasiricus		No details available.
Hyphydrus jaechi		No details available.
Laccophilus baturitiensis	B	No details available.
Microdytes elgae	B	No details available.
Hydraenidae		
Hydraena concinna		No details available.
H. feuerborni		No details available.
H. squalida		No details available.
Limnebius angustipennis		No details available.
L. feuerborni	B	No details available.
L. javanus		No details available.
L. leachi		No details available.
L. thienemanni		No details available.
Elmidae		
Ancyronyx acaroides	B	No details available.
Potamophilus javanicus		No details available.
P. orientalis		No details available.
P. perplexus		No details available.
Dryopidae		
Elmomorphus castaneus		No details available.
E. javanicus		No details available.
E. substriatus		No details available.
Sostea pilula		No details available
S. sodalis		No details available.
Psephenidae		
Eubrianax basipennis		No details available.
E. javanus		No details available.
E. major		No details available.
Microeubria jaechi		No details available.
Spioneubria jaechi		No details available.

Source: M. Balke, O. Biström, M.A. Jäch, P. Mazzoldi (pers. comm.)

known until there is further collection and revision, but it is likely that at least some of these species have already become extinct given that they tend to be inhabitants of lowland forested streams.

Major Publications

No comprehensive guide or listing exists, but the main agricultural zoology books (Dammerman 1929; Kalshoven 1981) allow the common species of beetles to be identified. In addition, the well-illustrated *Common Malayan Beetles* (Tung 1983) covers many of the larger, common species found in Java and Bali.

FLAT BUGS

Flat bugs (Aradidae) comprise some 1,800 described species known from temperate and tropical regions, although the latter contain 90% of the species. They are 2-30 mm long, are members of the sap-sucking Heteroptera, and their mouthparts are as long as their bodies. Unusually, virtually all their sustenance comes from fungal juices, and they are found primarily on fungus-laden rotting wood in forests or under bark. Most species have functional wings, but some forest species have reduced wings or no wings at all. Most species have developed cryptic camouflage to prevent potential predators seeing them on the bark. Although many of the species are known just from the type specimens, it is likely that species with reduced or no wings are endemic because they have such limited powers of dispersal. This amounts to eight of the 31 species known from Java and Bali (table 5.20), one of which repesents an endemic genus (fig. 5.32). A further 10 winged species are known only from Java and Bali, but it would not be surprising if these turned up on Sumatra or Borneo where the aradid fauna is similar to but richer than the fauna on Java and Bali.

EARWIGS

The earwigs (Dermaptera) are easily recognized by the pair of 'forceps' at the end of a flat, narrow, black or brown body. Most species are nocturnal predators on other insects, which they grip with the 'forceps', and can be very useful in the biological control of pests. They are found in a wide range of habitats, although they are most common where conditions are moist.

One of the more unusual species, *Xeniaria jacobsoni* (fig. 5.33) lives in

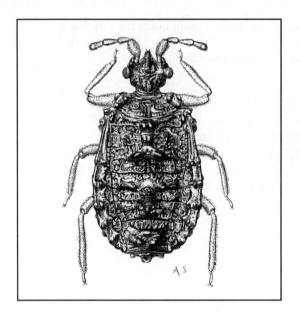

Figure 5.32. *Cremastaptera tuberculata*, an endemic genus of flat bug known only from a single female caught on Mt. Gede-Pangrango. Note the embellishments on the back that may prevent predators seeing the bug on rotting bark.

After Usinger and Matsuda 1959

Table 5.20. Endemic wingless and short-winged flat bugs of Java and Bali.

Carventinae			
Apteraradus bloeti	W		Wingless species known from a single specimen caught on Mt. Gede-Pangrango.
Indiaradus crinatus		E	Wingless species known from a single specimen caught on Madura.
Mezirinae			
Chelonocoris javanensis			Wingless species; no further information available.
C. kormilevi	W		Wingless species known from two specimens caught near Jampang Kulon and Mt. Malang.
Cremastaptera tuberculata	W		Endemic genus of a wingless species from a single specimen caught on Mt. Gede-Pangrango over 100 years ago.
Dimorphacantha usingeri	W		Short-winged species known from a single specimen caught on Mt. Megamendung.
Mastigocoris usingeri	W		Short-winged species known only from the type series caught near Bogor 90 years ago.
Scironocoris baliensis		B	Short-winged species known from a single specimen caught near Lake Buyan.

Source: E. Heiss (pers. comm.)

Figure 5.33. *Xeniaria jacobsoni,* a Javan earwig known from caves in Java and Malaya.

By A.J. Whitten

Table 5.21. Endemic earwigs of Java and Bali.

Pygidicranidae
Diplatys dohrni		No information available.
D. javanicus		No information available.
D. jacobsoni		No information available.
Schizodiplatys bensoni		No information available.
Cranopyia daemeli		No information available.
C. javana		No information available.
Mucrocranopygia horsfieldi		No information available.
M. vicina		No information available.
Echinosoma affine		No information available.
Gonolabis acuta		Also known from Bali.
G. kirbyi		No information available.
Antisolabis dammermani		No information available.
A. javana		No information available.
Parisolabis renschi		No information available.

Labiduridae
Gonolabidura javana		No information available.
Forcipula vanheurni		No information available.
Tomopygia abnormis		No information available.

Labiidae
Pseudovostox myrmecus		No information available.
Chaetospania bellator	B	No information available.
C. ferox	B	No information available.
C. javana		No information available.
C. thienemanni		No information available.
Spongovostox suspectus		No information available.

Chelisocidae
Hamaxas semiluteus		No information available.
Proreus horsfieldi		No information available.

Forficulidae
Cosmiella javana		No information available.
Eparchus globus		No information available.
Opisthocosmia vigilans		No information available.
Pterygida ferraria		No information available.

Source: Steinmann (1989)

caves where it is associated with certain species of insectivorous bats, and is adapted to life in dark caves by having greatly reduced eyes (Burr 1912). It spends most of its time on the carpets of guano or on the cave walls, and probably feeds on other insects (Medway 1958; Nakata and Maa 1974). The only known locality for this species was 'Gua Lawa' at Babakan near Banyumas, which was probably one of the sea caves on the south of Mt. Selok near Cilacap. A search for the earwig in those caves in 1993 proved unsuccessful, but it was found in Gua Petruk to the east.

Of nearly 2,000 species of earwigs known worldwide, 118 are known from Java and Bali of which 29 (25%) appear to be endemic to Java, and two to just Bali (table 5.21). They are not, as with many other groups, widely collected, and so the true distributions and their habits are poorly known (G. Cassis pers. comm.).

GRASSHOPPERS

Fauna

Grasshoppers are common components of ecosystems where some grass is

Table 5.22. Endemic species of grasshoppers from three Acridid subfamilies.

Acridinae		
Epacromiacris javana		No details available.
Phaeloba fumosa	W C E B	Commonly found in grass fields, forest clearings, and sawah.
Catantopinae		
Althaemenes javanica	W	
Anacridium javanicum		No details available.
Baliacris maculata	B	
Bibracte cristulata		No details available.
B. maculata		No details available.
Circocephalus microptera		No details available.
Eyprepocnemis javana		Possibly associated with alang-alang.
Javanacris montana	W	Known only from the slopes of Mt. Gede at 1,000 m.
Lepatacris javanica		No details available.
Lyrolophus javanus		No details available.
Meltripata javana		No details available.
Pareuthymia javana		No details available.
Racilidea exigua		Known only from 1,400 m.
Traulia javana		No details available.
Valanga chloropus		No details available.
Oxynae		
Chitaura lucida	W C	Found in forest clearings and hillside meadows between 1,400-1,800 m.
Stolzia javana		No details available.

Source: Willemse (1955, 1957, 1967); Hollis (1975); Dirsh (1975); K. Monk (pers. comm.)

present, although some species with short wings can be found in forests. Grasses are the primary food of some species, but probably the majority of species in Java and Bali eat the leaves of shrubs and trees. Some species may be dependent on just one or very few species of plant (Preston-Mafham 1990). There are two easily distinguished families, the Tettigonidae with long antennae which includes some minor pest species, and the Acrididae with short antennae which includes major pests on rice such as the green *Oxya* spp., and the swarm-forming migratory locust *Locusta migratoria*.

Endemic Species

Lists of endemic grasshopper species are difficult to compile (F. Willemse pers. comm.), but it has proved possible for three of the subfamiles of acridids found on Java and Bali. This shows that 20 (40%) of the 50 species present are apparently endemic (table 5.22) (p. 204).

Major Publications

No comprehensive guide or listing exists, but the main agricultural zoology books (Dammerman 1929; Kalshoven 1981) enable the most common species to be identified.

STICK INSECTS AND LEAF INSECTS

Fauna

Stick insects (Heteronemiidae, Phasmatidae, Bacillidae, and Pseudophasmatidae) and leaf insects (Phylliidae) are grouped into the order Phasmida, which is at its most diverse in Southeast Asia. Stick insects are characterised by their very long thorax which results in a considerable separation of their legs. This together with their generally cryptic green or brown colour make them look just like twigs. Some species even have 'thorns', 'spines', or 'leaflets' on their legs and body to complete the disguise (fig. 5.34). When disturbed, some species will pull their legs close to their bodies and feign death, whereas others will flee, startling any potential predator with brightly-coloured wings. Leaf insects have their front wings (and sometimes their legs) modified into elaborate leaf shapes complete with veins and tears. All phasmids are vegetarian, and all those in Asia are active only at night as a defence against predation.

It is not just stick insects themselves that mimic plants, for their eggs

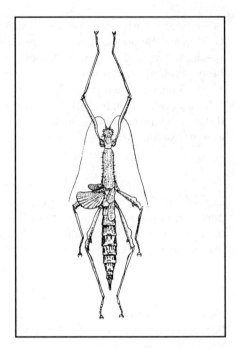

Figure 5.34. One of the stick insects endemic to Java, *Centema spinosissimum,* illustrating the spines employed to effect camouflage and its vestigial wings.

After Redtenbacher 1908

also have a remarkable resemblance to certain seeds. Some stick insects bury their eggs, but most flick them away as they emerge. Ants find them and take them back to their nests because, it has been suggested, the eggs are mistaken for certain plant seeds that have a fat-rich and knob-like appendage which is eaten by the ants. Such seeds are produced by species in a wide variety of plant families, and by all types of plants: trees, shrubs, herbs, climbers, and parasites (Beattie 1985). Interestingly, stick insect eggs have a similar appendage, the function of which had long puzzled zoologists. It seems that it is advantageous for a stick insect egg to be taken below ground, first because this reduces the chances of it being infested by parasitic wasps, and second because being on the surface for the long incubation period (up to three years) lays it open to desiccation or predation. It might be thought that a newly-hatched, soft, and frail stick insect would be easy prey for the ants, but ant colonies tend to move every few months, and by not hatching for a long time, young stick insects are likely to emerge into an ant-free environment (Hughes and Westoby 1993).

Endemic Species

Some 125 species are known from Java and Bali and 71 (57%) appear to be endemic (table 5.23), although it should be realized that the taxonomy of

Table 5.23. Endemic stick- and leaf-insects (Phasmida) of Java and Bali.

Heteronemiidae				
Lonchodinae				
Carausius exsul				No details available.
C. irregulariterlobatus	W			Known only from Cibodas.
C. proximus				No details available.
C. strumosus				No details available.
C. virgo		C	E	No further details available.
Lonchodes bryanti				No details available.
L. severum	W			No further details available.
L. spurcum	W			Known only from Mt. Gede.
Myronides reductus	W			No further details available.
Prisomera phyllopa				No details available.
Staelonchodes obstrictus			E	Known only from the Malang area.
Necrosciinae				
Acacus parvulus				No details available.
Anarchodes atrophicus				Possibly also occurs in Brunei.
Asceles adspirans	W			Known only from around Bogor.
A. rusticus				
Calvisia leopoldi			B	No further details available.
C. spurcata				Known only from the Malang area.
Centema adaequatum				No details available.
C. obliteratum				No details available.
C. recessum				No details available.
C. spinosissimum				No details available.
C. zehnteri				No details available.
Cylindomena acuminata				Known only from the Tengger Mountains.
C. scalprifera				No details available.
Diacanthoidea vittata	W			No further details available.
Diardia dentata			E	No further details available.
D. palliata			E	No details available.
Lamarchinus mirabilis				No details available.
Lopaphus pedestris				No details available.
L. transiens				No details available.
Marmessoidea quadrisignata				No details available.
Necroscia chlorotica				No details available.
N. fasciata				No details available.
N. horsfieldii				No details available.
N. pirithous				No details available.
N. redempta				No details available.
N. vittata				No details available.
Orthonecroscia fumata				No details available.
Orxines macklottii	W			Known only from around Bogor but not found for many years, although it has been cultured in Europe since 1940. Its possible occurrence in Sabah needs to be confirmed.
Paramyronides crishna				No details available.
Parasiphyloidea aenea				No details available.
P. exigua	W			Known only from Cibodas and Mt. Gede-Pangrango.
Pseudodiacantha obscura				No details available.

Table continues

Table 5.23. *(Continued.)* Endemic stick- and leaf-insects (Phasmida) of Java and Bali.

Scionecra analis			No details available.
Sipyloidea lutea		E	Known only from the Tengger Mountains.
Trachythorax unicolor	W		No further details available.
Pachymorphinae			
Ramulus javanus			No details available.
Phasmatidae			
Phasmatinae			
Baculum bidentatum			No details available.
B. dubiosum		E	Known only from the Tengger Mountains.
B. eminens	W		No further details available.
B. fissicornis			No details available.
B. ingerens			No details available.
B. irregularis	W		No details available.
B. lobipes		E	No further details available.
B. mediocris			No further details available.
B. modestum		E	No details available.
B. verecundum		E	No further details available.
B. warsbergi			No further details available.
Eurycnema beauvoisi			No details available.
E. versirubra			No details available.
Eucarcharus inversus	W		Known only from around Tugu.
Pharnacia heros			No details available.
P. semilunaris			No details available.
Platycraninae			
Echetlus jacobsoni			No details available.
Bacteriinae			
Ophicrania vittata			No details available.
Bacillidae			
Heteropteryginae			
Datames cylindripes			No details available.
Pseudophasmatidae			
Aschiphasmatinae			
Presbistus darnis			No details available.
P. servilleanus			No details available.
P. viridimarginatus			No details available.
Pseudophasmatinae			
Prisopus draco			No details available.
Phylliidae			
Phyllium jacobsoni			No details available.

Sources: Bragg (1992, pers. comm.); P.D. Brock (pers. comm.)

some species is by no means clear, phasmids are not often collected and, when they are, the locality information is not always reliable (P.E. Bragg and P.D. Brock pers. comm.). Many of the endemic species are known from single specimens or single collections, often with no locality beyond 'Java'. It is more than likely that new species await discovery and also that some of the supposed 'endemic' species will one day be collected in Sumatra and perhaps elsewhere.

Major Publications

There is no published account of this group for Java, but one is being prepared for Peninsular Malaysia which will be of use at generic level at least (Brock in press).

DAMSELFLIES AND DRAGONFLIES

Fauna

The damselflies (suborder Zygoptera) and dragonflies (suborder Anisoptera) of the order Odonata are well-known members of aquatic communities. The former rest with their wings held together above the abdomen, while the latter rest with their wings apart. About 150 odonate species are recorded from Java. They are found from sea level to over 3,000 m in a wide range of natural and man-made habitats. Although some of the species are quite strong fliers, and are relatively widespread in common habitats within the region, many are small, slender-bodied, and restricted (at least in their larval stages) to very specific habitats, and consequently have a relatively discrete and limited distribution. Around rice fields, *Crocothemis servilia* dragonflies with their bright red and flattened bodies, and *Neurothemis terminata* with their reddish brown body and wings and clear wing tips, are well known. On lotus ponds in Bali the white-bodied *Zyxomma obtusum* can often be seen flying up and down looking for prey. Dragonfly larvae are quite short, broad, and flat with no external gills; damselfly larvae are rather elongated with three posterior gills. They can be found in both standing and running water where they are fierce predators on small fish and tadpoles, and are sometimes regarded as pests in fish ponds. Conversely, they will also prey on mosquito larvae and have been used as biological control organisms in Bangkok where bins and buckets of water for household use, both potential mosquito breeding sites, were stocked with these rapacious little animals.

Lieftinck (1934) stated that with the alarming loss of forest and the drainage of swamps, the loss of many odonates is to be expected. A number of the little-known species were collected near Jeruklegi, 12 km north of Cilacap on the south coast, but this whole area has since been converted to rice fields.

Table 5.24. Endemic dragonflies and damselflies from Java and Bali.

ZYGOPTERA - Damselflies		
Chlorocyphidae		
Rhynocypha fenestrata	W C E B	Found throughout, from sea level to 1,000 m.
R. heterostigma	W C	Wells, spring-fed marshes, and streams in dense forest, 600-1,600 m.
Lestidae		
Lestes praecellens	W	Forest marshes and weedy ponds in coastal areas; a rare species.
Megapodagrionidae		
Rhinagrion tricolor	E	Deep ravines in dense primary forest, 1,000-1,300 m. Males rest on branches of trees fallen into water. Females oviposit in moss-covered logs and boulders in mid-stream.
Platystictidae		
Drepanosticta gazella	W C	Tiny, inconspicuous species found in small streams in forest from 100-1,500 m.
D. siebersi	E	Known only from the Tengger Mts. 1,700 m, apparently very rare.
D. spatulifera	C	Known only from the southern slopes of Mt. Slamet, 700-800 m.
D. bartelsi	W	Seepages in dark ravines in forest near the south coast. Known only from the type series.
D. sundana	W C E	Runnels and streams in shady surroundings in primary forest, secondary forest, and bamboo groves up to 900 m.
Protoneuridae		
Prodasineura delicatula	W C	Rocky streams in forested areas, up to 500 m.
Platycnemididae		
Coeliccia lieftincki	W C	Runnels in forest marshes up to 900 m.
Coenagrionidae		
Pseudagrion nigrofasciatum	C E	No further details available.
Aciagrion fasciculare	W	Tiny and inconspicuous, found in marshy lakes near dense forest at about 900 m.
ANISOPTERA - Dragonflies		
Corduliidae		
Macromia septima	W	Known from a few specimens caught in the first thirty years of this century in the hills west of Bogor. Possibly also found in North Vietnam.
M. jucunda	W	Known only from the male type found dead in Kebun Raya Bogor on 2 February 1954.
M. erato	W	Most records from Jasinga, but apparently rare.
Prosordulia karnyi	C E	Possibly also present in Sumatra.
Procordulia papandajensis	W	Known only from Mts. Gede and Papandayan.
Gomphidae		
Gomphidia javanica	W C E	Very local, near slow-flowing streams in forests and shady bamboo groves, 100-600 m.
Megalogomphus junghuhni		No locality known for this large species, represented by a single female specimen.

Table 5.24. *(Continued.)* Endemic dragonflies and damselflies from Java and Bali.

Burmagomphus inscriptus	C E	Small species known only from three specimens caught in ravines cut though montane and sub-montane dense forests.
B. javicus	W	Known from a single pair, one of which was from 600 m on Mt. Cisuru, Jampang Tengah.
Onychogomphus banteng	W	Apparently very rare, known from a single male caught near Cibadak on Mt. Salak, probably from shady mountain streams.
O. geometricus	W C E	Found near fast-running streams in forested areas up to 1,200 m.
Paragomphus reinwardtii	W C E B	Found near woodland streams with mineral bottoms.
Aeschnidae		
Gynacantha stenoptera		Represented by a single male caught in an unknown part of Java by an unknown collector last century.

Source: Lieftinck (1934); J. van Tol (pers. comm.)

Endemic Species

Twenty six species (18%) are endemic to Java, some widespread or locally common, and some known from a very few specimens (table 5.24). As with other insect groups it is likely that some of these may yet be found in the fast-disappearing forests of southern Sumatra, and that further collecting would reveal new species (Lieftinck 1934; p. 204).

Major Publications

An annotated list is provided by Lieftinck (1934), but this does not include means of identification and is somewhat out of date. Unfortunately, the often bright colours and patterns of these large insects are lost shortly after death, and so identification tends to be based on morphological features, particularly of the sex organs. It would be possible, however, to produce a coloured identification guide to live animals covering most of the genera if not species.

SPRINGTAILS

Fauna

Springtails (Collembola) are primitive insects rarely exceeding 5 mm in length. They are found in a wide variety of moist habitats such as soil,

Table 5.25. Endemic springtails of Java and Bali.

Hypogastruridae

Hypogastrura consanguinea	C	No further details available.

Neanuridae

Ceeratimeria lognicornis		No details available.
Lobella garuda		No details available.
L. gedehensis		No details available.
L. soemarwotoi		No details available.
L. wayang		No details available.
Oudemansia coerulea		No details available.
Protanura kräpelini		No details available.
Pseudachorutes javanicus		No details available.
Siamanura promadinae	E	Recently discovered in Baluran National Park.

Onichiuridae

Onychiurus bogoriensis	W	No further details available.

Isotomidae

Folsomia octoculata			No details available.
Isotomurus tricuspis	W	B	No further details available.
Subisotoma sundana		B	No further details available.

Entomobryidae

Acrocyrtus javanus	W	No further details available.
A. eurylabialis	C	No further details available.
A. merapicus	C	No further details available.
Alloscopus teracanthus		No details available.
Callyntrura batik	C	No further details available.
C. villosa		No further details available.
C. longicornis		No details available.
C. quadrimaculata	C	No further details available.
Dicranocentrus javanus	W C	No further details available.
Dycranocentroides villosus		No details available.
Entomobrya proxima		No details available.
Lepidosinella armata		No details available.
Lepidosira sundana	C	No further details available.
Microparonella annulicornis		No details available.
M. nigrofasciata		No details available.
Salina fasciatus		No details available.
S. insignis		No details available.
S. javana		No details available.
Seira marginatus		No details available.
S. schaefferi		No details available.
Willowsia hyalina		No details available.

Chypodestidae

Table continues

Table 5.25. *(Continued.)* Endemic springtails of Java and Bali.

Cyphoderus javanus	W	No further details available.
C. javanicus		No details available.
Cephalophilus yayukae		No details available.
Orchesellidae		
Heteromurus yosiius		No details available.
Sminthuridae		
Corynephoria jacobsoni		No details available.
Katianna coeruleocephala		No details available.
Ptenothrix gracilicornis		No details available.
P. pulchellus		No details available.

Source: Handschin (1926, 1928); Suhardjono (1989); Yoshii and Suhardjono (1989); Suhardjono and Deharveng (1992); R. Yoshii (pers. comm.)

decaying vegetable matter, under bark, among leaves, in ant or termite nests (mainly *Cyphoderus*), in caves, on the surface of freshwater, and even in situations where they are submerged twice a day by tides. In terms of numbers they are among the most important arthropod groups and may number over 20 million in a hectare of forest (Stork 1988). They are detritivores and a number of species are important in rice field ecosystems (B. Settle pers. comm.). Most species have a jumping organ attached to the fourth abdominal segment which, when moved rapidly downwards, causes the animal to spring into the air. Most species are dull-coloured, but others have a bright metallic sheen, and still others have bold patterns (fig. 5.35).

Endemic Species

Some 73 collembolan species are now known from Java and Bali of which 41 (56%) have only been collected on Java and/or Bali (table 5.25) (Suhardjono 1989; R. Yoshii pers. comm.). Given the large number of new species found in recent years, the relatively few workers in the field, and the concentration of collecting on Java, it is likely that these figures will alter substantially in the future.

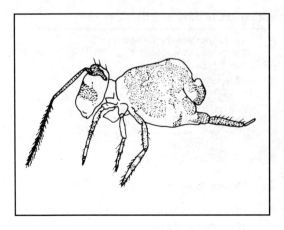

Figure 5.35. The attractively marked endemic springtail *Ptenothrix pulchellus.*
After Handschin 1928

SPIDERS AND SCORPIONS

Spiders are very adaptable animals, being found from mountaintops to air-conditioned offices, deep caves, and sandy shores. All spiders are carnivorous, catching living prey, especially insects. Spiders, particularly in the tropics, are little known, and the 30,000 species described to date throughout the world is only a small proportion of those that exist. Unfortunately, even if they were being collected, there is a singular dearth of spider specialists able to identify or describe species. Very few collections of spiders have been made, and distribution information, if it exists in more detail than just 'Java', is probably unreliable.

From the available data, however, it seems that the spider fauna of West Java (as represented by the species found in Cibodas and Ujung Kulon) is quite similar to that in Sumatra, Borneo, and Peninsular Malaysia. East Java and Bali have a less clear relationship. The spider fauna of open or disturbed areas is quite distinct from that of the shady forests and generally comprises relatively widespread species. The primary forest species (and those from cave systems) generally seem to be endemic to quite small areas. It is certain that many genera, and even more species on Java and Bali, have yet to be described. Unfortunately it is equally certain that many species are already extinct (C. Deeleman pers. comm.).

In the western world spiders tend to be feared and disliked, but among the Islamic inhabitants of Java they are respected because of the spider

Figure 5.36. The bird-eating or tarantula spider *Selenocosmia javanensis.*

By F. Kopstein with permission of the Tropen Instituut, Amsterdam

whose web concealed Mohammad when he was hiding in a cave from his irreligious Quraisj enemies. This respect is appropriate because spiders do a great service to people by ensnaring, deceiving, and jumping on pestilential insects that damage crops. This message is now getting through, and in East Java roadside signs showing enormous spiders implore drivers, farmers and passers-by to protect them as part of the effort to develop integrated pest management (p. 575).

Scorpions are close relatives of spiders, and about 350 species have been described worldwide. Ten species are known from Java (the number known from Bali is not certain), and all are widespread in Asia and beyond. They are mainly nocturnal, hiding under logs and rocks and in crevices during the day. They feed on insects. The related whip scorpions (Uropygi) and whip spiders (Amblypygi) have two species each in Java; at least one of the former seems to be endemic (p. 555).

Notable Species

The largest spiders are the hairy bird-eating or tarantula spiders of the Theraphosidae, of which three species are known, including one very rare endemic one *Selenocosmia raciborskii* (Kulczynski 1908). A more common relative is the yellow-kneed tarantula *S. javanensis* (fig. 5.36) the legs of which can span over 12 cm. When disturbed they rear up their bodies and lift their first two pairs of legs to reveal their heavy, curved jaws. This species is one of those which is gaining popularity in Europe and North America as a house pet.

Mention should also be made of some of the useful rice field spiders that prey on insect pests (p. 579). The most voracious appears to be *Lycosa pseudoannulata*, a wolf spider which lives among the rice tillers and jumps on adult stem borer moths, planthoppers, and leafhoppers. Each spider will claim 5-15 prey each day. The lynx spider *Oxyopes javanus* lives at the top of the rice plants where it sits concealed until its prey, mainly small moths, comes within range. The web-spinning orb spiders are represented by *Argiope catenulata*, which constructs four conspicuous zig-zag 'stabilimenta' at right angles to each other around the centre of the web. These serve to advertize the web to birds which might otherwise fly through it. The sticky strands would be a problem for the bird, and the spider would lose precious protein which it would normally recycle by eating it prior to

Table 5.26. Endemic freshwater crabs of Java and Bali.

Sundathelphusiidae					
Terrathelphusa kuhli	W	C			Known from Mt. Halimun, Sancang, Gede, and Baturaden; semiterrestrial, found near streams. The carapace and legs are a distinctive purplish red colour (fig. 5.37).
T. modesta	W				Known from Ujung Kulon, Banyumas, Tangerang; semiterrestrial, minor pest on tobacco.
Parathelphusidae					
Parathelphusa convexa	W	C	E	B	Semiterrestrial, streams and paddy fields.
P. bogorensis	W	C	E	B	Widespread. Specimens from Bali indicate that they may represent a separate species.
P. sp.		C			Known only from Bawean Island.
Grapsidae					
Geosesarma nodulifera	W				Known only from around Bogor.
G. sp.	W				Found in 1994 in Ujung Kulon National Park.
G. rouxi		C			Beautiful, purplish crab with red claws and bright yellow eyes, common in padi fields, drains, and even gardens.
Sesarmoides jacobsoni					Known only from a cave in the Mt. Sewu area; last caught in 1911.
S. emdi				B	Discovered in 1993 in a cave on Nusa Penida.
Potamidae					
Malayopotamon granulatum	W				1,000-2,000 m, aquatic, found in fast-flowing streams (fig. 5.37).
M. javanense	W				Known from lowlands, sawah, slightly polluted streams, fish ponds.

Sources: Ihle (1912); Pesta (1930); LBN-LIPI (1979); Ng (1988, pers. comm.); Stewart (1994); Ng and Whitten (1995)

building another web.

The most frequently encountered scorpion is the relatively small, light brown, spotted scorpion *Isometrus maculatus* which lives in houses. The larger black wood scorpion *Hormorus australasiae* is common under loose bark anywhere there are trees, and in the same habitat one might also find the large (15 cm excluding the claws) and dark blue-green *Heterometrus longimanus*. All scorpions can inflict a painful sting from the needle-like tip of the tail, and of the above species the spotted scorpion is the most painful. Young scorpions are born looking like miniature adults, and when still young they are carried on the mother's back.

Major Publications

Identification of forest spiders is well nigh impossible, but books are available that allow species of urban and agricultural habitats to be identified (Koh 1989; Shepard et al. 1987). There is an illustrated paper on the scorpions of Java including a key to the 10 species encountered (Kopstein 1926).

FRESHWATER CRABS

Eleven freshwater crab species are known from Java and Bali (table 5.26). All but three of the species are endemic to Java, two are found also on Bali, and one appears to be confined to Nusa Penida.

Notable Species

The common sawah crab *Parathelphusa convexa*, and to a lesser extent *P. bogoriensis* and in the hills *M. javanense*, are locally troublesome because they feed on nursery beds and newly transplanted rice, and because they burrow deep into the bunds thereby reducing their strength and effectiveness. Egrets and other water birds feed on the young crabs, but the best control against *P. convexa* is apparently still the traditional driving of flocks of ducks into the fields after harvest (LBN-LIPI 1980; Kalshoven 1981).

Major Publications

No illustrated text exists for the freshwater crabs of Java and Bali, but *Parathelphusa, Geosesarma,* and *Sesarmoides* are dealt with in Ng (1988).

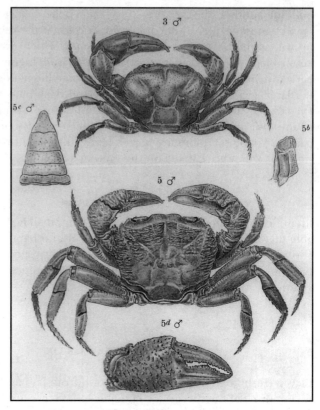

Figure 5.37. *Terrathelphusa kuhlii* (above) and *Malay-opotamon granulatum*.
After de Man 1892

TERRESTRIAL AND FRESHWATER MOLLUSCS

Fauna

This section concerns the terrestrial and freshwater molluscs including the slugs with no, or only an internal, shell (Rathousiidae, Vaginulidae, Arionidae, Limacidae) and the bivalves. Most of the nearly 190 species known from Java and Bali are snails with a coiled shell, generally turning to the right. Some species have a plate or operculum on the foot which is used to close the aperture when, for example, conditions become too dry. Some shells are high and thin, others very small and flat; some have up to 16 whorls, while others have only one. Most species are small, the smallest being less than 1 mm long. The largest and most familiar species is the introduced giant African land snail (p. 299), and the rapidly-spreading large golden snail, originally from South America, is also a major pest (p. 301).

The mollusc fauna of Java, not surprisingly, has fewer species than its larger island neighbours, and is characterized by the absence or near-absence of many genera known elsewhere on the Sunda Shelf (J. Vermeulen pers. comm.). The density of species is, however, considerably higher on Java, perhaps as a result of a greater collecting intensity. The number of known species in Java, Sumatra, and Borneo is shown in table 5.27.

Endemic Species

To date, 65 species (31%) of land and freshwater molluscs appear to be endemic to Java alone, 11 to Java and Bali, and 10 to Bali (including Nusa Penida) (table 5.28). This compares with 41% endemism for Sumatra (A. van Bruggen pers. comm.). Of 60 endemic land species found in West Java, no less than 40 (67%) are found in Mt. Gede-Pangrango National Park, and 12 of those appear to be restricted to it–probably reflecting the 'Mt. Gede effect' (p. 204) (fig. 5.38). Five (31%) of the bivalves, and three (3%) of the freshwater snails are endemic (fig. 5.39). Java has relatively few endemic species when compared with Borneo or the Malay Peninsula (not enough collecting has been conducted on Sumatra for any conclusion to be drawn), but certain genera such as *Philalanka, Landouria,* and *Acrophaeausa* seem to have evolved substantial numbers of endemics on Java but remained somewhat homogenous on neighbouring land masses.

Notable Species

A few species of molluscs are agricultural pests, but the major ones (giant African snail *Achatina fulica* (p. 299), *Lamallaxis gracilis, Bradybaena similaris,*

Table 5.27. Numbers of terrestrial and freshwater snail species known from Java, Sumatra, and Borneo.

	No. terrestrial species	No. species/ 10,000 km²	No. freshwater species	No. species/ 10,000 km²	Total species	Total species/ 10,000 km²
Java	181	14	105	8	286	22
Sumatra	192	4	133	3	325	6
Borneo	430	6	62	1	492	7

Sources: adapted from A. van Bruggen (pers. comm.); J. Vermeulen (pers. comm.)

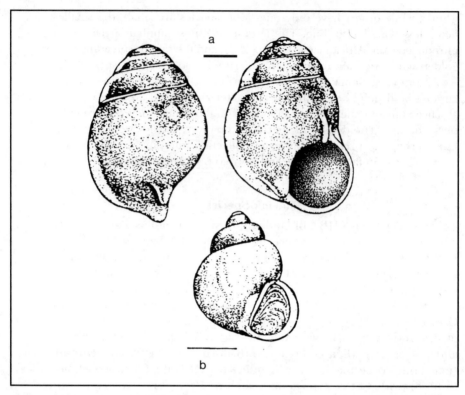

Figure 5.38. A - *Hydrocena laeviuscula, b - Pupina verbeeki,* two of the land snails known only from Mt. Gede. Scale bar indicates 1 mm.

After van Bentham Jutting 1948, 1950

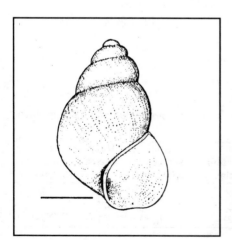

Figure 5.39. *Wattebledia insularum,* a freshwater snail endemic to Java. Scale bar indicates 1 mm.

After van Bentham Jutting 1956

Table 5.28. Endemic terrestrial and freshwater molluscs of Java.

PROSOBRANCHIA				
ARCHAEOGASTROPODA				
Anaglyphula whitteni			B	Found in 1992 above D. Buyan in volcanic soil under forest.
Hydrocenidae				
Hydrocena laevivscula	W			Known only from 3,000 m on Mt. Gede.
Helicinidae				
Helicina rollei		E		Known only from the Kangean Islands.
Sulfurina biconica	?W		B	Probably a Bali endemic, although it was once found in Bogor Botanical Gardens.
Cyclophoridae				
Alycaeus crenilabris	W			Known from various forested localities from 150-700 m.
A. reticulatus	W			Was known from a single specimen found in the Jampang hills at about 600 m, until recent collection was made at the side of the road near Padalarang.
Chamalycaeus fruhstorferi	W			Known from various forested localities from 200-1,500 m.
C. kessneri			B	Known only from Nusa Penida, and first found in 1992.
Cyclotus pendensis			B	Known only from Nusa Penida, and first found in 1992.
?Ditropis fruhstorferi	W			Known from a single specimen from Mt. Salak, unknown altitude.
Lagochilus convexum	W			Known from Mt. Cikurei at 1,500 m (and perhaps from near Pelabuhan Ratu).
L. humile	W			No locality known; not collected since about 1890.
L. macrophalum	W			Known only from Mt. Cikurei at 2,000 m, probably among damp leaf litter.
L. obliquistriatum	W	E		From various locations in W. Java and from Mt. Lamongan, from 200-700 m.
Pterocyclus sluiteri	W			Known from a single specimen from Mt. Gede, unknown altitude.
Pupinidae				
Pupina bipalatalis	W			Known from a number of localities from 200-700 m.
P. compacta	C	E		Known from Mt. Dieng at 2,000 m and Mt. Tengger at 1,200-1,500 m.
P. junghuhni	W	E		Known from numerous localities from sea level to 1,400 m.
P. verbeeki	W			Known only from Mt. Gede at 1,200 m.
Diplommatinidae				
Diplommatina auriculata	W	E	B	Known from many locations, 300-2,400 m.
D. cyclostoma	W C	E		Known from a number of localities including Dieng 2,000-2,900 m, but also at 100 m on Madura.
D. diplostoma		E	B	Known only from Mt. Ijen 1,600-1,850 m. and from Bali.

Table continues

Table 5.28. *(Continued.)* Endemic terrestrial and freshwater molluscs of Java.

D. heteroglypha		E		Known only from Mt. Ijen 1,600-1,850 m.
D. hortulana	W	E		Known from various locations, 150-1,350 m.
D. javana	W		B	Known from a few locations 500-2,700 m.
D. ornithorica	W			Known only from Mt. Cibodas or Bukit Kapur west of Bogor.
D. perpusilla	W			Known from Mt. Gede and Mt. Ciremai.
D. planicollis	W			Known only from Mt. Gede-Pangrango.
D. sulcicollis	W	E		Known only from Mt. Gede Pangrango and Mt. Lawu.
D. tetragonostoma	W			Known only from Mt. Gede 1,300-1,400 m.
D. sp.		E		First found in 1993 to the west of the Yang Plateau.
Opisthostoma uranoscopium	W			Known only from Mt. Cibodas or Bukit Kapur west of Bogor, and from Pangandaran.
Palaina gedeana	W	E	B	Known from mountains, 800-2,900 m.
P. nubigena	W			Known from mountaintops, 2,400-2,500 m.
P. vulcanicola			B	First found in 1993 in forest above Lake Buyan, 1,400 m.

MESOGASTROPODA

Thiaridae
Balanocochlis glandiformis

Freshwater. No information available.

Hydrobiidae
Wattebledia insularum   E

Freshwater. Known only from R. Gelapan and R. Pening S of Tulungagung, Kediri.

PULMONATA
STYLOMMATOPHORA

Helicarionidae
Inozonites imitator   W

Known from a single specimen from 1,000 m on Mt. Gede.

Ariophantidae

Asperitas waandersiana		E	B	Locally common in forest and more open situations.
Durgellina sp.		E		Found in 1993 to the west of the Yang Plateau, 1,700 m.
Elaphroconcha juvenilis			B	
E. patens		E		Known from a few locations in Tengger, Ijen, Kawi 1,100-1,800 m.
Lamprocystis gedeana	W			Known only from Mt. Gede 1,200-2,700 m.
Macrochlamys spiralifer			B	Found above D. Buyan in 1992.
Microcystina brunnescens				Bali and Nusa Penida
M. chionodiscus	W C		B	Found in surveys in 1992. In Bali known only from Nusa Penida.
M. subglobosa	W C	E		Known from a few locations around Mt. Gede, Mt. Tilu, Mt. Lawu, and Nongkojajar from 250-3,265 m.

Table continues

Table 5.28. *(Continued.)* Endemic terrestrial and freshwater molluscs of Java.

Microparmarion austeni	W			Slug-like with single-whorled shell. Known from a few locations 250-2,400 m.
Sasakina plesseni			B	Known only from tiny patches of forest on Nusa Penida (fig. 5.40).
Camaenidae				
Amphidromus winteri	W	C	E	Known from tree crops from sea level to 2,000 m.
A. banksi	W			Known only from P. Panaitan and P. Sertung (Krakatau).
A. javanicus	W			Known from tree crops 150-1,000 m.
Ganesella sphaerotrochus			B	Found in 1992 on the slopes of Mt. Abang, 1,700 m.
Landouria monticola	W	C	E	Known from three locations (Cikurei, Dieng, Kawi) 2,000-2,500 m.
Pseudopartula arborescens	W			Known only from P. Panaitan (two collections 1951 and 1992), and recently from Sertung (Krakatau).
Euconulidae				
Cuneuplecta macrostoma	W	E		Known from a variety of locations 2,000-2,400 m.
Liardetia viridula	W			Known from a few locations 2,400-2,900 m.
L. pisum	W			Known from a few locations 300-2,400 m.
L. reticulata	W			Known only from Mt. Gede-Pangrango 1,700-2,400 m.
L. platyconus	W	E	B	Known from numerous locations 1,000-3,000 m.
Achatinellidae				
Tornatellina periconspicua			B	Found in 1992 on the slopes of Mt. Abang, 1,700 m.
BASOMMATOPHORA				
Clausiliidae				
Acrophaedusa cornea	W	C	E	Known from a variety of locations including Madura and Nusa Barung Island.
A. fruhstorferi	W			Known only from Mt. Malabar and Mt. Cikurei, 1,900-2,300 m.
A. nubigena	W			Known only from Mt. Gede, Mt. Cikurei, and Mt. Malabar 1,300-2,900 m.
A. orientalis	W			Known only from Mt. Gede Pangrango, Telaga Warna, Puncak Pass, 1,000-1,700 m.
A. schepmani	W			Known from Mt. Gede and Mt. Jampang.
Pseudonenia fucosa	W			Known only from forests near Cipanas, 3 km from Taman Jaya on the edge of Ujung Kulon National Park.
P. javana	W	C	E	Known from a variety of locations including Nusa Barung, Madura, and Nusa Kambangan, 600-3,200 m.
Rathousiidae				
Atopos ouwensi	W			Known from Bogor Botanical Gardens, Mt. Guntur, and Mt. Masigit, 250-1,600 m.
Vertiginidae				
Gyliotrachela fruhstorferi	W			Known from a few locations on limestone 200-700 m.

Table continues

Table 5.28. *(Continued.)* Endemic terrestrial and freshwater molluscs of Java.

	W	C	E	B	
Pupisoma perpusillum	W				Known only from Mt. Gede Pangrango 2,400-3,300 m.
P. tiluanum	W				Known only from Mt. Gede and Mt. Tilu 1,700-2,100 m.
Enidae					
Coccoderma prillwitzi	W				Known only from Mt. Gede 1,400-1,800 m.
C. tenuilirata	W		E		Known only from Mt. Gede, Tengger, and Kawi, 1,500-1,500 m.
C. tenggerica			E	B	Known only from Mt. Tengger and Kintamani.
C. thrausta	W	C	E		Known from Mt. Gede, Trinil, Tengger, and Madura 75-1,500 m.
Endodontidae					
Pilsbrycharopa baliana				B	
P. setifera	W				Telaga Warna, Puncak Pass.
P. depressiopira			E	B	Found in 1993 in forests above Lake Buyan, 1,400 m, and west of the Yang Plateau, 1,700 m.
Zonotidae					
Geotrochus multicarinatus	W		E		Known from a number of locations in West Java and possibly Madura.
Trochomorpha concolor	W				Known from a few locations, 1,000-2,000 m.
Arionidae					
Meghimatium striatum	W		?	?	Known from a few locations, 1,000-2,000 m.
Achatinellidae					
Elasmias sp.	W				Telaga Warna, Puncak Pass.
LAMELLIBRANCHIA					
Sphaeriidae					
Pisidium javanum		C			Freshwater. Known only from Lake Cigombong, Dieng Plateau.
P. sundanum		C			Freshwater. Known only from Mt. Nukomo near Ngebel, 850 m.
Sphaerium javanum		C			Freshwater. Known only from R. Dolog, Dieng Plateau, 2,000 m.
Unionidae					
Contradens contradens	W	C	E		Freshwater. From lakes, ponds, and rivers. Remains found in Sampung cave (p. 202).
Elongaria orientalis	W	C	E		Freshwater. Widespread on Java and Madura.

Sources: Rensch (1931, 1932, 1934); van Bentham Jutting (1948, 1950, 1951, 1953); Loosjes (1953, 1963); Smith and Djajasasmita (1988); J. Vermeulen (1995, pers. comm.); Vermeulen and Whitten (in prep.)

Figure 5.40. *Sasakina plesseni,* a species of snail known only from Nusa Penida.
By A.J. Whitten

and *Subulina octona*) are not native and are distributed widely throughout the tropics as well as in greenhouses in temperate countries. Others, such as the slugs *Semperula maculata, Filicaulis bleekeri,* and *Parmarion pupillaris* (LBN-LIPI 1980), are as damaging as the snails, feeding on crops whose leaves should be harvested intact, such as tobacco and leafy vegetables, and ornamental plants such as orchids are attacked, too. Slugs will attack young rubber trees and also feed on the flowing latex from older trees. Unfortunately they sometimes get trapped in the collecting bowl, thus contaminating the product. A recent and most notable arrival is the golden or mulberry snail (after the shape and colour of its egg mass) which is discussed elsewhere (p. 301).

Exploitation

The introduced giant African snail apart, the only regularly consumed species are the apple snails *Pila* spp. and *Bellamya javanica,* both of which live in still, clear water with abundant aquatic plants. The large, introduced bivalve *Anodonta woodiana* (p. 300) is also consumed, and their empty shells can form large piles around Lake Bratan and Rawa Pening. There is a small trade in shells, but most collectors are interested in the larger, more colourful and bizarre marine species. However, the tree snails *Amphidromus* spp. are colourful and highly variable and have thus become collectors' items (Dharma 1992).

Distribution

Land snails favour moist habitats and avoid sunlight, either by being nocturnal or by living in shady places, because they face desiccation in hot and

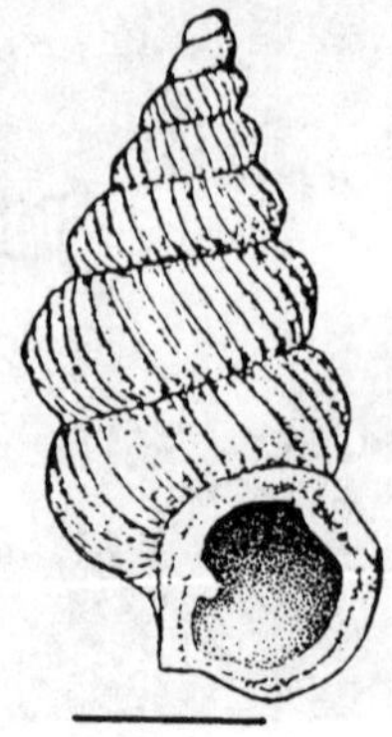

Figure 5.41. *Diplommatina ornithorica*, a snail known only from the now severely degraded Bukit Kapur limestone hill west of Bogor. Scale bar indicates 1 mm.
After van Bentham Jutting 1948

dry situations. Most snail species appear to be found in the regions of highest rainfall, where they also occur in their greatest abundance, but dry, vegetated areas can have a surprisingly rich fauna, and this is particularly true of the remaining forested areas of Nusa Penida. Such dry areas have attracted relatively few collectors. The other major factor influencing snail distribution is the presence of lime (calcium carbonate), without which the shells cannot be made. It is not surprising then, that many species and individuals are found in limestone areas, but volcanic deposits are not without lime, and so snails are by no means restricted to limestone areas.

All stated distributions of species must be regarded as provisional; for example, during a survey in 1991, a small bag of soil was collected from beneath a low vegetated cliff on Madura (south of Ketapang). This small sample contained the shells of 10 species of snail, two previously unknown from East Java, one thought to be confined to altitudes above 2,000 m, and one, *Opisthostoma javana*, thought to be possibly extinct since it was previously known from a single specimen from the degraded limestone Mt. Cibodas or Bukit Kapur hill, west of Bogor (p. 484); it is now also known from Borneo and Sulawesi (Vermeulen 1991). One other small species, *Opisthostoma uranoscopium*, which was thought to be confined to that hill, was found at Pangandaran in 1993. There remains a single species, *Diplommatina ornithorica* (fig. 5.41), still known only from Mt. Cibodas, but it now seems unlikely that it is genuinely restricted to that site.

Of the roughly 180 land snails and slugs native to Java, 157 species are so far known from West, 54 from Central, and 82 from East Java. Given the prevalence of limestone in East Java, one might expect a rich mollusc fauna with a high degree of endemism, but the long history of cultivation has probably wrought major damage to the fauna, although species adapted to arid climates may tolerate disturbance more than species from humid forest.

Data currently available suggest that there is an increase in the numbers, if not species, of molluscs with increasing altitude up to about 1,500 m. It seems that 52 snail species (17 of them endemic) from the lowlands and hills never reach beyond 1,500 m, and 21 species (13 of them endemic) from the mountaintops are never found below 1,500 m. Although 28 species are known from the highest mountain zone about 2,500 m, only two species, the hydrocenid *Hydrocena laeviuscula* and the endodontid *Philalanka micromphala,* both endemic, are not known outside this zone.

Major Publications

The only major works that allow access to the group are the papers by van Bentham Jutting (1948, 1950, 1951, [on terrestrial species], 1953 [on freshwater bivalves], 1956 [on freshwater gastropods]), the papers on Clausiliidae by Loosjes (1953, 1963), although the new books in the series *Siput dan kerang Indonesia* (*Indonesian Shells*) by Dharma (1988, 1992) have excellent photos of many of the larger species. The first volume comprises mainly sea shells with some freshwater species; the second covers mainly land snails.

DOMESTIC BREEDS

Java and Bali have a rich store of domestic breeds of the major livestock animals, most famous of which is the native Bali cattle, a domestic form of the banteng *Bos javanicus* (p. 220). While the loss of plant varieties as a result of modern plant breeding programmes is well known and germplasm is the subject of conservation programmes, the loss of variation among animals has received less attention. This is not to say that the desire to improve livelihoods through the introduction of new breeds from outside Indonesia is inappropriate, merely that every effort should be made to keep a broad genetic base of a variety of breeds as insurance against unforseen changing environmental circumstances and market requirements. While doubtless not complete, a number of declining breeds are described below.

Fat-tail Sheep

The fat-tail sheep (fig. 5.42) is a large, fecund, hair (rather than wool)

Figure 5.42. The fat-tail
sheep of East Java.
By A.J. Whitten

breed that is found mainly in East Java, and although probably introduced
from India or West Asia originally, it has evolved into an adapted local
breed since at least the 17th century (Nieuhof 1682). At the beginning of
the 19th century Java had just 5,000 registered sheep (compared with
402,000 buffalo) (Raffles 1817), but the number increased enormously
over the next century. The total number of fat-tail sheep existing today is
not known, although nearly one million are known from East Java. Most of
the pure fat-tails are found in Sumenep, Situbondo, Probolinggo, and
Pasuruan, mainly on the dry coastal plains (Soetranggono 1991). In the last
few decades, however, numbers have declined dramatically, partly because
of the export of the breed to elsewhere in Indonesia and partly because of
the introduction of new breeds from outside Indonesia such as Suffolks
and dormers.

Various ministerial and gubernatorial instructions have shipped numbers of fat-tail sheep outside East Java to prevent genetic erosion, but since
documentation has not been undertaken, the usefulness of these efforts is
limited. There is every reason to believe that the fat-tail sheep would
respond well to a development programme emphasizing its use in production systems integrating tree cropping and sheep (for manure, meat,
leather, and live export to the Middle East and elsewhere). There is at present no reason to believe that hybridization with other breeds would
improve the breed, although careful selection should be performed to
satisfy the various types of market demand (Iniguez 1991).

Kacang Goat

This relatively small goat with erect ears is gradually being lost and replaced with the larger but less fecund *etawah* goat with floppy ears from the middle east and India. The kacang may be smaller, but it matures earlier (six months against twelve), and has lower embryonic and perinatal losses (Wodzicka-Tomaszewska et al. 1993a, b). Its smaller size may be an advantage to small-scale livestock owners.

Bali Pig

These hung-back, low-bellied, short pigs have become quite scarce, except around Klungkung, because of the introduction of British breeds such as Yorkshire, saddleback, Berkshire, and landrace. The newly introduced pigs are much larger, but they require special feeding and are less fecund than the local breed.

Kampung Chickens

There are a host of chicken varieties running around Java and Bali villages, most of them quite small (about 1 kg body weight against 2.5-3 kg for broiler chickens). They are not prolific breeders, producing only 30-36 eggs in three distinct clutches through the year rather than the almost continuous egg production found in broiler chickens, they do not grow as fast as broiler birds, and they convert their food to meat less efficiently. Despite all these apparent disadvantages, kampung chickens command a premium price because of their superior flavour and because their 'firm' flesh does not fall to bits when cooked in traditional styles. Efforts have been made to develop intensive commercial operations using kampung chickens, but none has yet been successful. Meanwhile attention needs to be paid to the conservation of genetic variation within the kampung chicken populations.

Tengger Dog

At the end of the last century attention was drawn to the existence of a peculiar dog found only in the Tengger highlands (fig. 5.43). The Tengger dog had a fleece as thick as a sheep's, and was reddish brown with a dark stripe from the nose along the back to the tip of the plumed tail. Another dark streak formed a collar. The erect ears were very dark but the feet were light, and the belly, tail, and 'buttocks' were white (Kohlbrugge 1896). Its disposition was courageous and friendly, quite unlike that of the common village dog, and this was so admired by the Dutch that many were taken to the lowlands. Unfortunately, their thick coats and unfamiliarity with lowland diseases took their toll, and by 1896 only a few males were known to remain. All that remains of this dog today is two skeletons and half a skin, all in the Leiden Museum. The bones show several differences from other domestic dogs (Jentink 1897). There were also some peculiar small, long-

Figure 5.43. The extinct Tengger dog.
After Jentink 1897

haired, brown dogs found only on the Dieng Plateau; these too were nearing extinction by the end of the last century (Jentink 1897) but, together with the Kintamani dog from Bali, have been maintained by enthusiasts (Rokhman 1990).

INTRODUCED SPECIES

Buffalo

As was seen from the lists of prehistoric mammals (p. 202), the wild water buffalo *Bubalus arnee* used to live on Java. This was the ancestor of the domestic species *B. bubalis* which was probably domesticated by about 2000 B.C. in southwest Asia (Cockrill 1984). The wild water buffalo is described by Lekagul and McNeely (1977) as:

> a truly magnificent beast, much larger in all proportions, better coordinated, quicker-moving, and considerably more aggressive. The horns curve in much the same manner as in the domestic breed, but have a much wider spread, with the horn tips of a bull in its prime quite

sharp and slender. In color, the wild water buffalo looks like the domestic variety, except the stockings below the hock and knees are grayish or dirty white.

It is known that the wild and domesticated species will hybridize. This begs the questions of how close the buffaloes in Java and Bali are to the wild stock, whether some of today's buffalo are descended from Javan wild stock, when the wild species became extinct, and when the domesticated species was introduced. In this regard it is interesting that Dammerman (1934) notes:

- the use of the buffalo originates from a much earlier period than the arrival of Indian Hindus at the start of the first few centuries of the first millennium. This is evidenced by the large number of names by which the buffalo is known throughout the islands and its distinction from ordinary cattle;
- the Indian-influenced stone reliefs around Borobudur temple do have a few buffalo depicted, but in the very rare case that a plough is shown, it is being pulled by a hump-backed zebu cattle, a breed now very rarely kept; and
- there is a tradition in West Java that it was the first Hindu king of Paja-jaran that introduced the idea of using buffaloes to pull ploughs. Per-haps because of this, he (and his son) were accorded the title of 'male buffalo' or 'leader of the herd' (Temminck 1846), possibly indicating that the beasts were known as wild herds, not just as domes-ticated draught animals. It has also been recorded that, at about the same time, buffalo came to people's service out of the forest of their own free will (Schlegel and Müller in Temminck 1839).

Thus, while there is doubt and no case can actually be proven, it may be that wild buffalo did persist in Java into the historical period, that the domestic buffalo were introduced into Java, and that there may have been some inter-breeding between the two stocks. There is no evidence, however, that the buf-falo in areas such as Baluran are any more than feral domestic stock.

Molluscs

The largest (< 130 mm) and most familiar snail species on Java and Bali, the nocturnal, brown, and white striped giant African land snail *Achatina fulica* (fig. 5.44) was introduced in 1933 to Indonesia from India and Africa, although its original home was Madagascar. It is now a considerable pest, able to feed even on prickly pear cactus *Opuntia nigrans* and sap-filled *Euphorbia*, and also on dung and carrion. It is never found deep inside intact native forest although it can be found around the edges. These are highly adaptable and tolerant animals, and their large shells can be produced even in urban areas where the soils have little lime because they ingest calcium by rasping off mortar and whitewash. Where conditions

Figure 5.44. The giant African land snail.
With permission of Kompas

are favourable, they can grow to over 10 cm in just six months. Once detested as a food, they were eaten during the Japanese Occupation and are now cultured for export (Whendrato and Madyana 1989). The increase in exports has been remarkable: from just seven tons in 1981 to 3,648 tons in 1989. Most snails are produced around Kediri (East Java) and are used locally as medicine, livestock food, and as saté or crisps. The flesh is relatively low in fat and calories. Its relative *A. variegatus* has been introduced for cultivation but does not yet seem to have escaped.

Other introduced and pestilential snails include the ubiquitous *Brady-baena similaris* which was already in the Bogor area in 1928 (Rensch 1933), the African *Rachis zonulata* which is found in drier areas, the ubiquitous *Huttonella bicolor, Subulina octona, Gastrocopta servilis,* and *Lamellaxis gra-cilis,* the last of which was first collected in Indonesia in 1848 (LBN-LIPI 1980). Other exotic snails can be found in gardens and plantations but have not yet escaped into the wider environment (J. Vermeulen pers. comm.). In addition, the bivalve *Anodonta woodiana* is now widespread. This is a native of China and Taiwan, and was introduced to Java just over a decade ago (Djajasasmita 1982) and has spread elsewhere in the archipelago along with the introduced fish tilapia *Oreochromis* on which it encysts

Figure 5.45. Golden snail.
With permission of Kompas

during the glochidial larva stage (D. Dudgeon pers. comm.). *A. woodiana* is
the largest freshwater mussel recorded from Indonesia: it can reach 27 cm
long, 13 cm high, and 61 cm wide and is of an olive to dark green colour
(Djajasasmita 1982).

The golden snail *Pomacea canaliculatus* (fig. 5.45) was introduced into
the Philippines from its native South America in 1980 as a commercial venture to produce 'escargot' (snail meat) for the export market. By 1988 the
snail was considered the second most serious rice pest after the brown
planthopper. The snail feeds on newly transplanted rice and can destroy
50%-80% of the potential rice crop. It will also eat corn, citrus, and ornamental plants (Baldia and Pantastico 1991; Anderson 1993). This snail
was first found wild in Indonesia in 1987 (Marwoto 1988). Golden snails are
very similar to the native apple snails *Pila* (Djajasasmita 1987), but they have
a thinner, more translucent shell, a horny operculum, are more fecund,
and have young that develop more quickly, taking barely three months
before breeding. This pestilential snail lays conspicuous, reddish pink egg
masses about 20 cm above water level (giving it its alternative name of mul-

berry snail). Each mass contains 400-700 eggs (Marwoto 1988). Initially they were known only from the northern part of East Java, but by 1992 they were found west to Kebumen in Central Java (R.M. Marwoto pers. comm.). By 1994 they were widespread across Java in ricefields and other wet areas. Other species of introduced golden snail can be found in rice fields, but none is as damaging (Suryadi 1992).

Various chemical and mechanical controls are available but none, when used alone, is adequate. Molluscicides offer only temporary relief and the chemicals are in any case hazardous for humans, livestock, fish, and other animals. Part of the problem is that these snails can close themselves off from unsuitable conditions by pulling their horny operculum into the shell, and when the rice fields are drained, they can survive at least six months in a dormant condition within the soil. Integrated pest management (p. 575) offers better potential control, and the following options are available (Rondon and Sumangil 1991):

- handpicking of adults and egg clusters;
- construction of small canals or depressed strips into which the snails congregate when the field is drained and from which they can be collected and killed;
- using older seedlings for transplanting;
- using stakes to attract snails for egg deposition;
- using baits of kangkung, papaya, cassava, and taro leaves to facilitate collection;
- using traps or screens on water outlets;
- using ducks to eat the snails (rats probably have a more significant effect, but no one would suggest encouraging these);
- using dried tobacco leaves and other botanical preparations as deterrents and molluscicides;
- harvesting them for food; and
- applying commercial molluscicides only as and when necessary and only at the prescribed concentrations.

Aquaculture Introductions

The Directorate General of Fisheries (DGF) is satisfying the Indonesian Government's push for economic development by developing new export-quality aquaculture resources. These are generally exotic species or strains.

Bullfrogs and other amphibians. Frogs (like turtles and certain crabs) are *makruh* to Muslims, that is their consumption is not regarded as a sin, but to abstain from eating them will result in blessings and rewards. Thus the domestic consumption of frogs' legs is primarily found among non-Muslims. Traditionally, three native Javan frogs have been popular as food,

Rana macrodon, R. cancrivora, and *R. limnocharis.* Collecting these was subject to the vagaries of weather and crop harvest and it was not possible to supply consistent numbers or desired sizes to customers. It was logical to try caged cultivation of these species but it proved difficult to make an enterprise economically viable. In 1980, however, the giant American bullfrog *Rana catesbiana* was introduced and it has been much promoted. By 1990, Indonesian exports of frogs' legs reached nearly 8,000 tons, produced mostly in Java, with the primary importers being Belgium/Luxemburg, France, and Singapore. Indonesia's current top position in the export of frogs' legs used to be held by India until, in 1987, India decided that the income from exports was not sufficient to offset the problems of increasing insect populations in marshy frog habitat (Negroni and Farina 1993).

The bullfrog was originally imported by DGF in 1985 as an export-quality commodity. The imported strain was already domesticated, and growth rates and final size were adequate for culturing. There were, however, two major difficulties that had to be overcome: feeding and disease. The bullfrog's natural food consists of live fish and earthworms, and the problems associated with collecting live food and its expense made it unattractive until, in 1989, a specially formulated food pellet was developed. This solved practical feeding problems, but its cost put the culture on an economically shaky footing. The second problem, disease, could be solved by culturing bullfrogs in spring water rather than irrigation canals or other sources of water, but this limits culture to those few areas with adequate supplies of spring water.

It is very disturbing that no attention was drawn to the potential dangers of bullfrogs escaping into rivers and lakes, and no preliminary study has been made on its likely effects (e.g. Susanto 1989). Given this frog's well known ability to cover up to 2 m in a single leap, it has considerable propensity to escape from captivity if care is not taken. Indeed, specimens have already been found away from but in the vicinity of breeding ponds south of Jakarta and on mountain slopes north of Sukabumi (F.B. Yuwono pers. comm.). It may be, of course, that the environmental conditions are not conducive to explosive growth of feral populations, but this should have been investigated beforehand. Attention should be given to the domestication and selection of local species to replace the American bullfrog as a culture species. Any possible future genetic deterioration of the domesticated strain could easily be solved by introducing genes from the wild populations.

Of potential importance was the discovery in a pond near Sukabumi a few years ago of two giant salamanders *Andrias* (formerly *Megalobatrachus*) sp. native to Japan and China (D. Yuwono pers. comm.). There is also a record of a giant salamander *A. japonicus* (fig. 5.46) being released into the Bogor Botanic Gardens in 1969 (Boeadi pers. comm.). These enormous, carnivorous beasts can grow up to 150 cm in total length, but small ones are

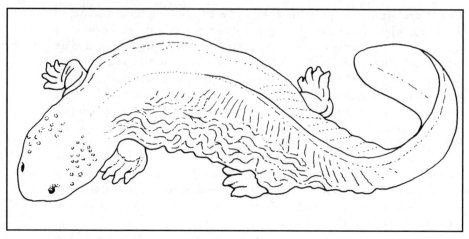

Figure 5.46. Giant salamander *Andrias* sp.

sometimes imported as aquarium curiosities. They require cool, pure running water, but should these ever become established, the impact on fish ponds in upland areas could be considerable.

Fishes. A great many fish species have been introduced into the major water bodies of Java and Bali. These are of two types, those brought to Indonesia as potential food fish, and those brought as ornamental fishes.

Probably the most abundant freshwater fish on Java and Bali is the introduced guppy *Poecilia reticulata*. This is a native of the West Indies, Venezuela, and Guyana (Courtenay and Meefe 1990), but it has been introduced deliberately and accidentally into many countries (Welcomme 1988; Arthington and Lloyd 1990). It probably arrived in Java shortly before the Second World War, as an aquarium fish. Certainly it is not mentioned in the relevant volume of *The Fishes of the Indo-Australian Archipelago* (Weber and de Beaufort 1922). While accepting that assessing the impact of introduced species without baseline data is subjective, the considered opinion of experts in this field is that the carnivorous guppy has had negative impacts on native fishes and perhaps other aquatic organisms. In particular, it appears to compete with native close relatives which in Java and Bali are the tinhead *Aplocheilus panchax* and rice fish *Oryzias javanicus*. In Jakarta huge numbers of guppies, but very few tinheads, can be found in the drains, but whether this is the result of competition is not clear. In any case, in the Indonesian situation, it would be hard to judge that the presence of guppies was important in ecological terms, given the much more potential ecological effects of pollution, overfishing, and river works. Other aquarium fishes now found in the rivers and lakes of Java and Bali include platies *Xiphophorus maculatus,* swordtails *Xiphophorus helleri,* mollies *Poecilia sphenops,* and blue acara cichlids *Aequidens pulcher* (Kottelat et al. 1993).

One of the most important groups of introduced food fish is the tilapia *Oreochromis* spp. The first species of tilapia introduced to Java was mujair *Oreochromis mossambica* which appeared in 1930s, perhaps through the escape of aquarium stock or pond fish. It became economically important during the Second World War, during which time the coastal areas were off-limits to milkfish farmers for military reasons, and so the people turned to tilapia for food (Ling 1977). In 1969 the faster growing, black-and-red nila *O. nilotica* was introduced (Kottelat et al. 1993) partly because *O. mossambica* populations were becoming stunted over time (Welcomme 1988). Many stocks of the former species were then dumped into waterways, where they are now exceedingly common. Both species are omnivorous and not averse to eating small fish, and many native species have probably suffered as a result. The new red tilapia from Taiwan is a hybrid between the above two species, and is becoming very popular.

The effects of careless fish introductions is perhaps best seen in Lake Bratan, Bali which was the only locality of the endemic *Rasbora baliensis* before our surveys. This lake has had at least nine fish species deliberately or accidentally introduced to it since Independence (table 5.39). In 1990 recreation fishermen (there is no commercial fishery on this small lake) started noticing wounds (necroses) caused by bacteria appearing on the fish, and by 1992 almost all fish of interest to them seemed to be so affected or carrying *Lernaea* parasites. This is the combined result of introducing unhealthy fish stock, and of cumulative stress caused perhaps by pesticide loads originating from the market gardening areas to the west of

Table 5.39. Fishes deliberately and accidentally introduced to Lake Bratan.

Guppy	*Poecilia reticulata*
Molly	*Poecilia* sp.
Soft-finned carp	*Osteochilus hasseltii*
Snakehead	*Channa striata*
Tilapia	*Oreochromis mossambica*
Common carp	*Cyprinus carpio*
Striped rasbora	*Rasbora lateristriata*
Two-spot barb	*Puntius binotatus*
Grass carp	*Ctenopharyngodon idellus*
Walking catfish	*Clarias batrachus*
Dumbo catfish	*Clarias gariepinus*

Source: Green et al. (1978); M. Kottelat (pers. comm.)

the lake. The lake has no outflow except perhaps through to springs on the outer slopes. Considering this, it is remarkable that the endemic *Rasbora baliensis* has survived until now.

One of the more frightening potential introductions is the piranha *Serrasalmus* spp. from the Amazon and nearby areas. The habits of these voracious carnivores are sometimes exaggerated, but they still pose considerable dangers for humans and animals. Although a Ministerial decree in 1982 banned their import, it is said that some do already exist in Indonesia in private collections (probably in Jakarta or West Java). They can be confused quite easily with the pacu *Colossoma mitrei*, a close relative which is already legally imported as an aquarium fish and as a candidate for aquaculture. It is hoped that the ban on piranha can be effective and enforced, and that before the existing fish die, their monetary and status value will prevent them from being tipped accidentally or deliberately into a river or other body of water.

Catfish. The large African catfish *Clarias gariepinus* was first 'adopted' by a private company in 1986. It was claimed to be of export quality and to command a high price in international markets. This particular population was of very good quality, already domesticated, and had high fecundity and growth rates compared to the most common local catfish *C. batrachus*. The imported individuals were first bred in January 1987, and were distributed soon after to local farmers. Despite the hopes, the last five years have proved to be unsuccessful for the following reasons:
- international prices dropped and this species has not yet been exported;
- there is no local market due to taste preference for local catfish;
- although initially of high quality, the strain was soon degraded as a result of poor broodstock management on the part of farmers (new stock has now been imported); and
- it was found that disease resistance was low.

The local catfish *C. batrachus* is not considered suitable as an export commodity because it is smaller, has relatively low growth and fecundity, and is not established in the international market. However, if local catfish were domesticated and selected for high fecundity, growth rates, and fry survival, the advantages over culturing the African species are numerous (V. Brzeski pers. comm.):
- as local consumers prefer the taste of the local catfish to the African catfish, a local market is assured with a potential international market which could out-compete the African catfish;
- this species is naturally adapted to Indonesian environmental conditions, thus would be much easier to culture;
- local farmers, already familiar with the species, would be more willing to adopt it and less extension training would be necessary;

- in the case of genetic degradation or development of low disease resistance, there is a very large pool of genes in the wild population to draw on to improve the domesticated strains and thus no need to import new stock as was the case with the African catfish;
- because of regional variations of this catfish species, several domesticated strains could be developed independently, each one maximizing efficient use of resources in a specific environmental and cultural region; and
- the potential dangers of *C. gariepinus* becoming a serious, exotic carnivorous pest will be avoided.

Mangrove Crab. The latest example of introduced commodity species is the mangrove crab *Scylla serrata* brought in from China in April 1992. Although the literature sources classify it as a brackish-water species, the private company that initially imported this strain claims it is a herbivore which can be cultured in freshwater. This is of interest to DGF as it could potentially be used to control the water hyacinth *Eichhornia crassipes* population which is a problem in reservoirs (pgs. 183, 453). From China, this crab is already exported to Japan, Hong Kong, and Taiwan. However, the crab is close to being carnivore, and the imported individuals died in the freshwater tanks provided.

    S. serrata occurs naturally in mangrove forests throughout Indonesia. Some individuals have been found to migrate upriver into freshwater. Instead of importing species of uncertain quality, DGF would be more successful in the long run if it collected mangrove crabs found in Indonesia and selected individuals tolerant to freshwater. This would not only lead to conservation of local crab strains but it would also empower DGF with more control in the development of this potential aquaculture commodity (V. Brzeski pers. comm.).

Future Developments. There are several advantages to importing species intended for aquaculture development when they have already proved themselves elsewhere (V. Brzeski pers. comm.):

- the animals imported have already been domesticated, thus enabling culture to proceed without delay;
- there is already an established international market for the species or strains that DGF imports—thus assuring Indonesian farmers an export commodity and saving time and resources spent on marketing, research, and development strategies; and
- the individuals imported have already undergone selection to suit international market demands.

Although the introduction of such animals satisfies one of the Indonesian government's priorities, namely the development of export commodities, it can be in conflict with the stated desires for the conservation and development of Indonesia's natural resources. These two goals need not be in conflict since they can work together to supply Indonesia with export-quality commodities and, at the same time, conserve Indonesia's biological diversity by developing the vast supply of Indonesian indigenous species. Although importing an established export commodity such as the African catfish may initially be economically rewarding, developing a local species (i.e. domesticating, selecting, and marketing) is more rewarding and more efficient in the long term. Such proposals may require some initial investments (from animal breeders and geneticists, marketing analysts and economists), but they should prove to be beneficial and, in the long run, more economically viable and sustainable (V. Brzeski pers. comm.).

Conclusion

It is clear that animal introductions are rarely beneficial and more often deleterious. Even though large numbers of species have already taken up residence, and it is unlikely that a pestilential creature would be introduced deliberately, there should surely be much stricter controls on the import of alien species. It is worth noting that many richer countries (notably neighbouring Australia) do all in their power to prevent alien species from arriving on their soil (Janzen 1987a).

Chapter Six

Human Background

Prehistory

A layman reading the literature on the remains of prehistoric, human-like apes found on Java could be excused for believing that a large community of species was present; the names *Anthropopithecus erectus*, *Meganthropus palaeojavanicus*, *Pithecanthropus duboisii*, *Hemanthropus*, *Javanthropus soloensis*, *Sinanthropus*, *Hylobates giganteus*, *Homo modjokertensis*, *Homo dubius*, *Homo trinilis*, *Homo primigenius*, and *Homo erectus*, among others, are all mentioned. The many names given to bone fragments was largely a failure to appreciate that the natural variation exhibited by prehistoric skeletons was no less than that of modern skeletons, particularly with regard to differences between males and females (Groves 1989).

It now seems to be agreed widely that all the Javan material (mostly from a 8 x 10 km area around Sangiran, 10 km north of Solo in Central Java) actually belongs to a single, variable species, *Homo erectus*, whose remains have also been found in Africa and Asia (fig. 6.1) (Pope 1988; Groves 1989). Suggestions that the remains of *H. erectus* in Central Java date from nearly two million years ago (Ninkovich and Burckle 1978), were eclipsed by the view that this species probably did not arrive in Java until about 1.25 million years ago at the very earliest, and most likely not more than 800,000-500,000 years ago, about a million years after it appeared in Africa (Pope 1983). It was felt that earlier estimates of age perhaps did not take account of the devastating effects of volcanic eruptions, tectonic movements, mudflows, and floods that all disturb a neat ordering of geological strata. New analyses have shown, however, that the volcanic pumice around the fossils is indeed nearly 2 million years old (Swisher et al. 1994). If this accurately reflects the age of the fossils (see de Vos and Sondaar 1994), then they are only very slightly younger than the oldest *H. erectus* fossils found in Africa. Whatever the chronology, Java is the furthest east these early humans are known to have been reached, and to have gone beyond here would have meant travelling across water. They appear to have survived and thrived, remaining unchanged for several hundred thousand years (Bartstra 1983). The migration to Java coincided with a relatively cool period when the sea levels were lower, and this would have made migration

Figure 6.1. Impression of *Homo erectus.*

By C. Raab, with permission of the Museon, 's-Gravenhage

to islands easier for people, animals, and plants.

Although it is generally agreed, on the basis of fossil remains found to date, that human-like animals arose in Africa, there are a variety of theories concerning the later stages of human development. The remains found in Java (and also those in China) are of singular importance in this debate. One contention is known as the 'Out of Africa' hypothesis in which it is proposed that modern humans *H. sapiens* arose relatively recently in Africa, from where the species spread rapidly, displacing archaic humans such as Neanderthals and *H. erectus*. An opposing view, the multiregional hypothesis, is that modern humans appeared more or less simultaneously around the world evolving from *H. erectus*. If the first view is correct, then the *H. erectus* remains in Central and East Java have no direct relationship with modern people. Recent genetic research into the make-up and variation of DNA in the nuclei and mitochondria of living humans suggests that modern humans could well have originated in Africa, perhaps as recently as 200,000 years ago (Wainscoat 1986; Cann et al. 1987; Lewin 1993). An important objection to the displacement theory, however, is that it seems unlikely that potentially colonizing *H. sapiens* populations would have had such outstandingly better life strategies than those of the resident *H. erectus* populations that displacement would have resulted. Indeed, nothing in the remains found of prehistoric humans in Asia indicates any sudden introduction of superior life strategies (Pope 1983).

Although no tools associated with *H. erectus* have yet been found, stone tools from later are known. Interestingly, the 'cleavers' and 'hand axes' (termed Acheulean tools) found in African and south Asian *H. erectus* sites, are absent from Java and Southeast Asia where the 'tool kits' are dominated by chopping tools. One interpretation of this is that a non-stone and biodegradable material was very important to the *H. erectus* in this area in order to complete their 'tool kit'. An excellent candidate for this material would be bamboo, which can be fashioned into almost anything these early Javan people could have wanted except chopping tools (Pope 1985, 1989; see also Hutterer 1985), although this does not exclude a whole range of other forest materials being used in addition.

Stone axes have been found, however, near to Pacitan on the south coast of Central Java (van Lohuizen-de Leeuw 1979). The cave-pitted limestone hills of Pacitan encircle a gently shelving and safe bay that would have provided ample and varied protein for an early population. Further detailed investigations in the area are almost certain to turn up more interesting artefacts and further clues to the ways in which these people lived. The finding of stone hand axes is not of itself necessarily evidence of the discovery of very old artefacts, however, since we have no idea how recently groups of people may have used stone tools. In any case, the Pacitan tools were made using highly advanced flaking techniques and cannot be older than 50,000 years. Possibly Wajak Man, early *Homo sapiens* who lived at the end of the Pleistocene and early Holocene, was responsible. The tools of this people reflect their environment (poor lateritic soils on the fringe of a karst area), and they may well have lived a semi-settled existence with a fringe form of Hoabinhian culture (Bartstra 1983). In addition to the tools found in Central Java, tools made from glass-like obsidian (excellent for making sharp edges) have been found around the former Bandung Lake and at another former lake 50 km to the southeast at Leles, near the small lake of Situ Cangkuang (Subagus 1979). The first artefacts, stone tools, definitely associated with the robust Wajak Man were recently described and, in the absence of pottery remains, point to a mesolithic culture from 10,000 to 5,000 years ago. Associated with the human remains, are mammal bones among which is the tapir *Tapirus indicus*, which is now extinct in Java (p. 200) (Storm 1992).

Java's original fully-human *H. sapiens* were probably australoids, similar to the Australian aborigines of today who came from Indonesia some 40,000 years ago. The australoids were probably later ousted by immigrants from Southeast Asia with more advanced cultural and ecological adaptations living as sophisticated hunter-gatherers. None of these people have obvious descendants in Java or Bali, but they live on as the Anak Dalam or Kubu of central Sumatra, and the people of eastern Indonesia. Then, some 3,000-5,000 years ago, another wave of immigrants arrived, this time cultivators known as the proto-Malays. Their descendants can be found among the Mentawai islanders of West Sumatra, the Tenggerese of East Java, the Dayak of Kalimantan, and the Sasak of Lombok. Finally, the Austronesians or the deutero-Malays, originally from Taiwan and southern China arrived by sea in Malaysia, Philippines, and Indonesia and probably arrived in Java and Bali by about 1,000-3,000 years ago (Bellwood 1984). Their descendants now dominate western Indonesia, by virtue of their knowledge of swidden and rice cultivation, irrigation (perhaps), pottery, and advanced and highly polished stone tools (Koentjaraningrat 1985). One famous site for these is around the village called Tugu just north of Bandung (Bennett and Bennett 1980). This group of people by no means stopped in Indonesia, for they sailed on in outrigger canoes and settled or colonized islands as far away as New Zealand, Easter Island, Hawaii, and Madagascar.

As permanent settlements began to be established, possibly around 2,500 years ago, so a spiritual life is believed to have developed. This was probably based on ancestor worship, and stone (megalithic) structures associated with burial are known from many places such as Wonosari and Pakuaman in Java, and Margatengah and Gilimanuk in Bali (Soejono 1984). The burial customs became most elaborate during the oldest metal age just over 2,000 years ago. An excellent area for bronze age burial objects is around Kuningan, West Java. The most characteristic and beautiful objects made in the Bronze Age were finely ornamented kettledrums, some imported from mainland Asia, but others, called *moko*, were made in Java and Bali, even into historic times. Bronze objects found at Gilimanuk (to the east of the road between the bus and ferry terminals) have been dated to about 7,000 years ago. Ornate and varied pottery objects have also been found throughout. All this indicates established social organization and stratification, religious conceptions, as well as skilled workmanship (Soejono 1984; Kempers 1988; McConnell and Glover 1988).

Other objects are very hard to interpret since essentially nothing is known about them. One example is the one-kilometre-long and strategically-placed fortification formed of earthworks with ditches and walls on a mountain pass in what is now the Argasari estate between Pacet and Santosa, south of Bandung. There is also a large but overgrown megalithic pyramid at Cikakak, Pelabuhan Ratu, and many ancient stone graves for which local myths are currently the only information (Bennett and Bennett 1980).

Effects on, and of, People

Between 22,000 and 18,000 years ago, Sundaland was exposed and the coastline was relatively short (less than half the present total length). Some of the rivers were extremely long, draining large areas (p. 118). It is assumed that communities of people were few and small, many of them living in the interior riverine areas, particularly near caves. Caves provide shelter, warmth, a certain degree of security, and they were focal points of activity. Bays near forested headlands with caves at the mouth of a river would have been prime sites for habitation; such sites are rare but can be found around Pacitan. The people may have rafted across stretches of water but, with only limited navigational skills, they are unlikely to have gone far. Any exploitation of the coasts would have been limited to the intertidal zone (Dunn and Dunn 1977).

During the times of low sea level the climate was cooler and drier and there would have been a greater extent of relatively open deciduous forest than is present now. In such habitats, a great deal of leafy food would have been within a neck's stretch of the 10 large herbivores (rhinos, wild cattle, and deer) and five browsing elephants that may well have still existed in Java at that time. Research from elsewhere in the world has suggested that human hunters caused the extinction of many species of large vertebrate during this period. The critical factor appears to have been a certain threshold density of population which, once exceeded, brought once-hunted but relatively stable populations of prey to extinction, perhaps in the space of only a few decades. Climatic variations are sometimes implicated, but this seems unlikely since the species had already successfully withstood thousands of years of climatic change. In either case, it is intriguing how and why the rhino and banteng have persisted.

The people, who were part-time hunters, would have been in competition with a dozen or so large carnivores (against just the leopard and wild dog of today, and the tiger of yesterday). These large carnivores were not just competitors because they were certainly predators on the early humans themselves, just as they had been on their ancestors. Whatever weapons the people possessed would have been of little use against a large and hungry carnivore intent on devouring them. Indeed, the invention of weapons was probably as much due to defence against such attacks as improving hunting techniques or inter-group warfare.

Between 18,000 and 9,000 years ago the climate warmed and rain forest developed again. As this occurred, so the fauna would have changed, with reduced densities of large herbivores. Human hunting would have shifted to smaller, less conspicuous game such as monkeys and mouse deer, and women would have likely increased their time devoted to collecting plant food and setting the stage for agriculture (McNeely and Wachtel 1988). By 9,000 years ago Java and Bali would have both been islands, with the Sundaland shoreline only 10-15 m below present levels. The human population would have grown, and a greater proportion would now be living on the coasts. Cave sites would still have been preferred, but some settlements would have been in open sites. Some form of boats probably allowed the people to explore offshore reefs and other coastal areas, but not to cross open seas. By 5,000 years ago the sea level would have exceeded and then returned to present levels. Certain sites would have become foci of activity and seafaring would have become both effective and competent as boat and navigational technology improved. As a result, exploitation of pelagic and benthic species could proceed (Dunn and Dunn 1977).

It was at about this time that Sampung Cave was occupied, and the remains from there, although not systematically documented, tell us much about the people's way of life, while raising interesting questions. In the top layer of the deposits were pot sherds, stone mortars, polished stone axes, and other utensils of a sort associated with the late Neolithic cultures. In the next layer down there were neither stone implements nor pot sherds, but there were scrapers and spatulas made from horn and bone. These may have been used to clean skins for clothes, although it is just as likely that bark cloth was worn. In the next layer down are stone tools including winged arrowheads, mortars and pot sherds. Interestingly, the bulk of the bones from the hunted animals are in the middle layer from which the hunting weapons are absent, and it is not clear why. Mammals were a major food, but the people's diet also included fish, molluscs, and fruit, some or all of which were cooked over fireplaces. The people also used animal parts as ornaments–palm civet teeth with holes drilled though them and pieces of mother-of-pearl have been found (van den Brink 1983).

These early people would have used spears and cord snares, set traps and camouflaged pits to catch their prey, and the large ground-living species such as wild boar, deer, and banteng were, not surprisingly, the most common prey. Interestingly, only remains of the head and legs have been found–the horns, ribs, backbone and trunk were not brought into or left in the cave. This applies also to the other mammals recorded. The skulls and long bones appear to have been smashed to extract the brain and marrow respectively. The complete absence of banteng horns or horn fragments, is hard to understand. After the banteng, the next most favourite (or simply most frequent) prey species appear to have been wild boar, deer, and then mouse deer. It is again strange that the endemic Javan pig *Sus verrucosus* is entirely unrepresented. The relative numbers of the different animal species found in the cave deposits unfortunately cannot be used to judge whether the hunters were selective or whether they simply caught and ate anything they could catch. This is largely because early excavations were not carried out with the rigorous techniques used in recent years, and it is likely that only a proportion of the large bones were collected and many of the smaller fragments were overlooked (van den Brink 1983).

Fires were probably used to clear vegetation and hence to improve visibility and increase herbage for the grazers. These fires were not necessarily extensive. When these early people were able to fell trees, however, they modified habitats, but this would have required great effort. It is possible to conclude, then, that the activities of the early horticulturalists did not involve large-scale forest clearance, and cultivation probably took place in small, fixed plots next to their dwellings (Bellwood 1980, 1985). Iron, introduced about 1,500 years ago, rather than stone axes, probably gave people their first efficient means of felling large trees, and it was at about that time that the process of forest clearance truly began. Some of the indigenous wildlife may have benefited from the greater areas of relatively succulent secondary growth in the 'edge habitats', but this would also have exposed them to greater hunting pressure.

It seems that the inhabitants of Sampung and Wajak were successful enough at exploiting wild forest produce not to need true agriculture until long after other peoples of the region had fully entered the Neolithic culture. There is no sign of agriculture or domestication of animals in any of the cave remains from Java. We do not know if these early people had begun to experiment with agriculture or animal husbandry as there are no archaeological remains to offer evidence. Indications of rice in cave deposits are generally clear enough and have been found in caves in Sulawesi dating from 6,000 years ago (Glover 1985). Settled agriculture and domesticated animals (dogs and pigs) probably made their appearance about 4,000 years ago, when the first major indications of forest disturbance are found. Even so, they would have had very little impact on the forest. Indeed, some animals such as deer, Javan rhinoceros, and banteng, would have benefited from limited clearance because they favour the succulent plants growing at the forest edge. These were all prey species, so the benefits to the beasts may not have been long-lived. Findings from other sites in Southeast Asia suggest that by 3,000 years ago people would have brought into cultivation plants such as candlenut *Aleurites moluccana* (Euph.), betel nut *Areca* (Palm.), kenari *Canarium* (Burs.), cucumber *Cucumis* (Cucu.), gourds *Lagenaria* (Cucu.), and betel pepper *Piper* (Pipe.) (Glover 1979). The people would also have dispersed trees by picking fruits in one place and discarding or voiding their seeds in another. An early form of agriculture would have been the accidental or deliberate sowing of tree seeds in the same area, thereby reducing the distance between fruit trees.

HISTORY

History, defined as past events recorded in writing, began on Java about two thousand years ago when two-way contact was established between its small coastal kingdoms and India (table 6.1). The Indians brought their version of Hinduism and Buddhism to Java (Koentjaraningrat 1985), while Bali was not affected until some years later. The most impressive historical remains must be the great Hindu-Buddhist shrines and temples of Central Java begun in the late eighth century. Borobudur, the greatest shrine of them all, is remarkable for all the carvings of 'ordinary' village scenes including the smoking of rats out of burrows in a rice field, marching elephants, fishing with cast nets, and ploughing fields (Lubis 1990; Miksic 1991). Most of these temples were built in what must have been an incredible one hundred-year period of intense creative energy and very high cost as acts of very conspicuous consumption (Christie 1983). Later, many smaller shrines and temples were also built, such as on Bukit Cinta, now a park above Rawa Pening, and sacred objects were carved and placed in spiritually significant locations. For example, statues were placed on the highest peak of Mt. Salak (Hartmann 1939), and numerous temples were built on Mt. Penanggungan over 500 years up to the 15th century, when the religious complexes on the peak of Mt. Argapura were also constructed. Many artefacts, such as statues, gold plates, and weapons have been found, such as around Rancakasumba and Ciparay on the Bandung plain where the Citarum was regarded as a sacred river in what would been the Galuh Kingdom during the period of Hindu influence from 500-1500 A.D. (Bennett and Bennett 1980).

Balinese culture is the result of at least eleven centuries of mixing indigenous elements with Hindu and Buddhist elements from Java. Javanese influence became very strong in the 14th century during Majapahit's golden age, and inscriptions were written in Javanese, not Balinese. Later, while the Muslim kingdom of eastern Java and the colonial powers had some influence, the culture was subject to very little interference until the beginning of this century. In 1906, the mass suicide of the major unsubjugated king, his family, relatives, court, and their families in the face of Dutch soldiers began another period of isolation as the result of a horrified response from Europe. To some extent, and although considerable changes have been wrought this century, the culture of Bali today is a window into the culture of Java (and indeed elsewhere in Southeast Asia) yesterday. Geertz (1980) has argued that the royal governments of Bali were concerned more with spectacle than with power and administration. Thus the many, varied, costly, and long ceremonies were expressions of the power of the gods and their earthly representatives (the kings) and are argued to have been ends in themselves, and the kingdoms merely devices for the enactment of mass ritual. While this is attractive as a hypothesis, it must be remembered that the various Balinese kingdoms often vied to advance themselves to the detriment (if not extinction) of the others and that elaborate ceremonies were only possible where there was relative confidence, peace, and wealth, that is power.

Environmental and Cultural Decline

During the Hindu period, grants of substantial areas of forested land were given by the kings to members of their family, court, priesthood, and

Table 6.1. Major historical events in Java and Bali.

	WEST	CENTRAL	EAST	BALI
Pre-700	**c.450:** King Purnawarman inscriptions in Jakarta, Bogor, part of Tarumanagara Kingdom along Citarum River.	**3rd century BC** merchants sailing to India where Java known as Yavadwipa (Barley Island) rich in grain and gold. Taking gold, camphor, gaharu, sandalwood, cloves, nutmeg, and pepper. **Early centuries AD:** much trade with India and mainland SE Asia. **c. 250:** trade started with China. **413** Chinese Buddhist monk stopped in Java, and few years later a Kashmiri prince. **449** China recognized three rulers in Indonesia, one of whom over Kingdom in Java known as He-hio-dan = Tarumanagara ? Then other kingdoms had contact.		Graves in W. Bali from several centuries BC-herding, farming people using bronze (some iron) to make implements and jewellery. Also stone sarcophagi, shaped like turtles, in mountains. Stone seats, altars, and big stones, connected with veneration of ancestor spirits that form core of pre-Javanized religious practices. Dongson-type 'Moon of Pejeng' kettledrum made, largest in SE Asia. Romano-Indian rouletted-ware pottery arrives from 1st century AD.
700		**695:** Dan-dan kingdom (Hindu) sends envoy to China. Inscription on slopes of Merbabu (Tuk Mas). **c.690-770:** Dieng temples built. **640-818:** Most important kingdom said to be Holing, probably in Central Java. Exported tortoise shell, yellow and white gold, rhino, and elephants. Buddhist. Mataram kingdom prosperous, with active agriculture and trade. Many temples. Queens sometimes reigned, large-scale war rare. Rivers main transport. Also ferries but overland travel not developed until 17th century. Borobudur has reliefs of boats. Buddhism in a small elite, Hinduism more adaptable to traditional religion. Caste systems not adopted so no great conflict.	Bits of East Java were perpheral parts of the kingdom of the Kedu and Prambanan plains.	

Table 6.1. (Continued.) Major historical events in Java and Bali.

	WEST	CENTRAL	EAST	BALI
		732: King Sanjaya inscription on Mt. Wukir 10 km from Borobudur. Sanjaya later said to be Mataram Prince. Panangkaran (his successor) became vassal of Sailendra which was supreme over Mataram.	**760:** Dinoyo inscription.	**732:** Bali said to have been conquered by King Sanjaya from Central Java but no proof.
		c.780-860: Candi Kalasan, Borobudur, Mendut, Loro Jonggrang built.		
800		**818:** Shepo becomes most important kingdom.		
		832: Sanjaya absorbs Sailendra dynasty by marriage.	Possibly even here or before Hindu-Buddhist kings were diverting Brantas waters to develop elaborate irrigation systems. Then they constructed temples and monuments.	
		860: Loro Jonggrang complete.		
		835-840: Balaputra (of Sailendra) rebelled, lost, went to Sumatra to take over Buddhist Sriwijaya empire, Mataram's foe.		
				882: Stone and copper inscriptions start here on statues, bronzes, caves, rock-cut temples, bathing places, close to rivers, ravines, volcanic mountains, and peaks.
900		**919:** No more inscriptions or monuments after this date here. No one knows why. Suggestions are volcanic disasters, epidemics, and Sriwijaya invasions.		

Table 6.1. Major historical events in Java and Bali.

	WEST	CENTRAL	EAST	BALI
900			**929:** King Sindok set up palace in Brantas valley, SW of Surabaya. After this many kingdoms emerged and became centre of Javanese culture for many centuries. Sindok very active and respected. Succeeded by daughter. **977:** Jalatunda. East Java invades Sumatra.	Close and peaceful bonds with E. Java, especially with Kediri (10th century–1222).
1000			**1016:** rebellion and capital destroyed. Surviving Crown Prince, Airlangga, son of Udayana then 16 years old, lived as hermit for four years. He then became ruler of a small coastal kingdom near Surabaya. By the time he was 49 (1049), he had recovered his ancestral domain. On his death he divided realm into two: Malang (Janggala) and SW Brantas near Kediri. Kediri became the dominant kingdom on Java until end of next century.	King Sindok's great-granddaughter married Bali king, establishing the root of Javanese influence in Bali culture. Their son was Udayana, father of Airlangga who ruled east Java and Bali. Cliff temples of Tampaksiring built. **1049:** Airlangga dies and Bali virtually independent again.

Table 6.1. *(Continued.)* .Major historical events in Java and Bali.

WEST	CENTRAL	EAST	BALI
1100		Up to 1200 many inscriptions indicating Kediri to be a thriving kingdom. But no records from Malang, Chinese records indicate that Java was one of the most prosperous countries in the region, but there are very few archaeological remains.	
		1222: Angrok, possibly an ambitious and brutal commoner, acquired power over the Malang area and unified the former kingdom of Airlangga. This new kingdom called Singasari, after his palace site just north of Malang. These were turbulent years. Several of the most beautiful E. Java temples (e.g., Kidal) are memorials to murdered rulers. By 1275, Singasari held power over major centres in Malay Peninsula, Sumatra and Java. Final ruler was Kertanegara (commemorated by the huge Buddha near the centre of Surabaya). He rejected Chinese demands to submit to Kublai Khan.	**1284:** Bali vanquished by Kertanegara; not sure whether colonized, but Javanese influences appeared in Balinese art from now on.
		1292: Kertanegara died. The following year Mongol fleet arrived, helped Kartanegara's son-in-law, Vijaya, to become ruler in civil war.	**1292:** Bali independent again.
		1293: Vijaya founded new kingdom, Majapahit ('Bitter Gourd') with capital at Trowulan, largest city in Indonesia before colonial days. Great centre of art and culture.	

Table 6.1. (Continued.) Major historical events in Java and Bali.

	WEST	CENTRAL	EAST	BALI
1300	Banten already a seaport of ancient Buddhist Sunda kingdom with capital Pajajaran near Bogor.		**1331-1364:** Gajah Mada Prime Minister. **1350-1389:** Hayam Wuruk ruled. During HW's reign, power claimed over area larger than present Indonesia. Kingdom short-lived once Hayam Wuruk died. Political splits and fights. **1360's:** Javanese Islamic converts in Majapahit heartlands.	**1334:** colonized by Majapahit's Prime Minister (Gajah Mada). Gajah Mada and nobles (aryas) went to subdue Bali's cruel king, and the island was divided between a Javanese vassal king and some of the aryas. New king's capital at Samprangan (Klungkung), Javanese court and culture. System of four castes started. Satriya ruled from Samprangan. Wesya were land-holding nobles. Those who did not participate fled to mountains and known as Bali Aga or Bali Mula. Mother Temple of Besakih begun.
1400	Maybe Cirebon became first Islamic city-state. Prince of Pajajaran made harbour master; he stopped sending tributes to now-Hindu capital and fought off their attacks. Sunda Kelapa important coastal outpost for inland kingdom of Pajajaran, competing for share of regional spice trade.		**1478:** Majapahit fell after Holy War led by Muslim ruler of Demak, although a part was extant when the Portuguese arrived in 1511. Most of Mt. Penanggungan temples built between 1400-1500.	**1460:** Gelgel (or Klungkung) became capital. **1480:** Possible immigration of Hindu-Buddhists from Java after Majapahit fall.

Table 6.1. *(Continued.)* Major historical events in Java and Bali.

	WEST	CENTRAL	EAST	BALI
1500	**1525:** Sunda Kelapa invaded by Demak. Sunan Mount Jati helped found Banten and Cirebon. Both flourished after 1546 demise of Demak. Banten became capital of new Islamic trading kingdom.	Demak the most prominent of the Islamic trading states. Sunan Mount Jati (one of the nine wali or venerated apostles of Islam on Java) helped its growth. Demak's influence reached S. Kalimantan, but failed to conquer Blambangan Kingdom in East Java. This contributed to its fall in 1546.	Giri (near Gresik) founded by another wali, Sunan Giri, from which Islam was taken to Lombok, Sulawesi, Kalimantan, Moluccas, and Ternate.	Bali conquers Lombok.
	1533: an inscribed stone in Bogor records ascendancy of King Surawisesa of Pajajaran, a powerful Hindu king who was alternately a vassal and rival to Majapahit rulers in East Java. Some scholars claim this was written 200 years earlier.	**1575:** Senopati founds Mataram kingdom - buried (1601) at Kota Gede.	Blambangan Kingdom invaded by Demak but repulsed. This, Java's last Hindu kingdom, Blambangan, ruled the eastern tip, capital Banyuwangi. Kudus and Giri eventually swamped by Mataram.	
	1578: Pajajaran fell			
	c.1590: Banten was the largest city in Southeast Asia and one of most famous ports in the world. Its population was about the same as Amsterdam's. European merchants establishing factories.			

Table 6.1. *(Continued.)* Major historical events in Java and Bali.

	WEST	CENTRAL	EAST	BALI
1600	**1603:** Dutch East India Company and British establish trading posts at Banten. **1611:** Dutch East India Company trading post at Jayakarta, named Batavia in 1619. **1628-29:** Sultan Agung attacked Batavia. **1651-82:** Agung's rule and Banten revived after 30 years of Dutch blockading Banten. Agung hired British and others to man his trading vessels. He constructed irrigation systems to increase rice production.	**1613-1646:** Sultan Agung, Senopati's grandson, caused Mataram to expand taking Tuban, Gresik, Surabaya, and Pasuruan. Introduced Javanese Islamic calendar. Buried at highest point of Imogiri. **1614** capital moved from Kota Gede to Kerta. Extended kingdom to most of Central and East Java. After his death the capital moved to Plered and Kartasura. **1671:** Agung's successor, Mangkurat I (who was hostile to Islam), died after a rebellion by Islamic leaders, and a Madurese prince helped by the Dutch who enthroned Mangkurat II. He moved the capital of Mataram to Kartasura west of Solo. Various wars ensued.	Blambangan attacked by Muslim/Mataram empire, and in 1639, after a two-year war, many Hindus lead away to slavery to prevent the growth of Balinese influence there.	Gelgel declines. Blambangan conquered in 1697 and made into a colony. Bali itself comprised several small autonomous districts. Descendants of Aryas became increasingly independent and formed separate kingdoms, but recognized Dewa Agung of Klungkung as the overlord.

Table 6.1. (Continued.) Major historical events in Java and Bali.

	WEST	CENTRAL	EAST	BALI
1700	1740: Chinese tradesmen, some specializing in sugar and arak, regarded by Dutch as a threat and deported to Sri Lanka. Tensions led to revolt, leading to death of thousands of Chinese prisoners. This spread along north coast and to Central Java where.... After this Chinese all had to live in Glodok outside and to SW of city walls. Chinese then bought from the Dutch the monopolies to deal in opium, run markets, and collect tolls. 1789: first coffee gardens on Tangkuban Prahu (Bandung).	....Pakubuwana sided with Chinese. Dutch fought and won, and were ceded entire 2.5 km of north coast and all of Java east of Pasuruan. New capital founded at Solo. 1755: After another war of succession, Mataram divided into two–Surakarta and Yogyakarta. 1757: Mangkunegaraan principality formed, dependent on Surakarta.	1767-77: Dutch fought long and bitter wars to effect the succession of eastern Java. Stamped out last Hindu kingdom (Blambangan) and depopulation of large areas, because of cholera and massive Ijen eruption in 1817. Osing (Hindu descendants) still present.	Karangasem most powerful kingdom, a position maintained through 19th century. By 19th century, Bali divided into nine great 'realms'.

Table 6.1. *(Continued.)* Major historical events in Java and Bali.

	WEST	CENTRAL	EAST	BALI
1800		**1811-16:** Raffles directs provisional British administration.		
		1815: Sultan of Banten abdicated and Banten became fishing village.		
		1808-11: Grote Postweg (highway) linking Bandung, Cirebon, Batavia. Netherlands assumes sovereignty.		
		1825-30: Java War against Diponegoro.		
		1846-49: Dutch expeditions to Bali.		
		1850s: Widespread development of plantations and large-scale agriculture. Expansion of money economy and consequent disruption of closed agrarian communities.		
		1879: Agrarian Law safeguarding indigenous rights to cultivated land.		
1900		**1900:** Ethical policy introduced.		
		1908: Bali conquered after suicides of resisting royal houses.		
		1914-1918: Nationalist sentiments increase.		
		1936: Indonesian language recognized by People's Council.		
		1945: Indonesia declares Independence.		

Figure 6.2. Candi Trowulan, possibly part of the capital of the Majapahit kingdom in the 14th century.
With permission of Kompas

army, on the condition that they were cleared and developed (*mbabad wana*). These areas developed into vassal states dependent on the main kingdom both economically and politically (Koentjaraningrat 1985). Some of the opened land was converted to agriculture, but apparently extensive areas were regularly burned to allow the rulers and their officers to hunt deer (p. 682). Some communities in the 14th century were exempted from the payment of taxes by virtue of their looking after (i.e., encouraging) alang-alang grasslands (Pigeaud 1962; Christie 1983). There were royal prerogatives on hunting, the use of alang-alang for thatch, and also the use of teak or *jati* trees ('jati' probably being an early Javanese word for valuable or durable) and turtles' eggs (Pigeaud 1962).

The capital of the Majapahit empire established in the 14th century has never been found, but evidence from early writings suggest it may have been in or near Trowulan, southwest of Surabaya (Edwin and Arifin 1988) (fig. 6.2). It has been suggested that the demise of the Majapahit empire may have been due, in part, to environmental reasons. Majapahit officials cleared large areas of land for homes and for timber, and the indiscriminate felling of trees led to damaging floods and the silting up of irrigation canals. This was aggravated by the practice of digging up the soil to make bricks, which continues today. The sedimentation was made worse still by lava and ash from the volcanoes to the south, with Mt. Kelud the most likely

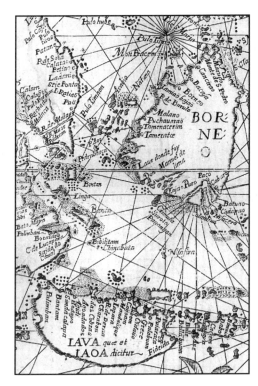

Figure 6.3. A map of Java and Bali from 400 years ago to illustrate that at this time Java was known primarily for its north-coast ports.

After van Linschoten 1598

candidate, and serious food shortages resulted. These problems were compounded by political instability and rivalry, however, and the empire crumbled in the early 15th century (Edwin and Arifin 1988).

In the 16th century the only part of Java that was known to the outside world was the north coast, in particular its ports (fig. 6.3). Banten, 75 km west of Jakarta, was the most successful and cosmopolitan of these, serving as an entrepôt for goods brought from China, India and Europe, and also as an exporter of highly valued local produce such as pepper and gold. The success continued into the 17th and 18th centuries when Dutch influence was growing, and at the peak of Banten's glory, ships from all over the world docked in the mouth of the Cibanten River, and boats could sail upstream to the agricultural hinterland.

So great were the rewards for the pepper traders that farmers in the sloping, well-drained soil of the Banten and Lampung areas were forced to grow a certain number of pepper vines per family, and to sell only to the Sultan's monopoly. As a result the farmers had to buy in rice from outside. Logically, the most intensive pepper-growing areas were along the Cibanten River, but soil erosion and garbage thrown into the lower reaches of the river aggravated natural rates of sedimentation, as did the digging of clay from riverbanks for building bricks. Access to the interior became more and more difficult, the productivity of the pepper fields decreased, and coastal sedimentation prevented the ships from getting closer than one kilometre to the port. As the quality of the environment declined, so did Banten's role as an international port and centre of commerce, a process hastened by the Dutch who had been trying to divert trade to their own centre at Batavia since the late 17th century. Settlement expansion at Banten continued, but this caused the Cibanten water quality to decrease still further, and in 1787 local people and Dutch soldiers died from the contamination. Batavia, later Jakarta, took over (Nurhadi and Azul 1987).

The blame for the fall of Banten should not be directed towards the pepper farmers or the clay diggers, but rather towards the economic system and the greed of the rulers whose policies and actions overstretched the environment. There is no evidence, however, that there was any awareness of the cause and effect that looking back is now dramatically clear. This contrasts with the present situation in which we have the understanding and knowledge necessary to prevent ecological degradation, impoverishment, and decay.

History of Deforestation

Java. Virtually all of Java and Bali would once have had some form of forest cover with the type determined by altitude, seasonality, and to some extent by soils. People have lived on Java and perhaps Bali for rather less than one million years (p. 309), and human impact on the forests and the flora probably began as soon as cutting tools and fire were available (p. 314). The first major loss of natural forest probably occurred after teak was introduced during the early years of Hindu contact (200-400 A.D.) (p. 317). By 1000 A.D. there may have been as much as 1.5 million ha of teak (with the loss of an equivalent area of lowland forest on volcanic, alluvial, and limestone soils). The forests are believed to have been managed, and in about 900 A.D .the post of 'Hunting Master' is mentioned in Javanese writings; it is believed this position was also concerned with forestry activities.

It seems that primary forest was not cleared and planted directly with teak, rather food crops were grown on cleared land, and only when the soil was exhausted was teak planted. Since teak prefers drier climates, the planting was concentrated in Central and East Java (Altona 1924, 1926). The litter on the floor of a teak forest is highly inflammable, aggravated by the dry climates in which it grows, but the adult trees are fire resistant; if other tree species are present, they, or their saplings, are likely to perish in the fires, and this will lead to dominance of the teak.

By the time the Hindu-Buddhist temples of Central Java were being built (p. 318), appreciable areas of forest had probably been cleared on the alluvial coastal plains. Mountains were ascended by Hindu pilgrims, and it has been suggested that the fire-climax *Casuarina junghuhniana* forests circling most of East Java's mountains may have had their origin in fires accidentally started by such pilgrims centuries ago (Smiet 1990b). Irrigated rice culture was introduced over 1,000 years ago and was probably confined to the lower slopes of the numerous volcanoes and limestone hills.

Before Dutch control began, teak was used for building the indigenous sailing fleet (Boomgaard 1988, 1991). During the early days of Dutch influence, teak was much exploited for a variety of purposes, but not in a sustainable fashion. Little extra land was put under this tree because it seemed to be so abundant. Major change began in 1830 when the Dutch administration imposed the '*cultuurstelsel*' which forced farmers to grow export crops among the food crops on communal ground (often forest), in a *tumpangsari*-type system (p. 619). Instead of having enough food and some surplus and goods for sale, the people had to grow cash crops at the expense of food to satisfy the desires of people halfway round the globe for certain commodities such as cloves, sugar, coffee, and tea. The human population grew rapidly (p. 27) and the land became crowded, forcing farmers to develop ever more intensive forms of agriculture, and inhibiting the development of indigenous capitalists and urban entrepreneurs, a process known as 'agricultural involution' (Geertz 1963). The production of indigo and sugar required considerable quantities of fuelwood, which further depleted the mixed and teak forests. Junghuhn (1854) identified overexploitation for fuelwood and conversion to coffee plantations as the primary causes of deforestation in the uplands. He had travelled all over Java, and he found that some mountains were clear of forest from the lowlands to the summit (e.g., Merbabu, Sindoro, and Sumbing). Fires were regularly started in the montane forests of East Java to encourage grassland for hunting (Smiet 1990b). To balance this, Junghuhn also described large areas of undisturbed lowland and montane forests. The qualitative descriptions of the Javan landscape in the last century also give vivid impressions of just how much things have changed. For example, the Pengelengan plateau south of Bandung and between Mts. Malabar, Tilu, and Wayang was forested until the middle of last century when the flatter areas were cleared for plantations of coffee and tea. In 1880 a traveller reported large herds of banteng and packs of wild dogs on the forested southern slopes of Mt. Malabar (Forbes 1885, 1989), and coffee plantations had to have ditches and fences to protect the crops from the depredations of rhinos (Bennett and Bennett 1980).

By the time the cultuurstelsel was abolished in 1870, and new private companies were allowed to manage plantations, more than 300,000 ha of Java was under coffee. Since coffee grows best in a seasonal climate between

1,000-1,700 m, the remaining upland forests in East and Central Java came under heavy, relentless pressure from these companies. Around 1885, however, coffee leaf blight began its ruinous advance through Java, and over the next 50 years (during which Brazilian competition increased) the area under coffee was reduced to 98,000 ha even though a resistant strain was introduced in 1900 (Purseglove 1968). The lifting of the cultuurstelsel also had the effect that communal village lands under coffee were converted to dryland farming. The occupation of these upland areas, at least in Central and East Java, led to the first complaints from those in the lowlands of diminishing water flow in the rivers. Erosion was ño great problem because the human population was not so great and the dryland farming alternated with fallow periods. The northern half of the island, comprising malaria-infested alluvial coastal plains, remained uncultivated because of the technical problems of irrigating such flat land. Between the 1850s and World War I, however, virtually all these lands were brought under cultivation.

Before 1850, deforestation occurred without Government directive (or concern), but over the following 80 years conversion of forest lands to agriculture and plantations was vigorously promoted. Diminishing forest areas led to diminishing wildlife with some species becoming 'pests' as they impinged on human activities in their desperate search for food and a place to live. It seems that between 1898 and 1937, some 22,000 km^2 of natural forest were lost, and one of the major uses of timber was in the building of the extensive railway network. Control of forest conversion began to be exercised between 1928 and 1937 after which, despite relatively high population growth, the loss of forest has not been in any way commensurate. This is partly the result of firm government resolve with clear enforcement in some areas, and partly the limited access and extreme steepness of much of the remaining forest land, which made it unsuitable for agriculture (Smiet 1990b), and largely a measure of just how little forest remains. There is gradual attrition and marked degradation of almost all remaining forests (e.g., Smiet et al. 1989). Many of the activities causing the continued loss of quality and quantity are traditional, but the intensity is now unsustainable and therefore not consistent with government policy.

Tracking the changing forest areas is far from easy (Durand 1989), at least in part because certain of the older maps are not particularly reliable for assessing forest cover. For example the 1912 map of forest areas in Java and Bali produced by the Batavia Topographical Office records no forest on the Ujung Kulon or Blambangan peninsulas. However, it seems that just before World War II about 23% of Java still had a cover of natural forest (Seidensticker 1987). During World War II there was widespread and uncontrolled deforestation, and afterwards many of the remaining suitable tracts of forest were converted to teak plantations. More forest was lost

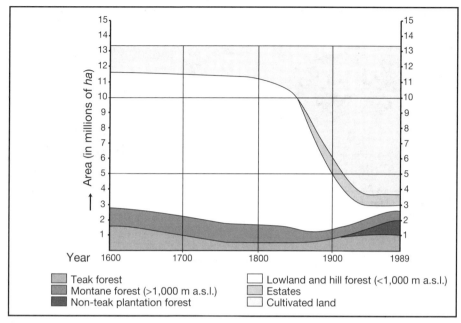

Figure 6.4. Decrease of forest land and changes of land use on Java, 1600 to 1990.
After Smiet 1990b

during the difficult years of the 1940s and 1950s, and by 1973 the natural forests covered only 11% of Java (Donner 1987). By 1990 this had dwindled still further to a mere 7% or 0.96 million ha (FAO 1990) (figs. 6.4-7). This more than 90% decrease in forest cover is entirely attributable to human - impacts of volcanic eruptions, earthquakes, and strong winds are negligible (Smiet 1990b). In both proportional and absolute terms it is the species-rich lowland forest that has suffered most. Most of the relatively large remaining lowland forest areas is almost all in three distinct areas of little agricultural value but major conservation value: Ujung Kulon (where the Krakatau explosion of 1883 wiped out human settlements), Baluran (where there is a shortage of water), and Alas Purwo (with poor soils and frequent water shortages).

The pattern, if not the time scale, of forest loss on Java is similar to that of the loss in England, the progress of which over the last 5,000 years can be described with some confidence (fig. 6.8). Thus 80% of the original forest (wildwood) cover had been lost by two thousand years ago, and for the last 700 years less than 10% has remained. Severe fragmentation of forests has also occurred, such that today 83% of the ancient woodland sites

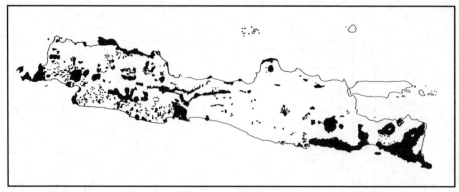

Figure 6.5. Natural forest cover on Java in 1891.

After Koorders 1912

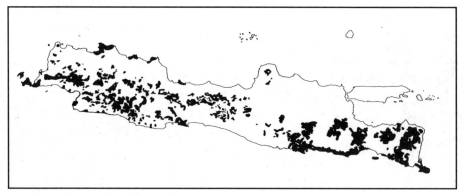

Figure 6.6. Natural forest cover on Java in about 1963.

After van Steenis and Schippers-Lammerste 1965

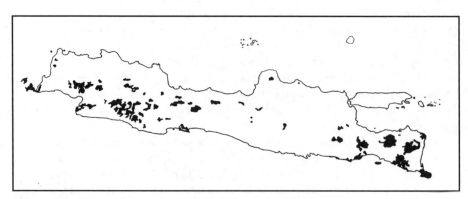

Figure 6.7. Natural forest cover on Java in about 1987.

After RePPProT 1989

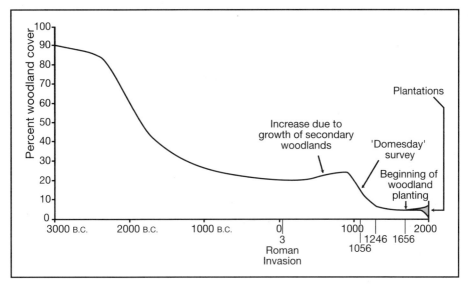

Figure 6.8. Decrease in forest land and changes of land use in England over the last 5,000 years.

After J.W. Spencer pers. comm.

(sites that have had forest cover since at least 1600) are less than 20 ha in extent, and only 2% are over 100 ha (Spencer and Kirby 1992). This is of even greater concern on Java and Bali because the biological impact of forest loss here is far greater than in Britain because of the enormously greater biological diversity. Sadly, many of the remaining forest areas in Java and Bali still receive inadequate protection, and the relentless degradation of Javan forests has been reported by many over the last 25 years (Soerianegara 1970; Soeriaatmadja and Rahman 1972; Wind and Soesilo 1978; Lembaga Ekologi Unpad 1980; FAO 1980, 1982; Rosanto and Priatna 1982; MacKinnon et al. 1982, Harun and Tantra 1984; van der Mijn 1988; Yusuf et al. 1988; Smiet et al. 1989; Manaerts et al. 1990; Smiet 1990, 1992).

The most detailed and comprehensive analyses of forest loss concerns the Kali Konto sub-watershed west of Malang (Smiet 1992). Almost all (93%) of the forests in the Kali Konto area show the effects of profound and long-term impact of human activities. About 8% of the forest is or has been subject to frequent burning with resulting *Casuarina* (Casu.) and *Engelhardia* (Jugl.) communities. Tree cutting is the major form of disturbance, and in over half the original area this has led to the formation of grassy shrubland. The 7% that remains relatively undisturbed is remote and difficult to reach (table 6.2).

Firewood collecting is a potent factor in forest loss on Java, and has been so since the start of this century (Ross 1985; Gillis 1988). Partly in response to this, the Government instituted a heavy subsidy on kerosene that was hoped would address income redistribution and forest conservation. While ostensibly a rational measure, it missed the point that kerosene was used for lighting, not cooking, in almost all households, at least in part because few families (15%) owned stoves. This subsidy cost over US$5 billion in the three years prior to 1981 representing about $80,000 for each forested hectare 'protected' from cutting - an enormous contrast to what can be spent by the Department of Forestry. One alternative was substituting liquefied petroleum gas (of which Indonesia has an abundance) for firewood. There would again be a problem of stoves, but the 12 million rural Javan households could all have been provided with a stove in 1980 for a one-time outlay of US$650 million, with a subsequent saving of billions of dollars (Gillis 1988).

Bali. Change came later to Bali. Without European influence, irrigated rice agriculture was primarily on the southern, wet slopes of the central volcanoes. Dutch-promoted cultivation for export was concentrated on the northern, dry slopes only from the late 1880s. Wholesale colonial control could not be wrought until about 1910 but, even then, progress in road building and other developments was slow. The loss of forest during this century has been due largely to the introduction of coffee, clove, and coconut plantations and also to increased demands for quicklime, bricks and tiles all of which require fuel wood for their manufacture (fig. 6.9) (Kanwil. Kehutanan Bali 1991). In addition, the eruptions of Mts. Agung and Batur destroyed areas of forest around their peaks (Foley 1987).

ETHNIC GROUPS

The island of Java has just two native ethnic groups, the Javanese and Sun-

Table 6.2. Areas of six forest disturbance categories in the Kali Konto basin, East Java.

Class	Area (ha)	% of total area
Virtually undisturbed	312	2.4
Slightly disturbed	603	4.7
Moderately disturbed	1,683	13.0
Heavily disturbed	2,407	18.6
Affected by burning	1,077	8.3
Converted to shrubland	6,860	53.0
Total	12,942	100.0

Source: Smiet (1992).

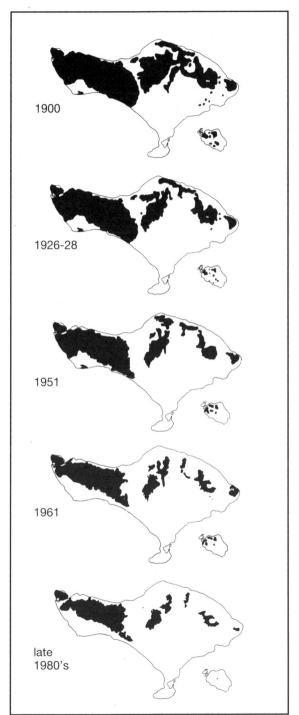

1900

1926-28

1951

1961

late
1980's

Figure 6.9. Forest cover on Bali 1900 (estimated from early maps and literature), 1926-28, 1951, 1961, and late 1980s.

After US Army Map Service 1943 copied from earlier 1926 Dutch maps; after Sandy 1964; after Sandy 1964; after RePPProT 1989

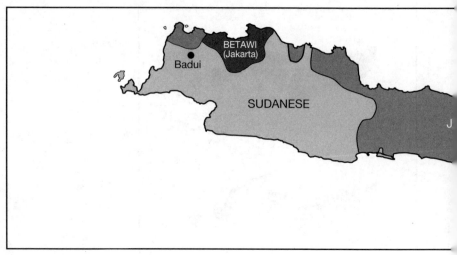

Figure 6.10. Distribution of the main ethnic groups on Java.
After Hefner 1991

danese. Two relict groups also persist, the Baduy derived from the Sundanese, and the Tenggerese from the Javanese. In addition, since the 18th century many of the Madurese from the neighbouring eastern island of Madura have migrated to East Java (fig. 6.10). These five groups together with the Balinese are described in more detail below. To these could be added the Betawi or Jakarta people, but they are identified more by the use of the Betawi language rather than by cultural background or traditions.

Javanese

Although the Javanese represent a single ethnic group, by convention they are generally divided into three areas. The kejawen area, extending from Banyumas to Blitar, is the heartland of what most people regard as the 'real' Javanese culture. Here, surrounded by beautiful sawah made fertile by a series of active volcanoes, are the homes of aristocratic dynasties that have developed and encouraged a wide range of artistic endeavours for over 1,000 years. The royal domination has generated almost an obsession with status and rank, and different levels of language are used to elevate, put down, or befriend the person being spoken to. The majority religion is Islam, although two distinct strands are encountered: the orthodox, and those who incorporate Javanese customs, rites, and mysticism.

The second area, known as pesisir, stretches across the northern plain

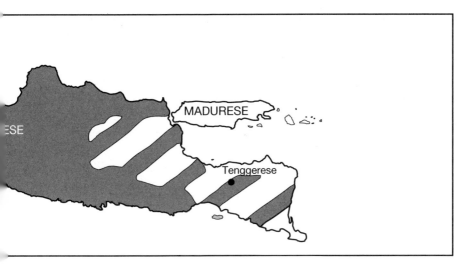

from Cirebon to Surabaya and Pasuruan. The Javanese of the pesisir are urban rather than rural, and perhaps not as refined or genteel as their southern kin, having had much more to do with trade than with the arts. The character differences even extend to clothing: with subdued almost sombre, colours in the royal courts to brazen, primary colours in the north. They are uniformly orthodox in the form of Islam practiced, and the small traders have taken this with them both through Java and to the rest of the archipelago.

The third Javanese area is Blambangan in the extreme east of Java. This was a small Hindu principality which resisted repeated attempts to force acceptance of Islam for 200 years after the fall of Majapahit. The warring period, dramatic volcanic activity, and disease had together decimated the population of Blambangan by the beginning of last century. The native Islamized people of Blambangan, known as Osing, had some unique cultural traditions, but these have been largely diluted by immigrants from Madura in the north and kejawen in the south. Only among the small Tengger group (p. 339) are some of the old ways of Blambangan preserved (Hefner 1985, 1990, 1991; Dove 1993).

Sundanese

The Sundanese live almost exclusively in West Java and are the second most numerous ethnic group in the country. Their language is more closely related to the trading languages of Minang (W. Sumatran) and Malay than to the neighbouring Javanese, but they have adopted the levels of speech that reflect actual or perceived rank, or respect. Other Javanese 'inventions' such as gamelan music, shadow plays, certain dance types, and terraced irrigated rice cultivation, have also been absorbed to the west and adapted. Today the Sunda culture differs from the Javanese by being more clearly Islamic and by paying less attention to social hierarchy. Sundanese are perhaps more independent by nature because, even last century, they were predominantly dry-field shifting cultivators in the mountainous, forested interior, living in small villages. By being so dispersed they were less susceptible to authoritarian, central control (Tjondronegoro 1984). Foci of traditional Sundanese culture can be found around Kampung Naga (fig. 6.11), and among the Kasepuhan people living in and around Mt. Halimun National Park (Adimihardja 1992; Adimihardja and Iskandar 1993).

Balinese

The history of the Balinese is related, as one might expect to that of the Javanese. There has been contact for over 1,000 years, and periods each of peace, war, colonial subjugation, and wielder of colonial power have taken their various turns. There is a story (without corroborating evidence) that it was the removal of a cruel Balinese king by the famous Majapahit general Gajah Mada in the 14th century which led to the introduction of a Javanese court and culture, including a society separated into four castes (Baum 1937; Covarrubias 1937; Eiseman 1989; Geertz 1991). Some groups of the native Balinese, known as the Bali Aga, objected to these changes and have kept themselves separate from mainstream Bali culture ever since.

The Hindu culture has changed and adapted over time, but the adherence to religion and its practices is still great. The round of offerings and temple ceremonies dominates the lives of most Balinese, and on most days there are traffic jams in some part of urban, modern Denpasar as followers process to a nearby temple. Music, dance, drama, shadow plays, weaving, painting, and carving, as art forms are still strong, and proficiency in at least one area is still pursued by the young (Vickers 1989; Eiseman 1990; Jensen and Suryani 1992).

Although sawah dominates the most popular views of Bali, only 20% of the island is given over to this land use. Nearly 30% comprises non-irrigated fields, and 17% is covered by gardens of cash crops such as coffee (RePP-ProT 1989).

Madurese

There are about six million Madurese people, but only half of them live on

Figure 6.11. Kampung Naga, a traditional Sundanese village.
By A.J. Whitten

Madura or the Sapudi and Kangean Islands to the east. The remainder live in the urban centres of Surabaya, Malang, and in the northeast part of East Java. More than 80% of those living on Madura live in the coastal and inland villages, where they engage in dryland agriculture, trade, cargo sailing, fishing, and salt making. The only major commercial crops grown are tobacco and various fruits.

Given the close proximity of Madura to the rest of East Java, it is not surprising that the histories, languages, traditions, and lifestyles of their ruling classes were closely related. Despite the similarities at the top of the social hierarchies, however, the 'ordinary' Madurese and Javanese are markedly different. For example, the Madurese have a much stricter moral code, and are far more brazen than the rather subdued Javanese, particularly when family honour is threatened, and a significant percentage live in extended families which gives their members a great deal of security. Life for many island Madurese is still dominated by the traditional races of paired bulls pulling a sledge and jockey, and to own a pair of good racing bulls is seen as one of life's ideals, despite the costs incurred of using special masseurs, herbalists, and magicians to improve the chances of success.

Tenggerese

High on the slopes of the massive Mt. Semeru and Tengger caldera within which squats the ever-popular Bromo volcano, live the 40,000 Tengger people (Hefner 1985), one of the smallest ethnic groups in Indonesia, revered for their freedom from dishonesty, jealousy, and quarrels (MacLeish 1971) (fig. 6.12). These are descendants of the Hindu Blambangan principality which held out against the rise of Islam until the late 18th century. Most of the people converted, but the Tenggerese forswore Islam, finding defence and security in the rugged and remote mountain slopes, to form distinctive *mandala* communities (see below). Many of these people today, descended mainly from commoners, retain their Hindu beliefs but, unlike the people in Bali, they have no court, literature or other rich traditions of art, although there are important objects such as bronze bells and a ruined water temple on the north slopes of the Tengger massif. They do, however, have rich folk beliefs, and their most famous ceremony is the annual Kasodo on the crater rim of Mt. Bromo which is now witnessed by visitors from all over the world. Their folk origins are also probably the reason they do not use levels of language that depend on the status of the person being addressed and, at least until recently, have been concerned with equality, democracy, and community (Hefner 1990).

There have been some critical changes over the last 30 years as excessive and inappropriate cultivation of the steep slopes have led to dramatic erosion. This has resulted in the reduction of the average farm size and a shift away from subsistence crops, notably maize, and towards cash crops such as cabbages, carrots, potatoes, and onions for the urban masses, which has aggravated the problems of soil erosion (p. 144) (Hefner 1990).

Baduy

The Baduy number only about 5,000 living in the rugged and mountainous south of Banten regency in West Java, just 150 km from Jakarta. Their population size in no way reflects their significance, the interest taken in them, or the degree of frustration felt by government officials charged with encouraging them to develop. The Sundanese living nearby hold them in great respect and indeed fear, because the Baduy are said to be masters of magic and to be watched over by phantom tigers.

A Baduy village has scarcely changed since written observations were first made, for it is believed that what was good enough for their ancestors should be good enough for them. Thus modern accoutrements of living are banned, and their agriculture is classic swidden cultivation (Iskandar 1992), which has all but disappeared elsewhere in Java although, because of land pressures from inside and outside their area, it is practiced on much steeper slopes now than it was in the past. The Baduy people probably arose from one of the ascetic religious communities or *mandala* that lived at sacred locations such as mountaintops or river sources during the Hindu-Buddhist period. These communities were charged to maintain the imperilled balance between the earthly and spiritual worlds, and thus

Figure 6.12. Tenggerese farmer returning to his village from market across the desolate Sand Sea of the Tengger Caldera.
By A.J. Whitten

had to lead a very regimented and pure life, and their importance and independence were reflected in their freedom from taxation. Changing their way of life would threaten the natural harmony with the consequent threat of floods, landslides, and other disasters befalling their kin in the lowlands. Because of their crucial task they maintained close ties with the princes who depended on harmony, peace, and tributes for their existence (Bakels 1988). In the middle of the 17th century, the son of a Baduy leader left the forests and sought shelter at the palace of the Sultan of Banten. This boy grew close to the sultan, eventually marrying one of his daughters, and thereby founded the aristocratic and influential Djajadiningrat family line (Epton 1956).

The heavy burden felt by the Baduy for the integrity of a principality has now been transferred to a feeling of responsibility for the entire nation. This is reflected in their brazen behaviour of walking barefoot to Jakarta to make representation directly to Cabinet Ministers, rather than using the more usual government channels (van Tricht 1929; Epton 1956; Purnomohadi 1985; Djoewisno 1987; Garna 1987; Bakels 1988). After direct representation to the President (their reigning 'prince'), their 5,000 ha (just 7 x 7 km) territory has now been marked with concrete boundary posts to protect it against the encroaching modern world, and they are primarily responsible for the forest and other resources within it. In this respect the Baduy has achieved what no other minority ethnic group has even been allowed to consider elsewhere in the archipelago. The territory is a fraction of the area the Baduy used to use, however, and so the population and land use must be kept in check if there is not to be a spiral of degradation.

Growing Modernism and Consumerism

There is a sense in which the different cultures of Java are becoming swamped given the pervasive and inevitable homogenization that accompanies mass education and mass media. Most of Java's middle-class urbanites in fact have little interest in their traditional cultures as is evidenced by the great popularity and cultural poverty of the independent television channels. Traditional dances and plays are living arts in remote villages, but are simply 'shows' in most cities, competing unfavourably with rock or *dangdut* concerts. Exceptions to this are, of course Yogyakarta and Surakarta. Attention is, in the main, attracted towards globalized consumerism, little of which probably serves the purchasers or the environment in significant ways. This general trend is not (yet) rife in Bali, however, where the whole range of arts receives much more participation and patronage, and where matters spiritual make greater demands on people's time.

Part C

Ecosystems

An ecosystem is composed of interdependent and interacting ecological components set in, or proceeding towards, a dynamic balance. The human component is just as important as the other components, and it has the capacity to alter both the structure and balance of an ecosystem. The result does not necessarily maintain or replace the capacity or function of the original ecosystem, however, and this is the root of ecological problems. Some of the basic ecological information that follows relies heavily on *The Ecology of Sumatra* (Anwar et al. 1984; Whitten et al. 1987c) and *The Ecology of Sulawesi* (Whitten et al. 1987a, b) where it is applicable to the situation prevailing in Java and Bali. Other information specific to Java and Bali is woven into this.

Java and Bali could become a model of the integration of ecology and economics to undergird sustainable development, a model which could be used as a template for other islands. If that model cannot be built, the tragedy will be mourned across the country. To avoid this, an understanding of ecological benefits and constraints must become more widely understood. This chapter thus examines the components and the interactions within the major natural and artificial ecosystems on Java and Bali.

Chapter Seven

Coastal Ecosystems

The ecosystems dealt with in this chapter are coral reefs, seagrass meadows, and open sandy and rocky shores. Mangrove forests are covered in the next chapter. Some of these are dealt with in greater detail in Tomascik et al. (in prep.).

CORAL REEFS

Coral reefs are of considerable importance because they protect coasts from erosion, and because many of the larger animals and plants associated with them have major direct and indirect economic value. Coral reefs are enormously rich in species which interact with each other to form extremely complex and little-understood communities. Their exquisite beauty can be marketed and has made them extremely, and increasingly, valuable to the tourist industry; indeed, Bali and the Thousand Islands promote the excitement, wonder, and fulfilment of diving or snorkelling over reefs as one of their major attractions for domestic and foreign tourists. The diving off Bali is world class, and thousands of people, including many internationally respected experts, go there each year simply to dive, having no interest in other aspects of Bali or its culture (Muller 1992).

Structure and Formation

Coral reefs are made up of the compacted and cemented skeletons and skeletal sediment of sedentary organisms which are then smothered by other organisms (p. 349). The outermost layer of a coral reef is living tissue comprising primarily scleractinian (hard) corals and algae with tissues impregnated with limestone. Coral reefs form in warm water (generally above 22° C) that is relatively clear and illuminated by the sun, and has a near-to-normal seawater salinity. Reefs are poor or absent around coasts near large rivermouths because they are intolerant of lowered salinity and high sediment loads.

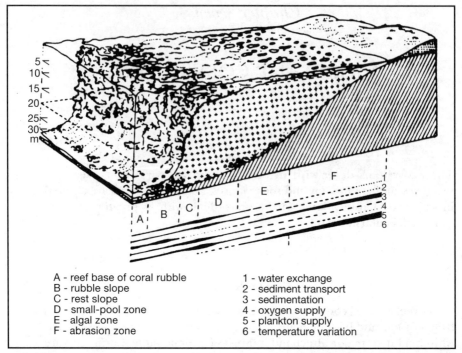

A - reef base of coral rubble
B - rubble slope
C - rest slope
D - small-pool zone
E - algal zone
F - abrasion zone

1 - water exchange
2 - sediment transport
3 - sedimentation
4 - oxygen supply
5 - plankton supply
6 - temperature variation

Figure 7.1. The zones of, and changes across, a typical fringing reef which grows away from the land. Zones D, E, and F are together sometimes known as the reef flat, reef platform, or lagoon.

After Barnes and Hughes 1982

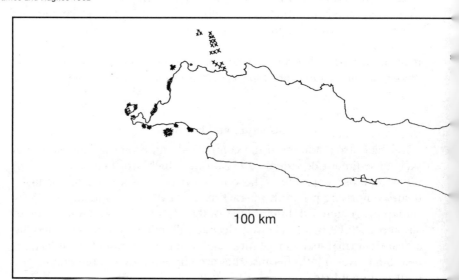

100 km

Figure 7.2. Coral reefs around Java and Bali.

Adapted from Salm and Halim 1984

Since coral is dependent on light, the depth to which it can grow increases with its distance from major land masses where water is more turbid. This intolerance of turbid or brackish water results in corals growing gradually away from the land, leaving shallow lagoons behind them floored with sand composed of broken, dead coral skeletons. The different structures of the steep reef edge and the gently sloping reef flat or reef platform, result in physical changes across the reefs (fig. 7.1).

The coral reefs around Java and Bali (fig. 7.2) are of three types: fringing reefs, barrier reefs, and patch reefs. Fringing reefs are formed close to the shore on rocky coastlines. If water depth remains the same or decreases over time, growth of the reef is entirely seawards, vertical limits being set by the level of low spring tides and the depth to which light can penetrate. Between the shore and the crest of a fringing reef there is usually a shallow reef flat where coral growth is poorer because of reduced water circulation, higher sediment load, and slightly lower salinity due to run-off from the shore. Barrier reefs occur where the coast has subsided, and only the seaward reefs have been able to continue their growth. Barrier reefs are therefore separated from the shore by a lagoon. Patch reefs grow on isolated patches of substrate none, of which is exposed above the sea.

Surveys of a number of coral reefs around Bali Barat National Park were conducted in 1980. The reefs close to shallow, mangrove-lined and muddy bays were relatively depauperate, with poor reef structure as a result of the more turbid water. The reefs examined showed a range of physical profiles,

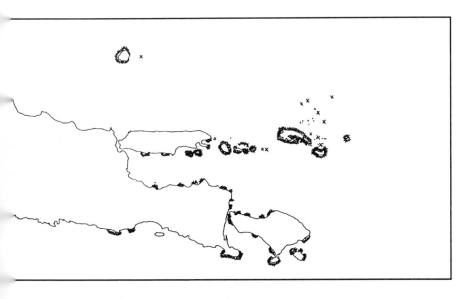

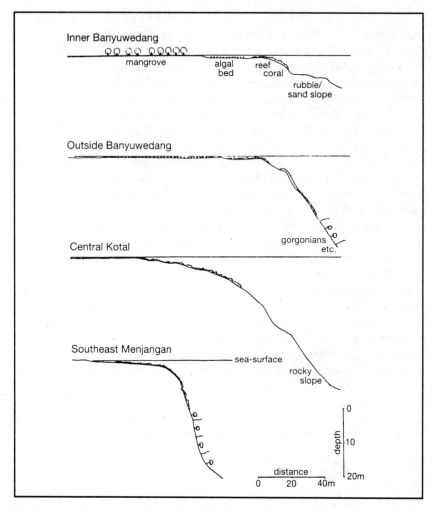

Figure 7.3. Physical profiles of four reef areas in Bali Barat National Park.
After Polunin et al. 1983

from gentle slopes with surge channels, to steep and unindented slopes indicating prevailing gentle water movements (fig. 7.3). The direction of the prevailing wind is very important in determining the shapes and compositions of coral reefs (Umbgrove 1928a, b). The degree of exposure, depth of water and low tide, sediment load, temperature and salinity changes, and other physical factors control the type and diversity of benthic communities. This was first recognized 60 years ago in the Thousand Islands (Verwey 1931b), and has subsequently been confirmed for many other reefs in Indonesia and around the world (Sukarno 1975; Kusyanto 1979).

Various types of coral survey have been conducted around Java and Bali and many of these are listed in Appendix 3.

Fauna

Invertebrates. Although the corals are the most obvious organisms of a coral reef, the living 'skin' of a coral reef comprises a wide range of sessile and mobile invertebrates from at least eight phyla (table 7.1).

Coral. An individual coral animal starts life as a small planktonic larva which comes to rest on a suitable substrate and metamorphoses into a polyp. This polyp begins dividing and forming genetically identical polyps next to it, and this process continues until the coral dies. Within the 'skin' of the polyps are small, yellowish brown granules, which are small photosynthetic organisms from the phylum Dinophyta (dinoflagellates), other members of which are found in the marine phytoplankton. Only a single species, *Symbiodinium microadriaticum*, is known to live symbiotically with corals. It absorbs waste products produced by the host polyp, converting the phosphates and nitrates into protein and, with energy from the sun, the

Table 7.1. Simplified classification of the major invertebrates occurring in the benthos of coral reefs. The phylum Cnidaria used to be known as Coelenterata.

Phylum	Class	Order	Vernacular name
Sarcodina	Granuloreticulosea	Foraminiferida	forams, foraminifera
Porifera			sponges
Cnidaria	Hydrozoa	Rhizostomeae	jellyfish
	Cubozoa	Cubomedusae	sea wasps
	Alcyonaria	Gorgonacea	sea whips, sea fans, sea feathers
		Alcyonacea	soft corals
	Zoantharia	Scleractinia	true corals
		Actinaria	sea anemones
	Ceriantipatharia	Antipatharia	black or thorn corals
Annelida	Polychaeta		marine worms
Mollusca	Bivalvia		clams, cockles, mussels
	Gastropoda		limpets, snails, sea slugs
Bryozoa	Gymnolaemata		marine bryozoans
Echinodermata	Crinoidea		sea lilies and featherstars
	Asteroidea		starfishes
	Ophiuroidea		brittlestars and basketstars
	Echinoidea		sea urchins
	Holothuroidea		sea cucumbers
Chordata	Ascidiacea		sea squirts

Classification from Barnes (1984)

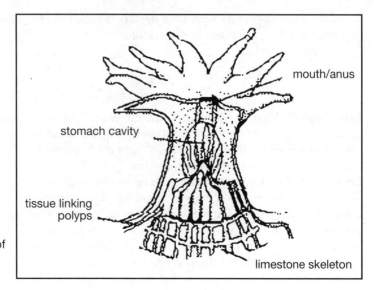

Figure 7.4.
Cross section of
a coral polyp.
After Henrey 1982

carbon dioxide into carbohydrates. Their waste product, oxygen, is used in turn by the polyp for its respiration. Coral polyps secrete their external skeleton of limestone from their bases but each polyp is connected to its neighbours by strands of tissue (fig. 7.4).

Corals mature after three to eight years and tend to breed seasonally and often simultaneously in a certain area. It used to be thought that most scleractinian corals were viviparous, that is brooding fertilized eggs within the polyp, often releasing larvae throughout the year. Careful observations in northeast Australia revealed, however, that most corals release gametes simultaneously at night a few days after full moons (Harrison et al. 1984). Why should corals breed simultaneously? If breeding were not simultaneous, there would be great risks involved in releasing eggs or sperm, both of which have limited periods of viability, into the sea. The chances of egg meeting sperm if spawning were random would be very low. In addition, if only a few eggs were in the water at any given time, it is likely that a relatively large proportion of the eggs would be eaten both by opportunistic or by specialist predators. Thus simultaneous spawning gives the maximum opportunity for successful fertilization, minimum chance of any particular egg, fertilized or not, being eaten, and prevents any animal specializing on coral eggs as food.

The first major study of the coral ecology in the Thousand Islands was in 1927 (Umbgrove 1928b). It was found that the composition of the coral fauna depends on the nature of the bottom substance (coral sand or shingle), water depth, wave movement, and so forth. In addition, the relatively sheltered southern reefs of P. Pari had a greater diversity of corals than the exposed northern reefs. Also, the different parts of the reefs supported a different fauna. On a shallow reef flat with a sandy bottom, *Montipora ramosa* with *Psammocora* corals, *Halimeda* green algae, and abundant sea cucumbers could be expected. On shingle walls, red algae thrive and many branched corals (e.g., *Hydrophora cresa*, *Millepora dichotoma*) with *Acropora* species at the top where they were exposed at low tides. On the outer slope of the shingle wall, *Montipora foliosa* is the characteristic coral (Umbgrove 1928b). *Acropora* is tolerant of heavy wave surge and also has a high growth rate, the ability to grow from fragments and tends to occur in high energy areas together with encrusting and branching *Montipora*, *Porites*, *Pavona*, and faviid species. Even so, each *Acropora* shows distinct habitat preferences (Brown et al. 1983).

Giant Clams. Giant clams (Tridacnidae) are bivalve molluscs which live either loose on the sea bottom or embedded in rocks or living coral. Like other bivalves, they filter plankton and other organic matter from the water, but they also have *S. microadriaticum* in their mantle lobes (the soft parts visible when open normally). Probably because of these symbiotic organisms, giant clams are the fastest-growing bivalve molluscs, growing up to 8 cm per year. Giant clams are hermaphrodite, and when both sets of sex organs are mature, sperm, and then later eggs are released into the water. The entire population of a clam species on a reef will do this at the same time which maximizes the chances of fertilization occurring. The fewer adult clams present, the less likely it is that sufficient fertilization will occur, and it seems probable that there is a density of clams below which no successful reproduction will occur, even if adults are still present (Wells et al. 1983).

Echinoderms. A ubiquitous though varied phylum of coral animals is the echinoderms ('spiny skin'), comprising the classes of sea lilies or crinoids, sea urchins, starfish, brittlestars, and sea cucumbers, all of which exhibit five-rayed symmetry, and have no head or brain. Nearly 90 species have been found in different habitats of sand flats, seagrass beds, algal beds, and the outer reef of P. Pari in the Thousand Islands, the most abundant group (in terms of species) being the brittlestars with 35 species, and the richest habitat the outer reef wall with 73 (Indriati 1984), which also has the most species of crustaceans and molluscs (Romimohtarto and Moosa 1977). The most common species of echinoderm on the sand flats of P. Pari is the blotchy grey starfish *Archaster typicus* (Darsono et al. 1978), and where there are broken pieces of coral among the sand the brittlestar *Ophiocoma scolopendrina* can be abundant. Where sand abuts against an algal area, there were relatively few species, but the long-spined, black sea urchin *Diadema setosum*, the sand-covered sea cucumber *Holothuria atra*, and the mottled brown sea cucumber *Bohadschia marmorata* are abundant (the last two being found also on the reef slope) (Heryanto 1984). In the seagrass beds the knobbly, orange starfish *Protoreaster nodosus* and the solitary grey-spined sea urchin *Echinothrix diadema* are most common, as they are on Sanur Beach, Bali (Whitten and Whitten 1993a), and five species of sea cucumber are found.

Although most starfish are predatory, their diets vary between species, resulting in different distributions. For example, *Protoreaster nodosus* grazes the algae growing on the leaves of seagrass, the bright blue *Linckia laevigata* feeds mainly on the mucus produced by corals, the crown-of-thorns starfish *Acanthaster planci* eats the top layer of coral containing the living coral polyps, the 'armless' *Culcita novaeguineae* eats sponges and coral polyps, while many of the species on the reef slope attack molluscs and brittlestars (Aziz 1981).

Reef Fish. There is a greater density of fish species on a reef than in any other place in the sea, with 100-200 species present in a single hectare. The colours, patterns, and shapes are breathtaking, and the relative ease of observation and the beauty of the subjects make study of their diverse behaviour and ecology extremely rewarding. The variations across a reef in biological parameters is also clear (fig. 7.5), with the greatest number of species being found over the terrace and reef slope. Two field guides are now available to assist in reef fish identification (Myers 1991; Kuiter 1992).

Certain fish are among the most dangerous animals around a reef. Among the fish this group would include are sharks (generally small except in deeper water), stingrays (Dasyatidae), moray eels (Muraenidae), lionfish (Scorpaenidae), and stonefish (Synanceiidae). These last two could not be more different: lionfish flounce around the reef in gaudy colours, whereas the drab, grotesque stonefish lie still with their large, warty heads and mouths almost invisible among the sand and corals. Spines of the dorsal fin inject a poison which can be fatal to humans.

Sea Snakes. The other dangerous vertebrates are the sea snakes but most are rarely seen. One, the grey-and-black striped *Laticauda colubrina*, however, is confined to coral reefs and the beaches where it lays its eggs. The bites of some sea snakes can be extremely dangerous, but *L. colubrina* is reluctant to bite, even when handled, and has a venom of low toxicity. This is not known to most of the tourists to Tanah Lot, Bali, who are persuaded, for a fee, to view 'dangerous' *L. colubrina* in a small, seashore cave.

Algae

Although a coral reef may appear to be dominated by animals, plants can

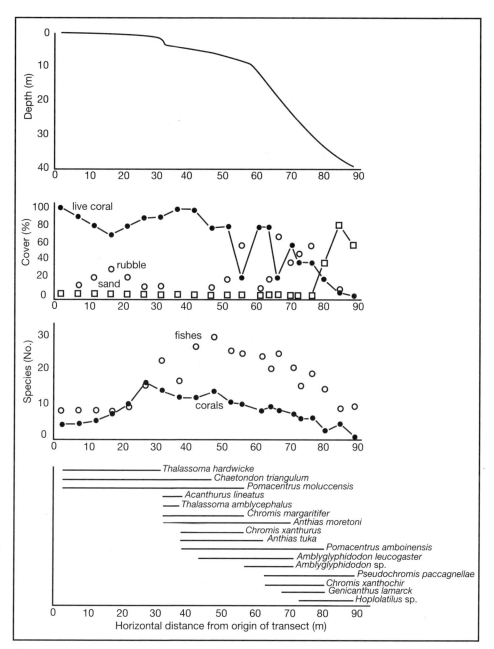

Figure 7.5. Details of, and transect across, the reef on the northwest of Menjangan Island showing (a) physical profile, (b) percentage cover by live coral, and rubble and sand, (c) number of conspicuous fish and hard coral species, and (d) zonation of some fish species.

After Polunin et al. 1983

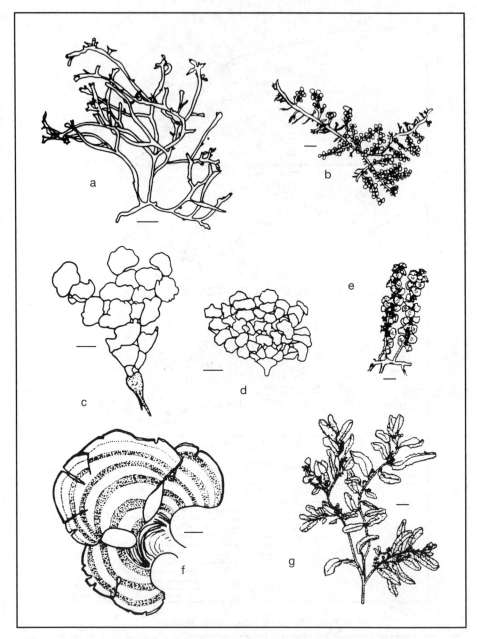

Figure 7.6. Common coral-reef algae. Red algae: a - *Gracilaria lichenoides* (Grac.); Green algae: b - *Caulerpa racemosa* (Caul.), c - *Halimeda tuna* (Codi), d - *H. opuntia;* Brown algae: e - *Turbinaria conoides* (Fuca.), f - *Padina gymnospora* (Dict.), g - *Sargassum polycystum* (Sarg.). Scale bars indicate 1 cm.

After Teo and Wee 1983

in fact constitute 75% of the biomass in an area of reef. These plants include very fine, filamentous green algae growing in dense mats or turfs, larger seaweeds, and coralline algae. Large seaweeds show distinct patterns in their distribution; red algae are found on inner reef flats and outer edges of coral reefs, brown algae such as *Sargassum* usually occur throughout the reef flat, and green algae tend to be found in intertidal zones (fig. 7.6). Coralline algae are red algae that are calcified, that is they deposit limestone in their cells and are consequently extremely hard.

Some 22 species of red, 15 species of green algae, and 7 of brown algae have been recorded on algal ridges at Tanjung Benoa, southern Bali, with 600 g (combined) per square metre, more than half of it red (Sulistijo and Atmadja 1980). Algae such as the red *Gracilaria lichenoides, Hypnea musciformis* and the green *Caulerpa racemosa* and *Halimeda opuntia* have been collected traditionally by local people and this is now a major commercial enterprise (p. 646). The algal community of the Penanjung beach of Pangandaran also comprised about 50 species, but without any species of economic value (Atmadja and Sulistijo 1980). Some algae have considerable economic value and this is discussed elsewhere (p. 646).

The presence of sedentary animals seems to be inversely related to the presence of algae. For example, it appears that filamentous algae interfere with the feeding structures of erect, sedentary bryozoans. In places exposed to grazing by parrotfish and surgeonfish, the presence of algae around sedentary animals such as sea squirts exposes the latter to damage from those fish species that scrape algae off rocks. In this way fish 'weed out' sedentary animals from algal beds (Day 1983).

Productivity and Plankton

The gross primary productivity of coral reefs is high, about 3,000-7,000 g carbon/m^2/year (equivalent to 30-70 tons/ha/year). This is balanced by very high respiration, however, so that the net primary productivity is about 300-1,000 g carbon/m^2/year (Mann 1982). Even so, coral reefs are about 20 times as productive as the open sea where net primary productivity is only about 20-40 g carbon/m^2/year.

Coral reefs support many fish, coral, and other invertebrates, but early studies of zooplankton densities showed that the density of plankton drifting over coral reefs from the sea was too low to support the reef organisms. In fact, certain zooplankton leave the reef at night and return before dawn (Alldredge and King 1977). Another major source of zooplankton is found close to the coral. These plankton rise from their refuges at dusk and retreat at dawn and thus represent the most important food for the night-emerging corals. So, most of the zooplankton are not drifting in from the sea but are resident among the coral, resulting in most of the nutrients being recycled within the reef ecosystem. It is important to note that neither of the major sources of plankton would be detected by the common methods of plankton sampling such as net towing (Birkeland 1984).

Exploitation and Management

Molluscs. The coral reefs of Java and Bali used to harbour commercially valuable species such as giant clams, including the world's largest shelled mollusc *Tridacna gigas,* which can weigh over 200 kg, 60 kg of which is the living tissue. This and other large and beautiful shells such as the triton *Charonia tritonis,* helmet shell *Cassis cornuta,* mother-of-pearl *Trochus niloticus,* green snail *Turbo marmoratus,* and the cephalopod mollusc known as nautilus *Nautilus pompilius* from deeper waters, have been protected by Indonesian law since 1987 because of overexploitation for their meat and shells, which in the case of the clams were used for floor tiles and to satisfy the demands of the ornamental shell trade for whole shells or pieces of mother of pearl. Between 1976 and 1979 Indonesia exported an annual average of 1,511 tons of *Trochus niloticus* and 100 tons of *Turbo marmoratus* for mother-of-pearl, mostly bound for Japan and Singapore.

Giant clams have been exploited traditionally for a range of materials such as meat, lime, tiles, and as ornaments. Unfortunately, as a result of reef destruction and demand for clams far exceeding their ability to maintain their populations, they are now extremely rare. In order to collect clams (except *Hippopus hippopus*), part of the coral reef must be destroyed, because they grow into the coral rock which also grows around them. The two largest giant clams *T. gigas* and *T. derasa* are now probably extinct around Java and Bali and the rest of western Indonesia, and the other species are rare, at least at commercially viable sizes. In 1983 a survey of giant clams was made around Karimun Jawa of the remaining giant clam population to determine how large an impact their exploitation for floor tiles was having (1,160 tons of shells were taken from the islands to Jepara in 1982), and whether marine culture by 'seeding' the reef was a viable proposition, in order to support the trade in shells and also as a source of seafood for export (Brown and Muskanofola 1985). It appeared that the dead shells were found in sandy areas and their extraction caused little damage to the coral, unlike the situation in the Seribu Islands (Salm 1987). In fact, shells of clams still living have much thinner shells than the old ones, and it seems that the resource being collected represents the remains of a once large and healthy giant clam population dating from at least 70 years ago. The surveys found virtually none of the three large species (*T. gigas, T. derasa,* and *H. hippopus*). *T. maxima* was the most common, and the most collected.

Although giant clams are hermaphrodite, they require cross-fertilization of eggs, and the probability of this occurring decreases with increasing distance between breeding-sized adults. There is a very high mortality (>99%) from the larvae to adult stages. Despite the paper protection afforded to them, all these protected molluscs have been, until recently, widely available in souvenir shops, particularly in Bali.

Giant clams could be managed as a commercial resource, though at lower levels of exploitation than have prevailed to date. The adults suffer very low mortality, have no predators (save people), the young grow rapidly and are easy to maintain. They attach themselves to the substrate with byssal threads but if removed carefully will reattach themselves to a new substrate. Clams can produce a billion eggs at a single spawning, more than any other creature known, and their rate of development is extremely rapid. Clams (which start off male) can reach maturity after only five years, becoming female and mature again a few years later. A number of Pacific nations are already cultivating the clams for their meat, and there appear to be no major biological or technical problems. Options for proper management are the reseeding of reefs with clam larvae, the bringing together of adults to establish breeding colonies, and the development of clam mariculture. All of these options could have positive benefits for local human populations. Some experimental clam farms have experienced problems with filamentous algae growing over the shells, but this can be controlled by developing mixed mariculture with juvenile mother-of-pearl shells *Trochus niloticus* which graze on algae.

Now that Karimun Jawa is a marine park, it would be possible to reseed the barren reefs with spat of the larger species to act as an additional tourist attraction since the large, iridescent blue or green 'lips' and the sheer size of these animals are quite remarkable. The technology supporting hatching, cultivation, and reseeding is already available (Brown and Muskanofola 1985). Reseeding in non-protected areas for meat and shells is being contemplated.

Reef Fishes. Reef fish, like reef invertebrates, are objects of trade. Indeed most of the marine ornamental fish traded on the domestic and international markets are associated with reefs, and they are caught at many locations around the coasts of Java and Bali, including national parks. The Indonesian region has more potential ornamental marine fish species than any other region, perhaps more than 250 species. The fish are caught in hand nets either at night when the fishermen can blind them with a bright light, or by stunning them with cyanide. With care this does not kill the fish, although the reef invertebrates are less tolerant. Most of the fish caught are sent to Jakarta for export. The impact on the wild populations of this collecting is hard to assess precisely, but intensively collected areas such as the Sanur flats have a rather depauperate fauna. Of the fish caught, about 70% survive to be sold, and only half of these live more than six

Figure 7.7. The capture of reef fishes is important to some coastal communities, but the practice at its current levels is not sustainable, and is wasteful and destructive.
With permission of Kompas

months (Anon. 1986) (fig. 7.7). The regulation of the trade needs to be encouraged, and enthusiasts should consider the unnecessary wastefulness of their hobby.

Artificial Reefs. Reefs have long been recognized as important to offshore fisheries. It has also long been known that fish congregate around wrecks and other structures in the sea which provide a three-dimensional habitat, often with a large surface area on which a succession of life forms can become established, providing food for fish and other organisms. At the end of the 18th century, Japanese fishermen began deliberately dropping heavy, wooden structures offshore in order to attract fishes. Since then, various items have been used as artificial reefs in southeast Asia, such as large concrete blocks, tyres, steel pipes, cars, oil rigs, boats, and, off Jakarta, pedicabs confiscated by the city government. Successful artificial reefs are deliberately constructed to provide appropriate habitat for target species– they are not just dumped in the sea to fall scattered on the seabed. The placement of artificial reefs is also extremely important if they are to result in an overall increase in harvestable productivity; a study in the Philippines, for example, has found that they must be at least 1 km away from natural reefs, near an alternative food source (e.g., seagrass beds), on flat or gently sloping sand in water with good visibility at depths of 15-25 m, protected from excessive wave action, and accessible to fishermen (White et al. 1990).

The only artificial reefs in Indonesia are small areas off Jakarta, but about one-fifth of Japan's coastline has artificial reefs, and in the Philippines some 50,000 tyres were dumped between 1990 and 1995 to make artificial reefs. Car tyres are popular because of their long life-span, insignificant cost, ease of shipping and handling, and their many crevices and surfaces, although they do have to be weighted down to counteract their buoyancy. After five years 15% of the tyre reef can be covered with 30 coral species, and after five more years 40% of the surface can be covered (Gomez et al. 1982).

The objectives of these apparently beneficial junk yards can be one or all of the following (White et al. 1990):

- to concentrate organisms to allow more efficient fishing;
- to protect small or juvenile organisms and nursery areas from destructive fishing gear;
- to increase the natural productivity of an area by providing new vertical niches for permanently attached or sessile organisms to stimulate the establishment of associated food chains; and
- to simulate natural reefs and create habitats for desired target species.

Although not true in all cases, artificial reefs seem to harbour fewer species of fish and other organisms than natural reefs, but a greater biomass and greater density of individuals, particularly where the structure of an artificial reef is complex. Real economic gains only occur, however, when artificial reefs enable capture of fishes that could not have been caught elsewhere for the same or less cost. For example, artificial reefs can be economic assets when fish are concentrated, resulting in less use of labour and fuel, and lower risk (Bohnsack and Sutherland 1985).

Critics of artificial reefs claim that in certain circumstances, such as when overfishing is occurring, they can be detrimental to fish stocks since they simply attract rather than directly produce fishes. This concentration makes them more susceptible to capture. Thus apparent increases in fish yields are simply the result of reducing the standing stock of target species. Where this does not damage the ability of the breeding population to sustain the harvest, there is no problem. In any case, production rather than attraction will not be possible until the reef organisms have colonized and grown, and it seems that this takes at least a decade. Indonesian experience with constructing different types of artificial reefs for specific fishing purposes, and managing the fishermen who will exploit the fish around them, is limited, but there is much to learn from Thailand and the Philippines. Even so, the basic ecological understanding of how they function is very limited and needs to be improved through solid action-orientated ecological research (White et al. 1990).

Protection and Management. It has been demonstrated elsewhere that when reefs are protected from fishing, the associated animal communities change. For example, the abundance and average sizes of many larger carnivorous fishes increase, and similar patterns probably exist for other species which had been targeted by fishermen. However, it has not yet been demonstrated that reserved areas serve adjacent areas by forming a supply of recruits that will enhance or sustain catches (Roberts and Polunin 1991).

A number of terrestrial conservation areas have significant stretches of reefs within their existing (or proposed) boundaries, such as Ujung Kulon, Pangandaran, Bawean, Baluran, Alas Purwo, and Bali Barat. A decade or more ago visitors to Pangandaran walked on and thus damaged the reef flats despite prohibitions to the contrary (even though at that time it was not a gazetted conservation area), fishes were caught live to supply the aquarium trade, and shells and coral were collected to sell to tourists (Kvalvågnaes and Halim 1979a). All these activities deleteriously affected the very values sought by most of the visitors. Increased protection with better awareness and education were achieved, together with the gazetting of a 500 m wide band of coast which included all the reefs.

While there may be differences in degree, all these coastal conservation areas experience similar problems related to their management (Ritchie 1991):

- the large number and severity of impacts threatening to reduce tourist appeal, fisheries production, and availability of other products;
- the almost total lack of marine management;
- the lack of staff with marine expertise;
- the absence of marine biology training for staff or boatmen;
- the low budget priority given to marine areas; and
- the general lack of base line information against which changes, if any, can be assessed.

P. Menjangan in Bali Barat National Park now has a qualified marine biologist associated with it (as proposed 10 years ago), and hopefully it will be possible to establish long-term monitoring and research programmes to evaluate changes in marine resources. This is important because the high-quality reef front, reef terrace, drop-off, and reef slope around the island draw large numbers of divers and snorklers every day. In the past, some boatmen and visitors have treated the reef and its inhabitants with little respect, and this has caused considerable damage. The Park authorities have taken the step of distributing leaflets, sponsored by EMDI, setting out codes of behaviour for the boatmen and the divers (table 7.2) to try to reduce the impact of the popular dive spots around the island.

Table 7.2. Codes of practice for boatmen and visitors to coral reef areas as used in Bali Barat National Park.

CODE OF PRACTICE FOR BOATMEN

Mandatory requirements

Boatmen are required to have:

1. A current recognized certificate of proficiency in Medical First Aid.

Boatmen are required to:

2. Ensure the safety and comfort of visitors in the coral areas and to identify and warn them of common dangers (e.g., marine life that stings, bites, is poisonous or otherwise harmful or dangerous) and to advise them how best avoid such contacts.
3. Have at least one other crewman on board on any visit to a coral area.
4. Proceed at not more than 5 knots (8 kph), over any coral reef flat, reef front, reef terrace, drop-off, and reef slope for at least 100 m seaward of the drop-off (the look-out zone).
5. Maintain, while moving in the look-out zone:
   a. From the helm, special watch ahead and abeam (both sides).
   b. From the foredeck or not more than 3 m from the fair-lead, special watch ahead and abeam.
6. Ensure that no part of his boat contacts any part of the coral reef at any state of the tide, sea or wind except in life, or vessel, threatening situations.
7. Never anchor on any living coral or allow his anchor or anchor wrap to come into contact with any living coral or any other marine life.
8. Always use any mooring provided.
9. When a mooring is not provided, anchoring must be carried out on sand using a proper sand anchor so that no part of the anchor, chain, or wrap can contact any living coral or marine life at any state of the tide, sea, or wind.
10. As a last resort anchoring is permitted on dead coral areas, e.g., the reef flat, in such a way that no part of the anchor, chain, or wrap contacts living coral or marine life.
11. Avoid discharge of any fuel, oil, bilge water, or rubbish in the look-out zone or anywhere in the marine protected area.
12. Warn passengers not to discharge any rubbish from the boat.
13. If there is no guard or guide on board, the boatman must inform his passengers of the status of the coral area, of the codes of practice, and of their obligations.
14. Not allow any marine life, living or dead, or any object or natural feature of the marine protected area to be brought on board his boat or conveyed by it.
15. Not allow any fishing or collecting gear to be carried on or used from his boat except with express written permission from PHPA, or in an area where fishing or collecting is allowed.
16. Refrain from entering, or conveying visitors for the purpose of entering, any area that PHPA has declared to be prohibited to divers or visitors.
17. Carry and display on board the boatmen's, and visitors' code of practice.

CODE OF BEHAVIOUR FOR VISITORS

Visitors to coral areas are required to:

1. Follow this code of behaviour the functions of which are to ensure maximum protection for the coral reefs of the national park marine areas, and the safety and comfort of visitors to them.
2. Follow advice and instructions from the boatman and guide or guard on board,

Visitors to coral areas are not allowed to:

3. Fish for, trap, snare, move, remove from the water, or roughly handle or damage or kill any marine life.
4. Damage, alter, move, or remove any object or natural feature.
5. Sit, stand, or rest on any living coral anywhere on the coral reef except in a life threatening situation.
6. Throw or discharge any rubbish of any sort into the sea within the national park marine protected area boundaries.
7. Land or go ashore anywhere other than at a place designated for landing and going a shore.

Threats

Reefs in Java and Bali are threatened by coral mining, fish bombing, pollution, sedimentation, and tourism (Ongkosongo 1986). When even quite minor and short-term changes are made to the environment, the members of a complex community, not used to such disturbance, may exhibit some initial resistance (an ability to avoid change), but low resilience (the speed with which the community can return to its original state) after the change has occurred (p. 22). Coral reefs, then, may be able to withstand just so much stress before the living community completely collapses.

Coral Reef Mining. Coral is mined for a variety of purposes such as white-wash (paint), road fill, decorative walls, and sawn temple bricks (fig. 7.8). The mining itself tends to be undertaken by otherwise jobless people who get little in return for unpleasant work (Subani 1982). Unfortunately, the removal of the coral heads and the coral reef crest results in coastal erosion, and this has been known for years (Hardenberg 1939; Umbgrove 1947; Verstappen 1953, 1954; Ongkosongo 1986; Subani 1982, Kaniawati 1985). Coral mining not only has negative physical effects on the shoreline, it also adversely affects the whole range of reef life including species of economic value (Subani 1982; Putra 1992).

Along the west-facing coast of Java the most damaging human activity is the manual and mechanical harvesting of coral fragments and blocks from the reef flats and beaches, and the trucking away of large quantities of sand and gravel for construction and cement making in Serang, Cilegon, and Pandeglang. Both activities are entirely unsustainable at current levels of exploitation, and serious onshore erosion is resulting in the loss of residential and agricultural land.

The situation has, perhaps, improved in many areas over recent years. For example, in 1987 there were over 200 lime kilns burning coral rubble in just Badung, southern Bali, and all the 400 kilns on Bali at that time were processing nearly 200,000 m^3 each year (Subani 1982). There are now only a handful left. One of the most important reasons for this, however, is that virtually all coral blocks accessible to the crowbar gangs on the reef flat have been removed already, and the local government regulations appear to have been merely incidental. Even so, it is encouraging to see that there are definite signs of recovery as far out as the reef slope off the popular Sanur Beach, Bali (T. Tomascik pers. comm.). A challenge will be to prevent a resurgence of mining as the coral grows.

Unlawful Fishing. Another common cause of reef destruction is the use of explosives to catch fish. Explosives are very efficient, but they kill fish indiscriminately without regard to food value, and more fish probably sink and decay than float to the surface for collection (Burbridge and Maragos 1985). This is because the major site of internal damage is the

Figure 7.8. The mining of coral causes coastal erosion and a decrease in the sea's productivity.
With the permission of Kompas

swim bladder, a sac of gas used to control buoyancy, and when this is rup-tured, many fish lose their buoyancy and sink as they die. Bottom-dwelling fish, with poorly-developed swim bladders, are less likely to be killed than other species (Wright 1982). The effects of underwater explosions on invertebrates has not been studied, but it is known that divers can suffer perforated ear drums if diving while coral is being blasted. Fishing using explosives was outlawed in 1920 partly because of the damage it caused to coral, and the mining of coral was prohibited in West Java in 1940 and in Bali in 1973. Despite the regulations, the use of explosives continues, par-ticularly in remoter areas. Fishermen who blast for fish must have some access to individuals who are legally entitled to store and use the explosives, which is illegal. Tracing and convicting the offenders is difficult because they are generous in giving away surplus fish to the villages in which they stay, and their friends in high places will help the cover to avoid their own detection.

Blast fishing is primarily for food fish, but another serious catching method, cyaniding, is directed primarily to ornamentals. In theory this is harmless to the fish if used correctly. Some expensive fish are very difficult to catch without it, and since it affects a small area, any fatalities will not have a large impact. Cyanide is fatal, however, even in very small concentrations, to many sedentary reef organisms. This should be more widely known to more aquarium enthusiasts.

The coral itself may recover from blasting (Parrish 1980), but it would take many decades before the effects of blasting are no longer obvious to even a casual diver. The coral rubble and the algae growing on it have a relatively low productivity, and the biomass of plankton over such areas is little greater than over plain sand (Porter et al. 1977). When considering different impacts on coral, it is worth remembering that most coral has a radial growth rate of only 1-10 mm/year (Soegiarto and Polunin 1980), although in certain conditions it can grow at 3-10 cm/year (Verwey 1931a), or faster (T. Tomascik pers. comm.).

Sedimentation. The most insidious cause of coral death is probably suffocation by sediment, either in suspension in the water, or settled on the coral polyps. In both cases light is prevented from reaching the photosynthetic organisms in the coral skin, and the suspended matter may also overburden the filter- feeding polyps with inorganic material. Thus, any plan concerning coral in coastal zone management must include considerations of the sources of sediment: uplands subject to bad farming practice, deforestation, overgrazed pastures, and public works projects that expose bare soil to the full force of the rain. To incorporate such concerns in regional planning and to act when necessary, requires dedicated and co-ordinated government action.

Tourism. Tourists and the operators who facilitate their visits can cause long-lasting damage to coral reef areas, through accidental damage to delicate coral structures, deliberate souvenir hunting, and unthinking walking on coral heads. Codes of practice to help prevent these problems are described above.

Exceptional Sea Warming. The issue of rising global temperatures is discussed earlier (p. 72). In a number of areas during the abnormal conditions associated with the El Niño of 1983, the strongest for perhaps 200 years, seawater temperatures rose as much as 5° C above normal. Shortly afterwards scientists noticed that some corals appeared bleached (Suharsono 1984). This bleaching is caused by a lowered concentration of photosynthetic pigments in the dinoflagellate within the skin of the coral which makes the white skeleton easier to see. Such bleaching can be induced experimentally by raising or lowering temperature, by diluting the seawater, by high sedimentation, or by pollution (Glynn 1991). Around P. Pari there was little evidence of bleaching on the reef flat or lagoons, perhaps because the corals growing in these shallow waters are adapted to a wide range of temperature, but many were affected on the outer reef slope down to about 9 m. Below this fewer and fewer were affected (Hoek-

Figure 7.9. Coral reefs, and other natural ecosystems, need to be the subject of long-term monitoring in order to detect the effects of environmental changes.

With permission of A. Ibrahim

sema 1991). It should be remembered that the dinoflagellates provide only up to two-thirds of the corals' nutrients, and so even when they are out of action, the coral does not necessarily die. Indeed, five years after that El Niño 50% of the previously-bleached corals on a reef in the Thousand Islands were apparently healthy again (Brown and Suharsono 1990). This event demonstrated the importance of long-term monitoring of natural ecosystems so that changes can be studied and interpreted correctly (fig. 7.9).

Effects of Coral Destruction. Destruction of coral reefs removes a natural means of protecting the shoreline from erosion. Although at first sight this may be difficult to understand, coral reefs play a very significant role in protecting coasts from erosion. In deep water the height of an ocean wave is one-twentieth of its length. When a wave approaches shallow water, however, the drag of the sea bed shortens the wave length and the wave is forced up into a peak. This sharp-topped wave breaks when its height is in the ratio of 3:4 with the water depth. Thus a wave 30 cm high will break in water 40 cm deep. It is for this reason that waves can be seen breaking (and have much of their energy dissipated) some distance offshore above the reef slope. In places where the reef is broken or blown apart for the collection of hard core for roads and other constructions, the coral 'lip' is

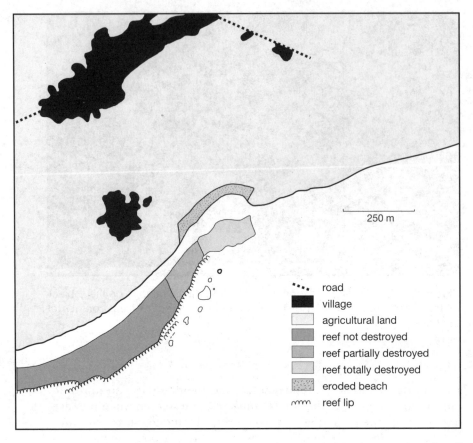

Figure 7.10. Beach erosion resulting from destroying fringing coral reefs off Sengkidu, near Candi Dasa, Bali. Note that the situation has deteriorated since this study despite the construction of groynes.

After Praseno and Sukarno 1977

removed, and the concave nature of the reef lagoon is changed into a sloping shore. The waves now break nearer the shore, and the wave energy is now not just effective in creating water turbulence but also scours the coastline (fig. 7.10-11). An example of this is Sanur beach, Bali, which is much narrower now than it was 20 years ago. The killing of coral reefs also greatly reduces the productivity of the coastal waters with the result that the production of reef-related fisheries is certain to fall.

Figure 7.11. Coastal abrasion as a result of coral reef destruction along the north coast of Java.
With permission of Kompas

Conservation

It is clear that coral reefs suffer much disturbance and face varied threats. In response to this situation a National Strategy on Coral Reef Conservation and Management was agreed to sustain coral reef ecosystems so that they may be utilized optimally for the present and future generations, with the expectation that the national income and the welfare of local communities will improve (KLH 1992c). A great many Acts, Government Regulations, Presidential Decrees, Ministerial Decrees, and other lesser instructions have been issued over the years, and if these were followed, the problems would diminish significantly. There is still room, however, for increasing community awareness of and participation in coral reef conservation, and for devising cost-effective management strategies and plans for the most important areas. Responsibility for coral reef conservation is held by the Department of Forestry, and there is a case for discussing whether this is its most appropriate home (p. 845).

SEAGRASS MEADOWS

Seagrass meadows are undervalued ecosystems which play an important role in coast stability and the support of coastal fisheries.

Original and Present Distribution

In lagoons or shallowly sloping sand beaches can be found 'meadows' of seagrass (fig. 7.12). They are fast growing and play an important role in binding shallow sediments against erosion. The nature of the substrate and its depth are the two most important factors in determining the presence of the species, but there is also a general change of species with distance from the shore. The meadows are best developed around calm, sandy beaches off the north coast of Java, at certain points along the southwest coast, and in the Sanur lagoon on Bali. Seagrass meadows in certain areas have been converted to commercial algae farms, and meadows are lost as substrate composition and suspended sediment concentration change as a result of coral reef destruction .

Vegetation

More than 10 seagrass species belonging to two families (Hydrocharitaceae and Potamogetonaceae) are found around Java and Bali (den Hartog 1957, 1970) (fig. 7.13), and a simple key can be found elsewhere (Whitten et al. 1987b, c). The largest and one of the most common species is *Enhalus acoroides* (Hydrc.), which has strap-like leaves up to a metre long and nearly 2 cm wide, and whose rootstock beneath the sand is covered with stiff, black hairs.

Seagrasses are among the very few higher plants that thrive entirely in sea water. *Halophila* and *Thalassia* are pollinated within the water, but *Enhalus* has flowers that reach the surface where they are pollinated by insects or the wind. The flowering coincides with low spring tides around the full moon. The male flowers break off from their stalks and float around in groups, bumping into the still-attached female flowers, attracted by surface tension to their waxy, water-repellent coating. The male flower floats next to the female flower as the tide rises, but when it rises beyond the length of the female flower stalk, surface tension pulls the female petals together, trapping the male flowers between them. The seeds are heavier than water, and so dispersal distances are probably relatively short and germination is rapid (den Hartog 1957). Green turtles and dugongs, which eat seagrass leaves and rhizomes respectively, probably assist the dispersal of seeds by ingesting and later defaecating them, although the dugong has been hunted to extinction around the Java and Bali coasts, and green turtles are increasingly rare (pgs. 244, 756).

Figure 7.12. A typical seagrass meadow of *Thalassia* and *Halodule* seagrasses with, in the foreground, some *Padina* algae. Animals shown are a *Synapta* sea cucumber, a *Echinometra* sea urchin, an *Archaster typicus* starfish, a *Holothuria* sea cucumber, and a *Diadema setosum* sea urchin.

After Göltenboth et al. in prep

Among the sea grasses can sometimes be found distinctive algae which are more generally associated with coral reefs (p. 352).

Productivity. It is frequently noted by bathers that a beach adjacent to seagrass meadows is fringed by floating and stranded, rotting seagrass leaves. These are regarded as 'dirty' and undesirable, and early morning walkers on Sanur beach on Bali will find hotel employees busily raking and burying these unsightly but harmless ecosystem products. Although a seemingly considerable quantity of decomposing seagrass is washed up on the beach, the majority is 'exported' into estuarine and offshore ecosystems. Indeed, seagrass meadows are highly productive (a fact probably best appreciated by those who clear up the leaves each morning) with a net primary productivity of up to 20 t/ha/year, and a standing stock or biomass of leaves and stems of some 4 (0.8-7) t/ha. At least a third of the primary productivity in reasonably open meadows may be accounted for by the epiphytic algae rather than the seagrasses themselves. The algae fail to thrive where

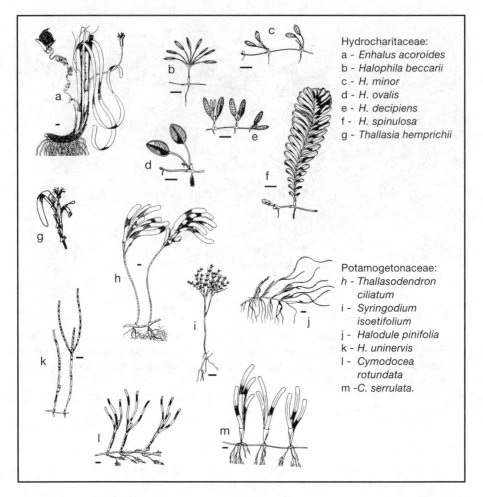

Hydrocharitaceae:
a - *Enhalus acoroides*
b - *Halophila beccarii*
c - *H. minor*
d - *H. ovalis*
e - *H. decipiens*
f - *H. spinulosa*
g - *Thallasia hemprichii*

Potamogetonaceae:
h - *Thallasodendron
    ciliatum*
i - *Syringodium
    isoetifolium*
j - *Halodule pinifolia*
k - *H. uninervis*
l - *Cymodocea
    rotundata*
m - *C. serrulata.*

Figure 7.13. Seagrasses found around Java and Bali. Scale bars indicate 1 cm.
After den Hartog 1957, 1970

the seagrass meadows are dense and less light is available. These epiphytic algae are grazed intensively, especially at night, by a range of invertebrates (Kitting et al. 1984).

Very little, perhaps only 5%, of the seagrass production is consumed directly, the majority entering the food chain when it is decomposing, when bacteria have begun to digest cellulose, and when there is a considerable population of microorganisms present. Among the few organisms that eat live seagrass leaves are green turtles, and sea urchins with cellulose-digesting bacteria in their guts.

Fauna

The seagrass meadows are habitat for many molluscs including the large and much overexploited *Lambis* snail with its seven or so projections, and species of *Strombus*. There are also sea cucumbers such as the long and corrugated *Synapta*, and the black and sand-encrusted *Holothuria atra;* nearly 25,000 sea cucumbers were found in a hectare of seagrass meadow in the Thousand Islands (Aryono 1987). In addition, various seaslugs or nudibranchs, as well as cauliflower starfish *Culcita* and the brown-knobbed, sand-coloured starfish *Archaster* can be found (Atmadja 1977, Whitten and Whitten 1993). In the Thousand Islands some 78 fish species were found in the seagrass meadows (Hutomo and Martosewojo 1977), but to what extent these are dependent on the meadows is uncertain. However, it is clear from snorkelling over the meadows that many young fish are present, such as dense shoals of black-and-white striped catfish *Plotosus anguillaris,* and the seagrass provides both shelter and a habitat that is not conducive to larger, predatory species. Studies in Banten Bay found that habitat diversity of seagrass meadows, rather simply the abundance of seagrass, was the main determinant of fish abundance (Hutomo 1985).

Uses and Management

The large seagrass meadow off Sanur beach, Bali, was studied during the writing of this book to ascertain its biological and physical importance. It was found to be of critical significance to the replenishment and stability of the beach system as well as supporting offshore ecosystems through a detritus-based food chain. It is dependent on the fringing coral reef which provides the shelter for the lagoon, and it serves the coral reef by stabilizing lagoon sediments, reducing turbidity. The lagoon does not simply stabilize the beach, it also actually produces the sand most of which comprises the tetrahedral tests or shells from species of benthic foraminifera, a form of aquatic protozoan. Thus the continued existence of each system depends on the well-being of the other (T. Tomascik pers. comm.).

There are clearly good ecological reasons for disturbing Sanur beach as little as possible, and certainly for not importing sand from other beaches, as was being suggested in 1992. This would have a finer sediment and could smother and disturb the seagrasses and their forams. Engineering solutions should give way to improving the ecological integrity of the lagoon, particularly by the lowering of phosphate and nitrate concentrations. Luckily the effects of human activities on the beach are minimized by the strong north-to-south current that effectively flushes away polluted water. If the seagrasses are removed, or the conditions made unacceptable, the production of forams would decline, and the loss of beach sand would exceed its production. The influential hotel trade should understand that they are dependent on an intact ecosystem for the maintenance of their beach.

As seagrass meadows are trodden on, dredged, smothered, and cut up as a result of development and recreation activities, so it is certain that the energy inputs into coastal waters will be reduced, that planktonic primary producers will replace benthic ones, that beach morphology will change because the sand is no longer bound, and that structural and biological diversity will be lost as bare, shifting sand predominates. If the main causes of seagrass destruction can be controlled, then they are likely to recolonize former haunts, and it has been demonstrated successfully in the USA that seagrass meadows can be restored by transplanting as and when the damaging activities are stopped or at least controlled (Thorhaug 1985), as is being increasingly shown with mangroves (p. 407).

An Indonesian Strategy for Seagrass Conservation and Management has been produced by a team coordinated by the State Ministry for Environment, which intends to implement it with the cooperation of other arms of government.

BEACHES

Open Area Communities

Most animals of sediment beaches rarely emerge onto the surface, and these are known collectively as the infauna. Those that spend some time on the surface such as crabs and snails are known as the epifauna. Most of the epifauna are large (macrofauna) but the infauna can be grouped into the microfauna or protozoans, the meiofauna (defined as animals able to pass through a 0.6 mm mesh sieve but retained by a 0.05 mm mesh sieve), and the large, conspicuous macrofauna such as bivalve molluscs or large worms. These are all resident animals, although their larvae may have originated elsewhere. Sandy beaches also receive visitors such as waders and other shorebirds (p. 233) and turtles (pgs. 244, 756). An interesting myth is that insects are rare in such coastal habitats. While certainly true for temperate regions (whence come most of the textbooks), nearly half the insect orders are represented in beach and shallow marine habitats. An unfortunate consequence of the myth is that marine biologists do not study insects, and entomologists cease their work at the high-tide mark (Howarth 1991).

The meiofauna, being too small to move the grains, generally comprises elongate, wriggling creatures. Among the most common animals are nematode worms, of which hundreds of thousands may occupy the top few centimetres in a single square metre of beach. These worms are an important food source for larger animals. The worms and most other members of the meiofauna are most common in fine sediments since coarse sands dry out too quickly and very fine sands are too easily deoxygenated. The environment inhabited by the meiofaunal infauna is less rich in sediments comprising a mixture of grain sizes, probably because the weak waves which result in such sediments are incompatible with the animals in some way. The ability to move through the sand is crucial to beach animals in order to avoid excessive wave action, surface predators, high temperatures, and desiccation.

The beach ecosystems are unusual in that the common plant-herbivore-predator structure of food chains is absent. The only 'plants' available are diatoms and bacteria, and predation on or in the sediment is difficult. Predators are found primarily among the epifauna, such as birds, certain mudskippers, insects, polychaete worms, and snails. Thus the majority of animals either filter plankton from the seawater (suspension feeders) or suck organic deposits and microorganisms off the sediment surface or sort out edible particles after ingesting sediment (deposit feeders), although the distinction between these is not always clear. The amount of organic material on or in the sediment is generally greater in the finer sediments because these contain higher concentrations of organic carbon and nitrogen (protein in bacteria), and so feeders on these deposits flourish here. Feeders on the sediments suspended in the water are more common lower down the beach, where they are covered by water for longer and so are able to feed for longer.

All these animals have an effect on the environment within which they live. Burrows increase the depth to which oxygen can penetrate, and the digging of them brings lower sediments to the surface; suspension feeders deposit faecal pellets on the sediment surface which become a food source for deposit feeders; the action of deposit feeders can re-suspend deposits in the sea water. These suspended particles can, however, clog up the filters of suspension feeders, so deposit feeders often predominate in very fine sediments while suspension feeders predominate in coarser sediments (Brafield 1978).

Animals living in finer sediments may, when the tide is out, experience oxygen deficiency. Those with burrows opening into surface pools, and those that can utilize atmospheric oxygen, will have few problems. Most, however, have to rely on oxygen trapped in the pore spaces, which is at low concentrations at the best of times. Some animals reduce their metabolic rate below normal levels at low tide, some create currents of water past their gills or body by waving cilia or fine hairs, whereas others move towards the surface as soon as oxygen levels fall below a certain threshold. Others are able to withstand low oxygen concentrations because of the structure of their respiratory pigments, such as haemoglobin, in their body fluids which have exceptional affinity for oxygen. Some air breathers, such as fiddler crabs *Uca*, have no problems at low tide, and are able to respire anaerobically for short periods when they are in their burrows and the sea is covering the sediment.

Small *Ocypode* crabs, prawns, fish larvae, polychaete worms, and small bivalves are among the most important foods for shorebirds, and the distribution of these foods between beaches is very uneven. Differences in the mudflat fauna can really only be determined by direct investigation (Swennen and Marteijn 1985), although the birds can probably identify a good beach by physical clues, such as casts of mud thrown up by suspension feeders, and swimming movements of small crustaceans (Pienkowski 1983). Some birds also use tactile rather than visual clues and have sensitive beak tips which can sense prey underground. Sandpipers, one group of partially tactile feeders, may avoid sandy mud because the sand grains are very similar in size to the worms upon which they feed (0.5-1 mm) (Quammen 1982). Where a suitable prey is present, density is the most important factor in attracting birds, followed by prey size, prey depth, and the penetrability of the substrate (Myers et al. 1980). The biomass density of prey varies widely, from 1-37 g/m^2 for a number of Javan sites, and is lower around Java than, for example, in the Malacca Strait. One possible reason for this is the large fluctuation in salinities as a result of very different river flows in the wet and dry seasons in eastern Java (Erftemeijer and Swennen 1990).

In addition to the waders, other common large birds of the coast include the white-bellied sea eagle *Haliaeetus leucogaster,* the Brahminy kite *Haliastur indus,* and the osprey *Pandion halietus,* all of which fish in the shallow waters and scavenge food along undisturbed beaches. There are also various storks, herons, egrets, and ducks seen around the shore and roosting and nesting in mangrove forests.

Sandy Beach Vegetation

Sandy beach vegetation forms a narrow fringe of characteristic plants above the maximum level reached by high tides. There are three vegetation formations of sandy beaches: that beneath the half-metre high sea wall, that upon it, and that on coastal dunes. The papers by Backer (1917-22) describe the flora and ecology of these formations.

- *Pes-caprae* **formation** - This is found immediately behind the reach of the highest tides and gets its name from the dominant purple-flowered, 'goat-foot'-leaved creeper *Ipomoea pes-caprae* (Conv.) (fig. 7.14). It can be seen in many areas along the south coast of Java. Indeed most of the plants are creepers producing runners with deep roots

Figure 7.14.
Ipomaea pes-caprae at the top of the beach in Leuweung Sancang Reserve, West Java. Behind the *Ipomaea* can be seen the white-flowered spider lily *Crinum asiaticum,* and the bush *Scaevola taccada.*
By A.J. Whitten

that reach fresh water, bind the soil, and trap organic matter that is exploited by animals and plants. Other dominant plants are *Canavalia* (Legu.), *Vigna* (Legu.), the spiky *Spinifex littoreus* (Gram.), *Thuarea involuta* (Gram.), *Ischaemum muticum* (Gram.), and *Euphorbia atoto* (Euph.).

- *Barringtonia* **formation** (fig. 7.15) - This is named after a frequent but not ubiquitous tree *Barringtonia asiatica* (Lecy.) (fig. 7.16) (Schimper 1903; van Steenis 1958; Kartawinata 1965). Other trees can also be common, such as the yellow-sapped *Calophyllum inophyllum* (Gutt.) which is virtually ubiquitous, the stilt-rooted *Pandanus tectorius* (Pand.), *Morinda citrifolia* (Rubi.), *Sterculia foetida* (Ster.) (fig. 7.17), *Terminalia catappa* (Comb.), *Cycas rumphii* (Cyca.), *Erythrina variegata*

Figure 7.15. Beach forest on P. Peucang, Ujung Kulon National Park, comprising *Pandanus tectorius, Calophyllum inophyllum, Hibiscus tiliaceus,* and *Thespesia populnea.*
By A.J. Whitten

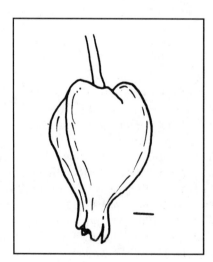

Figure 7.16. Characteristic fruit of *Barringtonia asiatica.*
After Whitten et al. 1987b

Figure 7.17. The characteristic
bark of *Sterculia foetida.*

*By J.G.B. Beumée, with permission of the
Rijksherbarium/Hortus Botanicus, Leiden
University*

(Legu.), *Hibiscus tiliaceus* (Malv.), *Thespesia populnea* (Malv.), and
Scaevola taccada (Good.). Magnificent, long stretches of this formation can be seen along parts of the south coast of Java, notably in
Ujung Kulon and Alas Purwo National Parks. In places where the sea
is abrading the sea wall, these trees are undercut, leaning into the
sea and resting on their major boughs. In wetter areas epiphytes
such as ferns and orchids are found, even where they are occasionally splashed with surf.

- **Dunes** - Dunes are mounds or ridges formed by sand being blown by
  the wind. They are not abundant on Java or Bali, but botanical
  descriptions are available for those on the north coast of Madura
  (Backer 1917-22), Parang Tritis near Yogyakarta (Lütjeharms 1973),
  in southeast Java, near Puger (fig. 7.18; Booberg 1929), and near the
  shallow lakes southeast of Lumajang (Kooiman 1938). One of the
  most characteristic plants is the tough grass *Spinifex littoreus*, which
  has massive, globular fruiting heads which roll along beaches when
  they are ripe and fall from the parent. The extensive dune system

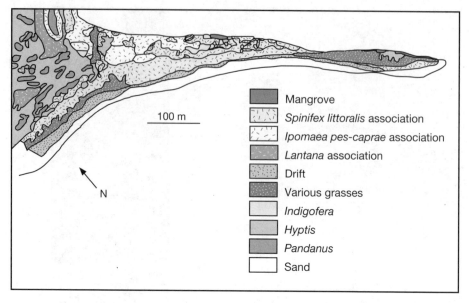

Mangrove	
Spinifex littoralis association	
Ipomaea pes-caprae association	
Lantana association	
Drift	
Various grasses	
Indigofera	
Hyptis	
Pandanus	
Sand	

100 m

N

Figure 7.18. Vegetation map of the Puger dunes made 70 years ago. By early 1994 the entire area had been converted to cow peas *Vigna unguiculata*. Scale bar indicates 100 m.

Simplified after Booberg 1929

near Selopeng on the northeast coast of Madura used to have numbers of gnarled, bonsai-like *Casuarina equisetifolia,* but these so attracted the interest of horticulturalists that none now remain on the dunes (fig. 7.19). Some can be seen, however, in the middle of roundabouts in Sumenep.

Fauna. A number of terrestrial animals frequent vegetated sandy beaches, which provide a valuable source of salt (fig. 7.20). Most large herbivores are dependent on sources of salt in order to maintain the physiological balance of their body fluids. Banteng, however, clearly do not seem to be so restricted. Other animals which appear to be quite common in the beach forests are monitor lizards, wild boar, fruit bats, and lutung, but no animals are restricted to these areas. Lack of access to salt has probably been the main cause, with hunting, of the demise of the vast herds of deer *Cervus timorensis* once found on the Yang Plateau (p. 807), and perhaps of inland populations of rhinoceros.

Three species of turtles regularly visit Java and Bali's beaches to nest

Figure 7.19. The extensive dune system near Selopeng, northeast Madura.
By A.J. Whitten

Figure 7.20. Most large herbivores, such as these Javan deer on the beach in Ujung Kulon National Park, need access to a source of salt.
By Gerald Cubitt

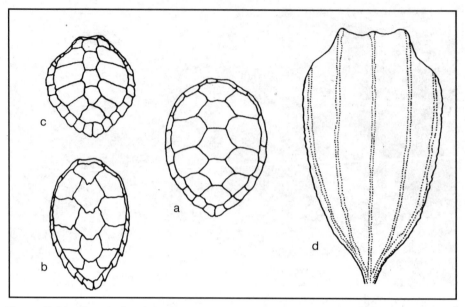

Figure 7.21. The shells of the four species of turtles which visit the beaches of Java and Bali to breed. a - green turtle *Chelonia mydas*, b - hawksbill turtle *Eretmochelys imbricata*, c - olive ridley turtle *Lepidochelys olivacea*, d - leatherback turtle *Dermochelys coriacea*.

(figs. 7.21-22), although they do so in many fewer numbers now than in the past. This demise and options for their management are discussed elsewhere (p. 756). Among the smaller animals, one species that seems to be widespread in beach forests is the mottled *Pythia pantherina* snail. This has a narrow and toothed aperture which looks designed to keep the snail itself inside, but in fact its body slips through it perfectly.

Uses and Management. Attempts to recreate a tamed version of beach forest are undertaken by most seaside hotels, although they stop short of encouraging the sand-binding *I. pes-caprae*. Even so, those with a fringe of trees seem to experience less erosion than others. For example, the Carita Beach Hotel has taken care not to damage the beach vegetation, and trees and shrubs have been planted in initially bare locations. Despite intensive human use and the annual strong westerly winds, there is little erosion. Stark comparisons can be made by looking south from that beach to properties where the vegetation has not been cared for–the beach is narrower, steeper, and the back of the beach is eroding (Joongjai and Rosengren 1983).

Figure 7.22. Turtle tracks leading down to the sea in Ujung Kulon National Park.

By A. Hoogerwerf, with permission of the National Museum for Natural History, Leiden

Most of the famous dunes at Parang Tritis south of Yogyakarta are now grossly disturbed, not just because of industrial developments, but because of tourist pressure (Singagerda 1991).

Seabirds

The seabirds seen around Java and Bali are boobies (Sulidae), frigate-birds (Fregatidae), tropicbirds (Phaetonidae), and terns and noddies (Laridae) (fig. 7.23-24). They are well known to fishermen who observe their diving behaviour to aid them in finding large shoals of fish. The delicate terns and noddies can take off easily and so nest directly on the ground, but the others are masters of soaring flight, and so they prefer to take off from some metres above the sea. Unlike wading birds, seabirds spend months or even years at sea without returning to land. They tend to nest in

Figure 7.23. The two species of frigatebirds seen around Java and Bali. Left pair: male and female great frigatebird *Fregata minor,* right pair: male and female lesser frigatebird *F. ariel.*

After King et al. 1975

Figure 7.24. Bridled terns *Sterna anaethetus* on P. Gundul in Karimunjawa in 1953. There is no recent information on the numbers still visiting the island.

By A. Hoogerwerf, with permission of the National Museum of Natural History, Amsterdam

large colonies, often on cliffs or small islands where there are fewer rats to predate on eggs or young, and less disturbance from people (fig. 7.25, table 7.3). Some islands appear to be roosting rather than nesting sites, but the pattern seen today is the consequence of human interference both direct and by the introduction of rats and other predators onto even the smallest islands. As a result of this, the seabird populations in Indonesia are experiencing a serious decline. For example, boobies are now seen in small groups or singly, but in the past they occurred in large flocks–there is a photograph of 275 boobies flying off the southeast of Nusa Barung (Appelman and Siccama 1939).

The habit of nesting in large colonies is disadvantageous because disturbance can have so serious an effect, but it has evolved for at least four important reasons: for ease of pair formation (most seabirds are solitary or live in small groups and range over vast distances when they are not breeding), to take advantage of a limited availability of suitable sites, for defence against predators (there is less risk on any one individual becoming the prey), and for the information shared concerning the locations of the abundant food necessary for feeding young birds.

Table 7.3. Seabirds roosting (R), nesting (N), or extinct (Ex) on or off Java and Bali. Sl - brown booby *Sula leucogaster*, Fa - lesser frigatebird *Fregata ariel*, Fm - great frigatebird *Fregata minor*, Am - white-capped noddy *Anous minutus*, As - brown noddy *A. stolidus*, Sta - little tern *Sterna albifrons*, Stal - bridled tern *S. anaethetus*, Stb - great crested tern *S. bergii*, Sts - black-naped tern *S. sumatrana*, Pl - white-tailed tropicbird *Phaeton lepturus*, Pr - red-tailed tropic bird *P. rubricauda*.

	Sl	Fa	Fm	Am	As	Sta	Stal	Stb	Sts	Pl	Pr
Thousand Islands						NR			NR		
Panaitan									NR		
Ujung Kulon	N?R					N?R?			NR		
Cikepuh islets						NR?			NR		
Karimunjawa - Gundul				NR	NR	NR	NR	NR	NR		
Bawean islets						NR			NR		Ex
Rongkop									NR	NR	
Nusa Barung	N?R?									N?R	
Bali Barat	R		R								
Uluwatu	N?R	R?								N?R	
Nusa Penida	NR									NR	

Sources: de Korte (1984); Salm and Halim (1984); de Korte and Silvius (1990)

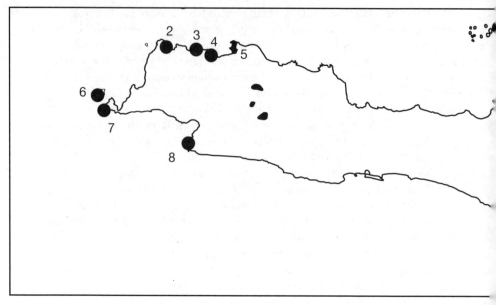

Figure 7.25. Nesting and roosting sites of seabirds around Java and Bali. 1 - Thousand Islands, 2 - P. Dua, 3 - P. Rambut, 4 - Muara Angke, 5 - Muara Gembong, 6 - P. Panaitan, 7 - Ujung Kulon, 8 - Cikepuh islets, 9 - Karimunjawa-P. Gundul, 10 -

Rocky Shores

The southern coasts of Java and Bali have some magnificent, tall cliffs, and many less dramatic rocky shores. These are buffeted by very strong waves and so only those animals and plants able to withstand such extreme conditions can survive. In some places the waves have cut caves and clefts into the rocks, and these provide roosts for bats and swiftlets. Lower down in the splash zone, rock-hopping blenny fish hold on to the rocks using their ventral fins like a sucker, scraping the thin film of algae off the rocks with their teeth. Large grapsid crabs are also a common sight. In the intertidal zone a series of communities can be distinguished, each of which requires a different ratio of exposure and wetting, although some species have a wide tolerance and are found throughout. The plant life is not particularly rich, the irregularly-shaped sea lettuce *Ulva* being the most characteristic, but there is a fair range of molluscs, including primitive chitons (fig. 7.26), and small crustaceans. The zoning is the outcome of the partitioning of resources between potentially competing species. The factors determining the upper limits tend to be physiological, but for the lower limits tend to be ecological and behavioural (Barnes and Hughes 1982).

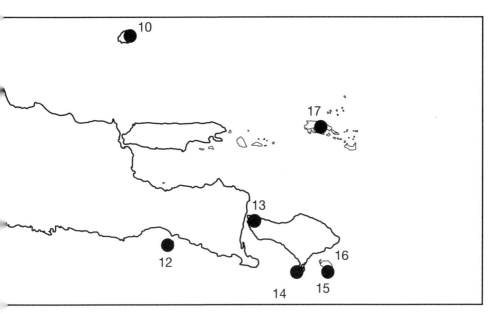

Bawean islets, 11 - Rongkop, 12 - Nusa Barung, 13 - Bali Barat, 14 - Ulu Watu, 15 - South Bali, 16 - Nusa Penida, 17 - Kangean islets.

After Salm and Halim 1984

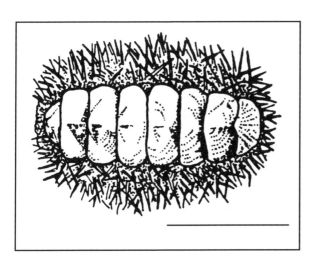

Figure 7.26. A chiton *Acanthopleura,* a primitive mollusc found on rocky shores. Scale bar indicates 1 cm.

After Whitten et al. 1987b

Chapter Eight

Mangrove Forest

The functions and products of mangrove forests are many and diverse. The transport of leaves and other debris out of these forests seems to provide an important source of nutrients for the inshore and neighbouring estuarine ecosystems, although there is not necessarily a direct correlation between amount of mangrove forest and productivity. Mangrove forests also act as nursery grounds for young fish, shrimp, and other organisms. In both these respects they can clearly play an important role in supporting coastal fisheries. Their physical presence also protects certain coasts from the destructive energy of waves. Among their products, the timber of some mangrove trees, notably *Rhizophora,* is in high demand for building, scaffolding, and the production of rayon, but it has never been harvested in a sustainable, commercial fashion on Java or Bali, and there is no large site remaining where this could now be achieved. Other mangrove forest products include fuelwood, wood for other uses, chemicals for tanning and dyeing, oils, green manure, and nipa palms *Nypa fruticans,* which have considerable potential in the production of alcohol (Mercer and Hamilton 1984). Mangroves also seem to give some protection against viruses attacking shrimps (Ami 1993).

Mangrove forests are the best-studied vegetation type in Indonesia (Kartawinata 1990), and a list of mangrove studies and surveys conducted on Java and Bali is provided in Appendix 5.

DISTRIBUTION

There were once quite extensive mangrove forests on Java, particularly along the north coast and around Segara Anakan (Fig. 8.1). In 1981 there were about 500 km^2 (Sukardjo 1990), which had declined to about 338 km^2 by about 1987, over half of which was around Segara Anakan in south Central Java. On Bali there were about 5 km^2 of mangrove forests in 1987 (RePPProT 1989). Since 1987 there does not seem to have been any let up in the loss of mangrove forest, but current figures are not available. In any case, mapping their extent is complicated by the fact that the

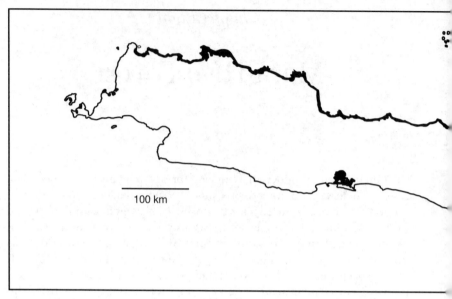

Figure 8.1. Original distribution of extensive mangrove forest on Java and Bali. Derived from KNAG (1938) showing coastal swampy areas. Note that many important coastal fringes of mangrove forest do not show at this scale.

remaining areas vary from luxuriant and relatively undisturbed (e.g., on Panaitan Island) to scrubby wetland. Thus, as with other forest types, the loss of area is only part of the problem because there has also been major degradation with accompanying decreases in mangrove richness, diversity, and structure.

The remaining mangrove forests are in theory managed either by the Department of Forestry (in a variety of conservation areas) or by Perhutani, but the remnants are still being replaced by industrial or real estate schemes or brackish-water ponds, irrespective of any management prescriptions. A 'green belt' policy is generally supposed to operate, leaving a minimum of 50 m of mangrove forest between the shore and any development, but this has yet to be widely implemented. Under the green belt legislation (Presidential Instruction No. 32 1990), three options are available:

- the area can be left alone either because there is no mangrove habitat, or because there is an extensive area of mangrove on a prograding coastline, or because local management of the mangrove does not threaten the mangrove vegetation;
- the area can be gazetted as some form of protected area with the potential to preserve the mangrove habitat and allow natural regen-

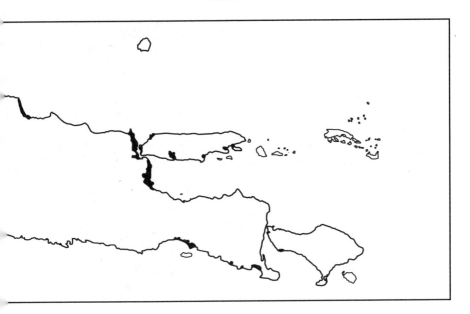

eration processes to restore the mangrove habitat; and

* the area can be reforested.

The third option is clearly the most intensive and expensive, and it is important that the critical questions of whether the planted trees will grow, and what benefits are expected, are answered before a project commitment is made. In any case, the green belt's ecological justification and application in coastal resort development is controversial (Sloan and Sugandhy 1994).

The two largest mangrove forests remaining in Java and Bali are Segara Anakan, opposite Nusa Kambangan, and Segoro Anak in Grajagan Bay (fig. 8.2), but there are also pockets along the south and north Java coast, around Ujung Kulon, and Benoa Bay in southern Bali (Sukardjo 1984a). The last of these is being developed and extended as a Grand Forest Park. Other major areas used to be found from Cilegon to Indramayu, around Surabaya, and Perencak Bay in Bali, but these have been almost entirely lost through overexploitation for firewood and conversion to brackish-water fish and shrimp ponds.

Figure 8.2. Mangrove-fringed lagoon in Segoro Anak, Alas Purwo National Park.

By A.J. Whitten

VEGETATION

Some 35 plant species have so far been recorded in the mangrove forests of Java and Bali, but the species composition of the forest depends on soil type, tide characteristics, distance from the sea, and human disturbance (Sukardjo 1990) (table 8.1). The most important plant family in mangrove forests is the Rhizophoraceae with *Rhizophora, Bruguiera,* and *Ceriops,* followed by Avicenniaceae (*Avicennia*) and Sonneratiaceae (*Sonneratia*) (fig. 8.3-4). Only one mangrove tree is found on purely sandy substrates and that is *Rhizophora stylosa,* although *Sonneratia* spp. can sometimes be found growing on sand-covered, dead coral reefs. There are few differences between the forests in the wetter and drier climatic zones of Java and Bali, except that epiphytes are less common in the latter. The landward fringe of a mangrove forest would have *Lumnitzera racemosa* (Comb.), *Intsia bijuga* (Legu.), pandans (Pand.), nipah *Nypa fruticans* (Palm.), and the spiny-trunked *Oncosperma tigillaria* (Palm.) but, in Java and Bali, this type of vegetation has generally been converted to some form of agriculture or fisheries production. It can still be seen, however, in Segoro Anak and in Ujung Kulon National Park. A cross-section through mangrove forest in

Segoro Anak is shown in fig. 8.5.

Many different types of mangrove communities have been identified in Southeast Asia (e.g., Sukardjo 1990), but given the considerable variation in macro- and micro-environmental conditions there are probably no real 'types' which have any regional meaning or use. At any one location, a zonation of species and communities can be discerned as environmental gradients are followed, but the patterns are neither constant nor particularly precise (Watson 1928; de Haan 1931). The reason for this is that the zones are formed as a result of a complex set of processes including geomorphological changes, physiological differences among the species, and different levels of seed predation (Anwar et al. 1984; Whitten et al. 1987a, b, c), the individual effects of which are virtually impossible to tease out, and which, it could be argued, are not terribly relevant in the context of Java and Bali where human activities are probably more potent than any of the above natural processes. This is not to say, of course, that species do not

Table 8.1. Major plant species (by family) found in the mangrove forests of Java and Bali

Trees		Shrubs and Herbs	
Apocynaceae	*Cerbera manghas*	Acanthaceae	*Acanthus ebracteatus*
Bignoniaceae	*Dolichandrone spathacea*		*A. illicifolius*
Combretaceae	*Lumnitzera littorea*		*A. volubilis*
	L. racemosa	Leguminosae	*Derris heterophylla*
Euphorbiaceae	*Excoecaria agallocha*	Lythraceae	*Pemphis acidula*
Meliaceae	*Xylocarpus granatum*	Palmae	*Nypa fruticans*
	X. moluccensis	Rubiaceae	*Scyphiphora hydrophyllacea*
Myrsinaceae	*Aegiceras corniculatum*	Rutaceae	*Merope angulata*
Myrtaceae	*Osbornia octandra*		
Rhizophoraceae	*Bruguiera cylindrica*		
	B. gymnorrhiza	**Fern**	
	B. parviflora		
	B. sexangula	Dennstaedtiaceae	*Acrostichum aureum*
	Ceriops decandra		
	C. tagal		
	Kandelia candel		
	Rhizophora apiculata		
	R. mucronata		
	R. stylosa		
Sonneratiaceae	*Sonneratia alba*		
	S. caseolaris		
	S. ovata		
Sterculiaceae	*Heritiera littoralis*		
Verbenaceae	*Avicennia alba*		
	A. marina		
	A. officinalis		

Source: Sukardjo (1990). Reprinted with permission of Kluwer Academic Publishers.

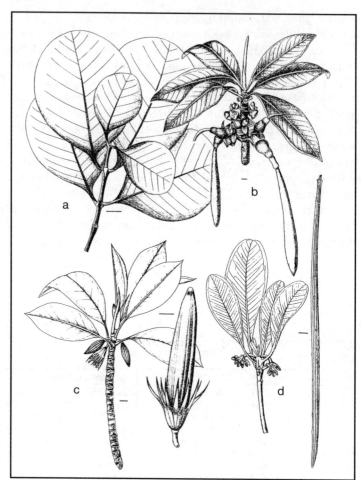

Figure 8.3.
Common mangrove trees. a - *Sonneratia alba,* b - *Rhizophora apiculata,* c - *Bruguiera sexangula,* d - *Ceriops tagal.* Scale bars indicate 1 cm.

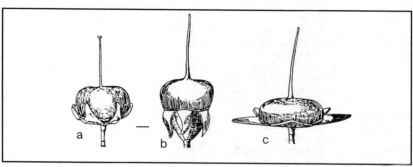

Figure 8.4. Fruits of the three species of *Sonneratia* found in mangrove forests of Java and Bali. a - *S. ovata,* b - *S. alba,* c - *S. caseolaris.* Scale bar indicates 1 cm.

After Koorders and Valeton 1914

Figure 8.5. Cross section through mangrove forest in Segoro Anak. a - *Avicennia officianalis*, b - *Rhizophora mucronata*, c - *Bruguiera gymnorrhiza*, d - *Xylocarpus granatus*, e - *X. moluccensis*, f - *B. cylindrica*, g - *Ceriops tagal*, h - *Excoecaria agallocha*, i - *Lumnitzera racemosa*, j - *Acrostichum aureum*, k - *Acanthus ilicifolius*. Vertical scale bar indicates 25 m.

After Sulistiadi 1986

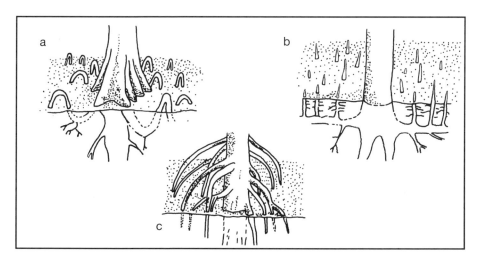

Figure 8.6. Different types of rooting systems in mangrove trees. a - knee roots as found in *Bruguiera* spp., b - spike roots as found in *Sonneratia* spp., *Avicennia* spp., and sometimes *Xylocarpus moluccensis*, c - stilt roots as found in *Rhizophora* spp.

After Whitten et al. 1987b

have clear habitat preferences, for *Avicennia* tends to be found on soft mud, *Sonneratia* is most commonly found in seaward areas on compact soils, and *Rhizophora* lies behind them. Interestingly, although these trees are associated exclusively with brackish soils and inundation, they are not restricted to these conditions and will grow successfully in fresh water in botanic gardens and in the swamp forest of Rawa Danau (p. 773). This suggests that they successfully outcompete other trees in the coastal areas, but rarely have the competitive edge in inland areas.

Because the soil below mangrove forest is water logged and devoid of oxygen, peculiar root systems have developed in most of the mangrove trees in regularly inundated areas to allow for gases to be exchanged above the soil (fig. 8.6). These aerial roots are known as pneumatophores or 'breathing roots', and the stilt roots of *Rhizophora* may serve the additional function of preventing the growth of seedlings close to an existing tree. In *Sonneratia* and *Avicennia* (and sometimes *Xylocarpus moluccensis*) the spike roots are side shoots off the underground roots, whereas in *Bruguiera* the underground root itself loops in and out of the soil, forming 'knees' above the soil. All these aerial roots have large openings or lenticels through which gases can pass. *Ceriops* trees do not have specialized roots, but the bark has large lenticels to assist in gas exchange. If these lenticels become covered, with oil from a spill for example, the plants will very soon suffocate and die.

Mangrove forests are distinctly different from all other forest types in the region in having virtually no climbing or understorey plants (Ding Hou 1958; fig. 8.7). This is probably partly because regular tidal flooding acts against the establishment of herbs and climbers, but other mechanisms may yet be determined (Janzen 1985; Corlett 1986).

Heavily degraded mangrove areas are characterized by the mangrove fern *Acrostichum aureum* (Denn.) (fig. 8.8), sea holly *Acanthus ilicifolius* (Acan.), *Pluchea indica* (Comp.), *Premna obtusifolia* (Verb.), and *Clerodendron inerme* (Verb.). In some of the disturbed areas, such as parts of the Gembong and Cimanuk deltas, all that is now left is isolated, large trees standing among weeds such as *Panicum repens, Paspalum commersonii, P. vaginatum, Sporobolus* sp. (all Gram.) (Sukardjo 1984b; BScC 1992). The dykes around intensive tambaks have very little in the way of vegetation except for a poor herbaceous cover of weedy *Suaeda, Arthrocnemum indicum,* and *Tecticornia cinerea* (all Chen.), and succulent-leaved *Sesuvium* and *Trianthema* (both Aizo.). In traditional and new silvo-cultural systems, *Rhizophora* trees are commonly seen.

Reproduction

The flowers of Rhizophoraceae are pollinated in a number of ways. In *Bruguiera,* for example, the pollen explodes out of the anthers when dis-

Figure 8.7. Coastal fringe of mangrove forest with *Sonneratia* and *Rhizophora,* east of Besuki, East Java, to show the typical lack of understorey plants.

By A. Jeswiet, with permission of the Rijksherbarium/Hortus Botanicus, Leiden University

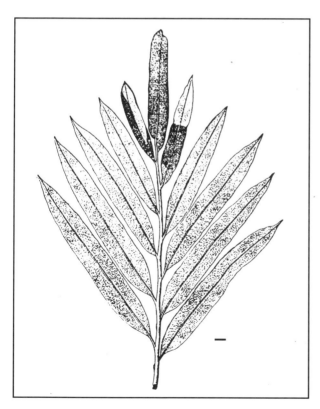

Figure 8.8. The mangrove fern *Acrostichum aureum,* commonly seen in degraded mangrove. Scale bar indicates 1 cm.

After Backer and Posthumus 1939

turbed by nectar-seeking sunbirds *Nectarinia* in large-flowered species, and by butterflies and other insects in small-flowered species. *Ceriops tagal* has a similar strategy but is pollinated at night by moths. *Rhizophora* flowers are probably wind pollinated but bees may also be involved. The flowers of *Sonneratia* are feathery and reminiscent of rose apples *Eugenia* and eucalypts *Eucalyptus* and, like many of those, are pollinated by fruit bats, particularly the small long-tongued fruit bat *Macroglossus* spp. and the medium-sized and closely-related cave fruit bat *Eonycteris spelaea*. The latter is the species that roosts in thousands around the temple at Goa Lawah in Klungkung, Bali. It feeds primarily on nectar, and has been found to travel at least 40 km to a good food source (Start and Marshall 1975). The mangroves in Benoa Bay are well within this range.

There are at least some mangrove trees fruiting or flowering in most months but, as in inland forests, there tends to be a peak of flowering during the dry season, and ripe fruits are dropped during rainfall peaks. The seasonal patterns are, in part, influenced by the abundance of insects which act both as pollinators and predators on seeds, but are obscured because the fruit of different species take different periods to develop and ripen. For *Avicennia* the time required is only 2-6 months, but for *Ceriops* it requires about 15 months (Christensen and Wium-Anderson 1979; Wium-Anderson and Christensen 1978; Wium-Anderson 1981; Duke et al. 1984).

The familiar long, green, dagger-shaped 'fruit' of *Rhizophora* trees is in fact an already germinated seed (fig. 8.9). The knobbly, long, green part is a root, and when this falls from the parent plant, the young first leaves jut out of the top; the fruit case from which the root became detached remains on the tree for a while before also falling. The root may penetrate the mud below the parent, or it may be swept away and washed up in a new location.

Dynamics

The mass of the above-ground parts of undisturbed mangrove forests throughout Southeast Asia and Australia appears to be in the range of 100-250 t/ha (Ong et al. 1980a, 1985; Boto et al. 1984), although the mass of a mangrove forest which had been subject to sustainable silvicultural management in Peninsular Malaysia over 80 years is about 300 t/ha (Ong et al. 1980b). The total production of 'litter' (leaves, twigs, branches, flowers, fruit, etc., falling to the forest floor) is 7-14 t/ha/year, similar to or higher than figures obtained for most lowland forests (Sukardjo and Yamada 1992). The productivity is similar, however, to young or pioneer forest on relatively young, nutrient-rich river terraces. In both cases the trees grow, reproduce, and die relatively fast (Jimenez et al. 1985).

The variation in net primary production in mangrove forests has been ascribed to the availability of phosphorus (Boto et al. 1984), and this is

Figure 8.9. *Rhizophora* seeds at different stages of germination.

By P. Arens, with permission of the Rijksherbarium/Hortus Botanicus, Leiden University

consistent with the view that both nitrogen and phosphorus are limiting in coastal environments (Rhyther and Dunstan 1971). Given that the mangrove areas of Segara Anakan and Benoa are enriched from outfalls of rubbish and river-borne effluent probably with high levels of both phosphorus and nitrogen, the net primary production might be expected to be higher there than, say, in Segoro Anak.

Probably less than 10% of the production of mangrove leaves is consumed in the form of living leaves by herbivorous animals (Johnstone 1981), the remainder entering the ecosystem as detritus (dead organic matter). The mangrove forest on the west of Prapat Agung, West Bali, is said to lose most of its leaves in the dry season, but the ground does not become covered with leaves (Soepadmo 1961). This is because much of the leaf and other litter that falls onto the floor of the mangrove forest is gathered and pulled into burrows by the crabs, the great abundance of

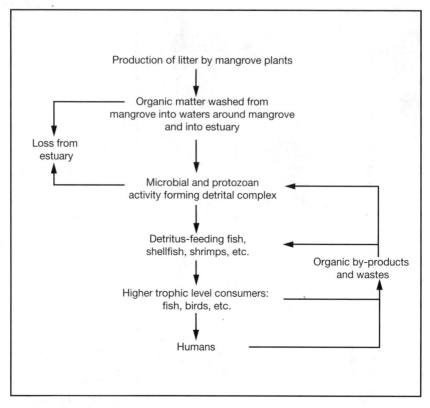

Figure 8.10. Major pathways of energy flow in a mangrove-fringed estuary.
After Saenger et al. 1981

which is only really appreciated if one waits quietly for a while, allowing them to emerge from their holes. Up to 90% of the mangrove leaves are either eaten or buried by crabs within three weeks of falling and enter the system again as excreted detritus which is enriched by fungi, bacteria, and algae growing on and within it. If crabs are prevented from reaching the leaves, the same degree of decomposition does not result for nearly six months (Ong et al. 1980a).

The detritus is eaten by a host of small animals such as zooplankton, prawns, other crabs, and small fish, which in turn are eaten by carnivorous animals, mainly fish. Many of these organisms are of major commercial importance, and the most important of them are the penaeid prawns, the juvenile stages of which live among, and in front of, mangrove vegetation.

In extensive prawn culture systems these are caught in coastal waters and raised in ponds. It has been established that mangrove detritus is also utilized by creatures that are not strictly part of the mangrove ecosystem, such as the commercially important cockle *Anadara granosa,* oyster *Crassostrea, Acetes* shrimps (used in making the fermented *terasi* paste used widely in cooking), swimming crabs *Scylla serrata,* and many fish such as mullet *Mugil* spp., milkfish *Chanos chanos,* and giant perch or barramundi *Lates calcarifer* (Polunin 1983; Rondelli et al. 1984). Thus this detritus is extremely important in the productivity of mangrove and adjacent ecosystems (fig. 8.10), though its importance to offshore systems has yet to be demonstrated. Unfortunately, the existing nature reserves are far too small to wholly serve this purpose.

FAUNA

The fauna of mangrove forests can be divided conveniently into two groups: the essentially aquatic component of crabs, snails, worms, bivalves and others which depend directly upon the sea in various ways, and the terrestrial, often 'visiting', component including insects, spiders, snakes, lizards, rats, monkey, and birds, which do not depend directly upon the sea. The animals mentioned below represent only a fraction of the total mangrove fauna but serve to illustrate that there are many different habitats within the mangrove forest, each of which supports a distinct community of animals, and that a proportion of the fauna is more or less restricted to mangroves. Faunal zonations within mangrove forest are effectively three-dimensional. The vertical component is shown, for example, by the worms and crabs burrowing into the mud, some snails living on the mud surface, and other snails and barnacles living in the trees. A time dimension can also be observed as animals climb up and down trees, or emerge and hide in the mud, in time with the tidal inundation.

The Challenges of Living in Mangroves

In comparison with other shore fauna, the mangrove forest fauna is heavily dominated by many species of crabs and snails. There are few types of worms, and very few bivalve molluscs, coelenterates (sea anemones), echinoderms (starfish, sea urchins), and so forth. (fig. 8.11). The dominance of crabs and snails may be ascribed to their successful adaptations to the peculiar conditions of mangrove forests. Animals living on the soil of the inter tidal area are subjected to long periods when they are not covered by the sea. At the mean high water level of spring tides, for example, the soil is left exposed for about 270 days per year, and for up to 25 consecutive

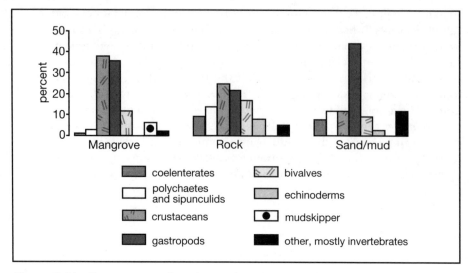

Figure 8.11. Percentages of total aquatic animal taxa recorded at three types of shore (omitting microscopic forms). Note that crustaceans and gastropod molluscs account for 74% of the total on the mangrove shore, and that these data, while not based on complete lists or on equally exhaustive surveys, serve to illustrate the major differences.

Data from Berry (1972)

days; at the inner margin of the mangrove forest the soil is left exposed for about 320 days per year, and for up to 30 consecutive days. Most marine animals cannot tolerate this because:

- they quickly dry out;
- most cannot respire in air;
- many can only feed on water-borne food; and
- many must release spermatozoa, eggs, and larvae into seawater.

Crabs and snails, however, have:

- impervious exoskeletons/shells to restrict water loss;
- many can breathe air;
- many feed on microorganisms or organic materials;
- many climb into the trees to find these foods; and
- they have internal fertilization and protect their eggs and early developing young in capsules or brood pouches.

The soil in mangrove forests is subjected to salinities ranging from nearly zero to 50%. During spring tides, salinities approximate those of tidal seawater, that is about 27-32%. During neap tides, however, when the landward mangrove soil is not covered by the sea for days at a time, rainfall may reduce salinity to less than 15%. Conversely, evaporation without rainfall can increase salinity to more than 32%. Most marine animals are only able to withstand very minor variations in salinity such as are experienced in the open sea because they cannot regulate the salt/water balance of their body fluids except within quite narrow limits. Many crabs are able to cope with wide salinity variation by controlling the balance in their bodies, whereas many snails can allow the salt concentration of their body tissues to vary without ill effect. These adaptations thus allow crabs and snails to predominate in the mangrove fauna.

Mammals

Ground-living mammals, such as wild boar, mouse deer, wild cats, rats, and lizards only venture into the landward edge of mangrove from inland areas for brief periods, because the soil is often too soft, and burrowing is not possible. The two mammals seen most often in mangrove are the Javan lutung *Semnopithecus auratus* (fig. 8.12) and the long-tailed macaque *Macaca fascicularis*. These both live in the canopy, eating leaves, shoots, and fruit, although the macaques also come to the ground and eat crabs, peanut worms, and small vertebrates. Faecal analyses have also shown that both species eat the nectar-rich flowers of *Sonneratia* (Lim and Sasekumar 1979). At Muara Gembong the lutung ate 60% leaves (with relatively high protein concentrations) and 40% fruit, while the macaque consumed 50% fruits (with relatively high carbohydrate), with lesser amounts of leaves, bark, roots, insects, and other animals. Both monkey species are widely distributed, and their diet differs in different habitats, but the basic difference in food preference remains, even in captivity. Although the two types of monkey at Muara Gembong do eat some of the same foods, it seems that at least in some cases, the same plant parts are eaten at different stages of development, for example, lutung tend to eat younger fruit than the macaques (Supriatna et al. 1989).

Leaves are a poor source of food for most mammals because the cellulose molecules are very long and cannot be broken down without special digestive adaptations. Leaf monkeys, such as the lutung, have a highly sacculated stomach, like a cow's, in which special bacteria ferment and break down cellulose into smaller, digestible molecules. In addition, many tree leaves are protected from herbivores by defence compounds, the most common of which are fibre (lignin) and phenols, particularly tannins. In a normal acidic digestive tract (the content of a human stomach are generally at pH 3), tannins will bind strongly onto any protein, both food and enzymes, thereby hindering the digestive process. In a leaf monkey's complex stomach, however, the pH is probably maintained between 5.0 and 6.7 (Bauchop 1978), and in such conditions not only are the fermentation

Figure 8.12. Javan lutung in a remnant patch of mangrove forest at Muara Gembong.

By Biological Sciences Club, with permission

bacteria able to thrive but the attachment of tannins to proteins is weaker. Both lutungs and long-tailed macaques have been observed to eat bark from the twigs of *Bruguiera* (Lim and Sasekumar 1979), which have considerable quantities of tannins–perhaps 30% dry weight. For the macaques, which have a simple stomach, this may be eaten as a stomach purgative.

Another way in which these apparently similar animals differ is in the way they treat fruit seeds. Any fruit eater is faced with the problem of what to do with the seeds which are often relatively hard, large, and indigestible. Macaques spit out seeds greater than 3-4 mm in diameter after removing their flesh, while leaf monkeys chew and swallow the seeds, which are thus destroyed (Corlett and Lucas 1990). The seed spitting of the macaques has the advantage that unnecessary food bulk entering the stomach is reduced, but it does mean that even small fruits must be processed one or a few at a time before the seeds are spat out, which seems to conflict with the need for a frugivore dependent on low-nutrient fruit flesh to maximize efficiency in handling its food. A macaque gets round this problem by using pouches in its cheeks into which excess fruits can be crammed during a feeding bout to be processed at leisure when searching for another patch of suitable food (Corlett and Lucas 1990). The lutung eats slower than a macaque because of its complicated digestive system, but many of the seeds it eats and digests have high nutritional value as compensation (Davies et al. 1988).

Among the other mammals, the small-clawed otter *Aonyx cinerea*, and the fishing cat *Felis viverrina* would once have been common residents of mangrove forests, but Java has been largely robbed of these animals as a result of the great reduction in the area of this forest. The status of these species is currently being investigated (Melisch et al. 1994a, b). Fruit bats, particularly the flying fox *Pteropus vampyrus,* roost in the mangrove forest canopy, and other bats such as the cave fruit bat *Eonycteris spelaea* and the long-tongued fruit bat *Macroglossus sobrinus* are very important in the pollination of the flowers of *Sonneratia* (p. 210) (Start and Marshall 1975).

Birds

Although the larger waterbirds are the bird species most generally associated with mangroves, there is also a distinct community of smaller birds including cuckoos, woodpeckers, warblers, and sunbirds which spend their lives in this forest type. Indeed, about 170 species of birds have been recorded from Javan mangrove forests, although this total includes occasional visitors and species seen on the periphery of the forest. Two 'typical' mangrove species, the mangrove blue flycatcher *Cyornis rufigastra* and mangrove whistler *Pachycephala cinerea,* appear to be rather uncommon. Mangrove forests are potentially an attractive bird habitat–the high productivity of the ecosystem ensures an abundance of small animal prey for kingfishers and shrikes, many of the flowers have sweet nectar which attracts both birds directly and indirectly through the insects attracted, and large, dead trees attract wood-boring or hole-nesting species (van Balen 1989d).

The different ways in which mangroves are used by birds have been listed (Nisbet 1968; van Balen 1989d):

- species confined exclusively to mangroves. In Java only six species belong to this category;
- species nesting in mangroves but feeding elsewhere. This group mainly comprises waterbirds;
- coastal species for which mangrove is their major habitat. Though the species belonging to this group are sometimes found in other habitats, they are considered as characteristic of the mangroves. Three species belong to this group;
- species in Java which move daily (or seasonally) to and from mangroves. Though the significance of periodic visits by frugivorous pigeons and other birds is not fully understood, it is believed that the mangrove trees are used as roosting sites with fruiting trees in the neighbourhood, though some pigeons eat shoots and buds of mangrove trees;
- northern (and southern) migrants which use mangroves as well as lowland areas. In Java only a few insectivorous birds and a kingfisher belong to this group, totalling seven species;
- aerial species such as swallows and swiftlets which feed over mangroves as well as over land. They only nest in the area if suitable substrates are available (buildings, bridges, etc.);

Figure 8.13. The lesser adjutant is a rare bird found in Segara Anakan.
By A. Hoogerwerf, with permission of the Tropen Instituut, Amsterdam

- species of open woodland which are also found in mangroves;
- species of rural and urban areas which are also found in mangroves. This is the largest group, with 21 species, and includes the most common birds in Javan towns and villages; and
- species of rain forest birds which are also found in mangroves. In Java, this group comprises just four species which are probably stragglers or occur only along the peripheries and are not permanent residents of the mangroves.

The extensive (though diminishing) mangrove forests and associated mudflats of the Segara Anakan lagoon near Cilacap support at least 85 species of birds, including 160-180 non-breeding milky storks *Mycteria cinerea*, an internationally endangered species (p. 779), and at least 25 of the rare lesser adjutants *Leptoptilos javanicus* (fig. 8.13). The rare endemic Javan coucal *Centropus nigrorufus* is also found (Erftemeijer et al. 1988). The mangrove forests appear to contain at least 18 bird species which live permanently in the mangroves (table 8.2), more than any other mangrove area in Java or Bali. Of the 18 species, ten are also common in urban and suburban areas, and six in open woodland, sometimes far from mangrove forests. Two of the species of mangrove birds, greater goldenback woodpecker *Chrysocolaptes lucidus* and great tit *Parus major*, which in Malaysia are strictly confined to mangroves, are, in Java, found in a much wider range of habitats, possibly due to a longer history of widespread cultivation in

Table 8.2. Birds species living permanently, though not necessarily exclusively, in the mangrove forests of Segara Anakan.

	1	2	3	4	5
Javan coucal					
Centropus sinensis	√				
Stork-billed kingfisher					
Pelargopsis capensis		√			
Small blue kingfisher					
Alcedo caerulescens		√			
White-collared kingfisher					
Halcyon chloris					√
Pacific swallow					
Hirundo tahitica			√		
Common iora					
Aegithina tiphia					√
Racket-tailed treepie					
Crypsirina temia				√	
Slender-billed crow					
Corvus enca				√	
Great tit					
Parus major					√
Striped tit-babbler					
Macronous gularis				√	
Flyeater					
Gerygone sulphurea					√
Ashy tailorbird					
Orthotomus sepium					√
Bar-winged prinia					
Prinia familiaris					√
Pied fantail					
Rhipidura javanica					√
Brown-throated sunbird					
Anthreptes malacensis					√
Olive-backed sunbird					
Nectarinia jugularis					√
Purple-throated sunbird					
Nectarinia sperata				√	
Copper-throated sunbird					
Nectarinia calcostetha				√	

Source: van Balen (1989d)
1 = Found exclusively in mangroves
2 = Found mainly in mangroves
3 = Found above mangroves
4 = Found in open woodlands and mangroves
5 = Found in rural and urban areas as well as mangroves

Table 8.3. 'Pairs' of similar bird species found in and around Javan mangroves.

Mangrove		Surrounding habitats	
Javan coucal	*Centropus nigrorufus*	Greater coucal	*Centropus sinensis*
Mangrove blue flycatcher	*Cyornis rufigastra*	Hill blue flycatcher	*Cyornis banyumas*
Javan white-eye	*Zosterops flava*	Oriental white-eye	*Zosterops palpebrosa*

Source: van Balen (1989d)

coastal areas, or possibly to the smaller absolute number of bird species in Java which allows certain species greater latitude for niche expansion. Most of the mangrove forest birds are insectivorous or nectarivorous, with those eating seeds or fruit rather uncommon. Some birds are relatively confined to mangroves because there is a similar, congeneric species outside those areas against which they cannot compete successfully (and vice versa) (van Balen 1989d). As the mangroves diminish in quantity and quality, so the former set of species will eventually disappear.

Reptiles and Amphibians

Mangroves are inhabited by a variety of reptiles such as the monitor lizard *Varanus salvator,* the common skink *Mabuya multifasciata,* and snakes such as the arboreal common cat snake *Boiga dendrophila* (fig. 8.14) and reticulated python *Python reticulatus.* The cattle snake *Elaphe radiata* is found on the ground, and dog-faced water snake *Cerberus rhynchops, Acrochordus granulatus,* puff-faced water snake *Homalopsis buccata,* and crab-eating water snake *Fordonia leucobalia* are found in the shallow water, where they feed on small fishes (Supriatna 1982). Most snakes seen in or near mangroves are not in fact sea snakes, as those live primarily in open water. The largest animal that could potentially live in the mangrove forests is the estuarine crocodile *Crocodylus porosus.* At the beginning of this century they were said to still be common along the coasts of Java (Cabaton 1911), but persecution has reduced their numbers to a very low level, and it is likely that large specimens (they can exceed 9 m) no longer exist around Java and Bali (p. 246). They are able to swim long distances, however, and so it is not impossible that relatively large individuals could arrive from Sumatra or Kalimantan, but even here their numbers are much reduced.

The mangrove frog *Rana cancrivora* is exceptional among amphibians in being able to live and breed in weakly saline water. The tadpoles are more resistant to salt than the adults, and metamorphosis into adults will only occur after considerable dilution of the salty water (MacNae 1968).

Insects and Spiders

Insects are probably the most significant devourers of mangrove plant material, even taking into account the crabs and molluscs. Where the mangrove forests are sufficiently extensive, such consumption is likely to be controlled naturally. As the mangrove forests are split up into smaller and less diverse blocks, however, insect attacks that affect the growth and survival of the trees are likely to become more frequent. This has been observed near Medan, North Sumatra (Whitten and Damanik 1986). Unfortunately, so little is known about 'natural' levels of herbivory that determining or predicting cause/effect relationships in cases of pollution,

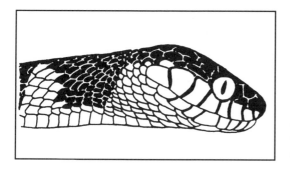

Figure 8.14. Common cat snake
Boiga dendrophila is a common
snake in mangrove forests.
After Tweedie 1983

or deterioration of soil or water quality in an environmental assessment, is singularly difficult. A detailed discussion of mangrove insects, together with colour photographs of the main species found in Singapore, where the mangrove communities are very similar to those found in Java and Bali, is available (Murphy 1990). The most conspicuous of the insects are probably mosquitoes, but only one of these, *Anopheles sundaicus*, carries malaria (p. 657). The larvae of this species can live in water with a salinity of 13%. Species of *Aedes* mosquitoes (p. 658) have been seen feeding on mudskippers, but they are also attracted to humans.

Spiders are abundant in mangroves but their presence is poorly documented (Murphy 1990), although a start is being made (Koh 1991).

Uses, Management, and the Future

Uses

Despite the rapid rate of mangrove attrition, it is clear that mangroves are widely believed to be useful, and this is shown by the many planting initiatives. The first programme for the replanting of mangrove trees in Java and Bali was instituted during the Dutch colonial period (Schnepper 1933), but the first modern plantings were conducted by Perhutani in 1965 near Cilacap, east of Segara Anakan. Since then other areas have been planted around Java, Madura and Bali by Perhutani, provincial Forestry offices, and local village groups. In some cases this has been done with the aim of supporting local fisheries and providing a timber crop, but elsewhere the primary reason is to save recurring expenses of protecting shore roads from wave erosion (Soesilotomo 1993). All areas with replanting programmes face major problems of boundary consensus and demarcation, land tenure, and conflicts of opinion over the desired, traditional, and proposed land use. In certain areas of West Java, Perhutani is working with farmers as part of its Social Forestry Programme to rehabilitate mangrove forests witha view to developing silvo-fisheries (for milkfish, wild shrimp, windu shrimp, and crabs) and timber production (Effendi 1990; Perhutani 1992a; Sastroamidjojo 1993).

It is worth examining the claimed ecological uses or roles of mangrove forest because the situation is not as straightforward as is sometimes made out. First, there is the role of maintaining the nutrient and energy export from the land to the sea, upon which many offshore fisheries depend (e.g., Morton 1990). The quantity of material exported will obviously differ from place to place, and if there are plans to replant an area of mangrove with the intention of assisting offshore fisheries it needs to be demonstrated that the planted mangrove forest will indeed contribute litter and nutrients in quantities which will be judged significant to those entering the sea from rivers bringing material from upstream. It is therefore not inevitable that a wider fringe of mangrove will benefit fisheries.

Second, it is widely recognized that mangroves protect shorelines from erosion. The evidence for this role is equivocal, however. Mangroves actually have limited capacity to withstand the erosion forces of wind, waves, or tidal currents, and on all but low-energy eroding shoreline, planted mangroves seedlings are often swept away by the sea unless special care is taken. An exception is *Rhizophora stylosa,* which does sometimes occur on wave-swept shorelines on stable coralline substrates. On prograding coastlines with a fringe of *Avicennia* or *Sonneratia,* however, replanting may be unnecessary because adequate and suitable propagules are produced naturally.

Third, it is clear that the conservation of biological diversity is enhanced by large areas of mangrove, and without them certain species such as fishing cat *Felis viverriana* or mangrove trees of the landward fringe face local extinction. The presence of humans is not necessarily detrimental to all members of the community, however, because over 25 species of mainly migratory wading birds can be seen on the mudflats bordering the mangroves in busy Benoa harbour, Bali, and the mudflats beyond the extensive tambak of Sidoarjo south of Surabaya are of international importance for this group of birds. In traditional extensive tambak areas, the tree-planted bunds can support breeding colonies of herons and egrets (if not too disturbed), and the aquatic fauna is rich. Such areas may not rate as highly as large areas of mangrove forests, but they do act as corridors for movement between more natural mangrove areas, and they are certainly more benign than intensive tambak areas.

Fourth, while few areas of mangrove have captured the interest of tourists, this is more of a public relations problem than a failing of the mangroves themselves. There is a major development planned in Benoa Bay, close to the international airport in Bali, however, where natural and planted mangrove will form the centre of a Grand Forest Park with tourist facilities such as campsites, walkways, a laboratory, bird observatory, arboretum, library, tambak exhibits, boats for hire, and souvenir shops. This will certainly be the mangrove with the highest profile in the country.

The conclusion must be that all the common arguments for conserving mangrove forests based on their uses are valid, but that not every argument can be applied with equal force to each mangrove area. The conservation of Java's mangroves, or rather the lack of it, makes an interesting case study. Knowledge of their ecology and calls for their conservation have been almost entirely indigenous, even if this is part of an international awareness of mangrove values. Every four years a national mangrove seminar is organized by the Indonesian Institute of Sciences, Department of Forestry, and Perhutani in different parts of Indonesia, and there are fine words from senior government officers about the importance of mangroves and the need to understand and conserve them. At the end of the seminar, resolutions are drafted calling for the protection of this important ecosystem. At the same time, licences have been granted for intensive brackish-water ponds and other developments that involve the clearance of mangrove forests, and many of these are in western Java, close to the offices and homes of many senior government officials. Unlike some other ecosystems, the economic arguments for the conservation of mangrove forests have been very clear—remove them and the nursery function for important fishes and prawns, the ecological support for coastal fisheries, the 'cheap' protection of coastlines, and the source of a wide range of forest produce all disappear. These were not disputed, at least not publicly. Decision makers are forever calling on ecologists to justify conservation moves on economic grounds, and mangroves were the shining example on Java where this could be done. Despite all this the boundaries of the mangrove forests continued to recede year by year.

The twist in this history is that prawns are now raised and bred in captivity, and numerous hatcheries support the capital-intensive prawn ponds found around Java and Bali (p. 641). Thus the days of mangrove forests being needed to maintain the ponds' supplies of young shrimp are nearly past. Not only that, but work at the Bogor Agricultural University has shown that milkfish can now be induced to spawn in tanks. Although this has not yet reached a commercial scale, the sight of people pushing large, triangular nets through the coastal shallows to collect the *nener* larvae of milkfish may yet be numbered (fig. 8.15). So, some of the economic arguments of conserving mangroves seem to be less convincing than they were, and this poses the question of whether conservation arguments should not in fact be placed on a level above the changing winds and vagaries of economics (p. 77).

Figure 8.15. Fisherfolk off the north coast of Madura catching milkfish larvae, and a girl carefully removing the individual larvae from the catch.
With the permission of Kompas

Reserves

Relatively extensive but heavily exploited mangrove forest is found in Alas Purwo National Park, but the most secure natural mangrove areas along the north coast of Java are the two small bird islands of D. Dua and R.

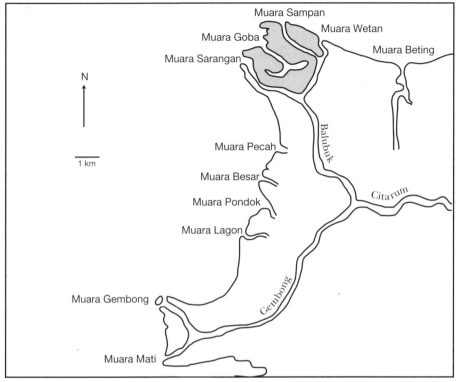

Figure 8.16. Area (shaded) of Muara Gembong proposed for conservation in 1980. Most of the forest has now been felled. Scale bar indicates 1 km.

After Djaja and Hadi 1979

Rambut. The colonies of birds which roost and nest there are threatened unfortunately by residential developments on and around their mainland feeding areas. For P. Rambut these are along the highway to the Soekarno-Hatta International Airport. The mainland reserves of Muara Angke and Cimanuk either need to be planned and managed seriously as 'urban parks' (for which they have considerable potential) or written off.

A detailed and reasoned proposal for a 1,200 ha reserve at Muara Gembong has been made (Djaja and Hadi 1979; Djaja et al. 1982; fig. 8.16). In 1979 there were 25-35 families living in the proposed area. Their presence was illegal but it was proposed that their move from the area be conducted with due care and consideration, and that they be given priority in obtaining employment to safeguard the reserve. In a report commissioned for this book to examine the present status of the area, it is reported that in the intervening 13 years virtually all the mangrove has been lost save narrow belts to the west and north of the islands, and there are now hundreds of families established (Syahminudin 1992). It may be advizable, though regrettable, to shelve the proposals for a reserve.

The relentless loss of mangrove on Java and Bali, in a country that was one of the main protagonists in the drafting of the UN Convention on Biological Diversity, requires that energetic and appropriate management be afforded to all the remaining and potential areas of mangrove forest. It is probably too late to execute the surveys, inventories, and evaluations (Sukarkjo 1990) that could fine-tune planning decisions, and the priority now is to demonstrate to developers that mangroves are not unproductive wastelands, but that they can add to the amenity value and quality of an area. It may also be possible to save unnecessary expenditure of money if developments are designed with the mangroves rather than against them, as is being attempted in a small way at the International Commerce Centre occupying the old Kemayoran Airport site in northeast Jakarta and in the Kapuknaga area just west of Jakarta (Ligtvoel et al., 1996).

Given that so much natural mangrove forest has been lost, the question arises as to how valuable the so-called 'degraded' mangrove systems are, such as those in the Solo-Brantas region. This 'degraded' epithet can certainly be challenged because the vegetation of the tambak area has evolved and exists in a highly dynamic environment where tall and species-rich forests are unlikely ever to have been supported. This is not to deny that extensive areas of mangrove forest have been felled to construct the tambak ponds, but it is not clear what the net effect of this pattern of land use has had on the stature, species composition and extent of the mangrove that would be present in the absence of the tambak (Davie and Sumardja in prep.). The mangrove fringe would never have been particularly broad, unlike certain areas of eastern Sumatra (Anwar et al. 1984; Whitten et al. 1987c). It may therefore be the case that the mangrove-type vegetation (natural fringe plus planted tambak) might be many times greater precisely because the tambak farmers have been managing the coastline. It is certainly unfortunate that the skilled tambak farmers in the Solo-Brantas region perceive their economic success as being due to having escaped the effects of industrial and other pollution in the nearly urban and industrial conurbations, rather than because their government understands their form of resource management and is committed to its continued sustainability.

Government concern about the fate of mangroves needs to be demonstrated now by halting the conversion of mangroves (with or without traditional tambak) to industrial and real estate uses, and by promoting their conservation and sustainable management.

Chapter Nine

Lakes and Rivers

Freshwater is an essential element of life. Its obvious value is in the daily uses of drinking and bathing, but it is also harnessed to produce electricity, used in industrial processes, diverted to irrigate crops, and utilized by animals and plants. Despite the economic importance of freshwater ecosystems to people throughout the tropics, their nutrient cycles, carrying capacities, and ecological limits are not well understood. The study of freshwaters extends beyond the lakes and rivers into the drainage basins above, features of which strongly influence the chemical composition of the drainage water. The life within the water is both an indicator of the health of a freshwater body, and a means by which quality is actually improved through biological processes.

BACKGROUND AND STATISTICS

The primary agent of lake formation in Java and Bali has been volcanic activity either by forming craters that fill with water, or blocking rivers with lava or other debris. Many crater lakes, such as the four main lakes on Bali (fig. 9.1), have no surface outlet, but they do feed underground springs. Once formed, a lake basin is doomed to 'die', following an assured sequence of stages from youth to senescence, at which time the lake has become full of sediment, although the process is generally measured in millennia (Cole 1983). In the past, Java has had some substantial lakes, such as that covering the area in which Bandung now sits (pgs. 96, 148), and the original Rawa Pening before it became a peat swamp and then a reservoir (p. 417). Now, however, Java has no natural lakes over 5 km^2 in area, while Bali has Lake Batur, which covers over 16 km^2. Java does, however, have a large number of small lakes, as is illustrated by those found around Mt. Lamongan (fig. 9.2). Since the New Order Government arose in 1966, numerous reservoirs have been created for irrigation and electricity generation, and for the sake of this book these are regarded as lakes. The largest lakes are thus Gajahmungkur (Wonogiri) (90 km^2), Jatiluhur (83 km^2) and Cirata (62 km^2), and Saguling (53 km^2). In terms of volume, however, Jatiluhur is four times as large as Gajahmungkur, and Batur is nearly

413

Figure 9.1. Lake Tamblingan, the smallest of the four main crater lakes on Bali.
With permission of Kompas

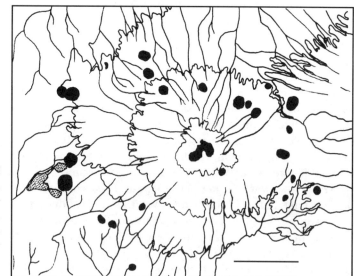

Figure 9.2. Small lakes around Mt. Lamongan. Contours at 300, 500, and 1,000 m. Scale bar indicates 1 km.
After Ruttner 1931

as large as Saguling despite having only 30% of the surface area (fig. 3.30; table 9.1). It should be noted that the area of lakes and rivers change with the seasons, and the depth of some reservoirs can vary regularly by as much as 20 m (Hardjamulia et al. 1988). Indeed, in very dry years large reservoirs can nearly dry up and, in the Gajahmungkur reservoir, this exposes the towns and villages that were submerged years before (Edwin 1988b).

The larger rivers of the level plains are not naturally fixed in their

Table 9.1. Characteristics of the major lakes of Java and Bali. * - total for the various lakes, na - not available.

	Max. area km^2	Max. depth m	Volume m^3 x 10^6	Altitude m
Cacaban	na	na	86	na
Gajahmungkur	90.0	20	736	na
Jatiluhur	83.0	95	2,970	107
Cirata	62.0	na	2,165	na
Saguling	53.4	na	982	na
Kedung Ombo	46.0	na	723	na
Sempor	29.0	na	52	na
Rawa Pening	25.0	11	52	463
Batur	16.1	88	815	1,034
Karangkates	15.0	na	343	na
Wadaslintang	14.6	na	443	na
Malahayu	9.3	25	60	na
Bening	5.7	na	37	na
Pacal	4.5	na	41	na
Darma	4.0	na	4	na
Selorejo	4.0	na	62	na
Bratan	3.9	23	49	1,239
Dieng lakes	3.8*	88	na	2,000
Wlingi	3.8	na	24	na
Wonorejo	3.8	na	122	na
Buyan	3.7	70	116	1,217
Lahor	2.6	na	37	na
Prijetan	2.2	na	10	na
Ranu Grati (Klindungan)	1.9	134	na	10
Telaga Ngebel	1.5	46	na	730
Tamblingan	1.2	40	27	1,200
Lamongan	0.5	30	na	240
Ranu Pakis	0.5	156	na	205
Ranu Bedali	0.4	11	na	150
Telaga Pasir (Sarangan)	0.3	28	na	1,290
Telaga Pasir	0.2	23	3	1,286

Sources: Ruttner (1931); Lehmusluoto and Mahbub (1989); Hardjamulia and Suwignyo (1988)

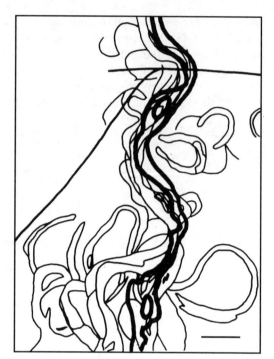

Figure 9.3. Changes in the course of the lower Serayu River, showing the present course in bold, and earlier courses in narrower lines. Scale bar indicates 1 km.

After Meijerink 1977; RePPProT 1989

course, and there is evidence of many earlier channels (fig. 9.3; van Setten van der Meer 1979) when seasonal floods occurred. This movement and unpredictable change is disliked by water engineers and causes significant economic losses, and so river courses have been straightened and their banks raised to allow water to flow faster to the sea, thereby reducing the risk of floods (p. 112).

Some rivers have very distinctive physical features, many of which are well-known tourist sights. The most obvious are waterfalls such as those at Cimahi, Curug Cisarua, Cibeureum on Mt. Gede-Pangrango, Grojogan Sewu near Solo, Baung near Purwodadi, Madakaripura near Probolinggo, Kakek Bodo near Tretes, Cuban Rondo near Malang/Batu, and Gitgit in Bali. Some waterfalls, however, appear to be virtually unknown yet are still spectacular, such as the huge but unmeasured waterfalls near Pronojiwo on the southern slopes of Mt. Semeru. In limestone areas, rivers can disappear into underground caves, termed swallow holes, for some of their length, and one of the largest of these in Java must be Sangiang Tikoro (the 'throat') adjacent to the small village of Cipanas near Rajamandala west of Bandung. Here the large Citarum River bifurcates around an island, and one half flows into a cave (Bennett and Bennett 1980). Nearby, the Citarum cataract or Curug Jompongan north of Soreang is possibly the largest series of waterfalls in Java in terms of volume if not height, but the crashing water stirs up thick rafts of foam, an unpleasant smell, and numerous old sandals and other refuse.

Surveys

The first surveys of Java and Bali's freshwaters were performed over a period of 10 months between 1928 and 1929 during the German Sunda Limnology Expedition which visited Sumatra, Java, and Bali. The results, written by over 100 specialists, total nearly 8,000 pages with over 3,000 figures, and were published in the journal *Archiv für Hydrobiologie (Supplement)* between 1931 and 1958. Most of the papers are taxonomic (over 1,100 new species of animals and plants were described), and this gives a wonderful base for modern studies. The results of the physical studies are summarized by Ruttner (1931, 1932). In West Java the lakes investigated by the expedition were the swampy lakes of Cigombang and Sindanglaya, in Central Java Telaga Pasir and Telaga Ngebel, in East Java the small lakes of Lamongan, Pakis, Bedali, and Klindungan, and on Bali the relatively large crater lakes of Bratan and Batur (fig. 9.4). Since then, many studies have been made of lakes and reservoirs, rather fewer of rivers, and most of these focus on fisheries. In many cases, however, the studies have not taken account of the dynamic nature of freshwater ecosystems, and this is discussed further below.

No area of freshwater in Java, or indeed in Indonesia, is better studied ecologically than Rawa Pening (fig. 9.5), thanks to the long-term interest of Satya Wacana Christian University in nearby Salatiga. Rawa Pening is Indonesia's oldest reservoir, originally a peat swamp (p. 487) (Polak 1951), but blocked in 1916 and again, after enlargement, in 1939. Many physical, chemical, biological, and sociological studies have been made there since 1975, and there is an excellent base of information from which informed management decisions can be made. The current information about Rawa Pening is summarized in a booklet (Göltenboth and Timotius 1992), and a bibliography of work there up to 1991 is available (Göltenboth 1992). One of the most important parts of Rawa Pening is the Muncul estuary, through which good-quality water enters the lake and where many of the lake's fish spawn. Great concern was expressed by the local scientific community some years ago when an indigo processing plant was being built on its banks. The force of sound argument eventually led to its closure (Kristyanto 1991). Other well-studied lakes are Saguling (e.g. Costa-Pierce and Soemarwoto 1990b), and Bojongsari or Situ Sawangan (Nontji and Hartoto 1989).

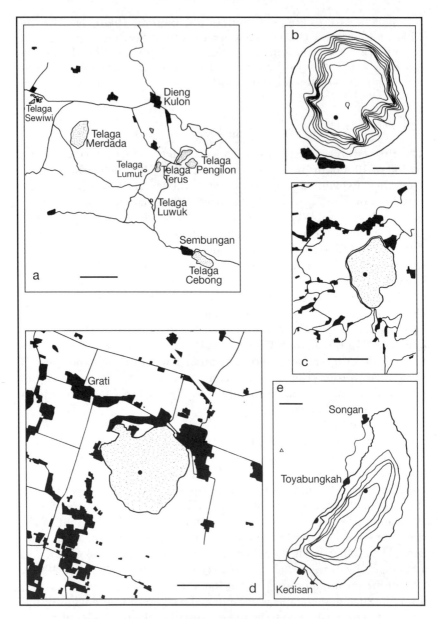

Figure 9.4. Some of the Java and Bali lakes studied by the German Sunda Limnological Expedition in 1928-29. a - Dieng Lakes, b - Ranu Pakis showing contours every 20 m and a disk over the deepest part (155 m), c - Ranu Ngebel with disk over the deepest part (46 m), d - Ranu Grati with a disk over the deepest part (134 m), e - Batur with contours every 20 m and a disk over the deepest part (88 m). Scale bars indicate 1 km except for b, where it indicates 100 m.

After Ruttner 1931

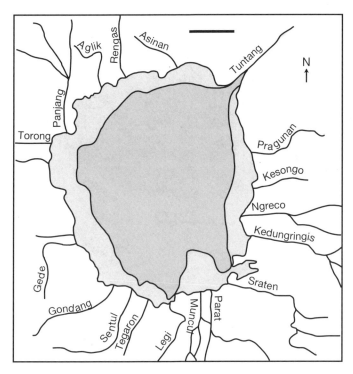

Figure 9.5. Rawa Pening, the best-studied freshwater body in Indonesia. Inner line is the dry season shore, the outer is the normal peak shore. Scale bar indicates 1 km.

After Göltenboth 1979

WATER

Water Inputs

The ultimate source of all freshwater is rain, but this is just one stage in the water cycle (fig. 9.6). Water flows along three main pathways: overland, under the soil surface but above the water table, and in the groundwater. Whichever path is travelled, the water is subject to a variety of physical and chemical forces, including gravity, and eventually collects in a depression, river, lake, or ocean. The amount of rain reaching the ground below any vegetation cover depends on the rainfall intensity and duration, and the configuration and density of leaves. Some of the rain water that reaches the ground later evaporates from the stomata or leaf openings of plants (transpiration), as well as from the soil, puddles, rivers, lakes, plant stems, trunks, and leaf surfaces (evaporation). Thus, only about 60-80% of the rainfall actually enters rivers and lakes in the streamflow (p. 137) (Wiersum 1975; Whitmore 1984).

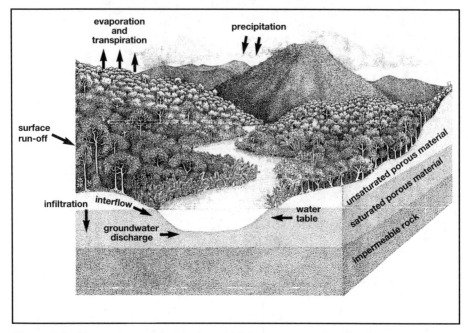

Figure 9.6. The various pathways by which water moves in a catchment.

After Göltenboth et al. in prep.

The response of a river to rain depends on the size, topography, geology, and soil conditions of the catchment area above the river. Small rivers typically display a quick rise and fall of water levels after rain, whereas larger rivers are slower to respond to rainfall, and the magnitude of the eventual response is less. After rain has fallen, groundwater may not appear as riverflow for hours, days, or even weeks. Delayed groundwater flow can therefore sustain low-water flows in rivers or lakes through long, dry periods (p. 141).

Water Chemistry

The chemistry of a river or lake reflects the complex interactions of rain water with soil, rock, plants, and climate. Rain is not pure water–it contains:

- low but measurable concentrations of dissolved gases, particularly oxygen, nitrogen, and carbon dioxide;
- positively charged ions (cations) such as hydrogen, sodium, potassium, calcium, magnesium, and trace elements; and

- negatively-charged ions (anions), such as sulphates, carbonates, nitrates, chlorides, and phosphates.

As the water percolates through or runs off a catchment area, it changes chemically due to the leaching of substances from the soil and rocks. These infiltrate into the soil and are stored temporarily in soil water or are attached to the surface of soil particles. This attachment or adsorption depends on the soil's capacity to exchange cations, which in turn is dependent on its pH and the amount of clay, humus, or organic matter. The chemical nature of clay gives it ten times the capacity of sand, to exchange cations. Water added to soil can cause some of the adsorbed cations to diffuse away from the exchange surface and into solution. The ease of replacement from the complex is approximately: sodium>potassium>magnesium>calcium>hydrogen; that is, sodium is more readily displaced than potassium, and hydrogen least readily of all. Leaching of cations out of the soil system is facilitated by mobile anions, particularly bicarbonate. Bicarbonate is a product of soil organism respiration, and it is also a common constituent of limestone bedrock. Thus, leached soil water reaching rivers through the groundwater should contain large quantities of magnesium, calcium, bicarbonate, sulphate, and have a pH of 7 or more (Freeze and Cherry 1979). Freshwater has remarkable buffering capacity, that is, it maintains its pH within certain close limits despite the input of liquids with a different composition such as rain water or factory effluent. The system can, of course, be overloaded.

PHYSICAL CHARACTERISTICS OF LAKES

Most lakes can be viewed as slow-moving rivers in which the river bed has become very wide and very deep. For example, the water in Rawa Pening is completely changed about every month (Göltenboth and Kristyanto 1987). Many of the same species of animals and plants live in both lakes and rivers, and many of the adaptations they require are also the same. So different are the physical regimes of lakes and rivers, however, that their behaviour requires separate attention. These are dealt with in detail (Anwar et al. 1984; Whitten et al. 1987a, b, c), and summarized here.

One of the most important aspects of lakes to appreciate is the way in which many characteristics vary with depth (table 9.2), and this is explained below.

Temperature and Oxygen

In the same way that air temperature decreases with altitude (p. 499), so the water in high-altitude lakes is cooler than the water in lowland lakes (fig. 9.7).

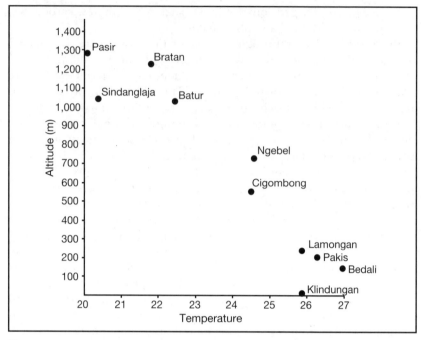

Figure 9.7. Water temperatures at the bottom of some lakes at different altitudes in Java and Bali.

After Ruttner 1931

Table 9.2. Characteristics of water at the surface and at the bottom (15 m) of Gajahmungkur reservoir.

		surface	bottom
Suspended solids	mg/l	73-157	175-490
Temperature	•C	24-27	20-24
pH	-	7.0-8.0	6.0-7.0
Dissolved oxygen	mg/l	1.9-6.2	0.4-2.5
Free carbon dioxide	mg/l	2-9	11-18
Phytoplankton density	no./l	315-546	42-105
Phytoplankton diversity	-	1.5-2.2	0.4-0.8
Zooplankton density	no./l	42-526	63-525
Zooplankton diversity	-	0.5-1.3	0.0-1.2

Source: Utomo (1983)

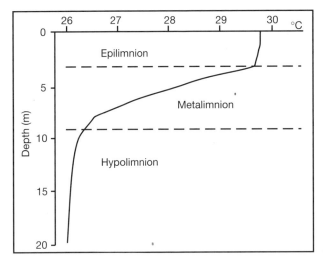

Figure 9.8. The three principal density layers of a lake illustrated by temperature readings from Lake Lamongan.

After Ruttner 1931

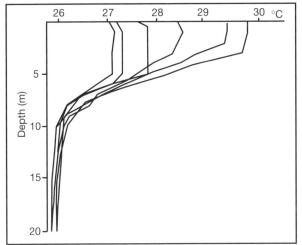

Figure 9.9. Profile of temperature through Lake Lamongan on six days at different times to show the variation in surface readings.

After Ruttner 1931

Temperature is most constant at the bottom of the lake because, as the sun warms the water surface, its density decreases, and these density differences within the water column produce a layering effect (fig. 9.8). The warmest and highest layer, the epilimnion, experiences fluctuations within and between days, although these are less than those experienced on land because of the higher thermal capacity of water. The narrower temperature range of water than air is due to water's superior capacity to retain heat. The metalimnion is an intermediate layer through which water temperature usually drops rapidly, and this gradient is termed the thermocline. The cooler layer below is referred to as the hypolimnion (fig. 9.9).

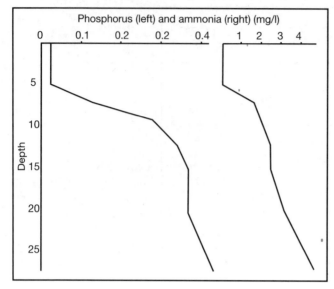

Figure 9.10.
Profile of phosphorus and ammonia concentrations (mg/l) in Lake Lamongan. Nutrient loss from the surface starts to be compensated for at about 5 m by the nutrient 'rain'.
After Ruttner 1931

Dissolved oxygen concentrations are highest in the epilimnion where photosynthetic activity of plants is greatest, and lowest in the hypolimnion where decomposition of the faeces, corpses, and other organic matter falling to the bottom consumes rather than produces oxygen. Oxygen concentrations at the bottom of a lake may be reduced to zero: this has been recorded at 15 m in Telaga Pasir, 25 m in Lake Lamongan, 35 m in Telaga Ngebel, 50 m in Ranu Pakis, and 90 m in Lake Klindungan (Ruttner 1931).

Nutrients

In tropical lakes, chemical processes proceed four to nine times faster because they are 15-20° C warmer than temperate ones. Carbon dioxide and other solutes are thus released relatively quickly, and in the deeper lakes much of the settling organic matter is probably broken down before it reaches the lake bed. The hypolimnion experiences a nutrient gain due to the rain of organic matter from above. The high concentrations of phosphate and ammonium in the hypolimnions of some lakes have led to the suggestion that it is these deep waters, rather than surface layers, that should be used for irrigation. A profile of nutrient concentration with depth can reveal approximately where nutrient gain, expressed as a higher concentration of dissolved materials, starts, although this varies over time and from lake to lake (fig. 9.10). Figures for the chemical composition of

Java and Bali lakes from various studies indicate a large degree of variation, but to what extent this represents genuine differences or simply reflects different methods of analysis is not known. Almost all measurements are from only the surface layers (probably 1 m or less), but authors do not always give details of the depth at which samples were taken.

Stability

Stability of tropical lakes is poorly understood, but may be very important. In general, warm waters have a greater resistance to mixing than cooler waters, and slow mixing may continue all the year round. Overturns, in which the hypolimnion is brought to the surface during a period of windy, cold weather when the surface waters cool, are not unknown, however, and the low oxygen concentrations of water from the bottom of a lake can kill fish when it comes to the surface (Green et al. 1976). Greatest stability is found in steeply walled, deep lakes with small surface areas. Indeed, the relationship between surface area, thermocline, and stability can be quantified roughly as follows (Ruttner 1931):

Area	1	:	100	: 1,000
Depth of thermocline	1	:	3	: 6
Stability (0-20 m)	50	:	10	: 1

Thus Lake Batur, which has ten times the area of Lake Tamblingan, would have a thermocline twice as deep, and require only one-tenth of the wind strength to mix the top 20 m.

Variations

Many parameters, such as oxygen and carbon dioxide concentrations, temperature, salinity, pH, phosphate, nitrate, plankton abundance, and productivity, vary through the day and between seasons and among sampling stations (fig. 9.11) (Nontji 1984), and so readings made on a single day are no more than a snapshot of the situation and provide little useful information. Far better to devise a programme of measurement that will identify diurnal, weekly, monthly, and yearly variations. This is perhaps the greatest lesson to be learned from the Rawa Pening studies. For example, the average dissolved oxygen concentration in the surface water is 3.4 mg/l (about 44% saturation), but individual studies have also reported average values from 0.4-7.6. The differences are likely due to the unpredictable movement of the water hyacinth beds, the proximity and growth of the submerged plant *Hydrilla verticillata*, the input of cold mountain water from some of the rivers, and an underwater spring of deoxygenated water. Selecting a single sampling station at random does not help, as was discovered by one set of researchers in 1979 on Rawa Pening whose chosen sampling station was close to a volcanic emission of hydrogen sulphide

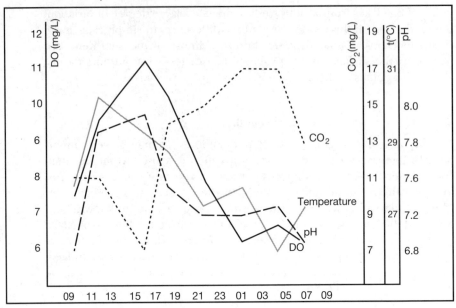

Figure 9.11. Variations over 24 hours in dissolved oxygen, carbon dioxide, temperature, and pH at a single sampling point in Rawa Pening.

After Göltenboth 1979

which depleted the water of oxygen (Göltenboth and Kristyanto 1979b). This oxygen-depleting emission was still active in November 1992.

Physical Characteristics of Rivers

Rivers vary in their velocity, depth, and substrate composition across and along their length, and these are also affected over time in response to the intensity of rainfall.

Discharge

The volume of water flowing through a cross section of river per unit time is called the discharge, which is the product of the mean velocity of the water and the cross-sectional area. It may seem strange, but the average velocity of a river is actually lower in the steep headwater regions than it is in the lowlands. In headwater streams, shallow, fast-flowing stretches alternate with slow-flowing deep pools. Discharge also increases as a river flows downhill, both because of increasing velocity and the additions to flow both from tributaries and from water entering from runoff and interflow. Headwater rivers respond very quickly to short, intensive rainfall, whereas the response of wide lowland rivers is much slower.

Shear Stress

Shear stress is the result of fast-flowing water moving past slower-flowing water or a stationary surface. Velocity is greater in the main body of flow just under the surface than at the river bed or at the sides, and mean velocity is generally found at 60% of the total depth (Townsend 1980). Turbulent flow moves particles and organisms on or near the river bed, and these are lifted up, pushed or bounced along, bounced into other rocks, or thrown into faster water layers, only to come to rest again on the river bed if sufficiently heavy. Shear stress on the stream bed is correlated with slope and water depth. Slopes decreases faster than depth increases, and so greater shear stress will be experienced by benthic organisms in the headwaters than in the lower reaches.

River Bed Particle Size

Rivers are capable of carrying the smallest clay fractions and also large boulders. The particles that travel along the river bed will be those that are larger than the flowing water is able to carry away. The particles within the water column are termed the suspended load, while the larger particles that are rolled along the river bed are termed the bed load. As shear stress decreases downriver, so the average particle size of the suspended load decreases, the larger particles having fallen to the river bed. Thus the average particle size on the river bed also decreases downriver, although the amount and type of suspended sediment is related to the type and exposure of sediment sources, such as slopes, riverbanks, and roads, as well as to discharge.

Temperature, Dissolved Oxygen, and Nutrients

Altitude, rain, exposure, water sources and velocity, and ambient temperatures are the major factors which influence river-water temperatures. Temperature layering, as occurs in lakes (p. 421), does not occur in rivers because the water is in continuous motion and depth seldom exceeds 2 m except during spates. River temperatures can respond quickly, however, to changes in ambient air temperatures, particularly if the river is shallow and slow. Night time temperatures of water remain warmer than air temperatures due to the warming effects of surrounding earth and groundwater seepage, and to the fact that the thermal capacity of water is much greater than that of air. Rivers at higher altitudes are cooler due to the lower mean daily air and soil temperatures.

Temperature is important in determining water quality. As temperature rises, the rate of chemical reactions increases, oxygen solubility decreases, and the rate at which oxygen is consumed through chemical and biological oxidation of organic compounds increases. Thus, a river suddenly exposed to increased sunlight, due to the felling of riverbank trees for example, will experience a dramatic change in water quality associated with the increased water temperature. In addition to a lower dissolved oxygen concentration, the warmer water is conducive to the growth of aquatic bacteria, that may be pathogenic to certain fish, and will probably alter the metabolic activity of most aquatic organisms. For a given quantity of organic matter, rivers that are shallow and slow-moving will have lower levels of dissolved oxygen and higher BOD levels than rivers that are deep and fast-moving. Rivers at high altitudes generally have higher oxygen levels than those in the lowlands (Chye and Furtado 1982) because turbulent water facilitates the mixing of layers and keeps water temperature low and constant across and through a river; oxygen may also be replenished from bubbles formed in rapids, waterfalls, or waves. Oxygen concentrations of headwater streams are generally about 6.5-7.5 ppm, although sluggish rivers may have a concentration as low as 4 ppm.

Under natural conditions, lowland rivers have low concentrations of chemical ions, weakly acid pH, low alkalinity (and therefore low buffering capacity), low BOD, and low concentrations of nutrients, especially phosphate and nitrate. These conditions will be altered to varying extents by human activities.

PLANTS

Phytoplankton

Phytoplankton are those microscopic plants that float or drift near the water surface, and are the major photosynthetic producers in freshwater. In addition to different types of green algae, phytoplankton include various flagellates, diatoms, and blue-green 'algae' (actually a group of bacteria). They have many unusual shapes, with spines, horns, and hairs, which were originally thought to increase surface area and thus help in buoyancy, but are now believed to be related to absorption of nutrients and defence against herbivory as well (Cole 1983).

The phytoplankton lead a precarious existence. They are readily preyed upon by fish and zooplankton, their habitats may be destroyed by desiccation or incoming floodwaters, or they may sink to the bottom of the lake where they perish due to lack of light. These potential problems have favoured the selection of rapid reproduction by simple cell division, which can occur every few hours or days (Moss 1980). In rivers, substantial phytoplankton development can occur only when river flow is slow or a river is long and not too turbid. The diversity and abundance of phytoplankton are also related to the diversity and abundance of microhabitats and surfaces. For example, phytoplankton in Rawa Pening is most abundant in and around the *Hydrilla* areas and least frequent beneath the mats of water hyacinth. The phytoplankton is most abundant in the wet season and least in the dry, and the dominant species changes from a diatom *Synedra* to a desmid *Closterium infractum,* respectively (Göltenboth and Kristyanto 1979a).

In the same way that physical conditions of lakes and rivers change through the day and between seasons, so it is clear that some plankton studies often give no more than a momentary impression of the situation (Mulyadi 1985), and are of no great use for management proposals.

Fungi, Bacteria, and Blue-Green Algae

The major decomposers in freshwater are bacteria and aquatic fungi. The dominant group of fungi is the Hyphomycetes which colonizes faeces, and dead plant and animal material. Knowledge of the biology and ecology of the decomposers is confounded, however, by their minute size and the difficulty of meaningful sampling. Bacteria are generally dominant among aquatic weeds or macrophytes where, along with algae, they live as epiphytes. The blue-green 'algae' or cyanobacteria are present in an enormous range of habitats including hot springs hotter than 55° C (Kullberg 1982). They have evolved the use of chlorophyll 'a', the most important photosynthetic pigment in higher plants, and are thus, along with green algae, primary producers of oxygen in freshwater habitats. They are also capable of fixing nitrogen (Cole 1983). Overgrowth or blooms of blue-green algae, however, produce toxins dangerous to fish and other animals. A bloom of the blue-green 'alga' *Microcystis* was reported from Jatiluhur in the early 1970s (Soemarwoto 1973), and a similar bloom occurred in Rawa Pening in 1980 (Basmi 1981; Putra 1987) as a consequence of water enrichment through erosion, fertilizer use, and a dense human population.

Macrophytes

Macrophytes are divisible into five groups:
- those that grow on wet banks and are frequently flooded;
- those that are rooted beneath the water but project above the water surface (emergents);

- those that are rooted beneath the water but whose leaves float on the water surface;
- those that are rooted and grow beneath the water; and
- those that float.

The submerged and floating freshwater plants recorded from Java and Bali (and other parts of Indonesia), are listed by Giesen (1991), and sources of species lists for specific areas are given in Tjitrosoedirdjo and Widjaja (1991). Interestingly, few species of indigenous macrophytes have adapted to rice fields (Backer 1912, 1913, 1914), and the reservoirs formed by damming water courses are generally very poor in species, though this is due partly to their steep sides that are not conducive to plant growth.

Large areas of macrophytes in various types of communities would once have been found in the low northern plains from Serang to Cirebon, but these, and the interesting, small areas around Jakarta, such as Rawa Tembaga (Edeling 1870) and at Rawa Bening (Kediri, East Java) (Coert 1934), have long since been drained and converted to sawah. Major areas remain, however, in Rawa Danau and Rawa Pening.

Macrophytes of a sort also occur in the splash zone around the base of waterfalls. Behind the falls themselves, the rocks have a slippery covering of green algae, on the fringes of which grow mosses and perhaps filmy ferns with their single-cell thick fronds. Away from the wettest areas are sappy plants such as *Begonia* (Bego.), *Impatiens* (Bals.), *Cyrtandra* (Gesn.), and *Elatostema* (Urti.). At high altitude deep moss cushions can be found closest to the water, followed by shrubs, tree ferns, other ferns, and orchids (van Steenis and Schippers-Lammerste 1965).

A special group of emergent macrophytes are the rheophytes: plants which grow only in rock, gravel, or sand river-beds. These plants must be able to resist being swept away in sudden floods, and consequently are well-rooted and generally have leaves which are relatively long and thin. Java has relatively few rheophyte species, but *Homonoia riparia* (Euph.) (sometimes planted as an erosion retardant), and *Elaeocarpus grandiflorus* (Elae.) are examples, reminiscent of willows (van Steenis and Schippers-Lammerste 1965). In relatively sluggish streams *Ficus racemosa* (Mora.) and *Salix tetrasperma* (Sali.) (recorded only in Central and East Java) may be found. A remarkable plant found throughout Java but only in clear, swift streams is *Cladopus nymani* (fig. 9.12) (Podos.). This tiny macrophyte grows to only about 4 mm high, remains sterile during the rainy season, and flowers when the water level falls (van Steenis 1949b). Other very small water plants include the floating *Lemna, Spirodela,* and minute *Wolffia* (all Lemn.) which can be very common in still water. *Wolffia* contains the world's smallest plants; *W. globosa* measures just 0.4 mm high and 0.5 x 0.2 mm in length and width (van der Plas 1971).

Larger macrophytes provide refuges from predators, roosting and nesting sites for birds, food for humans, green manure, cattle feed, deco-

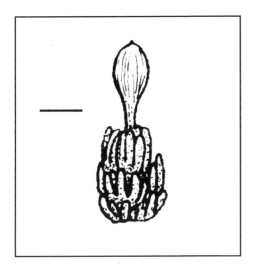

Figure 9.12. The tiny macrophyte *Cladopus nymani* which is found on rock in clear streams. Scale bar indicates 1 mm.

After van Steenis 1949b

rations, and substrates for algae, bryozoans (Vorstman 1928a, b), diatoms and rotifers, and the greater the diversity of macrophytes, the greater the diversity of invertebrates (Scheffer et al. 1984). Macrophytes generally have many non-benthic macroinvertebrates associated with them, but are not themselves eaten by invertebrates. At first sight this is peculiar since the leaf cuticles are thin and few species possess spines or hairs that might dissuade invertebrate herbivores from eating them. It has been suggested that the plants simply are not very nutritious, but in fact they contain as much protein as high-quality forage crops. The probable reason macrophytes are not eaten is rather because many of them appear to have significant quantities of a wide range of defensive chemicals (alkaloids) in their leaves (Ostrofsky and Zettler 1986).

Zonation. The shallow region close to shore, up to about 1 m depth, forms the primary habitat of emergent macrophytes. These plants utilize the resources of both aquatic and terrestrial environments: there is a very rich nutrient supply in the sediments in which they are rooted, and they have an advantage over submerged plants in that they have direct access to light, oxygen, and carbon dioxide. The productivity of emergent macrophytes in eutrophic lakes is consequently very high. Between 1 and 3 m depth, plants with rhizomes rooted to the ground but with leaves floating on the surface can be found: an example is the water lily *Nymphaea* (Nymp.). Deeper

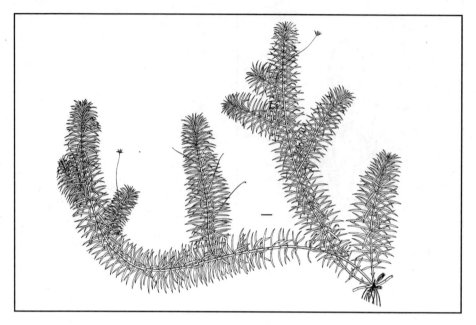

Figure 9.13. *Hydrilla verticillata,* a common macrophyte which has become a weed in some areas. Scale bar indicates 1 cm.

After Kostermans et al. 1987

still, up to about 10 m depth in clear water, are found the submerged macrophytes. This zonation is ultimately the result of the physiological problems faced by the aquatic plants caused by the limited diffusion of carbon dioxide and oxygen, relatively low light levels, and high water pressure. This pressure prevents vascular plants from growing below about 10 m, but mosses and algae can grow much deeper if there is adequate light. Oxygen tolerance is achieved by having up to 60% of the tissue volume occupied by air spaces (Moss 1980). For example, one-third of the tissue volume of the common *Ceratophyllum demersum* (Cera.) consists of air spaces. Light is absorbed quickly by water, so such submerged plants receive only a small proportion of the light reaching the water surface. Submerged leaves assume a morphology similar to terrestrial shade plants; their leaves are thin and contain many chloroplasts, thereby maximizing the use of available light energy. *Hydrilla verticillata* (Hydr.) (fig. 9.13), a widespread submerged macrophyte, for example, is able to maintain a high photosynthetic rate even at low light intensities (Finlayson et al. 1957b).

Floating plants can be found on water of all depths, and their distribution is determined primarily by currents and wind.

Productivity and Decomposition

The productivity of most submerged plant communities is about five times lower than that of emergent, floating, or terrestrial macrophyte communities (18 tons/ha/year against 75-100 tons/ha/year) (Sutton 1985) due to the reflection of light from the water surface and suspended particles in the water, lower rates of gas transmission, epiphytic algae and protozoans that shade the leaf surfaces, and a general absence of extensive root systems. As a result, emergent rooted plants will quickly overgrow submerged ones where conditions are suitable. Productivity varies considerably depending on environmental conditions: thus, in sunny pools algae has been found growing six to seven times faster than in shaded pools, and this influences the density of algae-feeding fish (Power 1983).

ANIMALS

The community of animals associated with lakes and rivers can be rich and diverse (fig. 9.14), and the animals can be categorized roughly by the microhabitat they occupy, thus:

- the neuston comprises those animals living on or immediately below the water surface, supported by surface tension, such as pond skaters and mosquito larvae;
- the nekton comprises swimming animals, such as fishes and water beetles;
- the zooplankton comprises animals that drift in the water or swim weakly, such as small crustaceans, rotifers, and protozoans; and
- the benthos comprises those animals associated with the river or lake bed and other surfaces, such as many insect larvae, molluscs, prawns, crabs, and non-swimming protozoans;

by their size, thus:

- microinvertebrates which are invisible to the naked eye; and
- macroinvertebrates which are visible to the naked eye;

or by the way they feed, thus:

- shredders which feed on large units of plant material;
- collectors which feed on loose organic particles either on the river bed or free in the water;
- grazers which feed on attached algae, rotifers, and bacteria; and
- carnivores which eat other animals either by scavenging or by killing them;

or by taxonomic group.

A useful illustrated guide to the freshwater animals is available (Ng 1991b). Information is provided below on some members of the major habitat groups.

Figure 9.14. Animals and important plants of Rawa Pening. Plants: 1 - *Rhynchospora corymbosa*, 2 - *Ceratophyllum demersum*, 3 - phytoplankton, 4 - *Eichhornia crassipes*, 5 - *Nymphoides indica*, 6 - *Hydrilla verticillata;* Animals: a - *Rana erythraea*, b - *Pipistrellus*, c - *Collocalia linchi*, d - *Urothemis abbotti*, e - *Lonchura molucca*, f - *Halcyon cyanoventris*, g - eggs of *Pila* snails, h - *Belostoma indica*, i - *Macrobrachium rosenbergi*, j - *Channa striatus*, k - *Aplocheilus panchax*, l - zooplankton, m - *Clarias batrachus*, n - *Chironomus* larvae, o - *Trichogaster pectoralis*, p - *Culex* larvae, q - hydra, r - *Spongilla lacustris*, s - *Corbicula javanica*, t - *Caridinia laevis*, u - *Pila polita*, v - *Chaoborus* larva, w - *Anodonta woodiana*.
After Göltenboth et al. in prep.

Neuston

The neuston is a relatively minor component of the freshwater system. Groups of shiny black whirlygig beetles (Gyrinidae) skim around the surface of quiet water and dive into the water if they are disturbed. They are peculiar in that each eye is completely divided into two, giving the impression of having four eyes, so that they can see clearly above and below the water at the same time. They are predators and scavengers, feeding on small animals on the water surface. Pond skaters or water striders (Gerridae) skate on their long legs. The hind legs are used for steering, the middle for propulsion, and the shorter front pair for catching prey and grooming. They are capable of skating across quite rapidly moving water, maintaining their position relative to the shore. Other neuston members include the water measurer (Hydrometridae) which walks slowly across water around emergent vegetation, the small ripple bugs (Veliidae) which are predators on brown planthoppers which drop onto the water surface, in flooded rice fields, recently emerged mosquitoes resting on the water before flying away, and mosquito and other fly larvae which rest with their breathing tubes projecting into the surface film.

Nekton

Insects. The Hemiptera (insects with a stout proboscis pointing backwards beneath the head) are represented by two families in the nekton: the backswimmers (Notonectidae and Pleidae) which swim upside down, and the water boatmen (Corixidae) which swim the 'right' way up (fig. 9.15). Both have long hind legs which are used as oars. They are important predators on mosquitoes (p. 657), although the water boatmen also eat plant material. Both types of bug will fly readily, and this allows them to move away from unsuitable habitats and to colonize new areas. Other insects of the nekton include the diving beetles (Dytiscidae), some species of which have large (60 mm) and voracious larvae (fig. 9.16). Both adults and larvae feed on tadpoles and small fishes. Some people around the shores of Rawa Lamongan eat water beetles *Hydrophilus* after removal of their wings and hard covering (Green et al. 1976).

Fishes. Java has about 130 species of native freshwater fishes (p. 251), but little is known of their abundance, status, migratory habits, or ecological role. Even when detailed analyses of their diet are performed (fig. 9.17), they are generally no more than snapshots because diets change with age, over time, and between lakes or rivers. River fish used to be caught, but as various factors have taken their toll (p. 718), and as pond fish have become more available, so their significance has decreased. Bali has relatively few species of fishes (p. 251). The Balinese are not as fond of fish as food as are the people on Java, particularly the Sundanese, and the ichthyological paucity is of no concern to most inhabitants. Thus initiatives to introduce pond fish for food are not overwhelmingly successful, except among residents from Java. The fishes caught for food in Java's lakes are listed in Table 9.3, and an indication is given of their movements. Their absolute and relative abundances in reservoirs are controlled by stocking programmes from the Fisheries Services.

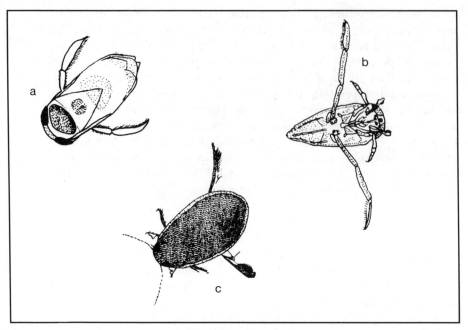

Figure 9.15. Insect members of the nekton: water boatman (a), backswimmer (b), and adult water beetle (c).

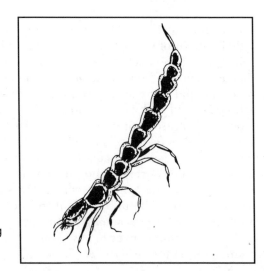

Figure 9.16. Larva of a diving beetle.

After Mudjiman 1986

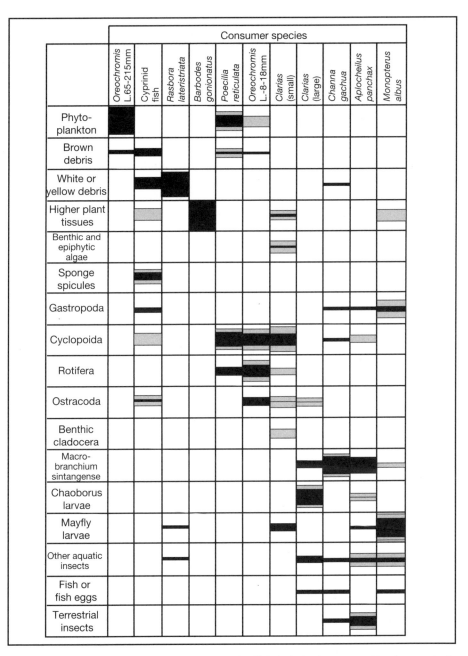

Figure 9.17. Results of stomach analyses for fish in Lake Lamongan to demonstrate the different habits and ecological separation. The depths of the black blocks show the percentage of stomachs with each food as main contents, and the depth of the shaded blocks show the percentages of stomachs in which food was present.

After Green et al. 1976

Many of the freshwater fishes eaten by people are introduced species, and probably every water body on Java has had one or both of carp *Cyprinus carpio* and tilapia *Oreochromis nilotica* introduced to it. Even the 7 ha and 75 m deep Telaga Cebong at 2,000 m on the remote Dieng Plateau has these fish, although they are not very productive (Budiman et al. 1979). Other fishes have been introduced for specific purposes, such as the grass carp *Ctenopharyngodon idella*, which is unable to breed in Indonesian lowland waters, to control the growth of aquatic plants (Göltenboth and Timotius 1992).

Probably the most abundant fish on Java and Bali is the guppy *Poecilia reticulata*, accidentally introduced from aquaria a few decades ago. It is now to be found in virtually every ditch, stream, canal, river, lake, reservoir, swamp, and rice field. A clue to the success of the guppy may be its remarkable ability to exploit different environments, varying its diet from benthos to plankton to surface prey. For example, analysis of their gut contents revealed that in Lake Buyan the guppies were eating the dominant zoo-

Table 9.3. Major reservoir fishes and their habits. JL - Jatiluhur, RP - Rawa Pening, L - Lahor, sl - standard length, dom - dominant species.

	Max. sl	Diet	Breeding behaviour
Barbodes balleroides	300	omnivorous on detritus, plant matter, and plankton (JL).	spawn in shallow areas that have been exposed during the dry season from August to November.
Barbodes gonionotus	330	herbivorous but will eat plankton (JL).	spawn in shallow areas that have been exposed during the dry season from August to November.
Hampala macrolepidota	700	crustaceans, insects, fishes (JL).	swim upstream to spawn at beginning of rainy season.
Mystacoleucas marginatus	200	omnivorous (L).	restricted range.
Osteochilus hasseltii	320	phytoplankton, periplankton (RP).	
Mystus nemurus	480	crustaceans, fish, insects (JL).	spawn in upper parts of JL or in Citarum between October and March.
Clarias batrachus	400	*Macrobrachium* prawns (dom.), insects (RP).	
Pangasius djambal	600	crustaceans, insects, molluscs, worms, carrion, rotifers, algae, small fishes.	spawn in deeper parts of upper reaches of JL at beginning of rainy season.
Oreochromis nilotica	300	detritus, plankton.	spawn at the end of the dry season from July to September.
Channa striata	900	insects (dom.), *Macrobrachium* prawns (RP).	

Sources: Kartamihardja (1977); Rahardjo (1977); Mahan et al. (1978); Darmawiredja (1979); Sastrawibawa (1979); Sutardjo (1980); Hardjamulia et al. (1988)

Figure 9.18. South American catfish such as these are one of many species of ornamental fishes bred in rice fields and open tanks which accidentally enter rivers and lakes.

With permission of Kompas

plankton and not benthic animals, which appear to be scarce around the shore of that lake. In contrast the guppies from neighbouring Lakes Bratan and Batur ate mainly benthos and a fair quantity of organic debris. Their guts also had remains of animals which would have been taken near the surface (Green et al. 1978).

In and around Jakarta and Bogor ornamental fish farms which breed fishes for the aquarium trade have been a major source of recent alien escapees which now have established populations (p. 304). These include sucker fish *Hyposarcus* sp. (fig. 9.18), cichlasoma *Cichlasoma nigrofasciatum*, flag cichlid *C. festivum*, black tetra *Gymnocorimbus ternetzi*, rainbow shark *Labeo erythrurus*, and seven-banded cichlid *Tilapia mariae* (Nasution 1990; Kottelat et al. 1993). More are to be expected and they will spread to other areas.

Zooplankton. The zooplankton are the smallest and most numerous animals of lakes and some rivers, with sizes ranging from 0.2-5.0 mm (Moss 1980). Zooplankton are primarily represented by four groups:
- protozoa are microscopic, one-celled animals, some of which have no fixed shape and glide across surfaces like animated bean bags,

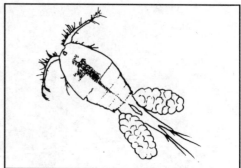

Figure 9.19. Typical copepod crustacean.

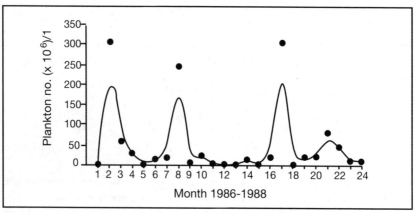

Figure 9.20. Changes in the plankton numbers in the Saguling reservoir, off-shore of Selacau, over a period of two years.

After Soemarwoto et al. 1990

while others use long hairs or flagella to propel themselves through the water. Some groups have chlorophyll within their bodies and so manufacture their own food, like plants;

- rotifers or wheel animals which have two rings of hairs around the mouth and a 'foot' for anchorage and movement;
- water fleas are small crustaceans which have long appendages for drawing food towards their mouth, and which can form dense, pink-coloured shoals in still water. The pink colour comes from haemo-globin within the transparent body. Cladoceran water fleas have the head and appendages exposed, whereas the ostracod water fleas are more or less enclosed in two shells like a bivalve mollusc; and

- copepods are also crustaceans, but are recognizable by the two egg pouches on either side of their body (fig. 9.19).

These animals all prey on phytoplankton or smaller zooplankton as well as ingesting detritus. Certain phytoplankton can travel unharmed through the digestive tracts of zooplankton and in fact absorb nutrients during their passage (Cole 1983).

Zooplankton have been studied in a number of locations on Java, but in general insufficient attention has been given to the diurnal and longer variations in abundance and diversity for any useful conclusions to be drawn. A recent exception is a study in Saguling over a period of two years (fig. 9.20). The reasons suggested for the changes were the high concentration of nutrients flowing into the reservoir from Bandung during the dry seasons, causing major blooms, and overturns of the water column due to the onset of the rainy season (Soemarwoto et al. 1990).

Benthos

The benthos is an important group of organisms which is richest in relatively undisturbed situations (Rondo 1982). For example, the benthic organisms in two streams arising in Mt. Tilu Nature Reserve were sampled at 1,200 m above sea level as they ran through the adjacent Gambung Plantation Research Station, south of Bandung. The benthic invertebrate community was rich, with 35 species found, 24 of which were met in both rivers (Jamidy 1984). Poor but natural communities can also be found in the sulphur streams of the Dieng Plateau, where just one species of fairly abundant chironomid fly is found (Boon 1977).

Crustaceans. Benthic crustaceans include the detritus-feeding and carnivorous crabs and prawns, the latter of which are sometimes caught to eat. Most are quite small, but a native giant *Macrobrachium rosenbergi* prawn with a head-to-tail length of 56 cm and a weight of 510 g has been caught in Rawa Pening (Göltenboth and Timotius 1992). One of the most common and widespread prawn species is *Macrobrachium pilimanus* (Fidhiany 1983; Asnawati 1990), which favours habitats with pebbles, stones, or rocks beneath or behind which they can shelter, although they are found in both strongly flowing as well as stagnant water. There are also small filter-feeding prawns such as the atyidid *Caridina*, which have tufts of hairs at the tip of their delicate pincers. When opened, these form a net to catch detritus and small organisms which are then eaten.

Caddisflies. Ubiquitous and interesting members of the benthos in relatively undisturbed rivers are caddis-fly larvae, the adults being moth-like insects, often dull brown in colour. The larvae of some groups construct a silken tube camouflaged with grains of sand or other material which they carry

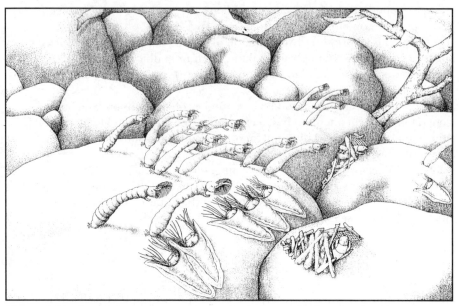

Figure 9.21. Invertebrate community on the bottom of a fast-flowing river showing two net-building caddisfly larvae, and the larvae (erect) and pupae (prone) of black flies *Simulium*. The current is flowing from right to left.

After Göltenboth et al. in prep.

around with them. These larvae feed on plants. Another group, the Hydropsychoidea, spin silken nets of different mesh sizes which trap detritus and drifting invertebrates on which they feed. Larger mesh sizes tend to be found in the headwaters and smaller sizes in downstream regions, appropriate to the differences in the sizes of suspended matter (Townsend 1980). Species may also be distributed differentially across a stream: the rigid nets of certain species predominate in the fast-flowing riffles, and flimsy and weaker nets of other species are more often found in the slower-flowing water of pools.

The degree to which certain species are able to exploit opportunities afforded by humans and their development activities never ceases to amaze and, to some, annoy. For example, the barrier screen in front of the water intake for the Tuntang dam, north of Rawa Pening, has encrustations formed from the larval tubes of the hydropsychoid caddisfly *Amphipsyche meridiana* (fig. 9.21). These blocks of material can become quite large, and have to be removed periodically so as not to impede the flow of water. From the point of view of the caddisfly the barrier represents an excellent substrate. The larval tubes (made of small mineral particles) are at such a high density on the barrier that they have to cooperate, albeit passively, and use a single, large net for feeding because many individual nets would

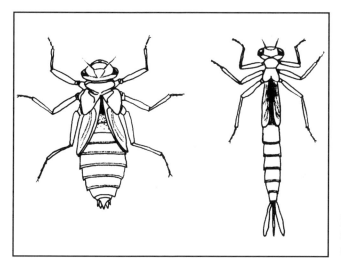

Figure 9.22. Larvae of typical damselfly and dragonfly.

interfere with each other. Interestingly, the mesh launched by the larvae from the barriers is three times larger than the mesh of nets spun by the same species in the river below the dam, and the food caught differs (Boon 1979, 1984). Indeed, the same species lives very differently on the screen and in the river downstream.

Dragonflies and Damselflies. The predatory nymphs of dragonflies and damselflies (Odonata) are wholly aquatic and are common in the benthos of both standing and flowing water (fig. 9.22). They move slowly but have a very efficient hooked 'lip' which is thrust forwards to catch prey. Their most common prey are tadpoles and small fishes, and they can become pests where young fish are being raised. When a nymph is fully grown, it crawls out of the water on the stem of a plant, and the adult emerges through a crack in the back of the nymph's skin. Dragonflies rest with their wings apart, whereas damselflies rest with their wings together over their backs. Normally both nymphs and adults are present all year round, but after prolonged rains and high water their populations are much reduced, partly because they have been washed downstream and partly because the substrate upon which the larvae rest prior to metamorphosis and emergence as adults are inaccessible or submerged.

A community of dragonflies and damselflies was studied for four years along the Cibarangbang, a small tributary of the Cidurian flowing north out of what is now Halimun National Park and under the Bogor-Rangkasbitung road, seven kilometres west of Jasinga (Lieftinck 1950). At the time of the study (the late 1930s) the stream was still forested on either side of

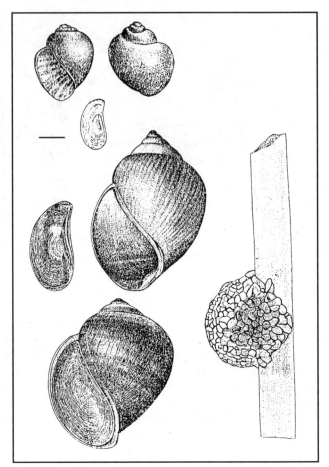

Figure 9.23. Apple snails *Pila scutata* (above), *P. polita* (middle), and *P. ampullacea* (below) with a typical egg mass of *P. ampullacea*. Scale bar indicates 1 cm.

After van Bentham Jutting 1956

the road. The odonate fauna was much richer (32 species) than that found in nearby areas in deforested agricultural areas. At least nine habitat types occupied by the nymphs could be identified, with some species found in four of these, but with the majority restricted to just one.

Molluscs. Freshwater molluscs are among the better-known groups of benthic animals, and they have been eaten by people since prehistoric times (van Bentham Jutting 1932), and in some areas, such as around Segara Anakan, the apple snails *Pila* spp. are still popular (fig. 9.23). A total of 16 bivalve molluscs and 88 snails are known to be native in Java's freshwaters (van Bentham Jutting 1953, 1956). From the distribution of species, it appears that the species on Java and Bali may be divided into two categories:

- those that appear to be endemic to the island or even a single lake, such as *Gyraulus terrascrae* and the bivalves *Pisidium javanum* and *Sphaerium javanus* from the Dieng Plateau; and
- those that are widely distributed species, such as the thiarid snails *Brotia testudinaria, Thiara scabra,* and *Melanoides tuberculata,* and the endemic bivalve *Elongaria orientalis.*

The common thiarid snails are found in lakes, irrigation ditches, and similar habitats, some from Africa to the Pacific islands. Not only do these snails give birth to fully formed young, but they reproduce without the eggs being fertilized (parthenogenetically). The eggs and larvae are retained in the brood pouch until they hatch. Populations of *Melanoides tuberculata* have very few males and these do not appear to function sexually. All individuals are therefore genetically identical, and this lack of evolutionary potential might be thought to doom the snail to extinction. It would seem, however, that they have all the adaptive potential required since they are extremely effective colonizers. One particular adaptation favouring colonization is the presence in the snail's brood pouch of young of all ages: eggs, larvae, and small snails, the last of which can be released throughout the year or when environmental conditions are favourable (Dudgeon 1986).

The generally large, air-breathing pulmonate snails, such as the widespread *Lymnaea rubiginosa,* generally breed frequently and abundantly, and this appears to lead to a reduced adult life span. These snails may thus be regarded as 'r-selected', and this probably reflects the often temporary nature of their habitats (pools, rice fields, lake fringes). The more common prosobranch snails (with gills, not lungs) such as the apple snails *Pila* spp. and the thiarids, on the other hand, tend to have frequent broods of relatively few young (sometimes born alive rather than hatching from an egg laid in water), and have longer adult life spans. These snails seem to be K-selected. The bearing of live young (viviparity) is clearly advantageous for a river-dwelling snail since planktonic larvae would be carried downstream and out to sea. It is not surprising, then, that the viviparous snail family Thiaridae dominates the headwaters and middle reaches of rivers in the Old World tropics (Dudgeon 1986).

In general, snails are most common where macrophytes are most dense (Koswara 1985; Suratiningsih 1986). In addition, some species are found more commonly over sand, and others over mud (table 9.4) (Suratiningsih 1986). It is often stated that snails feed by grazing algae from stones and plants, but this oversimplifies a situation in which different species prefer different foods, and many ingest predominantly inorganic material and detritus (Dudgeon and Yipp 1983).

The muscle tissue of the bivalves *Corbicula javanica* and the introduced *Anodonta woodiana,* found in the muddy northeast corner of Rawa Pening, is infested with trematode larvae, but thorough cooking prevents their transmission to humans (Göltenboth and Timotius 1992). In addition, three of the 15 gastropod snail species found in Rawa Pening carry the larvae (cercaria) of human parasitic echinostome worms (Suratiningsih 1986). A small industry exists around the lake for grinding the shells of both these species to make lime for chewing with betel nut.

Turtles. Five species of terrapins (Emyidae) and three soft-shelled turtles (Trionychidae) live in the rivers of Java and Bali, although some also spend considerable periods on dry land. One of the most common species is the herbivorous box turtle *Cuora amboinensis* of ponds and sawah, which has two hinges on its lower shell that allow the animal to cut itself off completely from the outside world. The black pond terrapin *Siebenrockiella crassicollis* is carnivorous, but generally confines its attention to frogs and snails. Whereas terrapins reach only 20-30 cm in length, the spotted soft-shelled turtle *Amyda cartilaginea,* which is found in muddy rivers, ponds, and swamps, can reach over 70 cm (fig. 9.24), and the much rarer giant

Table 9.4. Habitat preferences of aquatic snails in and around Rawa Pening. Ec = *Eichhornia crassipes,* Hv = *Hydrilla verticillata,* - = none, + = few, ++ = many, [i] = introduced species.

		Swamp			River		Sawah	
	surface	Ec	Hv	mud	mud	sand	grass	soil
VIVIPARIIDAE								
Bellamya javanica	-	-	-	-	-	-	-	+
AMPULLARIIDAE								
Pomacea paludosa [i]	+	+	+	+	-	-	-	-
Pila ampullacea	+	+	+	+	-	-	-	-
Pila scutata	+	+	+	+	-	-	-	-
Pila polita	+	+	+	+	-	-	-	-
THIARIIDAE								
Brotia testudinaria	-	-	-	-	++	+	-	-
Brotia testudinata	-	-	-	-	++	+	-	-
Melanoides tuberculata	-	-	-	-	+	++	-	-
Melanoides granifera	-	-	-	-	+	++	-	-
LYMNAEIDAE								
Lymnaea rubiginosa	+	-	-	+	-	-	+	+
HYDROBIIDAE								
Hydrobia sp. [i]	+	-	-	+	-	-	-	-
Amnicola sp. [i]	+	-	-	+	-	-	-	-
PLANORBIIDAE								
Segmentina sp. [i]	-	-	-	-	-	-	++	++
Drepanotrema sp. [i]	-	-	-	-	-	-	++	++
Armigerus sp. [i]	-	-	-	-	-	-	++	++

Source: Suratiningsih (1986) with updated nomenclature

Figure 9.24. Spotted soft-shelled turtle *Amyda cartilaginea,* which lays its eggs in riverbanks.

By F. Kopstein, with permission of the Tropen Instituut, Amsterdam

soft-shelled turtle *Pelochelys bibroni* can reach 120 cm in length (de Rooij 1915). The former species has a skin-covered shell, fully-webbed feet, and a long, tubular snout. It lies buried in mud from which it breathes simply by extending its long neck until its nostrils break the water surface. It is a sit-and-wait predator, biding time until a fish or frog passes within reach. These turtles can inflict a painful bite, and the only way to avoid this when picking one up is to grab the shell behind the rear legs.

BIOTIC PATTERNS

Lakes

The distribution of plants and animals in a lake is determined principally by the physical conditions, particularly the layers (p. 421). Differences in plankton abundance not only occur between the epi-, meta-, and hypolimnion, but considerable variation also exists within the epilimnion itself because of the differences in water density and viscosity, night-time cooling, turbulence, temperature, light intensity, and time of day. The feeding pattern of zooplankton in also important (Davis 1955).

Deep benthic animals either have to be able to cope with very little oxygen (such as the red, haemoglobin-filled chironomid fly larvae) or with no oxygen, and usually no light (such as anaerobic fungi and bacteria which derive their energy from dead organic matter and are the chief agents in the process of decay). Fish are not found in this environment, but those organisms that are adapted to it have few predators.

Rivers

As a river flows around a bend, so the water in the inner curve tends to be shallow and slow-running over stones, and the water in the outer curve is deep and fast-flowing over a muddy substrate. These patterns obviously influence the distribution of animals and plants, and the more a river meanders, the more diverse will be the life within it. As a river changes along its length in velocity, depth, turbidity, temperature, and slope, so the abundance and diversity of plant and animal life in it will change. In the headwaters the main aquatic plants are species of algae which attach themselves to rocks. As the river deepens and shear stress becomes less, so species of macrophytes will increase in importance. Gradually the river becomes more turbid and light penetration is reduced, and phytoplankton take over as the major group of plants. Phytoplankton are virtually absent in the headwater streams because they have no chance to develop here, and zooplankton are only found still lower down or in lakes where they are not swept away. There are no native fishes on Bali above about 500 m altitude, probably because they are intolerant of the low temperatures. The fishes that reach the highest altitudes on Bali are all gobies, including the beautiful *Stiphodon elegans,* which has a shimmering blue tail (Kottelat and Whitten in prep.). Above 500 m clean rivers have a rich invertebrate life with damselfly and dragonfly larvae, water cockroaches, water stick insects, whirlygig beetles, and tadpoles.

It might be thought that the many weirs found along the rivers block the distribution of river life. In fact, a surprising array of species can cope, either by taking the long route via the irrigated rice fields where the inclines are more gradual (as do various fishes), or by creeping up the concrete faces (as do snails, prawns, crabs, and gobies). Another creeper is the bottom-dwelling sucking catfish *Glyptothorax platypogon.* Its body is flat, offering little resistance to strong currents, and it also has loose lengths of skin between its pectoral fins which act as an adhesive organ (Kottelat et al. 1993). A study in the Cisadane River found that this species favoured fast-flowing water and a substrate with stones of diameter greater than 10 cm. It feeds most frequently on various aquatic insects and larvae such as mayflies, mosquitoes and midges, beetles, and caddisflies (Rachmatika 1987).

ENERGY

Importance of a Riparian Fringe

The major source of energy driving freshwater ecosystems is the material that drops, or is washed, into it from outside, such as dead leaves, wood, and fruit. If tall vegetation around a river is removed, plant matter fails to accumulate in sufficient quantities with the result that many organisms such as fish, crabs, prawns, snails, not to mention much smaller animals, will not survive even if other conditions are suitable (Dudgeon 1983). Some rivers without a riparian fringe will have fish, but few species are represented, and those present tend to be introduced or widely distributed. Native species with narrow niche requirements will tend to be lost. In addition to energy problems, the lack of shading caused by the absence of a riparian fringe along relatively narrow channels will result in higher water temperatures. An increase in water temperature from 25° C to 30° C will reduce the amount of oxygen the water can carry by 10%. This is likely to have marked impacts, particularly on animal communities, and most markedly on native species.

The way in which energy flows through a lake has been described for a number of sites in Java and Bali, although the importance of the different pathways can change markedly between seasons, and even among the same organisms between different habitats (Winemiller 1990). An example is shown on the following page (fig. 9.25).

MANAGEMENT OF MACROPHYTES

Dense growths of macrophytes, usually of introduced species, are generally found in waters that are naturally or artificially eutrophic, such as newly filled reservoirs. In the absence of their natural controls these plants are very adaptable, and have high potential rates of sexual and vegetative reproduction that are hard to control using biological, mechanical, or chemical means. It is for these reasons that they have spread far and wide around the world. Water hyacinth is probably the worst offender, although the water fern *Salvinia molesta* is also important (figs. 9.25-27; table 9.5). There are good reasons for not wanting large masses of macrophytes in a

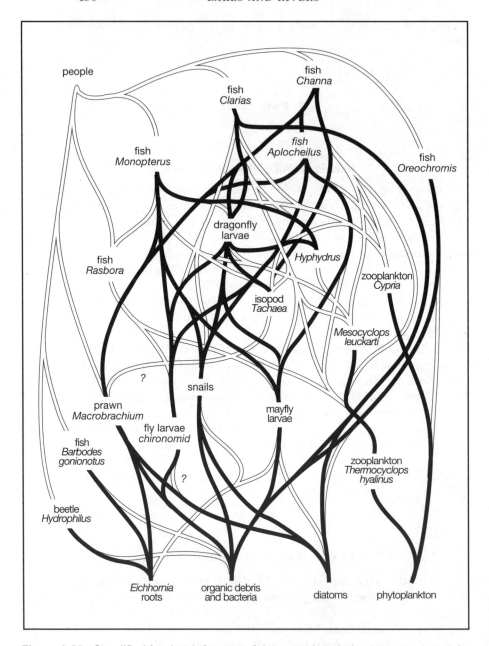

Figure 9.25. Simplified food web for part of the water hyacinth root-mat community in Ranu Lamongan. Linkages are based on gut contents of animals. Black lines indicate a major food of the consumer. Question marks indicate indirect evidence only. See also Göltenboth (1982); Göltenboth and Kristyanto (1987).

After Green et al. 1976

Figure 9.26. View of Rawa Pening showing extensive areas of water hyacinth. Eradication is not necessary because the plant actually benefits fisheries to a certain extent.

By A.J. Whitten

Figure 9.27. Gathering of water hyacinth in the Pluit reservoir, Jakarta.

With permission of Kompas

lake. For example:
- the accumulation of organic matter on the bottom reduces the water volume of the lake;
- the rate of evapotranspiration from floating plants is about twice the evaporation rate of open water, resulting in unnecessary water loss;
- floating masses change shape and position, sometimes blocking the outlet, or interfering with fishing activities; and
- the floating masses serve as breeding grounds for mosquitoes, rats, snails, and other vectors of plant, animal, and human disease;
- extensive growths of certain macrophyte species may endanger and cause the local extinction of native species of submerged macrophytes; and
- the submerged species *Hydrilla verticillata* can grow so thick that water transport and fishes can experience difficulty moving through the water (van Steenis and Ruttner 1933; Giesen 1991).

It is important to note, however, that by no means do all lakes have a macrophyte problem. Jatiluhur and Cirata reservoirs are water bodies where aquatic weeds are not a problem, partly because of the shape and depth of the reservoirs' bottom, and partly because the wave action on their surface is relatively severe (Tjitrosoedirdjo and Widjaja 1991). Second, in some places water hyacinth is not regarded as a weed at all, but as an important component of the fisheries system. At Ranu Lamongan, for example, floating mats of it are held between the shore and wooden pontoons. The water hyacinth itself is not eaten to any great extent by the fish,

Table 9.5. Some of the open waters in Java with major growths of macrophytes.

Location	Provinces	Important species
Rawa Danau	West Java	*Phragmites karka*
Eichhornia crassipes		
Lake Curug	West Java	*Salvinia molesta*
E. crassipes		
Hydrilla verticillata		
Saguling reservoir	West Java	*E. crassipes*
S. molesta		
Rawa Pening	Central Java	*E. crassipes*
H. verticillata		
Salvinia cucculata		
S. molesta		
Wlingi reservoir	East Java	*E. crassipes*
Bureng reservoir	East Java	*E. crassipes*
S. molesta		
Surabaya River	East Java	*E. crassipes*
Ipomoea aquatica		
I. fistulosa		

Source: various sources in Tjitrosoedirdjo and Widjaja (1991)

Figure 9.28. *Salvinia molesta,* a floating fern which causes major problems in many water bodies. Growth habit in crowded (left) and uncrowded (right) conditions.
After Kostermans et al. 1987

but its roots are coated with diatoms and brown organic debris, and many animals, ranging from dragonfly larvae to amoebae, at the base of the complex food web, feed on these (Institute of Ecology 1980; Göltenboth and Kristyanto 1987). It is unfortunate that most macrophytes are referred to as 'weeds', a pejorative word that has been defined as 'a plant where you don't want it', and efforts are sometimes made at control before estab-lishing that it is really necessary.

Water hyacinth was brought to Indonesia in November 1894 (Backer 1936), when it was grown in Bogor Botanical Gardens (p. 183). By 1931 Rawa Pening was completely covered by it (Göltenboth and Kristyanto 1987). Efforts to control water hyacinth and submerged macrophytes at Rawa Pening by manual and mechanical means have been made on and off by a variety of government agencies (Tjitrosoedirdjo and Widjaja 1991). In the most recent intensive efforts, in 1988 and 1989, 14 and 31 ha were cleared, respectively (Tjitrosoedirdjo 1991). There is currently a Japanese-built weed-gathering boat working on the lake, but it appears to be making no significant headway against the plants' growth. *Salvinia molesta* arrived in

Rawa Pening only in 1987, about 35 years after its introduction, adding just one more problem to the management of the lake .

By the time Saguling reservoir was first filled in 1985, there were already 2 ha of water hyacinth. Part of it was removed manually, but the remainder increased to 6 ha the following year. Over half of that was removed, but two years later the area covered by this prolific plant was 30 ha. Half of that was cleared using 750 workers, but the work force was not motivated to remove all the plants because the yearly manual clearance ensured supplementary income.

Control by Use

Macrophytes do have many uses, and considerable efforts were expended 10-15 years ago to demonstrate their economic value in the production of mushroom culture, fish food, biogas (600 L of biogas can be obtained from a single kilogram of dried water hyacinth, enough to bring 15 L of water to the boil), fertilizer, livestock foods, handicrafts, and coarse but eco-logically friendly paper (Soerjani 1976; Lembaga Ekologi 1979; Widyanto 1979). Water hyacinth can also be used to accumulate heavy metals dis-solved in water, thereby reducing their concentrations. Unfortunately, none of the schemes really succeeded because local use did not translate into economic success or effective control.

Local uses, such as the incorporation of water hyacinth as green manure into rice fields around Rawa Pening, are a major reason that these plants are not more of a problem, although most people would like to see less (Göltenboth and Kristyanto 1987). Other macrophytes in the same lake are also harvested. *Hydrilla,* which when dry contains 16% protein, is fed to pigs, and the *Salvinia* is mixed with peat from the lake to make a compost which is taken to the Dieng Plateau for mushroom culture. The mush-rooms are exported to the USA. The leaves of one species, *Ipomaea aquatica,* are widely eaten, but there is a limit to how much can be marketed and consumed.

Chemical and Biological Control

Control of macrophytes is a complex problem. For example, grass carp have been used successfully against a major growth of *Hydrilla verticillata* and less abundant weeds in the eutrophic artificial 1 ha lake Situ Leutik on the new Darmaga campus of Bogor Agricultural University. The lake became turbid, however, because of the large quantity of faecal wastes which the fish produced. This increased eutrophication stimulated phyto-plankton growth and unaesthetic turbidity, which developed into a 'bloom', using up oxygen and suffocating the fish. A balance needed to be sought between plant eaters and algal feeders such as tilapia, *Barbodes gonionotus,* and wild fighting fish *Betta* sp. (Wijaya et al. 1991).

Various chemical controls have been tried in laboratories and in the field (2,4 D has been used against water hyacinth), but none has been used widely or consistently on a large scale. Biological control once seemed a promising avenue for the control of water hyacinth, but none of the insects, fungi, or fish has been particularly successful. One problem has been finding an organism which is sufficiently specific. For example, a number of insects (grasshoppers, moths, and aphids) attack water hyacinth, but they also feed on economically important crops (Tjitrosoedirdjo and Widjaja 1991). One of these insects, an American weevil *Neochetina eichhorniae,* was brought to Indonesia in 1975 to control water hyacinth, but it was neither particularly potent nor very specific, having spread to arrowroot *Canna edulis,* lengkuas *Alpinia,* and ginger *Zingiber* sp. (M. Soerjani pers. comm.). As a result of these disappointments the last decade has seen little work on the biological control of aquatic weeds. Successful biological control of *Salvinia molesta* has been achieved in several countries by the introduction of a *Cyrtobagous* weevil which typically reduces the area covered by the weed by 99% within a year (Room 1990), but this has not been adopted. Rather, the current practice on infested Javan water bodies is to use manual and mechanical methods to collect the plants, and then tip them over the tops of dam walls to be washed away (or to become someone else's problem). Most of the mechanical control programmes against macrophytes have been judged unsuccessful, but this is often because the programmes were halted. Such programmes are indeed highly unlikely to lead to eradication (even if eradication were the declared goal), and so they must be properly planned and then applied consistently, either year after year, or every few years (Tjitrosoedirdjo and Widjaja 1991).

IMPACTS OF DEVELOPMENT

As stated earlier, one of the great benefits of the detailed work of the German Sunda Expedition, is that it allows present data to be compared with a baseline. For example:

- examination of the high-altitude Telaga Pasir at Sarangan, East Java, in 1981 (Ibkar-Kramadibrata and Oey 1987) found that the biological, physical, and chemical conditions were very similar to those found over 50 years earlier (Ruttner 1931);
- Ranu Lamongan was studied in detail in 1928 (Ruttner 1931) and in 1974 (Green et al. 1976). Despite a radical change in vegetation around the lake in the intervening years, and a dramatic increase in the human population, the chemical characteristics, phytoplankton, macrophytes, and the fauna appear to have remained much as they

were, except for four introduced species of food fish;

- at Rawa Pening, there were close similarities in the physical and chemical characteristics found in 1978 and 1949, although the nitrate, nitrite, and ammonium concentrations were rather low in the late 1970s (Kristyanto and Göltenboth 1979). Warnings were given that the lake could become eutrophic (Hardjosuwarno et al. 1974), and this has happened (Göltenboth and Kristyanto 1987); and

- two of Bali's lakes, Batur and Bratan, were visited by the German team in 1929, by British and Indonesian scientists in 1974 (Green et al. 1978), and by Finnish and Indonesian scientists in 1977 (Lehmusluoto and Mahbub 1989). It appears that the lakes have changed little over this period in their physical and chemical characteristics, although the effect of human activities on Lake Bratan are beginning to show with a slight increase in pH.

These results show that surprisingly little change has occurred in the lakes because lakes are large in volume and continually flushing. Where the water body is small or slow-flowing, or the inputs of pollutants high relative to the volume of the water, the impacts will be greater. Impacts will also be felt in lakes where there is little movement of the water, such as in the 'arms' of Saguling. The aquaculture potential here could be lost if aquatic pollution flowing from the Bandung-Cimahi-Padalarang area worsens. Already in 1989 the concentration of nitrates, sulphides, and ammonia were too high, and dissolved oxygen at night too low, for optimum aquaculture conditions.

Water pollution problems generally increase with distance from the headwaters (de Iongh and Ruhyat 1974), but this does not necessarily mean that the headwaters are in a good condition. For example, ecological surveys of the upper reaches of the Ciliwung found that even above the areas of major industrial and domestic activity, the river and its tributaries were biologically derelict, with few species and few individuals. The impoverished fish fauna was thought to be due to a number of factors: reported and probably unreported fishing by tipping large quantities of DDT into the water upstream, overfishing using electricity and underwater bombs, high levels of suspended sediment from erosion, excessive use of agricultural pesticides, and extraction of sand, gravel, and rocks (Sabar et al. 1984). Even so, quality can improve along a river because the activity of organisms purifies the water (fig. 9.29). Inadequate attention, perhaps, is given to this aspect of water management.

Poisons, Bombs, and Electric Shocks

The most common traditional fish poison is 'tuba' made from the roots of various species of the climber *Derris* (Legu.) in which the active chemical is

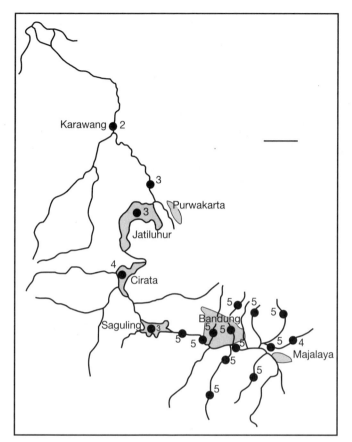

Figure 9.29. Water quality along the Citarum River and its tributaries to demonstrate that natural processes will ameliorate even serious pollution. 2 - moderately polluted, 3 - heavily polluted, 4 - very heavily polluted, 5 - incredibly polluted. Scale bar indicates 10 km.

After BP3U 1991

rotenone. Other poisons, typically commercial insecticides, are also commonly used. Poisons, bombs, and electric currents are very effective, and kill or debilitate large numbers of fish without regard to size or edibility. The effects of explosives are devastating, as described for coral reefs (p. 362). Apart from any other arguments, these forms of fishing are selfish, needlessly destructive and, in the case of poisonous insecticides, can cause illness in people who buy the fishes, unaware of their origins. As a result, many urban and rural rivers have virtually no fish left in them (fig. 9.30).

Forest Clearance and Land Management

From an ecological view, the accumulation of sediments, and the progressive depth reduction this causes in lakes, is entirely natural (Hardjosuwarno et al. 1974). It just happens to be in conflict with those development plans

Figure 9.30. Rivers in Central Jakarta have virtually no fishes, but for some people it is worth seeing what can be caught.

With permission of Antara

that insist that ecosystems should be kept in a state of eternal youth. Soil washed into reservoirs from the surrounding hills reduces their volume or storage capacity, and thus their useful life and return on the enormous investments (Hardjosuwarno et al. 1974; Soerjani 1976). Many of the sedimentation problems in the major reservoirs are caused by inappropriate land-use practices that cause soil erosion (p. 134), and these are best alleviated by controlling the cultivation patterns used by farmers in the watershed–for example, by discouraging the growing of cassava, and encouraging the planting of useful trees and vetiver grass (p. 144).

Dams

Dams are built primarily to stimulate economic growth by electricity generation, and to increase food production through irrigation. There are many negative aspects concerning large dams and reservoirs, such as lack of sustainability, negative environmental and social impacts, disruption of migratory freshwater organisms, and the poor and inequitable distribution of benefits. Huge investments and debts have been undertaken to create these infrastructures, but perhaps not enough attention has been devoted to increasing the efficiency of energy generation, transmission, distribution, and end use, which are the best ways to increase energy supply with minimal cost and environmental impact. Even so, power from hydroelectric schemes is environmentally preferable to power from diesel or coal stations. For example, the Mrica dam, 100 km southwest of Semarang, is the largest in Central Java and by producing about 580 GWH annually, saves the burning of about 290,000 tons of oil each year. Despite the relative advantages of hydroelectricity generation, every effort should still be made to move towards means of electricity generation such as wave and solar power which can have less serious environmental impacts.

Reservoirs and Socio-ecological Impacts

One of the most recent reservoirs to be created was Saguling. Attempts were made to learn from past mistakes, making thorough analyses of potential impacts, and ensuring that the dam should be an agent of development, with the displaced people regarded as a resource rather than a liability (Costa-Pierce et al. 1988; fig. 9.31). It was proposed to develop fisheries to benefit the displaced people, whose Sundanese culture had a deep-rooted fisheries interest. At the time, floating net culture of carp was becoming popular and these proliferated as the reservoir filled. Early on it was only the moderately rich farmers who could afford the rafts, nets, and fish food, but schemes were developed so that poor, displaced farmers could benefit, too (Soemarwoto 1989).

Unfortunately, ecologically and sociologically equitable projects sometimes do not survive in the 'real' world. The successful floating fish cages of Saguling and Cirata reservoirs became attractive to urban capitalists, thus displacing the local people for the second time: the first being physical displacement, the second economic, thereby reducing many of the inhabitants of the area to the level of labourers. Only if the people organize themselves into cooperatives and/or the fisheries authorities give licenses preferentially to the local people (and act on untruthful declarations of ownership) will the full social benefits be realized (Soemarwoto 1989). The capture fisheries on the waters of Saguling had been estimated to produce 45 kg/ha/year. Although productivity grew rapidly, it soon decreased markedly and stabilized at about 1-2 kg/ha/year. Part of the reason for this was overfishing, and part was the dominance of the fish community by the predatory *Hampala macrolepidota*. To improve the yield of capture fisheries, stock of *O. mossambicus* and *O. nilotica* were introduced, licenses were issued to individual fishermen, and it was proposed to introduce the freshwater sardine *Clupeichthys aesarnensis* from Thailand because it would exploit a pelagic and currently unoccupied niche in the reservoir (p. 637) (Costa-Pierce and Soemarwoto 1990a).

There may be good ecological studies on the effect of creating reservoirs, but we have not found them. The closest is a study of the condition

Figure 9.31. A disadvantage of large reservoirs is that good-quality rice fields are lost, such as here at Neglasari, now submerged under the Saguling reservoir.
With permission of Kompas

of the freshwater fishes in the area that became the Wonogiri reservoir, which was monitored for two years after it began filling in late 1981. It is the first two years after closure of the dam that are regarded as the most critical in terms of ecosystem change because it is during this period that terrestrial organic material decays, and the physical characteristics of a river change to those of a lake. Among the early changes found as the river became a lake were a decreasing abundance of snakeheads *Channa striata* and catfish *Macrones* sp. in the south. Populations of most species increased initially, then crashed, but exactly what the mechanisms of these changes were is unknown. By 1982 *Barbodes gonionotus* (omnivore) and *O. hasseltii* (plankton feeder) were most abundant, at least in part because of major stocking of the former (nearly 90,000 fry were introduced in 1981-82) (Praptohardiyo and Muluk 1982). It is also known that of the 20 native fish species found in that part of the Citarum that became the Jatiluhur reservoir prior to completion of the dam, only eight grow and breed well in the reservoir. A ninth species, *Labeo chrysophekadion,* was caught in reasonable numbers but since then declined dramatically for no known reason (Krismono and Atmadja 1983; Sarwita 1983; Hardjamulia et al. 1988).

Sediment Mining

One of the most obvious and widespread activities creating negative impacts on riverine ecology is the large-scale collection of river stones and sand for road and other building projects. The disturbance this causes lifts large quantities of sediment and makes the river channel more prone to erosion. This was studied in the Ciapus River, a tributary of the Cisadane near to Bogor, West Java. In general it was found that effects were felt upstream, downstream, and in the immediate vicinity where the riverbed is broadened due to the use of heavy trucks and equipment. Regulations and guidelines concerning the activities were completely ignored. The small Ciapus was carrying an average of nearly 216,000 tons of sediment each day into the Cisadane (itself carrying just 12,255 tons up to the confluence), and it is estimated that the effects of this enormous extra load would be experienced 11-17 km downstream (Sundjoto 1983).

Regulations covering extraction of river sediments do include some environmental safeguards and make it clear that the license holders are themselves responsible to some extent. However, implementation and supervision are not as tight as one might hope. Part of the problem is inadequate manpower and budgets, but at least as significant is a lack of understanding of the importance of the safeguards by the police, judiciary, local people, extraction company personnel, and local government employees (Laksono 1985).

Chapter Ten

Lowland Forests

Tropical lowland forests are among the world's richest and most diverse ecosystems, providing not only unequalled aesthetic inspiration, but also a wide range of economic benefits, from sustainable supplies of high-quality timber to watershed protection, and from useful chemicals to cane. It is thus unfortunate that so much of Java and Bali's lowland forests has been lost, and it is difficult, and in some ways pointless, to describe what is essentially an historical situation (p. 328, figs. 4.24, 10.1). There are remnant areas, however, and for the sake of encouraging some interest in and understanding of these, the descriptions in this chapter are provided. The remaining mangrove forests are dealt with in chapter 8. The details of forest dynamics, such as succession, regeneration, nutrient cycling and partitioning, are described elsewhere (Whitmore 1984, 1990; MacKinnon 1986, 1992; Whitten et al. 1987a, b, c; MacKinnon et al. 1996).

RAIN FORESTS

Rain forests are the luxuriant, rich, and stately forests or jungles of legend that still clothed much of Java at the beginning of the last century, but which are now confined to isolated or disturbed remnants. Many species have probably been lost by the agency of man, since more than 100 plant genera found in southern Sumatra do not occur in Java–too many to explain by island biogeography alone (van Steenis and Schippers-Lammerste 1965). The major loss of forest has skewed our perception of certain plants. For example, although the important timber tree *Altingia excelsa* (Hama.) is generally regarded as a mountain plant, it grows best at low altitudes (from 200 m), and it is seen now only at one end of its altitudinal spectrum. It is disturbing that some of the –largest blocks of rain forest, such as most of Nusa Kambangan (fig. 10.2), Mt. Wayang (p. 793), and Lebakharjo (p. 802) are not gazetted conservation areas.

Figure 10.1. A stand of *Dipterocarpus* forest near Trawas, photographed in 1930. This and many similar forests no longer exist.

Photographer unknown, with permission of the Rijksherbarium/Hortus Botanicus, Leiden University

Figure 10.2. View over part of the eastern end of Nusa Kambangan to show the dense forest and the small islands of Karang Bolong and Wijayakusuma where the culturally important flower *Pisonia grandis* is said to occur (p. 175). Although taken 70 years ago, the situation today is not much changed.

By H.J. Lam, with permission of the Rijksherbarium/Hortus Botanicus, Leiden University

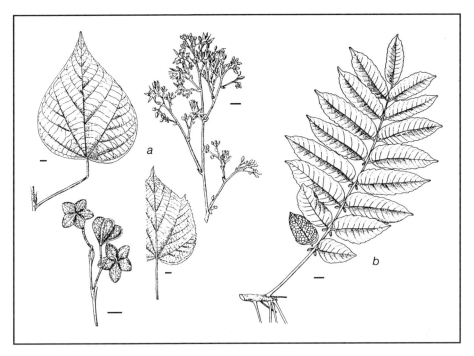

Figure 10.3. Two trees characteristic of semi-evergreen rain forest: a-*Kleinhovia hospita,* and b-*Garuga floribunda.* Scale bars indicate 1 cm.

a-after Leenhouts 1956; b-after Koorders and Valeton 1915

Vegetation

Two types of rain forest are recognized in this book: evergreen and semi-evergreen, and their natural distributions are shown elsewhere (p. 192). The difference between the two is that the semi-evergreen grows in slightly more seasonal areas, with two to four dry months each year, and is characterized by the presence of *Garuga floribunda* (Burs.), *Pterospermum diversifolium* (Ster.), and *P. acerifolium* in primary forest, and of *Kleinhovia hospita* (Ster.) in secondary areas (fig. 10.3). Up to about half the tree species present are deciduous.

Java. Unlike the majority of the lowland forests of Borneo and Sumatra where the dominant trees are dipterocarps, most of the lowland rain forests of Java have no dominant species or family, and there is such variability in species composition that no 'typical' mix of species can be given. However, four species are common to (although not necessarily abundant in) the four main areas of remnant lowland rain forest in West Java: Peucang, Dungus Iwul, Yanlappa, and Sukawayana. The species are *Artocarpus elasticus*

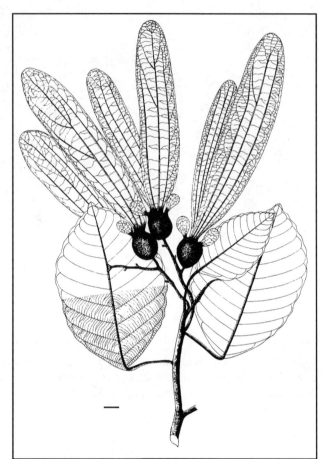

Figure 10.4. *Diptero-carpus hasseltii* domi-nates the vegetation in Sangeh Reserve, Bali, but the trees are probably the result of planting or selection. Scale bar indicates 1 cm.

After Whitmore et al. 1989

Table 10.1. Comparisons of lowland rain-forest tree composition and structure between six areas of lowland rain forest. dbh = diameter at breast height.

	Plot area (ha)	Species total	No. trees/ha (>10 cm dbh)	Average basal area
Panaitan Island	0.3	82	453	0.04
Peucang Island	36	330	534	0.14
Ujung Kulon	1.85	128	403	0.04
Dungus Iwul	?	?	420	?
Yanlappa	32	275	523	0.09
Meru Betiri	?	146	207-437	?

Sources: Kartawinata et al. (1985); Soejono (1992)

(Mora.), *Dysoxylum caulostachyum* (Meli.), langsat *Lansium domesticum* (Meli.), and *Planchonia valida* (Lecy.), and Kartawinata (1975) has asserted that Javan lowland rain forests are characterized as *Artocarpus elastica-Planchonia valida* forests. Readers needing more detail should refer to the site specific studies listed in Appendix 2, and to the descriptions of individual protected areas (chap. 20). The number of trees and species per unit area in Javan forest differs between areas (table 10.1) and is rather lower than found elsewhere in the region.

Bali. The general comments about the composition of Javan rain forests apply also to Bali. Very little remains of the Bali rain forests below about 500 m, although some of the deep, steep, and inaccessible ravines cut into the soft volcanic tuff by swift rivers still have remnants of lowland forest species clinging to them. Species composition is variable. The forest in the nature reserve around the famous Sangeh monkey temple is a nearly pure stand of the native *Dipterocarpus hasseltii* (Dipt.) (fig. 10.4), although there are some 50 other flowering plants present. These are described, illustrated, and keyed in a useful report by Tantra (1982). There are over 550 trees with trunk diameters over 60 cm, the largest reaching 2 m. Although it is regarded locally as a remnant of natural forest, there are few botanical indications of this (de Voogd 1937a; Soepadmo 1961, 1965; Meijer 1976; Tantra 1982). The largest of the trees may once have been part of a forest, but the remainder would have been planted or selectively preserved.

There is no consistent forest composition in the planned extension to Bali Barat National Park, but on parts of the southern slopes the most common large tree is *Planchonia valida* (Lecy.), the leaves of which turn a deep red before falling. Other common trees include *Palaquium javense* (Sapo.), *Duabanga moluccana* (Sonn.), *Meliosma ferruginosa* (Sabi.), and *Pterospermum javanicum* (Soepadmo 1961) (fig. 10.5), but around the Pura Luhur temple on the southern slopes of Mt. Batukau the most common tree is a species of *Tabernaemontana* (Apoc.).

Fauna

The majority of animals native to Java and Bali would have been found in the rain forests, although such forests do not support large populations of most species because the resources available to them are both scattered and uncommon (Kikkawa and Dwyer 1992). By far the best and most detailed account of the vertebrate communities concerns Ujung Kulon (Hoogerwerf 1970), but its relevance is much wider. In many respects there have been few advances in the understanding of general mammal ecology since those studies, and people with a serious interest should obtain a copy. The role of animals in seed dispersal, seed predation, pollination, leaf

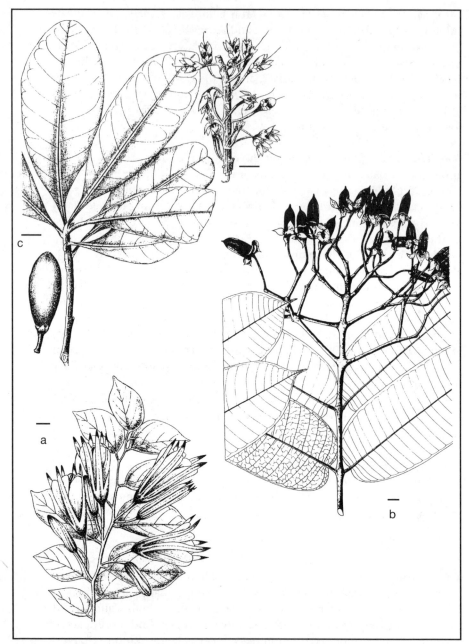

Figure 10.5. Three trees common in the lowland rain forests of West Bali: a - *Ptero-spermum javanicum*, b - *Duabanga moluccana*, and c - *Palaquium javense*. Scale bars indicate 1 cm.

a-after Koorders and Valeton 1915; b-after Whitmore et al. 1989; c-after Koorders and Valeton 1918

eating, decomposition, and other important functions has been discussed elsewhere (e.g., Anwar et al. 1984; Whitten et al. 1987a, b, c), and it is almost certain that their diminished populations jeopardize the sustainability of disturbed forests.

Effects of Disturbance

When lowland or hill forest is cut or otherwise disturbed, or when gardens or orchards on former forest areas are abandoned, various successions of temporary vegetation types will occur, each successive one having more similarity to the natural forest than the last, assuming further disturbance does not occur. Disturbances such as landslips, storm damage, volcanic eruptions, changing riverbanks, and earthquakes are entirely natural, and there are plants adapted to exploit the ecological opportunities these bring. A return to the original forest composition and structure can take decades or centuries depending on the degree of disturbance, the extent of the area affected, and the proximity of forest from which seeds can be dispersed.

Where fires are not prevalent, a secondary growth will develop comprising widespread pioneer or 'nomad' trees such as: *Mallotus* (Euph.), *Macaranga* (Euph.), *Homalanthus* (Euph.), *Trema* (Ulma.), *Gironniera* (Ulma.), *Pipturus* (Urti.), *Vitex* (Verb.), *Piper aduncum* (Pipe.) (p. 185), *Melochia umbellata* (Ster.), *Grewia* (Tili.), *Adinandra* (Thea.), *Ficus septica* (Mora.), and *F. padana*. These types of tree share certain characteristics such as: small seed size, rapid growth, short life, and shade intolerance. Many are also quite succulent, such as *Brugmansia suaveolens* (Sola.), *Leucosyke diversifolia* (Urti.), and *Villebrunea rubescens* (Urti.), and these are often cut to provide fodder for cattle and goats (Sukardjo 1982a). In addition, one can sometimes find fruit trees deliberately planted in the young growth such as kedondong *Spondias* (Anac.), rambutan *Nephelium* (Sapi.), durian *Durio* (Bomb.), and jackfruit *Artocarpus* (Mora.). Herbaceous plants are dominated by large gingers, wild bananas, and aroids. In some areas, climbers such as *Mikania* (Comp.) (p. 185), morning glory *Ipomaea* spp. (Conv.), and *Thunbergia* (Acan.) may smother the young secondary growth and delay or prevent the growth of young trees.

Sure signs of previous disturbance within an area of forest are groves of bamboo (van Steenis and Schippers-Lammerste 1965; Hommel 1987). These produce a thick litter which, together with possible allelopathic effects, prevent other plants from growing. Such bamboo forests can be seen in Ujung Kulon, and in the northern part of the coastal plain the forests are also characterized by many-stemmed trees reminiscent of a European coppice, a type apparently unique in the Malesian region. The most common tree in these areas is the hard-wooded, fire-resistant pioneer *Ardisia humilis* (Myrs.) which, it is supposed, was cut close to the ground

Figure 10.6. Mature *Ardisia humilis* trees in the semi-evergreen rain forest of Ujung Kulon National Park. Their characteristic shape probably results from people cutting them while clearing land for cultivation before the eruption of Krakatau in 1883, after which they grew undisturbed.

By Gerald Cubitt, with permission

when the grazing savannas in this part of the peninsula were maintained by cut and burn. When this practice was abandoned, pioneer trees were able to grow (fig. 10.6; Hommel 1987). Where the soil is poor and the land used as pasture, then unpalatable or spiny foliage of *Melastoma* (Mela.), *Tetracera* (Dill.), and *Lantana* (Verb.) can predominate.

DECIDUOUS FORESTS

The remaining areas of deciduous forests on Java and Bali occur where there are four or more dry months. Two major types are recognized in this book: moist deciduous forest where annual rainfall is 1,500-4,000 mm and the dry season four to six months long, and dry deciduous forest where the rainfall is less than 1,500 mm and there are more than six dry months. During the dry periods the soils of both types dry out, deep cracks develop,

Figure 10.7. Deep cracks appear in the soil of Baluran National Park during the long dry season.

By A.J. Whitten

and evaporation exceeds precipitation (fig. 10.7). In contrast with trees in the less seasonal areas, the trees here have to develop very deep root systems, and annual plants and those with underground tubers are common. Those who have walked through this forest will remember that many of the plants are spiny, but their apparent dominance may be caused by people and grazing animals cutting or eating species which are less unpleasant to deal with (van Steenis and Schippers-Lammerste 1965). Where fire and herbivores are major ecological factors, the dry deciduous forest has formed a savanna vegetation with a more or less continuous grass layer and a discontinuous tree layer (Boulière and Hadley 1983; Cole 1986; Walker 1987).

Vegetation

Typically the forest comprises a simple, lightly closed community of trees few of which ever exceed 25 m in height, and containing a large proportion of deciduous or leaf-shedding species. Deciduous forest contains relatively few species found in the lowland rain forest, and one of the few emergents in certain areas is *Salmalia malabarica* (Bomb.) (fig. 10.8). During the Hindu period in East Java, these used to be planted as boundary markers.

Figure 10.8. *Salmalia malabarica,* a common tree in the deciduous forests of Java and Bali. Scale bar indicates 1 cm.

After Prawira et al. 1972

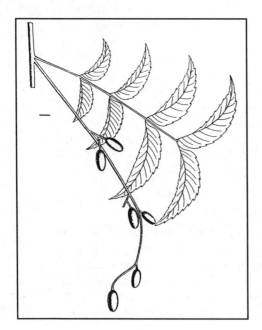

Figure 10.9. The neem tree, *Azadirachta indica,* is an evergreen species of deciduous forest. Scale bar indicates 1 cm.

After Prawira et al. 1972

This can be seen in the dry hills east of Yogyakarta where, at the start of the dry season, they lose their leaves and shortly thereafter sport great masses of red flowers (Beumée 1929). One of the trees that keeps its leaves is the exotic but widely naturalized neem *Azadirachta indica* (Meli.) (fig. 10.9). This has had minor uses in Java and Bali (Heyne 1987), but in India its use and potency as a pesticide attracted the interest of international drug companies. In 1993 a stable form of the active pesticide compound azadirachtin was produced and patented in the USA, much to the annoyance of the Indians (Pearce 1993b).

Probably the most conspicuous plant indicators of a climate where deciduous forest would be found are the two large palms, *Borassus flabellifer* and *Corypha utan,* since they are quite resistant to fires. They are easy to confuse, but *Borassus* has blue-green or greyish leaves, a smooth trunk with a diameter similar to that of a coconut palm, and no large spines on the leaf stalks. *Corypha* is generally more massive, its leaves are nearly 2 m in diameter, twice as large as *Borassus,* and it regularly produces small fruits up to 5 cm diameter compared with the massive, 18 cm fruits of *Borassus. Corypha* can be identified from a distance by the lines formed by the persistent leaf bases spiralling up the trunk. When it is 50-70 years of age, *Corypha* flowers (Whitmore 1970), and *C. utan* has the second-largest inflorescence of any flowering plant, with the flowering branches projecting nearly 4.5 m above the top of the palm and setting hundreds of thousands of fruit. After fruiting, the tree dies. Other good indicators, often seen growing at the roadside, are *Calotropis gigantea* (Ascl.), with large, woolly leaves and pale purple flowers which are sometimes candied for food, and the introduced prickly pear cactus *Opuntia nigrans* (Cact.).

Among the more common deciduous trees in the eastern part of Java and on Bali not already mentioned above are *Homalium tomentosum* (Flac.) (the source of 'Moulmein lancewood'), *Albizia lebbekoides, Acacia leucophloea, A. tomentosa, Bauhinia malabarica, Cassia fistula* (all Legu.), *Dillenia pentagyna* (Dill.), *Tetrameles nudiflora* (Dati.), *Ailanthus integrifolia* (Sima.), and *Phyllanthus emblica* (Euph.) (fig. 10.10). Among the herbs, many are confined to deciduous forest such as *Turraea pubescens* (Meli.) with ant-dispersed seeds, *Helicteres angustifolia* (Ster.), and *Glinus lotoides* (Moll.). Many grasses are also so confined, but the generally troublesome alang-alang *Imperata cylindrica* is uncommon, substituted by *Andropogon amboinicus* or *Sorghum nitidum.*

The deciduous vegetation, even in and around teak stands, shows very clear seasonality, and the contrasts are best seen at the end of the dry season. Some dormant trees, such as *Acacia leucophloea,* start to come into leaf before rain falls on the forest itself, possibly as a result of a rise in the water table caused by rains in the hills. Bees become active as they are attracted to the sweet smells of *Garuga floribunda* and *Adenanthera microsperma* (Legu.) flowers among their leafless branches. The flower buds of

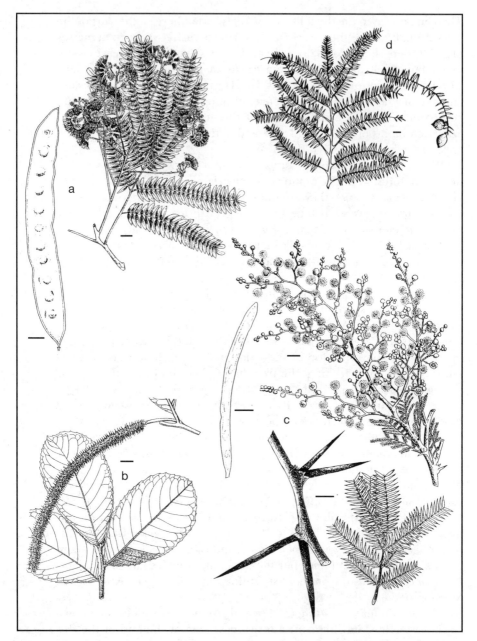

Figure 10.10. Four trees common in deciduous forest: a - *Albizia lebbekoides*, b - *Homalium tomentosum*, c - *Acacia leucophloea*, and d - *Phyllanthus emblica*. Scale bars indicate 1 cm.

a-after Koorders and Valeton 1913; b-after Koorders and Valeton 1914; c-after Prawira et al. 1972; d-after Whitmore 1972a

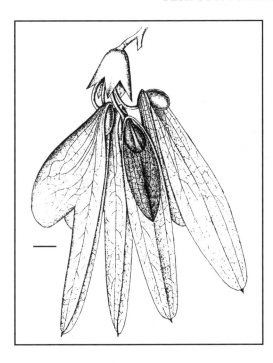

Figure 10.11. Distinctive fruit of *Pterocymbium javanicum*. Scale bar indicates 1 cm.

After Koorders and Valeton 1914

Pterocymbium javanicum (Ster.) are visible for some months as balls hanging below the branches before they break out into dark red, bell-like flowers (fig. 10.11). Once the rains fall, then within a week there is a riot of leaf and flower colour from the ground to the canopy. On the forest floor lilies, such as the yellow-flowered *Curligo orchioides* (Lili.) and the larger *Crinum asiaticum* (Lili.) with long, white petals and broad leaves, use up reserves stored in their deep underground bulbs. One might also find the endemic, purple-flowered ground orchid *Nervilia campestris,* whose leaves and flowers appear at different times. Also at this time the small shrubs such as *Munronia javanica* (Meli.) and *Tabernaemontana pauciflora* (Apoc.) excel themselves with shows of flowers (de Voogd 1927).

Indramayu. The fire-influenced deciduous vegetation closest to Jakarta was in Indramayu, in areas that are now irrigated. Here there were massive termite mounds 2 m tall and 5 m across. Typical trees were *Schleichera oleosa* (Sapi.), *Grewia* (Tili.) and *Schoutenia ovata* (Tili.), *Phyllanthus emblica, Vitex, Morinda tinctoria* (Rubi.), *Acacia leucophloea, Albizia,* and fire-resistant *Dillenia* (van Steenis 1936b).

Figure 10.12. Dry deciduous forest in Baluran National Park, showing the flat-topped crowns of *Acacia leucophloea.*
By A.J. Whitten

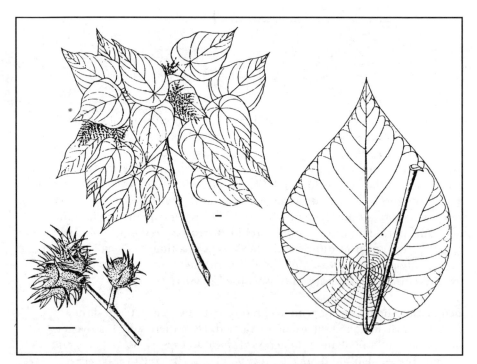

Figure 10.13. *Macaranga tanarius.* Scale bar indicates 1 cm.
After Koorders and Valeton 1913

Figure 10.14. *Ziziphus jujuba* may have been brought from India centuries ago because of its highly edible fruit.

By E.W. Clason, with permission of the Rijksherbarium/Hortus Botanicus, Leiden University

Madura. The vegetation of Madura was described by Beumée (1929), but even 60 years ago most forest elements had already been lost. Remnants appear to exist on the rocky cliffs east of Pagantenan, but they have not been investigated.

Baluran. Most of the vegetation of this national park (p. 817) comprises dry deciduous forest with evergreen *Ziziphus rotundifolia* (Rham.), *Albizia lebbekoides* (Legu.), *Phyllanthus emblica* (Euph.), *Sterculia foetida* (Ster.), *Tamarindus indica* (Legu.), and *Azadirachta indica* (Legu.), and deciduous *Schoutenia ovata* (Tili.), *Kleinhovia hospita* (Ster.), and *Flacourtia indica* (Flac.), as well as the large palms *Corypha utan* and *Borassus flabellifer.* Above 400 m on the mountain the forest is rather more evergreen, and *Drypetes ovalis* (Euph.) and *Homalium foetidum* (Flac.) can be found (Appelman 1937; Partomihardjo and Mirmanto 1986). In the thorny, fire-influenced savannas there are grasses with isolated *Acacia leucophloea* (Legu.), *Ziziphus rotundifolia, Schleichera oleosa* (Sapi.), and *Corypha utan* palms.

Eastern Bali. The forests of eastern Bali are extremely dry because much of the rainfall percolates rapidly through the unstable, ash-cinder soil. The river beds are generally dry. Most of the larger trees are felled by villagers and there are abundant pioneers such as *Macaranga tanarius* (fig. 10.13), *Mallotus paniculatus, Trema orientalis* and occasional *Manglietia glauca* (Magn.) in the mountains which is much favoured for carving, *Dendrocnide stimulans* (Urti.), and *Engelhardia* (Jugl.) with tri-lobed, winged seeds (fig. 11.16).

West Bali. The coastal savanna forest in northwest Bali is characterized by the flat-topped crowns of spreading *Acacia* trees and by numerous *Borassus* palms. The lower parts of the trees and shrubs are charred as a result of the regular fires. The grassy areas have about ten species of grass and numerous small herbaceous legumes such as species of *Crotalaria* and *Indigofera.* Among the trees are the widespread *Albizia lebbekoides, Cordia obliqua* (Bora.), *Antidesma ghaesemblica* (Euph.), *Phyllanthus emblica* (Euph.), *Schoutenia ovata* (Tili.), and *Ziziphus jujuba* (Rham.) (fig. 10.14). *Z. jujuba* may possibly have been brought to Java and Bali from India many centuries ago because of its highly edible fruit, its tanning properties, and its pro-duction of lac (p. 622). Its Sanskrit, Sundanese, and Javanese names are certainly all very similar. In from the coast this savanna grades into a shrubby forest. The species here are typical of seasonally dry areas: most have thorns, furry or waxy leaves, and are deciduous, for example, *Capparis micrantha* (Capp.), *Glycosmis* sp. (Ruta.), *Strychnos* sp. (Loga.), *Uvaria* sp. (Anno.), *Albizia lebbekoides, Garuga floribunda, Pterocymbium javanicum,* and *Fagara rhetsa* (Soepadmo 1961; Meijer 1976).

The most unusual forest in Bali is the single-species stand of *Manilkara kauki* (Sapo.) (fig. 10.15) forest on the Prapat Agung peninsula of Bali Barat National Park, although some *Doryxylon spinosum* (Euph.), *Schoutenia ovata* (Tili.), and *Polyaulax cylindrocarpa* (Anno.) also occur (Daryono 1985b). This forest can be seen from the Java-Bali ferry as rounded, grey crowns. Mature tropical forests dominated by a single species of tree in the main canopy are not unusual, although the proportion of the total forest area that they cover is insignificant. Other examples in Indonesia are iron-wood *Eusideroxylon zwageri* (Laur.) in Borneo and Sumatra (Anwar et al. 1984; Whitten et al. 1987c; Peluso 1992b; MacKinnon et al. 1994a, b; Whitten in prep.), *Dryobalanops aromatica* (Dipt.) in Malaya and Sumatra, and certain mangrove species and *Casuarina* throughout the region. Where the forest is self-sustaining with young trees regenerating successfully below the canopy formed by their parents, the single-species dominance is a result of very harsh soils or otherwise extreme environments excluding potential competitors. In benign areas, however, the dominant species is generally a superior competitor and/or is particularly tolerant to stresses such as shade, and is most likely to be found where the pool of available species contains few late-succession species with similar life-history traits (Hart 1990).

This is not the case for the *M. kauki* forest. It is not a regional forest type, and it is interesting that on the next hill to the southeast, over the main road, there is no *M. kauki* to be found. Conversely, many of the species found on there are not found on Prapat Agung, such as lancewood *Homalium tomentosum* (Flac.), *Berrya cordifolia* (Tili.), *Diospyros buxifolia* (Eben.), and *Cratoxylon formosum* (Gutt.) (Soepadmo 1961). This forest seems to fall into the category of single species, reflecting dominance

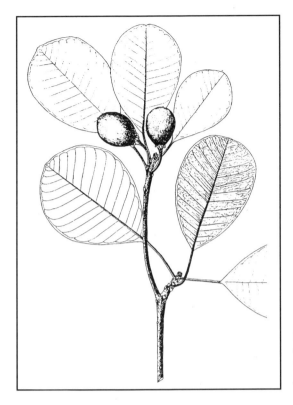

Figure 10.15. *Manilkara kauki* occurs in a pure stand in Bali Barat National Park..
After Prawira et al. 1972

occurring years ago during early forest succession. This dominance is likely to occur when there is the massive establishment of a single, efficiently dispersing, and rapidly growing species that was fortunate in being the first species to become established. In these situations, however, the dominance lasts only a single generation (Hart 1990). The successional nature of this forest is further illustrated by the fact that in the mid-1930s the appearance of the *M. kauki* forest was most unlike that of today, resembling more a temperate woodland (de Voogd 1937c), and it used to be more common than now, dominating the vegetation.

Thus, the *M. kauki* trees on the Prapat Agung peninsula probably represent mature members of an early successional stage. The succession would have begun after some major disruption to the area, most probably fire during a very dry period (aggravated by being on limestone), or strong winds. The hypothesis is that the fire left some *M. kauki* trees alive, the fruit from which was able to be dispersed into the forest areas. The fruit, tasting not unlike its domesticated cousin from South America *Manilkara zapota* (whose sticky sap was used as chewing gum by the Aztecs), is eaten and dispersed by monkeys, flying foxes, and pigs. Interestingly, enumeration of the

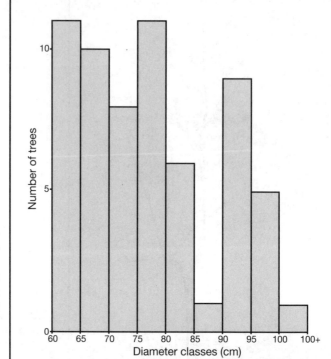

Figure 10.16. Histogram of trunk diameter classes of *M. kauki* trees in a 5.8 ha plot on the Prapat Agung peninsula, Bali Barat National Park.

From data in Taman Nasional Bali Barat 1989

Figure 10.17. Large *Manilkara kauki* tree on the Prapat Agung peninsula, Bali.

By A.J. Whitten

healthy trees with trunks over 60 cm diameter indicated that most trees with trunks 75-79 cm, very few with trunks 80-90 cm wide, and another age class with trunks 90-94 cm wide (fig. 10.16). This may indicate that the largest trees are the older generation of seed trees from which most of the present generation is derived (fig. 10.17). Certainly when one is within the forest it is striking that almost all the *M. kauki* trees are about the same size and the largest trees do not form part of the continuum. Recognizing that the existence of the pure stand of *M. kauki* is an interesting but temporary natural phenomenon should influence its management. *M. kauki* is a valuable tree because its wood is very highly valued for carvings. Early regulations covering the cutting of the trees on the Prapat Agung peninsula only allowed for those of 30 cm trunk diameter and over to be cut. However, since almost all the *M. kauki* fell into this category, their exploitation radically changed the forest on the southern slopes (Soepadmo 1961). It seems now, however, that there is a desire to intervene and maintain the single-species dominance artificially by opening the canopy, reducing the density of potential competing species, and protecting the existing adult and young trees from human and animal disturbance (Taman Nasional Bali Barat 1989).

Fauna

Nothing has been published specifically on the fauna of deciduous forests, but the species would clearly have to be able to withstand the long dry season, either by some behavioural or physiological adaptation or by moving away from the area. An indication of what the mammal ecology of Baluran might once have been like can in found in papers on deciduous forest in Thailand (Rabinowitz and Walker 1991; Walker and Rabinowitz 1992; Srikosamatara 1993). Given the harshness of the climate, some quite surprising animals can be found in Baluran National Park, such as snails which withdraw inside their shells and hide in cracks out of the sun. During the peak of the rainy season, the chorus of frogs and toads around pools of water can be deafening, although it is not known how they survive the long dry season. Although generally dry and harsh, the deciduous forest takes on quite a different face at the beginning of the wet season when young leaves and flowers appear–not unlike the change from winter to spring in temperate regions (de Voogd 1927). During the wet season the dormant animals become active again. For example, the dung beetles are scarcely seen at the peak of the drought, with the result that the dung of the deer, buffalo, and banteng is not recycled. When the rains come, however, they become very active, and Java deer faeces can be removed within 24 hours and those of barking deer in only eight (S. Hedges and M. Tyson pers. comm.).

Effects of Disturbance

No primary lowland deciduous forest still exists in Java and Bali, or indeed anywhere else in Indonesia, because it is so susceptible to fire during the dry season (van Steenis 1957; Whitmore 1984). Thus none of the grasslands or savannas are natural but are instead the result of a wide range of people's activities on deciduous forests. In fact, no vegetation type on Java and Bali has suffered more than deciduous forest (van Steenis and Schippers Lammerste 1965). Some of the influences are known to be ancient: for example, old irrigation canal systems and a pagoda have been found in parts of what is now the forested area of Baluran National Park (Junghuhn 1854), possibly dating back to when the Hindu majority in eastern Java were being conquered by the Islamic kingdoms from the west (p. 323). The area was later more or less depopulated and returned to nature because of disease, famine, and major volcanic activity from nearby Mt. Ijen (Appelman 1937).

FOREST AND OTHER VEGETATION ON LIMESTONE

The limestone areas of Java and Bali (fig. 3.5) comprise rocks of various origins, but they share many of the same physical characteristics. The limestone erodes into characteristic topographies, and both major forms of karst landscape typical of the humid tropics are found in Java and Bali: conical hill karst, such as found in the Mt. Sewu area and on Nusa Penida southeast of Bali (fig. 10.18), and the tower karst of Padalarang near Bandung, although this is hardly recognizable after the intense limestone mining in the area. These hills were formed probably by streams flowing in cracks between blocks of uplifted limestone. The soils are, not surprisingly, often richer in bases, particularly calcium and magnesium, with a higher cation exchange capacity than soils in similar situations on different parent materials. On moderate slopes and hollows, clay-rich, leached, brownish red latosols are formed (Burnham 1984).

Vegetation

Virtually all forest on limestone has been lost, and that which remains is mostly in areas bearing moist deciduous forest. It is therefore difficult to give an account of the range of vegetation occurring in different areas. Information from elsewhere in the region can be used, however, to piece together the likely original conditions.

Forests on limestone generally have a similar total basal area of trees (sum of cross-sectional area of trunks measured at breast height) to other types of lowland forests, but when they occur on steeper slopes and rocky

Figure 10.18. Cone karst landscape on Nusa Penida.
By A.J. Whitten

hilltops, the shallow soil conditions can seriously affect tree growth. Compared with forests on deeper soils, forests on limestone generally have few tree species (Crowther 1982; Proctor et al. 1983a, b), although the total number of plant species is probably not dissimilar to other forest types. The relative paucity of tree species probably arises because some trees of lowland forest cannot tolerate the high calcium levels in the soil. The different tolerance of trees to calcium and the unique physical habitat of limestone makes the composition of these forests rather different from other lowland forest, giving rise to a specific community of trees. On steep limestone cliffs with bare rock faces, clefts and shelves, a distinctive herbaceous flora occurs. When a severe dry season occurs the tolerance of these plants may be exceeded, and they may be forced to survive as annual plants; that is, they must complete their life cycle during the wet season so that their seeds can germinate when the following drought ends.

Despite being fairly well studied, there are apparently no plants on Java that are specific to limestone (van Steenis 1931; van der Pijl 1933), a striking contrast to the situation in Peninsular Malaysia where 21% of the limestone flora is restricted to that habitat, generally on isolated tower karst hills, and half of those are endemic to the peninsula (e.g., Chin 1977, 1983).

Mt. Cibodas. The best-studied limestone site from the ecological point of view is Mt. Cibodas near Ciampea west of Bogor, a long coral reef raised to above 350 m above sea level. The hill is the only known site for a species of snail (p. 294), although it is possible this persists elsewhere. Sixty years ago the forest was untouched (van Steenis 1931); the hill had six Hindu statues at the top, and there were probably proscriptions against disturbing the surrounding slopes. The beliefs giving protection have eroded with time and Mt. Cibodas is now virtually treeless and the rock is being used for cement, white wash, and road fill. The forest used to have numerous *Dipterocarpus hasseltii* (Dipt.), *Stelechocarpus burahol* (Anno.), and a number of *Diospyros* (Eben.) species, some of which provide ebony wood for carving, but no single species was dominant (van Steenis 1931). The fruit of *S. burahol* in Central Java was so sought after that only the aristocracy were permitted to eat it, and the tree was to be grown only in the royal palaces. It was chosen as the official plant of the Special Area of Yogyakarta in 1993. In West Java the Sundanese did not even bother to cultivate it in their villages (Heyne 1917, 1987).

Nusa Barung. The limestone island of Nusa Barung off the south coast of East Java is likewise not dominated by any particular genera or species although locally *Diospyros maritima* is common on the flat lowlands and *Croton tiglium* (Euph.) on hills, although *Pisonia grandis* (Nyct.) (p. 175) and *Kleinhovia hospita* (Ster.) are locally common (Jacobs 1958).

Padalarang. The vegetation of Karang Panganten, one of the Padalarang hills, was studied over 60 years ago, but even by this time the forest had already been lost. The herbaceous plants and scrubs were still quite varied, and could be classified into five ecological groups:

- shade-loving herbs with small root systems growing in pockets of moist humus on small shelves in cracks and crevices. During the dry season they shrivel and die. Notable examples are violet-flowered *Rhynchoglossum obliquum* (Gesn.), the blue-flowered *Cyanotis cristata* (Comm.), and the brown, fleshy-leaved *Peperomia laevifolia* (Pipe.);
- epiphytes attached to bare rock such as the ferns *Drynaria quercifolia* (Polyp.) and the smaller *D. rigidula* (fig. 10.19). Both form 'nests' from the long leaves produced, and various organic and mineral matter fall into these. 'Roots' grow into this nutritious, decomposing mass, allowing these plants to live in seemingly very inhospitable habitats. Another fern found in this category *Antrophyum reticulatum* (Adia.) has quite fleshy leaves, and it was found to lose 63% of its weight in water and yet survive. Conversely an *Elatostema* (Urti.) from the first group lost only 20% of its weight before dying;
- succulents are not part of the native flora, but various euphorbs,

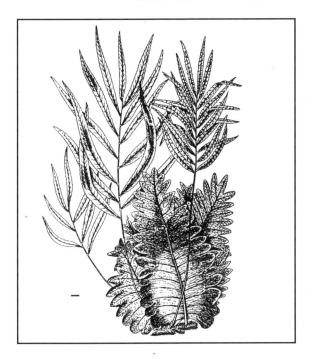

Figure 10.19. *Drynaria rigidula*. Scale bar indicates 1 cm.
After Backer and Posthumus 1939

Agave (Agav.), and cactus such as the prickly pear *Opuntia nigrans* (Cact.) may be found;

- other herbs become dormant during the dry season, such as voodoo lily *Amorphophallus* (Arac.), which has large, deep tubers. Other plants adopt alternative strategies. The fleshy aroid *Remusatia vivipara* (Arac.) produces, as its name suggests, tiny bulbils or miniature plants on what seem at first glance to be erect flowering stems. These drop off and await the rains before growing. Meanwhile the herb *Pellionia* (Urti.) forms swellings in the leaf axils. As the weather becomes drier, so the plant breaks up and the swellings lie waiting for the rains before growing as a new plant; and

- drought-adapted trees and shrubs of various sorts, typically with small leaves, often deciduous (particularly the larger trees), and armed with thorns, such as *Orophea hexandra* (Anno.). The largest tree is *Firmiana malayana* (Ster.) which, while still leafless, produces masses of orange flowers. Also found growing here was *Pisonia grandis,* normally found on remote islands (p. 175). It produces sticky fruits which are dispersed by birds, but it has been known for birds to be so covered with the fruits that they cannot fly and, as a result, die (van der Pijl 1933).

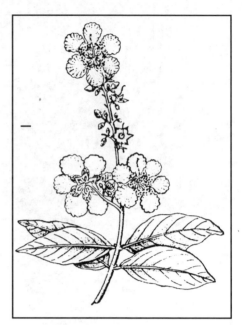

Figure 10.20. *Lagerstroemia speciosa.*
Scale bar indicates 1 cm.
After Koorders and Valeton 1913

Nusa Penida. The island of Nusa Penida has virtually no vestige of original forest left, most of it being dominated by cultivated land and scrub comprising various legumes, typical of open, strongly seasonal areas (de Voogd 1937b). There are, however, a few small, sacred patches of largely secondary forest around some of the temples, and a 10 ha remnant of primary forest is found just west of the highest peak, Bukit Mundi, where, again, it surrounds a temple. This was investigated for this book in early 1994, and it was found to be dominated by *Buchannia arborescens* (Anac.), and to be home to probably all the island's endemic snails, some of which can be found in great abundance.

Fauna

No vertebrates are restricted in their distribution to the surface of limestone hills, but some species of snails are because they need calcium to form their shells (p. 294). Many animals are more or less restricted to the caves within them (chap. 12).

Effects of Disturbance

It is patently obvious from the situation on Java and Bali that forest and other vegetation on limestone is easily destroyed in the process of limestone quarrying. The secondary vegetation in these limestone areas is dominated by the introduced *Chromolaena* (Comp.) and *Lantana* (Verb.), the dense thickets of which probably hold back the succession process by some years. Trees found in the succession include *Homalanthus, Lagerstroemia* (Lyth.) (fig. 10.20), *Pterospermum, Kleinhovia,* and *Villebrunea.* Given their 'aggressiveness' in competing with *Chromolaena*, some of these trees could reasonably be used in programmes for the reclamation of weed-dominated lands on limestone soils.

SWAMP FORESTS

Freshwater swamp forest grows where there is occasional inundation of mineral-rich freshwater of pH 6 and above, and where the water level fluctuates such that drying of the soil surface occurs periodically. Freshwater swamps are normally found on riverine alluvium but also occur on alluvium deposited in lakes such as was once the case in Rawa Pening. The alluvial soils are more fertile than those on adjacent slopes and have great agricultural potential when drained. The soils are very variable but are generally young and therefore do not have clearly differentiated horizons. Soil animals such as termites and earthworms, which normally take organic matter into the soil, are not tolerant of the periodic waterlogged, anaerobic conditions and so there is little mixing of soils. The agricultural potential of these swamp soils has meant that freshwater swamp forests have been greatly reduced by human activities.

Swamp forest also grows in periodically inundated conditions where there is peat (soil with more than 65% organic matter) at least 50 cm deep. The extensive peat lands of Sumatra and Borneo are mainly of a deep, raised, rain-fed (ombrogenous) type which has very acid (pH 3) drainage water, but in Java the only type recorded occurs in shallow layers in depressions (topogenous). Plants growing on this can extract their nutrients from the peat itself, from mineral soil beneath the peat, from river water, plant remains, and rain. Peat accumulation in topogenous swamps is relatively slow, and the peat and drainage water are only slightly acid (pH 5) and nutrients are relatively abundant.

Swamp forests have all but disappeared from Java and were probably never present on Bali (fig. 10.21). It has been estimated that they once covered 72,000 ha of Java, and that 7,700 ha remain, of which only 2,600 ha are in conservation areas (Scott 1989). They were generally formed in depressions, such as the craters of ancient volcanoes and behind river-

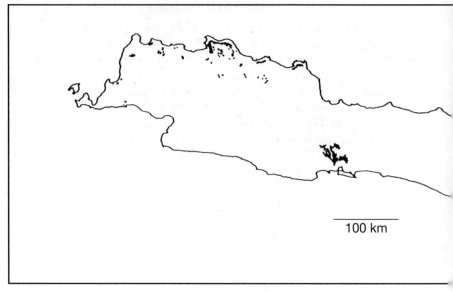

100 km

Figure 10.21. Likely natural distribution of swamp forest on Java.

Drawn from inland swampy areas shown in KNAG (1938)

banks, but most of these have long since been converted to rice fields. Examples are:

- Java's only peatswamp of Lakbok near the southern boundary between West and Central Java;
- Rawa Danau Reserve between Labuan and Serang;
- Rawa Bojong near Pameungpeuk;
- Rawa Bening near Tulungagung (IPB 1986b);
- the Siradayah swamp forest near Cicadas between Bogor and Jakarta as described by van Steenis (1934c);
- behind the coastal forests in Cikepuh Nature Reserve;
- around Rawa Pening near Ambarawa in Central Java (Polak 1951); and
- near Jatiroto west of Jember (fig. 10.22).

Of these, Rawa Danau is the only significant area of swamp forest remaining, the others having been drained for the cultivation of irrigated rice.

Vegetation

Swamp vegetation varies according to the wide variation in its soils and the availability of free water. The forests are relatively poor in species and this is a reflection of the inability of most dryland forest plants to tolerate

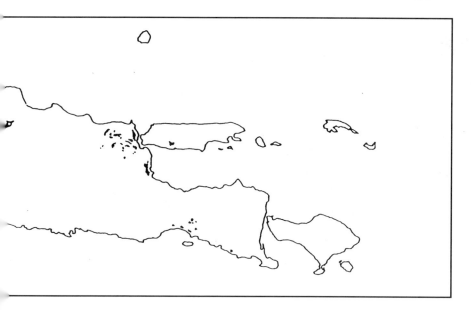

inundated conditions. The few species that appear to be restricted to swamps are generally able to live in drier conditions, but are unable to compete successfully. The fringes of the swamp near open water are grassy but, where the ground is firmer, palms and pandans can be common. As the soils become drier, a forest very similar to normal dryland lowland forest is generally found. The relative instability of the soil, by virtue of its periodic inundation with water, is probably one reason that supportive structures associated with trunks, such as long, winding buttresses and stilt roots, are common, and certain species have accessory breathing structures such as pneumatophores (Corner 1978).

Rawa Danau. The largest area of freshwater swamp remaining in Java and Bali is Rawa Danau, Banten, in West Java, which exists (even in its grossly disturbed state) only because it was made a nature reserve in 1921. The original forest was described by Endert (1932) and was dominated by large trees of *Elaeocarpus macrocerus* (Elae.), *Alstonia spathulata* (Apoc.) (fig. 10.23), *Ficus retusa* (fig. 10.24), *Mangifera gedebe* (Anac.), *Lagerstroemia*, *Horsfieldia irya* (Myri.), *Barringtonia racemosa* (Lecy.), and *Gluta renghas* (Anac.). This last mentioned tree can usually be recognized by the black stains of dried sap on the bark, dippled, scaly bark, and spirally arranged leaves. It has a pinkish brown, warty, puckered fruit, the seed in which can apparently be eaten only after roasting (*Gluta* is in the same family as the cashew

Figure 10.22. Sites of four Javan swamp forests: a - Rawa Bojong, photographed in 1941, showing large *Nauclea coadunata* trees, b - Rawa Bening in 1927 to show that even at this time no swamp forest remained, c - tiny Siradayah swamp forest persists not far from the Jagorawi highway, d - Jatiroto swamp forest, near Jember, as it appeared in 1912.

a-photographer unknown, with permission of the Rijksherbarium/Hortus Botanicus, Leiden University; b-photographer unknown, with permission of the Rijksherbarium/Hortus Botanicus, Leiden University; c-by A.J. Whitten; d-by F. Jeswiet, with permission of the Rijksherbarium/Hortus Botanicus, Leiden University

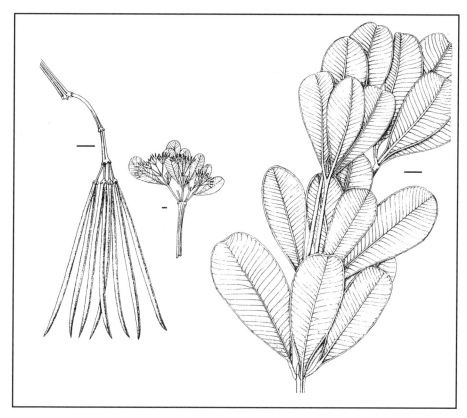

Figure 10.23. *Alstonia spathulata*. Scale bars indicate 1 cm.

After Koorders and Valeton 1918

Figure 10.24. *Ficus retusa*. Scale bar indicates 1 cm.

After Koorders and Valeton 1918

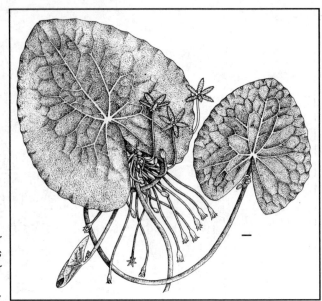

Figure 10.25. Water gentian *Nymphoides indica*. Scale bar indicates 1 cm.

After Kostermans et al. 1987

Anacardium occidentale). Otherwise its parts should be regarded as poisonous and it is unwise even to shelter under these trees during rain, since the resin washed from the leaves and branches can cause severe skin irritation (Burkill 1966, Ding Hou 1978). This species used to grow in Rawa Bojong, west of Pameungpeuk and at Subah on the north coast of Central Java (see photos in Ding Hou (1978)).

Lakbok. The largest areas of peatswamp forest in Java and possibly the largest group of topogenous peatswamps in Indonesia (Silvius et al. 1987) were to the west of the lower reaches of the Citanduy River, south of Lakbok, Ciamis. In the middle of last century Junghuhn remarked on the striking contrasts in the wildlife and vegetation of Rawa Lakbok.

> In one part of the year the waters are alive with swarms of many large and tasty fishes, and the surface is covered with countless numbers of tree ducks *Dendrocygna javanica* and other water birds. Crocodiles lurk on the banks, and pelicans and many types of herons wade about in the mud. Some months later, in April or May, when the water level begins to fall, thousands of fishes are stranded along the shore. Tigers and wild dogs arrive and compete for these with birds of prey. Some months later, in the middle of the dry season, the bottom of the lake is transformed into a meadow covered with dense and tall grass among which deer and wild

boar gambol, even when stalked by tigers and leopards (Junghuhn 1854).

Drainage of these swamps for rice fields began in 1924, but in the late 1940s there was still some 3,000 ha of swamp remaining (Polak 1949). A general survey conducted in 1990 revealed that major changes had occurred in the area (Indrawan 1990), and a subsequent survey conducted for this book in 1992 found that remnant *Nauclea* (Rubi.) and *Neonauclea* (Rubi.) trees were all that remained, and that only Rawa Jangraga (Kec. Padaherang), measuring just 500 x 600 m and with peat up to 6 m deep, had not yet been used for rice culture. It has been recommended as a nature reserve (Budi-Asmoro 1992). In the other remaining swampy areas there is a mixture of rice culture with swamp grasses, lotus, rushes, water hyacinth, and water gentian (fig. 10.25). Local people hunt birds, fish, and gather medicinal plants, and peat for burning. There are still fishes in the swamp waters, but these are of common and/or introduced species, and in at least one area 'fishermen' use potassium cyanide. Good numbers of waterbirds of both migrant and resident species can be seen on the open waters, probably because similar habitats are now very rare. Among the birds is the pheasant-tailed jacana *Hydrophasianus chirurgus*, a migratory wader which is rare in Java (Indrawan 1990, 1991).

Siradayah. A very small area (<1 ha) of holy, freshwater swamp forest not far from the Jagorawi highway between Bekasi and Bogor was described sixty years ago as a 'botanical jewel' (van Steenis 1934c). Rather remarkably it is still standing, largely because it is regarded as sacred, but the sawah around it is being replaced by a housing estate. The local proscriptions on entering the forest still seem to be strong. The main tree is *Elaeocarpus littoralis,* which grows to 30 m tall, and its bent, knee roots cover the forest floor. Other abundant trees are *Eugenia operculata* (Myrt.), *Horsfieldia glabra,* and *Glochidion glomerulatum* (Euph.) with red-and-white pencil roots. The undergrowth is dominated by the spiny-stemmed aroid *Cyrtosperma merkusii* (Arac.).

Other. A small inland swamp forest known as Rawa Tembaga is said to persist just north of the toll road between Karawang and Cikampek. Swamp forests behind mangrove forests have virtually all been converted to agricultural or fisheries uses. In places, such as the estuary of the Citanduy, the oil-giving aromatic *Melaleuca cajuputi* (Myrt.) is reported in similar situations to those where it is found in southeast Sumatra and the southern coast of Kalimantan. It has been lost from the degraded and much-reduced Muara Angke Reserve (Indrawan and Sunarto 1992; Indrawan et al. 1993) (p. 411, figs. 10.26-27). Under natural conditions such forests tend to be dominated by *Ficus retusa* and *Nauclea coadunata,* with abundant rattan and the wet-loving climbing fern *Stenochlaena palustris* (Blec.) with its edible,

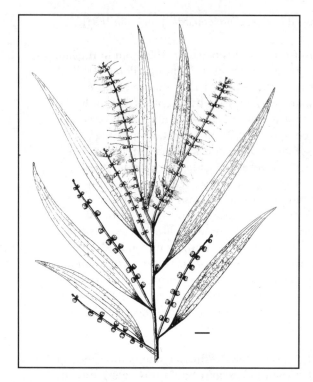

Figure 10.26. *Melaleuca cajuputi.* Scale bar indicates 1 cm.

After Whitmore et al. 1989

Figure 10.27. *Melaleuca* forest at Muara Angke in the early part of this century. It has now been lost.

By C.N.A. de Voogd, with permission of the Rijksherbarium/- Hortus Botanicus, Leiden University

red, young leaves (said to be especially beneficial if eaten for breakfast during the fasting month) (van Steenis and Schippers-Lammerste 1965).

Fauna

The fauna of swamp forests has elements from dryland ecosystems and from aquatic ecosystems and, as the quotation from Junghuhn above shows, the intensity of use by different species will vary according to the time of year and the quantity of water in the swamp. Little seems to have been written on the fauna of swamp forests except for Rawa Danau, and this is described elsewhere (p. 773).

Chapter Eleven

Mountains

When one considers how little lowland forest is left, and how even its remaining fragments are becoming impoverished and reduced in area, it is clear that montane forests are very significant as sites for the conservation of much (if not all) of Java and Bali's biological diversity. This is because many species are confined to montane areas (Chap. 5), and some lowland species are able to live in the lower montane zones. Mountains are also culturally significant (p. 672). Before Hinduism arrived in Java, the forested mountaintops were believed to be the abode of the gods and of the spirits of the dead, and for many people the remote, windswept summits have not lost their significance. It is interesting that Mt. Mahameru, the centre of the Hindu cosmos, was believed to be a high mountain surrounded by four or eight lower peaks. The shape of Mt. Penanggungan, south of Surabaya, was long ago realized to conform to this description, and dozens of temples, shrines, and sanctuaries were built there, 81 of which can still be identified (Edwin 1988e) (fig. 11.1).

THE MOUNTAINS OF JAVA AND BALI

There are few parts of Java and Bali from which, on a relatively clear day, at least one mountain cannot be seen (figs. 11.2-3). Early navigators found Java by its mountains, and Sir Francis Drake was drawn by the sight of Mt. Slamet to anchor at Cilacap in 1580 during his circumnavigation of the globe (Turner 1988). All the mountains are volcanoes, although some are older and more eroded than others. Large areas of montane forest have been lost and much is degraded, but the natural mountain ecosystems are relatively intact and offer some of the best prospects of conserving a sample of the biological diversity of Java and Bali. The mountains have been the focus of a good deal of research, early examples of great depth being Docters van Leeuwen (1933) and van Steenis (1972), and enough is probably known to execute ecologically appropriate management.

Figure 11.1. Mt. Penanggungan, south of Surabaya, conforms to the description of the holy Mt. Mahameru.

By Luchtvaart Bandoeng, with permission of Tropen Instituut, Amsterdam

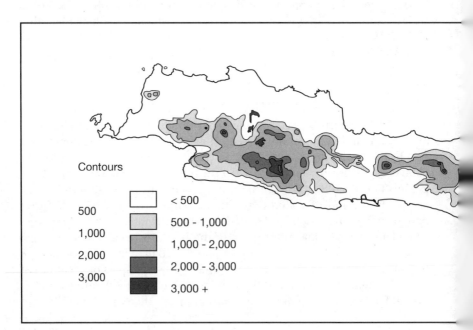

Figure 11.2. Contour map of Java and Bali.

PHYSICAL CONDITIONS

Temperature

Heat is lost quickly from high altitudes at night, causing the daily temperature range there to be as much as 15° C to 20° C. As the surfaces of
plants, soil, or rocks cool down at night, so the air around them also cools.
This, being heavier than the warmer air, will become progressively colder
if there is no slope down which it can flow, for example in hollows and valleys. Frosts are most likely to occur in flat-bottomed hollows called frost
pockets on clear, calm nights when long-wave radiation is lost to the skies.
The fastest rate of cooling occurs on nonconductive surfaces such as dead
plant material and dry soil. The gradual decrease in temperature with
increasing altitude, known as the lapse rate, is generally accepted to be
about 0.6° C/100 m, but this depends on factors such as cloud cover, time
of day, and amount of water vapour in the air.

Low temperatures are the main reason the tops of mountains have not
been cleared for agriculture. Above 2,000 m it is regularly less than 10° C at
night, and occasional local frosts can interfere with crops. A temperature of

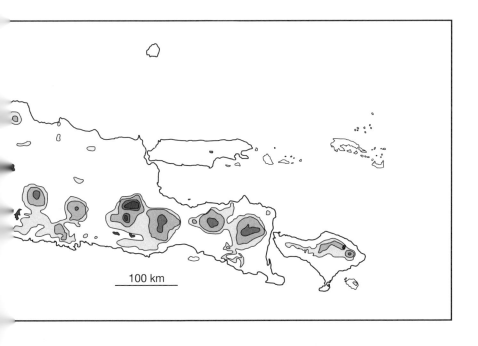

100 km

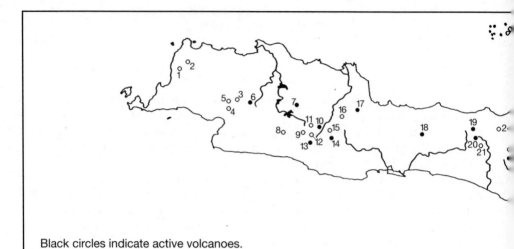

Black circles indicate active volcanoes.

Figure 11.3. Major mountains of Java and Bali and their heights. 1 - Pulasari 1,346 m, 2 - Karang 1,778 m, 3 - Salak 2,211 m, 4 - Halimun 1,744 m, 5 - Ciparabakti 1,525 m, 6 - Gede-Pangrango 3,019 m, 7 - Tangkubanprahu 2,076 m, 8 - Patuha 2,434 m, 9 - Wayang-Windu 2,182 m, 10 - Guntur 2,249 m, 11 - Malabar 2,350 m, 12 - Kenang 2,608 m, 13 - Papandayan 2,622 m, 14 - Galunggung 2,168 m, 15 - Telaga Bodas 2,241 m, 16 - Cakrabuana 1,721 m, 17 - Ciremai 3,078 m, 18 - Slamet 3,432 m, 19 - Perahu 2,565 m, 20 - Sindoro 3,136 m, 21 - Sumbing 3,371 m, 22 - Merbabu 3,142 m,

-10° C has even been recorded at 2,000 m a few centimetres above the soil in a hollow during the dry season. This frost is a potent ecological factor in high plateaux which have been cleared or repeatedly burned (for hunting) (p. 536) because it will hinder the growth of young tree seedlings. This, together with grazing by deer attracted by the grass, such as on the Yang Plateau (p. 807), will keep the turf low, which again favours the formation of frosts and the gradual eradication of tree seedlings. Thus, once deforested, such plateaux remain dominated by grasses (van Steenis 1972).

Relative Humidity

The percentage saturation of a mass of air increases as its temperature falls. The dew point (the temperature at which condensation occurs and clouds or drops of dew form) at different altitudes depends therefore on the temperature and the initial moisture content of the air. The forests of higher altitudes experience a very high relative humidity, particularly at

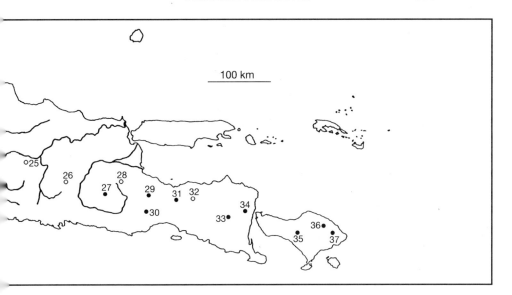

23 - Merapi 2,911 m, 24 - Ungaran 2,050 m, 25 - Lawu 3,265 m, 26 - Liman-Wilis 2,563 m, 27 - Kelud 1,731 m, 28 - Arjuna-Welirang 3,339 m, 29 - Tengger 2,775 m, 30 - Semeru 3,676 m, 31 - Lamongan 1,671 m, 32 - Argapura 3,088 m, 33 - Raung 3,332 m, 34 - Merapi-Ijen 2,800 m, 35 - Batukau 2,276 m, 36 - Batur 1,717 m, 37 - Agung 3,014 m.

After KNAG 1938

night when the temperature falls, and the dew point is frequently passed so that water condenses on the leaves. During dry spells at higher altitudes above the main cloud layer (above about 2,500 m), however, the relative humidity may be only 20% during the day, leading to wide daily extremes. Further detail can be found in van Steenis (1972).

Clouds

Clouds originate where ascending air reaches its dew point and where there are the necessary dust or other particles for the water vapour to condense upon. Once water droplets have been formed, they tend to act as 'seeds' and the droplets grow as they bounce around in the clouds. During the wettest months, the slopes and peaks of high mountains can be enveloped in clouds for days on end. During the drier months, however, when the air is not saturated with water vapour, it is common for a belt of clouds to form around a mountain, often at about 2,000 m, although a high peak may be seen above this.

Rainfall

Rainfall is generally higher on the side of a mountain facing the prevailing wind, but there appear to be no guiding principles relating altitude to rainfall, except that rainfall on mountain slopes up to 1,500 m will generally be greater than on the surrounding lowlands. At the cloud level, however, rainfall measurements are not especially useful for ecological studies because much of the water used by plants is in the form of water droplets in the clouds which adhere to the surfaces they touch. Studies on Mt. Pangrango in West Java, showed that the summit received more frequent rain than the slopes, but the total was less. There is, however, great variation between years: the summit of Mt. Pangrango has in one year received only 13 mm of rain between July and September but 460 mm during the same period in another year (van Steenis 1972).

Soils

The dominant soil type on Java and Bali's mountains are dystrandepts, which are slightly weathered volcanic ash soils with a low base saturation and a thick, black topsoil. The nature of mountain soils changes with increasing altitude, becoming more acid and poorer in mineral nutrients, and this is particularly the case where acid peat is present. Peat forms in wetter places, in the cloud belt or upper montane zone, for example, where decomposition processes are generally slower. Ridges, knolls, and summits only receive water from the atmosphere, so their soils are continually leached because soil and water cannot trickle down to them. The soil in these places is therefore drier and more nutrient-poor than soil in valleys or the lower slopes (Burnham 1984). Differences in climate are the major factors influencing soil formation up a mountain, although steepness of slope and openness of the vegetation cover are also important. Low temperatures slow down the processes of soil formation, reduced evapotranspiration causes less movement of water through the soil, chemical reactions occur 2-3 times slower with every 10° C drop in temperature, and the reduced abundance of soil organisms means that biological processes affecting soil formation are also slower. With increasing altitude the soil is less well formed, and the roots of the plants growing in it are shallower. The differences in the soil obviously have their effects on the vegetation, but whether the soil is more important than direct climatic effects and the changes in communities of disperser animals is not known.

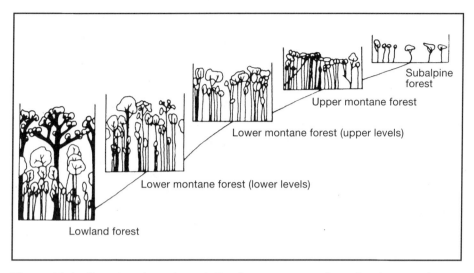

Figure 11.4. Cross sections through five forest types to show the decrease in tree height and simplification in structure with an increase in altitude.

Adapted from Robbins and Wyatt-Smith 1964

VEGETATION ZONATION

The forests on mountain slopes below about 1,200 m are very similar to other lowland forests. With increasing altitude, however, the trees become rather shorter and less massive, and epiphytes such as orchids become more common (fig. 11.4). This is lower montane forest, and plots above Cibodas were found to have the equivalent of 280-586 trees with trunks of 10 cm diameter or more in a hectare (Meijer 1959c; Rollet and Sukardjo 1976; Yamada 1976a). Further up, there is abundant moss, the tree canopy gradually becomes more uniform, the trees become shorter, eventually looking squat and gnarled, and leaves are small and relatively thick. This is upper montane forest, and in another plot above Cibodas, over 1,500 rather small trees/ha were found (Yamada 1976a). The thicker leaves characteristic of montane habitats (Grubb 1977; Tanner and Kapos 1982) may be an adaptation to resist undesirable penetration by fungi, bacteria, moss, and lichens, some of which grow on leaves. The high humidity provides ideal conditions for the growth of these plants, and they probably represent a greater threat to leaf life in the frequently cloud-swathed upper montane and subalpine forests than does insect damage, the major factor in the lowlands (Grubb 1977).

Beyond the upper montane forest is the subalpine forest of yet smaller trees (up to nearly 4,000/ha) with smaller leaves, and branches covered

Figure 11.5. *Vaccinium varingaefolium* blown into shape on Mt. Papandayan. Beneath the bush is the fern *Histiopteris incisa* (Denn.).

By C.G.G.J. van Steenis, with permission of the Rijksherbarium/ Hortus Botanicus, Leiden University

with epiphytic lichens but virtually no orchids (table 11.1). Valleys and other depressions in the subalpine zone are generally devoid of trees, and it appears that no trees have adapted to both high-altitude and water-logged conditions, particularly where frosts sometimes occur. Lack of trees on mountain slopes, however, is the result of hunters having set fires in order to facilitate the hunting of deer and other quarry. Here and else-where in this zone, there is instead a covering of shrubs, colourful herbs, and tough grasses. Some subalpine plants have woolly hairs on their leaves and stems, and these have been variously attributed to the ability to protect the plants against high temperatures (Lee and Lowry 1980), intense ultra-violet radiation (Mani 1980), and frost/freezing (Smith 1970). They may play a role in all three. Other high-altitude plants provide their growing buds and perennial parts with thermal insulation by retaining old leaves, and by having tufted branches and persistent scales. Such protection is nec-essary because it is more of a drain to a subalpine plant to replace a leaf than to plants growing at lower altitudes.

Some montane trees take on permanent windswept shapes caused by the strong winds experienced at higher altitudes (fig. 11.5). On Mt. Papan-dayan, for example, some slopes of the mountain experience the same wind direction irrespective of the monsoon (fig. 11.6) (van Steenis 1935b). In Java, gnarled 'elfin' forest can generally be found from about 2,000 m. It appears that the frequency of cloud or fog (or both) is the most important factor determining the stature of montane forest (Grubb and Whitmore

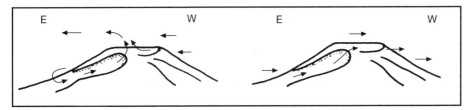

Figure 11.6. Wind directions over Mt. Papandayan during the west monsoon (left) and east monsoon (right) to show that in some parts of the mountain the wind direction stays the same.

After van Steenis 1935b

Table 11.1. Characteristics of four types of forest found on mountains. The most useful characters shown in bold type. *- the leaf-size classes refer to a classification of leaves devised by Raunkier (1934) and modified by Webb (1959). The definitions are: mesophyll: 4,500-18,225 mm^2, notophyll - 2,025-4,500 mm^2, microphyll - 225-2,025 mm^2, nanophyll - less than 225 mm^2. An approximate measure of leaf area is 2/3 (width x length). See figure 11.7 for leaf sizes.

	Lowland forest	Lower montane forest	Upper montane forest	Subalpine forest
Canopy height (m)	25-45	15-33	1.5-18	1.5-9
Height of emergents (m)	67	45	26	15
Leaf-size class*	**mesophyll**	**notophyll or mesophyll**	**microphyll**	**nanophyll**
Tree buttresses	**common and large**	**uncommon and small**	usually absent	absent
Trees with flowers on trunk or main branches	common	rare	absent	absent
Compound leaves	**abundant**	present	rare	absent
Leaf drip tips	**abundant**	present or common	rare or absent	absent
Large climbers	**abundant**	usually absent	absent	absent
Creepers	usually abundant	common or abundant	very rare	absent
Epiphytes (orchids and ferns)	common	abundant	common	**very rare**
Epiphytes (moss, lichen, liverworts)	present	present or common	**usually abundant**	abundant

Sources: Whitmore (1984); Grubb (1977)

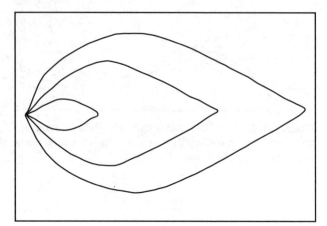

Figure 11.7. Hypothetical leaves of 4,500, 2,025, and 225 mm² to assist assessment of leaf class.

After Whitten et al. 1987b

1966), and many reasons have been suggested for this. The most recent hypothesis is that the stunting is caused by low rates of photosynthesis and transpiration, possibly related to a reduced supply of nitrogen in the soil at higher altitudes (Bruijnzeel et al. 1993).

Montane forests cause the same difficulties as lowland forests when one tries to classify and delimit the different types (Seifriz 1923). The altitude, volcanic activity, height of prevailing cloud cover, soil nutrient status, and particularly the degree of seasonality are major factors influencing the vegetation. Some of these are at work across Java and Bali, giving rise to the occurrence of, for example, *Altingia excelsa* (Hama.), *Podocarpus (Dacrycarpus) imbricatus, Podocarpus neriifolius* (both Podo.), and *Schima wallichii* (Thea.) as characteristic trees in the montane forests of West Java, but the absence of *A. excelsa* and *S. wallichii* on Mt. Slamet and mountains further to the east. Conversely, *Casuarina junghuhniana* (Casu.) (fig. 11.8) is not found west of Mt. Lawu but dominates the mountain forests to the east where long dry seasons and fires are common (fig. 11.9) (van Steenis 1972). Although different trees dominate in different areas, these do not really represent forest 'types' but rather biological responses to past events (such as landslips, volcanic ash falls, and fires). For example, *D. imbricatus* and *C. junghuhniana* are pioneer species and will not regenerate under a closed canopy (van Steenis 1972; Darmatin 1978). The dominance will change to a more mixed character in time, assuming further 'events' do not occur, although this may be measured in centuries rather than decades.

The physical conditions experienced at high altitudes vary from frost pockets (p. 500), to the edge of mountain streams or waterfalls, to bare soil. Clearly, the height above sea level alone will give little idea of the vegetation

Figure 11.8. *Casuarina junghuhniana.*
Scale bar indicates 1 cm.
After Koorders 1913

Figure 11.9. *Casuarina* forest on the
steep slopes of Mt. Agung, Bali.
By A.J. Whitten

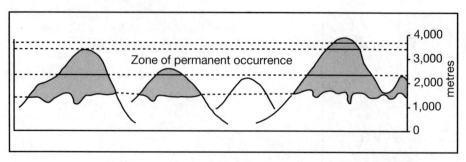

Figure 11.10. Zone of permanent occurrence with zones of temporary occurrence above and below it, showing the distribution of a hypothetical species (shaded areas).

After van Steenis 1972

to be expected. High-altitude plants found in relatively low locations may not flower or may have infertile flowers, and so to maintain a population, they are dependent on seeds or spores being dispersed from above. A similar but opposite mechanism occurs up the mountain, whereby sterile or nonflowering plants are found higher above the species' normal range, for example, around warm volcanic fumaroles (van Steenis 1936). Thus, for a given species there is a 'zone of permanent occurrence' which supplies seeds or spores to adjacent zones of temporary occurrence. For this reason, a relatively small mountain may lack a plant species which is found at the same altitude on a higher mountain, which also has a zone of permanent occurrence (fig. 11.10) (van Steenis 1972). Finally, the composition of plant communities can also differ significantly between mountains, or even neighbouring ridges, as a result of quite minor differences in aspect or age of soil. For example, *Paraserianthes lophantha* (Legu.), *Myrica javanica* (Myric.) (fig. 11.11), and *Gaultheria fragrantissima* (Eric.) are all on Mt. Gede, but absent from its sister Mt. Pangrango, with older soils, despite these peaks being separated by only a few kilometres and being joined by a saddle not far below (van Steenis 1972).

Given the above complications, it is not surprising that it is difficult to apply a single system to every mountain. Even so, the one below, based on Whitmore (1984) for southeast Asian forests, appears intuitively reasonable and is useful in the field in most situations:

lowland forests	0-1,200 m
lower montane forest	1,200-1,800 m
upper montane forest	1,800-3,000 m
subalpine forest	above 3,000 m

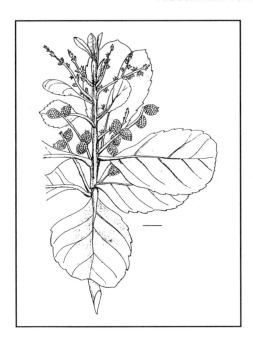

Figure 11.11. *Myrica javanica.* The male plants have long catkins. Scale bar indicates 1 cm.
After Koorders and Valeton 1914

The altitudes are by no means the same in every locality, because zone compression occurs such that dwarfed, moss forest is found at the top of even quite small coastal mountains exposed directly to moisture-laden sea breezes, but the apparent similarity of these forests to true upper montane forest is due to similar physical conditions (steep slopes, high winds, and a relatively low cloud level), rather than floristic similarities. Examples are Mt. Payung at the west of Ujung Kulon, and Mt. Tinggi on Bawean Island, where stunted 'montane' forest can be found at just 400 m (Hommel 1987; Bruijnzeel et al. 1993). These low-altitude species have grown to look like the trees one finds at high altitudes because of the relatively harsh conditions.

The demarcation between the lowland and the montane zones at 1,000 m is largely floristic, with major families more or less confined to the lowland zone (table 11.2). Conversely, many genera, if not families, are confined to the zones above 1,000 m, and many of these feature in the vegetation of northern and southern temperate regions such as *Anemone* (Ranu.), *Aster* (Comp.), *Berberis* (Berb.), *Galium* (Rubi.), *Gaultheria, Lonicera* (Capr.), *Primula* (Prim.) (fig. 11.12), *Ranunculus* (Ranu.), *Rhododendron* (Eric.) (fig. 11.13), *Veronica* (Scro.), and *Viola* (Viol.). The lower boundary of upper montane forest is generally quite abrupt elsewhere in southeast Asia, but on Java there is a gradual change, probably because the volcanic

Figure 11.12. *Primula pro-lifera* at 2,800 m on the Yang Plateau.

By J.G. Kooiman, with permission of the Tropen Instituut, Amsterdam

Figure 11.13. *Rhododendron javan-icum.* Scale bar indicates 1 cm.

After Koorders 1912

Table 11.2. Lowland plant families represented only by certain species or specially adapted genera between 1,000-2,000 m.

Annonaceae	Connaraceae	Menispermaceae	Sapindaceae
Apocynaceae	Cucurbitaceae	Myristicaceae	Thymelaceae
Araceae	Euphobiaceae	Palmae	Vitaceae
Asclepiadaceae	Leguminosae	Rhamnaceae	Zingiberaceae

Sources: Seifriz (1923); van Steenis (1972); Sukardjo (1982a)

soils are relatively nutrient-rich (Whitmore 1984). The demarcation between the montane and subalpine zones is partly floristic, but it is also clearly marked by physiognomic changes, and changes in the relative abundance of species (Yamada 1990); the forest below has a relatively high canopy with more than one layer; that above has a lower, denser, thinner-stemmed, lighter forest with only a single-layered canopy covering a ground layer of herbs and small shrubs (van Steenis 1972).

FLORA

Lower Montane Forest

On Java the lower montane forests are dominated in terms of abundance by the oaks *Lithocarpus* and *Quercus*, chestnuts *Castanopsis*, and numerous species of laurels (Fagaceae and Lauraceae, respectively), but the Magnoliaceae, Hamamelidaceae, and Podocarpaceae are also well represented (van Steenis 1972; Sukardjo 1978; Mukhtar and Pratiwi 1991). Although the oaks are thought of as montane trees, all of them can also be found in lowland forests, albeit in lower numbers (Soepadmo 1972). The composition of the lower montane forests of Bali is strikingly different from that in western Indonesia, because the oaks and chestnuts (Fagaceae) are missing. For comparison, five species of Fagaceae are known from East Java, 14 from Central, and 18 from West (Backer and Bakhuizen van den Brink 1963). The 18 species of oaks on Java can be distinguished by the size and shape of their acorns (fig. 11.14). These are relatively large and heavy fruit which are poorly adapted for long-distance dispersal (Soepadmo 1972). Pigs eat the seeds, and those which pass through the gut without being chewed or otherwise damaged may remain viable. Squirrels predate on the seeds, but they also take them out of the parent crown and store and hide some of them in the ground or in other 'safe' places (Becker et al. 1985). A proportion of these are not reclaimed and subsequently germinate.

Other trees of the main canopy of montane forest include:
- *Acer laurinum* (Acer.) (fig. 11.15), a deciduous tree which flowers and has its leaf flush as soon as the canopy is bare, and which has white lower-leaf surfaces;
- *Engelhardia spicata* (Jugl.) (fig. 11.16), garlands of three-lobed bracts attached to a small hairy nut;
- *Schima wallichii* (Thea.) (fig. 11.17) with five large, white petals, one of which is cup-shaped; and
- *Weinmannia blumei* (Cuno.) (fig. 11.18) with yellow flowers;

and at least 30 species of montane forest trees have commercial value (van

Figure 11.14. Acorns of Javan oaks *Lithocarpus* and *Quercus*. a - *L. conocarpus;* b - *L. crassinervis;* c - *L. daphnoideus;* d - *L. elegans;* e - *L. indutus;* f - *L. javensis;* g - *L. korthalsii;* h - *L. kostermansii;* i - *L. pallidus;* j - *L. platycarpus;* k - *L. pseudomoluccus;* l - *L. rotundatus;* m - *L. sundaicus;* n - *Q. gemelliflora;* o - *Q. lineata;* p - *Q. oidocarpa.* Scale bar indicates 1 cm.

After Koorders and Valeton 1913

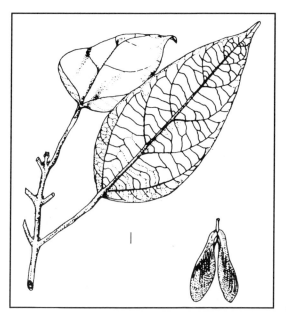

Figure 11.15. *Acer laurinum.*
Scale bar indicates 1 cm.
After Whitmore 1972b

Figure 11.16. *Engelhardia spicata.* The related *E. serrata* has similar bracted fruit but serrated leaves and is found only in western Java, Scale bars indicate 1 cm.

After Koorders and Valeton 1918

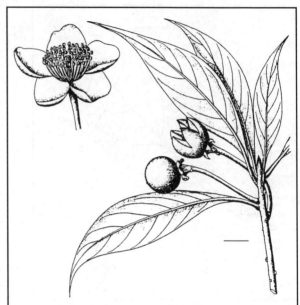

Figure 11.17.
Schima wallichii. Scale bar
indicates 1 cm.

After Koorders and Valeton 1915

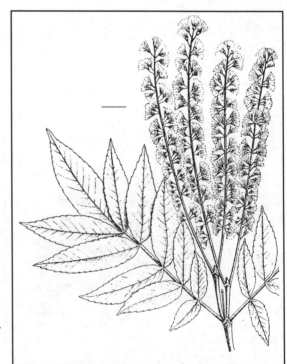

Figure 11.18.
Weinmannia blumei. Scale bar
indicates 1 cm.

After Koorders and Valeton 1913

Figure 11.19. *Altingia excelsa* showing a twig with a young fruit-head and a young male inflorescence. Scale bar indicates 1 cm.
After Vink 1957

Steenis 1972; Sukardjo 1982a).

The few emergents that are found tend to be *Altingia excelsa* (fig. 11.19) or *Podocarpus* spp. both of which can reach 60 m tall. *A. excelsa* is a mighty forest tree, and the most massive specimens are probably several centuries old. It has been much exploited for its attractive wood and is now used as a plantation species. Although once found down to 200 m (Backer and Bakhuizen van der Brink 1963), none can now be found growing naturally below about 1,500 m. *Podocarpus* is a confusing genus comprising two types of tree: the feathery-leaved *Podocarpus* (better referred to as *Dacrycarpus*) *imbricatus*, and the long-leaved species that are generally identified as *P. ner-iifolius* (fig. 11.20). The taxonomy of these trees has been revised (de Laubenfels 1988), but the characters used for distinguishing the long-leaved species are so minor and inconsistent that they have little utility in the field.

Supreme and most fire-resistant among the very few species of pioneer trees is the mountain *Casuarina junghuhniana*, which is found east of Mt. Lawu, on the border of Central and East Java. Unlike most pioneers, it is long-lived and can reach great size (up to 45 m), but in common with most pioneers, its seeds can grow only in light conditions and when in contact with mineral soil or ash. In other words, it will only germinate on

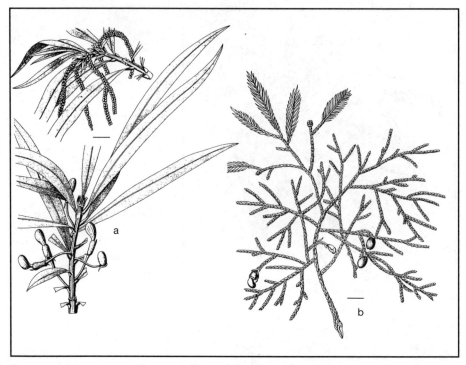

Figure 11.20. *Dacrycarpus imbricatus* (a); and *Podocarpus 'neriifolius'* (b). Scale bars indicate 1 cm.
After Koorders and Valeton 1915

bare soil and not within existing forest. It would thus be expected on sandy, rocky, or gravelly places and on scree (fig. 11.21). Beneath the growing trees is a dry tinder of 'needles' and branches, and when fires sweep through, tender, young plants are killed. At altitudes over 1,400 m it can occur over wide areas as the only tree, and can be found over 3,000 m. Where fires have not been persistent, however, the *Casuarina* forest has matured, other species have grown up, and it is just one element in a mixed mountain forest. This can be seen above Bali's Eka Karya Botanical Gardens, on the slopes of Mt. Batukau just to the west (Soepadmo 1961; Tantra and Anggana 1977), and on the slopes of Mt. Kawi (K. Kartawinata pers. comm.).

If the thick-barked *Casuarina* do suffer in a severe fire, however, they regenerate from burned stumps and below-ground roots. Grand, tall *Casuarina* forests, such as around Mt. Argapura, are a wonderful sight but, as with all pioneer species, unless fire rages or volcanic action occurs, they will be

Figure 11.21. During the violent eruption of Mt. Semeru in 1918, falls of hot ash stripped the branches of *Casuarina* trees. Many trunks remained erect on the ridges, however, and sprouted again.

By W. Roepke, with permission of the Tropen Instituut, Amsterdam

replaced in due course by other species. *Casuarina* is far more abundant now than in the past because it has colonized fire-generated grassland on land once covered with oak-laurel and ericaceous forest. Beneath the tree canopy are a variety of shrubs, including the large green *Euphorbia javanica* (Euph.), which can form dense carpets and, where the soil has been disturbed, the fearfully armed nettle *Girardinia palmata* (Urti.) (fig. 11.22). The sharp, turgid stinging spines of this nettle can be over 0.5 cm long, and it is remarkable that male Javan deer are reported to eat this plant on the Yang Plateau during the breeding season (van Steenis 1972).

Tree ferns, mainly *Cyathea* (Cyat.), some growing to 15 m, are a common component of montane (particularly lower montane) areas in both disturbed and primary forests. Tree-fern trunks can grow vertically one metre in 15 years but, as with palm trees, some years elapse between the first growth of a young tree fern and the start of the vertical growth of its trunk (Tanner 1983). At present 15 species of tree ferns are known from Java, almost all of which are found on mountains, and half of which are

Figure 11.22. An unpleasant component of the vegetation at the eastern tip of Java is *Girardinia palmata,* the poisonous spines of which can inflict a painful sting.
By A.J. Whitten

restricted to West Java, although none is endemic. The species are distinguished by the form of the reproductive parts, and by the shape of the fine, brown scales found covering the young, coiled leaves and the bases of older leaves (Backer and Posthumus 1939; Holttum 1963; Piggott and Piggott 1988). The tree ferns of Cibodas are described by Meijer (1955). Another notable fern of lower montane forest is *Aglaomorpha heraclea,* which is a huge nest-building epiphyte with fronds up to 3.5 m long. The nest sometimes becomes so weighed down with old leaves and other litter that it falls to the ground, where it continues to grow.

Strobilanthes (Acan.) is a particularly interesting shrub found in montane forests, because almost all the various species synchronize their flowering and fruiting, which occurs only once before they die, though each species appears to have a different rhythm, ranging from 5-12 years. In West Java *Strobilanthes cernua* (fig. 11.23) is a very common constituent of the forest floor from 750-2,100 m and has a nine-year cycle that has been followed without exception on Mt. Gede since 1902. The last mass flowering was in 1992 when there was a beautiful show. After the flowering, the entire population dies, leaving the forest floor peculiarly empty for a while before the seedlings grow up, again simultaneously (van Steenis 1940b, 1942). The ecological purpose of simultaneous fruiting is probably to lessen predation on the plants' seeds. By producing ripe seeds rarely and at more or less the

Figure 11.23. *Strobilanthes cernua* in Mt. Gede-Pangrango National Park during the mass flowering of 1992.

By A.J. Whitten

same time, no seed predator can become specialized in exploiting the resource, populations of the seed predator cannot build up, and the massive production of seeds satiates the predators' appetites so that most can germinate successfully.

A single hectare of the *Altingia*-dominated lower montane forest in Mt. Gede-Pangrango National Park, just above the Cibodas Botanical Gardens, was enumerated 40 years ago, and 78 tree species (from 283 trees), 30 lianes, 10 climbers, 100 epiphytes, and 73 herbs were found (Meijer 1959c) (fig. 11.24). This total compares with the total of 870 species of flowering plants and 150 ferns estimated for the entire Mt. Gede-Pangrango National Park (van Steenis 1972). The forest, as with many other tropical forests, was characterized by the high frequency of rare species. Only one tree *Villebrunnea rubescens* (Urti.) was present with more than 30 individuals, whereas 23 of the total 74 species were represented by a single specimen with a trunk larger than 10 cm diameter (table 11.3). Almost all species of tree in this lower montane forest had young individuals growing beneath them or in the vicinity, and it has been estimated that the maximum age of trees here is 200-250 years with a mean of 130 years, and with 0.77-0.83 dying each year (Baas Becking 1948). This plot was affected by the very strong winds of 1984, and it would be very interesting to enumerate the plot again. Another plot just 15 minutes walk from, and 100 m above, that

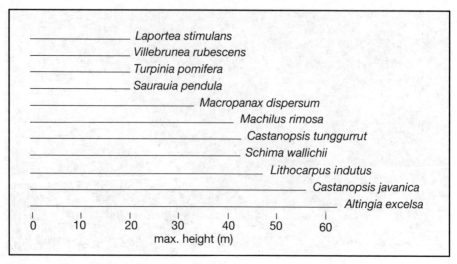

Figure 11.24. Relative heights of trees in a hectare plot of lower montane forest above Cibodas Botanical Gardens.

After Meijer 1959c

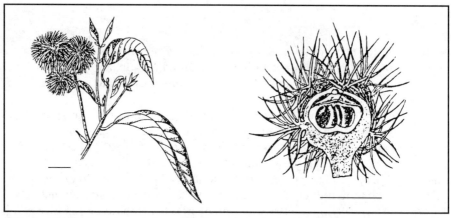

Figure 11.25. *Castanopsis javanica.* Scale bars indicate 1 cm.

After Koorders 1913

Table 11.3. The number of species represented by different numbers of individuals in a hectare of lower montane forest.

	Number of Individuals					
	1	2-5	6-10	11-20	21-30	31-40
Number of species	23	26	4	4	3	1

Source: Meijer (1959c)

used by Meijer was selected for part of Yamada's studies. This was sited in the lower regions of the *Podocarpus*-dominated forest, but *A. excelsa* was still present. Yamada's hectare plot had 57 species dominated by *Schima wallichii* and *Castanopsis javanica* (fig. 11.25), followed by *Persea rimosa* (Laur.), *Lithocarpus pseudomoluccus,* and *Vernonia arborea* (Comp.).

Upper Montane Forest

One of the most obvious characteristics of the wetter parts of the upper montane forest is the enormous quantity of *Aerobryum* moss covering the ground and adorning every twig, branch, and bough up to 2-3 m above the ground. Above the cloud zone the most common epiphyte is the beard 'moss' *Usnea.* This is in fact a lichen, a composite plant composed of a complex, spongy framework of fungal mycelium strands within which are held algal cells. The algae photosynthesize, and the fungus feeds on a proportion of the food produced. People exploit its astringent properties for curing intestinal troubles, and it has been found also that the usnic acid found in the tissues of *Usnea* has antibiotic properties against tuberculosis (Burkill 1966).

Many species of trees found in the lower montane forest persist into the upper montane forest, such as the *Dacrycarpus imbricatus,* which forms a girdle around the southern slopes of Mt. Ciremai and can be found in a unique single-species stand between 2,400-2,700 m. Although now regarded as a tree of high altitudes, this is probably a reflection of its overexploitation lower down because of the value of its timber. Most characteristic of the upper montane forest are trees and shrubs of the Ericaceae, such as the large and colourfully flowered *Rhododendron,* bilberries *Vaccinium,* and wintergreen *Gaultheria.* The leaves, flowers, and large, black berries of *Gaultheria* taste strongly of wintergreen oil (methyl-salicylate), a substance which is often applied to alleviate the symptoms of rheumatism. *Rhododendron* has very small seeds, each of which has a pair of wings which is presumably an adaptation for dispersal by wind. The fleshy fruits of the other species are eaten by birds (particularly thrushes and white-eyes), and mammals (Sleumer 1967). Some of these plants are found lower down the mountains, but as epiphytes rather than being rooted in the ground. The forest floor of the upper montane forest has a rich shrub and herb flora, and in the open areas many beautiful herbs can be found.

Yamada (1990) found on Mt. Gede that, although the occurrence of species in the upper montane forest did not show any particular zonation, the contribution to the total basal area by the main tree species indicated a floral transition at about 2,300 m. In a plot at this altitude he found 33 species of higher plants: 10 trees, 11 shrubs, 6 ferns, 3 herbs, and one each of tree ferns, lianas, and orchids. The trees *Myrsine affinis* (Mysi.) and *Acronodia punctata* (Elae.) dominated the plot. *M. affinis* also dominated

Figure 11.26. Large white-flowered edelweiss on Mt. Sumbing about sixty years ago.

By W.M. Docters van Leeuwen, with permission of the Tropen Instituut, Amsterdam

among the 24 species present in the plot at 3,000 m where there were 12 species of shrubs, five herbs, four ferns, and one each of orchid and liana (Yamada 1976a).

On Mt. Agung, the most abundant species at about 2,000 m were *Myrica javanica, Weinmannia fraxinea* (Cuno.), *Saurauia distosoma* (Acti.), and *Viburnum coriacea* (Capr.) (Meijer 1976).

Subalpine Forest

Subalpine forest has only one layer of trees which is rather poor in species, many of them belonging to the Ericaceae, such as bilberries *Vaccinium* and rhododendrons *Rhododendron.* On Mt. Gede the dominant species in terms of basal area are *Myrsine affinis, Eurya obovata* (Thea.), and *Symplocos sessifolia* (Symp.), and elsewhere other common plants include species of

Figure 11.27. *Anaphalis viscida* standing 5 m tall on the Yang Plateau. Note man.
By A.J. Whitten

Rapanea (Myrs.), *Leptospermum flavescens* (Myrt.), and *Myrica javanica*. The herbaceous ground flora has many temperate elements, with gentians *Gentiana* (Gent.), primrose *Primula* (Prim.), buttercup *Ranunculus* (Ranu.), strawberries *Potentilla*, barberry *Berberis*, and the endemic cow-parsleys *Pimpinella javana* and *P. pruatjan* (Umbe.). Terrestrial orchids are relatively common and epiphytic ones relatively rare, although mosses cover branches, trunks, ground, and rocks.

Best known of the characteristic plants of the subalpine zone is the edelweiss *Anaphalis javanica* (Comp.) (p. 175) (fig. 11.26). It is, in fact, a secondary species which appears after subalpine vegetation is burned or otherwise destroyed, and its advance in the alun-alun behind Mt. Gede has been at the expense of *Vaccinium* and *Rapanea*. It dislikes shade, and even in strong sunlight grows slowly; one particular seedling is reported to have grown just 20 cm in 13 years (Docters van Leeuwen 1933). Thus the tall specimens with stems 15 cm thick would have been well over a century old. The largest specimen mentioned by him was a giant 6 m tall with a trunk 50cm in diameter found on Mt. Sumbing. We have not been able to find even any photographs showing really large specimens, and it seems that they are unlikely ever to be found again. A close relative, *A. viscida*, with green rather than woolly leaves, is found on the Yang Plateau where some

Figure 11.28. Montane *Festuca nubigena* grassland surrounding a montane swamp in an old crater on Mt. Lawu at 3,200 m.

By W.M. Docters van Leeuwen, with permission of the Rijksherbarium/Hortus Botanicus, Leiden University

reach 5 m tall (fig. 11.27).

The woody components of the subalpine flora are in general poorly developed around most of the high summits because they were cleared when triangulation pillars were being erected during the Dutch period. The pillars were used for survey work and this required good visibility all round. Slow plant growth and occasional fires have conspired to hinder regeneration.

Swamps

Montane swamps occur across Java, but are not known from Bali. They are formed in former craters or where a lava stream or lahar cut off a valley, and the best sites were, and perhaps still are, on the Dieng Plateau, which was entirely forested when visited by Junghuhn (1854). Such swamps often have *Sphagnum* moss as a major peripheral or floating plant, but sedges (Cype.), rushes (Junc.), *Xyris* (Xyri.), and *Eriocaulon* (Erio.) are common. A list of characteristic montane marsh plants is given in van Steenis (1972). Another excellent sets of swamps of different types are found on Mt. Patuha. The little stream draining the Tegal Alun-alun on Mt. Papandayan has many marsh plants and is the only known site in Java of the wild 'straw-berry' *Potentilla polyphylla* (Rosa.), which may be the rarest plant on Java (van Steenis 1972). Perhaps the most beautiful montane marsh and forest-fringed lake is Taman Hidup above Bremi on the west of the Yang Plateau. As swamps 'mature', they tend to develop into *Festuca* grasslands (fig. 11.28).

BIOMASS AND PRODUCTIVITY

The trends in biomass and productivity change, from lowland to upper montane forests, are as follows (Grubb 1974):

- biomass decreases proportionately less than height of plants, resulting in shorter, stockier trees in the upper montane forest;
- production of woody parts declines from about 3-6 t/ha/year to about 1 t/ha/year;
- production of litter, particularly leaf litter, decreases proportion-ately much less than biomass or production of woody parts;
- biomass of leaves decreases proportionately much less than total biomass;

Table 11.4. Leaf-fall types among the main species of trees and shrubs in the four seasons for the three layers of forest.

	Dry season	First half of rainy season	Middle of rainy season	Latter half of rainy season	All times
1st layer	Lithocarpus rotundatus	Cinnamomum sintoc	Castanopsis javanica	Engelhardia spicata	Vernonia arborea
	Platea latifolia	Litsea resinosa		Lithocarpus indutus	
	Schima wallichii			Lithocarpus pseudomoluccus	
2nd layer	Litsea mappacea		Polyosma integrifolia	Decaspermum fruticosum	
	Persea rimosa			Flacourtia rukam	
	Polyosma ilicifolia				
	Laplacea integerrima	Laplacea integerrima			
3rd layer	Strobilanthes cernua	Meliosma nervosa	Castanopsis argentea	Lithocarpus elegans	Ardisia fuliginosa
		Symplocos fasciculata		Saurauia blumiana	Saurauia pendula
	Villebrunea rubescens	Villebrunea rubescens			Saurauia reinwardtiana
	Prunus arborea		Prunus arborea		Turpinia sphaerocarpa

Source: Yamada (1976b)

- mean life span of leaves increases only slightly;
- the leaves become thicker and harder and the leaf area/g decreases from up to 130 to about 80 cm^2/g.

Thus, with increasing altitude, total production decreases, but plants invest a greater proportion of their production in making leaves, although these more 'expensive' leaves may last no longer than the 'cheap' leaves of the lowlands.

The two studies of litter production in Javan montane forests found that the annual rate of total litter production in lower montane forest was between 6,000 and 7,500 kg/ha/year, about 75% of which was leaves (Yamada 1976b; Bruijnzeel 1984), which is similar to results from lowland forest in Peninsular Malaysia (Lim 1978). Leaves fall throughout the year, but the main peak is during the dry season. Branch fall is greatest during storms, which occur most often during the wetter seasons.

Leaf-, fruit-, and flower-fall occur in all months in the montane forest and there is no clear general pattern. When individual species are examined, however, some patterns can be discerned (table 11.4). Some species, mainly in the lowest layer, lose leaves throughout the year, while species in the highest layer, exposed to all the changes in climatic conditions, seem to show the greatest seasonality (Yamada 1976b).

MINERAL CYCLING

There are two main classes of mineral cycling in a forest. First, there is the rapid cycling of minerals in the fallen leaves and twigs, and in the rain falling through the forest canopy to the floor. Second, there is the much slower cycling of minerals held in the tissues of large woody parts of trees.

Lower quantities of minerals cycle through leaves and woody parts in montane forests than in lowland forests. This is because the concentration of nutrients in leaves of montane plants is about half that of their counterparts in lowland forest, and because of the relatively low rate of production of woody parts. The concentrations of minerals in the montane leaf litter are proportionately even less than in lowland litter because montane plants absorb into their permanent tissue about half of their leaf minerals before shedding them, whereas lowland plants absorb only about one-quarter of their leaf minerals.

The major external input of minerals appears to be from the rain (Edwards 1982), which has been contaminated by ash from fires (Ungemach 1969) and volcanic eruptions. Not all the rain falling over a forest reaches the floor because it evaporates, is intercepted by epiphytes and other leaves, or is absorbed by bark, and root mats. When the rain is light, little may reach the forest floor, and even in the heaviest rain about

25% is intercepted (Edwards 1982). The rain that does complete its journey to the ground will have also picked up minerals from animal droppings, plant exudates, and humus around epiphyte roots and will be far more mineral-rich than rain falling in a clearing.

As indicated above, leaf litter does not have the same mineral content as the living leaves because of absorption by the plant. Some minerals, such as calcium and magnesium, do not seem to be reabsorbed, while others such as nitrogen and phosphorus are. It is possible that the minerals which are absorbed are relatively scarce and limiting in the soil.

ANIMALS AND THEIR ZONATION

Although a number of animals are restricted to high mountains, most tropical animals are unaccustomed and unadapted to cold temperatures. This is particularly true for invertebrates, fishes, amphibians, and reptiles, which draw their body warmth from their surroundings. The reptiles, at least, raise their temperatures by basking in the sun, and as long as nighttime temperatures do not reduce body temperatures below a critical minimum, they can live successfully in relatively cold climates. No native fishes are found in the montane zone, but the endemic toads *Leptophryne cruenata* and *Philautus pallidipes* have been found up to 2,500 m and at the summit on Mt. Pangrango, respectively (van Kampen 1923), and a small, unidentified toad has been reported from 2,500 m next to Rawa Kombolo on Mt. Semeru (Docters van Leeuwen 1933). A number of snakes are known from over 2,000 m (de Rooij 1917), but no analysis has been performed on the meagre data.

For birds and mammals it is availability of food rather than temperature which limits distribution, although some species may travel through seemingly inhospitable habitat to reach suitable areas. The larger mammals, such as rhinos, banteng, Javan deer, and tiger, used to be found occasionally over 3,000 m, but there are no remaining populations that can extend to such altitudes.

Invertebrates

The number of invertebrates declines with altitude, and this is most conspicuous among the ants and termites which are absent in the soils of the high peaks but ubiquitous in the lowlands. Biomass, however, can be high on vegetated peaks, even high ones such as Mt. Pangrango where large carabid and staphylinid beetle larvae take over from termites as the major detritivores. Other detritivores found in such locations are bug and fly larvae, thrips, millipedes, and even earthworms (Docters van Leeuwen

1933). On the soil surface and vegetation quite a number of small snails can be found, most of them endemic (p. 287). Lower down, in the montane forests from Mt. Gede to Mt. Slamet, the semi-terrestrial crab *Terrathelphusa kuhli* is one of the larger detritivore organisms. At high altitudes, centipedes and spiders take over from ants as the major soil-based predators (Docters van Leeuwen 1933; Collins 1980).

When searching for flying and other above-ground invertebrates on overcast days, the fauna may seem more rare than it actually is, but on sunny days there is a great apparent abundance of insects, even if it does comprise large numbers of rather few species. One of the few insects that seems not to be influenced by the weather is the thick-haired, black bumble bee *Bombus rufipes,* which is also the most important pollinator of high-mountain flowers. It is found also down to about 1,300 m.

The most dangerous of the mountain invertebrates are the social wasps *Vespa velutina* which nest in the bare parts of craters against rocks and in the ground. A severe case of stinging may result in a high fever, and the best defence is simply to run away (Docters van Leeuwen 1933; van Steenis 1972).

It is apparent that invertebrate communities differ between and among mountains even at the same altitude. Although this has not been quantified in natural forest, in planted pine forests in Central Java the ground fauna among three different locations on the slopes of Mts. Merbabu, Merapi, and Sindoro was strikingly different in composition and relative abundance. Indeed, only four of over 40 species were found in all three locations (Gunadi and Notosoedarmo 1988).

Casuarina trees support a very pauce fauna, possibly because of their tough, modified leaves and twigs, but, growing in a natural monoculture, they are susceptible to pest outbreaks. Some of these have been recorded. For example, hairy lasiocampidid caterpillars *Voracia casuariniphaga* and caterpillars of the geometrid moth *Abraxus contradens* defoliated the trees in some areas in 1921, although the trees did not die (Kalshoven 1953). The litter of *Casuarina* forests can support huge numbers of St Mark's fly (bibionid) larvae, just below the surface.

Table 11.5. Number of bird species found in different altitudinal zones on Java.

Species not found above 800 m	187
Species living between sea level and 1,500 m	233
Species living between 1,000 and 3,000 m	134
Species living exclusively at sea level	98
Species living from sea level to mountaintops	72

Source: Hoogerwerf (1948a)

Figure 11.29. Damage done to *Engelhardia* trees by deer *Cervus timorensis* at 2,100 m on the slopes of Mt. Argapura.

By C.G.G.J. van Steenis, with permission of the Rijksherbarium/Hortus Botanicus, Leiden University

Birds

Bird life at high altitudes is scarce. The altitudinal distribution of Javan birds has been analysed (Hoogerwerf 1948a), and this revealed that 420 species of birds are found between sea level and 800 m, compared with about 300 species between 800 m and 2,000 m (p. 236). By examining the maximum height at which species have been found, it seems that there is a major boundary at about 1,300-1,600 m, where 105 species reach their highest altitude. Many of Java's birds are found in mountain forests, and some are confined to it (table 11.5). The most conspicuous species above 3,000 m are the rather tame Sunda whistling thrush *Turdus poliocephalus* with a conspicuous, yellow bill and a bubbling call, and the red jungle fowl *Gallus gallus*. Also quite common are the endemic white-flanked sunbird *Aethopyga eximia,* and the grey-throated darkeye *Lophozosterops javanicus,* both of which eat insects and take nectar from *Rhododendron* and *Vaccinium* flowers. At the peak of Mt. Gede it is often possible to see the endemic volcano swiftlet *Aerodramus vulcanorum.*

Mammals

For the larger, terrestrial species of mammals, mountains represent no particular obstacle. Such species would include leopard and other cats, deer (fig. 11.29), and wild boar. For the arboreal species, however, problems are

Figure 11.30. Yellow-throated marten *Mustela flavigula*. Its total length is about 80 cm.

After Payne et al. 1985

encountered. The gibbons and two species of leaf monkeys are rarely found above 1,250 m, although the lutung monkey extends highest, and the surili monkey the next highest. As described on p. 503, montane forest becomes increasingly gnarled, short, and mossy with altitude and the ability of the gibbons to travel, swinging below boughs and branches, would decrease significantly. In addition, the productivity of the montane forests are lower than that of lowland forests, and the species composition changes, being dominated by chestnuts and conifers which offer no food to these animals. The upper limit for the gibbon is lower than for the leaf monkeys because gibbons have more fruit in their diet. It is clear that, unless sizeable areas of lowland forest below 1,200 m are conserved, the endemic gibbon and surili will head inexorably towards extinction.

Although mammals generate their own heat, heat loss poses enormous problems for the smaller species because their surface area is very large compared with their volume, and their heat loss must be compensated for by consuming large quantities of high-energy food. It is therefore surprising that the smallest native Javan mammals, the shrews, are found at high altitudes. Shrews are almost entirely carnivorous and eat small insects, snails, earthworms, millipedes, centipedes, and spiders; resources which are scattered, cannot be stored and whose abundance is often unpredictable. Shrews therefore have to spend a great deal of time searching for food, an activity which in itself uses up energy, and they consume about their own body weight in food each day, often in their underground burrows, having

Figure 11.31. Stink badger *Mydaus javensis*. Its total length is about 45 cm.
After Payne et al. 1985

to be active almost the whole day and night, although they rest briefly every few hours. If more than a few hours passes without a meal a shrew will starve to death. Larger shrews, such as *Crocidura maxi* (head and body 70 mm) and *C. fuliginosa* (head and body 95 mm), can presumably survive hunger better than the smallest one *C. monticola* (head and body about 60 mm), and their response to food shortage is also different: the smaller species increase and larger ones decrease their foraging activity (Hanski 1985).

The resident predatory mammals include the yellow-necked marten *Mustela flavigula* (fig. 11.30), which preys on small birds. It is an accomplished climber, and few nests with their eggs and nestlings would be safe. Another predator is the Javan skunk or stink badger *Mydaus javensis* (fig. 11.31), which feeds at night on insects and earthworms. It is notable for its ability to spray evil-smelling drops of liquid out of its anus which cause nausea. Almost as remarkable is the fact that the leaves of one of the ferns found in the mountains *Didymochlaena truncatula* (Aspl.) smell almost identical, and this odour is particularly strong in hot weather (Docters van Leeuwen 1933; Backer and Posthumus 1939).

A number of rat species are found in the montane and subalpine zones, and it is interesting that the majority of these are endemic species (table 11.6); indeed, only one of them appears to venture below 1,000 m. The rat confined to the highest altitudes, the woolly mountain rat *Niviventer lepturus,* has a thick coat as a defence against the cold. In contrast to the commensal rat species of towns and fields, which have a high reproductive output (p. 652) and relatively low inherent survivorship, the montane rats of the much more stable forest environment probably have much smaller litters of just one or two young which have a high chance of survival.

It has been suggested that some small mammals seem to walk upwards from the forest into the highest parts of a mountain before dying. Whether they die of the cold, or whether they are sick before the venture upwards, is not clear. In the case of flying squirrels, which normally climb to perches from which they can glide, it may be that they walk up an increasingly barren mountain in the forlorn hope of finding a better take off point (Gisius 1930; van der Veen 1936; van den Bosch 1938; van Steenis 1972). It may be, however, that dead animals are simply easier to see in the sparse undergrowth or bare soil of the highest altitudes, and more often seen because flesh takes longer to decay where the temperatures are lower and there are fewer decomposing organisms.

Table 11.6. Native species of Javan rats and mice shown in order of altitudinal range (m). [e] - endemic species.

		0 - 500	500 - 1,000	1,000 - 1,500	1,500 - 2,000	2,000 - 2,500	2,500 - 3,000
Polynesian rat	Rattus exulans	x	x	x	x	x	x
Woolly mountain rat [e]	Niviventer lepturus				x	x	x
Javan volcano mouse [e]	Mus vulcani					x	
Javan mountain spiny rat [e]	Maxomys bartelsi			x	x	x	
Fawn-coloured mouse	Mus cervicolor			x	x		
Javan monkey rat [e]	Pithecheir melanurus			x	x		
Jayan giant rat [e]	Sundamys maxi			x			
Javan flat-nailed rat [e]	Kadarsanomys sodyi		x	x			
Hill tree rat	Niviventer bukit	x	x	x	x		
Large bandicoot rat	Bandicota indica	x	x	x			
Pencil-tailed tree rat	Chiropodomys gliroides	x	x	x			
Dark-tailed tree rat	Niviventer cremoriventer	x	x	x			
Ricefield rat	Rattus argentiventer	x	x	x			
Roof rat	Rattus rattus	x	x	x			
Long-tailed giant rat	Leopoldamys sabanus	x	x				
Spiny rat	Maxomys surifer	x	x				
Wood rat	Rattus tiomanicus	x	x				

Sources: van der Zon (1979); Corbet and Hill (1992); Boeadi (pers. comm.)

EFFECTS OF DISTURBANCE

Volcanoes

Although people are a potent factor in the destruction of montane forests, it is their cumulative effects over time that draw attention. Volcanoes, on the other hand, are rarely active, but when they are, the physical changes are dramatic with hot and cold lahars, lava streams, and smothering clouds of hot ash. Of these the hard lava streams, such as on Mt. Batur (de Voogd 1940) (fig. 11.32) and Mt. Guntur (van der Pijl 1939), and the old one on the Ijen Plateau (Rejengan), are the slowest to develop new vegetation. It may take centuries before lava has been sufficiently weathered to support closed forest but, if there is sufficient moisture, the first plants can be found almost as soon as the lava cools down. Early colonizers of ash slopes include grasses, *Vaccinium*, *Gaultheria*, and *Rhododendron*, and the fern *Gleichenia* (Glei.) (fig. 11.33) is common where landslips have occurred. An account of succession on the lahar deposits of Mt. Kelud is available (Clason 1935).

Ash is the major volcanic product, however, and even a thin layer deposited far from the erupting volcano can interfere with living leaves by adhering to them and preventing photosynthesis. If rain falls soon, it will wash off the dust and the plants will survive, but without rain they will die. When Mt. Agung erupted violently in February/March 1963 (pgs. 97, 685), all the plants around the Balinese mother temple of Besakih on its southern slopes were smothered with ash and killed. A botanical survey in October 1963 found only three species of plants were growing among a dead landscape. These were the elder *Sambucus javanica* (Capr.), the grass *Eleusine indica* (Gram.), and the common purple-head herb *Ageratum conyzoides* (Comp.). One year after the peak of the eruptions only 10% of the ground around Besakih had a green cover of grasses, herbs, shrubs, and trees – mainly new growth, although some supposedly 'dead' plants were sprouting new leaves – and a total of 83 plant species were found. The remaining 90% of the ground was still bare "as if the area had been cemented" (Dilmy 1965).

'Sand seas' of unvegetated ash can be found on the summit of Mts. Sumbing, Merapi-Ijen, and Slamet, but by far the best known is in the caldera of Mt. Tengger (Schröter 1929). Near Mt. Bromo, in the southeast of that caldera, the sand is virtually bare because it receives most of the fresh ash ejected from Bromo. In the 'Rujak' area, however, there are grasses, herbs, and short stands of *Casuarina*. The sand is virtually sterile and does not hold water, so it is difficult for plants to become established. Some tussock- or mat-forming plants, such as the grasses *Calamagrostis*, *Imperata*, and *Festuca*, as well as *Polygonum chinense* (Poly.), the white, hairy-

Figure 11.32. Regeneration of vegetation on hard lava streams, such as here on Mt. Batur, can take centuries.

By A.J. Whitten

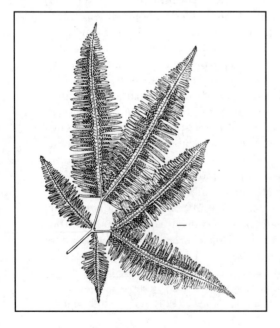

Figure 11.33. *Gleichenia linearis.* Scale bar indicates 1 cm.

After Backer and Posthumus 1939

Figure 11.34. *Vaccinium varingae-folium.* Scale bar indicates 1 cm.
After Koorders 1912

flowered endemic *Styphelia javanica* (Epac.), and some *Carex* (Cype.), can eventually take hold, gathering organic material within their mass, and forming small dunes around themselves. Some ash areas on steep parts of volcanoes generally lack vegetation, and this is because they do not retain moisture, they are sterile, and any germinating seedlings are buried as the ash and developing soil creeps downhill.

Around active craters or sulphur vents the soil is usually rocky, very pervious, sterile, acid, and lacking in organic matter. The atmosphere is generally a choking mixture of sulphur dioxide and hydrogen sulphide, mixed with smaller quantities of chlorine, carbon monoxide, and nitrous oxide. In addition, the ground is frequently heated from beneath. These are conditions which are hardly favourable to plants, but it is amazing how tolerant some can be. Those closest to the sulphur vents tend to be dwarfed, prostrate, and to grow extremely slowly (von Faber 1927). The closest species are often *Vaccinium varingaefolium* (fig. 11.34), *Rhododendron retusum,* and the fern *Selliguea feei* (Polyp.). Lichens may be found closer still, and some blue-green algae are able to live in hot, muddy, sulphurous pools (Kullberg 1982). If the noxious gases are not too concentrated, then plants more typical of lower altitudes may be found because the vicinity of fumaroles is akin to a natural hot-house (van Steenis 1935a, 1936c).

People and Fire

The most ubiquitous, intransigent, intractable, and insidious disturbance in the mountains is caused by people cutting and burning the vegetation (fig. 11.35). Some of this is virtually traditional, such as the firing of high slopes and plateaux to open up the vegetation to encourage deer for future hunting. It is likely, however, that many fires are now started by recreational mountain walkers. Most of these people live in lowland environments where plants grow and recover extremely rapidly. It is important, therefore, that people who climb mountains for recreation or study should realize just how long-lasting the effects of disturbance can be in a much less dynamic environment (fig. 11.36).

Lower montane forests are not easily ignited since they are generally wet or moist and there are no piles of easily dried material that could act as tinder. As a result, none of their trees have any fire-resistant characters. In upper montane and subalpine forests, however, periods of drought do occur, and the oil-rich leaves of *Rhododendron, Vaccinium, and Gaultheria* can catch alight even in the rain. In addition, people are able, through successive attempts to burn an area, to increase the abundance of grasses and sedges, the dead leaves of which serve as tinder (Burger 1930; van Steenis 1972). Above 2,000 m the ground cover after a fire tends to comprise ferns such as *Gleichenia linearis, Dipteris conjugata* (Diptd.), and *Dicranopteris* sp. (Glei.) (Sukardjo 1982a), but when subalpine vegetation is burned, regeneration proceeds very slowly. The fire kills the above-ground parts of shrubs, but suckers often grow from their bases, albeit slowly. The soil surface exposed by the fire remains unvegetated for years, and very few seedlings of the surrounding plants are found. Dead stems of plants stand for many years, and even when they fall, they decompose slowly because of the low temperatures and the absence (or extreme paucity) of most of the lowland decomposers.

The higher altitudes have fire-resistant trees such as *C. junghuhniana*, but only one, the small *Paraserianthes lophantha*, is positively encouraged by burning, and it is sometimes found as a girdle around active volcanic cones. The seeds of this species germinate only after fire has cracked open their hard case, and as a result, in any given area the trees will all be of the same age. This tree is very common and the young seed pods are a favoured vegetable, a substitute for petai *Parkia* (Legu.). People set fire to such areas every five years or so in order to start another 'crop' cycle.

Disturbance is also caused by the introduction of exotic species. In Mt. Gede-Pangrango National Park the understorey, particularly the gaps, of the lower montane forest is becoming swamped with the orange-flowered, small (<5 m) tree *Cestrum auranticum* (Sola.), introduced from Central America (Bruggeman 1927a, b; Meijer 1954c; R. Beckwith pers. comm.). Also, the large-leaved passionfruit *Passiflora edulis* (Pass.), intro-

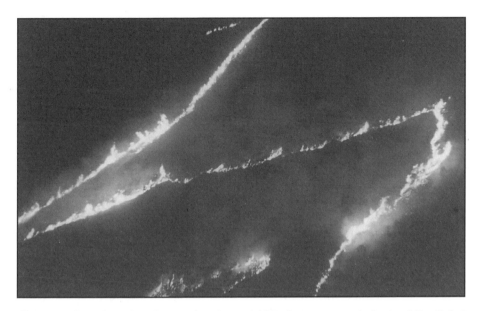

Figure 11.35. Trees burning on the slopes of Mt. Argawayang, between Mts. Kelud and Arjuna, in 1940.

By M. Sigg, with permission of the Rijksherbarium/Hortus Botanicus, Leiden University

Figure 11.36. Even relatively remote summits, as here on Mt. Arjuna, do not escape the damage of 'nature lovers'.

With permission of Antara

duced from South America for its sweet fruit, is smothering the canopy. Monkeys eat the flesh and seeds (R. Beckwith pers. comm.), and local people collect the fruit to sell. It is particularly common (or at least most visible) in the gaps caused by the very high winds of October 1984.

Succession

The succession of plants in montane areas up to 2,000 m altitude is quite similar to that lower down. Tree falls and other moderate gaps are colonized by pioneer species such as *Macaranga rhizinoides, Homalanthus populneus, H. giganteus* (in East Java and Bali), *Mallotus, Glochidion rubrum* (all Euph.), *Trema orientalis, Parasponia parviflora* (both Ulma.), *Dendrocnide stimulans* (Urti.), *Weinmannia blumei, Rapanea hasselti, Persea rimosa* and *Pinanga* palms (Soepadmo 1961; van Steenis 1972). Large figs *Ficus* spp. (Mora.) can also be common.

Abandoned highland fields are soon dominated by *Chromolaena inulifolium* (Comp.) and *Lantana camara* (Verb.), and these form a protective cover for the soil and for young tree seedlings. Where the area is subjected to repeated burning, herbs such as white-leaved 'edelweiss' *Anaphalis longifolia* and bracken fern *Pteridium aquilinum* (Denn.) increase in abundance. With further burning, alang-alang grass *Imperata cylindrica* becomes dominant. In moist places the wild sugarcane *Saccharum spontaneum* (Gram.) can become very common.

Chapter Twelve

Caves

Caves are simple natural ecosystems of great value for understanding ecological interrelationships, for regulating and purifying water, for economic products such as fertilizer and birds' nests, as well as for their own intrinsic interest. They have advantages over many other ecosystems in terms of potential for research, both theoretical and applied, in that the boundaries are discrete, and most of the species inhabiting them can be studied easily and manipulated in the cave or laboratory. Caves may have small entrances opening into large chambers, or massive entrances behind which the cave is only penetrable for a short distance. Some caves have rivers flowing through them, and are called active, and others are more or less dry, having been formed in the past, and these are called nonactive or fossils. Caves of sorts were created during the Japanese occupation when tunnels were dug into hillsides. These have many of the same features as natural caves and provide habitats suitable for some cave organisms.

The characteristic features of a cave are its definable limits, enclosed nature, low light levels, and the comparative stability of temperature, relative humidity, and air flow. Variations in these characteristics between caves creates a surprisingly wide range of habitats which determine the type and number of animals that can inhabit a cave. Habitats associated with caves are shown in figure 12.1, and include soil and litter in limestone forest, the superficial underground compartment (the very small cracks, passages, and chambers which are found between the soil and caves), speleothems (such as stalactites), cave streams, sump zone, and cave-floor habitats.

DISTRIBUTION AND FEATURES

There are about 1,000 caves in Java and Bali, some 200 of which have now been mapped (R.K.T. Ko pers. comm.). The vast majority of the caves are formed in limestone areas, and most of these are in the 1,000 km^2 area of Mt. Sewu east of Yogyakarta (fig. 12.2), and in the smaller area around the Karangbolong Hills, 100 km to the west of Yogyakarta. There are also a few

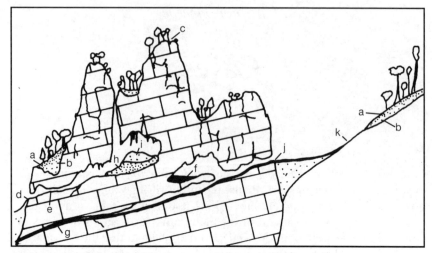

Figure 12.1. Habitats associated with limestone caves. a - soil and litter in forest; b - superficial underground compartment; c - soils; d - resurgence; e - underground stream; f - pools inside cave; g - sump; h - speleothems, clay, etc.; i - guano; j - river sink; k - river.

After Ko 1986

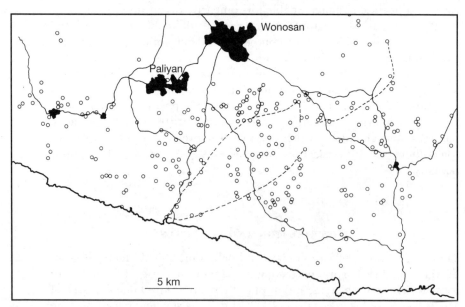

Figure 12.2. Major caves in the Mt. Sewu area.

After Waltham 1983

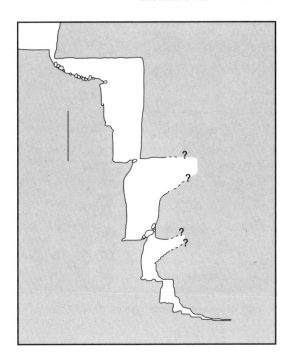

Figure 12.3. Cross section through Luweng Ngepoh, the deepest cave in Java. Scale bar indicates 40 m.

After Alkantan and Lambertus 1986

lava caves. In the spectacular cone karst limestone area of Mt. Sewu no less than 261 caves have been catalogued within the boundaries of the Special Region of Yogyakarta, one of which is over 4 km long (Waltham 1983; Quinif and Dupuis 1984; Willis et al. 1984; Sautereau de Chaffe 1985), and a further 55 in its eastern extension west of Pacitan. One of these, Luweng Ngepoh, at 236 m is the deepest cave in Java (fig. 12.3) (Alkantan and Lambertus 1986), and another, Luweng Jaran, was explored and mapped for 11 km in 1984 (fig. 12.4) (Stoddard 1985, in press). By 1993, however, no less than 25 km of passages were known. This makes it the longest known cave in Indonesia (R.K.T. Ko pers. comm.). It has large passages and chambers with exquisitely beautiful 'decorations' of stalactites hanging from the ceiling, and stalagmites and 'walls' growing upwards from the floor, with streams and small lakes.

The Karangbolong massif west of Yogyakarta also has a good number of interesting caves, but these are not yet fully documented, although one of the caves has been mapped for over 3 km (Waagner et al. 1983; Sautereau de Chaffe 1985). The caves here have formed in three distinct layers. There are short fossil caves in the cone tops of the area, partially fossil caves (e.g., Petruk Cave), and long, active caves (e.g., Barat Cave with a 7.5 km

Figure 12.4. Luweng Munung Plente, part of Luweng Jaran, the longest known cave system in Indonesia.

By S. Stoddard, with permission

passage) (Koschier et al. 1983). Another, Macan Cave, has Java's largest chamber, reached through a small passage, which has an area equivalent to 1.5 football pitches. Bali has rather few caves except those dug by human hand (e.g., Gajah Cave) (Kusch 1978), short 'shelters', and wave-cut coastal caves (Rachman 1990). On Nusa Penida to the southeast of the main island, however, there are a few substantial caves in the extensive cone karst (Kusch 1980; Rachman 1990).

Most of the caves mentioned above are more or less horizontal, but in some areas almost all the caves are vertical. Examples are the caves on the degraded Cibodas hill near Ciampea (HIKESPI 1983), and the hills just north of the Jatiluhur reservoir (Ko 1983a).

A few of the caves are developed for tourism to varying degrees (table 12.1), and others have potential tourism value, although for a number it will be only real enthusiasts who will want to penetrate the gloom (e.g., Wyeth and Boothroyd 1990). Caving and speleology are becoming increasingly popular in Indonesia, largely due to the enormous energy of Dr. R.K.T. Ko of the Federation of Indonesian Speleological Activities (IIIKESPI) (P.O. Box 55, Bogor), where great emphasis is placed on correct

Figure 12.5. Caving and speleology are exciting and beneficial activities if the necessary precautions are taken.
source unknown

and safe techniques and search-and-rescue skills (fig. 12.5). One is struck by the respect that the caves engender in the enthusiasts and researchers. Regrettably it is very difficult to get decision makers to reach the same level of emotional involvement because few of them are prepared to enter this unknown underground world.

Caves have a special place in Javanese culture (Wahyono 1982), and many caves are well known for their tradition of worship and meditation (fig. 12.7). Some of these have become foci of pilgrimages or even tourist attractions, while others, such as in Alas Purwo National Park, are occupied by ascetic hermits. In addition, almost every cave seems to have some legend or tale associated with it (e.g., Hartono 1979; Purnomo 1981; Hidayat et al. 1982).

Table 12.1. Some of the tourist caves in Java and Bali.

	Location	Notes
West Java		
Sanghyang Sirah	A day's walk from Tamanjaya on the south coast of Ujung Kulon.	A cave for meditation.
Lalai	Near Pelabuhan Ratu.	Domestic and foreign tourists come to marvel at 2 million wrinkle-lipped bats *Tadarida plicata* flying out at dusk. The cave mouth is protected by fences of barbed wire to protect the valuable guano deposits from unlicensed collectors.
Cimenteng and Cigudeg	Near Jasinga.	Recently developed with cave mouths in the form of dinosaurs.
Donan	At Kalipucang, near Pangandaran.	Up to 1982 every visitor was given an insectivorous bat as a souvenir (R.K.T. Ko pers. comm.).
Pawon	Near Padalarang, Bandung.	
Pemijahan	South of Tasikmalaya.	Where Haji Abdulmuchyi, one of the Kyai who brought Islam to W. Java is said to have come to meditate and to use a direct link to Mecca. On pre-scribed days thousands of Muslim pilgrims go here to pray.
Central Java		
Seplawan	On the west of the Menoreh Hills near Purworejo.	A cave of considerable proportions with a 1 km main passage where golden statues and other ancient items were found in 1979 after a local villager had a vision of their location. There are also old kawi inscriptions (Soekatno 1980; Darsoprajitno and Handokotresno 1981; Purnomo 1981).
Jatijajar	South of the road between Kebumen and Banyumas, on the west of the Gombong Hills.	Highly developed with steps, banis-ters and embellishments of historical dioramas, painted dragons and dinosaurs (Bachri 1978), and there are plans to build a swimming pool and hotel complex. This cave is very popular and thousands of domestic tourists flock here at weekends. HIKESPI members have now mapped over 3 km of caves with many beau-tiful waterfalls beyond the limits reached by tourists.
Petruk	Seven km south of Jatijajar Cave.	A beautiful cave of international significance (Ko 1983c), 'developed' sensitively as an interesting, educa-tional, and natural tourism experience by Perhutani with HIKESPI advice; well worth a visit (fig. 12.6).

table continues

Table 12.1. *(Continued.)* Some of the tourist caves in Java and Bali.

	Location	Notes
Karangbolong	On the coast southwest of Kebumen.	These tall, wave-cut caves are famous for their birds' nests and the collectors who scale ladders to prise them from the cave walls. Demonstrations of what is involved in collecting birds' nests can be seen at Karangbolong beach. The main cave cannot be reached except by boat, however, and then only by licensed and 'spiritually prepared' individuals.
Lawah	On the eastern slopes of Mt. Slamet.	An unusual lava cave (fig. 12.7) (Darsoprajitno and Ermadi 1979; Hartono 1979).
Maria	Just west of Tawangmangu on the slopes of Mt. Lawu.	
Mt. Selok	Wave-cut caves east of Cilacap.	Some of the caves are believed to be spiritually important, and are used for prayer and meditation (Ko 1983c). Neighbouring caves have some biological interest.
Yogyakarta		
Selarong	On the south coast of Yogyakarta.	Cliff-top cave where the greatest Indonesian freedom fighter, Prince Diponegoro, sheltered and meditated.
Kiskendo	30 km west of Yogyakarta on the east of the Menoreh Hills.	Traditional site of the fight in the Ramayana epic between the two monkey kings and the demon king (Darsoprajitno and Astawa 1982).
East Java		
Pongangan	Five km from Gresik, near Surabaya.	Millions of bats which once attracted visitors have shrunk in number but still offer a fine wildlife spectacle.
Semar	On the west of the Tengger caldera.	A meditation cave (Prihantono 1981).
Tetes	Near Lumajang.	
Bali		
Lawah	Klungkung.	An important temple is sited around and within a cave occupied by thousands of cave fruit bats *Eonycteris spelaea*.
Giri Putri	Karangsari, Suana, Nusa Penida.	A sizeable cave with impressive whip scorpions, cave crickets, bats, and an endemic crab. Guarded well by the surrounding community, which has prevented and erased graffiti.

Figure 12.6. Large entrance to the beautiful Petruk Cave, Central Java.

By A.J. Whitten

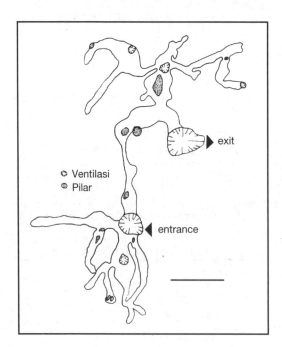

○ Ventilasi
◉ Pilar

Figure 12.7. Plan of Lawah Cave, one of Java's few lava caves open to tourists, situated due east of the summit of Mt. Slamet. Scale bar indicates 100 m. Stippled blocks indicate pillars, rayed blocks indicate openings.

After Darsoprajitno and Erimadi 1979

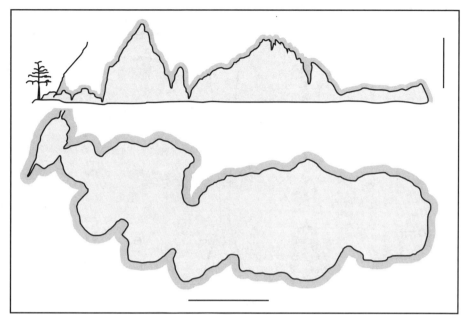

Figure 12.8. Cross section through and plan of the meditation cave of Tuyo Ngrembes (Imogiri). Vertical and horizontal scale bars indicate 10 m.
After Suryakusuma 1980

CAVE FORMATION

Caves in limestone areas are formed by rain water, which contains carbon dioxide absorbed from the atmosphere and is therefore slightly acid. This weak acid dissolves calcium carbonate (the main constituent of limestone) and forms channels which, in time, achieve the dimensions of caves, often with a stream running through them.

Two of the commonest features within a cave are stalactites and stalagmites, which together with other cave decorations are known as speleothems. They are made of calcium carbonate with various impurities (the cause of the wide range of pale colours found) and are formed by the repeated evaporation of water containing calcium carbonate, leaving thin layers of mineral deposits (fig. 12.9). Evaporation in caves is very slow because there is no solar radiation to excite the water molecules, air movement is absent or minimal, and the air is virtually saturated with water vapour. This explains why speleothems take so long to grow. It has been suggested that small stalactites commonly grow in length by only 0.2 mm per year.

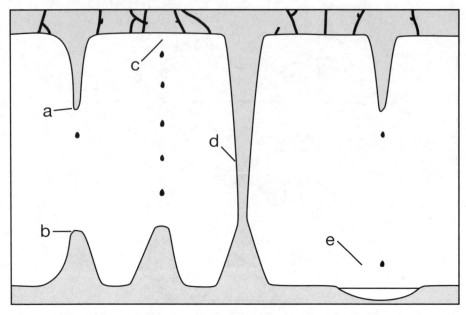

Figure 12.9. The formation of stalactites and stalagmites. a - water evaporates from drips, precipitating calcium carbonate and impurities thereby forming stalactites; b - water evaporates, also precipitating calcium carbonate but forming a squatter stalagmite; c - where water drips quickly, no stalactite is formed; d - stalactite and stalagmite eventually join to form a single column; e - where the drips from a stalactite fall (or fell) into a river, no stalagmite is formed.
After Whitten et al. 1987b

There are many types of speleothems in addition to the well-known stalactites and stalagmites. Water dripping onto the cave floor may form rings similar to the shapes formed momentarily when a raindrop falls into a puddle; spheres known as 'pearls' may form under special conditions; calcium carbonate may precipitate out of water as it flows over cave walls and rocks, thereby forming 'frozen waterfalls', often of beautiful colours because of the impurities (fig. 12.10). Lastly, on floors once covered with shallow water, slow evaporation may result in the formation of coral-like spikes, fans, and glistening crystals. All speleothems act to increase the surface area of a cave, and therefore also the living area available to the cave inhabitants.

Lava caves are formed when molten volcanic lava of different chemical compositions cools at different rates (Darsoprajitno and Ermadi 1979; Hartono 1979; Sautereau de Chaffe 1985).

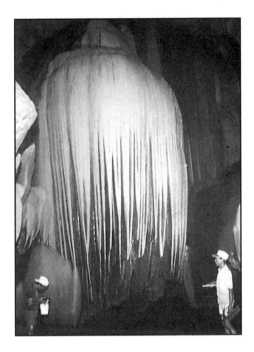

Figure 12.10. 'Frozen waterfall' in Lawah Cave.
With permission of Kompas

HYDROLOGY

Karstic limestone areas are characterized by the rapid and substantial infiltration of water into the soil and rock, by limited overland flow which rarely reaches streams, and by limited throughflow or lateral movement of water through the soil (Jennings 1971). As explained above, the water penetrates and enlarges small cracks in the rock, emerging eventually as springs. Thus the whole of the limestone part of the Karangbolong or Gombong Hills is an enormous water storage tank (Darsoprajitno 1985), the land use on which should be managed carefully. It is bizarre that Cilacap, less than 50 km to the west, should be experiencing water shortages for its growing domestic and industrial users when this supply of clean water is so close. But the water and Cilacap lie in different kabupatens. Elsewhere advantage is taken of the high-quality water. For example, at Semanu on the north of Mt. Sewu, a dam has been built deep

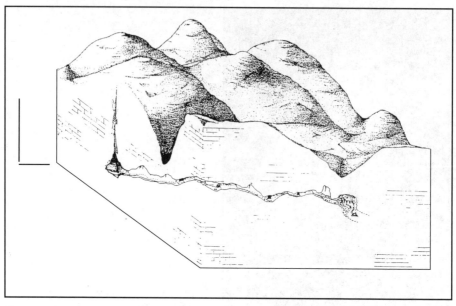

Figure 12.11. Bribin Cave near Semanu; some 500 m from the entrance is a dam. Vertical and horizontal scale bars both indicate 60 m.
After MPA-UGM 1992

within a cave (fig. 12.11).

The chemical composition of water seeping into a cave depends on the capacity of the percolating water to dissolve rock and sediment, the rate at which minerals dissolve, and the rate and nature of deposition of calcium carbonate by evaporation. Water seeping from caves shows considerable variation in chemical composition determined by the above factors and whether it has come into contact with guano.

TEMPERATURE, HUMIDITY, AND CARBON DIOXIDE

The insulating role of the walls and roofs of caves effectively buffers the rel- atively wide, daily variations in temperature and humidity of the outside world. Conditions thus remain fairly stable day-to-day, especially deep within a cave, but there are still seasonal changes which can greatly alter conditions in caves. For example, during a rainy period the humidity and amount of free water within a cave tend to increase. Air movement is also buffered by the cave walls, but still occurs as air is drawn out of the cave during the day when the air outside is warmer and lighter. This air movement follows a regular pattern, but leaves pockets of stagnant air in deep caves where spiders can weave delicate and complex webs, and preserves pockets of high humidity. Under such stable conditions the minute disturbance of air caused by the approach of predators or prey can be detected. In these deeper areas, the concentration of carbon dioxide increases if there is no inflow of air except from the cave mouth. It has been suggested that the lower metabolic rate of some cave invertebrates may be a physiological response to this high carbon dioxide concentration (Howarth 1983).

EFFECTS OF DARKNESS

The cave environment can be divided into three zones according to the degree of darkness and other physical conditions:

- the twilight zone near the cave entrance where light and temperature vary, and in which a large and varied fauna can be found:
- the middle zone of complete darkness but variable temperature, in which a number of common species live, some of which make sorties to the outside; and
- the dark zone, where a hand can no longer be seen even five minutes after all lights have been extinguished, is of almost constant temperature and is home to obligate, cave-adapted species (Poulson and White 1969). Since light is essential for photosynthesis, green plants are not found in the dark parts of caves. Plant roots can penetrate fissures leading to a cave, and these can commonly be seen attached to or hanging from the cave roof. The most important effect of this virtual total exclusion of green plants is to make all cave dwellers dependent on material brought in from the outside, and to exclude all animals that feed directly on the above-ground parts of green plants.

Most animals have clearly defined daily cycles of activity, being most active either at night (nocturnal), during the day (diurnal), or around dawn and dusk (crepuscular). Such cycles are obviously associated with daylight and darkness and thus might be thought to be absent in a cave. There are, however, certain events which may impose a daily rhythm on cave inhabitants. The most important of these is the departure and subsequent return of the bats and swiftlets. In their absence, food is not available for the free-living ectoparasites in the roosts, and there is a halt to the rain of fresh faeces from the roof (Bullock 1966). In addition, air draughts probably have a daily rhythm caused by heating and cooling of air outside

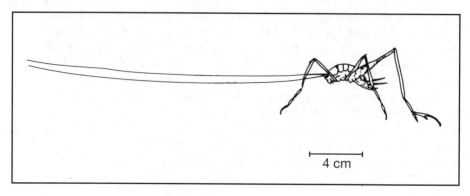

4 cm

Figure 12.12. A cave cricket; note the extremely long antennae.
After Whitten et al. 1987b

the cave. Therefore, although data are lacking, it is probable that a daily rhythm does exist in a cave.

The departure of the bats is an event which may continue for two hours after the first bat flies out of the cave. The different timing and pattern of flight activity of bat species in a single cave have been interpreted in different ways, such as avoidance of competition for food, avoidance of predators, and optimizing energy budgets (Fenton 1979; Erkert 1982). It should be stressed, however, that bats are not out of their caves from dusk to dawn, nor are they are necessarily roosting or asleep while in the cave. Instead, even if not disturbed by voices or carelessly aimed flashlights, individuals or groups can often be seen flying around the caves, and there appears to be a constant, high-pitched chatter during the day. It might be thought that the different activity periods of bats and swiftlets would represent temporal partitioning of a common food resource. In fact, swiftlets feed mainly on small, wasp-like insects (Hymenoptera) (Medway 1962; Hails and Amiruddin 1981), whereas insectivorous bats concentrate on various moths and beetles (Yalden and Morris 1975; Gaisler 1979; Fenton 1983).

In total darkness, a cave-dweller becomes reliant on senses other than sight to detect food or enemies. This is not peculiar to cave animals, however, because many nocturnal and cryptic animals found only above ground, and animals living in soil, also depend almost exclusively on hearing, smell, and touch (Bullock 1966). Some, for instance, have very long appendages, such as the legs of scutigerid centipedes and the antennae of cave crickets (fig. 12.12). The relative scarcity of food in the deepest parts of caves has led to the few animals that live there being able to withstand long periods of starvation, to gorge when food is available, and

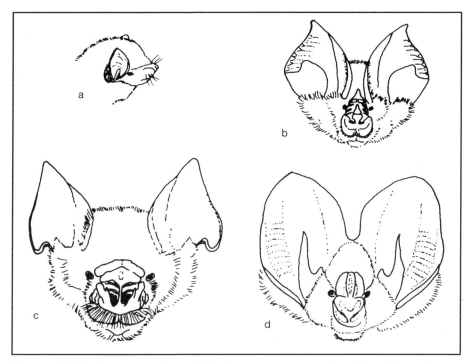

Figure 12.13. Representatives of the four most common families of insectivorous bats. a - *Miniopterus schreibersii* (Vespertilionidae), b - *Hipposideros diadema* (Hipposideridae), c - *Rhinolophus arcuatus* (Rhinolophidae), d - *Megaderma spasma* (Megadermatidae).

After Payne et al. (1985)

to store large amounts of fat (Howarth 1983).

Many bats and swiftlets have developed the ability to echolocate; that is, to produce a sound and interpret the echoes which reflect back from solid objects to give a 'picture' of the surroundings. Rain disturbs echolocating bats because very humid air absorbs high-frequency sounds, and the raindrops confuse the echoes received by the bats. Different families (fig. 12.13) use different systems of echo-location, and some can detect objects as small as 1 mm across. It used to be thought that bats using echo-location had no difficulty catching their insect prey, but it now appears that some moths can detect echo-locating bats from 40 m away and before the bat has detected the moth, using 'ears' on their chests, abdomens, or mouths. Lacewings, the adults of ant-lions (Neuroptera), have 'ears' on their wings, and those that have been artificially deafened have a 40%

greater chance of being caught by a bat than those which could still hear (Fenton 1983). Some moths have developed the ability to utter clicks that confuse the bat, while others have a variety of behavioural responses which make it difficult for the bats to predict their flight pattern. Some bats in their turn do not keep their echo-locating system 'switched-on' continuously so as to give as little warning as possible to the moths, while others emit frequencies above the hearing threshold of moths (Fenton and Fullard 1981; Fenton 1983). Whereas the bats mentioned above catch flying insects, false vampires (Megadermatidae) feed by picking lizards, frogs, and small rodents off the ground, insects off leaves, or fish from the surface of water. They have also been known to eat other bats (Medway 1967).

The only fruit bats to echo-locate, the cave-dwelling rousette bats *Rousettus*, use a low-frequency (1.5-5.5 kHz) tongue-click like swiftlets, which is audible to humans and reminiscent of a wooden rattle (Yalden and Morris 1975; Fenton and Fullard 1981). For this bat and the swiftlets, the echoes enable them to detect large objects or rock walls such that they are able to navigate, nest, and breed within a totally dark cave, but the system is not as accurate as those which to enable bats to catch insects at night (Medway 1969).

CHARACTERISTIC ANIMALS

Cave animals can be divided into three ecological groups:
- troglobites or obligate cave species unable to survive outside the cave environment;
- troglophiles or facultative species that live and reproduce in caves but that are also found in similar dark, humid microhabitats outside the cave (e.g., beetles and other insects);
- trogloxenes or species that regularly enter caves for refuge but normally return to the outside environment to feed (e.g., bats and swiftlets).
- some other species which wander into caves accidentally but cannot survive there (Howarth 1983).

Available information suggests that the troglobitic Javan cave fauna comprises a pale, long-legged crab *Sesarmoides jacobsoni* (which has not been collected since 1911) (Ihle 1912; Jacobson 1912), a white prawn with small eyes and reduced cornea, *Macrobrachium poeti* (Holthuis 1984), and possibly a blind fish, a barb known as *Puntius microps* (Jacobson 1912), which is now protected by Indonesian law. It seems that this fish may be synonymous with *Puntius binotatus* notwithstanding the fact that some have small eyes, some reduced eyes, some no eyes, and some eyes on one side only (Kottelat

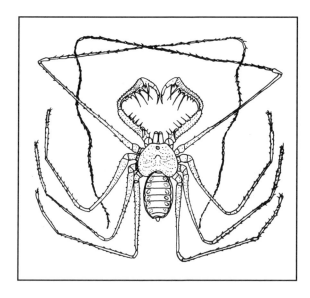

Figure 12.14. *Stygophrynus dammermani*, an endemic whip scorpion from caves around Bogor.
After Roewer 1928

et al. 1993). A possible troglobitic crab *Sesarmoides emdi* was found in a cave on Nusa Penida in 1993 (Ng and Whitten 1995).

Troglophilic animals include the endemic whip scorpion *Stygophrynus dammermani*, known from a few caves in West Java (Roewer 1928), and although some of its original sites have been destroyed, it is probably still present. The body is only 11 mm wide, but the greatest reach of the first pair of legs is about 33 cm (fig. 12.14). Other abundant troglophiles are crickets *Rhaphidophora dammermani* and *R. dehaan* (e.g., Boeadi 1980; Nugroho 1986), and cockroaches. In suitable conditions, cockroaches can form living carpets on the cave floor; on fermenting dung in the Ngerong river cave in the Tuban area, members of HIKESPI counted 100 cockroaches per square metre (Ko 1986). Aquatic troglophiles include the common prawn *Macrobrachium pilimanus* and the crab *Parathelphusa convexa*, both of which are also found in surface waters (Holthuis 1984).

The major group of trogloxenes is the bats. The majority of bat species in Java roost in trees during the day, with only a few species roosting regularly in caves. Some of these cave-roosting species can be found in huge numbers, such as the wrinkle-lipped bat *Tadarida plicata*, which is found in the hundreds of thousands at Lalai Cave at Pelabuhan Ratu and in Gresik. Other abundant cave-roosting bats are *Hipposideros bicolor*, *Rhinolophus pusillus*, and the nectar- and pollen-eating cave fruit bat *Eonycteris spelaea*, whose most famous roost is at the temple site of Lawah Cave in Klungkung, Bali. Here, in the absence of any disruption to their lives from the worshippers or tourists, these bats are remarkably tame and can be approached to within less than a metre.

All cave dwellers are ultimately dependent for food on material brought into the cave from outside. Some animals feed on plant roots attached to the cave roof, wood and other material washed in during floods (if the cave has a river running through it), or the organic matter percolating through from the surface. The major providers of food, however, are the bats and swiftlets which roost and breed in the cave but feed outside. Bats and swiftlets supply food through their:

- faeces, collectively known as 'guano', which various animals known as coprophages feed on, and which are a source of nourishment to fungi and bacteria;
- parasites, both internal and external, which in turn provide food for predators;
- moulted hair and feathers and shed pieces of skin;
- progeny which may be susceptible to different predators and parasites; and
- corpses which form a source of food for various necrophages or corpse-feeding organisms.

Almost all animals provide food for others in these five ways, but within the cave ecosystem, in the absence of green plants, these are the only major sources of food. The bats themselves roost on the roof and walls, and form the primary basis of one community; their faeces and dead bodies fall to the cave floor and form the basis of another. Thus there is a distinct division of the animals into a roof community and a floor community (Bullock 1966).

Roof Community

The roof community includes bats and swiftlets as well as all those animals that feed on or parasitize them. Over half of the insectivorous bat species, and three or four of the fruit bat species, probably use caves as permanent or temporary roosts. Cave-roosting bat species differ in their preference for certain conditions. Some, such as the dawn fruit bat *Eonycteris spelaea* are found in chambers near to the cave mouth. Some have wings that allow them to manoeuvre in tight spaces such that they are found roosting in narrow crevices or 'chimneys' (Goodwin 1979). Others, such as the long-fingered bats *Miniopterus* tend to be found in the dark zone.

Bats are hosts to parasites, some internal, and many external, which bite their hosts to suck blood. Some, such as the spider-like wingless nycteribiid bat flies, live almost their entire lives on bats (Marshall 1971), while others, such as streblid bat flies, bed bugs (Cimicidae), and chigger mites (Trombiculidae), spend only part of their life cycle on bats (fig. 12.15). It has been suggested that the tropical bed bug *Cimex hemipterus* may have begun

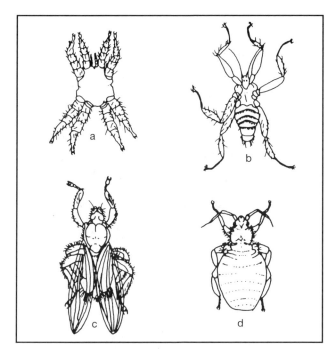

Figure 12.15. Common parasites of bats. a - mite; b - nycteribiid bat fly; c - streblid bat fly; d - bed bug.
After Fenton 1983

its association with humans when they used caves as shelters. Within a species of host, however, not all individuals are necessarily crawling with these parasites. A detailed study of one species of nycteribiid bat-fly on its major host showed that male bats were more frequently infested than females, but even the males were mostly not infested, and those that were mostly had only one individual parasite. The reason for this low parasite population is not fully understood, but it may be because of the relatively low density of bats in the abundant roosting sites (Marshall 1971). There is some evidence to suggest that animals on low-protein diets carry larger populations of ectoparasites (Nelson 1984), and this may be due to the host being less able to scratch the ectoparasites out of its fur.

It is frequently supposed that bats carry rabies. This is certainly true in Central and South America, but there is only a single report of rabies in a bat from Asia – a dog-faced fruit bat *Cynopterus brachyotis* in Thailand (Hill and Smith 1984).

Swiftlets are well known for the protein-rich saliva nests, made by certain species, which are gathered and sold for use in soups and other dishes

Figure 12.16. Swiftlets *Aerodramus fuciphagus* produce commercially valuable nests in caves.

After Coomans de Ruiter 1953

(p. 234; fig. 12.16). The first Chinese records of these nests date back to the Tang dynasty (618-907 A.D.), and Dutch colonial records make mention of the trade as early as 1625. During the 18th and 19th centuries 3-4 tons (representing 300,000-400,000 nests) were harvested annually from the sea caves of Karangbolong alone (Ko 1983b). The annual yield from all four caves at Karangbolong now is only about 100 kg, worth some Rp35 million. Clearly there is scope for improvement in the management prescriptions. The enterprise is currently managed by the local government, but the sanctity of the caves and the harvesting is perceived to be very strong among the local people, and proper conduct must be adhered to.

Figure 12.17. The black toad *Bufo asper* is a major predator on large arthropods on the cave floor; Gua Ratu, Nusa Kambangan.
By A.J. Whitten

Floor Community

The organic matter on the floor of most dry caves is composed largely of material formed from waste products and bodies of animals. This guano has a low level of carbon, a moderately high concentration of nitrogen, a very low carbon:nitrogen ratio, and an extremely high level of phosphorus. Not surprisingly, the guano is in great demand as fertilizer, and a large company on Madura exports it all over the archipelago.

On the cave floor the coprophages and necrophages predominate. It is often difficult to distinguish between them because, whereas a few animals are exclusively necrophagous, many of the coprophages will include dead bats or swiftlets in their diet. The majority of cave-dwelling bats are insect-eating, and the faeces they produce are hard and dry and readily exploited by coprophages such as woodlice, caterpillars of *Tinea* moths which carry a cocoon around with them, and flies and beetles, although the primary decomposers are bacteria. The faeces produced by the few species of fruit-eating bats that roost in caves, however, are soft and rich in carbohydrates and not generally utilized by coprophages. One of the most common predators on the cave floor is the large (<26 cm) toad *Bufo asper*, which is almost always seen singly (fig. 12.17).

The various relationships described above can be drawn as a food web (fig. 12.18). As the understanding of a cave ecosystem grows, so the web becomes more complex. The animals at the base of a food web are relatively abundant, while those at the top are relatively few in number, with a

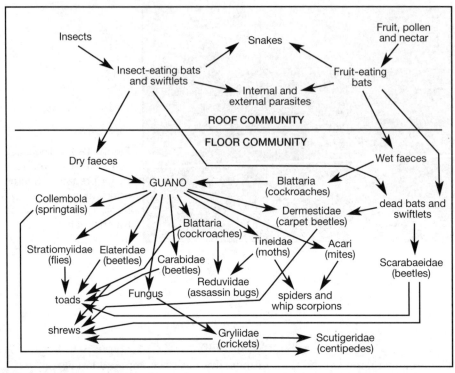

Figure 12.18. Simplified food web of a Javan cave.
Modified after Whitten et al. 1987b

progressive decrease between the two extremes. This is known as the 'pyramid of numbers' and is found in almost all ecosystems. Exceptions are systems including parasites and hyper-parasites (parasites of parasites), in which case the pyramid is inverted.

DIFFERENCES WITHIN AND BETWEEN CAVES

It is clear from the descriptions above that, although at first sight a cave appears to be a fairly uniform habitat, this is certainly not the case (Boeadi 1980; Notowinarto et al. 1986). One of the chief factors causing variation in the cave habitat is the distribution of bats, the main producers of guano. Physical factors also vary from place to place in the cave. For example, during a rainy period, standing water will accumulate in one part of a cave but not in another, and some parts are subject to air movement, while others are not.

Conditions between caves clearly differ far more than conditions within a cave, or even within a single area of limestone, and even when comparing caves of broadly similar conditions, the differences are striking.

EFFECTS OF DISTURBANCE

Visitors

Bats are sensitive to disturbance and, when caves are visited during daylight hours, strong lights and loud noises should be avoided in the darker chambers. Some of the bats that pollinate flowers of commercial fruit trees and other plants roost in caves, and if they abandon a disturbed cave and leave the area, this could cause considerable financial and ecological loss (Suyanto 1984). Those involved with developing caves as tourist sites should consider in detail the ecological impacts of the visitors.

While most visitors to caves respect their intrinsic interest and do not deliberately damage or deface their features, there are those who seem to have an irrepressible desire to paint or carve their names in conspicuous places. In some cases arms of government have joined with enthusiasts from HIKESPI to remove this visual pollution (Soeroto 1985), but the cleaning action can itself be damaging. In addition, wanton damage to and removal of ancient speleothems is not unknown. For example, remarkable and rare features such as caves 'pearls' were found in the first exploration of Luweng Jaran, but soon afterwards they were removed (R.K.T. Ko pers. comm.).

Mining

Caves are threatened by two forms of mining: for limestone and for the nutrient-rich guano derived from the huge quantities of swiftlet and bat droppings falling onto the cave floor (Hidayat et al. 1982). The limestone mining is obviously the most destructive since the very fabric of the caves is dynamited or otherwise broken up and removed. The vibrations caused by blasting necessary to break up the rock can cause shock waves that break stalagmites and stalactites, or cause thin cave roofs to collapse. Limestone is vital, of course, for construction projects and for the agricultural productivity of the marginal acid soils of the outer islands, but to avoid unnecessary negative impacts of quarrying, it is important that the blasting be conducted with care and informed understanding. The removal of guano is generally conducted under local licenses, and particularly rich caves have barriers of barbed wire in front of them to prevent guano theft–guano sells at about US\$15/ton. The digging for guano results in temporary disturbance but is not necessarily seriously destructive in the long-term. For example, some of the small caves seen in central Madura which had been fully exploited for guano by the major guano company on the island still have good populations of bats and swiftlets roosting within them.

Pongangan Cave near Gresik, close to Surabaya, has been well known for the huge numbers of wrinkle-lipped bats *Tadarida plicata* bats which swarm out every night between about 5 and 6 p.m. The limestone area around the cave had more or less been razed to the ground by 1987, with the rock being sold to the PT. Semen Gresik cement works. The company mining the limestone was about to start dynamiting Pongangan Cave when a newspaper reporter caught wind of it, and the issue was hotly debated in the open, not least because the tourist draw of the cave was falling as the number of bats fell. Finally, due to the intervention of former Minister Emil Salim, the mining company agreed not to proceed with its plans, because they were convinced by his arguments that the bats must be major agents of insect control. Unfortunately, the number of bats has continued to decline. This may be due to dynamiting damage in the cave system that has changed the microclimate within the parts of the cave occupied by the bats, to the effects of pesticides in the bats' food, to the destruction of vegetation around the cave mouth which allows sunlight to penetrate further into the cave, or to the increasingly hostile environment outside the cave. The whole area of Gresik is remarkably barren, with everything natural or semi-natural being flattened to make way for the new industrial estates. There is, of course, no *a priori* reason why industrial estates have to be devoid of vegetation, and working environments would be more pleasant if some thought were given to this.

Mining also has serious effects on the swiftlets and the harvesting of their valuable nests. On Cibodas Hill, for example, mining and unsustainable overexploitation of the nests have led to virtual desertion by the birds (HIKESPI 1983).

Chapter Thirteen

Rice and Maize

RICE

General

Rice is not just a food – it is a culture, a way of life, a precious grain believed by Hindus to have been specially created by Vishnu and cultivated after the instruction of the god Indra. Rice *Oryza sativa* is a crop of the tropics and subtropics, but within that area it is found from sea level to 2,500 m, in lakes and irrigated deserts, on sands and clays, and in a wide range of soils from pH 3-10, and from acutely deficient to clearly surplus mineral status. It is one of the very few crops (together with the aroids) that thrives in water-saturated soils, tolerant even of being submerged for part or all of its growth cycle. Major developments in the form of canal and dam construction have allowed rice to be grown beyond its 'natural' ecological limits. The wide variety of habitats, climates, cultures, and cultivation practices in southeast Asia has resulted in intense selection of rice varieties for thousands of years. In Indonesia alone it has been estimated there are more than 8,000 varieties (Bernsten et al. 1982), although there may be at least that many on Java alone (Fox 1991).

Rice fields nowadays dominate the landscape in many areas of Java and Bali, but it is likely that the wild relatives of domesticated rice originated in upper Burma, Thailand, and southwest China. The oldest evidence suggests that even by 7,000 years ago in the Yangzi delta in China, cultivated rice was an important food, but that, as a staple, it was superseded by tubers and millets. The growing of rice in marshy areas using buffalo to puddle the mud (and incidentally to fertilize the fields) was practiced before the cultivation of hill rice (Glover 1985), although this was not necessarily true for all people in all areas (Marschall 1988). This is supported by analyses of pollen from 4,000-5,000 years ago that may indicate large-scale and possibly permanent forest clearance or at least the type of shifting agriculture that leaves grassland maintained by regular burning (Flenley 1985). Unfortunately rice pollen cannot be distinguished from other grass pollens.

563

The carvings on the sides of the 9th century Borobudur temple in fact depict millet *Panicum miliaceum* or *Setaria italica* rather than rice, and this and Job's tears *Coix lachryma-jobi* (Gram.) were probably the original staples. No one knows for sure how or when rice actually arrived on Java and Bali. The cultivation of irrigated rice was possibly practiced before the period of Indian influence (van Setten van der Meer 1979), but some argue that it was introduced only a while before the establishment of the Majapahit Kingdom in the 13th century, from which it later spread to Malaya (Bray 1986). Whenever rice actually arrived, it stayed and people were reluctant to return to former staples once they had expended time and energy in building the rice terraces.

Rice is the staple food of half the world's human population and on average is the most productive of the major cereals. It is unique among cereals in being eaten after boiling alone, without being pounded into a flour or mush; it is also highly digestible, nutritious, and high-yielding. Over half of the rice grown in Java and Bali is in irrigated fields, a quarter in rainfed fields, and the rest in dry, upland fields. It is generally not realized that rice only tolerates rather than requires inundation, and the water serves to deter many species of weeds and to hold the soil particles together so that steep terraces do not collapse. Nitrogen-fixing blue-green algae growing naturally in the water enable repeated rice yields of 2 ton/ha to be obtained even without any fertilizers, and this has probably been the case in Bali and some parts of Java for nearly 800 years (Lansing 1987). In contrast to shifting agriculture or swiddening, which takes its toll on the soil and requires a 15-20 year fallow period for the soil to recover, irrigated rice can produce acceptable yields season after season for centuries. Chemical fertilizers are necessary for the high-yielding rice varieties, but for older varieties the water and soil are fertilized sufficiently through the agencies of nitrifying blue-green algae, rice straw, faeces of buffaloes working the fields, ducks foraging for food, and fish cultured in the field.

Cultivation Techniques

The hoeing, ploughing, and harrowing of the irrigated rice fields or sawah serves to change the inundated soil into mud. This 'puddling' facilitates the transplanting of the young rice tillers, improves the soil condition, reduces percolation of water through the soil, which increases the efficiency of water channelled onto or otherwise arriving in the field, and helps control weeds by cutting and burying them. This puddling uses one-third of the total water required by a rice crop, so farmers of rainfed fields have to wait until the rain has fallen before land preparation can begin. All rice farmers try to prevent their land becoming too dry since this makes land preparation much harder and hampers root development (Wirjahardja 1987).

New varieties of dry field rice currently in use mature after only 120 days and yield up to 2-3 ton/ha of milled rice in a season; local varieties yield 0.8-1 ton/ha. This does not compare well with the 4-6 ton/ha expected from irrigated fields, and for this reason the development of upland rice is not a government priority. In some areas the rainfall may be adequate for rainfed rice one year but not the next, and one cannot predict with certainty how much rain will fall. In Central Java a system called *gogo rancah* has developed to allow the farmers to hedge their bets. In this, seeds of inundation-tolerant varieties are broadcast or dibbled into terraced or bunded dry fields when the first rains fall. Water outlets are closed, and as much water as possible is let into the fields. If little further rain falls, then the rice plants will produce an acceptable crop. If the farmer feels that the rains are likely to continue, however, the seedlings are pulled out and distributed more evenly, or are even all pulled up while the field is tilled again prior to replanting in an orderly manner, after which an even better crop would be expected (Wirjahardja 1987).

A variation of gogo rancah is practiced near large rivers in some northern coastal areas of West Java and southern coastal areas of Central and East Java where water comes from floods rather than the rain. Tall rice varieties are grown which are tolerant of inundation and which, when broadcast rather than transplanted, mature after only 80-90 days. Improved varieties can yield as much as 7 tons/ha/year on these lands, and even local varieties can yield 3 ton/ha/year. These lands experience serious weed and tilling problems when the ground becomes dry, and it is difficult for the farmers to predict when the rivers will flood. All rainfed and dryland rice farmers practice different forms of multiple cropping using maize, cassava, and legumes in order to maximize the productivity of the land (Wirjahardja 1987).

The cultivation of irrigated rice has, for hundreds of years, been employing soil conservation techniques if only because, when contour planting on terraces is not used, the water runs away. The terraced lands are called *nyabuk gunung* in Javanese or *ngais gunung* in Sundanese. The first means the fields are like belts around the mountain, and *ngais* means carrying a child in a *selendang* (long batik cloth). Thus both terms have a sense of care and protection, and it is clear that if the protection is removed, soil will be lost, because of heavy rain and the steep slopes (Soemarwoto 1984a).

Upland rice may not be a major crop in terms of the area devoted to its production or cash returns, but it is a key crop in some upland areas because it is a desired commodity, is well adapted to growth during the rainy season, provides straw for animal forage and bedding, and can be used as mulch to improve the soil (McIntosh 1986). The yields are rather low, but then very little effort has been given to developing improved varieties, only one crop is produced each year, and generally only small quantities of pesticides or fertilizers are used (Soemarwoto 1987b). It is unlikely

to gain much support for development, however, because it does not provide good ground cover, unless incorporated into a multi- or sequential-
cropping system.

Rice Production and Consumption

Java has long been Indonesia's rice bowl. In 1968 it produced 65% of the
total rice harvest, and in 1986, after many developments in the outer
islands, it produced over 63%, and all this from less than 50% of the land
under rice. Given that large areas of new sawah (much of it of marginal
quality) have been established outside Java and significant areas of Java's
have been lost to other uses (p. 41), it is clear that Java has held its position
only by virtue of increasing efficiency of sawah use and higher yields (Fox
1991).

Java and Bali together have about 60% of the total irrigated ricefields in
Indonesia, and some 40% of this has technical irrigation and 15% has
semi-technical irrigation. About 20% of the irrigated sawah on Java is supplied with water from its major reservoirs (Bali has only one small reservoir
irrigating just 1,300 ha), which have been built and provided with control
equipment at considerable cost. Arguments have been made that the same
money could have been used for greater local benefit on different types of
schemes, but it is undeniable that the reservoirs have improved the cropping intensity and yields in the areas they irrigate. The benefits may be relatively short lived, however, or at least shorter than originally envisaged,
because the removal of forest and other soil-exposing practices in the
upper catchments have resulted in higher-than-expected rates of sedimentation in the reservoirs, thereby shortening their economic lives (RePP-
ProT 1989).

Rice is a status food, and to eat any other staple is regarded by some as
an indication of backwardness. Since this is the prevailing view of the dominant cultural group, so rice consumption increases while consumption of
other staple crops decreases, and the dependency on this single crop
becomes even more marked (p. 44). It is important that full support be
given to the ecologically appropriate production of other staple crops
such as maize on Madura and other parts of East Java, as well as the drier
parts of Bali (p. 584).

Reducing the emphasis on rice is as much a public relations and adver-
tising problem as any other, and *per capita* consumption has already
decreased in Thailand, Vietnam, and Japan. Diversifying the average diet in
Java and Bali would have the additional advantage of reducing the periods
over which fields were flooded, which in turn would reduce the production
of methane, one of the important greenhouse gases (Soemarwoto 1991a, b,
1992). These issues are explored further elsewhere (p. 45).

Subak

The farmers of Bali organize their sawah irrigation through subaks, which are autonomous and socio-religious cooperative user groups based on microwatersheds rather than village divisions, and organized through a system of water temples linked ultimately to the temple of Ulun Danau Batur. Such organization dates back to at least the 9th century (Suadnya 1990). The state of the whole ricefield ecosystem is controlled by the local subak organization, chaired by the *klian subak,* himself beneath broader watershed authority, which institutes fallow periods or determines planting times. The Balinese subak are unique, having no exact parallel in Java now or in ancient times (Wisseman 1981), although the Dharma Tirta in East Java and the Mitra Cai in West Java have similarities (I. N. Oka pers. comm.). The subak are remarkable not least for their irrigation tunnels which have been dug through hillsides for up to a kilometre (Lansing 1987).

Detailed analysis of the subaks has demonstrated that they encompass extremely sound resource management, resulting in one of the most stable and efficient farming systems in the world. The government started its involvement in 'improving' the subak system in 1925 when a number of weirs were built in the Ayung watershed, and since then subaks have been of two types: systems operated jointly between government and communities, and systems wholly under the control of the community. The basic difference is that in the first system the government has taken responsibility for the primary and secondary delivery of water. While this relief from burdensome responsibility at first glance seems a boon, it generates apathy towards, and lack of involvement in, the system as a whole (Sutawan et al. 1984). In 1979 the Asian Development Bank launched the Bali Irrigation Project (BIP) to improve the island's irrigation systems by widespread intervention in about 10% of the subaks. It was accepted that as a result of the $40 million of weirs and other infrastructure, and of greater centralized control over irrigation and planting regimes, the subak would lose some of their autonomy and traditions. At the time that BIP was being executed, Green Revolution rice varieties were being adopted (with great reluctance in some areas), continuous cropping was being encouraged, the traditional rice calendar was being abandoned (p. 686), coordination of cropping cycles eroded, and there was an upsurge of water shortages and pest problems. There were calls from the farmers to reinstate the authority of the subak, but the Bank's consultants were deaf to the arguments and recommended more conventional responses.

What finally convinced the decision makers of a fundamental flaw in the project was a computer simulation model made in 1988 that showed the vital and dynamic role of the water temples. Unfortunately, by then, it was too late to stop the spending of the Bank's loan funds, to undo the damage, and to repay the farmers' costs resulting from the flat-footed and

tunnel-visioned bureaucratic process, but the official Project Performance
Audit did at least admit that the project's high technology and bureaucratic
'solutions' had been the major factors behind the decreasing yields of
1982-85, that the costs had been high, and that the irrigated rice terraces of
Bali formed a complex artificial ecosystem which had been recognized
locally for centuries. As a result the temples have regained informal control
of cropping patterns over most of Bali (Lansing 1991).

Rice-Field Weeds

Rice farmers expend, on average, about one-fifth of their labour on
weeding (Wirjahardja 1987), even though many weeds are deterred by
the waterlogged soils. They do this because over 10% of the potential production of rice can be lost by competition from weeds. The weeds that grow
up with the rice are influenced to some extent by the management of the
crop, and the spacing of the rice is of major importance since closer
spacing will suppress weed growth, although if planted too closely, the
rice will limit its own growth. Not all of the 266 plant species regarded as
rice-field weeds (Kostermans et al. 1987) necessarily compete seriously
with the rice because some occupy distinctly different niches, and some are
encouraged and viewed as intercrops because they find local uses as food,
for example *Ipomoea aquatica* (Conv.), *Cyanotis cristata* (Comm.), *Limnocharis
flava* (Limn.), and *Marsilea crenata* (Mars.) (fig. 13.1), and forage such as
Panicum repens (Gram.). In addition, some of the fishes cultured in rice
fields (p. 637), such as tilapia *Oreochromis nilotica* and barbs *Barbodes
gonionotus,* feed on algae and other weeds and convert these into fertilizer
that is utilized by the rice plants (Grist 1978; Soerjani 1987). It is thus
simplistic, and for many reasons undesirable, to control weeds by blanket
application of herbicides. Far better to have a good understanding of each
plant's ecology and physiology in order to develop more subtle means of
control (p. 575).

The preparation of land for planting rice deters some of the potential
weeds. A successful weed has to survive periods of unfavourable conditions (wet or dry), to reproduce sexually rapidly or be capable of vegetative
reproduction from persistent parts, and to generate a potentially large
population to compensate for the high death rate caused by peoples'
attempts at control. The germination of many weed species is inhibited by
inundation (Pons et al. 1987), and the weed communities in upland and
lowland rice fields are consequently different; those found in the lowlands
are tolerant of inundation. Those that find germination difficult underwater also do not grow well underwater. Most of the species which reproduce vegetatively also do not fare well under the management regimes of
lowland rice culture, because they are killed by burial or submergence.
Notable exceptions are the clover-like fern *Marsilea crenata*, which grows

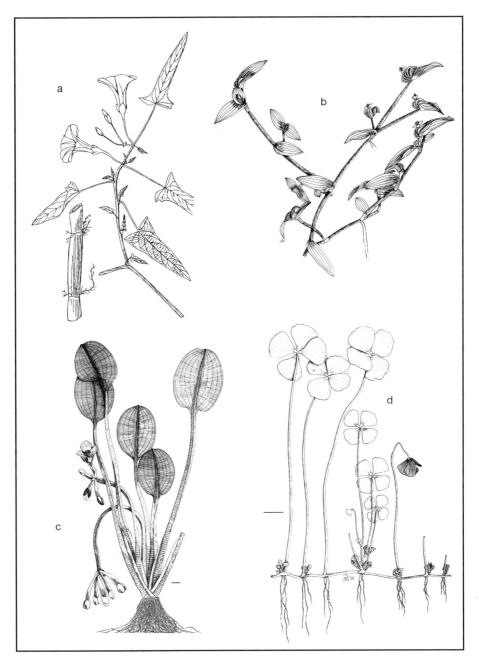

Figure 13.1. Useful weeds of rice fields: a - *Ipomoea aquatica,* b - *Cyanotis cristata,* c - *Limnocharis flava,* and d - *Marsilea crenata.* Scale bars indicate 1 cm.

After Kostermans et al. 1987

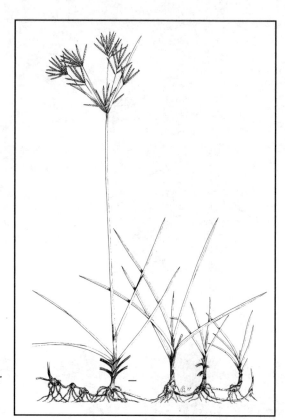

Figure 13.2. *Cyperus rotundus.*
Scale bar indicates 1 cm.
After Kostermans et al. 1987

very fast along creeping runners and is a serious competitor for nutrients in the first half of the rice plant's growth period (Kostermans et al. 1987). Weeds that reproduce vegetatively are in fact more important in upland rice, and the best-known examples are *Imperata cylindrica* (Gram.) and *Cyperus rotundus* (Cype.) (fig. 13.2), which reproduce rapidly when the rhizomes or tubers, respectively, are fragmented. The management of the rice field does, however, favour many weeds, for example, those which are stimulated by the application of nitrogen fertilizer, such as *M. crenata, Fimbristylis littoralis/miliacea* (Cype.), *Echinochloa crus-galli* (Cype.), *Ageratum conozoides* (Comp.), and *Cyperus rotundus*.

As might be expected, the earlier weeds can become established, the more successful they are. This is another major reason why transplanted rice suffers so much less from weeds than upland rice; the rice is well established by the time weeds have started to grow. Weeds that can establish within a rice field do not start to compete seriously for the first three to six weeks, depending on the species. Tall weeds such as *Echinochloa crus-galli*

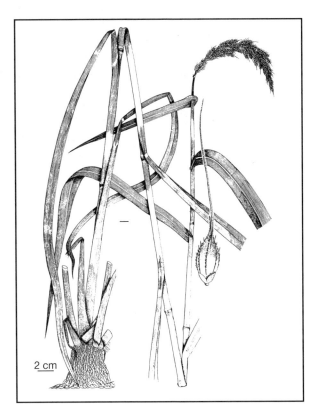

2 cm

Figure 13.3.
Echinochloa crus-galli.
Scale bar indicates 1 cm.

After Kostermans et al. 1987

(fig. 13.3) become competitive after three weeks when they start to shade the rice and reduce yields accordingly. Indeed, it is at the end of the three-week period that the most effectual weeding of the rice is conducted, unless *Marsilea* is present in large numbers, in which case a weeding after two weeks is necessary. Indeed, weeding is determined by a range of factors such as pest problems that may be aggravated by the presence of weeds. *E. crus-galli* is also remarkable in its growth rate; its seed weighs only one-twentieth of a grain of rice, but both produce the same dry weight of vegetable matter five to six weeks after germination. It is also stimulated more than rice by the application of nitrogen fertilizer. The problems can be reduced if fertilizer application is split or delayed, neither of which necessarily adversely affects crop yield (Kostermans et al. 1987). In the light of the above, it is not surprising that the most competitive rice cultivars have a large leaf area, a high tillering capacity, and are tall. Given that some of the more recently introduced high-yielding varieties are semi-dwarfs, it is clear that considerable effort has to be given to weed control if they are to fulfil their potential.

Green Revolution

Some decades ago widespread famine was seen as the inevitable fate of the increasing population of Java and Bali. The famines have not yet come, Java and Bali still produce a rice surplus, and the real price of rice has halved. The first high-yielding variety (HYV) of rice from the International Rice Research Institute (IRRI) was distributed in 1966, and to this can be traced the turnaround in fortunes that was dubbed the Green Revolution. The first variety distributed widely was a cross between an Indonesian variety and a dwarf Chinese species, and it had shorter and stronger stems that were able to support the heavier seed heads produced with applications of nitrogen fertilizer and 'protected' by pesticides.

The Green Revolution brought few if any benefits to upland populations, largely because its fundamental features of improved seed and larger quantities of fertilizers cannot be applied to upland areas because of the greater range of topographies, soil types, and a lack of dependable water supplies. Instead it has been suggested that adaptations of existing cropping systems which minimize farmers' risks should be advocated to generate, not a green revolution, but a green evolution that would likely be ecologically sustainable (Morooka and Mayrowani 1990).

The overall success of the rice intensification programmes, known as BIMAS (or 'mass guidance'), during the 1970s can be counted as one of the most impressive achievements of the post-1965 government, although the first programmes began in 1963 (Booth 1988). It involved major investments in new or rehabilitated irrigation systems, fertilizer plants, transport and storage networks, and the establishment of effective research facilities, extension services, and administrative bureaucracies, all with trained staff, as well as a rural banking network, and local cooperatives. The management of sawah has changed considerably over this period, notably in seed breeding and use, pest management, fertilizer use, and cultivation practices (Fox 1991). The crowning achievement has been a remarkable increase in rice production, increasing from about 11 million to 30 million tons (Fox 1991). As a result, Indonesia has moved from being the world's largest rice purchaser to the status (albeit precarious) of being self-sufficient (p. 41). Even so, the programme has not been without its problems.

For example, after the HYV were introduced, it was recognized that valuable local varieties were being lost and so the International Rice Research Institute in Los Baños, Philippines, set up a gene bank. Even so, it is not known how many varieties were lost before this initiative began (Soemarwoto 1983, 1985d, 1987b). Rice is a self-pollinating plant with, at the most, 2% outcrossing, but even this limited hybridization forms new 'varieties' around the edge of a field. This has long been recognized by farmers, with the result that seeds for the next harvest have traditionally been collected from the centre of a field. The government's encouragement for the growing of the relatively few high-yielding varieties (by 1975

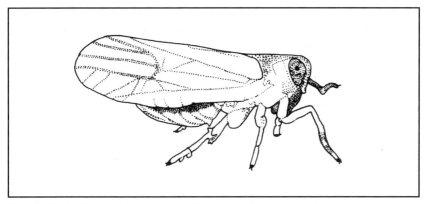

Figure 13.4. Brown planthopper *Nilaparvata lugens.*
After van Vreden and Ahmadzabidi 1986

half of the sawah were planted to just four varieties) led to increasing genetic uniformity which, predictably, opened the crop to disease and insect pests. This vulnerability was aggravated by increasing uses of pesticides and fertilizers, closer spacing of plants, and double or triple cropping each year without pause. The first signs of trouble appeared in 1974 with attacks by the brown planthopper *Nilaparvata lugens* (fig. 13.4), which damages the rice directly by sucking sap from the phloem of the leaf sheaths, and by introducing the 'grassy stunt' and 'ragged stunt' viruses. The viruses are a serious problem only when the pest populations are high, but the planthopper numbers proved difficult to control because planthoppers have:

- only about four weeks between generations;
- females which lay 100-300 eggs in the two-week laying period;
- males which can mate on the day after emergence and live for nearly a month;
- eggs which are lodged deep in the leaf sheath out of the reach of pesticides;
- tolerance to great crowding (up to 6,000 hoppers per rice hill);
- tremendous mobility, being able to fly for 10 hours;
- a complex life cycle that produces winged adults only every other generation; and
- voracious appetites which can turn a healthy, green plant to a withered, brown shadow in just two days. This condition is known as 'hopperburn' (Kenmore 1980; Gallagher 1984). There were traditional methods of dealing with these types of insects. On Bali, for

example, black and fire ants were treated with respect because it was understood that when the fields were drained, they would eat pest eggs (Foley 1984).

Four new IRRI varieties with the BPH-1 gene for resistance to the planthopper were introduced in 1975, just one year after the first outbreak, but these became susceptible after just 4-5 cropping seasons due to the development of a new form of the planthopper known as biotype-2 (Oka 1980, 1986). This happened in part because the 'new' varieties were quite similar genetically to their predecessors, and because of the enormous reproductive potential of the female planthopper. In the first four years of infestation, an estimated three million tons of rice (worth some US$500 million) was lost as a result of the planthopper. Then in 1977, four further rice varieties with the BPH-2 gene for resistance to biotype-2 became available. Two varieties were adopted widely, but resistance broke down in one of them in 1982 due to the evolution of another biotype (Oka 1983; Oka and Bahagiawati 1983). The remaining variety, IR36, remains resistant, probably due to the presence of a minor gene which helps to stabilize the function of the BPH-2 gene (I.N. Oka pers. comm.).

Meanwhile Indonesia's own rice breeding programme had produced some excellent varieties such as Cisadane, with good taste, even higher yields than IR36, and which fared well in wetter conditions (Soewito and Harahap 1984). These served to increase the number of varieties planted, and helped Indonesia to achieve rice self-sufficiency and surplus by 1984, although this was founded upon a genetic base even more uniform than that of the 1970s (Fox 1991). In 1985 the planthoppers were again causing serious damage to the rice crops, having evolved the ability to overcome the genetic resistance of almost all the varieties, and these increases in damage were generally preceded by increases in pesticide use. A new IRRI variety, IR64, appeared to be resistant to the insects and was very rapidly adopted throughout, causing Indonesia's rice production to be pivoted on a yet more narrow base, supported by still heavier use of government-promoted, highly subsidized pesticides.

It was at this time that ecologists' views began to be heard (Oka 1979, 1983). They had been intrigued that an insect which had been of such minor importance that it had not even been mentioned in the comprehensive *Agricultural Zoology of the Indonesian Archipelago* (Dammerman 1929c) should so quickly become of such significance that the Cabinet itself was holding meetings to discuss its control. It was explained that, under 'natural' conditions, over 100 predators, parasites, and diseases kept the planthopper numbers under control, neutralizing its tremendous reproductive potential, and that the broad-spectrum insecticides that had been used against the pests were even more damaging against the predatory spiders, beetles, dragonflies, and other insects. It was also explained that sublethal doses of insecticides caused the development of resistance and resurgence of the planthoppers, and that this was probably due to the stimulation of egg production, and to the destruction of natural enemies (I.N. Oka pers. comm.).

Integrated Pest Management

In late 1986 President Soeharto took a bold step and instituted a number of landmark ecological measures, the most radical of which was the ban on the use of 57 varieties of organophosphate chemicals on rice. These had been implicated in the resurgence or explosions of pest numbers due to the unintentioned demise of their predators. Never before had any country adopted ecological solutions to the problems of their major crop in such a sweeping manner. Even *Greenpeace*, not an organization known for its praise of Indonesia, has waxed lyrical about this integrated pest management (IPM) strategy (Treakle and Sacko 1989). The Presidential Instruction placed emphasis on promoting the use of an insect hormone analogue which prevents the larvae from developing into adults (p. 657), while the only conventional pesticides permitted were certain systemic carbamates, and these only when severe pest outbreaks occurred. These were supposed to affect only the pest insects attacking the plants, but it is now known that other insects, including parasitoids, drink from the plants, and that the chemicals drip from the leaf tips into the water surface where many important insect predators live.

But IPM is a broader concept than just making do with less chemicals (Flint and van den Bosch 1990). It ultimately has much to do with empowerment of farmers. Indonesian IPM farmers are taught to become responsible managers of their own and their community's fields, solving problems, researching, training others, promoting, and organizing. While it is vital that farmers learn to think and act with consideration for the complexities of ecological systems, understanding that some insects and other animals in the rice fields are decidedly beneficial and will reduce the numbers of the pest organisms, it is equally important that human and social processes and local organizational capacity for development are given full rein. There is still a need for appropriate extension services and for the interpretation of scientific advances, but the farmers should be confident enough to be able to adapt, tune, or reject recommendations depending on their own agroecological realities.

The presidential decree was effectively the start of Indonesia's world-leading position in IPM, replacing regular calendar spraying with a variety of biological and cultivation controls, and spraying only when defined levels of infestation were exceeded. In addition, within two years of the decree, Indonesia had removed all the pesticide subsidies. The benefits of the decree have been highly visible, and this demonstrates how development which works with the natural ecosystem rather than against it can achieve dramatic results:

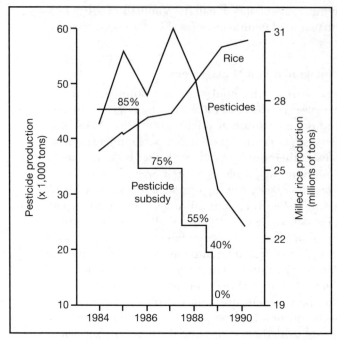

Figure 13.5.
Changes in rice yields,
pesticide production,
and pesticide subsidy.

With permission of Program PHT

- yields have continued to increase and have become more stable (fig. 13.5);
- farmers are saving money previously spent on pesticides;
- the government is saving $150 million annually on subsidies;
- the government has made direct savings of $1 billion;
- no serious outbreaks of brown planthopper have occurred;
- a limited outbreak of white stemborer in West Java, probably due to an overuse of carbamate insecticides, was dealt with by rapidly training 75,000 farmers (during three weeks of the fasting month) to deal with the problem using IPM methods;
- water quality has almost certainly improved;
- 500,000 farmers have been trained for one season, 7,000 farmers working as IPM trainers and 2,000 extension agents trained intensively for 15 months; and
- trained farmers pass on their knowledge and experience (Oka 1988, 1990, 1992, 1993.

It is no wonder, then, that governments throughout Asia have been watching the Indonesian programme closely, and are now embarking on their own. Clearly the otherwise successful implementation of the Green Revolution was marred by a serious error of judgement with regard to pesticides. Unlike inorganic fertilizers, for which a direct correlation with yields exists, there is still no evidence, after some 30 years, that the use of pesticides in rice fields increases yields. Rather, the evidence supports the thesis that insecticides have had a major destabilizing effect on overall yields, causing losses of millions of tons of rice, wasting billions of dollars in hard currency, and degrading the health and well-being of farmers and their environment. All this for the lack of understanding of how a rice field functions. The importance of ecology has perhaps never been so clear.

The high-yielding varieties of rice only give of their best when relatively large quantities of fertilizer are applied. Indeed, the use of urea has more than doubled since 1972 (Fox 1991), partly because of following recommendations, and partly because it has been so cheap – the farmers paid only 8-15% of the actual production costs in 1989 (Fox 1991). Also relatively unknown are the effects of the large quantities of triple superphosphates and potassium used, since the appropriate rates depend on their existing availability within the soil, and this itself depends on the soil type and the previous applications of fertilizers. Unfortunately the application rates used do not take these important factors into account with the probable result that more of these fertilizers were being introduced into the environment than was necessary. The recommendations are constantly being revised, however, and the much-expanded fertilizer production industry looks forward to satisfying export markets when domestic demand is reduced.

The problems of pesticide and fertilizer misuse will not be solved, however, until even more farmers have been trained and become confident to manage their fields, and until the banned pesticides stop being used on secondary crops on the same land where rice is grown, thereby eliminating useful predators and parasites. The continued availability of these pesticides also means that clandestine spraying on rice is very easy. It is clearly urgent to extend IPM to other crops.

Rice-Field Ecology

The ecology of rice fields in Indonesia is remarkably understudied as noted ten years ago (Anwar et al. 1984; Whitten et al. 1987a, b, c). In 1990 a bibliographic search was made for articles on rice yields and pesticide use published in the previous 30 years. Out of 1,356 references found, only seven mentioned the third trophic level comprising the natural enemies of the pests (W. Settle pers. comm.). This is a reflection of the 'pesticide company' view of rice-field ecology, a simplistic system involving just the rice and its pests. As is abundantly clear to anyone who cares to sit by a rice field for a short while, there is much more to the system than this. Tropical rice ecosystems are one of the most complex and stable agricultural ecosystems found anywhere in the world. The water surface is disturbed by

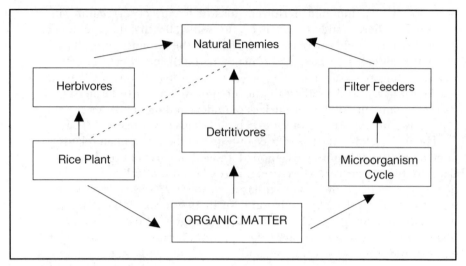

Figure 13.6. Three energy pathways supporting natural enemies of rice pests.
After Settle in prep.

the movements of mosquito larvae, the dragonflies hawk just above the rice and particularly around the bunds, and other creeping, crawling animals can be seen among the rice stems themselves. When it is considered that rice is the world's most important food crop, and that it grows in probably the most complex ecosystem of any crop, it is remarkable that only now is the plant being studied with an ecological perspective.

The conventional wisdom is that the pests' natural enemies (predators and parasitoids) attack either herbivores or other natural enemies. In this simplistic system, changes in the size of the enemies' populations would necessarily follow behind those of the pests. There are in fact two additional pathways of energy flow, both of which are independent of the herbivores (fig. 13.6). In the first, predators and parasitoids feed on organisms such as the larvae of beetles, flies, and springtails, and bacteria, which decompose organic material or detritus in the soil, most of which is derived from weeds, rice straw, and algae. In the second, the bacteria are fed upon by zooplankton which also feed on phytoplankton. The large zooplankton are fed upon by filter-feeding organisms such as chironomid midges and mosquitoes, which themselves become prey. In fields where insecticides are not used, the populations of filter feeders and detritivores build up early in the season, causing the populations of natural enemies to increase accordingly. These in turn act as a buffer against damaging increases in pest populations. The high early-season abundance of these 'neutral' populations provide an alternative food source for generalist predators that, in

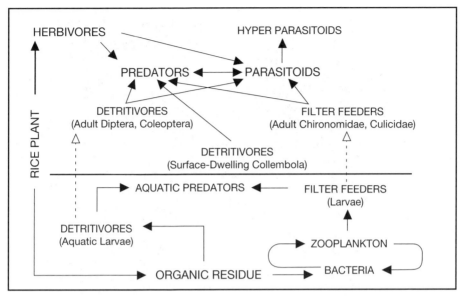

Figure 13.7. Simplified food web of a rice field.
After Settle in prep.

effect, uncouples predator populations from a dependence on pest population dynamics. This allows the predator's population to increase in advance of the pests. If pesticides are used early in the season, then the population of natural enemies is low, pest populations are released from natural controls, and the phenomenon of 'pesticide resurgence' occurs (Settle et al. in press). Linkages between these systems can be extended to form a simplified rice-field food web (fig. 13.7).

As a result of the Indonesian IPM programme, over 500,000 farmers have learned that spiders are among the most voracious predators on rice hoppers and other small rice pests, and spider statues can be found at the roadside in certain areas of East Java. Of the various species, the wolf spiders *Pardosa pseudoannulata* are the best performers, and are easily recognizable, having a fork-shaped mark on their back. They are highly mobile animals and colonize rice fields early in the development of the crop and can control the prey before damaging levels are reached. The female lives for three to four months and can lay 200-400 eggs during that time. After the spiderlings hatch, as many as 80 of them can be seen riding on her back. Wolf spiders do not spin webs but hunt their prey directly among the base of the plants, jumping over the water when disturbed. The spiderlings eat planthoppers and leafhopper nymphs, and the adults also eat stem

Figure 13.8. Wolf spider devouring a brown planthopper.
With permission of IRRI

borer moths. Each individual spider will eat 5-15 prey each day (fig. 13.8) (Shepard et al. 1987, 1991). IPM farmers know these statistics, not because they have been told them by an extension agent, but because they have all made these observations by housing planthoppers in cages – with and without spiders. The foundation of the Indonesian IPM programme is that farmers learn every important concept by doing experiments and making observations themselves (figs. 13.9-10).

Wolf spiders are just one of a wide range of spiders and insects, including beetles, damselflies, earwigs, pond skaters, ripple bugs, ants, and wasps, that prey directly on the pests of rice (Shepard et al. 1987, 1991). Of the approximately 750 species of arthropods collected in rice fields by IPM researchers, some 65% are predators and parasites, 20% are detritivores and filter-feeding species, and 15% are herbivores (Settle et al. in press). In addition to all these there are also numerous pathogenic fungi and viruses.

Upland rice is attacked by a greater variety of insect pests than lowland

Figure 13.9. Farmers learning about integrated pest management by observing pests, predators, and plant health.

With permission of Program PHT

Figure 13.10. Farmers discussing their observations in the rice fields with trainers in integrated pest management.

With permission of Program PHT

rice, primarily because of the larger numbers of soil pests. Since rice is a lowland plant, the upland varieties do not suffer from specialist pests; instead, the pests are rather catholic, with long life cycles and effective means of dispersion (Litsinger et al. 1986). This, together with the relatively low-yielding ability of upland rice, gives no encouragement to farmer or government to invest in control measures. The control of pests should stress conservation of natural enemies. The main constraint to upland rice is, however, the fungus rice blast *Piricularia oryzae* (Oka 1986).

The largest predators feeding on insect rice pests are probably wrinkle-lipped bats *Tadarida plicata*, which are relatively common in Java, being found in very large numbers in old buildings and caves (pgs. 544, 562). A study of the bats' stomach contents in West Java found that each bat caught about one-third of its body weight (14-20 g) of insects per night. Considering that a single cave can be a roost for 200,000 bats, more than one ton of insects (30-40 million individuals) would be consumed each night. Most of these insects are moths which form the great bulk of harmful pests in crops and plantations. Thus this one species of bat (and Java has 50 insectivorous species) must be a significant force in pest control (Boeadi et al. 1983).

Certain birds probably play a role in controlling insect numbers, and the more complex (layered) the vegetation is around the crops, the more predatory birds can be supported. Also, even simple measures like placing bamboo poles around the fields to act as perches and look-out posts enable birds such as drongos, treeswifts, wood swallows, and falconets to hunt more efficiently and thereby to help the farmers (van Balen 1989c). Exactly how beneficial they are has yet to be assessed. The same applies to two of the most conspicuous birds in some rice field areas: the Javan pond heron

Table 13.1. The top five most important food items (in terms of number, volume, and frequency) in the diets of four frog species found in rice fields near Bogor, West Java.

		Rana erythraea	*Rana limnocharis*	*Hylarana chalconata*	*Ooeidozyga lima*
Acridiidae	Grasshoppers	1	-	-	3
Oligochaeta	Earthworms	-	-	1	-
Araneidae	Spiders	-	4	-	5
Gryllidae	Crickets	-	2	3	-
Isoptera	Termites	2	1	-	-
Pentatomidae	Bugs	-	-	-	4
Scarabaeidae	Beetles	5	5	-	-
Formicidae	Ants	3	3	2	1
Diptera	Flies	-	-	4	-
Tabanidae	Horseflies	-	-	-	2
Lepidoptera larvae	Butterfly and moth larvae	4	-	5	

Source: Atmowidjojo and Boeadi (1986)

Ardeola speciosa and cattle egret *Bubulcus ibis.* The herons forage in rice fields mainly just after they have been ploughed or planted, and eat mainly dragonfly and waterbeetle larvae, molecrickets, and spiders. The egrets are generally seen in wet rice fields while they are being ploughed, and on dry harvested rice fields, where they feed mainly on grasshoppers, crickets, and spiders. The egret is generally considered to be beneficial since their main prey damage crops and compete with cattle (Kalshoven 1981; Vermeulen and Spaans 1987). These rice pests include the short-horned grasshoppers *Oxya japonica* and *Stenocatantops splendens,* and the locust *Locusta migratoria.* The long-horned grasshoppers *Conocephalus* are also eaten, but this situation is complicated by the fact that some members of the genus are beneficial by the fact of their predeliction for the eggs of the pestilential stinkbug *Leptocorisa oratorius.* They also eat spiders but it is not known how this affects pest populations (Vermeulen and Spaans 1987). The mole cricket *Gryllotalpa* sp. is a pest because it lives among the rice roots, feeding on the lower stem and loosening the soil which causes the plants to wilt (Kalshoven 1981).

Rice field frogs (i.e., *Rana erythrina, R. limnocharis, Hylarana chalconata,* and *Ooeidozyga lima*) are also beneficial biological control agents (table 13.1). Some animals eaten by the frogs are predators themselves, but the vast majority are potential pests. Rice also suffers serious damage from rats and introduced golden snails, and these are discussed elsewhere (pgs. 301, 652).

The Future

Studying the present situation is instructive, but we must be concerned equally about the future, because to keep pace with population growth, the production of rice must increase. According to trends observed in other countries, this increase will have to exceed population growth because as socioeconomic conditions continue to improve, so the rice consumption *per capita* increases before decreasing as imported foods are adopted. Under the 1998 GATT agreement, Indonesia will import some rice, but the government's avowed intention is to maintain self-sufficiency. To achieve this, yields will have to increase and the variation in yields among years must be kept to a minimum. The evidence suggests that promoting the natural buffering afforded by an ecologically healthy and diverse rice field ecosystem offers the best guarantee of this. There are dreams of a 'super rice' with relatively few tillers but heavy panicles on even sturdier stems, more vigorous roots, genetic resistance to even more diseases, capable of being sown directly into the field thereby avoiding the transplanting stage, and producing 13-15 tons/ha (rather than the current maximum of 8-9 tons/ha). It can only be hoped that if a second Green Revolution occurs, then the lessons of the first, relating to the biological and social environments, will be heeded.

Specific breeding targets have been made possible because the rice genome has now been mapped with over 300 'landmarks' to help biotechnologists manipulate specific parts, and DNA from bacteria has been introduced successfully into rice cells which have replicated and been transmitted to offspring. In this respect development of rice is advancing faster than other crops, helped by the fact that it has a relatively small genome (only a tenth the size of maize). *In vitro* techniques now allow hybrids between cultivated varieties of *Oryza sativa* and other *Oryza* species to be developed, bringing in genes for drought resistance, pest resistance, and so forth.

Ultimately, however, it is the ecologically-minded farmers who know how to test, evaluate, and apply new technologies in their own situation who really matter, since no arm of government can provide precise guidance at the micro-level (Pingali et al. 1990). This will be good for rice productivity, biological diversity, ecosystem quality, stability, human dignity, and democracy.

MAIZE

General

Maize originated in South America, and it was introduced to Java in the 16th century by the Portuguese, who wanted to increase the agricultural economic base of Java and other islands. Its Indonesian name *jagung* is said to derive from the Indian word *jawa* or 'sorghum' and *agung* or 'great'. It fares poorly in aseasonal areas, but in the seasonal areas, particularly in eastern Java and eastern Bali, it produces good yields. Although the total area under maize over the last decade or more has been stable, yields have increased dramatically over this period because of inbreeding and hybridization programmes.

Maize is now the third most important cereal crop in the world after wheat and rice, most of it being consumed in the countries in which it is grown. It is used as a staple human food, as feed for livestock, and in industrial processes such as the production of oil and sugar. In areas where the climate and soils are suitable, maize is a popular crop because:
- it gives one of the highest rates of yield/hour worked;
- it provides nutrients in a compact form;
- it is easily transportable;
- the husks give protection against birds and rain;
- it is easy to harvest and to hull, and does not shatter;
- it stores well if properly dried;

- parts not consumed by humans are nutritious and can be fed to livestock;
- it can be harvested over a long period, first as immature cobs, and can be left standing in the field at maturity before harvesting; and
- cultivars with different maturing periods are available (Purseglove 1972).

There are a number of types of maize: pod corn, popcorn, dent maize, soft or flour maize, waxy maize, sweet corn, and flint maize, but only the last two types are grown in any quantity in Java and Bali. Sweet corn has been grown increasingly since the early 1980s in order to satisfy the demand from the wealthier urban dwellers, whereas flint corn is the staple crop of the drier regions. It is hardy, matures early, and is relatively resistant to insect attack. The tender, young pods can be eaten whole and these are particularly popular in West Java. The rather small grains of the mature pod make them suitable as a poultry food (Purseglove 1972). It has, on a dry-weight basis, about the same number of calories as wheat or rice, and its protein percentage is higher than that of rice but less than wheat. The grain is used as a livestock feed in richer nations, and this is increasingly the major use for the crop in Java and Bali, much of which is grown under contract to livestock feed companies. Maize stems are used as a forage crop, being served green or dried. The dry stems can also be used as livestock bedding, and the cobs for fuel. There is almost a stigma attached to its consumption as a human food, because rice is seen as the food of civilized populations. In villages the ground kernels are generally cooked with rice but this is hardly ever served in restaurants, even in maize areas of Madura and other parts of East Java.

Among the disadvantages of maize are that it can make a heavy drain on soil nutrients, and it has a high nitrogen requirement, such that nitrogen is often the limiting factor in maize production. This shows itself in reduced vigour, a pale colour, and yellowing or drying along the midribs in widening bands towards the leaf tips. Application of organic or inorganic fertilizers is virtually essential. In addition, since the crop leaves much of the ground uncovered, soil erosion and water losses can be severe unless appropriate action is taken.

Cropping Patterns

In areas where irrigated rice is grown, maize is grown as a subsidiary crop in association with legumes or vegetables following a wet-season rice crop (McIntosh et al. 1977), but it is more often grown under rain-fed upland conditions. Maize is grown in various cropping patterns: intercropping, multicropping, relay cropping, sequential planting, and mixed cropping (Aak 1993). These polycultures generally result in a lower incidence of pests (Risch et al. 1983). The two main patterns found are multicropped maize/upland rice/cassava, and maize/upland rice followed by maize alone. The former pattern is adapted to an extended single peak rainfall regime. Thus, quick-maturing maize is planted in the first part of the wet season and is interplanted shortly afterwards with upland rice; cassava is interplanted 4-6 weeks later. When rainfall is at its peak, all three crops are making heavy demands on soil water. Maize is harvested first, after about 100 days, rice second, and cassava continues to use the end-of-season rainfall and water stored in the soil until the peak of the dry season (McIntosh et al. 1977).

In one area at the eastern tip of Bali a cropping system can be found which uses maize with sweet potato. This is just one of three systems found in the area, each of which uses two staples; the others are sweet potato and dry rice, and sweet potato and cassava, each grown in a different ecological area. By cultivating two starch crops, the farmers are able to gain some insurance against crop failure (Poffenberger 1983), reflecting the general trend of subsistence farmers to insure the dependability and stability of farm productivity and to minimize risk rather than maximize profit (Scott 1976).

Pests and Diseases

There are numerous pests on maize, though only a few are significant. Pyralid and noctuid moth larvae known as stem borers such as *Ostrinia furnacalis* and *Sesamia inferens* are perhaps the most serious. The adult lays eggs on the leaves and the larvae then make extensive tunnels in the stem which can provide entry points for pathogens. Corn earworms *Helicoverpa armigera* damage the silk-like extensions of the female flowers and the maturing ears themselves. The other major pests are cutworms or noctuid moth larvae such as *Agrotis* spp., the evidence for which is neatly felled young plants. Another insect pest is the larva of the rice seedling fly *Atherigona exigua,* which can be troublesome when conditions are wet. It also attacks common grass weeds such as *Cynodon dactylon, Paspalum* spp., and *Panicum repens. P. repens* is one of the most ubiquitous and troublesome weeds, and it is found in rice, sugar cane, tea, irrigation systems and lake shores, and from sea level to 2,000 m (Siregar and Soemarwoto 1976). It is also a high-quality pasture grass, however, and the leaves and rhizomes are often cut for fodder. It is a native of India, and in Java it never sets seeds (Backer and Bakhuizen van den Brink 1963). Damage to maize plants can also be caused by pigs, rats, and squirrels.

Pest problems can be kept to a minimum by giving every possible opportunity to natural enemies, rotating crops, rapid removal and burning of affected parts, and clearing weeds (Aak 1993). It is interesting that maize is not normally sprayed against pests because the level of losses is acceptable. One wonders whether, had maize been regarded as an important crop, organized programmes of pest control using chemicals would have led to pest resistance, the death of predators, and pest buildups requiring greater applications of pesticides, as has been the case with rice.

Maize suffers occasionally from fungal blights, mildews and smuts, and viral stunts, but none is particularly serious, and all the new maize hybrids are resistant to downy mildew.

Chapter Fourteen

Plantation Forestry, Agroforestry, and Estates

Planted trees cover large areas of Java and Bali in the form of timber plantations, agroforestry, and estates of industrially important crops. These form the subjects of this chapter.

PLANTATION FORESTRY

There is no official commercial exploitation of natural forests on Java and Bali, and all official forestry activities are restricted to plantations. These are dominated by teak *Tectona grandis* (Verb.). There is very little biological distinction between plantation forestry (for timber, fuelwood, fibre, resin, and so forth from trees) and plantation agriculture (for latex, fruit, leaves, and so forth from trees and other crops), but they are considered separately because they come under the wings of different parts of the government administration. All the plantation forests, and the 500,000 ha of protection forests, on Java are managed by Perhutani, a state-owned enterprise. Plantation forestry in Indonesia is now a major government priority, partly in the hope of reducing pressures on the natural forests, partly to promote non-oil/gas exports, and partly to meet high and growing domestic demand for paper and other wood products. At present, most productive forest plantations in Indonesia are on Java. The percentage share of the national total is diminishing, however, as timber plantations are developed on the outer islands, albeit on marginal soils which have not proved themselves capable of supporting sustainable plantation forestry. The Ministry of Forestry has set the goal of 20,000,000 ha of timber plantations on Java by the year 2000, and this figure has more or less been reached already (table 14.1).

In ecological terms, plantations raise a number of issues, mainly because they are planted as monocultures, and these are addressed later (p. 598).

On Bali almost all the forests outside the conservation areas are managed by the provincial forestry service. Bali does, however, have nine 'traditionally managed' forests, the management of which is under the direction of village authorities. Six of them are just 5-20 ha in extent, and the largest, adjacent to the prosperous and much-visited village of Tenganan, near Candidasa, is 373 ha. The term 'forest' is perhaps a misnomer for all the sites because most of the trees do not derive from natural Bali forest. They are very interesting, however, as agroforestry areas for they are generally managed under very strict local laws (*awig-awig*), and the Tenganan village has received a Kalpataru environment award in recognition of its exceptional forest management. This village is not at all typical: for historical reasons this village owns much more land than most villages of its size, and it is relatively very wealthy because of tourist interest in its old village buildings and in its unique double ikat cloth. The 'forest' is managed sustainably, but any timber required by the villagers beyond what can be supplied from its own forest is simply bought from timber traders outside its territory.

Table 14.1. Areas of forest plantations on Java (ha). Potentially productive forests are areas covered with plantations of nondesignated species, such as mahogany in an area designated for teak. The term *conversion forest* was sometimes used for plantation forest which was to be cleared of one species and replanted with another (W. Saleh pers. comm.); outside Java it refers to natural forest destined to be converted to another land use.

	West Java	Central Java	East Java	Total
Plantations				
Productive plantations	177,521	331,201	376,455	885,177
Potentially productive	297,989	197,158	392,013	887,160
Limited production	47,558	45,192	67,570	160,320
Net plantation area	523,068	573,551	836,038	1,932,657
Infrastructure	7,358	5,635	3,865	16,858
Gross plantation area	530,426	579,186	839,903	1,949,515

Source: Perhutani (1990); W. Saleh (pers. comm.)

HISTORY OF FORESTRY MANAGEMENT

Java

As was explained earlier (p. 183), the extensive teak forests of Java almost certainly originated from trees brought from India by early Hindus. The teak trees were so numerous that they were cut without thought for sustainability. After the Dutch East India Company acquired large territories along the northern part of Java in the 18th century, it was agreed that all teak wood on Java belonged to the company, and it monopolized purchases and sales, although special arrangements were made for the Surakarta and Yogyakarta (Mataram) royal families. It was agreed that felled areas would be replanted, but this was generally ignored, although regeneration from stumps did occur. As a result, the quality of the stands and their area declined.

At the beginning of the 19th century, after Dutch East India Company control over Indonesian trade was assumed by the newly formed Netherlands government, some teak areas had been badly damaged due to over-exploitation. There were still extensive areas of teak forests, however, along the northern part of Java from Tegal to the east coast, and the Solo River near Surakarta and the Brantas River near Kediri both flowed through vast and beautiful teak forests. Dirk van Hogendorp, one-time Resident of Jepara, is quoted by Raffles (1817) as having written in about 1800:

> Batavians! Be amazed! Hear with wonder what I have to communicate. Our fleets are destroyed, our trade languishes, our navigation is going to ruin – we purchase with immense treasures the timber and other materials for ship-building from the northern powers, and yet on Java we leave warlike and mer-chantile 'squadrons' [i.e., teak stands] standing with their roots in the ground. Yes, the forests of Java have timber enough to build a respectable navy in a short time, besides as many merchant ships as we require ... In spite of all [the cutting] the forests of Java grow as fast as they are cut, and would be inexhaustible under good care and management.

Governor-General Daendels helped to reorganize forest management between 1808-11, but the British administration under Raffles returned to the old system.

While few people in the mid-19th century were concerned about the loss of primary forest or the conservation of species (pgs. 330, 723), the economic concern over the loss and degradation of teak forests grew, and a Forest Service was finally established with professional German foresters in Rembang in 1849, half a century after van Hogendorp's plea (Cordes 1881). A balance between felling and planting was sought, although the first management plan for an area of teak, based on the principle of sustainable yield, was not fully implemented until 1890 (Peluso 1992a), and management for the then entire 700,000 ha of teak on Java did not begin until 1934 (Boomgaard 1988).

In order to conserve the dwindling supplies of teak, a system of engaging farmers to assist with planting and tending in return for rights to cultivate crops around the seedlings for a few years was introduced from Burma in 1873. This was known as *taungya*, but became known as *tumpangsari* in Indonesia (p. 619). By 1881 it was found in all the plantation areas of Java, and was also used with mahogany which was introduced in 1880. Species of pines were first planted in 1924 near Bandung, and by 1935 it was clear that the Sumatran pine *P. merkusii* was the best, and this was grown on a large scale (Becking 1935).

Much of the resource base was lost during the periods of physical and social revolutions of the Japanese Occupation (Lundqvist 1955), the Independence struggle, and in the 17 years prior to the presidency of Soeharto (Peluso 1992a). The large trees being felled today were almost all planted in the Dutch colonial period, and although some of the teak trees now found in plantation areas are regarded as giants, there are probably few which could compete with the giants over 2 m in diameter which once graced the hills (Schuitemaker 1950). By the time Independence was won, the structure and philosophy of Indonesian forestry (including the ideal of sustainable forest management) had more or less been set in place, and by so doing, had criminalized nearly all traditional local uses of forestland and timber.

Bali

Forestry on Bali is of much less importance than on Java, and has a very short history. It was only after all organized opposition on Bali was finally subdued by the Dutch in 1908 that its forest areas began to be opened for dryland agriculture, rice fields, and plantations (p. 334), and plantations were relatively slow to get started. In the early 1960s some 5,000 ha of what is now Bali Barat National Park were cleared of forest by the Forest Service for the planting of teak, *Manilkara kauki* (Sapo.), and other valuable timber trees. There are just over 100,000 ha of forested land on Bali (18% of the island area). The agreed land use map of the island (TGHK) formalized a plan to have 176,700 ha of forestland (31% of the island area). Given that forest cover on Bali is expanding at only 150 ha/year, it is hard to see how the target total can be achieved.

M. kauki is the most important indigenous forestry species (p. 478). It is a short (less than 20 m) but stocky tree found in very seasonal lowlands such as Prapat Agung in west Bali and in Alas Purwo National Park (East Java), and has a distribution from Thailand to northern Australia (Hamzah 1977). It used to be found in many drier parts of Java and Bali (e.g.,

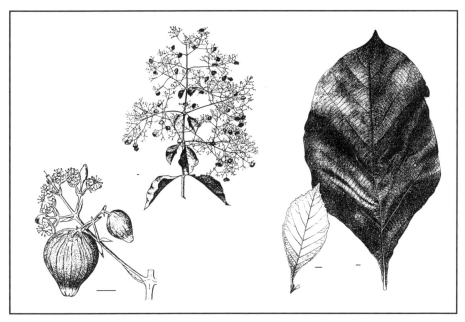

Figure 14.1. Teak *Tectona grandis.* Scale bars indicate 1 cm.
After Koorders and Valeton 1914

Schnepper 1934), but virtually all mature specimens have been felled, although seedlings are probably still present. *M. kauki* is a favourite wood of Balinese carvers, and the demand eventually resulted in attempts to establish plantations. Plantations started before independence were unsuccessful, and even the plantation established in 1956 at Sumber Kelapok near Prapat Agung was a failure. Trees planted nearly 20 years later have survived and can be seen close to the road at Sumber Kelapok (Kanwil. Kehutanan Bali 1991).

TEAK

The most important forestry tree on Java is teak *Tectona grandis* (Verb.) (fig. 14.1), and of the total 843,356 ha of productive plantations on Java, it accounts for 572,900 ha. Teak trees can grow up to 50 m tall with a diameter at breast height of 125 cm or more. The timber is highly durable, quite hard, yet easily worked. Teak grows best where there are 4-7 dry months, and between sea level and 700 m altitude. The tree produces large panicles of whitish flowers which open a short while after sunrise, and have an optimum pollination period around midday; each flower lasts only one day. Pollination is effected by insects, the most important being bees. Seeds develop to full size 50 days or so after pollination, but are ripe only after 120-200 days. The natural fruit dispersal agent is believed to be strong pre-monsoon winds. The ground litter of large leaves and branches decays slowly and suppresses the growth of other plants, but it does form an inflammable tinder. When occasional fires do sweep through, the teak trees remain undamaged, but other trees, if present, succumb and the soil surface is exposed to erosion.

Teak plantations are started with seeds sown where they will grow, and they are generally thinned every 10 years, with all the trees having been felled by 80 years. The present age structure of the plantations is such that 10% of the productive area of teak is more than 50 years of age (RePPProT 1990). The felling and skidding (dragging the trunks out of the forest) is performed using manual labour with buffalo. This is in marked contrast with the situation elsewhere in the country, where heavy machinery is used and serious erosion often results. A mature plantation can produce up to 200 m^3/ha.

A number of 'varieties' have been described from Indonesia but these appear to be derived from environmental rather than genetic roots, and a plantation area should plant mixtures of quality seeds. In fact, the genetic base of teak plantations in Java is very limited, and so it is being broadened in order to counter the attacks of boring insects and other pests, such as the larvae of pyralid moths, using seeds from other countries.

Major Pests

One of the major pests of teak is the termite *Neotermes tectonae* which hollows out branches and stems. The best controls seem to be to burn infected trees and to ensure that the schedules of tree thinning are adhered to (Kalshoven 1954; Intari 1991). Teak is naturally deciduous, but young leaves can be attacked by moth caterpillars such as the noctuid *Hyblaea puera*. It is interesting that a related species which defoliates teak in India is found also in Java but does not damage teak here. Young teak leaves occasionally suffer from plagues of nymphs and adults of acridiid grasshoppers *Valanga nigricornis*, especially after a succession of dry years (Kalshoven 1953). Another important pest in less seasonal areas is the wood-boring scolytine beetle *Xyleborus destruens,* which does not kill the trees but burrows into the wood and hence ruins the timber quality.

Secondary Species

There are a number of other plantation species which are suited to a variety of environmental conditions and for a variety of forestry purposes.

After teak, the next most common species is the Sumatran pine *Pinus merkusii* with 319,526 ha of existing plantations, followed by mahogany *Swietenia* spp. (Meli.) with 105,185 ha, rosewood *Dalbergia latifolia* (Legu.) with 45,864 ha in periodically dry areas, and *Altingia excelsa* (Hama.) and native kauri *Agathis dammara* (Arau.), occupying smaller areas. As might be expected from the climatic differences across the islands, there are species planted in East Java and Bali which are rarely seen in West Java, such as *Maesopsis eminii* (Rham.), *Manilkara kauki, Schleichera oleosa* (Sapi.), *Pterospermum javanicum* (Ster.), *Michelia velutina* (Magn.), and *Melaleuca leucodendron* (Myrt.), and species commonly planted in West Java which are not appropriate in East Java, such as *Altingia* and mahogany. In recent years, five times as large an area has been planted to nonteak species as to teak (Perhutani 1992b).

Pine plantations are generally found above 200 m altitude, and the trees grow best between 800-1,500 m. They are started using nursery-grown seedlings, and final felling occurs at 30-40 years with a timber production of up to 250 m^3/ha. In 1992 there were 298,000 ha under pine plantations in Java, roughly one-third of which was being tapped for resin. Resin has been tapped from pine trees since ancient times, being gathered in a similar way to rubber latex. Its distillation produces rosin and turpentine. Rosin is used in varnish, paint, ink, and paper industries, while turpentine is used in various pharmaceutical and perfumery industries. This industry employs (part-time) some 36,000 tappers (Mukerji 1989). *Eucalyptus urophylla* (Myrt.) is slowly replacing pine because its timber properties are similar, but it is more resistant to fires; it cannot be tapped for resin.

Pine plantings can be damaged by overgrowth of the introduced climber *Mikania macarantha* (Comp.) (p. 185), which can form such dense masses in the crowns that the tops of trees collapse under the weight.

Social Dimension

The problems facing plantation forests today, such as timber and fuelwood theft, grazing by cattle and other domestic stock, and deliberate fires, are identical to those at the start of the century (Altona 1924). Commercial production of timber is the primary objective of Perhutani, but it is impossible to ignore the gulf between the poverty of many of the people living around the plantations, and the relatively enormous value of teak and other timbers, or the other products that are needed by local people, notably fuelwood, charcoal, and construction timber. The temptations of living near teak are enormous, and many ingenious means have been used to steal timber, such as cutting large slices from the rear of the trees, away from the patrol road, to avoid detection for the longest possible period. Such damage hampers long-term planning and efficiency, and in response Perhutani began in 1973 to initiate a large number of programmes to

Figure 14.2. An example of tumpangsari, intercropping cassava with *Acacia mangium,* near Yanlappa, West Java.
By A.J. Whitten

involve local people and to increase their prosperity, many of them related to forms of tumpangsari (p. 619) (fig. 14.2) (Atmosoedardjo and Banyard 1978). In 1981 this was refined into a programme for developing rural social welfare. Efforts are made to upgrade the quality of the settlements of people living legally in their areas, and to provide training and employment for them. In 1986 the narrow view of tumpangsari was broadened into a social forestry programme whereby broader forms of agroforestry were embraced, particularly where land pressure was severe. These have aimed to increase the share of benefits obtained by local people from Perhutani enterprises, at the same time as improving plantation management. Multiple use of plantations is encouraged in order to share benefits with local people, using programmes for fuelwood production (p. 623), silk production (p. 623), beekeeping (p. 621), forest recreation, fruit trees, fish ponds, medicinal plants, and wildlife breeding.

In addition, since 1987, Perhutani has adopted the policy of giving farmers longer tenures on tumpangsari lands and rights over the collection of fruits from the fruit trees they were allowed to plant, in order to build mutual trust, to increase the income of the farmers around the plantations, and to improve the chances of the land being properly restored, rehabilitated, or reforested. While steps have been taken to address these issues, they have not solved all the problems of the farmers and landless at the plantation edge, and dialogues are being pursued and other reforms considered (Simbolon 1988; Peluso 1992a). Perhutani's own experience demonstrates that success in forest production systems depends on the degree to which those systems fit the availability of labour, the local ecological conditions, and the level and stability of income flows generated by them (Stoney and Bratamihardja 1990).

The emphasis of the programmes is on villagers adjacent to plantations rather than to natural forests, because most Perhutani staff work in and around plantations. Perhutani is providing useful models for future integrated conservation development programmes around conservation areas (p. 760), and it is important that they are applied soon, because there is no letup in the pressure of human populations on the boundaries of natural forests. Initial steps in this direction are being made in Baluran National Park (W. Saleh pers. comm.).

Reforestation

Planting of treeless areas within the forest estate is termed reforestation *(reboisasi)* and planting outside the forest estate is regreening *(penghijauan)* (p. 135). Degraded forestland was reforested beginning in the 1920s, and 95,000 ha were planted between 1920 and 1940 (an average of 4,750 ha/year). In contrast, between 1951 and 1985 an average of 20,000 ha/year of nonteak plantations were established in areas needing watershed protection using a range of species, but the main species used were the large-leaved mahogany *Swietenia macrophylla, Pinus merkusii,* and *Agathis dammara.* Thus grassland, shrubland, abandoned estates, and abandoned agricultural land were brought into productive use (Smiet 1990). Reforestation received central support during the first five-year plan (1969-73), and this was increased during Repelita III (1979-83). Large areas have been planted, but often using species which needed care and attention after planting, and which were of little use to the people. As a result, many young trees died, and some areas were cleared or burned by local people almost as soon as they were planted.

Reforestation is intended primarily to rehabilitate and protect land from erosion and burning, and only secondarily to grow timber (Coster 1938; Soemarwoto and Soemarwoto 1983; Soemarwoto 1991b). The tree species used tend to be fast-growing, resilient species such as pines, eucalypts, and acacias. Reforestation programmes have been hugely expensive, but the results have been disappointing because the soils into which the young trees are planted are generally markedly infertile as a result of erosion and burning, because of competition from alang-alang grass, and because there has been little incentive for the local people to look after the trees during their critical first few years. The programme has experienced many problems, and it was acknowledged in 1983 by the Minister of Forestry that the efforts in many areas had failed. Failure was generally caused by inadequate care of the saplings after planting, combined, especially in the major drought of 1982-83, with severe water stress. Looking back and with the benefit of hindsight, too much was made of the blanket benefits of reforestation (Soemarwoto 1991b), and even where tumpangsari models were used to assist the programme (Wiradinata and Husaeni 1987), primary attention was given to the cost effectiveness for Perhutani and only secondarily to the minimal, and temporary, welfare-improving potential for the farmers (Peluso 1992a). Success has been achieved, however, where there has been dialogue and good relations between Perhutani and local communities such as at Mt. Pinang (8 km from Serang in the direction of Cilegon), where Perhutani built a fence in 1991 to prevent raids on the farmers' crops by wild boar (Danaatmadja and Natawidjaja 1990; W. Saleh pers. comm.).

Ecological Concerns

It is conventional wisdom that simplicity in ecological systems leads to instability, and in many cases this appears to be true. Agricultural or forestry monocultures are thus perceived as ecologically inappropriate, and three aspects of this are considered below.

Table 14.2. Plantation and natural forests compared.

	Plantations	Natural forests
Richness	low	generally high, but poor natural forests are known
Diversity	low	high, but variable
Age and size class distributions	narrow	broad, after major disturbances the distribution can also be narrow
Canopy layers	one	many, except in early successional growth
Soil use	rooting space available	rooting space limited
Total production	low	high, but most of it is consumed by animals and decomposer organisms
Net primary productivity	very high, because they are designed to behave similarly to an early successional stage of forest growth	low, even zero or negative
Nutrient uptake	varies with the age of the stand, but not so efficient because there are few roots near the surface and few decomposition organisms	more or less constant, and efficient
Nutrient balance	major losses when trees are harvested	in equilibrium
Litter	tends to accumulate	tends to be broken down rapidly

Source: Evans (1992). With permission of Oxford University Press

Lowered Richness and Diversity

There are many ecological differences between plantation and natural forests, and these are listed in table 14.2. Plantations clearly lose in an ecological comparison, but broad conservation opportunities will be lost if the simple view that plantations are 'bad' and natural forests are 'good' is adopted.

Plantations are remarkably hostile to other organisms at some stages of their growth, but their monotony and uniformity does not persist throughout their entire development. While a crop is being established and when it is maturing, the more open conditions allow a greater variety of plants to grow, and a variety of scrubby habitats, gaps, edges, and decaying tree parts, gullies, streams, tracks, gaps, and firebreaks planted with different tree species can be found (Evans 1992). Even so, less than 40% of birds found in Java have ever been reported from teak forests, and only 14 species regularly nest and feed there, none of them endemic (Sody 1953), and reasons for this include:

- plantations are structurally simple;
- plantations lack holes (used as nest sites) because trees are not allowed to grow old;
- food is less abundant and less diverse; and
- birds may not be adapted to feed on whatever insect species are present (Recher and Rohan-Jones 1978).

Plantations can be used, however, by a wide range of large animals such as banteng, leopards, wild boar, deer, peafowl, and monkeys, because the relative lack of human presence means that they represent safe corridors between blocks of natural forest, and extensions to foraging ranges, but they do not provide much beyond shelter. These features make plantations effective buffer zones around conservation areas, and this should be exploited as a model for the management of such areas (p. 760). Plantation managers, particularly those whose area abuts onto natural forest, need to review their plantation and determine what means can be found to increase habitat diversity, such as not clearing the noncommercial vegetation along streams and rivers. Plantations can never fill the same role as natural forest, but there is no reason to abandon them on an ecological scrap heap. Indeed, where plantations replace critical grasslands, then ecosystem complexity is increased.

Susceptibility to Pests and Diseases

A large area covered by a single or very dominant tree species represents an enormous and effectively limitless food resource for a pest or pathogen adapted to exploit this species. A buildup of pest populations is further favoured by the close proximity or even direct contact of leaves, twigs, and roots of neighbouring trees. In addition, unlike short-season crops grown in a rotation, plantation trees are grown on a single site for decades, thereby allowing long-term growth of pest and disease populations. Since most of the plantation species are exotic, they owe some of their success to not having the natural, adapted pests and pathogens (or, indeed, their natural enemies). Evidence suggests, however, that this apparent immunity to problems does not last indefinitely (Evans 1992). When serious pest and disease problems do occur, however, they are generally associated more with the intensive management practices than with the monoculture itself. Thus site clearance, nursery conditions, site selection, means of establishment, and thinning and pruning methods, are all relevant. It can also be argued that protection from pests and diseases is best achieved in single-species plantations because identification of any problem is likely to be faster, trees are felled before they become sick and a threat to their neighbours, and care is exercised in the selection of planting material (Gibson 1980).

So, although it is often asserted that monocultures lead to increased risks of disease and pest attack, the evidence is pauce (Gibson and Jones 1977). Indeed, there have been very few disastrous pest or disease outbreaks in plantations anywhere in the world. The problem is that 'diversity' is not just the sum of species present in an area, but in a pest situation it is more important to assess which part of the total diversity impinges on the pest's life, its food, and its predators. Unless this is done, the total diversity could be increased without having any affect on the pest 'problem'. It is clear that an understanding of the problem will not be achieved without sound ecological study.

Long-term Productivity

Trees are long-lived organisms, and some teak trees standing now would have been growing when forest management was first being discussed on Java. In most areas of the tropics the expectations and projections of yields are therefore based mainly on the first plantings. Unfortunately, there are so many confounding variables, such as climate changes, changing biological and silvicultural conditions, genetic changes, and differences in trace-element content of soils among sites, that to generalize about long-term productivity is virtually impossible. Loss of productivity over the long term is probably a serious problem for plantations on the marginal soils found in many parts of the Outer Islands, whereas in Java the occasional volcanic top dressings may alleviate the problem. Reduced growth in second rotation teak was observed earlier this century, but this may have been due mainly to the effects of erosion when the soil was exposed after the litter was burned. Research elsewhere suggests that erosion under teak can amount to many tons/ha/year of topsoil (Bell 1973), and this should influence the choice of site for teak trees. In general, for plantation tree species grown in the tropics there is little evidence so far of long-term productivity problems provided attention is paid to conserving organic matter and minimizing weed competition during the establishment phase (J. Evans pers. comm.).

AGROFORESTRY

Definition and Characteristics

Agroforestry was being practiced in Java and Bali long before the term was originated by development experts who thought they had discovered something new. While not an easy term to define, it covers both agriculture that incorporates tree planting for fuel or fruit, and forestry plantations that incorporate arable food crops. A useful definition of agroforestry is: a system of permanent land use compatible with local cultural practices

Table 14.3. Agroforestry systems of Java and Bali.

Cultivation system	Subsystem	Indigenous system	Forest Service system
Rotational cropping (of annual and perennial crops)	with natural tree fallow	Shifting cultivation (*ladang, huma*) (almost extinct on Java)	
	with planted tree fallow	Wonosobo system: *Acacia*-tobacco and vegetables	Taungya system (*tumpangsari*)
Intercropping (annuals and short-lived perennials in spaces between trees)		Scattered trees on or beside agricultural fields (*tegal-pekarangan*)	Regreening (*penghijauan*)
Mixed cropping (low perennials in spaces between high perennials)	trees and agri-cultural crops	Mixed garden (*kebun campuran*)	
	trees and forage crops for stall feeding		Timber trees and forage system
	trees, fodder crops, and green manure crops	Several local systems	
Multistoreyed cropping (combination of long and short duration crops simultaneously)	in housing compound	Home garden (*pekarangan*)	
	outside housing compound	Mixed garden (*kebun campuran*) Forest garden (*talun*)	
Mixed farming (crop growing and animal husbandry)	in housing compound	Home garden (*pekarangan*)	
	outside housing compound	Several local systems (e.g., grazing in coconut plantations)	

Source: Wiersum (1980)

and ecological conditions, by which both annual and perennial crops are cultivated simultaneously or in rotation, often in several layers, in such a way that sustained multiple-purpose production is possible under the beneficial effect of the improved edaphic and microclimate conditions provided by simulated forest. Such systems may also include animal husbandry (Wiersum 1980; Torquebiau 1992). The different major types of agroforestry are shown in table 14.3, and the subject is reviewed succinctly elsewhere (Wiersum 1981; Tejwani and Lai 1992).

Despite the many types of system that can be found they often display similar functional characteristics (Wiersum 1980):

- perennial crops dominate the system such that there is a relatively high proportion of nutrients in the vegetation rather than the soil. So long as the perennial cover is not grossly disturbed and the proportion of annuals remains low, the soil is well protected from exhaustion, leaching, and erosion;
- there are at least two layers of vegetation, which allows for improved exploitation of water, nutrients, and light. This can lead to symbiotic synergisms (beneficial) or allelopathic reactions (detrimental). Unfortunately most food crops require high light levels and so are generally unsuitable in agroforestry systems; and
- there are relatively high numbers of species, as well as high structural and spatial diversity, which spreads the risks of loss through pests and disease and provides daily yields of at least some crops.
- resilience and resistance (p. 22) are high, at least in traditional systems, but they may decline under certain management regimes.

Uses

The many products of agroforestry include the following categories (RePP-ProT 1989):

- **Fuelwood** - this is often in short supply even if the situation is not critical, and in many areas it is taken 'free' from whatever forest areas are left in the vicinity, thereby accelerating the relentless forest degradation. The fuelwood is partly for domestic uses and partly for small-scale industrial uses such as tile- and brickmaking. Unfortunately but understandably, until the fuelwood supply becomes critical, farmers prefer to use land to increase the production of major food crops – only when the critical point comes does the realization dawn that even fast-growing trees take years to grow. Certainly, plantation estates and other responsible large-scale land users around natural forest areas have a moral responsibility to plant fuelwood at least on small, unutilised plots or even on core land. The excuse that they are not involved with forest protection is inadequate in today's world. Elsewhere it may be that tree-growing pro-

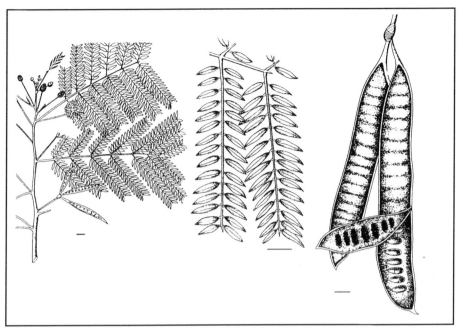

Figure 14.3. Lamtoro *Leucaena leucocephala.* Scale bars indicate 1 cm.
After Koorders and Valeton 1914

grammes for fuelwood should be directed at the women and women's organizations since it is they who bear the brunt of finding, carrying, and then using the wood for cooking.

- **Increased food production** - Perhutani has developed agreements with farmers by which food crops and food trees are planted beneath the rows of teak and other plantation trees. This cooperation between Perhutani and local communities is ideally of mutual benefit because the poor and often landless farmers are given a stake in the long-term future of the area, and the state-owned enterprise enjoys cheap protection and management of its investment.
- **Site amelioration** - where the soil is erodible and/or of poor quality, the planting of deep-rooting leguminous trees that can be used for green manure to build up topsoil fertility is an attractive option.
- **Economic buffer** - dependency on a single crop is risky both because of pests and diseases and because of variations in market prices. The more diverse a farmer's biological sources of revenue, the less such problems will affect him. Agroforestry is popular in densely populated areas because with good management, cooperation, and

environmental awareness, it allows land that is too steep or too degraded for conventional agricultural systems to be kept in appropriate and potentially sustainable production and to absorb manpower, for example in terracing.

- **Fodder** - fodder trees are important in seasonally dry areas to supplement pasture. Some systems rotate maize (a major food crop in eastern Java) with *Gliricidia* and *Sesbania*, or lamtoro *Leucaena leucocephala* (all Legu.) the branches of which are used to fatten cattle (fig. 14.3) (NAS 1977).

The types of agroforestry system chosen by farmers is determined in part by geomorphological aspects of the land corresponding to agroecological zones, and also by a household's access to the land. For example, in the dry and relatively poor Gunung Kidul area of Yogyakarta, three major agroecological zones were recognized: Batur Agung in which fuelwood for Yogyakarta was a major crop, Gunung Sewu where systems were more subsistence-based, and Ledok Wanasari with mainly fodder-based systems (Tolboom 1991) corresponding to three major agroforestry categories: tree production, crop production, and livestock production. As for land, households with little land tend to produce as many different crops as those with more land, and grow trees close together which are for subsistence purposes only. As might be expected, the production of staple requirements is regarded as more important than producing a marketable crop. The trees are also seen as a savings account, however, that can be turned into cash when there is crop failure or a special need, such as a wedding or funeral (van Dijk 1987).

Even in clearly degraded areas, such as Gunung Kidul, there is no widespread understanding of soil erosion, despite the projects of governmental and nongovernmental bodies to support the construction or repair of terraces or regreening, and the information provided on preventing soil erosion through contour ploughing and mulching. From the farmer's point of view these activities are expensive in terms of time and effort when crop production is actually stable or even increasing due to the widespread application of artificial fertilizers. The degree to which tree-planting is embraced also depends very much on whether the trees have a clear function, such as fruit or timber production, in addition to protecting the soil against erosion (Mouwen 1990). Many species of trees are grown, but most project packages include forestry species such as *Gliricidia*, *Calliandra*, *Sesbania*, and other legumes, together with teak and mahogany; but there is little socioeconomic data indicating that farmers actually want these trees (Mackie 1988).

One of the most widely planted species a decade ago, particularly in drier areas, was the 'miracle tree' *lamtoro*, the planting of which reached fever pitch as its benefits of controlling erosion, producing stock food, as well as producing firewood, even in drought-prone areas, were being realized. A

Figure 14.4. Lamtoro trees defoliated by jumping lice, and the ladybirds introduced to control them (above).
With permission of Kompas

voice of moderation was raised by Otto Soemarwoto, who foresaw the eco-logical, economic, and social risks of such massive adoption of the tree. He observed that lamtoro could grow under a certain range of conditions and that plantings that were not within this optimum range risked failure. Under optimum conditions, he opined that fast-growing lamtoro could become a serious pest. He also reminded people that the high evapotran-spiration of lamtoro would aggravate water shortages in some areas. Prophetically, he urged people to remember that, although the tree was claimed to be disease and pest resistant, these were not eternal states and that changes in such 'labels' could occur rapidly. He proposed careful monitoring of pests and diseases and of natural enemies, controlled use of pesticides, the maximization of variety in the planting of species and vari-eties, and the accordance of selection for any single type of resistance. Finally he implored planners to be sure they understood precisely why they wanted to plant lamtoro and then work to meet that goal, rather than try to make the tree do everything in all locations (Soemarwoto 1982).

In 1986 this efficient, nitrogen-fixing tree began suffering from depre-dations of the jumping plant louse *Heterosylla cubana*. Introduction of its natural enemies, the predatory coccinelid ladybird *Curinus coeruleus* from Hawai'i (fig. 14.4), and the parasitic *Psyllaephagus jaseeni* from Thailand successfully controlled the louse (Oka and Bahagiawati 1988; Mangoendi-hardjo 1989). A note of caution should be sounded here regarding such apparently successful and 'benign' biological control. The reports of problems are limited and few, but this does not necessarily confirm the general assertion that biological control is environmentally safe. The extinctions of nearly 100 species have been reported as a direct or indirect consequence of biological control programmes around the world (Howarth 1991). Some of these species were indeed the target organisms, but many were not, and most of the extinctions were discovered circumstantially with hindsight, or simply by luck. In this regard it is relevant that the ladybird introduced to control the jumping louse on lamtoro also attacked various coccid scale insects on coffee, clove, and coconut (Mangoendihardjo 1989). These additional prey were noted because they were on economic crops. No one in fact has any idea what the total spectrum of ladybird prey is, and which beneficial insects it may include.

Shifting Cultivation

Whereas the present population density on Java and Bali actually requires that the dominant agricultural system be as intensive as possible and make the best use of the large available labour force, that was not the case in the past. In the days of the early kingdoms when population density was far lower, the combination of intensive agriculture, population concentration, and proscription of forest-based alternatives supported, indeed were essential to, the survival of the ruling classes who extracted financial and material benefits from the people to gain profit and power. During the colonial era the same bias against extensive agricultural systems could be seen. The view prevails today, unconsciously perhaps, against the few remaining shifting cultivators such as the Baduy and Kasepuhan people in West Java (Iskandar 1992; Suharti and Alrasyid 1992; Adimihardja and Iskandar 1993) even though, under certain conditions, notably low population density, extensive agriculture can be shown to be a rational use of land and labour (Soemarwoto 1987b, 1985d). The enormous efforts expended to resettle and develop the Baduy (and other groups in the Outer Islands) have often met with varying degrees of failure, the blame for which is generally laid at the door of the 'strength of traditions' (used in a pejorative sense), or inadequacies of the government servants charged with executing the programme. The possibility that their *huma* or swiddening is actually more economically attractive to the practitioners is not entertained (Dove 1985).

The colonial government delimited small areas of West Java that could be used for shifting cultivation, but their small size automatically rendered them unsustainable. At about this time terraced wet-field rice cultivation was introduced from Central Java. A number of rice varieties was grown

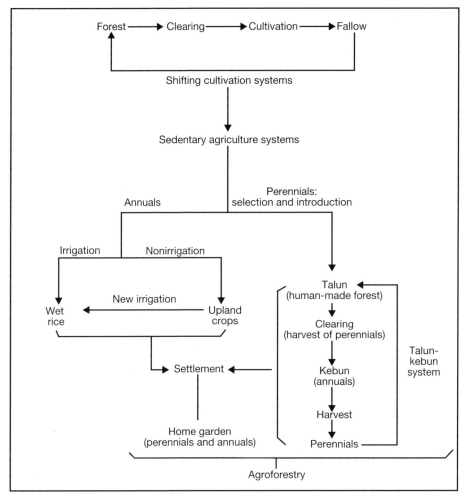

Figure 14.5. The evolution of agroecosystems in Java.

After Soemarwoto 1985d

and a diverse array of crops were grown on the fields during the fallow. As is clear, rice cultivation spread but the roots of shifting cultivation persisted and man-made forests, known as *talun-kebun,* arose around the settlements (Soemarwoto and Soemarwoto 1984).

The progression or evolution of agroecosystems from shifting cultivation to forms of sedentary cultivation can be described graphically (fig. 14.5).

Home Gardens

When flying over Java and Bali, it is striking that villages can be picked out not by their buildings but by their tree-dominated home gardens (Soemarwoto et al. 1976). These comprise many species at different stages of growth and different heights. The tallest trees are generally the coconuts, below which one finds a dark green canopy of fruit trees such as mango and rambutan, a layer of head-height food crops such as maize and cassava, and nearer the soil other vegetables such as taro, spices, and scrambling sweet potato. There are also climbers such as yam, passionfruit, and melons which can reach the tops of even the tallest trees. The number of species and cultivated varieties clearly mark these gardens as major and dynamic sources of potentially important genetic resources. With such variety something can be harvested almost daily either for consumption or for cash. In addition to food crops there are also medicinal plants, weather-indicating plants, and plants with supposed magical powers (Abdoellah 1990). Among the most striking additional benefits of a home garden is the coolness and shade making the compound a pleasant place to live (Soemarwoto and Soemarwoto 1984). Home gardens account for 17% (1.6 million ha) of Java's agricultural land, and this percentage has varied little during this century despite the great increase in agricultural land (Soemarwoto and Conway 1991).

Javan home gardens are an ancient agroecosystem, with their first records dating back to a charter of A.D. 840. A home garden is on land with definite boundaries which also encompass the house. In Central Java the home gardens tend to be around each of the houses, whereas in West Java, the houses tend to be very close and the gardens surround a cluster of houses (Soemarwoto and Soemarwoto 1984). In both types, however, the garden comprises a mixture of annual and perennial crops distributed in

Table 14.4. Differences in output of three major agricultural systems.

	Rice fields	Upland fields	Home gardens
calories	high	low	low
protein	high	low	low
calcium	moderate	moderate	moderate
iron	moderate	moderate	moderate
riboflavin	moderate	moderate	moderate
vit. A	low	moderate	high
vit. C	low	moderate	high
return on labour	low	moderate	high
return on cash	low	moderate	high

Source: Abdullah and Marten (1983)

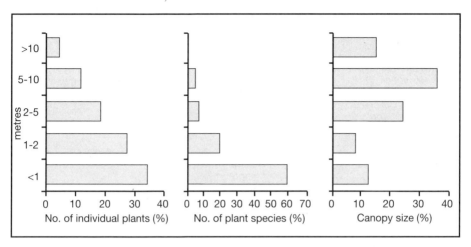

Figure 14.6. Number of individual plants and plant species, and size of canopy, in relation to plant height in home gardens.

After Karyono 1990

three-dimensional and temporal space. This maximizes production while still meeting the social and economic needs of the people in the form of cash income, carbohydrate, protein, minerals, and vitamins (table 14.4) (Terra 1953; Soemarwoto and Soemarwoto 1984; Soemarwoto 1987a, 1991a, b; Freeman and Fricke 1984; Marten 1990). It also mimics the hydrological, climatic, and genetic and soil conserving functions of a natural forest, although this is largely incidental (p. 675).

Studies on home gardens were begun in 1900 at the instigation of the Queen of Holland after it was acknowledged that the welfare of the Javanese was declining because of increasing population pressure and the costs of forced labour and service under colonial policies (Karyono 1990). The first detailed work, however, was not conducted until after the Second World War (Terra 1953, 1954), since when home gardens have become known internationally, largely through the work of Otto Soemarwoto and his staff and students from the Institute of Ecology, Pajajaran University, Bandung.

Composition

A single home garden may have within it some 50 species of plants, including 12 tree species, and the total number of species for a small hamlet may be as high as 200, including 64 tree species (fig. 14.6, table

Table 14.5. Plants found in a 500 m^2 home garden and their main function.

Cocos nucifera	Palm.	Coconut	Cash crop and spice
Coffea canephora	Rubi.	Coffee	Cash crop
Syzygium aromaticum	Myrt.	Clove	Cash crop
Maesopsis eminii	Rham.	Musizi	Construction material, firewood
Toona soereni	Meli.	Suren	Construction material, firewood
Ananas comosus	Brom.	Pineapple	Fruit
Artocarpus heterophyllus	Mora.	Jackfruit	Fruit
Carica papaya	Cari.	Papaya	Fruit
Citrus grandis	Ruta.	Pomelo	Fruit
C. nobilis	Ruta.	Tangerine	Fruit
Durio zibethinus	Bomb.	Durian	Fruit
Mangifera indica	Anac.	Mango	Fruit
Manilkara zapota	Sapo.	Sawo, sapodila	Fruit
Musa paradisiaea	Musa.	Banana	Fruit
Psidium guajava	Myrt.	Guava	Fruit
Syzygium aquaeum	Myrt.	Rose apple	Fruit
Curcuma domestica	Zing.	Turmeric	Medicine and spice
Kaempferia galanga	Zing.	Kencur	Medicine and spice
Languas galanga	Zing.	Laja	Medicine and spice
Sericocalyx crispus	Acan.	Keci beling	Medicine and ornamental
Zingiber officinale	Zing.	Ginger	Medicine and spice
Canna edulis	Cann.	Arrowroot	Snack or supplementary staple food
Colocasia esculenta	Arac.	Taro, cocoyam	Snack or supplementary staple food
Dioscorea alata	Dios.	White yam	Snack or supplementary staple food
Ipomoea batatas	Conv.	Sweet potato, kumara	Snack or supplementary staple food
Xanthosoma atrovirens	Arac.	Talas padang, tania	Snack or supplementary staple food
Manihot esculenta	Euph.	Cassava	Snack or supplementary staple food and vegetable
Amaranthus hybridus	Amar.	Bayam, 'spinach'	Vegetable
Capsicum annuum	Sola.	Bird pepper	Vegetable
C. frutescens	Sola.	Chili pepper	Vegetable
Cucurbita moschata	Cucu.	Pumpkin, squash	Vegetable
Dolichos lablab	Legu.	Lablab, hyacinth bean	Vegetable and green manure
Gnetum gnemon	Gnet.	Melinjo	Vegetable
Leucaena leucocephala	Legu.	Petai cina	Vegetable
Ocimum basilicum	Labi.	Basil	Vegetable
Parkia speciosa	Legu.	Petai	Vegetable and cash crop
Phaseolus vulgaris	Legu	Green beans	Vegetable
Pluchea indica	Comp.	Fleabane	Vegetable
Psophocarpus tetragonalobus	Legu.	Winged bean	Vegetable
Sauropus androgynus	Euph.	Katuk	Vegetable
Sechium edule	Cucu.	Vegetable pear, chayote	Vegetable
Solanum melongena	Sola.	Aubergine, egg plant	Vegetable
Vigna unguiculata	Legu.	Cow pea, long beans	Vegetable

Source: Karyono (1990)

14.5) (Penny and Ginting 1984a, b; Karyono 1990). Indeed, over 602 plant species (including 195 ornamentals) were found in 351 home gardens in the Citarum area (Karyono 1990). When one considers cultivated varieties, the numbers are still higher – as an example, 40 banana varieties have been identified in the Citarum River basin alone (Soemarwoto and Soemarwoto 1984). Home gardens therefore represent a significant reservoir of genetic resources (Sollart 1986). This is most true for traditional or remote gardens, and it is interesting that the number of plant species grown is significantly higher in villages away from rather than along a main road (table 14.6) (Soemarwoto et al. 1976). Where the number of plant species is high, the appearance of home gardens may appear to be a random accumulation of annuals and perennial plants with no discernible order. The farmers do not agree, however, and can explain the motivation and strategy behind the planting of particular plants in certain positions (Christanty 1982).

Among the trees, the most valuable are durian *Durio zibethinus* for its fruit and timber, *Pterospermum javanicum* (Ster.) for its timber, *Toona sinensis* for its timber and its role as a shade tree, nutmeg *Myristica fragrans* (Myri.), cinnamon *Cinnamomum burmannii* (Laur.), robusta coffee *Coffea canephora*, and cashew *Anacardium occidentale* (Anac.), the last of which is particularly favoured in the drier regions. By no means do all plants grown have a utilitarian value, however, and part of the planting of home gardens, at least in less remote or more prosperous areas, is done to appeal to aesthetic senses, and this is evident from the number of ornamental plants present. Traditional home gardens may also have a variety of useful animals associated with them such as chickens, ducks, goats, sheep, cows, rabbits, and fish.

Many of the inputs which are provided for a home garden would be

Table 14.6. Number of plant species in home gardens in two West Java districts.

	Cinangka (far from main road)	Kramatwatu (near main road)
Ornamental plants	44	33
Food plants	10	6
Vegetables	17	6
Medical plants	10	6
Fruit plants	29	15
Spice plants	14	4
Plants for industrial products	4	2
Others	14	4

Source: Soemarwoto et al. (1976)

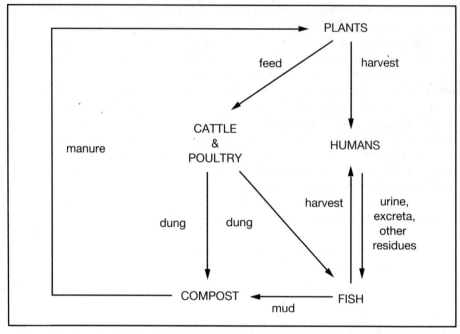

Figure 14.7. The traditional recycling system in West Javanese villages.

After Soemarwoto 1985b

regarded by some as waste, but the scraps and dung are perceived by the garden owners as simply a product to be used and converted into another product, and to protect the soil from nutrient depletion (fig. 14.7). Fish ponds, domesticated animals, plants, and people are integrated in a closed system, with kitchen and other human waste being fed to the fish. Dung, mud from the pond, and agricultural wastes are mixed and processed into compost to be used as fertilizer (Abdoellah 1990). In an ideal system food produced by the higher plants in the garden and the phytoplankton in the ponds is eaten by people either directly or indirectly by feeding the products to stock animals or fish, with surplus processed in a village home industry. Organic wastes are digested in a biogas unit to produce methane which is used to generate electricity to serve the village home industry. The sludge from the digester is composted to fertilize the gardens and the fish ponds. As an additional benefit, the digester eliminates the aesthetic and health problems of human excreta being used directly or deposited directly into rivers. In the above ideal situation, agriculture, animal husbandry, and industry are thoroughly integrated (Soemarwoto et al. 1976).

Economic Role

The home garden cannot, of course, compete with irrigated rice fields for monetary income, but since the produce from a home garden has such low production costs, the *net* income both on an absolute and a per hectare basis from a home garden can exceed that of a rice field (Soemarwoto and Soemarwoto 1984). For this reason, when people find themselves in a desperate situation, they prefer to sell a rice field or mixed garden rather than a home garden (Abdoellah 1990). Despite the clear importance of home gardens, extension officers are never trained in the development of home gardens, and one suggested reason for this (Dove 1990) is that their sheer complexity and harvesting makes them virtually impossible for anyone except the owner to exploit on a systematic and large-scale basis (although governments have taxed them in the past) (Penny and Ginting 1984a, b). Rice is the antithesis of this: a single species, harvested all at the same time, just once or twice each year. Thus it pays a farmer, particularly a relatively poor one, to shelter behind his home garden while belittling its importance. This situation arose over the centuries of tension between rural communities and central feudal courts, during which the strategy of emphasizing the sawah protected the farmers from being taxed on the apparently less valuable home gardens. One negative effect of this situation is that it is believed that the full potential of the home gardens is still not being realized. It is hard to judge, however, because the production of home gardens does not enter national statistics, and because much of the produce is not sold (Karyono 1990).

Change

Changes in home gardens are most visible around the more developed areas of Java and Bali, notably in the vicinity of Jakarta and Bogor. Traditionally this has been a wet-rice growing area with home gardens dominated by fruit trees for home consumption or intervillage markets. In the last 25 years, however, the increasing population densities and the demands of the urban populations for fresh produce and snacks have resulted in 75% of the sawah fields in the area being converted to vegetable and cassava cultivation (processed locally as tapioca) for national and international markets. In addition, more of the land is now in fewer hands, and home gardens have been transformed to conform to the rules of intensive, market-orientated production. Pieces of garden land and even individual trees are given monetary value and can be sold or mortgaged, such that modification of traditional gardens is favoured. Even so, traditional fruit

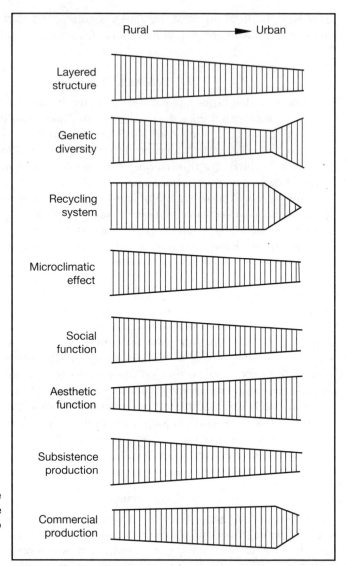

Figure 14.8.
Changes in the structure and function of the home garden from rural to urban settings.

After Soemarwoto 1985d

species and varieties are often grown alongside modern types, and farmers are able to respond to modern markets which provide livelihoods in the rural economy (Michon and Mary 1990). As the farmers increase their orientation to cash generation, however, their social responsibilities erode, traditional institutions based on mutual help are destroyed, trespass and theft are noticed, sanctions are sought, and traditional income sharing is

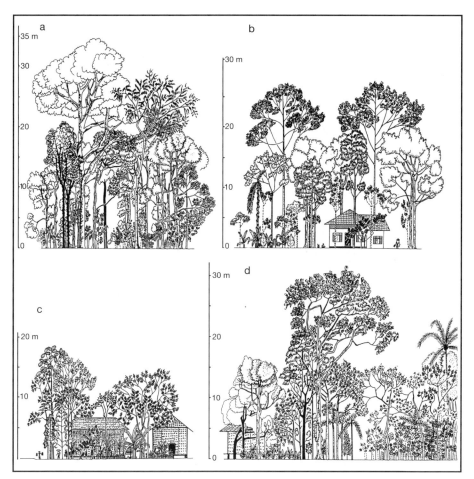

Figure 14.9. Changes in home gardens in West Java. a - traditional garden showing diverse, dense, and stratified vegetation away from the house; b - traditional garden with a new house showing reduced tree cover; c - more modern home garden with house, more open canopy, and many short-cycle species; d - renewed home garden with productive ground layer.

After Michon and Mary 1990

disturbed (Abdoellah 1990). By this stage also the garden is fenced, its loose, neighbourly social function lost, and the recycling function usurped (fig. 14.8) (Soemarwoto 1985d).

As these changes proceed, so the traditional home gardens (fig. 14.9a) have been divided up by the building of houses, and the individual plots have become smaller (fig. 14.9b). Light-demanding trees with high eco-

Figure 14.10. Intensified home garden in West Java with only species of high economic value.

After Michon and Mary 1990

Figure 14.11. Modern home garden in West Java with tall durian and petai over clove and nutmeg, and a ground cover of weeds.

After Michon and Mary 1990

nomic value have replaced the tall, traditional species, and the space in front of the house is given over to fruit, vegetables, and tubers for home consumption or market (fig. 14.9c). The demand for cash has also wrought changes, mainly the introduction of short-cycle species grown in the ground layer with produce for sale in markets. Plants introduced are shade-tolerant taro *Colocasia* and *Alocasia* (both Arac.) for their tubers and leaves, *Halopegia blumei* (Mara.) for its leaves used as packing material, coffee, and pineapple (fig. 14.9d). Changes in the tree composition have had the most marked effects. The tall trees are selectively felled, leaving only trees with the highest economic value, such as durian and petai, and other trees are introduced such as clove, nutmeg, and coffee, all of which can give good crops in five to seven years, and between which are grown bananas and papayas for market (fig. 14.10). Thus, the garden eventually matures and characteristically has a discontinuous tall canopy of durian and *Parkia*, a middle canopy of nutmegs and cloves, and a herbaceous layer of unutilized weeds (fig. 14.11). In extreme cases the gardens are given over to a single species – cloves, orchids, or cassava – and all the traditional features of diversity, complexity, multiple use, and stratification are lost (Christanty 1990; Kijne 1990; Michon and Mary 1990).

Another consequence of commercialization is the increasing demand for credit and other types of capital. This is transforming what was a low-input, low-yield, low-risk system into a high-input, high-yield, high-risk system. Income also becomes more seasonal and little is reinvested in the garden. Also, since the traditional vegetables and fruits are being displaced, and a higher percentage is marketed, so there tends to be less food for the family owning the garden. Lastly, equitability decreases as sharing is not compatible with the new commercial enterprise (Soemarwoto and Conway 1991).

Talun-kebun

Talun-kebun is a mixed garden outside the settlement area comprising perennials, the *talun*, integrated with an area of annual crops, the *kebun*. Wood is pruned or cut from the perennials: fruit trees, bamboo, *Albizia* , or similar trees. The leaves and small branches are burned, or mixed with dung and composted, with the compost used on the annual crops. Talun resembles a home garden but it has no house on it and is some distance away from the settlement, and the talun-kebun system is basically shifting cultivation in a man-made forest. In some areas the annual crops are dominated by *Dolichos lablab* (Legu.), an African shrub providing edible pods and seeds as well as green manure and fodder, while the perennials are dominated by large bamboos. As with home gardens, talun-kebun represent a largely unassessed on-farm repository of genetic material.

The cycle of talun-kebun cultivation is about six to eight years and begins with an area of cleared ground being planted with annual and perennial crops. After 18 months all crops will have been harvested and the canopies of the perennial trees closed. An area of talun will then be harvested and a new kebun prepared (Soemarwoto 1985a). The advantage of this system over traditional shifting cultivation is that virtually all species grown have a market value, or at least utility, and it is therefore able to support a higher human population density (table 14.7). Also, even though it is practiced in hilly terrain, erosion is slight because the kebun plots are relatively small and often surrounded by silt-trapping vegetation or litter. Also, although there is nutrient loss through the cutting and removal of wood and other products, the introduction of composted dung and ash helps to rejuvenate the soil.

The degree to which talun-kebun represents an ecological adaptation to steep and unstable slopes is graphically and horrifically demonstrated in the area south of Garut: fifty people were buried alive in October 1992 in a landslip in Sodonghilir which originated from a part of a hillside that had been converted from talun-kebun to rainfed terraced rice fields. The talun-kebun on either side were unscathed, but of the rice fields and the houses below them the rapid and tragic slip left no sign.

The factors that are tending to force change in home gardens are also having an effect on talun-kebun. The system tends to break down as the ratio of kebun to talun increases and the cycle is thus shortened. Also, diversity is being lost as economically attractive crops such as widely spaced mandarin and clove trees are planted; unfortunately this is often done on steep slopes, which is not good for soil conservation.

Table 14.7. Comparison between traditional shifting cultivation and talun-kebun system.

	Shifting cultivation	Talun-kebun system
Forest	natural	man-made
Land ownership	not well defined	well-defined
Clearing	to open a space for the garden	to harvest the planted trees and thereby create an opening for the kebun
Burning	almost all parts	only a small part
Mineral cycling	almost closed, few minerals exported, no import of minerals	open, large quantity of minerals exported, but balanced by import of minerals
Economy	subsistence	market economy
Population density	low	higher

Source: Soemarwoto and Soemarwoto (1984)

Tumpangsari

Tumpangsari is an agroforestry system in which annual crops are planted between rows of commercial timber trees for as many years as possible before the tree canopy closes and precludes the growth of the annuals. It was brought to Java in 1873 from Burma, where it is known as taungya. The basic system has been changing and developing ever since, with the integration of different tree and crop species and a wide range of combinations and densities. In the early 1970s tumpangsari models were intensified to increase the overall productivity of the forestland, and farmers were trained in the use of high-yielding varieties, better soil tillage, fertilizers, pesticides, and the proper adjustment of planting to rainfall. Tumpangsari agroforestry has increased both the growth of plantation trees (because of the fertilizer use) and the yield of food crops. The success in improving tree growth unfortunately resulted in more rapid closure of the canopy, and so caused the tumpangsari period to become shorter (from averages of 29 to 15 months). This in turn has led farmers to damage trees in order to maintain the cropping period. These problems have, in part, been overcome by increasing the tree spacing, but wide spacing encourages branching which benefits neither the quality of the timber nor the crops. There are exciting ecological potentials for improving tumpangsari systems by utilizing species interactions and species adaptabilities (Sukandi 1990).

Since 1984 Perhutani has been increasingly involved with 'full rotation agroforestry' in which farmers intercrop and harvest throughout the full rotation period of the primary tree species, rather than just during the first two years of establishment. This is made possible by increasing the tree spacing and permiting perennial crops such as fruit trees and multiple-use tree species to be grown to the extent that they can occupy 20% of the total stand. Early evaluation indicates major success in terms of yields, income, economic feasibility, harmony of relationships between farmers and foresters, and the skills of farmers (Bratamihardja 1990). Perhutani is also changing some of its guiding policies by increasing the area planted to short (less than 10 years) rotation species such as the hard and termite-resistant *Paraserianthes falcataria* (Legu.) from eastern Indonesia, *Gliricidia sepium* (Legu.) from Brazil, and *Gmelina arborea* (Verb.) from India, especially in densely populated areas facilitating some forms of tumpangsari. Tumpangsari has been used to establish 30,000-40,000 ha of teak, albizia, pine, mahogany, and agathis plantations per year in Java.

Various forms of tumpangsari are economically beneficial to Perhutani, but the farmer generally only obtains an adequate crop yield if he receives fertilizer. This implies that fully successful tumpangsari projects must be planned by both parties (Hutabarat 1991). Such problems, and those described earlier (p. 595), have not endeared Perhutani to the local communities adjacent to or on forestry land. The Social Forestry Programme of Perhutani began in 1986 to develop a community forestry management

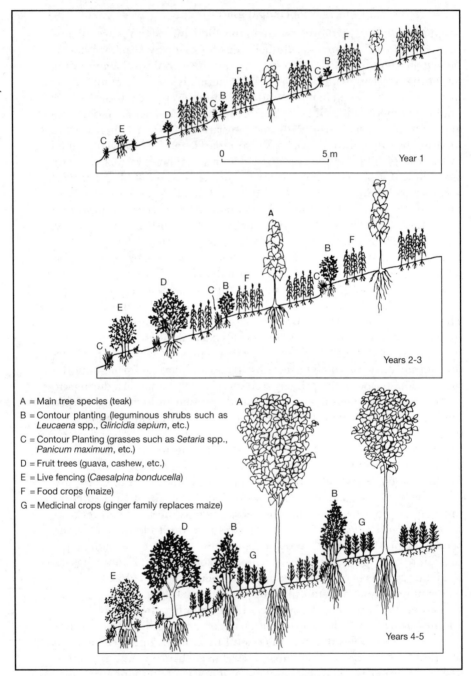

Figure 14.12. Changing vertical stratification in a Javanese agroforestry plot.

After Stoney and Bratamihardja 1990

system, and also to facilitate better relations between Perhutani and the target communities, and to establish agroforestry models that provide optimum yields for both Perhutani and the farmers (Bratamihardja 1988; Sukandi 1990). There is a large variety of models, and in addition to the standard timber and food species, fruit trees, fuel wood, hedge trees, and forage crops are also grown. Different combinations create different problems of plant competition (for light and nutrients), adaptation, and site conditions, but many of these can be dealt with after research into spacing, planting regimes, altered thinning and running regimes, fertilizer application, etc. It must also be remembered that the competitive relationships between the species change over time (fig. 14.12). Although there have been many major advances, the Social Forestry Programme has so far fallen short of fulfilling its primary objective (Seymour and Rutherford 1993).

Honey, Shellac, and Silk

Agroforestry also encompasses the cultivation of trees combined with some form of animal husbandry. A number of useful insects such as bees, for honey and wax, silkworms, and *Kerria* scale insects for shellac are bred to maximize production potential in the plantation ecosystems.

Honey. The indigenous honey bees *Apis indica, A. cerana,* and the sweat bees *Trigona* spp. have traditionally been kept in hollow logs on Java and Bali, and their honey used mainly as a local medicine. This activity has been encouraged by Perhutani for over twenty years, and is suitable in areas within 2 km of good nectar and pollen sources, such as forest, or extensive areas with mixed species of trees. Good nectar plants are coconut, mango, cashew, and guava, whereas rich sources of pollen are maize and lamtoro. In addition to the honey, the honeycombs are a useful crop themselves since they provide beeswax for further cash income. The bees themselves also provide an important service in terms of pollination.

The main problem with log hives is having to destroy all the combs to collect the produce, and so modern hives with comb frames are becoming popular. The European honeybee *Apis mellifera* has been used in agroforestry programmes to increase land productivity, increase farmers' income, and improve environmental quality (Kasno 1990). Fortunately, despite decades of attempts, this bee has demonstrated that it is not adapted to tropical climates, suffers from infestations of two mites *Varroa jacobsoni* and *Tropilaelaps clarae* to which the native bees have evolved defences, and its combs are persistently destroyed by the wax moths *Achroia grisella* and *Galleria mellonela* (Akratanakul 1986, 1987; FAO 1986). The

Figure 14.13. The hard, resinous lac secreted by scale insects.
By A.J. Whitten

ecological dangers of introducing alien species cannot be overemphasized.

Many people still regard the best honey as coming from the enormous, single-comb nests hanging beneath boughs in and around forest areas which belong to the relatively large and fierce *Apis dorsata*. This bee can grow up to 20 mm in length and is dark coloured except for the first three abdominal segments, which are dark yellow. It has been estimated that a single bee may effect 12,000 pollinations in a day (Dammerman 1929). This species can be very aggressive, and collecting its honey is exceedingly dangerous, but these factors simply add to the attractiveness of the honey (Akratanakul 1986).

The most successful bee projects have concerned the development of hive cultures of *Apis cerana* in coastal coconut areas, where they occur naturally in hollow trunks, heaps of broken shells, under house eaves, and in abandoned buildings. Since coconuts flower all year round, developing bees always have adequate forage, and there are generally other flowers in the vicinity. Management and protection of the bees also increases farm productivity.

Shellac. The only area where shellac is produced is in planted *Schleichera oleosa* forest east of Probolinggo, just behind the enormous, new Paiton power station. The scale bug *Kerria* (sometimes known as *Laccifera) lacca* was introduced from India in 1936, but it was not until 1956 that Perhutani devised a commercially viable means of cultivation. The insect population comprises almost entirely females which live stationary lives attached to branches by their mouthparts. Hundreds of red nymphs are produced, and the first instars are mobile and disperse. Later instars are stationary and they secrete a hard, resinous substance which forms a crust, or lac, around the branch of a host plant (fig. 14.13). Only 900 ha are currently used for shellac production, some having been felled to make way for the power station, although there are plans to establish new *S. oleosa* plantations near Madiun (A. Setiadi pers. comm.). The twigs with the crust are snipped off and taken to a processing factory where some are processed. The others are returned to be tied onto the trees for further infection by the first instar nymphs. After 21 days the second group are also taken to the factory. The shellac resin is used in paints, varnish, and similar products.

Silk. The culture of the silk moth *Bombix mori* by growing the mulberry tree *Morus* spp. (Mora.) has been much encouraged in order to reduce the quantity of silk imported into Indonesia, and to increase incomes in remote areas. The programme has been successful; Indonesian silk production was just 67 kg in 1963, but 8,695 kg in 1993. The mulberry leaves are fed to the silk moth caterpillars which are kept on trays in villagers' home for the eighteen-day feeding period before the caterpillars make a silk cocoon and pupate. Each cocoon weighs about 1.2 g, with 0.1 g silk with a length of about 1 km. In the past, poor management, pests, and diseases have caused problems but these have largely been overcome.

Rattan

Rattan can still be found growing in the remoter forests of Java and Bali, but gross and continuing overexploitation of the best species has denied future generations the option of continuing the harvest of wild rattan. There is, however, considerable potential for the development of rattan plantations on degraded lands, in line with the government's policy of continuing to export rattan even as wild stocks dwindle. Cultivated rattans can grow 15-20 m in their first 10 years, after which they are harvested. They grow well in old fields where the scattered trees can be used for support. The RePPProT (1989) review suggests that the Sewu-Lengkong Karstic Plateaux (pgs. 109, 110) could be rehabilitated using rattan, even if growth is slow, because once it is growing on the steep slopes, this natural barbed wire would deter other uses – human and animal.

Fuelwood

The supply of fuelwood need not be an issue in Java, although it tends to be presented as such, because the production of fuelwood from agroforestry areas can easily meet the annual demand of 90 million m³ (Smiet 1990a). Regional shortages do occur, but could be met by supplies from areas where there is a surplus. Thus, the problem is one of distribution, not of overall supply. This should be remembered when hearing that remote villages depend on natural forest for fuelwood (about 10% of fuelwood comes from forests), because the dependency and forest disturbance so caused can be eliminated by appropriate agroforestry programmes. Gathering of waste material is permitted both from plantation and protection forests, as well as at saw mills, but these have long since been inadequate. Local over-demand occurs particularly where wood is used to fuel furnaces which burn lime, such as at Padalarang, west of Bandung. It would be possible to install gas-fired furnaces, but price and competition preclude it.

Fuelwood species include the exotic *Leucaena leucocephala, L. glauca, Calliandra calothrysus, Gliricidia sepium, Acacia auriculiformis,* and *A. decurrens* (all Legu.), which provide pulp, firewood, charcoal, construction timber, and leaves rich in protein which can be used as cattle fodder or green manure. In order to produce fuelwood, these are planted either as an intercrop or beneath mature stands of trees. Probably the most popular fuelwood tree planted in agroforestry programmes is *C. calothrysus.* This is a small tree or tall shrub which was introduced from Central America in 1936 to shade coffee plantations. It also successfully reclaims degraded areas, and grows well beneath pine and other plantation trees. It also coppices well, but this means that it has no straight stem and so is awkward to carry. The leaves can be used as cattle fodder and its abundant red flowers are much visited by honey bees, and so it is sometimes planted for its nectar alone.

Other Non-timber Products

There is an enormous range of fruit, canes, oils, tannins, dyes, and bamboos in the forests of Java and other Indonesian islands which could be

Table 14.8. Areas (km²) of estate crops in Java and Bali.

	West Java	Central Java	East Java	Bali	Total
Sugar cane	386	425	1,508	7	2,326
Rubber	1,357	253	299	6	1,915
Coconut	633	313	219	457	1,622
Coffee	12	69	692	235	1,008
Tobacco	0	69	356	7	432
Tea	296	0	18	0	314

Source: RePPProT (1989)

introduced with profit into agroforestry programmes. These require the vision and budgets for research and development (Kartawinata 1990a).

ESTATES

Estates, or extensive areas of commercial agriculture, were introduced by the Dutch to produce crops for export. One of the earliest estate crops was quinine *Cinchona calisaya* (Rubi.), which was first brought to Indonesia in 1854 from the slopes of the Andes in South America where the Inca people used preparations of the bark to cure malarial fevers. Some 75 plants survived the journey and were held initially at Cibodas before being transplanted to the slopes of Mt. Malabar, near Bandung. By the outbreak of the Second World War about 90% of the world's quinine came from estates around Bandung (Cabaton 1911; Taylor 1945; Neill 1973). Subsequently, synthetic products have replaced natural quinine, and almost all the original estates have been turned over to tea production.

The most extensive estates in Java and Bali are of sugar cane, followed by rubber, coconut, coffee, tobacco, and tea, although their distribution across the provinces is by no means even (table 14.8): sugar cane is found mainly in East Java, rubber mainly in West Java. Apart from sugar and tobacco, which in some areas compete with rice for irrigated land, most are dryland crops.

Estates provide valuable foreign exchange and so, although they have major problems associated with them, their operation and goals have not changed markedly since Independence:
- they still serve urban dwellers rather than rural populations;
- they are subject to global fluctuations in prices;
- they are susceptible to disease because they concentrate on very few species or varieties;
- the processing factories are major sources of pollution;
- they support a low human population density within them; and
- since they do not employ very many people, they aggravate the land/people problems around them (Soemarwoto 1991a, b).

The outworkings of some of these problems are seen where the order and neatness of estates with their perennial crops contrasts sharply with the denuded forested slopes around them sown with annual crops, and the adjacent protected forests are used intensively for the collection of fuelwood and other nontimber forest products. Some of the problems would be relieved, perhaps, if estates were changed into cooperatives which share the profits more equitably than conventional business interests. Where the estates grow annual crops (tobacco and sugar cane), it is now common for the farmers to be allowed to grow those crops on their own land

Figure 14.14. Sugar cane being harvested near Pasuruan.
By A.J. Whitten

without a lease agreement, and to sell the produce for processing and marketing. This gives the farmers a better deal, but it creates problems of quantity and quality control (Soemarwoto 1991b).

Estate crops are grown as monocultures which, as explained earlier, are generally regarded as unsound from an ecological viewpoint. Whereas forest plantations and tree crops may have some redeeming ecological features (p. 598), the same cannot be said of annual crops. Ecologists are woefully ignorant of how the pests and their predators or parasites interact, and it would seem to be a major priority to follow the lead given by rice scientists (p. 577) and to examine the components and dynamics of these ecosystems.

Sugar Cane

Sugar cane plants are cultivars of different hybrids of *Saccharum* (Gram.) species, many of them indigenous. Sugar pressed from the canes was produced in Java as long ago as 1635 (Donner 1987), and by the mid-nineteenth century it had become an enormously profitable crop for the Dutch. At times "the whole east end of the island and the low, hot lands along the coast were green at their season with the giant grass" (Scidmore 1899). In the 1920s and early 1930s sugar was Indonesia's most valuable export. It lost its position partly because of the Great Depression, partly because of an increase in domestic consumption, and partly because the cultivators turned to other crops. It is still produced on the drier eastern lowland plains, largely by nuclear estates and by nearby smallholders under contract (fig. 14.14). Total national production from all estates and households in 1993 reached 2.5 million tonnes (52% from East Java), but this was not sufficient to meet the demand, which is somewhat over 3.2 million tons. Trials of the daisy-like South American stevia shrub *Stevia rebaudiana* (Comp.) are under way: the leaves produce crystals 200-300 times sweeter than cane sugar and contain fewer calories.

Sugar cane requires high temperatures, plenty of sunlight, abundant water supplied either from rain in excess of 1,500 mm annually or by irrigation, good drainage, and highly fertile soil, or generous fertilizer applications. It suffers from stem borers (moth larvae), and the tunnels they make provide access for disease organisms such as red rot fungus. Rats also eat the cane and promote the access of disease, although various snakes are generally present which help to control the rat numbers.

Rubber

Rubber *Hevea brasiliensis* (Euph.) is native to the Amazon basin, where its grows as an understorey tree in the lowland rain forest. The indigenous people used its latex before the European conquest to make bottles, shoes, and waterproof clothing, and ate its seeds, but only after long boiling or soaking to remove cyanide poisons. In Europe it was regarded as an interesting product, but not until 1770 did it actually have a use; to rub out pencil marks. Once a solvent was found, it was possible to make other items, and rubber tubes were first made in 1791. Unfortunately rubber items became sticky in the sun until the process of vulcanization (heating rubber with sulphur) was discovered by Goodyear in 1839. Suddenly rubber was a desirable product and the wild trees were extensively exploited. The demand increased still further during the American Civil War when waterproof clothing was needed, and again in 1888 when Dunlop discovered the pneumatic tyre (Purseglove 1968).

It was first suggested that rubber be grown in plantations in 1824, but nothing was done until 1870 when some seeds were taken to Asia to boost the world supply of latex. The seeds and seedlings proved to be reluctant travellers, and it was not until 1876 that the first live seedlings arrived in Sri Lanka, Singapore, and Java. Planters were reluctant to give the new crop room on their estates, but by the early 1900s it was established as an important plantation species. Interestingly, rubber plantations have never succeeded within its native range due to a leaf blight which, fortunately, has never reached Asia.

Most of the land under rubber in Java was cleared of lowland rain forest to make way for the crop. Rubber is tolerant of a wide range of soils, but it does best on deep, well-drained soils of pH 5-6, and so most rubber estates are found in the lowlands and hills of West Java. Java and Bali have 5% of the total area planted to rubber in Indonesia, but yields are low because many of the estates have old and unimproved stock, and traditional single-panel tapping methods are still used. As a result, Java and Bali contribute only 2% of the national production. Rubber is Indonesia's third-highest foreign exchange earner after oil/gas and timber, and Indonesia is the world's largest rubber producer (Dirjen Perkebunan 1993).

The simplicity of the rubber ecosystem with a single species of tree and a poor herb layer means that few larger animals can use it. Lutung monkeys and long-tailed macaques will enter plantations to eat leaves and tri-lobed fruit, but they are dependent on adjacent forest for most of their needs.

Rubber suffers from relatively few diseases, a notable exception being white root rot *Fomes lignosus,* and is little troubled by pests, although termites, rats, squirrels, wild pigs, and the giant African snail can cause problems (p. 299).

Coconut

Coconuts *Cocos nucifera* (Palm.) are wonderfully useful trees, being a source of oil, medicine, fibre, timber, leaves for thatch and mats, wood for utensils, and charcoal for fuel (Purseglove 1968). It is for the oil pressed from the dried endosperm or copra, that coconut palms are grown in estates. The best-quality oil is used for margarine and cooking oil; lower grades are used for soap, detergents, cosmetics, resins, wax, candles, and confectionery. With increasing competition from other vegetable oils, the coconut's importance as a plantation crop may decline. Nevertheless it will always be important in rural areas because it grows under conditions where other food crops often do not perform well and it produces many products used in local households. New varieties of hybrid dwarf coconuts are gradually replacing older palms in plantations since they fruit sooner. Until 1960 coconuts were the major source of vegetable oil in Indonesia, but soybean and oil palm are now more important (Purseglove 1968) and most of the coconut oil production is now for domestic use.

Coconuts probably arrived in Java naturally, their seeds buoyed on the sea from sources to the east. They can grow on a wide range of soils from coarse sand to clay, provided the soils have adequate drainage and aeration. They require regular rainfall for, although the leaves are designed to minimise water loss and can withstand drought, long dry spells impede palm growth and nut production. The tree's fine pinnate leaves and flexible stem allow it to withstand strong winds, which makes it a suitable tree for exposed coastlines.

A stand of mature coconut palms allows considerable light to reach the ground, and multicropping can be practised with plants of different heights and different nutrient and shade requirements all grown together. Coffee, cocoa, cloves, bananas, pineapple, ginger, beans, maize, and even rice can be grown beneath the coconut crop. Intercropping is valuable to the smallholder because it increases his crop and his income. Multicropping increases ground cover and thereby increases soil protection. It may also be an important factor for obtaining maximum returns from coconut plantations. Coconut production may also increase from this use of the land because the palms benefit from tilling, weeding, and fertilizer applied to the other crops.

More than 750 species of insect pests attack coconuts to some degree, with about one-fifth of these not known to attack other plants (Dammerman 1929c; Kalshoven 1981; Davis et al. 1985). Many insect larvae feed on coconut leaves, particularly when they are young and tender and within the 'protective' crown. For example, larvae of the rhinoceros beetle *Oryctes rhinoceros* tunnel through the unopened leaves, creating holes that appear as triangular gaps once the leaves unfold. This pest can be controlled biologically using a virus (Ohler 1984). The bright colours of the limacodid nettle caterpillars which eat the mature leaves warn predators of the unpleasant consequences of attack, and indeed the caterpillars bear barbed spines which release a painful toxin. Leaf-mining hispid beetles *Brontispa longissima* attack the coconut fronds but can be controlled biologically by introducing a parasitic eulophid wasp *Tetrastichus brontispae* which deposits its eggs in pupating beetle larvae. About 20 wasps emerge from a single beetle pupa (Davis et al. 1985). Orange and black weevils *Rhyncophorus ferrugineus* do not eat leaves but rather bore into the starch-rich coconut stem to lay their eggs.

Rats and plantain squirrels *Callosciurus notatus* live in the crowns of coconut trees and cause considerable damage to young leaves and nuts; they also eat the drying copra and can cause losses of 20% or more of the crop (Davis et al. 1985). One way to control rats is to encourage birds of prey such as owls in plantations (Whitten et al. 1987c). Wild boar *Sus scrofa* and the Javan pig *S. verrucosus* can be destructive in coconut plantations.

Chapter Fifteen

Aquaculture

Aquaculture on Java dates back to at least A.D. 1400, from which time a law code *Kutara Menawa* laid out punitive measures against those who stole fish from brackish or freshwater ponds (Schuster 1952). Whereas the brackish-water ponds were begun in East Java by outlawed convicts, the backyard ponds have simply evolved as an integral part of village life, often as part of the home garden (p. 608) as an expedient recognition that common carp, for example, grow one and a half times faster than do cattle or sheep. There are now numerous systems in operation using a wide variety and combination of fishes, prawns, molluscs, and algae, and these are summarized below (table 15.1). Marine aquaculture development for oysters, cockles, and mussels are beyond the scope of this book.

Brackishwater ponds or tambak cover the greatest area of all the aquaculture types and produce the greatest quantity of fish. The fastest-growing part of aquaculture is the floating cage culture, but this still produces less than 5% of that produced in tambak (table 15.2).

FRESHWATER PONDS

The most common fish cultured and bred in ponds for consumption are common carp *Cyprinus carpio*, Java carp *Barbodes gonionotus*, red-eyed carp *Puntius orphoides*, giant gourami *Osphronemus goramy*, kissing gourami *Helostoma temminckii*, Nile tilapia *Oreochromis nilotica*, red tilapia (a hybrid between the Nile and common mujair tilapia *O. mossambicus)*, nilem carp *Osteochilus hasseltii*, snakeskin gourami *Trichogaster pectoralis*, and walking catfish *Clarias batrachus* (Susanto 1990). Seven others are cultured, but the fry are gathered from rivers: river eel *Anguilla* spp., swamp eel *Monopterus albus*, jambal catfish *Pangasius djambal*, jelawat *Leptobarbus hoevenii*, sand goby *Oxyeleotris marmorata* (actually an eleotrid, not a goby), and *Barbodes belinka*. Invertebrates are also cultured, such as giant prawn *Macrobrachium rosenbergi*, *Pila* snails, and freshwater clam *Meretrix*.

There are interesting differences between the provinces in terms of the species favoured. For example, the common carp is by far the most-

Table 15.1. Aquaculture systems found on Java and Bali. Extensive systems are defined as having no feed or fertilizer inputs; semi-intensive systems as having some feed and/or fertilizer inputs; and intensive systems as being mainly reliant on external feed inputs. The possible consequences of the introduction of exotic species and breeds apply to all the systems listed here.

	Benefits	Environmental and social costs
EXTENSIVE		
Seaweed culture	Income; employment; foreign exchange.	May occupy formerly pristine reefs; rough-weather losses; market competition.
Coastal bivalve culture (mussels, oysters, clams, cockles)	Income; employment; foreign exchange; improved nutrition.	Public health risks and consumer resistance (microbial diseases, red tides, industrial pollution); rough-weather losses; seed shortages; market competition, especially for export produce; failure.
Coastal fishponds (mullets, milkfish, prawns, tilapias)	Income, employment, foreign exchange (prawns); improved nutrition.	Destruction of ecosystems, especially mangroves; increasingly noncompetitive with more intensive systems; non sustainable with high population growth.
Pen and cage culture in eutrophic waters and/or rich benthos (carps, catfish, milkfish, tilapias)	Income; employment; improved nutrition.	Exclusion of traditional fishermen; navigational hazards; management difficulties; wood consumption.
SEMI-INTENSIVE		
Fresh and brackishwater ponds (prawns, carps, catfish, milkfish, mullets, tilapias)	Income; employment; foreign exchange (prawns); improved nutrition.	Freshwater: health risks to farm workers from waterborne diseases. Brackishwater: salinization or acidification of soils and aquifers. Both: market competition, especially for export produce; feed and fertilizer availability/prices; conflicts/failure.
Integrated agriculture-aquaculture (rice-fish; livestock/poultry-fish; vegetables-fish and all combinations of these)	Income; employment; improved nutrition; synergistic interactions between crop, livestock, vegetable, and fish components; recycles on-farm residues and other cheap resources.	As freshwater above, plus possible consumer resistance to excreta-fed produce; toxic substances in livestock feeds (e.g., heavy metals) may accumulate in pond sediments and fish; pesticides may accumulate in fish, although rice-fish farmers tend to be careful with chemicals.
Sewage-fish culture (waste treatment ponds; latrine and septic tank wastes used as pond inputs; fish cages in wastewater channels)	Income; employment; improved nutrition; turns waste disposal liabilities into productive assets.	Possible health risks to farm workers, fish processors, and consumers; consumer resistance to produce.

Table continues

Table 15.1. *(Continued.)* Aquaculture systems found on Java and Bali. Extensive systems are defined as having no feed or fertilizer inputs; semi-intensive systems as having some feed and/or fertilizer inputs; and intensive systems as being mainly reliant on external feed inputs. The possible consequences of the introduction of exotic species and breeds apply to all the systems listed here.

	Benefits	Environmental and social costs
Cage and pen culture, especially in eutrophic waters or on rich benthos (carps, catfish, milkfish, tilapias)	Income; employment; improved nutrition.	As extensive cage and pen systems above.
INTENSIVE Freshwater, brackishwater, and marine ponds (prawns; fish, especially carnivores (catfish, snakeheads, groupers, seabass, etc.)	Income; employment; foreign exchange.	Effluents/drainage high in BOD and suspended solids; market competition, especially for export product; conflicts/failures.
Freshwater, brackishwater, and marine cage and pen culture	Income; foreign exchange (high-priced carnivores); a little employment.	Accumulation of anoxic sediments below cages due to faecal and waste-feed build-up; market competition, especially for export produce; social disruption; consumption of wood and other materials.
Other – raceways, silos, tanks, etc.	Income; foreign exchange; a little employment.	Effluents/drainage high in BOD and suspended solids; many location-specific problems.

Source: based on Pullin (1989)

Table 15.2. Fisheries production in 1990 from different sources and their water area.

	Inland open water	Tambak		Freshwater ponds		Cages	Sawah	
	tons	tons	ha	tons	ha	tons	tons	ha
Jakarta	0	78	447	424	106	0	310	167
West Java	4,438	55,881	39,034	67,945	15,951	3,493	33,933	55,094
Central Java	6,858	49,864	14,839	15,498	1,831	212	1,239	3,342
Yogyakarta	1,063	0	0	931	428	0	506	3,937
East Java	11,361	61,094	40,562	4,604	1,937	25	35,811	16,896
Bali	662	4,096	711	327	131	0	584	3,491

Source: Dit. Jen Perikanan (1992)

favoured species in West Java and Bali, Java carp the most-favoured in Central Java, and walking catfish the most-favoured in East Java (table 15.3).

The simplest cultivation systems serve household needs with low-value Java carp or Nile tilapia, or cash needs using common carp, which consistently sells for more than the other species save for giant gourami. The tilapia are sometimes regarded as pond pests partly because they excavate the pond base when nesting and can thus cause leakage, and partly because they have a much higher reproductive capacity and so can crowd out other cultured species. Because of this they are sometimes confined to stream channels where they are less damaging. Pond fish generally receive no prepared food, but will receive scrapings off plates and rice bran, and the pond will be fertilized by wastes falling from chicken cages, rabbit hutches, or human latrines built above them (p. 612). The common carp has to be managed rather more carefully than the others and has to be given a spawning substrate. Throughout the carp-growing world this palm-fibre mat is known by the Sundanese name *kakaban*.

The magnificent large native carp *Tor dourouensis* is kept in ponds fed by springs in the Kuningan and Sumedang areas, but they are regarded as sacred and may not be fished (p. 682). They used to be found in the surrounding waters, but this appears to be no longer the case. The life cycle, development, and growth of this species is hardly known, yet it may have considerable potential.

The ever-increasing interest in aquarium fishes has seen a concomitant increase in the number of ornamental fishes bred in large tanks or even rice fields. Indeed, one of the major exports of the mayoralty of South Jakarta is aquarium fishes, and it is local policy to increase the productivity of the fish farms and to improve the quality of the fishes. There are nearly one hundred species of native and exotic ornamental species which are cultured for the aquarium trade, and many of these have escaped into rivers, streams, and lakes (p. 439).

Table 15.3. Production (tons) in 1990 of the main cultured fish species by province.

	Common carp	Java carp	Mujair tilapia	Nile tilapia	Nilem	Giant gourami	Kissing gourami	Walking catfish
Jakarta	39	20	10	31	0	137	1	180
West Java	22,503	9,804	9,589	6,935	8,567	1,516	5,058	674
Central Java	526	6,711	3,310	421	3,368	490	116	418
Yogyakarta	188	221	39	0	0	164	0	178
East Java	507	1,148	982	157	0	401	1	1,157
Bali	180	4	16	74	0	25	0	25

CAGES

Floating net cages, each of which can hold hundreds of fish, were first tried in Indonesia on Jatiluhur in 1972. They have since become an increasingly popular form of culture technique, and some 600 have been built on Saguling alone (fig. 15.1). This popularity is due to the high returns, and a single cage with several thousand common carp can produce the equivalent of the annual protein needs for a family of five. The fish produced tend to be for the speciality market, however, since they are grown to 0.5-2 kg, rather than the 100-150 g size which is most commonly eaten in rural homes.

The practice of siting the cages in sheltered and often shallow bays or inlets has resulted in ecological problems. No institution concerned with licensing or fisheries development had the scientific competence to determine – or the weight of authority – to enforce the siting, density, size, or stocking of the cages based on the water movements, water depth, turbidity, and the water's natural ability to decompose large quantities of fish faeces and surplus food. The scale of losses are enormous: overcrowding of cages on Saguling resulted in 250 tons of fish being lost in late 1990 (Zerner 1992). Also, the cage culture on Cirata reservoir is currently excessively overcrowded, with resultant water quality problems: in mid-1992, 50 tons of fish died as a result of low water quality after an overturn or upwelling. This occurred after two days of cloudy weather which cooled down the surface waters, allowing the bottom waters that are artificially enriched with fisheries wastes to rise to the surface. Decomposition of the wastes in the relatively oxygen-rich top layer used up the available oxygen, causing the fish to suffocate (V. Brzeski pers. comm.). Government programmes are encouraging intensification of cage culture, but unless some control over siting is exercised, it is likely that basic ecological principles will undermine well-intentioned developments.

It should be pointed out, however, that the abundant experience with fish cages in temperate parts of the world cannot be applied directly to tropical reservoirs such as Saguling. Saguling has a much higher carrying capacity for fish cages than might be thought, and there is little or no loss of phosphorus to the open water from the fish food. This is probably due to a combination of three factors:

- relatively high water temperature causes high rates of decomposition;
- rapid decomposition of wastes and their utilization by bacteria and phytoplankton, notably *Microcystis,* which occurs in almost continuous blooms even without the fish cages because of the large amounts of organic wastes originating from Bandung; and
- the high biomass of fish efficiently assimilates the relatively low amount of phosphorus in the feed (1.0%) (Costa-Pierce and Roem 1990).

Figure 15.1. Floating cages holding carp on Saguling reservoir in 1993.
With permission of Kompas

Figure 15.2. Fish cages in the Cimula River near Tasikmalaya.
With permission of Kompas

Figure 15.3. Farmers harvesting fish from their rice fields.

· *With permission of* Kompas

It has been suggested that if a pelagic plankton-feeding fish such as the freshwater sardine *Clupeichthys aesarnensis* from northeast Thailand were introduced, the algal blooms would be held in check, the fisheries production of the open waters would increase, and the risk of cage fish problems would be reduced (Costa-Pierce and Soemarwoto 1990a).

Carp and other fish species are also held in bamboo cages in rivers and small water channels, and are fed village-made 'pellets' of rice bran, leaves, and soya (fig. 15.2). Good yields can be obtained using this method (Brown and Prayitno 1987).

RICE-FISH CULTIVATION

Farmers on Java and Bali have probably been raising fish in rice fields for as long as irrigated rice fields have been found, although initially the fish were wild and entered the rice fields naturally (Vaas 1947) (fig. 15.3). The practice decreased in importance as the numbers of wild fish declined as a result of deforestation, overfishing, fish disease, and pollution (p. 718). The current moves to reduce artificial fertilizers and pesticides will doubtless be beneficial, and there is active interest in increasing the complexity of rice field ecosystems by the introduction of ecologically appropriate, agriculturally acceptable, and economically useful fish to increase the productivity of the land using low-capital, low-risk techniques. Managing fish populations (and other aquatic life such as frogs, crabs, and prawns) is a natural partner to integrated pest management, and the hope is that full benefit will be taken of this (dela Cruz 1990; Horstkotte et al. 1992).

The integration of rice and fish culture also has benefits for weed control. The choice of fish species is critical, however, because the popular tilapia eats plankton rather than plants, and grass carp are as likely to eat the young rice as any weeds, and tend to languish in shallow, warm water. Again, there are clear differences between the provinces in the fish cultured: common carp dominating in West Java, and Java carp in East Java. Probably the most useful species for weed control are the milkfish *Chanos chanos* and common carp which damage shallow-rooted weeds while disturbing the bottom in search of benthic organisms. This activity also increases turbidity, and decreases light penetration, thereby slowing the growth of weeds. For overall benefits, however, an ecologically balanced mix of species utilizing a variety of niches provides both weed control and a good harvest of protein (Mudjiman 1986; Koesoemadinata and Costa-Pierce 1988; van Dam 1990; Cagauan 1991). Both the rice and fish seem to benefit from the presence of the other, and this is true virtually irrespective of size and arrangement of trenches dug in the field, so long as water is abundant. Three main systems exist for cultivating fish with rice:

- growing them for three to four weeks between rice crops *(penyelang);*
- growing in the field while the rice is growing *(minapadi);* and
- growing them as a single crop for two to three months after the second rice crop *(palawija).*

OPEN WATER

People living around the shores of Java's open water are ingenious in the ways they have devised to outwit aquatic animals, and on Rawa Pening alone, no less than 27 types of fishing gear are used to catch fish, prawns, mussels, and turtles (Göltenboth and Timotius 1992). The largest of these, seen also north of Segara Anakan and elsewhere, are the square *branjang* dip nets which are lowered slowly into the water from bamboo piers (fig. 15.4).

The yield of fish from Java's lakes varies enormously, from 15 to nearly 400 kg/ha, but not all bodies can aspire to high productivity because of differences in the nutrient status of the water, the shape of the bed, the abundance of macrophytes, the community of species present, and the altitude. As a general rule the fish yields from lakes and reservoirs with tilapias

Figure 15.4. Large branjang nets are often used to catch fishes in still water.
With permission of Kompas

is greater than from those without (Fernando and Holcik 1991).

Small to medium irrigation reservoirs receive the most attention for aquaculture development because they have relatively few weed problems, are reasonably rich in nutrients, and the water can be totally drained during the dry season, allowing easy and efficient harvesting (Hardjamulia and Suwignyo 1988). In large lakes with complex suite of species and interactions, there is a need to understand their ecological workings, such as the relative population sizes, the ratio of predators to prey, the location of spawning grounds, spawning seasons, the minimum breeding size of different species, effects of permitted and other fishing gear and fishing seasons, and the establishment of reserve areas to protect brood fish and their fry (Hardjamulia et al. 1988). This was done for the Saguling reservoir and, as mentioned above, it was found that there was ecological space for the introduction of a freshwater sardine from Thailand (Costa-Pierce and Soemarwoto 1990a).

Introductions do really need considerable care. For some, the knee-jerk reaction to the low productivity of some lakes has been to introduce exotic species of fishes. This has been the case in Bali where the high-altitude, oligotrophic lakes look, to the untrained eye, as though they should have more fish in them. In Lake Bratan this has worked to a certain extent, perhaps because of the increasing effects of human activities in and around the lake, but introductions of grass carp *Ctenopharyngodon idellus* and jambal catfish into the nearby nutrient-poor Lake Tamblingan failed. The fish introduced to Lake Bratan were not healthy, with the result that most fish caught now have parasites or sores.

Fish yields from Rawa Pening dropped from 548 to 18 kg/ha between 1972 and 1980, probably due to overfishing. Many of the fish caught in the increasingly small mesh nets were not yet mature, and the average length of the fish caught decreased (Carlander and Kastowo 1978; Göltenboth 1978). Since 1980 the annual yield has more or less recovered, possibly because of the introduction of the floating cages, the yields from which have reduced demand for fish caught from the open waters (Göltenboth and Kristyanto 1993).

PESTS AND DISEASES

Freshwater fisheries face a number of major problems: diseases and parasites, excessive predation on eggs, fry, and fish, competition from noneconomic species, genetic degradation of target species, loss and non-use of indigenous genetic resources, and water pollution from faecal and food wastes in intensive fisheries (Sukadi and Suseno 1991).

As aquaculture develops (for it has a much less developed technological base than agriculture), so an inevitable result will be the creation of better domesticated breeds. These will be dispatched from breeding centres and, inevitably, some will escape. This can lead to populations that displace or interbreed with wild stocks, disrupt natural habitats by causing the disappearance or proliferation of vegetation or by increasing turbidity (in the case of energetic benthos feeders), and can accidentally cause the introduction of pathogens, parasites, and predators. For example, back in 1932, serious epidemics of *Ichthyophthirus multifilis* occurred in West and Central Java, probably from imports of ornamental fish. Twenty years later an epidemic of *Myxobolus pyriformis,* an accidentally imported sporozoan parasite, killed thousands of young carp in Central Java and has continued to cause losses ever since because the cysts formed around the spores are resistant to normal chemical controls (Djajadiredja et al. 1983). The worst outbreak of fish disease, however, was caused by a bacterium *Aeromonas* sp., probably introduced with carp imports from Taiwan which killed 125 tons of carp and large numbers of walking catfish in Java during 1980 (Kabata 1985).

Another troublesome parasite is the crustacean *Lernaea cyprinacea* or anchor worm, which was accidentally introduced to Java from Japan in 1953. This attacks carp, kissing goramy, gurami, and other economically important fishes, and claimed up to 30% of hatchery production in the worst year, 1974. It is a damaging pest even at low levels, and it is probably

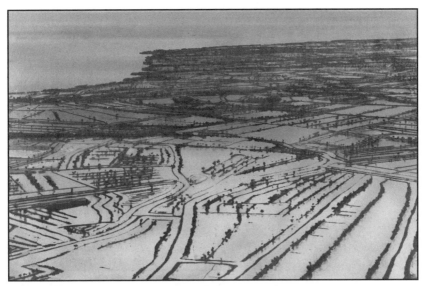

Figure 15.5. Extensive tambak near Krawang, West Java.
With permission of Kompas

the cause of the lesions that plague many of the fishes in Lake Bratan. Control methods have been developed, but farmers have been slow to adopt the necessary practices (Djajadiredja et al. 1983). A serious loss was experienced in 1990-91 when a genetically important caged collection of common carp in Jatiluhur was killed by an infestation of the fish louse *Argulus* (V. Brzeski pers. comm.). One of the latest new fish parasites is the larval glochidia of the freshwater mussel *Anodonta woodiana,* imported recently from Taiwan (p. 300) (Djajasasmita 1982). The glochidia do not seem to kill fish but they do stress them, making them more susceptible to diseases (Sachlan 1978).

BRACKISHWATER AQUACULTURE

Traditional brackishwater aquaculture or tambak comprised coastal ponds into which high-tide water was directed, and whatever animals happened to be washed in were grown and then harvested; the harvest was primarily milkfish, followed by prawns (Hannig 1988). The practice became more intensive as the ponds were stocked with larvae caught for the purpose, but such systems are still sustainable (fig. 15.5). For example, there is a large area of such tambak to the east of Surabaya and Sidoarjo. Here the ponds

Figure 15.6. Tambak banks stabilized by mangrove trees east of Sidoarjo.
By J. Davie

are separated by tall, steep bunds stabilized by *Avicennia* and other trees of mangrove forest (fig. 15.6). These trees provide a green mulch of leaves which fertilizes the ponds, and large trees are felled for timber. In places there are copses of up to six species of mangrove trees (Enex 1994), which support some major breeding colonies of herons, egrets, and similar birds. While it is undeniable that these water birds do take some of the fish and prawns from the ponds, they probably have a significant influence on the productivity of the ponds by virtue of their enriching faeces falling into the water. On the seaward mudflats there are major concentrations of migratory wading birds (fig. 15.7) (Erftemeijer and Djuharsa 1988). The mudflats are being extended by deposition of sediments from the Solo and Brantas Rivers, and these are being brought into the system by the pond farmers. In this way they are actually extending the area of mangrove, albeit rather pauce compared with natural mangrove forest.

Rapid development of tambak was triggered in 1980 by a Presidential decree which banned the use of offshore prawn trawlers. This obviously led to dramatic decreases in prawn production, and by means of compensation, great efforts were put into developing prawn aquaculture. The explosion in the area of prawn ponds was also due to the successful development of artificial spawning of the prawns, allowing hatchery-produced eggs to be used for monoculture production. Prawns command a high price in the

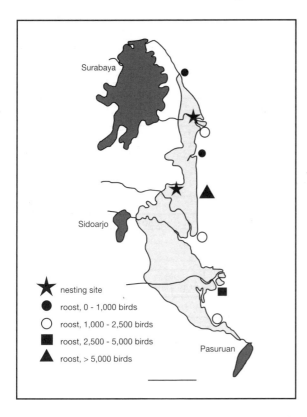

Figure 15.7. The tambak area (shaded) east of Sidoarjo, showing the sites of nesting colonies and major concentrations of migratory wading birds. Scale bar indicates 10 km.

Based on Erftemeijer and Djuharsa 1988; RePPProT 1989

Legend (in figure):
★ nesting site
● roost, 0 - 1,000 birds
○ roost, 1,000 - 2,500 birds
■ roost, 2,500 - 5,000 birds
▲ roost, > 5,000 birds

international market, where profits are high but the stumbling blocks many. Low-quality prawn exports from Indonesia to the U.S. resulted in a ban by that government in 1989, and these damaged and otherwise inferior animals were diverted to the domestic market. In addition, the widespread use of antibiotics in artificially crowded prawn ponds in order to check the spread of disease is having effects within the whole system since the antibiotics enter the natural system when the ponds are drained. Fish are also cultured, the principal species being the milkfish, which is raised from transparent larvae caught by fisherpeople pushing large nets around the coastal shallows (fig. 8.15). This fish primarily serves the domestic market.

Tambaks are in essence situated between sawah and mangroves, usually desperately impoverished ones, and the three systems are directly or indirectly linked by creeks or canals, and by a complex of human activities and interactions. The tambak ponds receive water from both their neighbouring ecosystems at different rates at different times, and there is considerable skill and understanding required to ensure that the relative

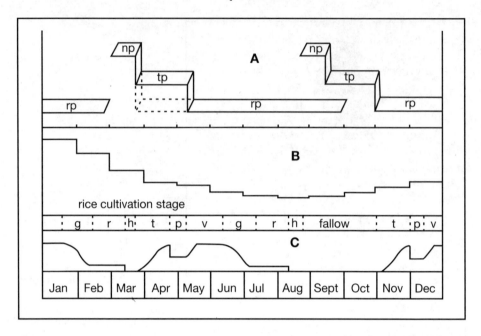

Fig. 15.8. Tambak cultivation pattern (**A**) in Karawang, West Java (np = nursing period; tp = transition period; rp = rearing period), and its relation to water resource conditions, rainfall (**B**), and graph of irrigation schedule (**C**) to support rice cultivation stages (t - tilling, p - planting, v - vegetative growth, g - generative growth, r - ripening, h - harvest).

After Muluk 1985

heights of the dykes and pond floors allow influx and outflow of water. The flow of fresh water is related to rainfall and, in most areas, the quantity of water released from major reservoir authorities. The flow of saline water is determined by predictable tidal patterns. The flow of fresh water is lowest during the dry season fallow, so water for the tambak during this period is almost exclusively taken from the sea (fig. 15.8). Different areas have different types of tambak managed under different systems which reflect adaptations to local environmental conditions over the last century or longer (Schuster 1952; Muluk 1985). What has not, and perhaps cannot, be adapted to is the pollution originating in the industrial zones that are replacing the coastal sawah. This can cause serious problems for the tambak owner, but seems very difficult to deal with.

Tambak can provide better incomes to owners, tenants, and labourers than rice fields. In addition a tambak near mangroves can produce crabs the sale of which can generate an annual income equivalent to nearly 500 kg of rice. This is welcomed by the tambak operators because the crabs compete with the fish and prawns (Muluk 1985).

Brackishwater ponds will only be successful where strict physical requirements are met (RePPProT 1989). It must be possible to control both the flow of brackish and fresh water, and to do this, the bottom of the ponds should be above the level of the lowest average low tide (to allow good drainage), but below the highest average high tide (to prevent continual inundation and possible escape of the stock). Some ponds that were built without due care and attention can be seen (such as in northwest Bali), and large areas which might seem suitable are not. For example, in Segara Anakan the soil is very porous and does not lend itself to the building of impoundments (IPB 1984). A great deal of knowledge and understanding of local conditions and constraints exist among the owners of extensive tambak, and this needs to be tapped (Muluk 1985).

Brackishwater ponds for intensive culture of prawns and fish have seemed an extremely good way of increasing income among coastal dwellers, even if most of the profits end up in Jakarta and other cities. However, it seems that in some cases the developers of the land are not as careful with it, or as concerned about sustainability, as are the traditional coastal dwellers, and badly managed intensive tambak can be ruined in as little as 18 months.

Some 90% of the fish caught within 10 km of the coasts in this region have food fragments from mangrove in their guts (Rondelli et al. 1984), and mangroves are important in the life cycles of many species of marine life, many of which are of economic value (though not necessarily to the developers of the fish ponds). The principles of a 'green corridor' of about 50 m width on the seaward side of a brackishwater pond development, and of a maximum of only 20% of the mangrove forest in any given area being converted to ponds are widely accepted, but these principles are not applied in any area of Java or Bali. In addition, it has not been demonstrated unequivocally that even this is sufficient to maintain all the important ecological processes that occur in a mangrove forest.

An important but often overlooked fact is that intensively managed brackishwater ponds require five times as much fresh water per hectare as sawah. In coastal, multi-landuse areas the water once used in the ponds must be regarded as contaminated because of the admixture of seawater and so is discharged to the sea without further use. Given the intense competition for scarce water resources in coastal areas of North Java, (p. 49) this is an important consideration. Of course, intensively managed ponds are more profitable than sawah. Extensive, traditionally managed ponds use less fresh water, and while showing lower returns per unit of fresh water, they not compete with sawah for water.

SEAWEED

Seaweed has been gathered from the coasts of Java and Bali for centuries. Traditionally, a wide range of seaweeds has been gathered, such as the feathery *Gelidium* sp. (Geli.) and *Hypnea* sp. (Rhod.), which are made into a snack eaten locally known in Bali as *jaja bulung*. The larger *Gracilaria gigas* (Grac.) was only gathered for sale to exporters. These unlikely green or red natural rock growths are the source of agar, which is used as a thickening agent in the foods of many countries, and eaten with flavouring as a jelly, especially in Asia. The most widespread uses, though, are in soft ice cream, toothpaste, jam, milkshakes, ice lollies and cottage cheese. Other seaweeds ,such as *Eucheuma* (Euch.), produce gums called carrageens which are similar to agar but are thicker and are used in the food and cosmetic industries, as well as in paint, and insecticides (Teo and Wee 1983). In fact, the number of uses for these natural gels were found to be so many that, during the 1960s, wild seaweed was gathered intensively from all around Indonesia. Japan produces the bulk of the world's edible seaweed where it is a favoured smallholder crop. This has now spread to Indonesia, notably on Bali.

The earliest experiments on Indonesian seaweed culture began in the Thousand Islands in 1967, but were not wholly successful. In 1978 a major Danish agar producer began new experiments in southern Bali, just south of the Nusa Dua tourist complex. The excellence of this site was reflected in the rapid growth of the alga *Eucheuma spinosum* (Soegiarto and Atmadja 1980), but the gel content was consequently low. The slower the growth, the higher the concentration of gel (Eiseman 1990). Foreign interest waned until success in the Philippines flooded the market; political practicalities there later led to Bali receiving attention again. A number of sites were evaluated but the best were around the offshore islands of Nusa Lembongan/Ceningan and the north coast of Nusa Penida, to which the first *E. spinosum* 'seedlings' were brought in 1984. Another species, the fatter bulung gede *E. cottoni*, has been brought in because it is easy to farm and is more disease resistant, but it grows more slowly and takes longer to dry. The farming of *E. cottoni* is a relatively attractive investment since, from a study of a 1 ha plot near Jungut Batu, Bali, the payback period is less than a year, it has an economic rate of return of 153%: it is relatively labour intensive, and requires little in the way of fertilizer, fuel, or chemicals (Firdausy and Tisdell 1991). *Gracilaria lichenoides* and *G. gigas* have potential in Bali as an additional crop from tambak ponds as in Taiwan, although the damage caused by wild fish and crabs has to be controlled (Sjafrie 1985; Sulistijo 1986).

The 'farm' itself comprises 2.5 x 5 m intertidal plots along which lengths of rope are tied (fig. 15.9). Seedlings are fastened onto these such that each plot contains about 400 seedlings. The crop is harvested every

Figure 15.9. A seaweed farm on Nusa Lembongan, Bali.
By A.J. Whitten

month, leaving some behind for the next crop, but there is much work between times weeding out a string-like alga known as *wit*, which has to be untangled from around the crop. Large mounds of wit can be seen on the shore. Unfortunately cattle will not eat it and it is too salty for mulch or compost.

Almost all the seaweed from Bali is exported to Denmark. Bali's exports in 1983 were 413 tons, 43 collected from the wild and 370 from farms. These were worth only US$15,000, compared with Bali's total 1983 exports of US$20 million, but only 169 ha of the potential 1,200 ha in Bali were being farmed. By 1991, however, Bali produced 9,654 tons, of which 2,171 tons were exported, worth US$1.3 million. This was less than half the success of 1990 because of the competition from resurrected Philippine farms (Daerah Tk. I Bali 1992). Still, barely one-fifth of the potential area is being farmed. Despite very low prices in 1991 (about Rp.250/kg dried seaweed), seaweed production is a popular addition to farmers' activities on these dry and highly seasonal islands. A family can tend about 0.25 ha, which produces about 20 tons of dried seaweed each year, bringing in at least Rp.5 million, a sizeable sum for subsistence coastal dwellers. In 1991 there were said to be over 2,500 Balinese families engaged in seaweed farming.

Chapter Sixteen

Urban Areas

Urban ecology is a term which has been given two distinct meanings. One concerns the way of life and interactions of living organisms within the urban environment, while the other views the urban ecosystem on a grander, almost organismic, scale and seeks to understand the predation of resources, their digestion, their excretion, and the circulation of their useful products through the organism. Such an urban organism can also be viewed as a colony of smaller organisms, and the allocation of energy among them, the competitive forces, and the relative rates of net and gross productivity can be studied. Some have likened cities to ecological black holes, attracting and absorbing material resources, human resources, and the productivity of a vast and scattered galaxial hinterland many times their own size (Rees 1993). While these views produce useful models and may be relevant to urban planning, it is urban and suburban ecology in the former sense which is examined in this chapter.

URBAN ECOLOGY STUDIES

Urban areas are excellent for ecological study. Such work may not rank as priority research, but someone who has looked in depth at a common urban species will be better equipped to study an organism in the context of some environmental issue than someone who has simply had book learning as her or his qualification. There is a surprising dearth of relevant material from Java and Bali (or our searches have failed to bring it to light), possibly because teachers fail to see the potential of the urban environment in education. There are texts available, however, which give many ideas of avenues of interest (Bornkamm and Lee 1982; Collins 1984; Hammond and King 1984; Smith 1984). They are, admittedly, written for a temperate audience, but an imaginative teacher will be able to adapt them. Ideas for urban ecology studies are also provided in the Sumatra and Sulawesi ecology books (Anwar et al. 1984; Whitten et al. 1987a, b, c).

Students and teachers should be encouraged to identify urban animals and plants since this is the foundation of any ecological investigation.

A handicap is the paucity of guides to the more common species but, if guides from other parts of the region are also used, the bulk of the organisms can be identified (e.g., Corner 1952; Piggott 1979; van Steenis 1981; SSC 1987; Piggott and Piggott 1988).

VEGETATION

Relatively few species of plants are found in urban areas, and they tend to be those species that are permitted, encouraged, or planted by the human inhabitants, although a fair range of native and introduced weeds can be found in untended corners. Many of the most common planted species are not native, and more efforts should definitely be made to plant more native species since few introduced plants are favoured by native animals. There would still be room for some of the magnificent exotic species that grace the streets of Asia, such as the rain tree *Albizia saman* (Legu.) from tropical America, the flamboyan *Delonix regia* (Legu.) from Madagascar, or the royal palm *Roystonea regia* (Palm.) from Cuba.

There is considerable horticultural expertise in the public and private sectors, although the main interest is in survival and aesthetics rather than in boosting the diversity of the environment. The choice of species is important because urban centres are quite hot and dry and the soils relatively infertile. For this reason legumes and *Casuarina* (Casu.) (whose root-nodule bacteria can fix nitrogen from the air), and species from seasonal climates such as the purple-flowered *Lagerstroemia flos-regina* (Lyth.) are commonly planted. Ecologists, on the other hand, are more concerned with the appropriateness of the plants in relation to other components of the environment.

Urban Forests

Urban forests are areas in excess of one-quarter hectare planted with closely spaced trees which are required to conform to urban design and to survive in a harsh environment. The World Health Organization has recommended a minimum of 9 m^2 of green space per person, and for Jakarta this would be the equivalent of 81 km^2, or 14% of its total area. The concept of urban forests has received increasing attention in the past few years, but as yet they are hard to identify. They do exist, however, and examples are Jakarta's Ragunan Zoo, Bogor's Botanical Garden, Bandung's Juanda Forest Park, and Surabaya's Zoo. Forest is in fact a misnomer for many areas which are planted with trees because these generally comprise shrubs, small trees, and other decorative plants (Ramlan and Iskandar 1987).

Urban forests are seen as ways of increasing the quality of life in cities. The trees create relatively cool and shady conditions where noise is dampened, there is a pleasing and restful view for the eye, habitats are provided for a range of wildlife (if allowed), and water infiltrates into the soil rather than running over asphalt and concrete surfaces into drains. Leaves can also trap dust particles from the air, but precisely how many trees would have to be planted in order to achieve significant dust reductions in the cities and towns of Java and Bali remains unclear. The different shapes and heights and colours of trees are also used to complement architectural features of buildings.

The ecological value of planted forests will increase with their degree of similarity to natural forests in terms of diversity, number of vegetation layers, and susceptibility to insects in order to support insectivorous mammals, birds, amphibians, and reptiles. Certain plants with demanding cultivation requirements or poisonous parts would not be favoured. Rare or expensive trees would also be inappropriate since vandalism is not unknown (Ramlan and Iskandar 1987).

BIRDS

The bird fauna of towns and cities is poorer than that of forested habitats, even degraded ones, and consists of very few dominant species. The paucity of the bird fauna is at least partly due to the predatory inclinations of young humans armed with catapults or even air guns, but ecological reasons include the lack of fruit suitable for birds, and relatively few insects able to utilize the 'foreign' trees, leading to less food being available for insectivorous or partially insectivorous birds. Most of the birds present are grain-eating ground-feeders rather than the bark- and canopy-insect feeders more typical of forest (Ward 1968; Yorke 1984).

Cities in Southeast Asia seem to have about 20-25 species of resident breeding birds (Anwar et al. 1984; Whitten et al. 1987a, b, c), about half of which are typical of coastal habitats, and a quarter of which are introduced or recent immigrants (Ward 1968). The proportion of nectar feeders, such as flowerpeckers *Dicaeum* and sunbirds *Aethopyga*, is very small, but this is the group which is most attractive to urban dwellers because of their small size and often brilliant colours. In addition, the flowers from which they feed tend to be large and showy. To attract these birds down to heights at which they can be seen, the wild-type red hibiscus *Hibiscus rosa-sinensis* (Malv.), *Ixora* (Rubi.), and *Calliandra* (Legu.) should be planted.

Restoration of Urban Bird Faunas

A rich and diverse bird fauna can be restored to large cities and towns. The keys are species and habitat diversity, and protection from being shot. For example, it has been found that the diversity of plants (and more particularly the types of food they provided) in villages around Bogor was directly related to the diversity of birds seen there (Farimansyah 1981; Alikodra and Amzu 1984). Thus, if it is desired to have more birds in towns and villages, it is essential that a good diversity of plant species is grown, using species which attract birds for different reasons (table 16.1). A programme to diversify the ornamental plantings in Singapore has been very successful, with particular attention having been paid to growing species which provide food at times of general food scarcity. Certain fig trees do this and they are also useful for the large numbers of pollinating small wasps which are fed upon by swiftlets (Bartels 1929). It is equally important that the wanton killing of birds be eliminated. The banning of all unlicensed air rifles (4.5 mm or larger) from March 1992 should have a beneficial effect, but only if it is enforced.

Several major Asian cities outside Indonesia are attempting to attract back those birds and insects that have long since given up on these centres of pollution and inappropriate habitat. Pollution, in some cases, is better than it was, and so the efforts that have been made in Singapore, and are being made in Kuala Lumpur, are concentrating on increasing the quantity and diversity of the native trees planted, increasing the complexity of at least part of the urban vegetation (by providing mosaics of rough grassland, bushes, and trees as favoured by many birds for nesting or feeding), and forming green corridors between the urban sprawl and remnant forest areas.

URBAN AND RURAL PESTS

This section concerns some of the animals on Java and Bali that have adapted to urban ecosystems, to the extent that people try to limit their numbers or even seek to exterminate them.

Rats

The most successful animals on Java and Bali are rats. This does not include all of the 21 rat species present because most live inoffensive lives in the forest and scrub. In fact, there are only four major problem species:

- the native house or roof rat *Rattus rattus* which took the terrifying Black Death around Europe in the Middle Ages;
- the native rice-field rat *Rattus argentiventer;*
- the native bandicoot *Bandicota indica* which grows as large as a rabbit

Table 16.1. List of plant species attractive to birds in Java and Bali for different purposes. Fr = fruits and seeds, Nc = nectar, In = insects and other arthropods, Wa = water in flowers/fruits, Ns = nesting, Ro = roosting. The species in bold are particularly valuable. [i] = introduced species. * Especially attractive to the palm swifts *Cypsiurus balasiensis* which attach their nest to the underside of the leaves.

		Utilization					
		Fr	Nc	In	Wa	Ns	Ro
TREES							
Albizia spp.	Legu.			x			
Antidesma tetrandum	Euph.	x					
Cananga odorata	Anno.	x					
Cordia obliqua	Bora.	x					
Erythrina variegata	Legu.		x				
Ficus spp.	Mora.	x				x	x
Flacourtia rukam	Flac.	x					
Gossampinus heptaphylla	Bomb.		x				
Muntingia calabura [i]	Elae.	x			x		
Sesbania grandiflora	Legu.		x	x			
Spathodea campanulata	Bign.		x		x		
Trema orientalis	Ulma.	x					
Vitex spp.	Verb.	x					
Ziziphus jujuba	Rham.	x					
PALMS							
Arenga pinnata	Palm.					x	
Caryota mitis	Palm.						
Corypha utan	Palm.	x					x
Livistona rotundifolia *	Palm.					x	
Oncosperma filamentosum	Palm.	x					
Ptychosperma macarthuri [i]	Palm.	x					
Roystonea regia [i]	Palm.	x				x	x
BAMBOOS	Gram.	x				x	
SHRUBS							
Ardisia sp.	Myrs.	x					
Bridelia lanceolata	Euph.	x					
Brucea amarissima	Sima.	x					
Bryonopsis laciniosa	Cucu.	x					
Clerodendrum serratum	Verb.	x					
Ehretia microphylla	Bora.	x					
Lantana camara	Verb.	x					
Laportea stimulans	Urti.	x					
Leea spp.	Leea.	x					
Melastoma malabathricum	Mela.	x					
Morus alba [i]	Mora.	x					
Rubus rosaefolium	Rosa.	x					
Sambucus canadensis [i]	Capr.	x					
Trichosanthes spp.	Cucu.	x					
Triphasia aurantiola	Ruta.	x					
FERNS							
Asplenium spp.	Aspl.					x	
ECONOMIC (CROP) PLANTS							
Andropogon sorghum	Gram.	x					
Areca catechu	Palm.					x	
Capsicum annuum/frutescens [i]	Sola.	x					
Carica papaya [i]	Cari.	x					
Cleome gynandra	Capp.	x					
Cocos nucifera	Palm.		x	x		x	
Curcuma longa	Zing.					x	
Lansium domesticum	Meli.	x		x			
Oryza sativa	Gram.	x				x	
Panicum viride	Gram.	x					
Pennisetum glaucum	Gram.	x					
Piper nigrum/aduncum [i]	Pipe.	x					
Saccharum officinarum	Gram.					x	
Salacca edulis	Palm.					x	

Sources: updated from Sody (1955); van Balen (1989a)

with long, thick, black hair; and

- the ground-dwelling and proficient-swimming brown rat *Rattus norvegicus*, originally from northeast China, which, so far in Java, is found only around the major ports where it has arrived on board ships. This species has displaced the roof rat as the dominant species in European cities, and it is possible the same may happen here.

In addition to the above species, the scrub rat *Rattus tiomanicus*, small house rat *Rattus exulans*, and rice mouse *Mus caroli* also cause damage, but to a much lesser degree.

Rats are despised partly because they rob people of so much food – both in the field and during storage. In addition they carry debilitating and potentially lethal diseases such as leptospirosis, murine typhus, scrub typhus, rabies, trichinosis, and bubonic plague, keep people awake with their incessant scratchings and squealings, and leave unpleasant urine and faeces in homes. They are also despised because they are so wretchedly difficult to control, seeming to outwit human endeavours with impunity. Their success as pests results from the following attributes:

- they are nocturnal;
- they are generally unobtrusive;
- they have an incredibly high reproductive rate (such that one adult pair could be responsible for 20 million rats births after just three years under 'ideal' conditions);
- they intimidate household cats and dogs which, when they do catch rats, generally limit themselves to the rather small *R. exulans*;
- they feed on foods that humans like;
- they have perpetually growing and razor-sharp teeth that make short work of most food storage containers; and
- they are protected from predators such as birds of prey by virtue of their close living with humans and the human persecution of those species.

Faced with all these advantages, it is perhaps not entirely surprising that even human ingenuity and technology have not come up with a way to reliably or continuously control rat numbers. People have tried traps, cages, anticoagulant and other poisons, dogs, cats, snakes, mongooses, gas, sterilization, and electrocution. Even unconventional approaches, such as the Central Java authorities demanding 25 rats' tails from a bridegroom before allowing him to marry, or allowing a married man to divorce (McNeely and Wachtel 1988), and encouraging industries that use rat meat for human or animal consumption, have been tried (fig. 16.1). In many parts of Java and Bali it has been traditional for people to join together in rice fields just after the harvest in order to dig out nests and club as many rats as possible (fig. 16.2). In Bali the rat corpses were cremated, but two living rats were released back into the field in the hope of gaining forgiveness from the spirits of the other rats (Frazer 1939). Thirty years ago on Bali the feeling

Figure 16.1. Rats being skinned prior to being processed as animal food.

With permission of Kompas

Figure 16.2. Result of a day's community rat killing.

With permission of Kompas

was that when a rat was killed, then four more would spring up where the first one died because the Chief Mouse had not received enough sacrifices and thus felt neglected. A ceremony was held at Besakih, the 'mother temple', to appease the Chief Mouse, after which poison was allowed to be distributed free by government agencies (Matthews 1983). Bali continues, however, to suffer the ravages of rats. Indeed, despite all these diverse efforts, the rats' capacity to reproduce is so enormous that a season or two after a major control effort the rat numbers and the problems would be as bad as before.

While extermination is an unrealistic goal, the control of rats is probably best effected by applying a variety of methods such as:

- reducing the available food sources;
- removing scrub and other weedy areas where they can hide;
- keeping bunds narrow and free of weeds;
- destroying nests and burrows;
- applying appropriate poisons; and
- giving maximum opportunity for natural predators to keep the rat populations in check. These predators include owls, diurnal birds of prey, crows, snakes, monitor lizards, civet cats, mongooses, and the small leopard cat *Felis bengalensis*. It is recognized that some of these also take chickens, but perhaps if farmers were fully aware of the predators' benefits, the prevailing attitude towards them, or the husbandry of the livestock, may change.

Cockroaches

There are about 3,500 species of cockroach known to science, most of which are inhabitants of forests where they live among leaf litter. They are instantly recognizable by their oval shape, long antennae, odour, and scuttering run. They have two pairs of wings, with the rather leathery fore pair overlapping across the abdomen. Although they can fly, most species do so rarely. Almost all of them are nocturnal and shun the light, although a few species are attracted to lamps. Eggs are laid in cases or purses which the females carry around on the tips of their abdomen for some weeks. Each purse holds 10-50 narrow eggs arranged in two alternate rows with the embryo heads facing the purse aperture. When they first emerge, the nymphs are worm-like but after they shed their first skin, they look like small versions of the adult. Some 5-12 moults later they become full-sized adults. This takes up to six months.

Cockroaches are omnivorous, eating human foodstuffs, refuse, faeces, book bindings, pastes, and glues, although dead animal material is probably their major food in the wild. They also contaminate foodstuffs with the characteristic smell from their liquid faeces. Although they do carry disease organisms on their legs, they are not associated with any particular disease and cannot be regarded as a serious vector. However, *Salmonella, Eschericha coli,* and *Staphylococcus* have been found in their guts and faeces, and so where a lavatory and kitchen are situated in close proximity, there is a risk of infection. Only one species is an agricultural pest, the black and thick-set *Leucophaea (Pycnoscelis) surinamensis,* which chews the roots of certain crops (Kalshoven 1981).

The most common species in houses in Java and Bali are *Periplaneta australasiae* and *P. americana* (which did not originate in either Australia or America). During the daylight hours they congregate in dark places such as drains and wells, and their flat shape enables them to squeeze into narrow crevices. The species of cockroaches found in houses and offices in Java and Bali are probably indigenous to Africa, but they have spread throughout the world, carried on ships and planes feeding on refuse, concealed in cracks and crevices.

Cockroaches are very aware of their surroundings and this makes them very hard to catch. Their antennae have organs that detect changes in air pressure (such as when a shoe fast approaches them), temperature, and moisture, and they have special receptors in their legs that are very sensitive to vibrations. Their natural enemies include egg parasites, hunting spiders, and the house shrew *Suncus murinus.* Numerous means of control have been tried, including many insecticides, but the use of these is complicated by the fact that their daytime haunts are usually difficult to treat thoroughly with sprays. Some control methods use baits to lure the cockroaches out of hiding. For some species it has proved possible to synthesize the female sex attractant or pheromone which lures the males from hiding where they can be killed with insecticides, eliminating the risks to nontarget species. A new form of control uses an insect growth-regulating hormone which acts on the juveniles, preventing their development into adults; it does not actually kill the cockroaches, which eventually die of 'old age' in their immature form. Another effective cockroach control does not require these complex chemicals, but uses balls made from flour, sugar, milk, and baking soda (sodium bicarbonate). The baking soda forms fatal bubbles of carbon dioxide in the insect's digestive tract which it is unable to remove by burping. This mixture is harmless to children and pets.

Mosquitoes

Mosquitoes are flies (Diptera) of the family Culicidae and all have a long, needle-shaped proboscis. There are about 3,000 species world-wide divided into three groups, two of which are medically important. These are the anophelines including *Anopheles,* which transmits malaria and filariasis, and the culicines including *Aedes, Culex, Mansonia,* and *Haemogogus,* which transmit viral filiarial diseases. The differences between the major genera are shown in figure 16.3.

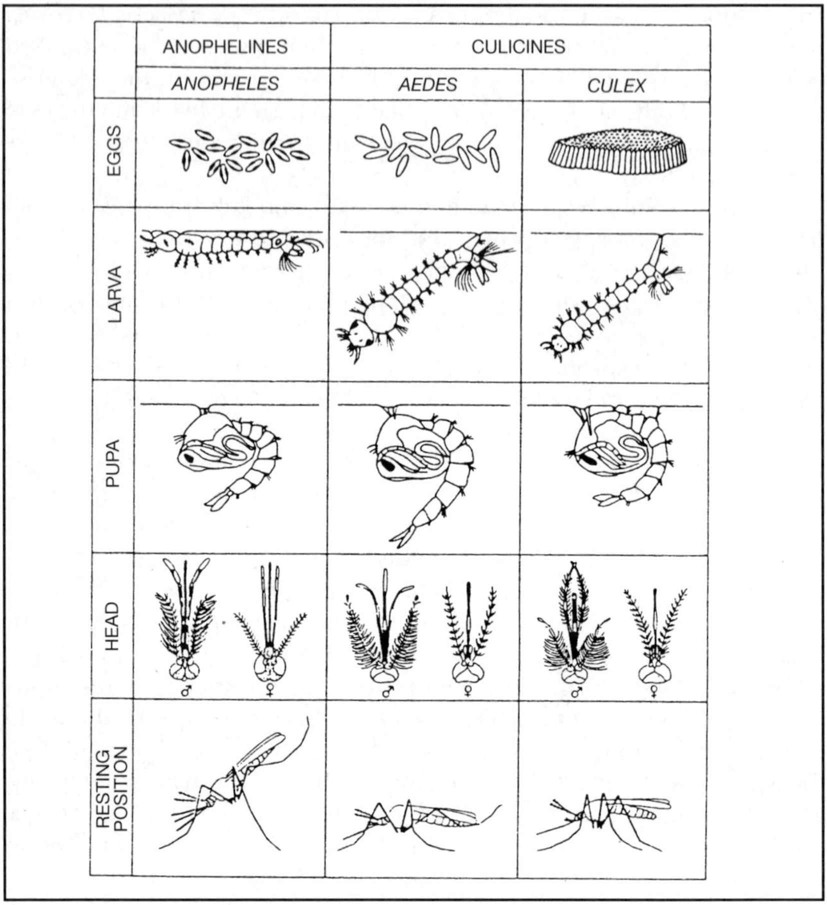

Figure 16.3. Differences between *Anopheles, Aedes, and Culex* mosquitoes at different stages of development.
After Bruce-Chwatt 1980

There are 180 species of mosquitoes known from Java, out of a world total of 2,960 (O'Connor and Sopa 1981), of which six *Anopheles* species have been implicated in the transmission of malaria, notably *A. aconitus* in rice-field areas, and *A. subpictus* and *A. sundaicus* in swampy and brackish areas (Ten Houten et al. 1980; Takken and Knols 1990). Worldwide only 25 of the 300 anopheline mosquitoes are significant vectors of malaria. The main texts for anopheline mosquitoes are Bonne-Wepster and Swellengrebel (1953) and Bonne-Wepster (1954).

Malaria has probably been with people from the days of *Homo erectus* (p. 309) and has spread from Southeast Asia to Africa where 90% of cases are now found. Malaria currently kills about three million people every year, causing acute illness in 100 million, and the malaria parasite is believed to infect about 500 million people every year. Drugs that were once fail-safe are now little more than useless, and in Southeast Asia resistance is advancing so rapidly that some strains may soon be declared untreatable. It used to plague western industrial countries but they are now largely free. Although malaria has been virtually eradicated from the larger cities on Java and Bali, where epidemics have not occurred for many years, it still occurs sporadically and is more common in rural areas.

To understand the success of mosquitoes and the prevalence of malaria, it is important to understand the salient features of the organisms involved. Malaria (and dengue) involve three organisms—the infective organism (a protozoan *Plasmodium* in malaria, a virus in dengue), the reservoir (man), and the host (mosquito). The malarial organism breeds within the stomach of the adult female mosquito, forming 'larvae' which develop in her salivary glands. The period required for this incubation depends greatly on temperature but is often roughly the same length as the mosquito life expectancy. As a result only those females that feed on infected human blood early on, and feed again on a non-immune human after incubation is complete, will actually transmit the disease. Thus the rate of transmission depends on:

- the number of adult female mosquitoes, which itself depends on the rate at which they emerge from their aquatic larvae, and their longevity;
- the frequency of feeding and the proportion of meals taken from the relevant host; and
- the incubation period of the disease organism.

When the female later feeds on a susceptible person, she will inject these 'larvae' into the human bloodstream, in which they migrate to the liver for further development. When the enlarged cells burst, the new form of larvae and foreign proteins issuing forth will cause the fever. In severe infections, up to 50,000 of these larvae can be found in each millilitre of blood. The normal *P. vivax* malaria causes recurrent fever, anaemia, and enlargement of the spleen, and while it does not have a high fatality rate, its debilitating nature can be critical for undernourished or otherwise unfit people. In most people with no immunity to malaria, infection with the malaria organism will produce acute illness (potentially fatal in the case of *P. falciparum*), and repeated infections will generally produce chronic illness.

Dengue in its common form is an unpleasant viral disease transmitted by the mosquitoes *Aedes aegypti* and *A. albopictus* and was first encountered in Indonesia in 1960. *Aedes aegypti* is a native of Ethiopia and neighbouring regions which came to Asia during the last century on large ships which had passed through the Suez Canal. *A. aegypti* is a household species which bites far more often in houses than *A. albopictus*, which is 25 times more likely to bite outside, than inside, a house. In the case of dengue, the first infection may not cause any obvious disease, and only later reinfections will produce chronic disease. When more than one type of virus is transmitted, the disease is known as dengue haemorrhagic fever, which is far more serious, and may be fatal.

Control and Ecology. Control of mosquitoes requires a strategy designed to hit one or a number of weak and accessible links in the disease cycle. The alternatives are:

- preventing mosquitoes feeding on man;
- reducing mosquito breeding by controlling potential breeding and other favoured environmental conditions;
- destroying mosquito larvae;
- destroying mosquito adults; and
- killing the infective organisms in people.

The control strategy can be aimed either against all anopheline mosquitoes, or selectively against the one or two vector species. The former strategy would be unwise because some harmless species compete successfully for resources against dangerous species, and the larvae of some species also prey on the larvae of other species, thus helping to control their numbers naturally (Adisoemarto 1977). The latter, focussed, approach must be based on a sound knowledge of the ecology and behaviour of the species.

The ecological preferences of larvae and pupae differ between mosquito species in terms of temperature, shade, water movement, water salinity, pollution, turbidity, microflora (food), and the absence of predators and parasitizing organisms. No one species seems to have the same preferences as another (table 16.2-4). For the adults important differences relate to mating and egg laying, behaviour, dispersion, biting behaviour, host preference, activity period, seasonality, longevity, and susceptibility to pesticides (Oomen et al. 1990). The preferences based on detailed literature searches are described elsewhere (Takken and Knols 1990). Species also differ in the way their populations tend to increase with the onset of the rains (Lee 1983). A word of caution should be exercised here to stress that since we know that mosquitoes have evolved resistance to drugs and insecticides, it may be that their ecological preferences have changed too, and this needs to be checked.

In the early days of control, efforts were made to prevent the larvae and pupae from breathing atmospheric air by spreading oil on the water surface to suffocate them, but this also prevented fish and other organisms from breathing, with obvious results. After World War II larvicides were introduced but they proved costly, labour-intensive and, more important, encouraged the development of resistance, and had potential for killing other organisms. DDT started to be used in Indonesia shortly after that war, primarily against *A. aconitus* in inland rice fields and against *A. sundaicus* in coastal areas. The frequency with which malarial parasites were found in people's blood dropped dramatically, until, in 1954, DDT resistance was found near Cirebon. Dieldrin was introduced and *A. sundaicus* was eliminated from the northern coast of Java, although there are still pockets found in the south. *A. aconitus* developed resistance to dieldrin in the late 1950s, although the problems of susceptibility differed between the three provinces (Atmosoedjono 1990). Perhaps related to this is the fact that malaria incidence has always been considerably higher in Central Java than in the other provinces in Java and Bali, though precisely why is not clear. DDT use started to be phased out in 1985 and there was a rise of 13,000 malaria cases in Java and Bali in 1987, which was attributed to a drastic decrease in indoor spraying of DDT. In 1990 the government decided that DDT should not be used for malaria control because there were dangers of the mosquitoes becoming resistant and because it did not break down in the environment. Its alternative (for Java and Bali) was fenetrothion but, because supplies of this are limited, DDT is still being used in the outer islands.

In 1986 the malaria incidence in Java and Bali was similar to that in 1963, but with the increase in the human population, this represents a far larger number of people. Adequate control is not being attained largely because of drug and insecticide resistance, but two other factors are judged to be very important, too. First, as a result of the 'malaria eradication' policy of 1955-69 in which attempts were made to eradicate mosquitoes without paying attention to their ecology and behaviour, there is a nationwide shortage of senior vector ecologists able to deal with the control problems in an ecological context. Second, since malaria is primarily a rural disease, it has not received the attention it might have done had it still afflicted large numbers of urban dwellers (Takken et al. 1990). As attention turns to the rural areas, and now that the authorities are aware that malaria is increasing world-wide, it is hoped that renewed, ecologically based, integrated approaches using drug therapy, vaccination, and vector control will spread. The first of these will clearly not succeed unless drugs from an entirely new group is found. There are technical and biological reasons why finding a malaria cure is fraught with problems, and very little research directed towards finding a cure for malaria is conducted around the world. Drug research is hugely expensive and is generally done by private companies with a profit motive. Their logical reasoning would be that producing a drug for sufferers, almost all of whom are poor, would not be a financial success. The earlier cures or prophylactics had ulterior motives: quinine helped European colonial powers to extend their influence, and chloroquine protected U.S. troops in Vietnam. There is, however, an exciting possibility that a malaria vaccine developed in Colombia, SPf66, may soon pass its final efficacy and safety tests and become widely available

Table 16.2. Natural and human-made breeding site preferences of *Anopheles* mosquitoes. The species shown here are those known as malaria vectors in Indonesia, with those important as vectors in Java shown in bold. √ = preferred habitat, • = preferred habitat, + = common habitat, + = rare habitat.

| | Large Water Bodies | | | | Human-made | | | |
| | Natural | | | | | | | |
	Lagoons	Lakes	Marshes	Slow rivers	Pits	Sewak	Fish ponds	Irrig. channels
A. aconitus		+	•	•				√
A. aitkeni							+	
A. annularis		+	•				√	√
A. balabacensis								
A. barbirostris			+	+			√	•
A. beazai			√					
A. flavirostris				+			+	+
A. kochi							+	+
A. maculatus		+		•			+	
A. minimus							+	•
A. nigerrimus	•	•	•				√	+
A. subpictus	• √		+			•	√	√
A. sundaicus							• √	•
A. tesselatus							√	•
A. umbrosus			√					•

Table continues

Table 16.2. *(Continued.)* Natural and human-made breeding site preferenced of Anopheles mosquitoes. The species shown here are those known as malaria vectors in Indonesia, with those important as vectors in Java shown in bold. √ = preferred habitat, ● = common habitat, + = rare habitat.

| | Small Water Bodies | | | | | | | | | | | |
| | Natural | | | Human-made | | | | | | | | |
	Small Streams	Seepage springs	Pools	Wells	Ground depressions	Over-flow water	Irrig. ditches	Pits	Wheel ruts	Hoof prints	Puddles	Empty cans, shells, etc.
A. aconitus												
A. aitkeni	√	√	●						●			
A. annularis							+	+				
A. balabacensis		●	●	●	●	+						
A. barbirostris	●						+					
A. beazai			√	●								
A. flavirostris	√		+			+	+					
A. kochi	+		●			●						+
A. maculatus	√	+					●				+	
A. minimus	√	+					√	+	●		+	
A. nigerrimus	+	+										
A. subpictus		+		+			●		+	+		
A. sundaicus		+		+					+			
A. tesselatus	+		●						+			
A. umbrosus			●				●					

Source: Takken and Knols (1990)

Table 16.3. Breeding site characteristics of malarial vectors. Species of *Anopheles* mosquitoes important as vectors in Java are shown in bold. √ = common, • = rare, ? = not known.

	Light intensity		Salinity		Turbidity		Movement		Vegetation	
	Sun	Shade	Brackish	Fresh	Clear	Turbid	Stagnant	Running	Present	Absent
A. aconitus	√	•		√	√		√	•	√	
A. aitkeni		√		√	√		√		√	
A. annularis	√	√		√	√		√	•	•	
A. balabacensis	√	√	•	√	√		•	•	•	•
A. barbirostris	•	√	•	√	?	?	•		?	?
A. beazai				√	•	?	•		?	
A. flavirostris			•	•	•	•	•	•	•	•
A. kochi	•	•		•	•	•	•	•	•	•
A. maculatus	√	+	√	•	•		√	√	√	√
A. minimus	•	√			•	•	•		√	•
A. nigerrimus	√		√		√	•	√		•	•
A. subpictus	√		√		√	•	√	•	•	•
A. sundaicus	√		•	•	•	•	•	•	√	•
A. tesselatus		√		•	•	•	√		√	•
A. umbrosus		√		•	√	•	√		?	?

Source: Takken and Knols (1990)

Table 16.4. Ecological characteristics of adult *Anopheles* mosquitoes in Java. The species important as malaria vectors in Java are shown in bold. √ = common habit, • = rare habit, √ = affirmative.

	Feeding habits		Resting habits		Biting habits		Insecticide resistance		
	Humans	Other animals	Outdoor	Indoor	Outdoor	Indoor	DDT	Dieldrin	HCH
A. aconitus	•	•	√		•	•			
A. aitkeni	•	√		√	√		√		√
A. annularis	√	•	•	√	•	•			
A. balabacensis	√	•	√	•	√	•			
A. barbirostris	√	•	√		?	?			
A. beazai	•		√		√	√			
A. flavirostris	√		√	√	√	√		√	
A. kochi	√	√			√	√			
A. maculatus	√		√	√	•	•			
A. minimus	√		√	√	•	•			
A. nigerrimus	√		•	√	√	•			
A. subpictus	•	√	√	•	•	•	√	√	
A. sundaicus	√		√	√	√	•	√	√	
A. tesselatus	•	•	•	√	√	•			
A. umbrosus	√		•	•	√	√			

Source: Takken and Knols (1990)

before the end of the century.

Meanwhile, an array of biological controls offers the best present option for malaria control. Between 1916 and 1938 a number of environmental methods of malaria control were tried successfully in Indonesia, but were forgotten as the use of insecticides and drugs became widespread (Takken 1990). They are coming back into favour, however, as chemical methods have failed to live up to their promises. Options include:

- larvae-eating fishes – the fishes most commonly used around the world for mosquito control have been the guppy *Poecilia reticulata* (which has not proved successful because of its relatively small size), and its relative the mosquito fish *Gambusia affinis,* which has been introduced from the USA and Mexico into many tropical countries with some success, though at some environmental cost (Courtenay and Meffe 1990). The mosquito fish has not, it seems, ever been introduced to Indonesia or neighbouring countries (Welcomme 1988). Among the native species, the tinhead *Aplocheilus panchax* is supposedly a predator on mosquito larvae, and the introduced *Oreochromis niloticus* (when young), grass carp, common carp, and gurami will feed on the larvae, too (Petr 1987). It would therefore seem unnecessary to consider any further introductions. Other good mosquito predators are dragonfly larvae (p. 443).

- mosquito pathogens – certain strains of bacteria produce toxins that are deadly to mosquito larvae but do not seem to adversely affect other organisms. These will become available soon and are said to be particularly useful in large water bodies.

- nematodes – these and other predators on larvae have potential, but none has been wholly proven in large-scale trials (Takken 1990).

- environmental measures – *A. aconitus* and *A. sundaicus* have been successfully controlled in the past using intermittent drying of their breeding sites and, in the case of the latter, by the replacement of natural vegetation and increasing the salinity of their breeding sites (Takken and Knols 1990). The other consequences of these actions should, of course, also be considered. A simple scheme would also be to establish a belt of dryland crops around a village, then people are protected from the mosquitoes breeding in rice-fields, because they do not fly very far. Other environmental controls are discussed by Birley (1991).

- botanical pesticides – there is a growing number of pesticides derived from the flowers of marigolds and chrysanthemums, which appear to be relatively benign in the environment. Control could be achieved either by planting the flowers around houses or by extracting the volatile chemicals and applying them as a spray.

The mosquitoes transmitting dengue have a much wider niche requirement than those transmitting malaria, with the result that this disease is still relatively common. These mosquitoes typically breed in domestic water tanks as well as in smaller containers, which should be emptied and cleaned weekly so that any larvae are unable to develop to adulthood. The adults rarely fly more than 100 m, so a community can look after itself as long as *everyone* takes the precautions necessary: one renegade household, and all the neighbours are at risk. In other Asian countries compliance was achieved using inspectors who were permitted to levy fines on households where larvae were found. Interestingly, outbreaks seem to occur every five years in Jakarta – 1968, 1973, 1978, 1983, 1988, and 1993 (though this one was mild), but the cause of this cycle is unknown.

DOGS AND RABIES

One of the most striking differences between urban settings in Java and Bali, is the presence of a large number of feral dogs wandering the streets in the latter. These hounds generally cower in the presence of humans and seem to get most of their nourishment from the ubiquitous small offerings of rice and other edible items found around shrines. Street dogs do exist in Javan cities, but an important difference between them and those in Bali is that none of those in Bali carry rabies. It is for this reason that there are signs at the ferry ports on Bali prohibiting the entry of dogs and monkeys (which also carry rabies) into the island. This should not result in complacency; in 1975 a rabid dog jumped ashore from a ship moored offshore of Manado in North Sulawesi, at that time free of rabies, and within a short while five people had died and a further 53 had received treatment in hospital (Koesharyono et al. 1985). On Java the situation varies considerably, with 10 of the 20 kabupatens in West Java suffering at least one rabies death each year, the last case in Central Java being in 1992, and East Java seeming to be clear.

The spread of rabies will not be wholly brought under control until the ecology and behaviour of the dogs that transmit this horrific viral disease is better understood. This requires not just a biological perspective, but also a sociological one that seeks to understands the relationship between people, society, and the dogs, and the positive and negative influences they have on each other. Features of society are the cause of the current rabies situation, and they also determine the choice of method of attack against the disease (Baran and Frith 1986). It is thus not possible to transfer control methods developed in industrialized countries where dogs are often cossetted family members.

Rabies in dogs was first noted in Java in 1889, and the first person died in 1894. There was one human victim every five days in 1977, and one every ten days in 1981 (Koesharyono et al. 1985). The domestic dog is the sole known reservoir host to the virus in Indonesia (Joseph et al. 1978); cats are also affected, but they have almost always been bitten by rabid dogs and they very rarely pass it on to other cats. To control rabies, the chain of transmission from infected to susceptible dogs must be broken, and this can be achieved in three ways:

- preventing the movement of dogs;
- killing all susceptible dogs; and
- immunizing the population by vaccination.

Given the loose relationship between most dogs and their owners (if any), it is impossible to control the movement of dogs. Killing susceptible dogs, using poisons or crack shooters from the army, for example, do not solve any of the long-term problems. The sudden vacating of dog territories, and the disruption of established inter-dog relationships as a result of a killing programme, results in the migration into the area of younger dogs and increased levels of canine violence as territories are fought over and the relative status among individuals established. In addition, the reduced density of dogs causes a temporary increase in available food, and this can increase the fecundity of female dogs and result in a higher survival rate for the puppies. The situation could be improved if the garbage and other food sources were eliminated at the same time, reducing the carrying capacity for dogs. Another problem is that owned, expensive, vaccinated, and purebred animals may accidentally be killed, to the obvious distress of their owners.

The only reasonable course is to immunize as many dogs as possible. For over ten years this has been the policy in Java, and the vaccinations have been given free. Despite this, rabies has continued to be somewhat of a problem, not least because the types of vaccines used give protection for only a few months (Hirayama 1990). Different percentages of actively immune dogs in a population have been suggested as minima for stopping the spread of the virus, but this depends to what extent the movement of the dogs is controlled. Certainly some dogs virtually never leave a house compound, whereas others wander homeless for their whole lives.

Part D

Conservation

No one would suggest that Java and Bali do not face some of the most challenging conservation problems:

- human attitudes tend to be more and more consumptive and to place value on things based on economic importance alone. The prevailing levels of concern indicate that problems such as providing enough space for greenbelts, care for threatened species of plants and animals, and participating in programmes to combat urban waste, are considered rather unimportant;
- economic success in Java and Bali encourages people from outside the islands to settle there, aggravating the problems of overcrowding, and putting pressure on the remaining natural areas, resulting in a loss of biological diversity. The problems are compounded by insufficient funds, lack of human resources, encroachment by land-hungry farmers, and exploitation by entrepreneurs; and
- conservation, commonly defined as the wise use of natural resources, is rarely seen in many of our lives, even for resources of immediate use such as water, plastics, and paper and, while this continues, there seems to little hope for conservation of genetic resources, species, and ecosystems, which seem rather more remote.

Solutions are complex and wide ranging, but the task of finding and implementing them would be easier if there were a genuine and deep desire to succeed among the politicians, decision makers, economists, scientists, technologists, and all Javanese and Balinese communities. To help the process, we have addressed the subjects of human attitudes, loss of biological diversity, the general state of conservation, and the best existing and potential conservation areas in the following four chapters.

Chapter Seventeen

Human Attitudes

Our lives and actions reflect the attitudes of our nation, our culture, and of ourselves as individuals. Traditional attitudes towards the environment evolve from years of experience, observation, and trial-and-error problem solving. The resultant interpretations of events and processes may not align with rational views, but they still have a validity and are thus worthy of incorporation and consideration in the development process (Wickham 1993). Some traditional attitudes towards the environment may disallow (deliberately or accidentally) activities which are damaging to the environment, and the erosion of these attitudes will cause environmental damage or degradation unless they are replaced.

HINDU BASIS

According to Hindu tradition, the gods were once distraught because the ambrosia which afforded them immortality, power, and control over the world could not be found after one of the universe's cycles of rebirths. The gods sought the help of Vishnu, the Supreme Deity out of whom all elements arise and back to whom they must return. He instructed them to obtain new supplies of ambrosia by churning the ocean using Mt. Meru as a stick and the giant naga serpent as a rope wound round the mountain with the gods at its tail and the demons at its head. To allow the mountain to swivel, Vishnu himself turned into a pivot at its base and so the sea was churned. As it did, some sacred items were brought to the surface, such as the tree of life or kalpataru, the moon, the goddess of prosperity, a white elephant (the mount of Brahma), and everything else that made life on earth worth living. From the very depths of the sea a single golden cup of ambrosia then emerged, and there ensued a ferocious argument between the gods and demons over whom should get the cup. Vishnu again intervened, this time by transforming himself into a seductress for whose favours the demons began to fight. While the demons were distracted, the gods drank the invigorating ambrosia and, so empowered, they banished the demons to the sea. They praised and worshipped Vishnu, and the world

671

found peace again (McNeely and Wachtel 1988).

Although Mt. Meru is believed by some to be in the Himalayas, legend has it that it is much closer to home. According to a 15th century Javanese manuscript, Java was once adrift in the ocean, tossed helplessly around by the waves. The gods decided to pin Java down by bringing Mt. Meru from India. Vishnu, transformed into a giant turtle, carried it on his back while Brahma, in the form of a long snake, wrapped himself around both to secure it. They set the mountain down on the first part of the island they came to, the west, but its weight caused the eastern end to tip up. The gods moved it to the east but it rested askew, so they severed part of it and placed it to the northwest. This fragment formed Mt. Pawitra, known now as Mt. Penanggungan (fig. 11.1), and the main Mt. Meru, the dwelling place of Shiva, is known today as Mt. Semeru (Edwin 1988e).

The influence of these stories can be found throughout Java and Bali in the *abangan* or modified pre-Islamic religion that is seen most clearly among the Hindu and Hindu-influenced people in Bali and East and Central Java, but also penetrates into Islam and Christianity (Geertz 1960; Koentjaraningrat 1985). Some of the items churned from the sea still have a place in many peoples' lives: the tree of life was chosen as the symbol for the annual Presidential awards for exceptional service to the environment; the Balinese represent Bali or the world (originally thought as one and the same thing) resting on a giant sea creature; the God Queen of the Southern Ocean remains real to many Javanese and Balinese, and the nights of the full moon have a special atmosphere noticed both by people and by packs of howling dogs. The Hindu epics, told in classical Javanese through the media of human actors or puppets in the Ramayana and Mahabarata, are still popular on stage, radio, and television.

About 1000 years ago the early Buddhist and later Hindu influences in Java merged and built on each other and on earlier beliefs in natural deities, spirits, ghosts, imps, nymphs, and devils (Supratmo 1945). The settled and productive rice cultivation on the wonderfully fertile soils supported accoutrements of temples, shrines, priesthoods and ceremonies, and temples were often sited where traditional beliefs held that land was sacred, on certain mountains, for example. The rulers were raised to the elevated position of gods on earth, but were expected to exercise their divine authority to ensure a mild, productive, and healthy environment. Success confirmed that the rulers were in contact with and supported by powerful animals, ancestors, and the forces that controlled droughts, floods, volcanoes, and pests.

ATTITUDES TO THE SEA

The traditional and prevailing view on Java and Bali is that the sea is an unhealthy, evil place, the 'repository of all human waste' (Hobart 1990). What is more, the southern sea is the domain of the powerful queen of the spirits, Ratu Kidul, mystical wife and patroness of the Mataram rulers represented today by the royal line of Yogyakarta's and Solo's sultans. Once each year, the day after the sultan's birthday, some of his nail and hair clippings and some flowers are buried in a special area of the beach at Parangkusuma, by the sand dunes of Parangtritis, south of Yogyakarta, and food and a full set of women's clothing are cast adrift, all as offerings to Ratu Kidul. As well as being a 'type' of the Hindu sea goddess, it is believed that Ratu Kidul was once a relative of an orphaned prince of the western Pajajaran Kingdom who was instructed by her to find the Majapahit Kingdom in East Java. He was told that, when his descendants established a kingdom near Mt. Merapi, she would marry each of its rulers.

This kingdom was (the second) Mataram, and Parangkusuma is believed to be the place where Prince Senapati, its founder, came ashore after spending three days and nights with Ratu Kidul learning of love, government, and war. Senapati later managed to defeat 5,000 troops from a neighbouring principality with just 800 of his own, partly by deception and partly by the opportune eruption of nearby Mt. Merbabu which induced the opposing general to retreat (Raffles 1817). This gave clear demonstration of his favoured status. Ratu Kidul is believed to have had communion with each of Senapati's descendants up to the present sultan, and the Water Palace or Taman Sari in Yogyakarta was built in 1758 by Sultan Hamengkubuwono I as a replica of her palace beneath the waves. Evidence of her immediate presence is believed to be gust of wind that slows to a breeze and leaves behind a powerful fragrance. At the coronation of Sultan Hamengkubuwono X in 1989 many of those present felt the wind and smelled the lingering fragrance (Suyenaga 1991).

Activities on the southern coast of Java are still influenced by the presence of Ratu Kidul. For example, swiftlet nests from coastal caves near Karangbolang are collected only at mystically determined times, in moderate quantities, and only after a buffalo's head has been offered to her. And the people who drown in the riptides, crosscurrents, and heavy surf that prevail along the south coast are believed to be recruited as servants in her kingdom in the waters that lead to a steep, 7,000 m deep trench 250 km offshore.

The Madurese people are unusual in Java and Bali in being at home on the sea, regarding it as a workplace and source of income, and they travel across it to all parts of the Indonesian archipelago. At the other extreme the southern Balinese of the exquisite and productive terraced landscapes regard the sea as the abode of demons and evil spirits. Balinese aversion to the sea is quite understandable, given the lack of natural harbours or other anchorage, fringing reefs that obstruct shipping, and cliffs which prevent access to the sea – and of course the turbulent Indian Ocean swell facing the populous southern, rice-growing areas. The Sundanese have never really had maritime inclinations, but in the sixteenth century, Javanese war fleets were sent out against the Portuguese at Malacca (de Graaf and Pigeaud 1974) and there was regular trading with other islands. The fishermen off the south coast of West Java, however, still retain a deep respect for the sea, and they were reportedly 'filled with remorse' when 400 dolphins were accidentally caught in nets set in shallow water in March 1992. Dolphins are believed to push boats ashore when they are in trouble, and the dead animals were buried rather than eaten.

The sea has always been regarded by the Javanese as a place of mystery, and at least one folktale tells of a prince who swam to the middle of the sea and obtained wisdom from a small being whom he met there (Sya'rani and Willoughby 1985). But since the arrival of the Dutch, the attention of the Javanese has increasingly been directed towards its productive land-based agriculture, and exploitation of the marine resources off its shores has been left to people hailing from other islands. One exception to this are the Osing Javanese of the Blambangan area who are distinguished fishermen, and who regard themselves as bolder and more skilled in their occupation than their relatives to the west (R. Hefner pers. comm.).

ATTITUDES TO THE FOREST

A similar antipathy is found towards the forests. Traditional attitudes hold these to be dark places entered only by woodcutters and inhabited by 'other' people – gnomes, elves, goblins, fairies, and witches. Among these beings are the malicious and teasing *gendruwo* and *wewe*, male and female forest spirits, and it is believed their scheming can result in illness if someone so much as walks through a forest (Boomgaard 1991). Indeed, there are many lingering beliefs in the effects on health and wealth of disturbing some element of the macrocosmos.

In Javanese tradition, forests are also the abode of wise men living as self-denying hermits from whom bold aristocrats can seek counsel and with whom they can practice meditation. Forest caves have been used as places to meditate and pray, as they still are in Alas Purwo National Park and Pangandaran Reserve, but he who enters the forest is figured as brave or foolhardy. Whichever he is, his way is often barred by obstacles, such as the stinging wasps, python, rhinoceros, beetles, and tiger encountered by Raden Tanjung, whose brave exploits made it possible for the Sundanese goddess Nyi Pohaci (more widely known as the rice goddess Dewi Sri) to change her body into the first weaving loom (Veldhuisen-Djajasoebrata 1988). The dark and unknown nature of the forest mirrors the dark and desperate straits of the person who has been forced into the forest. Old Javanese texts typically contrast the wild forests with 'garden' scenes, environments made or controlled by man which are safe and pleasant to be in.

In Majapahit lore, forests were dangerous places populated by evil spirits that could only be controlled and conquered by even more powerful good spirits represented by the kings themselves (Dove 1985). The official court chronicle of the Mataram kingdom emphasized the fundamental requirement to create cleared, cultivated, safe, useful, and controllable space from chaotic, wild, and fearful forests. This chronicle is entitled *Babad Tanah Jawi*, which can be translated either as 'The history of Java's land' or 'The clearing of Java's land' (Dove 1985). Fear is also the reason that the forests of Mt. Halimun in West Java, of Mt. Slamet in Central Java, of Mt. Penanggungan in East Java, and of Mt. Batukahu and Mt. Agung on Bali, are all facing attrition at lower rates than elsewhere. The 25 ha of forest on Mt. Pancar just 15 km northeast of Bogor, near Citeureup, has been left relatively undisturbed because there are holy graves within it; as a result, at least 60 bird species and a population of the Javan lutung monkey persist (Erftemeijer 1987).

Such views of forest are not peculiar to the Javanese. The report commissioned by British King Charles II to survey the status of forests in his realm expressed the opinion of the time that "it is natural for a man to feel an aweful and religious terror when placed in the centre of a thick wood" (Evelyn 1664; Thomas 1994). Part of that terror would have been due to superstitions and the folk stories of bears and wolves, by then long extinct, but part would also have been due to a sense of reverence for the atmospheric wildness (Crawford 1985), that sense which develops into environmental concern and respect.

A popular assumption is that the forest-like structure of home gardens is the result of deliberate planning to mimic natural forests (p. 609). In fact, forests do not have a high cultural value in Java and it is doubtful that its re-creation around peoples' houses would be encouraged (Soemarwoto 1987a). This is because forests are considered dangerous places where wild animals and evil spirits roam, and clearing forest for a settlement, *babat alas*, is considered to be a noble deed. Indeed, the term *babat alas* is still used to indicate pioneering activities of a praiseworthy nature, such as the founding of a university. A person would surely feel offended if his home garden were said to resemble a forest. It is much more likely, then, that the appearance of forest is an accidental result of people's efforts to fulfil many and varied kinds of subsistence needs in the absence of a market economy. In any case, the equating of home gardens with forest does both of them a disservice, and may indicate a certain unfamiliarity with the genuine article on behalf of some writers. Natural forest is not nearly so dark and dense as many home gardens.

ATTITUDES TO PLANTS

While the community of plants that makes up a forest may be feared,
many individual plant species have assumed considerable importance
because of their economic uses, their pestilential nature, their role in the
arts, or their curative properties.

One of the most common plants on Java and Bali is alang-alang *Imperata
cylindrica.* Alang-alang is regarded as a scourge by some, and a great deal of
money has been spent to control it, but it is not seen universally as a
problem. To the astonishment of many, alang-alang is actually cultivated as
a crop in Bali in order to provide thatching material which can last 10-15
years if the roof pitch is correct, and if unshaded and therefore allowed to
dry after rain. Also, among the traditional rural Javanese it is managed to
produce fodder for stock animals and for thatching. On Mt. Merapi the
people's way of life is still very much affected by alang-alang, all the more
so since the 1920s, when the upper slopes of the mountain were declared
protection forest. This reduced the area of arable land available to the
people, forcing the abandonment of shifting cultivation and the development of a more intensive system of semipermanent cultivation on the
lower slopes. With the passing of the fallow period, the fertility of soil was
improved by the application of cattle dung. Stall animals are fed on fodder
which is cut by hand from nearby fields and, during the dry season, from
the volcano's higher slopes, above the state forest, up to three hours walk
from the village. Some 50-60 kg are cut and carried by each household
every day. The most important component of this fodder is alang-alang
(Dove 1986). The swards of alang-alang are deliberately managed to produce the best-quality grass for the cattle by cutting, burning, manuring, and
cutting back the competing scrub. The people sometimes even plant this
grass (Dove 1986).

The topmost grasslands on Merapi are not cut because it is believed
they are the property of spirits who prohibit their use by humans so that
their horses can graze undisturbed. These spirits are just one part of a complex spirit world within the crater whose organization parallels that of the
royal court of Yogyakarta. Indeed, even the ownership and exploitation of
the mountain grassland has parallels (Dove 1986), for the historic Javanese
royalty maintained grasslands for hunting, grazing, and thatch for their
own use, and villagers' use was proscribed or at least regulated (Pigeaud
1962). These grasslands were perceived as having far more value than the
forests from which they were created (Dove 1985) and alang-alang became
a symbol of the aristocracy and was used in Hindu ritual (Dove 1986), as is
still the case in Bali (Rifai and Widjaja 1979) and in certain Javanese household traditions.

The traditional belief on Mt. Merapi is that alang-alang arose on the
mountain from the rotting bodies of winged termites that, for some reason,

failed to swarm out of their earth nest after the first rains of the rainy season. The termites themselves arose from deer fat buried on the slopes by one of the legendary apostles who brought Islam to Java (Dove 1986). Indeed, the Javanese word for deer, *menjangan*, appears to be related to the old Javanese word for alang-alang, *munja* (Pigeaud 1962).

The Javanese also have a strong tradition of using plants for a wide range of medicinal cures (p. 180). The number used is relatively small compared with the total flora, and most of the more commonly used species are grown around houses (p. 608). Traditional Balinese house compounds have over 30 species of plants within their walls, but most of these are cultivated for their roles in religious ceremonies (Suryadarma 1989, 1991, 1992; Sastrapradja 1990).

The upas was a tree from Java that gained notoriety in 1785 when a surgeon named Foersch, told that this lone tree stood in a desolate valley in Java and gave off such fumes from its poisonous gum that no living thing could approach and survive. The local people were said to try all types of expedients to obtain the gum in order to make their arrows and spears even more deadly (McMillan 1915). This tree was made famous by its mention in a poem by Erasmus Darwin (Darwin 1789), and by a large painting by Frank Danby, now in the Victoria and Albert Museum, London. In fact two natural phenomena had become mixed. First, there are 'death' valleys such as Pejagalan at Telaga Bodas where the poisonous volcanic fumes are sulphurous and so thick that neither people nor animals are able to cross. Likewise the Kawah Manuk, where birds, flying too low, are overcome and fall into the pool below (McMillan 1915). Second, there is a tree *Antiaris toxicaria* (Mora.) (fig. 17.1), which is well known for the toxic properties of its latex (Burkill 1966) and was used as a potent poison at the time of European expeditions to Java.

The most conspicuous village trees are the grand fig trees that dot the rural landscapes in Java, and particularly in Bali. The deciduous fig, without or with few aerial roots and long leaf stalks and leaf tips, is generally *Ficus religiosa* (Mora.), whereas the *Ficus benjaminica* is evergreen, has few if any aerial roots, and short leaf stalks but quite long tips (figs. 17.2-3). These are believed to be dwelling places of spirits, and in Bali altars piled high with offerings can be found without fail at their base. Wilful cutting of such a tree would be unthinkable, and its removal for road widening or some other perceived need would have to be preceded by protracted propitiation (Boomgaard 1991). At the foot of one such fig tree in Mojokerto, southeast of Surabaya, East Java, sits a 30 cm high statue of Ganesha the elephant god, which was originally placed there centuries ago by a pre-Islamic king of Mataram after a rest stop for his troops when he was repulsing the King of Madura. Only once was it moved, 15 km away, but it returned under its own steam, and the one person who is known to have shown physical disdain for it, a junior Dutch colonial officer, died within a week, and his

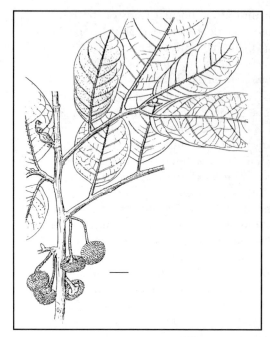

Figure 17.1. The fabled upas tree *Antiaris toxicaria*. Scale bar indicates 1 cm.
After Koorders and Valeton 1918

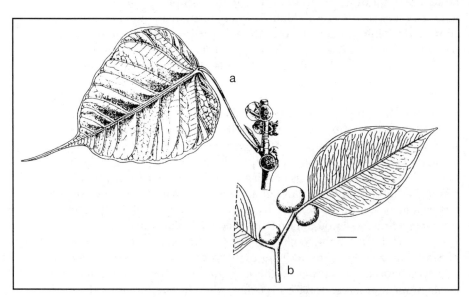

Figure 17.2. *Ficus religiosa* (a) and *Ficus benjaminica* (b). Scale bars indicate 1 cm.
After Koorders and Valeton 1918

Figure 17.3. Large *Ficus benjaminica* tree in Marga, Bali.

By P. Arens, with permission of the Rijksherbarium/Hortus Botanicus, Leiden University

entire family was dead by the end of the next year (Lubis 1969).

While no paintings from about 800 A.D. survive, some idea of the environment of those times can be seen in the Karmawibhangga reliefs at the foot of the Borobudur temple. Many of the plants can be identified and they tend to be those still common in and around villages eleven centuries later (Sarwono 1988) (fig. 17.4). In addition to the utilitarian trees, the tree of life or kalpataru is also depicted. This was adopted in 1979 as the logo of the State Ministry of Population and Environment. On the Borobudur this mythical tree is used to denote the beginning and end of the tales, similar to the *gunungan* used in Javanese shadow plays.

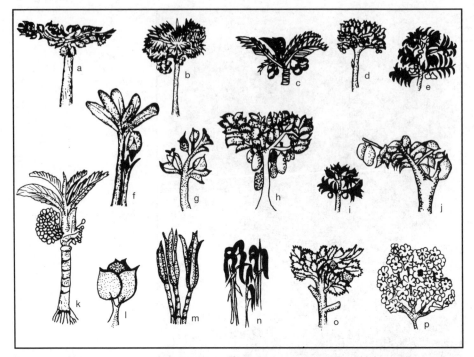

Figure 17.4. Representation of plants on the reliefs of Borobudur Temple. a - durian *Durio zibethinus,* b - lontar palm *Borassus flabellifer,* c - coconut *Cocos nucifera,* d - duku *Lansium domesticum,* e - rose apple *Eugenia* sp., f - banana *Musa* sp., g - kecubung *Datura metel,* h - jackfruit *Artocarpus heterophyllus,* i - mangosteen *Garcina mangostana,* j - mango *Mangifera indica,* k - betel palm *Areca catechu,* l - talas *Alocasia/Colocasia,* m - sugar cane *Saccharum officinarum,* n - millet *Setaria italica,* o - breadfruit Artocarpus communis, p - hibiscus *Hibiscus rosa-sinensis.*

After Sarwono 1988

ATTITUDES TO ANIMALS

It has been argued that the inclusion of animals as essential elements of dance, art, literature, and ceremonies demonstrates that people and nature are all part of the same whole, and that animals are often more appreciated by the gods than are people (McNeely and Wachtel 1988) (fig. 17.5). This is attractive and in line with the alluring doctrines of New Age philosophy and religion, but normal animals are scarcely to be seen in any of the important positions in Javanese or Balinese arts. Ganesha (the goblin-like, elephant-headed god), Barong (the 'well-meaning' but fearsome-looking

Figure 17.5. Detail of a Dodot (status) cloth made in Surakarta early this century. The animals and plants are all shown the same size, supposedly symbolizing the equal importance of all parts of nature.

After Veldhuisen-Djajasoebrata 1988

lion), Garuda (the humanoid, golden sun-bird, mount of Vishnu, enemy of Naga the serpent), Hanuman (the humanoid Monkey King who conquers Rawana for Rama), Raksasa (the devilish monster with long fangs), Naga (the dragon serpent), and hosts of other weird beasts are depicted and revered but this has not resulted in any noticeable respect for earthly animals or their habitats. Indeed, one of Bali's best-known folktales relating to tigers has the moral that tigers are best kept caged or as far from human contact as possible and should never be helped (Alibasah 1990). In addition, the Balinese, who regard themselves, and are often regarded by others, as among the most culturally advanced people on earth, seek to distance themselves from animals and less-cultured people by filing down the canine teeth of their children to ensure that the 'animal-like' fangs do not project beyond the incisors and premolars.

The fact that rhinos and tigers are said to have inhabited Mt. Meru, as depicted on the diamond-shaped gunungan of Javanese shadow plays, does not seem to have helped them. Where genuine creatures are respected, even cossetted, such as the large (<1 m), carp-like fish *Tor soro* in ancient fish ponds around Kuningan, West Java (Rachmatika 1991), the large swamp eels *Monopterus albus* in Toya Bubuh, Bali, or the bats in the trees surrounding the Alas Kedaton temple, Bali, it is out of fear of the hex that will fall on anyone who disturbs or otherwise maltreats them. Other examples of fear abound. It is what keeps the inhabitants of Buahan village, Gianyar, Bali, from collecting firewood or cutting timber in a small area of forest surrounding one of their temples. It is believed the forest is guarded by a mother worm about 2 m long and her daughter, who will attack anyone who attempts to disturb the forest or the temple (Wickham 1992, 1993). Fear of 40 days of bad luck prevents the hunting of pangolin (Neill 1973).

It is interesting that different groups of people have distinctly different views of and relationships to certain animals. For example, small house geckos are regarded as holy on Bali and their chattering call is believed to be uttered when true or important words are spoken. Conversely, the same animals are despised by Javanese Muslims because a similar small lizard gave away the hiding place of Mohammad to his enemies. He was not found, however, because a spider concealed him with its web, and so this small beast gains the Muslims' respect. Most westerners treat spiders with disdain bordering, for some, on fear and hate for no obvious reason. Likewise the bandicoot, a large type of rat, may not be killed on Bali because it is regarded as the magical incarnation of a Princess, but may be killed along with other rats on Java.

Interest in hunting wild animals has led, in some countries, to a concern about disappearing habitats and diminishing populations, but hunting in Javanese tradition is largely a thing of the past. Deer used to be a popular prey of Javanese hunters, and fire was a major weapon to flush the deer from cover, to drive them in a desired direction, to clear obstructing scrub, and to stimulate the growth of fresh, young leaf blades. This fire created and maintained large areas of grasslands, and without it, successional growth would have advanced. The deer is also a symbol of aristocracy, and mounted antlers are still to be found on houses of men of wealth or rank in the rural areas of Central Java. Indeed, a deer's head was buried as part of the ritual before the reconstruction of the Sultan of Solo's palace in 1985 (Dove 1986). Banteng were also a royal hunting quarry, and records of this for the Hindu Blambangan kings date back to 1365 (Hoogerwerf 1970). Prior to the arrival of Islam, pigs would have been a major hunting quarry. Although their habitats overlap with those of deer, pigs prefer forests and deer prefer open grassland. With proscriptions on the eating of pigs, so their value as game animals and the value of their forest habitat diminished. This was balanced by an increase in the cultural value of deer and grasslands, reinforcing the pre-Islamic bias against forests and for grasslands. This bias may even have promoted the adoption of Islam (Dove 1986). It should be added, however, that some devout Madurese Muslims in Pasuruan, at least, have hunted and eaten wild boar, hiring out their services to farmers wanting to rid their fields of these pests, believing that the wild boar are not pigs at all and therefore not proscribed (R. Hefner pers. comm.).

In Javanese tradition the essential attributes of the ideal male were a home (*wisma*) and a wife (*wanita*), both of which represent security and fulfilment, a dagger (*kris* or *curiga*) representing status and power, a horse (*turangga*) representing ease of communication and movement within society, and finally a bird (*kukila*) representing all of nature, without which the rest are meaningless (Mason 1992). This has led to the tradition of keeping birds in cages, the implications of which are discussed elsewhere (p. 232).

Petulu Egrets

One exceptional relationship between people and birds can be seen at Petulu Gunung village, north of Ubud, Bali. Here, chiefly along the main street, some 6,000 cattle egrets *Bulbulcus ibis* roost and nest, together with a few little egrets *Egretta garzetta*, pond herons *Ardeola speciosa,* night herons *Nycticorax nycticorax*, and a mobile population of intermediate egrets *Egretta intermedia* (fig. 17.6). The egrets are said to have arrived in 1965 after the excesses following the attempted coup. The villagers believed then and still strongly believe, that their arrival and continued presence are deeply significant (Kertonegoro 1987, Bawa et al. 1988; Boehmer 1992).

The peoples' concern for the birds is shown in several ways. They react swiftly to snakes or civet cats in the area that may take eggs or chicks; they will replace fallen birds in their nests or, failing that, will hand-rear them; no one may fell a tree in the village without agreement from the majority, and some villagers opposed the taking of even a tree branch as a specimen for identification (Noor 1992). The colony does not appear to be threatened by any activity around the nesting and roosting sites, but threats exist in the form of villagers from outside Petulu who, knowing nothing of the special nature of the birds, catch them for sale or for food, using nets or air rifles. Both *B. ibis and E. garzetta* are protected by Indonesian law (Noor 1992). Despite this, there is no reason to believe that the herons will leave Petulu so long as they remain undisturbed. This is also the key to

Figure 17.6. Some of the egrets roosting and nesting along the road in Petulu, Bali.

By A.J. Whitten

other large heronries such as on P. Panjang offshore from Jepara (Central Java) where Forestry and Tourism offices control visitors (Noor and Permana 1990), and in Semarang where a large colony is found next to a military housing complex (Purwoko 1991).

Tiger and Banteng

Two animals, the tiger and banteng, are regarded by the Javanese as epitomies of important but opposite human traits. Thus the tiger's power represents base human impulses (Lubis 1975, 1991), while the banteng represents the Javanese ideal of steadfastness and restraint, an animal at peace with and in control of itself. The two traits are both necessary but must be kept in balance for internal harmony to be maintained. Both animals find a place near the centre base of the gunungan, which represents Mt. Meru with animals and birds in the 'tree of life' growing on its slopes. Beneath the tree can be found a tiger and a banteng standing equally above a small house with a closed door, representing the mind of man. Peace of mind will only be attained, and the doors opened, if the opposing emotions the animals represent can be balanced and controlled.

Although today this tiger-banteng opposition is remote (perhaps unknown) to many Javanese, it was used as political allegory during the Dutch colonial period (unknown to the Dutch) in the form of gladiatorial fights between tiger and banteng in large cages. The animals had to be goaded to fight (the former with sticks and burning straw, the latter with boiling water and stinging nettles). To the Javanese, the tiger represented the Dutch (whose national symbol is a lion), who were deadly in firepower and quick to act, but lacked stamina. The banteng, meanwhile, represented the Javanese, being slow to anger but deadly when aroused. The result of the battles was unexpected (at least to the Dutch) because the banteng generally scored victorious. If the tiger did happen to survive the half-hour contest, it was removed and speared to death (Raffles 1817; McNeely and Wachtel 1988; Boomgaard 1994).

ATTITUDES TO NATURAL DISASTERS

When Mt. Tambora erupted on Sumbawa in 1815, the world's greatest volcanic event in historical times, Madura, 500 km to the west, was cloaked in darkness for three days, and the Javanese explained it as Ratu Kidul celebrating the marriage of one of her children. In former times the causes of natural disasters, such as volcanic eruptions, could not be explained with traditional knowledge other than as retribution by gods. With no direct means of controlling the events, the proper course was to make sacrificial offerings which, it was hoped, would placate the awesomely powerful deities.

In 1979, scores of animal sacrifices were made on the crater of Mt. Agung, the most holy and tallest mountain on Bali, abode of Siva, Brahma, Vishnu, and countless ancestral deities. This once-a-century rite (*Eka Dasa Rudra*) requires the sacrifice of one of every type of animal, and while this was not strictly achieved, the range of beasts tumbled into the bubbling crater was nonetheless impressive – among others, scorpions, leopard cat, crocodile, eagle, pigs, chickens, and buffalo. The rite had been planned for 1963, after a gap of several centuries, to atone for people's wickedness and excesses during World War II, the Indonesian Revolution, and the unholy exploitation of the mountain's sulphur, timber, and rock. In fact, just one month before the rite was to begin, the volcano erupted dramatically and killed over 1,000 people, some by suffocation with hot ash, and others by incineration with rolling, molten lava as they waited, patiently and clothed in ritual dress, within their temples (Matthews 1983). It was not until 16 years later that it was felt the Great Rite could be performed properly (Stuart-Fox 1982).

In 1982, after 64 years of being dormant, Mt. Galunggung became active. This dramatic event spawned, resurrected, or mixed a whole host of mysterious stories. For example, some local people believe that the eruption began immediately after three men, posing as hermits, had found the spiritually powerful *kris* daggers that had been hidden for safety on the mountain by King Galunggung after a defeat by other rival kings at some undefined time in the past. Even after the *kris* had been returned, the volcano continued to shake and belch. In penitence, hundreds of villagers made offerings, meditated, prayed, and sacrificed cattle and goats – but all to no avail. One leader had a revelation that the former king needed a tomb in which his soul could find rest, and the building of this was begun. However, a local senior official who lacked sympathy for this approach had the tomb destroyed. On the same day a jumbo jet nearly crashed after flying through the thick ash cloud (Tootell 1985).

ATTITUDES TO SEASONS

Farmers, particularly in Central Java, have been guided in their agricultural activities by the traditional *Pranatamangsa* ('arrangement of the seasons') calendar, the use of which can be traced back to before the time of the first Hindu arrivals. The calendar, which normally has 365 days each year, correlates cosmological, biological, climatological, and sociological aspects of agriculture, enabling the farmer to forecast likely seasonal deviations. The calendar is extremely complex – the 12 months, known both by their number and character, vary from 23 to 43 days in length and are arranged in seasons and periods reflecting expected environmental conditions. For example, the sixth month (9 November - 21 December) is given the characteristic *rasa mulya kasucen* ('a feeling of holiness') because the new green growth after the first rains stimulates a peaceful mind. The following month (22 December - 2 February) is known as *wisa kentiring maruta* ('flying poison blown by the wind') because floods and diseases can be expected. The fourth month, around September, is recognized as the driest month and has the rhyming name *sate sumber* 'drying spring' and likewise January, when there is the first surplus water after many months is known as *jan-ana-warih* 'just enough water' (Daldjoeni 1984).

Even now, great importance is given to the calendar and many of the taboos on planting and other practice are adhered to. In many cases the traditional practices do not conflict with advice from the extension services because the calendar advises appropriate activities at appropriate times. There were conflicts, however, when there was encouragement (and indeed the possibility) to plant three rice crops annually instead of the two allowed under the Pranatamangsa. There is now a sense of traditional justice that the third crop is actively discouraged to combat exacerbated problems of pests and disease.

The existence of seasons and variations from year to year tends to make the farmers optimistic, and the inevitability of success or failure means their fatalistic attitude gives them no fear for the future. Unfortunately the imposition of government schemes is sometimes met with a certain lack of enthusiasm and implementation problems, especially if activity is advocated during times when the farmers are supposed to be 'in despair' (Daldjoeni 1984). The failure of modern predictions and promises (e.g., of the extension services) to deliver on yields, pest protection, etc., increases farmers faith in the Pranatamangsa.

MODERN MYSTICISM

While many urban sophisticates would claim no lingering beliefs in the spirit world, and that the practices of the recent past, such as the presentation at village shrines of offerings to guardian spirits, are diminishing, it would be a mistake to underestimate the importance of traditional spirituality to probably the majority of the population of Java and Bali. Offerings of incense, food, and flowers can be found on many mountains, particularly those with special mystical significance.

Fear of a hex, and not landscape gardening, is the reason a large kapok tree *Ceiba pentandra* (Bomb.) stood for many years in the western segment of the Semanggi interchange between Jalan Sudirman and Jalan Gatot Subroto in Jakarta. When Indonesian and Russian engineers started construction work on the original cloverleaf junction in the early 1960s, they were informed by local people of the special spiritual powers of the tree that was scheduled to be felled. The news was greeted with both scepticism and alarm – but after two men (one Russian, one Indonesian) died while trying to bring it down, the alignment of the road was modified (Lubis 1969). A few years ago the tree was finally removed, without apparent problem, when the flyover was widened. Traditional beliefs have also had an influence on other development projects. For example, prior to work beginning on the Kamojang thermal electricity generating station in 1980, local people made offerings to the spiritual inhabitants of the fumarolic area to make amends for the human desecration of the area. In addition, a goat's blood was drained into one of the foundation shafts of the main building to ensure safety for the workers (Tootell 1985).

Despite Indonesia being the world's largest Islamic nation (distinct from an Islamic state) and Java alone having more Muslims than any Middle Eastern country, there are significant strands of traditional mysticism woven into the day-to-day lives of the people. This is true too for many of the 10% or so of the population who profess to be Christians.

Indeed, it has been said that Java is the only place in the world where one can approach almost any farmer in his field, inquire as to the ultimate nature of reality, and receive a thoughtful, mystical and extended reply (Geertz 1960). Many feel there is little conflict between their mystical and orthodox religious beliefs – it is held that the nine saints who spread Islam through Java in the 15th century had considerable mystical abilities to deflect daggers, make themselves invisible, and appear in two places at once (Hefner 1991). Indeed, *dukuns* (masters of the occult arts), are respected or feared members of the community, even when they describe themselves less pejoratively as 'traditional healers', and are consulted by many for a wide range of bodily and spiritual needs. There are strong movements to purify Islam by rejecting these practices, but the traditional beliefs are embedded deep within the people. The Central Javanese sultans were defenders and sponsors of Islam, but more as 'Allah's representatives on earth', for Hindu flavour can be found in small features such as the yellow 'mountain' of rice found at many celebration meals, which traditionally represents Mt. Meru.

MAJOR RELIGIONS

All the major religions appear to have something to say about our environmental problems (p. 836), but the teachings are only newly being emphasized, and have yet to make a significant mark on peoples' attitudes to nature.

SCENARIOS FOR THE FUTURE

It is very difficult not to be extremely pessimistic about the future of Java and Bali when one examines local, national, and international trends in environmental quality, trade, and economic 'progress'. The traditional beliefs are not serving, and probably never served, people and the environment particularly well. There are exceptions, however, such as the widespread traditional use in upland areas of green manures, watershed management for village springs, and so forth. It is impossible to be sure but it is conceivable that, had the agents of pollution, overconsumption, and forest destruction been available 1,000 years ago, they would have been used. There certainly seems to be little evidence of restraint.

Modern interpretations of the major religions all conclude that sound environmental management is an inevitable outcome of serious adherence to the faith. Yet there appears to be no trend among countries with a predominance of Muslim, Christian, Hindu, or other believers for a monopoly on successful sustainable development. Nor does the modern atheistic religion of economic advancement inevitably meet the real needs of people and the environment, and this is no better exemplified than by the near scuppering of the United Nations Conference on the Environment and Development in June 1992 by the powerful, economically dominant, developed countries which appeared unable or unwilling to practice what they preach in terms of self-restraint.

It can only be hoped that those with power will see the inevitable outcome of present trends and not just talk about solutions, but rather implement and enforce the necessary changes, even if those in the commercial environment with singular power and loud voices complain. If Java, Bali, and the rest of the world are to survive beyond the forthcoming nexus of issues, then it is almost inevitable that attitudes must change to result in:

- relative self-sufficiency with a new balance between extensive low technology and intensive but resource-conserving high technology – there is no need to return to the past;
- local responsibility resulting from a lessened need for a strong controlling central government, with those directly dependent on local resources being largely responsible for them; and
- a return to spirituality towards the godhead envisaged in the Pancasila. Others have proposed the deliberate strengthening of people's ties to spirits as a psychological tool to ensure increased harmony between people and nature (McNeely and Wachtel 1988). Apart from its manipulative overtones, this approach aims to subject people to the perpetual fears of offending invisible beings and forces. Instead, spirituality should be interpreted as a closeness and obedience to God and a resting in his peace.

Chapter Eighteen

Biogeography and the Loss of Biological Diversity

Less than a hundred years ago, rhinos still wandered on Mt. Slamet, tigers were found in most of the forest blocks on Java and Bali, rice fields were infested by dense flocks of Java sparrows, and the noisy calls of the Javan lapwing could be heard along the coastal plains. How things have changed. Java and Bali have already lost a considerable amount of biological diversity this century as a result of habitat loss (p. 328) and are poised to lose still more. Unfortunately, as the rate of loss increases, so the species that are being lost become less noticeable: the larger, predatory animals such as tigers are lost before the smaller, herbivorous ones (p. 701). It should be stressed that, although in this chapter we examine a few cases where the process of losing species can be traced or quantified, these are by no means the only cases. We hope that if decision makers can become familiar with the past and present situations, then they may make every effort to prevent any further irreversible losses. The pragmatic, economic, scientific, and moral reasons for doing this are discussed elsewhere (p. 62). Before examining some case studies from Java and Bali, we present some background to the distribution and abundance of biological diversity.

BIOLOGICAL DIVERSITY AND GEOGRAPHY

Every organism has a distribution related to its ecology, behaviour, physiology, ability to travel long distances, the other organisms living in the same area, and to the geological history and climate of the area in question. A further and important factor in determining distributions is chance. The distribution of an organism is generally bounded by unsuitable habitats, unsuitable climates, or the occurrence of a species against which it cannot compete successfully. Alternatively, a species may be actively dispersing, and the edge of its range may simply be the furthest point it has reached at that time. Clearly a barrier for one species is not necessarily a barrier for another, and a barrier for one life stage of an organism is not necessarily a barrier for a different life stage of the same species. For example, the larvae of many invertebrates, and the seeds of plants, are much more

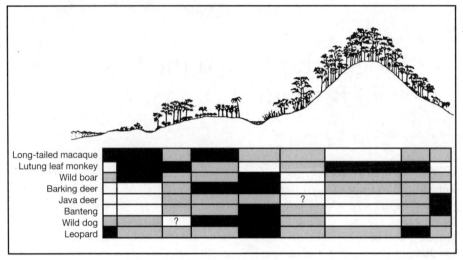

Figure 18.1. Cross section of Meru Betiri National Park showing the utilization of different habitats of Meru Betiri National Park by certain mammal species to illustrate the unevenness of their distributions. Black – high use, shaded – moderate use, white – low use.

After Seidensticker and Suryono 1980

mobile than the adults. Distributions can also vary over the course of a day, over seasons and, for migratory species, over a year.

Distributions can be described at various levels of definition: thus a bird might live in forest, but perhaps only in a few types of forest, at certain altitudes, avoiding forest near rivers, and only using the tops of the taller trees, and so forth. Equally, it would be correct to say that large animals such as banteng, deer, and hornbills live in Meru Betiri National Park, but closer examination shows that the actual distributions are more complex (fig. 18.1).

Niche and Niche Separation

It is important to appreciate some of the ways in which a community of organisms fits together, and this is best done perhaps by discussing species' ecological niches. The niche of a species can be thought of as the role that species has in a community (see Schoener 1989). This is best conceptualized as a multidimensional space, in which the coordinates are various ecological parameters such as diet, time of peak activity, and use of habitat. Species niches in simple communities are generally quite distinct, but as the community becomes more complex, so overlaps occur. For example, in

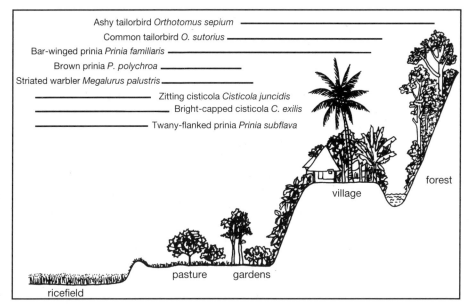

Figure 18.2. Habitat preferences, indicated with a bold line, of eight warbler species in rural areas in West Java.

After van Balen 1989c

the montane forests of West Java, up to 13 different insectivorous bird species feed together in a single flock. Although they overlap in their diet and general behaviour, each species uses the available food sources in different ways and thereby maximizes feeding efficiency and avoids total overlap (Erftemeijer 1987). Similarly, eight insectivorous warblers in another area of West Java were found to have different feeding strategies which minimized competition and maximized efficiency at exploiting available resources in a range of neighbouring habitats (fig. 18.2). Even when there was apparent overlap, finer preferences could be detected. For example, the golden-headed cisticola *Cisticola exilis* is generally found only in dry rice fields away from water, whereas the zitting cisticola *C. juncidis* favours wetter rice fields. Also, some species seek food in different vertical layers of the same habitat, such as the two common swiftlets on Java and Bali, the edible-nest swiftlet *Aerodramus fuciphagus,* and the slightly smaller linchi swiftlet *Collocalia linchi* which divide the air space, with the former found higher than the latter (van Balen 1989c).

Niche separation of two monkeys, the lutung *Semnopithecus auratus* and the long-tailed macaque *Macaca fascicularis* occupying the same forest has been studied in the mangroves at Muara Gembong, east of Jakarta (Supriatna et al. 1989). Some 60% of the lutung's diet comprised leaves (with rel-

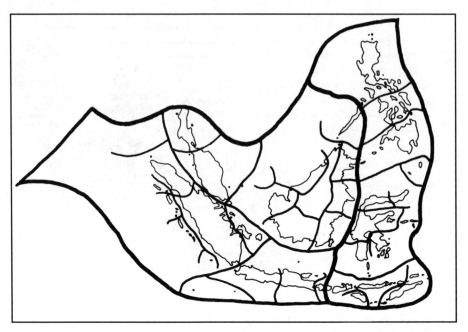

Figure 18.3. Biogeographic divisions and subdivisions of the eastern part of the Oriental or Indo-Malayan realm.

After MacKinnon and MacKinnon 1986

atively high protein concentrations) and 40% comprised fruit, while the macaque consumed 50% fruits (with relatively high carbohydrate) with lesser amounts of leaves, bark, roots, insects, and other animals. Although the two species of monkey at Muara Gembong do eat some of the same foods, it seems that in many cases the same plant parts are eaten at different stages of development – lutungs tend to eat younger fruit than the macaques (Supriatna et al. 1989). Both monkey species are widely distributed and their diet differs in different habitats – the lutung in Pangandaran and Mt. Gede-Pangrango eat considerably more fruit and flowers respectively than leaves (Kool 1992a, 1993; R. Beckwith 1995), but the basic difference in preference remains, even in captivity.

The Effects of Regional Geography

The species compositions of the fauna and flora of Java and Bali are unique, characterized by the presence of endemic species or subspecies

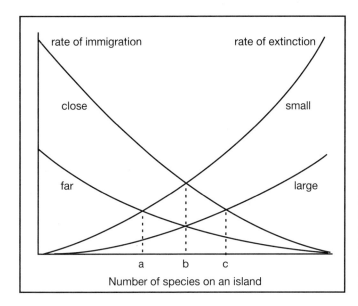

Figure 18.4. The relative number of species on (a) small, distant islands, (b) large, distant or small, close islands, and (c) large, close islands.

and the absence of species found in neighbouring areas. The biogeographical area known as the Oriental or Indo-Malayan Realm, stretching from India to Sulawesi, can be divided into regions with distinctive faunas and floras, one of which comprises Java and Bali with four subdivisions: western Java, eastern Java, Bali, and Christmas Island (governed by Australia) and the Cocos-Keeling Islands (an independent state) (fig. 18.3).

Species and Islands

It is understood intuitively by laymen and scientists alike that small islands support fewer species than large islands. For example, Bawean is a relatively small island with few types of habitat and it would be surprising if its biological diversity were not relatively poor compared with the Java mainland. The total number of species on an island represents a balance between the colonizing of the island by immigrant species and the extinction of existing species. The rate of colonization is clearly higher when an island is near the mainland because more species are likely to cross a relatively narrow sea gap. Also, the rate of extinction is clearly greater when an island is smaller because the population of any species will be smaller and the chance will be greater of disease and other detrimental events causing extinction. These relationships can be drawn graphically and represent the foundation of the Theory of Island Biogeography (fig. 18.4) (MacArthur

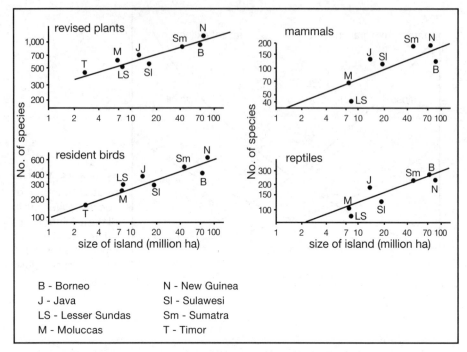

Figure 18.5. Relationship between number of species and island size for revised plants (those with a recent taxonomic review) and three groups of animals.
After MacKinnon et al. 1982

and Wilson 1967).

The relationship between island size and number of species is relatively constant for a given group of animals or plants, and in general reducing island area by a factor of 10, halves the number of species (fig. 18.5). Where an island supports fewer species than expected and so falls below the line, the reason may be that:

- the group is not sufficiently well known for all the species to have been found or recognized;
- equilibrium in species number has not yet been reached (where a volcanic island has been destroyed and is being recolonized);
- the island comprises a relatively restricted number of habitats, or habitats which do not support large numbers of species;
- the island is extremely remote and difficult to colonize.

Conversely, where an island supports more species than expected and so falls above the line, the reason may be that:

- more than the equilibrium number is present ('supersaturated'), and some species will in due course be lost. Such islands should exhibit their highest extinction rates soon after isolation (sea level rise) and progressively lower rates as the fauna is depleted;
- the island is peculiarly rich in habitat types; or
- the island is a centre of species radiation in a certain group.

It has to be remembered that species-area relationships hold only when an equilibrium has been reached, and the timing of this equilibrium cannot be predicted with any certainty. For some long-lived species one might expect it to take several centuries (Simberloff 1992) or, if the area is drastically too small, just months.

As described elsewhere, the sizes of Java, Bali, Borneo, and Sumatra have decreased and increased as the sea level has risen and fallen, alternately isolating them from and connecting them to the Asian mainland (p. 117). When Java was part of a larger land mass, it would have had more species than now and when the rising sea isolated it, it would have been 'supersaturated' (MacArthur 1972) with 'too many' species in relation to its new area. Such islands should exhibit their highest extinction rates soon after isolation (sea level rise), particularly among the larger species (p. 700), and progressively lower rates as the fauna is depleted. It is therefore not surprisingly that many of the species known to have lived on Java during the Pleistocene are now extinct (p. 199) whether or not people had a role in their demise. The related topic of transient diversity is discussed below.

Implications of Island Biogeographic Theory

As disturbance of natural ecosystems proceeds, so the relatively pristine habitat is divided into smaller and smaller pockets, or 'islands', which contain smaller populations of animals and plants dependent upon them. There are certain shapes of habitat which are theoretically better than others for maintaining biological diversity (Diamond and May 1976; Frankel and Soulé 1981), but as far as Java and Bali are concerned, the subject of design of conservation areas is largely one for academic debate. The question of what percentage of species will become extinct in different-shaped reserves in 50, 500, or 5,000 years' time is not of great relevance when it is by no means certain how much of the conservation areas will be intact in even 25 years' time. The shape of conservation areas does not generally conform to the theoretical ideal, because their design has been set primarily by expediency and patterns of human settlement. The major priority is simply maintaining the integrity of reserves.

During the last few years, attention has begun to focus on the rather more pragmatic problem of extinction. Extinction is now no longer viewed as an event, but rather as a process that can be observed and understood (p. 699). To avoid extinction, the population of a species must be large enough to maintain a sufficient proportion of its genetic variation to avoid the deleterious effects of inbreeding (p. 704). Thus the concept of 'minimum viable populations' has evolved, and this is of most concern for large, wide-ranging animals, and for keystone species of plants which provide an important or critical resource for a significant number of other species (p. 63). It has been ascertained that the effective size of an absolute minimum viable population is about 50 randomly mating adults living in a population with a one-to-one sex ratio. With an effective population of 50, however, it has been estimated that the genetic variation remaining after 100 generations will be, on average, only about one-third of the original, and the additional variance due to mutation would be negligible. It has therefore been proposed that an effective population of 500 randomly mating individuals is the minimum size of a population that would enable the species to adapt to environmental changes (Soulé 1987). It should also be remembered that the 50/500 minima are derived from theoretical considerations, and there are many apparently healthy captive populations derived from a very few animals. For example, there is the thriving herd of spotted deer *Axis axis* which has grown from the introduction of just three pairs at the beginning of last century to over 400 in the grounds of the Bogor presidential palace (Hasanuddin 1988b). Also, cultures of the Javan lichen stick insect *Orxines macklottii* have been maintained in Britain for 50 years from stock which was imported in the 1940s (Brock 1992). Nevertheless, for the largest animals on Java and Bali it is clear that there are good biological reasons for protecting their present populations, most of which are even now well below what would be regarded as the theoretical minimum for long-term survival.

It is clear that if one applies the 50/500 rule, then none of Java's protected areas can support the larger animals in populations which will remain genetically healthy in perpetuity. In the light of this, planners might seem justified in arguing that those lands should be transferred to alternative uses as has been tried in other countries (Simberloff 1988). When all the remaining areas are relatively small, the onus is on those responsible to manage the reserves so that the critical species are given the best chance of long-term survival. This does not yet apply to Java and Bali, but future management could include routine genetic and demographic monitoring as well as subsequent translocation and swapping of animals of some species from disjunct areas and possibly captive or semi-captive breeding. The costs of these options would be great but, as written by Soulé (1980), "there are no hopeless cases, only expensive cases." This, at least, should deter their adoption until proven necessary.

Edge Effects

An important consideration when calculating the size of an individual block of forest is that what may appear to us as an edge may not be recognized as that by the species dependent on primary forest. For example, many plants and smaller animals are greatly influenced by the internal microclimate of the forest, and deep-forest species are adapted to the relatively high humidity, low wind speed, small temperature variation, and low light levels, and are easily outcompeted by other species when conditions change. At the edge of a forest, dark and damp conditions clearly do not prevail, and it is surprising how far into the forest an 'edge effect' can be detected. The smaller the area of a forest 'island', the greater the proportion of edge-affected forest to deep forest, and proportionately less 'deep' forest remains.

In addition, the forest edge is occupied by non-forest species, many of them exotic, capable of producing large numbers of easily dispersed fruit or seeds which can spread into the forest, remaining as dormant seeds until a tree falls and conditions become conducive to their germination and growth. If they are aggressive and grow successfully in the gap then, again, the forest dynamics will have changed. Degradation accelerates and aggravates further degradation because forest trees are generally protected from the wind by neighbouring trees. When the canopy is disturbed, particularly near the edge, the shallow-rooted trees are subjected to the full force of the wind which may sometimes be sufficient to push them over.

EXTINCTION PROCESSES

Extinction is nothing new, because species have been lost throughout the history of the earth. Ecological factors involved in natural extinction in the past include:

- rising sea levels (p. 117) which would have isolated relatively small populations which were more prone to extinction;
- climatic warming that would have left relict populations of cold-loving species on mountains;
- habitat changes that would have disadvantaged species with large horns such as *Bubalus palaeokerabau* when savanna became rain forest; and
- population fluctuations (McNeely 1978).

What is changing is that in the recent past, and at present, almost all species have become less abundant solely due to the actions of humankind. The rate of mammal species extinctions has increased from about 0.01/century during most of the Pleistocene, to about 0.08 during the late Pleistocene, to 17 from 1600-1980. The higher rates were caused by neolithic hunters or climatic change and the hunting and commerce encouraged by European expansion. Ironically, people are the main cause of extinctions and also the only organisms with the ability to stem or eventually halt these losses (Myers 1990).

Periods of mass extinctions have dotted the earth's history and these mark boundaries in the geological time scale (p. 88). In the last 500 million years there have been six mass extinctions, the greatest of which occurred 240 million years ago and resulted in the loss of perhaps 95% of all marine organisms. These events can only be viewed as sudden when viewed against the age of the earth. For example, the loss of the dinosaurs 65 million years ago took about 2 million years. There seems to have been a lag of several million years before species diversity climbed up towards previous levels.

The causes of biological diversity loss and extinctions are listed elsewhere (p. 61), but it is clear that not all species are subject to the same extinction pressures, and the species most vulnerable to extinction are (McNeely 1978):

- those with poor dispersal and colonization abilities (such as many rain forest species);
- those found in only a small area;
- migratory species or those that require large areas for their continued survival;
- large species with low population densities confined to small islands of appropriate habitat;
- large predators persecuted by people;
- highly specialized forms intolerant of any but a narrow range of ecological conditions;
- those that compete directly or indirectly with people;
- those with commercial value; and
- those with a once large and contiguous range but now confined to small pockets of habitat.

Elephant, tapir, water buffalo, Bawean deer, orang-utan, siamang, jungle cat, and tiger, are the documented mammal extinctions on mainland Java over the past few thousand years. From the list above, these are exactly the species one would have predicted would go extinct first. They are the largest species, targets for hunting, and dependent on large areas of more or less natural forest. Smaller species are relatively less easy to hunt, and depend on proportionately smaller forest areas. The seven most-threatened mammal species on Java are listed below to show their propensity to decline and possible extinction (table 18.1).

The animal species found to be most prone to extinction at the Bogor Botanical Gardens were indeed those with a small initial population, and those that were rare or absent in the surrounding areas. Thus, although some species are able to breed and live in a small area like the Gardens, they cannot sustain themselves without the arrival of individuals from outside. This example demonstrates graphically that *small* reserves cannot function as species reservoirs but rather will tend to mirror the species composition of the surrounding depauperate areas (Diamond et al. 1987, 1990). In addition to small population sizes, the loss of species from the Gardens is also due to:

- habitat destruction, for irrigated rice and later for export crops grown for the Dutch colonial government. And today the expanding population continues to put pressure on remaining patches of natural vegetation around Bogor for firewood and short-term dryland agriculture;
- hunting and persecution of certain species such as birds of prey, cage birds (p. 232);
- heavy pesticide use affecting particularly those at the top of the food chains such as herons, egrets, and birds of prey; and
- 'trophic cascades', by which the disappearance of one species leads to the inevitable extinction of others. This is probably much more frequent than is realized, but inadequate and irregular surveys and inventories fail to identify this type of lost. One known example is the common koel *Eudynamis scolopacea,* a species of cuckoo which has been lost around Bogor because the crows whose nests it parasitizes have disappeared (Diamond et al. 1987).

When predicting extinctions, an important concept is that of populations or species being 'committed to extinction' (Simberloff 1986; Reid and Miller 1989) or joining the 'living dead' (Janzen 1986b, 1991), or subject to 'latent extinction' (Sutton and Collins 1991) even while they are living and breeding. Thus, before an equilibrium is reached in a reduced area of

Table 18.1. Globally threatened mammal species found on Java and their extinction propensity.

	1	2	3	4	5	6	7	8	9
Javan gibbon	√	-	-	-	-	√	-	-	√
Surili monkey	√	-	-	-	-	√	-	-	√
Wild dog	-	-	-	√	√	-	√	-	√
Leopard	-	-	-	√	√	-	√	-	√
Bawean deer	-	√	-	-	-	-	√	-	-
Javan rhino	√	-	-	√	-	√	√	√	-
Banteng	-	-	-	√	-	-	√	√	√

Key to headings:
1. Poor disperser   **4.** Large species, low density   **7.** Compete with people
2. Found in small area   **5.** Large predator   **8.** Commercial value
3. Migratory, need large area   **6.** Highly specialized   **9.** Fragmented distribution

Source: adapted from McNeely (1978)

habitat, certain species are destined to become extinct. An example could be the gibbons currently surviving in small areas of lowland forest on the slopes on Mt. Slamet. But there are such enormous gaps in our knowledge and understanding of the extinction process that prediction of the rate of population decline is nigh impossible (Heywood and Stuart 1992). Extinction is normally regarded as a lengthy process, but it is interesting that early papers on the now-extinct Javan lapwing did not stress that it was scarce or predict its imminent demise (p. 227).

In some cases, larger-than-expected numbers of species are found in small and isolated 'islands' of forest. This is termed 'supersaturation', and such a state is said to have transient diversity – a diversity that will not last. For example, the Bukit Timah Nature Reserve in the centre of Singapore protects the last 75 ha of primary rain forest on the island, though it is adjacent to about 2,000 ha of secondary forest. Within the reserve, one is still able to find 800-1,000 species of flowering plant (including more tree species than the whole of North America), 10,000 species of beetle, 200 species of ant, and 200 species of cockroach (Wee and Corlett 1986). Another example is Dungus Iwul, a 9 ha forest reserve about 30 km west of Bogor. It was gazetted in 1931 because of its striking abundance of *Orania sylvicola* palms, which are still the most common tree with seedlings thick in the forest floor (p. 783). There were other forest patches in the area at that time, but now it is at least 15 km from the next-closest block of forest and is immediately surrounded by rubber plantations of PTP XI. Amazingly, there is a group of endemic lutung monkeys in the forest. They venture into the rubber trees, but it seems unlikely that they have any contact with other groups. This is impressive, but the process of extinction has already claimed the larger and some of the smaller species which would once have roamed in both these areas. Since each species plays a role in the ecosystem, or at least interacts with other species, each loss will affect the total dynamic of the forest in some way. For example, a plant is no longer pollinated, a fruit is no longer passed through a digestive tract to stimulate germination, a herbivore is no longer preyed upon by a predator, a host-specific caterpillar is no longer able to feed, and the rate of decomposition of organic matter changes. These changes will generate further extinction in the areas (table 18.2).

Thus the relatively high diversity observed today in small areas of remnant forest is a transient phenomenon. Many species exhibit latent (hidden, dormant) extinction and for some species this is greater than for others. Thus the few gibbons remaining on Mt. Slamet probably have a greater latent extinction than, say, the species of abundant and fecund forest rats. The larger, rarer species have the greatest latent extinction because they are most at risk from random fluctuations in their populations or chance events such as disease, drought, and accidents. Extinctions may be balanced by immigration of species from other forest areas, but if these are also undergoing the same processes of attrition and degradation, this process cannot be relied upon to maintain diversity.

A practical problem in recording extinction is that it is very difficult to be absolutely sure that a rare species is extinct since nothing but a totally comprehensive, detailed, exhaustive survey would turn over every leaf or lift up every stone to confirm a species' absence. Where a search is incomplete, the apparent absence of a species may either be because the species is indeed absent, or because it was simply missed – hence the many student expeditions aiming to rediscover the Javan tiger. Thus, statements concerning extinction are probabilistic, and there is a relationship between the intensity of sampling, the rarity of a species, and the probability that the species will be encountered; details of this can be found elsewhere (McArdle 1990). An understanding of this is important if a researcher is trying to assess the probability of detecting a rare species, the number of samples that should be taken, or what rarity could be detected with a given sample size.

Table 18.2. Causes and effects of extinction of key species.

Ecological categories	Cause of extinction	Effect of extinction on biological diversity	Indirect effects
Large predators	Reduction in area of habitat, hunting.	Increased population density of herbivorous prey, increased competition amongst prey.	Degradation of habitat as a result of over-grazing and soil compression.
Large herbivores	Reduction in area of habitat, hunting.	Extinction of large predators, loss of previously regenerated habitats.	As below.
Key animal species (bees, butterflies, bats, birds)	Reduction in area of habitat, loss of regenerated habitats.	Reduced reproduction rate among plants of low population density due to loss of certain animals.	Extinction of plant species, extinction of specialized herbivores, extinction of specialized parasitoids and predators.
Key plant species	Commercial gathering, loss of previously regenerated habitats.	Starvation or emigration of pollinating species.	As above.

Source: Frankel and Soulé (1981)

Preservation of Genetic Variation

The assumption is frequently made that higher genetic variation improves the probability of survival of a population. Thus, in recent years, attention has been directed to determining the minimum effective sizes of a population that would avoid appreciable short-term reduction in fecundity, fertility, and offspring vitality (known as inbreeding depression), and of a population that would avoid long-term erosion of genetic variation in quantitative characters with a high likelihood of being inherited. Numbers frequently quoted, as above, for these two population sizes are 50 and 500, respectively, for medium to large vertebrates, although the low thousands is a conservative and perhaps more applicable figure. Even so, "anyone who applies the 'few thousand' estimate to a given species, citing this author as an authority, deserves all of the contempt that will be heaped on him or her" (Soulé 1987). Genuine minimum viable populations will vary considerably from species to species and could differ by two to three orders of magnitude (Soulé 1987). Conservation science has been rich in theories and hypotheses but not all have been particularly robust. A great deal has been learned, however, and the innate difficulties in an inexact science are being grappled with (Simberloff 1988; Caughley 1994).

The concern with maximal genetic variation is not just founded on the folk knowledge that inbreeding produces problems in man and his domesticates, but on the discovery that genetic variation is markedly reduced in a range of threatened species or small and isolated populations of a species (O'Brien et al. 1985; Vrijenhoek et al. 1985; Ledig 1987; Lesica et al. 1988; Gilbert et al. 1990), resulting in lowered fitness (Wildt et al. 1987; O'Brien and Evermann 1988; Quattro and Vrijenhoek 1989). Inbreeding depression can be observed in historically large, outcrossing populations that have been reduced rapidly to a small number of isolated populations for one reason or another. Such events are referred to as population 'bottlenecks'. There are no simple guidelines, however, and there is one report of genetic variation actually *increasing* following a population bottleneck (Carson and Wisotzkey 1989). Excessive outbreeding can in fact break up specialised gene combinations which have evolved in discrete local populations (Corbett and Southern 1977).

Conversely, when lethal, deleterious recessive genes are contributed by both parents and are therefore expressed, they will be eliminated, as occurs in inbreeding. Mutations which are only slightly deleterious or merely additive in their effects will not be lost so easily. It is true that many invertebrates and plants reproduce normally by sib mating or through self-fertilisation, and do not seem to exhibit lack of vigour, but the degree of inbreeding depression can sometimes be revealed indirectly by the loss of vigour that occurs when such inbred lines are crossed.

The avoidance of inbreeding depression by maximising genetic variation between individuals or subpopulations clearly creates a conservation conundrum since it would involve increasing the effective population size, either in reality or by translocating gametes (sperms or eggs) from one population to another. The latter could endanger the retention of the genetic basis of current adaptations to local conditions. Translocation (the deliberate moving of individuals from one area to another within the species' natural range) has sometimes been conducted without sufficient regard for other factors, thereby bringing the technique into disrepute in some quarters. However, selection forces in wild or reintroduced populations will continue to select against environmentally inappropriate genetic composition, and the introduction of small quantities of genetic material should not swamp the existing genetic variation in host populations.

Populations of some species can fall to a level below which recovery is not possible; for example, it may be difficult for an animal to locate a mate, or for pollen from one plant to reach the stigma of another. In fact, local extinctions generally occur through chance events either among the individuals in a population, or in the environment. Chance population events act because the individuals of a given age class or developmental stage have a range of probabilities of survival and reproduction. If it is assumed that these probabilities apply independently to each individual, then chance demographic events produce a variance around the mean of these probabilities which is inversely proportional to population size. Chance environmental events that produce changes in the probabilities of survival and reproduction will, however, affect all individuals of a given age or stage similarly and be more or less independent of population size. For this reason and because most populations fluctuate due to weather changes and the abundance of interacting species, environmental chance effects are generally agreed to dominate over demographic chance events in populations over about 100 individuals (Lande 1988).

Genetic Drift, Microspecies, and Conservation

Genetic drift (random fluctuation in gene frequencies of small, isolated populations not due to selection, mutation, or immigration) tends to lead to increased homozygosity or loss of genetic variation throughout the genome. Interpopulation crosses between the populations may then show complete or strong partial sterility barriers between them all, even if their places of origin are very close. Genetic differentiation between two isolated populations of a species is not surprising, and if the isolation has been caused or aggravated by human activity, an argument can be made for the introduction of genetic material from outside such populations. Care must be taken, however, because lack of molecular genetic information has led in some cases to the protection of 'species' that did not warrant the rank, or to insufficient protection or recognition of genetically distinct but morphologically similar populations (Alvise 1989).

Small reserves are more valuable for plants than for animals because the 'bank' of seeds in the soil can be viable for years, because vegetative reproduction is often an option, and because the individuals of many species have long life spans. Indeed, it has been found that small plant populations tend to lose heterozygosity more slowly than those in large populations and are more likely to persist for longer than might be expected. This is not to recommend small reserves or to be complacent about the degradation and demise of natural areas. Small populations should be avoided wherever possible since they are subject to the loss of rare alleles which may be very important for long-term survival, for example in terms of climatic change, and they are prone to extinction from random environmental fluctuations (Lesica and Allendorf 1992).

EXAMPLES OF BIOLOGICAL DIVERSITY LOSS FROM JAVA AND BALI

Tigers

The Bali tiger is known from just five skins and eight skulls, one set of which is in the Bogor Zoological Museum (Mazak et al. 1978). The Bali tiger was first described in 1912 and was probably extinct just 30 years later, although there is an unconfirmed record from 1963 when some villagers came across a tiger corpse in the forest. Five Bali tigers are known to have been killed in the first half of 1936 (de Voogd 1937d), but the last definite record is from 1937 when one was deliberately shot for the Bogor Zoological Museum, and that is now the only earthly evidence of the animal in Indonesia (de Iongh 1980). If one accepts an approximate average density in lowland rain forest of one adult tiger for every 15 km^2 (females have ranges of about 20 km, and the range of each adult male overlaps with several females), then even if *all* of the island were tiger habitat, only about 110 adult Bali tigers would have been alive at any one time. It is possible that occasional interchange of tigers across the Bali Strait occurred but there is no evidence to support this conjecture.

The Javan tiger, between the Sumatra and Bali subspecies in size, persisted a little longer. In the 1850s, Javan tigers were regarded as a nuisance in some urban areas, and in 1872 the price on the head of a man-killing Javan tiger in Tegal, Central Java, was about 3,000 guilders, and dozens were killed in an attempt to claim the reward. Even into this century they were not uncommon and were claiming hundreds of human victims each year, but the villagers waged no war against them because to do so, in their experience, was to court the destruction of their crops by herds of pigs (Cabaton 1911). However, the fanatical hunter Ledeboer claimed to have shot 100 between 1910 and 1940 (Anon. 1988). In addition, the tigers' plight was not helped by the persistent demand from makers of tiger and peacock Singabarong masks used in traditional *reog ponorogo* dance performances in Central and East Java (Edwin 1988d).

Until 1940 tigers were regularly observed (and shot) in the southern parts of West Java, with occasional individuals reaching Subang and Cibadak (figs. 18.6-7). These populations dwindled, and by the mid-1960s, Javan tigers were found only in Ujung Kulon, Leuweung Sancang, Baluran, and Meru Betiri Reserves, but the civil strife of this time caused armed groups of people to seek refuge in these areas. Tigers not killed as a result of this succumbed to anthrax or because the deer population plummeted (Seidensticker 1987). None of the forest areas remaining on Java by the middle of this century was prime tiger habitat, and they were being increasingly fragmented. It is clear that the Javan tiger became extinct because there was simply no more room (Seidensticker 1987). The same conclusion is valid for the Bali tiger, aggravated by the last few animals being promoted for sports hunting in the 1930s (Dressen 1937).

Surveys undertaken in 1976 by PHPA and World Wide Fund for Nature confirmed the presence of three tigers in Meru Betiri National Park, but found no evidence of breeding. The animals were not confining their activities to the Park, nor were they using all of the forested area available. Three animals remained in 1979. President Soeharto repeated the need to protect them, but this would have required the relocation of 5,000 plantation workers. Some politicians felt this action was too extreme for a handful of tigers, and conservation efforts became stuck in a quagmire (McNeely and Wachtel 1988). The necessary instructions to protect the tigers were eventually issued, but they were never quite executed, and by the mid-1980s the Javan tiger was no more than the symbol of the Army's Siliwangi Division in West Java, a hopeless quarry for student expeditions to seek, and as a symbol of base human impulses (Lubis 1975, 1991). Although Meru Betiri was their last refuge, it is not especially good tiger habitat, and they would not naturally have lived at a very high density, because the lower alluvial flood plains that would have supported large populations of prey (mainly deer) were converted to plantations soon after World War II (Seidensticker 1987). Eyewitness reports and tracks were reported from the forested southern slopes of Mt. Slamet in 1979 (Becking 1989), but with no repeat observations since then, there seems no hope of them having survived. Dating the extinction of such a metaphysically important animal as the tiger is difficult because people carry with them a lingering image of it (Seidensticker 1987), and it is not surprising that occasional reports of isolated individuals reach the newspapers, but

Figure 18.6. Last known photograph of a Javan tiger, taken in 1938 in Ujung Kulon National Park.

By A. Hoogerwerf, with permission of the National Museum of Natural History, Leiden

Figure 18.7. Tree used by tigers for scratching, near Sindangbarang. Photograph taken in 1936.

By M. Bartels, with permission of the Tropen Instituut, Amsterdam

these are almost certainly of the more adaptable leopard *Panthera pardus* (p. 214) (Bismark and Wiriosoepartho 1980), for which the vernacular name is very similar.

Although it has not been officially acknowledged, one can state without much fear of contradiction that the Javan tiger has been extinct for some years (Thohari 1988). It is not possible to point to any hard evidence of its existence from the last 15 years or so, despite many expeditions. Meru Betiri National Park is only 50 km^2, the area normally required for six or seven females and three males. A few more could be squeezed in if they took advantage of domestic stock around the Park, but there are no recent reports of livestock deaths due to tigers, and any such increase in density would not be sustainable. *If* there were one or two left, the Javan tiger would still essentially be extinct, certainly so in an ecological sense and evolutionarily. The dire circumstances of its cousin in Sumatra, where networks of hunters and agents gather in skins, serve as a reminder that it is not very hard to hunt a tiger – tie a hungry, bleating goat to a tree in the forest and in a few days your quarry will come. Given the millions of people within easy reach of the Meru Betiri National Park, the great publicity given to the Park as the 'tiger's last refuge in Java', the ineffectual nature of the guarding system, the high price of a pelt to make a mounted specimen or *Singabarong* mask for the *reog ponorogo* dance (p. 236), and the medicinal and monetary value of other bits and pieces, it is hard to believe that someone has not come in, probably some time ago, to remove the last specimens. We would be happy to be proved wrong – it is certainly easier to prove existence than extinction.

An English Comparison. It is worth noting that during the last 3,000 years in England, virtually all the natural forest has been lost (p. 331), and during this period most of the large wild animals have been lost too: the wild auroch cattle (before 750 B.C.), the wild boar (1260 A.D.), the beaver whose pelt was highly valued (before 1100 A.D.), and the wolf (by 1290 A.D. after a campaign of extermination by King Edward I) (Rackham 1986). Bears were lost before this. The numbers of wolves declined as the 'wildwood' was cleared and fragmented, aggravated by hunting. Just as there are few people on Java or Bali who would really welcome the return of the tiger, with all its implications, so there are few inhabitants of England who would encourage the return of the sheep-eating, swimming wolf.

Gibbons

A 1978 survey of remaining forest areas revealed that only 24 still had at least some gibbons (fig. 18.1), and that 14 non-mangrove patches had no gibbons, either because of hunting (four cases) or because the forest was

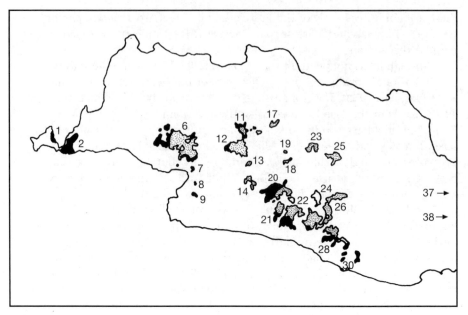

Figure 18.8. Forest areas occupied by Javan gibbons (black) in 1978. Forests above 1,500 m shown stippled.

Adapted from Kappeler 1984a

largely montane (six cases). In the other cases the reason was unknown (Kappeler 1984a). The hunting had been conducted both by local, traditional groups, and by weekend sport hunters from the towns, many of whom are undisciplined and do not differenciate between protected and unprotected species. The Javan gibbon has been protected by law since 1924.

Much has happened since 1978. Between 1992 and 1994 researchers from the Indonesian Foundation for the Advancement of Biological Sciences undertook a comprehensive resurvey of the gibbon in West and Central Java (Supriatna et al. 1992; Asquith et al. 1995). Only six forests in Java may not support populations larger than 100 individuals, meaning that there are now probably fewer than 3,000 Javan gibbons. Critical areas for the survuval of the species are Mt. Halimun and Ujung Kulon National Parks, Mt. Simpang Nature Reserve, Mt Salak Protection Forest and two large unprotected forests. One of these forests surrounds Mt. Kendang, south of Mt. Papandayan, and the other is around Mt. Wayang, south-west of Cikajang (referred to as Mt. Limbung in MacKinnon et al. 1982; and Kappeler 1984a). Since 1978, gibbons have been lost from at least seven areas. The attrition and degradation of lowland forests are clearly causing local extinctions of one of the larger Javan endemic mammals.

It is regrettable that one of the best areas of forest for gibbons (and much other endemic wildlife), Mt. Wayang (p. 793) is ungazetted, and there are rumours that it will be cleared for plantation forestry (Asquith et al. 1995).

Table 18.3. Forest areas with Javan gibbons present (or unconfirmed) in 1978 and 1993. Numbers refer to figure 18.8.

	No. of patches	Altitude range	Area (km²) (total/<1,500 m)	1978 Notes	Still present in 1993
1. Ujung Kulon	1	0-480		Few gibbons, unsuitable vegetation.	?
2. Mt. Honje	1	0-620	110/110		+
6. Mt. Halimun	3	400-1,929	400/380		+
7. Mt. Jayanti	1	300-687	5/5		+
8. Lengkong	1	400-758	10/10		+
9. Mt. Porang	1	400-600	5/5		+
11. Telaga Warna	1	300-1,700	80/75		-
12. Mt. Gede-Pangrango	1	500-3,019	180/95		+
13. Mt. Kencana	1	800-1,233	15/15		-
14. Mt. Malang	2	700-1,305	35/35		-
17. Sanggabuana	2	200-1,287	35/35	Unconfirmed.	+
18. Bojongpicung	1	400- 800	10/10	Unconfirmed.	+
19. Mt. Susuru	1	400	5/5		+
20. Mt. Masigit	3	900-2,078	205/135		-
21. Mt. Simpang	4	400-1,816	220/205		+
22. Mt. Tilu	1	1,400-2,140		Few gibbons, montane forest.	+
23. Mt. Tangkuban Perahu	1	1,500-2,081		Few gibbons, montane forest.	-
24. Mt. Malabar	2	1,000-2,321	50/5		-
25. Bukittunggal	1	1,400-1,833		Few gibbons, montane forest.	-
26. Mt. Papandayan	2	700-2,622	315/160		?
28. Mt. Wayang	2	500-1,813	140/130		+
30. Leuweung Sancang	5	0-700	45/45		+
37. Mt. Slamet	1	1,500-3,428		Few gibbons, montane forest.	+
38. Mt. Lawet	1	800-1,100	10/10		?

Source: Kappeler (1984a); Martarinza (pers. comm.); N. Asquith (pers. comm.)

Birds

It is possible for people untrained in ornithology to enter a forested area and see a good number of birds and conclude that nothing serious is amiss. It is fortunate, then, that good sets of data on birds from earlier this century are available for a number of locations in Java, and these allow a more objective assessment to be made. For example, comparison of the most recent checklist of birds from Mt. Gede-Pangrango National Park (Andrew 1985) with earlier works (Hoogerwerf 1949a), reveals that nine species of lowland forest and forest-edge birds have been lost in the intervening 35 years and that seven others are now much rarer (table 18.4). It is doubtful whether any of the species listed had ever been common in these forests, but on an island where there is so little forest, even a montane refuge (or lack of one) is of some significance. This further demonstrates the gradual erosion of biological diversity.

Likewise, between 1932 and 1952, 62 bird species bred in the Bogor Botanic Gardens, but by 1980-85, 20 had completely disappeared, four were close to local extinction, and the population of five other species had declined noticeably. Until about 50 years ago this 86 ha oasis had some more or less adjacent areas of lowland forest connecting to relatively extensive areas on the hills. The six species that increased their abundance over this time were the edible-nest swiftlet *Aerodramus fuciphagus*, the Asian palm-swift *Cypsiurus balasiensis*, ashy tailorbird *Orthotomus ruficeps*, great tit *Parus major*, scarlet-headed flowerpecker *Dicaeum trochileum*, and the eurasian tree sparrow *Passer montanus* (Diamond et al. 1987, 1990), all of which adapted well to the urban/suburban environment.

The loss of biological diversity among Java's bird fauna as a result of forest fragmentation has been the subject of an ongoing study (van Balen 1987b, in prep.). Nearly thirty forest areas, both protected and unpro-

Table 18.4. Species of birds lost from and rarer in the eastern part of Mt. Gede-Pangrango National Park between 1949 and 1985.

Lost		Rarer	
Indian cuckoo	*Cuculus micropterus*	Sunda pin-tailed pigeon	*Treron oxyura*
Banded bay cuckoo	*Cacomantis sonneratii*	Emerald pigeon	*Chalcophaps indica*
Chestnut-breasted malkoha	*Rhamphococcyx curvirostris*	Greater coucal	*Centropus sinensis*
Brown cuckoo-dove	*Macropygia phasianella*	Blue-eared barbet	*Megalaima australis*
Rhinoceros hornbill	*Buceros rhinoceros*	Checker-throated woodpecker	*Picus mentalis*
Banded pitta	*Pitta guajana*	Lesser cuckoo shrike	*Coracina fimbriata*
Large wood shrike	*Tephrodornis virgatus*	Orange-headed thrush	*Zoothera citrina*
Blue whistling thrush	*Myphonus caeruleus*		
Scarlet sunbird	*Aethopyga mystacalis*		

Source: Andrew (1985)

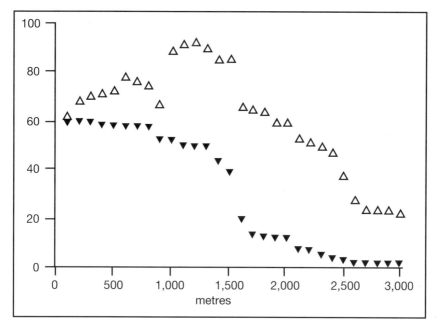

Figure 18.9. The number of obligate (open triangles) and facultative (closed triangles) forest birds found at different altitudes.

After Hoogerwerf 1948a; Sody 1956; van Balen 1987b

tected, large and small, and covering all altitudes from sea level to 3,000 m, have been surveyed. The birds observed can be divided into two categories:

- those which live exclusively in primary forest and those living in and at the edge of primary forest but never venturing far outside (obligate forest species); and
- those which live both in primary and secondary forest and even urban areas (facultative forest species).

The numbers of obligate and facultative forest species are related to altitude (fig. 18.9) in different ways – the first group shows a rise and fall, and the second a general decrease. One would have expected a general decrease for both categories because it is well known that lowland faunas are richer than those at higher altitudes. The observed rise and fall in the numbers of the obligate forest species almost certainly relates to the very small area of lowland forest remaining in Java, giving a much-reduced number of birds at low altitudes. The facultative species are indifferent to the loss of forest and so no loss is evident. A basic tenet of biogeography is that the larger an area of habitat, the greater the number of species present (p. 695). This is found among the forest birds of Java (fig. 18.10). Although

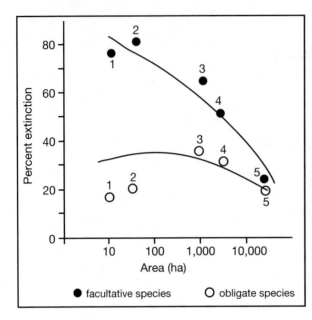

Figure 18.10. Percent of losses of forest bird species from presumed initial intact community in five West Javan forest areas.
1 - Dungus Iwul,
2 - Yanlappa,
3 - Ciogong,
4 - Leuweung Sancang,
5 - Ujung Kulon.
After van Balen 1987b

no forest birds are known to have become extinct in Java (but see below), some are now very rare and others have not been recorded for decades, such as the buff-rumped woodpecker *Meiglyptes tristis,* thick-billed flowerpecker *Dicaeum agile,* and yellow-eared spiderhunter *Arachnothera chrysogenys.* The key to stemming the loss of biological diversity is, again, conserving an adequate area of suitable habitat.

Among the birds that have experienced significant declines in rural West Java since 1950 are the Java sparrow *Padda oryzivora,* scaly-breasted munia *Lonchura punctulata,* white-bellied munia *Lonchura leucogastroides,* pied starling *Sturnus contra,* and Javan pond heron *Ardeola speciosa.* This is a surprising group since these are birds of rice fields and open habitats, both of which are widespread. Other rice-field birds such as white-collared kingfishers *Halcyon chloris,* white-breasted waterhen *Amaurornis phoenicurus,* purple swamphen *Porphyrio porphyrio,* and cinnamon bittern *Ixobrychus cinnamomeus* have not declined in the same way, although it is not clear why. The abundance and diversity of bird species in rural West Java is most closely linked with habitat complexity and structure, and different habitats had typical bird faunas (van Helvoort 1984), and this gives a clue to the means by which birds can be encouraged back into an area (p. 652).

The group of birds that has probably suffered most during the last century is the raptors or diurnal birds of prey – hawks, eagles, kestrels, buz-

Figure 18.11. The crested serpent-eagle appears to have suffered less than most other raptor species on Java, although the reason is not clear.

By A. Hoogerwerf, with permission of the Tropen Instituut, Amsterdam

zards, and kites. A survey was conducted in 1987 by a team of experienced observers, after which it was stated that, among over 50 countries where they had counted birds of prey, they had never encountered a place where raptors were so rare outside reserves as they were on Java (Thiollay and Meyburg 1988). Of those that could be loosely termed 'forest species', three species, including the relatively common crested serpent-eagle (fig. 18.11), are not found in primary forests, but four species appear to be more or less confined to primary lowland forests (table 18.5). These are all very local in occurrence and becoming rarer. Most notable of these is the endemic Javan hawk-eagle, Indonesia's national bird (p. 226), which may have a minimum home range size of some 20-30 km^2, and it is unlikely to survive in forest patches smaller than 20-100 km^2. As might be expected, the larger the size of a conservation area, the greater the number of raptor species encountered, and it seems that areas less than 300 km^2 in extent are unlikely to support viable populations of all the species (fig. 18.12) (Thiollay and Meyburg 1988).

Seven further resident species raptors are not tied to forest habitat, and these all seem to be more uncommon than a few decades ago. Indeed, after driving over 3,500 km through Java, the survey team had tallied just five raptors. Species which seem to have experienced major declines are the

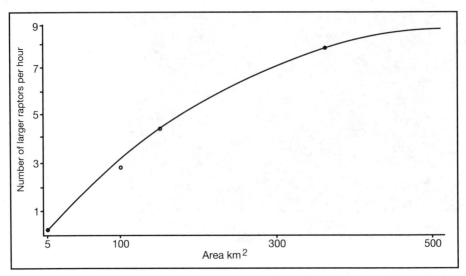

Figure 18.12. Relationship between the total frequency of forest raptors (mean number of birds, not including falcons, seen per hour), and the size of conservation areas (respectively Pangandaran, Baluran, Gede-Pangrango, Halimun, Meru Betiri). Note that for Baluran National Park, only the area of dense forest was considered.

Table 18.5. Habitat distribution of the forest raptors on Java, listed roughly in decreasing order of susceptibility to forest destruction.

	Primary forest	Edges and secondary forest	Open and degraded woodlands	Large clearings and plantations	Estimated home range size (km^2)
Javan hawk-eagle *Spizaetus bartelsi*	————————————				20-30
Chestnut-bellied hawk-eagle *Hieraaetus kienerii*	————————————				20-30
Besra sparrowhawk *Accipiter virgatus*	————————————				?
Crested goshawk *Accipiter trivirgatus*	———————————— – – –				?
Oriental honey-buzzard *Pernis ptilorhynchus*	—————————————————				50
Black eagle *Ictinaetus malayensis*	———————————————————————				20-30
Black-legged falconet *Microheirax fringillaris*		———————————————————			?
Crested hawk-eagle *Spizaetus cirrhatus*		———————————————————			10
Crested serpent eagle *Spilornis cheela*		———————————————————			5-10
Moluccan kestrel *Falco moluccensis*		———————————————————			?

Source: Thiollay and Meyburg (1988)

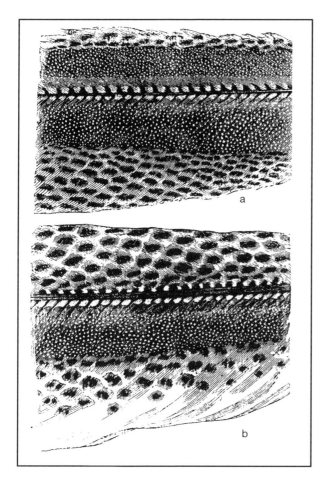

Figure 18.13. Part of a primary feather from the double-barred argus pheasant (a), and the great argus pheasant (b).
After Wood 1871a

black-shouldered kite *Elanus caeruleus,* which would be expected to be common in agricultural areas, the white-bellied sea-eagle *Haliaeetus leucogaster* of sea coasts, and the tan, black, and white Brahminy kite *Haliastur indus* of coastal and inland waters. These have been unable to adapt to the trapping, hunting, and agricultural practices of an increasingly dense human population (Thiollay and Meyburg 1988). The Brahminy kite is probably better adapted to human activity than any other raptor, and they can be numerous in many harbour cities feeding on refuse. They are not averse to stealing chickens and so farmers have felt no qualms at destroying their nests, felling their roosts, or hunting them. Although chosen by the Jakarta authorities as their official animal, they have disappeared from

Jakarta and just about everywhere else in Java in the space of a few decades (van Balen 1984; Erftemeijer and Djuharsa 1988; van Balen et al. 1993).

Before leaving birds, it is worth mentioning an enigmatic species, the double-barred argus pheasant *Argus bipunctatus*. This was described nearly a century ago (Wood 1871a, b) from part of a single feather from a feather collection of unknown origin bound for England and the fashion trade (fig. 18.13). It has been suggested that the feather in question originated in Java because the widespread great argus pheasant *A. argusianus* is absent there (Delacour 1951; Davison 1981). The argument has been made, however, that such a large bird is unlikely to have escaped the collecting activities of ornithologists, and that P. Tioman, off the east coast of Malaya, is at least as good a candidate (Davison 1983). This view perhaps overemphasizes the intensity of collection on Java, particularly in the southeast of the island. It is certainly possible that this large and possibly flightless bird (Davison 1983) was an inhabitant of the once-extensive lowland forests of Java, which has an impressive list of endemic forest birds. When most of the collecting was being conducted, it could well have already been extremely rare. It is interesting that a call similar to that of the great argus was heard in the barely surveyed lowland forests of Lebakharjo in East Java just a few years ago (van Balen 1989) (p. 802); could it have been the double-barred argus?

River Fishes

For the purposes of this book, extensive fish collections were made all over Bali and in selected major Javan rivers during 1991-92, the latter by the LIPI Centre of Biology Research and Development and most with the assistance of M. Kottelat. The status of the indigenous fish faunas in the Ciliwung, Citarum, and Brantas Rivers appears to be somewhat different. The Ciliwung, that passes through Bogor and Jakarta, is one of Java's most polluted and abused rivers and it is a tribute to the tolerance of some species in that 19 native species were found. However, 44 species (75%) reported before Independence and at least nine species recorded after Independence, including diadromous eels, large tambra carp *Tor* spp. (Rachmatika 1990), and various catfish, were not found and appear to be very rare or absent (S. Wirjoatmodjo pers. comm.). The surveys also revealed five native species not previously reported. Many of the species now 'missing' from the Ciliwung would have inhabited medium-sized, forested streams, intermediate in size between hill streams and slow-flowing lowland rivers. Such streams generally have a great diversity of habitats such as deep pools, riffles, and rapids, and a correspondingly high diversity of aquatic biodiversity (Kottelat 1996).

Somewhat fewer native species appear to have been lost from the Citarum (13 out of 28), perhaps because it is somewhat less polluted, although where it leaves Bandung, it appears to be lifeless. Some migratory species of fish may have been lost from the Citarum because of the Jatiluhur, Cirata, and Saguling dams. Seven species found had not been reported previously (S. Wirjoatmodjo pers. comm.).

Sixteen species previously reported from the Brantas were not found, but eight of the 34 species caught were new records for the basin (M. Kottelat pers. comm.). Inadequate time spent surveying is the most likely reason for the absence of certain species, but additional reasons may be:

- the loss through drainage of the Campurdarat/Tulungagung swamps near Blitar (where local informants claim seven species have disappeared);
- the completion of the Brantas dams, which appear to have had negative impacts on the catfish *Pangasius djambal* which generally migrates upstream to breed and was formerly of economic importance and now is decidedly rare; and
- a degree of doubt over identification for a few little-known species.

There are many other general reasons why species may have been lost or become uncommon. For example:

- it is well known that large quantities of agricultural chemicals have been and are used, particularly in irrigated rice-field areas, and while there are few studies on their ecological effects, it would seem that sensitive species would be adversely affected, particularly in downstream areas where irrigation water joins the mainstream. These chemicals tend to occur in major flushes within a watershed, when different stages of the rice cultivation cycle occur more or less simultaneously;
- the control of water whereby major flow changes are minimized is clearly good for people, but it can be anathema to certain species of fishes whose life cycles are controlled by major fluctuations in flow regimes. The weirs built to supply irrigation water to the rice fields are probably less significant than they seem since some fish, such as gobies (and prawns and some insects), can clamber up the wall, and the irrigation canals do provide a link between upstream and downstream;
- excessive use of fish poisons, both the traditional types such as rotenone (from *Derris* plants) which is relatively harmless to humans, and newer types such as 'Baygon' which is decidedly unsafe if ingested, explosives, as well as electric fishing and normal netting and trapping, have robbed most rivers of their fish populations. Poisons are said to be used regularly even in remote rivers;
- when forest is cleared from around a river, a number of changes will ensue:
  - first, there is a reduction in the available niches because fallen trees, overhanging roots, trailing vegetation, and accumulation

of leaf litter become fewer. This loss of complexity in turn will reduce the number of species a river can support;

- second, the loss of soil from the area will tend to increase, and so will the suspended sediment in the water which will adversely affect species adapted to clear-water conditions;

- third, the amount of allochthonous material (that originating from outside the river) will be reduced. Some species specialize on eating insects and other small animals, or fruit, falling from above; and

- fourth, and perhaps most significant, the temperature of the water will increase, with a consequent reduction in the concentration of dissolved oxygen and carrying capacity of the water for gill-breathing species;

• with most rivers receiving considerable quantities of human and other waste, their concentrations of nutrients has increased. In most cases, however, this has probably not had dramatic effects on river life, other than increasing their productivity; and

• there have been numerous introductions of a variety of exotic fish species (p. 304). It has been said that although many fish species have been introduced into the tropics, relatively few have had either a positive or negative impact (Fernando and Holcik 1991). It must be admitted, however, that a search for impacts among the native, non-commercial fish species has rarely been made, and deleterious effects are known (Arthington and Lloyd 1990; Courtenay and Meffe 1990).

THE FUTURE

Java and Bali have already witnessed a number of animal and plant extinctions and more are almost certain to follow. One of the most critical problems facing the conservation of biological diversity in Java and Bali, let alone the rest of Indonesia, is the dearth of expertise, indigenous or foreign, to identify, assess, and monitor species. As was found while compiling the lists in Chapters 4 and 5, the vertebrates and the more showy of the invertebrates are quite well covered (though in some cases by a very small number of individuals). For the majority of the invertebrates and for many plants, however, there appears to be no one with in-depth specialist knowledge, and even producing a simple list of species found on Java and Bali is impossible (p. 204). It is a serious matter that some of the little-known animals are from economically important groups.

Despite the extinctions, the total number of species known to live on Java and Bali is actually increasing, not decreasing. The reason for this is that new species of animals and plants are being discovered yearly (e.g., Killick et al. 1982; Goff 1983; Akhtar and Achmad 1985; Noordam and Lambers 1985; Reichart 1985; Schultz 1985; Noordam 1986; Suzuki 1986; Haines and Lynch 1987; Harrington 1988; Scheller 1988; Nässig and Suhardjono 1989; Okajima 1989; Wewalka and Bistrom 1989; Baehr 1990; Haitlinger 1990; Hippa 1990; Kramadibrata and Hedger 1990; Lienhard 1990; Lobl 1990; Yafuso and Okada 1990; Yamazaki 1990; Gilbert and Neuendorf 1991; Noordam 1991; Puthz 1991; Roth 1991; Vecker and Jordan 1991; Gibbons et al. 1992; Hasegawa et al. 1992; Mendes et al. 1992; Ricchiardi 1992; Suhardjono and Deharveng 1992; van Achtenberg 1992; Kitchener and Maryanto 1993, Aoki et al. 1994; Manohara et al. 1994; Watson and Kottelat 1994). During the course of writing this book, about a dozen new species of snail, a new fish, and a crab were discovered on Java and Bali.

It is hoped that in the future more concern will be evident for conserving the genetic variety of the agricultural base. This need not depend on high technology facilities or large annual budgets, because farmers themselves can play a major role.

Indonesia has ratified the U.N. Convention on Biological Diversity, and has thus committed itself to policies and actions which should serve to slow the loss of species in the future. The government has agreed (among a greater number of articles) to:

- develop a national strategy for the conservation and sustainable use of biological diversity;
- integrate the above strategy into relevant sectoral and cross-sectoral plans, programmes, and policies;
- identify important components of biological diversity: ecosystems, habitats, species, communities, genomes, and genes;
- monitor those components, paying particular attention to those requiring urgent conservation measures, and those offering the greatest potential for sustainable use;
- identify processes and activities which have adverse effects on those components;
- establish a system of protected areas to conserve biological diversity;
- develop guidelines for the selection, establishment, and management of those conservation areas where special conservation measures are required;
- promote environmentally sound and sustainable development in areas adjacent to conservation areas to improve the protection of those areas;
- rehabilitate and restore degraded ecosystems;
- prevent the introduction of, control or eradicate those alien species which threaten ecosystems, habitats, or species;
- endeavour to provide the conditions needed for compatibility between present uses and the conservation of biological diversity;

- respect, preserve, and maintain knowledge, innovations, and practices of indigenous and local communities embodying traditional lifestyles relevant to conservation, and allow, with consultation, the equitable sharing of the benefits arising from them;
- develop or maintain necessary legislation;
- support local populations which develop and implement remedial action in degraded areas where biological diversity has been reduced;
- encourage linkages between the government and private sectors in developing methods for sustainable use of biological resources;
- establish and maintain programmes for scientific and technical education;
- promote and encourage understanding of the importance of, and the measures required for, the conservation of biological diversity (UNEP 1992).

Many of these actions have already begun (some have been in operation for some years), but not until they are fully and wholeheartedly implemented will the future of biological diversity on Java and Bali be secure.

Chapter Nineteen

State of Conservation

Conservation is a process by which ecosystem preservation, management of biological diversity, environmental management, education, and conflict resolution are all integrated in a given area. Thus, in order to protect natural systems, it is essential to consider people living around them and to ensure that their food and income supplies are sustainable, that their environment does not degrade, and that there is a desire among the concerned parties to cooperate and resolve conflicts. The aim of conservation is therefore not akin to establishing museums, but rather about maintaining functioning, dynamic ecosystems as repositories of living components which live, compete, adapt, evolve, die, and decay, responding to environmental changes. The aim must be to maintain expanses of ecosystems which are large enough for natural processes to continue with minimum human intervention while maintaining all possible habitats and thus structural and taxonomic diversity. Where the areas are too small for this, active management intervention will probably be required.

From the beginning of this century concern has been increasingly expressed regarding the extent and condition of the forests and their wildlife, and there has been a growing awareness of the need to protect rare species. The first conservation area was established on the slopes of Mt. Gede in 1889 and from 1912 conservation issues were brought to the fore by the Netherlands Indies Society for Nature Protection, which was recognized officially in 1916 and given an advisory role in all matters relating to nature conservation. By 1940 most of the existing conservation areas on Java and Bali had been gazetted, but there was no management until 1967, when the Basic Forest Act gave a legal basis to those areas, and an office with responsibility for conservation was formed within Forestry. By this time some Reserves had already lost their forest.

RATIONALE

General

There is a wide variety of reasons for conserving areas of natural ecosystems, but they fall into two categories: social/moral, and economic. The social and moral reasons involve maintaining the quality of life, accepting a responsibility not to leave future generations less than we received or to deprive other species of existence, and national pride. Moral attitudes are discussed in detail elsewhere (p. 833). The economic reasons for, or benefits of, conservation are many but so are their costs (table 19.1), and these can be compared to assess the net present value or the benefit-cost ratio of a conservation area. This approach is fraught with problems, however, because not all the benefits or costs are present, and the burdens and advantages are not borne and appreciated by the same people. In response to this, one study has focussed on the distribution – local, national or regional, and global or transnational – of those costs and benefits (Wells 1992). This revealed that the overall economic benefits of conservation areas tend to be rather low at the local level, higher at the national/regional level, and potentially significant at the global level. Conversely, the overall economic costs tend to be significant nationally and insignificant globally. In addition, there are different types of costs and benefits at the three levels (table 19.2). Sorting out the empirical reasons for the imbalance between the three categories is extremely complicated, but the important point is that there are rather limited *economic* incentives for a country to spend its money protecting conservation areas (Wells 1992). If the economic books are required to balance, then the imbalances need to be dealt with through donor grants, increasing local and regional/national benefits from tourism, and integrated conservation development projects (Wells 1992).

In the 1970s the view that conservation was more to do with utilization than protection became prevalent, not just in Indonesia but also internationally. This view enabled some of the larger reserves to be developed into national parks, thereby allowing more people to appreciate the splendours of Java's natural heritage. Unfortunately, it now appears that unless a reserve can be utilized, it does not deserve a realistic (or any) budget for protection, with the result that encroachment proceeds unhindered. The 1990 Act No. 5 on the Conservation of Living Resources and their Ecosystems stresses the importance of utilization, e.g.:

> "Living resources and their ecosystems ... need to be managed and utilized sustainably, harmoniously, and in line with, as well as in a balanced way for the welfare of present and future generations of human beings in general and Indonesians in particular", "conservation efforts are

Table 19.1. Benefits and costs of conservation areas.

Benefits	Costs
Recreation/tourism	*Direct costs*
Stimulation of employment	Budget outlays
Rural development	Relocation
Watershed protection	Infrastructure development
Regulation of streamflows	Maintenance
Erosion control	Monitoring and research
Local flood reduction	Enforcement
Ecological processes	*Indirect costs*
Circulation and cleansing	Compensation for animal damage
of air and water	Opportunity costs
Production of organisms for	Income forgone from prohibition of timber,
commercial harvesting	mineral exploitation
Fixing and cycling of nutrients	Compensation for use restrictions
Soil formation	
Global life support	
Biological diversity	
Ecosystem diversity	
Gene resources	
Species protection	
Evolutionary processes	
Education and research	
Monitoring	
Awareness building	
Consumptive benefits	
Medicinal plants	
Fuelwood	
Game animals	
Nonconsumptive benefits	
Cultural/historical	
Aesthetic	
Spiritual	
Existence value	
Future values	
Option value	
Quasi-option value	

Source: Dixon and Sherman (1990)

Table 19.2. Comparison of conservation area benefits and costs on three spatial scales.

Scale	Potentially most significant benefits	Potentially most significant costs
Local	Consumptive benefits Recreation/tourism Future values	Opportunity costs
Regional/ National	Recreation/tourism Watershed values Future values	Various direct costs Opportunity costs
Global/ Transnational	Biological diversity Nonconsumptive benefits Ecological processes Education and research Future values	(Minimal costs)

Source: Wells (1992)

Figure 19.1. Part of Mt. Papandayan Reserve, West Java, photographed after a road was built along its edge to provide access to the Darajat geo-thermal electricity generating plant.
By R.E. Soeriaatmadja

necessary to promote the sustainable utilization of living resources and their ecosystems", and "Conservation ... shall mean the management of living resources whose wise utilization will ensure the maintenance and improvement of their value and variety."

This emphasis on utilization aside, the Act does, of course, recognize that all ecological elements are interdependent such that deterioration or extinction of one element may lead to damage of the ecosystem as a whole, and there is ample provision for conservation areas to be protected purely for their intrinsic value.

Table 19.3. Range-restricted bird species on Java and Bali.

	Range-restricted species confined to the specified area				Range-restricted species occurring in the specified area			
	Threat'd	Near threat'd	Total	Threat'd	Near threat'd	Total	Area (km²)	% area protected
Java and Bali mountains	3	2	18	5	5	33	18,000	<10
Java lowlands	4	-	6	5	-	7	16,000	<15

Source: Bibbey et al. (1992)

The Situation on Java and Bali

As a result of the radical land use changes during the last 150 years, the remnants of Java's natural ecosystems are small and scattered (fig. 4.24, 6.6). So little of the natural vegetation remains in Java and Bali (especially in lowland areas), that all extensive remaining patches are of the highest conservation importance. Many of the existing and proposed conservation areas are scenically magnificent, are naturally and culturally interesting, and have very high recreational and touristic value. In order to avoid or at least delay further extinctions and degradation, habitat management, wildlife management, visitor management, and boundary protection will have to be conducted both urgently and earnestly (fig. 19.1).

Java and Bali may not rank as 'hotspots' of biological diversity (Myers 1988), but they nevertheless have considerable numbers of endemic species (Chaps. 4 and 5). This was recognized in a major new study on assessing the world's most important areas for conservation (Bibbey et al. 1992) using the distribution of those bird species (the best-known taxonomic group) whose ranges were restricted to less than 50,000 km^2 (known as range-restricted species); this is about the area of East or West Java, Sri Lanka, Costa Rica, or Denmark. Indonesia was found to have more 'endemic bird areas' than any other country – 24 out of a global total of 221 - and two of them were the Java and Bali mountains (regarded as absolutely critical and the highest priority in Indonesia) and the Java lowlands (regarded as an urgent priority) (table 19.3).

CONSERVATION AREA SYSTEM

The most important task in conserving biological diversity is securing an adequate system of protected conservation areas. The first step in achieving this is an investigation of the existing system in terms of coverage, boundaries, status, threats, and needs for management (Salim 1992). The current situation requires rapid biological inventories and research into resource management techniques that meet the needs of conservation and local people. In this regard, regional land use planning must give at least as much weight to biological diversity conservation as to agricultural production (Reid and Miller 1989).

Areas felt to be of value to conservation are given legal protection by the government, and viewed together, they should form an integrated system of conservation areas wherein any activity is sustainable and conservation of biological diversity and ecosystem functioning are given paramount priority. The imperatives, or irreducible minimum requirements, for a conservation area system are the following (MacKinnon and MacKinnon 1986):

- within each biogeographical division, the main priority should be the establishment of major, large reserves selected to include a continuum of many habitat types, including, if possible, the richest examples of those habitats;
- smaller reserves should augment those major reserves by protecting additional habitat types or covering regional variants of habitats;
- small reserves should be included in the system to provide additional recreational, educational, or research facilities or to protect unique sites of special interest or beauty; and
- some small reserves should also be included to protect specific localized species or sites, such as nesting areas of important birds on islets, or caves known as roosting sites for exceptional communities of bats.

In 1979 the Food and Agriculture Organization of the United Nations began working with PHPA (the Directorate General of Forest Protection and Nature Conservation) to design a National Conservation Plan using objective methods. This resulted in many unprotected areas being proposed for inclusion in the conservation area system. It should be stressed that the areas proposed were the essential minimum required for the long-term conservation of biological diversity, not subjective desires without scientific basis. A methodology for ensuring that the above conditions were met was designed (MacKinnon et al. 1982), and this was further refined by IUCN (MacKinnon and MacKinnon 1986). In outline the methodology was as follows:

- maps were prepared of the original extent of each habitat type;
- areas of intact remaining habitat were identified from the latest imagery or photographs then available (much of which dated from the mid-1970s);
- the percentage of remaining habitat was taken as an index of threat;
- the boundaries of the existing and proposed reserves were plotted over the maps of forest area;
- the area of each extant habitat type within the reserves was measured. Degraded habitat was not measured since it supports few animals, but selectively logged and subsequently undisturbed forest was included;
- an index of the degree of protection afforded to each habitat type was calculated; and
- where a given habitat in a particular zone was clearly underrepresented, efforts were made to identify suitable areas for inclusion in the conservation area system.

In the IUCN refinement a simple scoring system was then employed to obtain a more objective evaluation of the effective contribution to overall conservation of major individual existing or proposed conservation areas. The highest-scoring existing or proposed reserves in each major habitat type within each biogeographical zone were regarded as Category A, representing areas of global importance. Nearly 80 areas in Indonesia are worthy of the 'A' rating, reflecting the diversity of the country's natural heritage, and these all deserve urgent attention regarding protection and management, particularly those not yet gazetted. Those in Java and Bali are listed in table 19.4.

Conservation Areas

There are a number of types of conservation areas divided into two main groups: sanctuary reserves (comprising strict nature reserve, and wildlife sanctuaries) and nature conservation areas (comprising national parks, grand forest parks, and nature recreation parks). The various definitions from the 1990 Act No. 5 concerning conservation are as follows:

- Sanctuary reserves – terrestrial or aquatic areas having sanctuary as their main function, preserving plant and animal biological diversity as well as an ecosystem which also acts as a life support system.
  - Strict nature reserves — a sanctuary having a characteristic set of plants, animals, and ecosystems, which must be protected and allowed to develop naturally; and
  - Wildlife sanctuary – a sanctuary having a high value of species diversity and/or a unique animal species where habitat management may be conducted in order to assure the species' continued existence.

Table 19.4. Conservation areas in Java and Bali awarded 'Category A' status in an IUCN analysis. Bold type indicates an ungazetted recommended or proposed conservation area. TN = Taman Nasional or National Park, TL = Taman Laut or Marine Park. * - the necessary extension to 770 km^2 has not yet been agreed.

		Area (km^2)	Status
Mt. Gede-Pangrango	W	150	TN
Mt. Halimun	W	400	TN
Ujung Kulon	W	786	TN
Thousand Islands	W	1,100	TL
Meru Betiri	C	500	TN
Segara Anakan	C	125	-
Baluran	E	250	TN
Bromo-Tenger	E	580	TN
Mt. Kawi-Kelud	E	776	-
Alas Purwo	E	620	TN
Bali Barat	B	190*	TN

Source: RePPProT (1990)

- Nature conservation areas – terrestrial or aquatic areas whose main functions are to preserve biological diversity and to provide a sustainable utilization of living resources and their ecosystems.
  - National park – an area which possesses native ecosystems and which is managed through a zoning system to facilitate research, science, education, breeding enhancement, recreation, and tourism purposes;
  - Grand forest park – an area intended to provide a variety of indigenous and/or introduced plants and animals for research, science, education, breeding enhancement, culture, recreation, and tourism purposes; and
  - Nature recreation park – an area mainly intended for recreation and tourism purposes.

In addition to the national categories, there are also Biosphere Reserves which are nominated to UNESCO by countries participating in the Man and Biosphere Programme. Such reserves are supposed to be globally significant, characteristic of the region, managed with a view to allowing people to be an integral component, established as a centre for monitoring, research, and education, and lastly a symbol of voluntary cooperation to conserve and use resources for the well-being of people everywhere. Mt. Gede-Pangrango National Park is Java's only UNESCO Biosphere Reserve (the others in Indonesia are Mt. Leuser, Siberut, Tanjung Puting, Komodo, and Lore Lindu).

The conservation areas that have been formally gazetted, recommended (by PHPA), or proposed (by nongovernmental bodies) on Java and Bali are listed, together with their scores for genetic value, socioeconomic justification, and management viability, their overall priority, IUCN score (where applicable), and notes or recommendations from the National Conservation Plan (tables 19.5-8). The extent of each area is difficult to determine for sure because the gazetted area does not necessarily tally with the official boundaries. All the areas are detailed in the appropriate volume of the Plan (MacKinnon et al. 1982), and it is disturbing that, even after more than a decade, many important areas for conservation do not yet have legal protection and suffer continual degradation, perhaps to the point where they have lost their conservation value. There is hope, however, that under the present administration of the Forestry Department, appropriate action may be taken.

A summary of protected areas in Java and Bali shows over 13 million ha of terrestrial conservation areas (table 19.9). This is not as reassuring as it

Table 19.5. Gazetted Reserves, West Java and DKI Jakarta. CA = Cagar Alam or Nature Reserve, SM = Suaka Margasatwa or Wildlife Reserve, TN = Taman Nasional or National Park, TW = Taman Wisata or Nature Recreation Park, HL = Hutan Lindung or Protection Forest. The notes are taken from the National Conservation Plan.

Area name	Gazetted area (ha)	Area as mapped (ha)	Genetic value	Socio-economic justification	Management feasibility	Overall priority	Score	Notes
TL Pulau Serbu	108,000	700	88	15	11	1		
TN. Ujung Kulon	76,100	62,200	904	15	10	1	383	
TN. Mt. Halimun	40,000	36,500	313	15	9	1	180	
TN. Cibodas Mt. Gede Pangrango	15,040	-	118	17	11	1	120	
CA. Mt. Simpang	15,000	16,000	165	12	10	2	62	
TB. Masigit Kareaumbi	12,420	14,200	98	11	12	2	24	Change status to SM.
SM. Cikepuh	8,127	6,500	-	12	10	2	16	Change status to CA and include Cibanteng and turtle beach.
CA. Mt. Tilu	8,000	8,600	36	12	9	2	32	
CA. Kawah Kamojang	7,500	3,500	42	10	9	3	40	+500 ha TW.
CA. Mt. Papandayan	6,623	7,100	36	11	11	2		+221 ha TW.
SM. Mt. Sawal	5,400	5,300	36	11	10	3	16	
CA. Mt. Burangrang	2,700	2,700	62	10	10	3		
CA. Rawa Danau	2,500	3,700	-	10	9	2		
CA. Leuweung Sancang	2,157	2,000	-	11	10	2		Enlarge to include Cipayabuh.
CA. Mt. Tukang Gede	1,700	1,000	-	10	7	2		
CA. Mt. Tangkuban Perahu (Bandung)	1,290	900	25	12	11	2	50	+370 ha TW.
TW. Mt.Tampomas	1,250	1,500	25	10	9	3		
CA. Bojong Larang	750	600	-	10	5	2		
TW. Curug Dago	590	600	-	10	4	3		
CA. Cibanteng	447	-	-	11	6	2		Add to CA Cikepuh.
CA. Penanjung Pangandaran	419	500	-	14	9	2		+37 ha TW.
CA. Telaga Warna	368	200	-	12	8	3		
CA. Telaga Bodas	263	300	-	12	8	2		+24 ha TW.

table continues

Table 19.5. *(Continued.)* Gazetted Reserves, West Java and DKI Jakarta. CA = Cagar Alam or Nature Reserve, SM = Suaka Margasatwa or Wildlife Reserve, TN = Taman Nasional or National Park, TW = Taman Wisata or Nature Recreation Park, HL = Hutan Lindung or Protection Forest. The notes are taken from the National Conservation Plan.

Area name	Gazetted area (ha)	Area as mapped (ha)	Genetic value	Socio-economic justification	Management feasibility	Overall priority	Score	Notes
TW. Cimanggu	154	200	-	11	6	3		Drop, forest destroyed, include in HL.
CA. Mt. Jagat	126	100	-	7	6	3		
CA. Rawa Cipanggang	125	-	-	-	-	-		Included in TN Gede-Pangrango.
TW. Situgunung	120	-	-	8	7	3		
TW. Carita	95	100	-	10	6	3		+65 ha TW.
CA. Telaga Patengan	86	-	-	10	6	3		Extend to 3,000 ha.
CA. Takokak	50	-	-	8	4	2		
TW. Jember	50	100	-	10	5	3		
CA. Sukawayang Pelabuhan Ratu	46	100	-	10	5	3		
CA. Pulau Dua	38	100	-	13	10	2		
CA. Tangkuban Perahu (Pelabuhan Ratu)	33	-	-	10	6	2		
CA. Yanlapa	32	100	-	10	6	3		
CA. Cadas Malang	21	-	-	7	4	3		Drop, tiny, damaged.
CA. P. Rambut	18	100	-	13	10	2		
CA. Nusa Gede Panjalu	16	200	-	6	4	3		Drop, tiny, destroyed.
CA. P. Bokar	15	100	-	11	10	2		
CA. Muara Angke	15	100	-	10	6	2		
TW. Linggar Jati	11	100	-	9	4	3		Drop, tiny, destroyed.
CA. Cigenteng Cipanji	10	-	-	6	4	3		Drop, destroyed.
CA. Dungus Iwul	9	-	-	7	6	3		
CA. Malabar	8	100	-	10	7	3		Extend to cover whole of Mt. Malabar.

table continues

Table 19.5. (*Continued.*) Gazetted Reserves, West Java and DKI Jakarta. CA = Cagar Alam or Nature Reserve, SM = Suaka Margasatwa or Wildlife Reserve, TN = Taman Nasional or National Park, TW = Taman Wisata or Nature Recreation Park, HL = Hutan Lindung or Protection Forest. The notes are taken from the National Conservation Plan.

Area name	Gazetted area (ha)	Area as mapped (ha)	Genetic value	Socio-economic justification	Management feasibility	Overall priority	Score	Notes
CA. Depok	6	100	-	8	4	3		Drop, tiny, damaged.
TW. Telaga Warna	5	-	-	-	-	-		
CA. Junghuhn	2	-	-	7	6	3		Drop, already in TN Gede-Pangrango.
CA. Arca Domas	2	-	-	8	6	3		
Recommended Reserves, West Java and DKI Jakarta.								
CAL. Ujung Kulon	19,200	-	-	-	-	-		
SM. Leuweung Sancang Cipatujah	1,156	1,000	108	11	8	2	-	
CA. Muara Cimanuk	1,000	-	156	11	7	2	12	
CA. Muara Gembong	855	400	166	11	7	2	15	
CAL. Pangandaran	470	1,100	-	-	-	-	-	
TW. Telaga Warna extension	25	-	-	-	-	-	-	
Proposed Reserves, West Java and DKI Jakarta.								
TB. Mt. Pangasaman	34,000	35,200	-	10	9	3	-	Low value.
SM. Mt. Kencana	25,000	-	211	11	12	2	50	
SM. Mt. Limbang	20,000	-	273	11	12	2	40	
CA. Waduk Gede/Jati Gede	10,500	12,700	-	8	9	3	-	Better as HL.
CA. Mt. Masigit	9,000	29,000	68	11	12	2	46	
CA. Tj. Sudari	8,200	4,700	125	11	7	2	6	
CA. Cikamurang	5,500	8,000	-	8	6	3	-	Drop.
SM. Muara Bobos	5,000	7,200	-	-	-	-	20	

table continues

Table 19.5. *(Continued.)* Gazetted Reserves, West Java and DKI Jakarta. CA = Cagar Alam or Nature Reserve, SM = Suaka Margasatwa or Wildlife Reserve, TN = Taman Nasional or National Park, TW = Taman Wisata or Nature Recreation Park, HL = Hutan Lindung or Protection Forest. The notes are taken from the National Conservation Plan.

Area name	Gazetted area (ha)	Area as mapped (ha)	Genetic value	Socio-economic justification	Management feasibility	Overall priority	Score	Notes
Proposed Reserves, West Java and DKI Jakarta *(continued).*								
SM. Pasir Alam	3,500	5,200	-	10	7	3	-	Possible TB.
CA. Mt. Karang	3,447	3,700	-	-	-	-	-	
CA. Cipatujuh	3,000	3,500	108	11	8	2	-	Manage with Leuweung Sancang.
CA. Sanggabuana	2,500	16,900	103	10	9	2	-	
CA. Cikencreng	2,500	2,000	212	11	9	2	-	
SM. Cimapang	1,500	4,500	-	10	9	2	-	Enlarge and link to Papandayan.
CA. Ciobong	1,500	3,400	-	7	8	3	-	Drop.
CA. Salatri	1,000	1,100	-	7	7	3	-	Drop.
CA. Mawuk	1,000	800	-	8	5	3	-	Drop.
CA. Muara Camal	1,000	-	-	-	-	-	-	
CA. Medang Kamulyan	500	400	-	-	-	-	-	
CA. Mt. Kendeng	500	1,300	-	-	-	-	-	

Sources: MacKinnon and MacKinnon (1986); RePPProT (1989)

Table 19.6. Gazetted Reserves, Central Java and Yogyakarta. CA = Cagar Alam or Nature Reserve, SM = Suaka Margasatwa or Wildlife Reserve, TN = Taman Nasional or National Park, TW = Taman Wisata or Nature Recreation Park. The notes are taken from the National Conservation Plan.

Area name	Official area (ha)	Area as mapped (ha)	Genetic value	Socio-economic justification	Management feasibility	Overall priority	Score	Notes
CAL. Karimun Jawa	111,625	4,200	-	11	12	2		
CA. Mt Celiring	1,379	2,100	118	11	10	2		
CA. Nusa Kambangan Barat	928	1,200	126	10	8	2		
CA. Plawang Turgo	198	100	-	10	7	3		
TW. Plawangan Turgo	131	100	-	-	-	-		
TW. Mt. Selok	126	100	-	10	5	3		
CA. UlolarangKecubing	71	100	-	10	5	3		
TW. Grojongan Sewu	64	100	-	12	5	3		
CA. Keling I/II/III	60	100	-	10	7	3		
CA. Pringombo I/II	58	100	-	?	?	3		Drop, damaged.
CA. Mt. Butak	45	100	-	10	6	3		
TW. TelogoWarno/Pengilon	39	100	-	10	5	3		
CA. Peson Subah I/II	30	100	-	7	3	3		Drop, tiny, damaged.
CA. Pagerwunung Darupa	30	-	-	?	?	3		Drop, tiny, damaged.
CA. Telogo Dringo	26	100	-	?	?	3		Not really CA quality.
CA. Bekutuk	25	100	-	9	5	3		Drop, tiny, damaged.
CA. Vak.53.Comal	24	-	-	?	?	3		Drop, tiny, damaged.
CA. Bantar Bolang	24	200	-	?	?	3		
CA. Telogo Sumurup	20	100	-	10	5	3		Drop, tiny.
CA. Telogo Banteng	18	100	-	?	?	3		
TW. Sumber Semen	17	100	-	10	5	3		
CA. Sub Vak 18c dan 19b	7	100	-	?	?	3		Drop, tiny, damaged.
TW. Tuk Songo	6	100	-	10	7	3		Extend or drop.

table continues

Table 19.6. *(Continued.)* Gazetted Reserves, Central Java and Yogyakarta. CA = Cagar Alam or Nature Reserve, SM = Suaka Margasatwa or Wildlife Reserve, TN = Taman Nasional or National Park, TW = Taman Wisata or Nature Recreation Park. The notes are taken from the National Conservation Plan.

Area name	Official area (ha)	Area as mapped (ha)	Genetic value	Socio-economic justification	Management feasibility	Overall priority	Score	Notes
CA. Sepakung G.Telomoyo	2	100	-	10	9	3		Extend or drop, tiny.
CA. Guci	2	100	-	?	?	3		?incl. in Mt. Slamet.
CA. Teluk Baron	2	100	-	?	?	3		Drop, tiny, damaged.
CA. Gebungan Mt. Ungaran	2	100	-	8	7	3		Extend or drop.
CA. Wijaya Kusuma	1	100	-	10	10	3		No info.
CA. Moga	1	100	-	?	?	3		Drop, tiny, damaged.
CA. Curug Bengkawak	1	100	-	?	?	3		Drop, tiny, damaged.
CA. Getas	1	100	-	?	?	3		Drop, tiny, damaged.
TW. Mt. Gamping	1	-	-	-	-	-		
CA. Karang Bolong	<1	100	-	10	10	3		No info.
CA. Mt. Batu Gamping	<1	-	-	-	-	-		
CA. Plelen	134	200	-	-	-	-		
TW. Plawangan Turgo extension	101	-	-	-	-	-		
TW. Curug Sewu	61	-	-	-	-	-		
CA. Lumbung Mas	43	100	-	-	-	-		
CA. Pasar Sore	31	100	-	-	-	-		
Proposed Reserves, Central Java and Yogyakarta.								
CA. Mt. Perahu	25,000	32,900	69	13	10	3		
CA. Nusa Kambangan extension	22,077	3,300	-	-	-	-		

table continues

Table 19.6. (Continued.) Gazetted Reserves, Central Java and Yogyakarta. CA = Cagar Alam or Nature Reserve, SM = Suaka Margasatwa or Wildlife Reserve, TN = Taman Nasional or National Park, TW = Taman Wisata or Nature Recreation Park. The notes are taken from the National Conservation Plan.

Area name	Official area (ha)	Area as mapped (ha)	Genetic value	Socio-economic justification	Management feasibility	Overall priority	Score	Notes
Proposed Reserves, Central Java and Yogyakarta (continued).								
CA. Mt. Lawu	21,000	25,400	66	13	10	3		
CA. Mt. Slamet	15,000	-	58	12	11	2	45	
CA. Merapi Merbabu	15,000	13,300	58	13	10	2	15	
CA. Pembarisan	13,000	12,100	251	11	10	2	39	
CA. Segara Anakan	12,500	7,300	-	-	-	-		
CA. Mt. Muria	12,000	12,900	207	12	10	2	36	
CA. Mt. Sumbing	10,000	5,800	152	11	9	2	22	
CA. Mt. Ungaran	5,500	5,600	40	11	9	3	16	
TW. Kaliurang	18	100	-	?	?	3		Link with Plawangan Turgo.
CA. Kembang	-	200	-	-	-	-		
CA. Jiwo/Bayat	-	100	-	-	-	-		

Sources: MacKinnon and MacKinnon (1986); RePPProT (1989)

Table 19.7. Gazetted Reserves, East Java. CA = Cagar Alam or Nature Reserve, SM = Suaka Margasatwa or Wildlife Reserve, TN = Taman Nasional or National Park, TW = Taman Wisata or Nature Recreation Park. The notes are taken from the National Conservation Plan.

Area name	Official area (ha)	Area as mapped (ha)	Genetic value	Socio-economic justification	Manage-ment feasibility	Overall priority	Score	Notes
TB. Maelang	70,000	-	526	10	11	2	120	Change to SM and combine with Ijen Raung.
TN. Alas Purwo	62,000	41,000	73	13	11	2	120	
TN. Meru Betiri	58,000	53,900	423	15	11	1	220	
TN. Bromo Tengger Semeru	57,000	53,200	211	15	11	2	171	
TN. Baluran	25,000	26,000	134	15	12	1	52	
SM. Dataran Tinggi Yang	14,145	12,700	141	14	11	2	225	Extend.
CA. Nusa Barung	6,100	8,000	37	13	13	2	25	
CA. Arjuno Lalijiwo	4,960	5,700	-	-	-	-		
SM. Bawean	3,831	-	180	12	11	2		
CA. Kawah Ijen Ungup-Ungup	2,468	1,900	49	14	11	2		Link to Raung Maelang.
CA. Ranu Kumbolo	1,340	-	17	10	7	3		Include in Bromo.
CA. P. Sempu	877	900	-	11	10	3		Join with Teluk Lenggosana.
CA. P. Bawean	725	5,100	-	-	-	-		
CA. Saobi (Kangean)	430	-	-	-	-	-		
TW. Ranu Darungan	380	-	-	9	5	3		Included in Bromo.
TW. Mt. Baung	195	-	-	12	5	3		
CA. Mt. Sigogor	190	400	-	11	6	3		Extend to Liman Wilis.
TW. Ranu Pane-Ranu Regulo	96	-	-	10	5	3		Included in Bromo.
TW. Kawah Ijen Merapi	92	4,300	-	-	-	-		
CA. Ceding	50	-	-	8	5	3		Join to Maelang or drop.
CA. Mt. Abang	50	-	-	10	4	3		Drop.
CA. Mt. Picis	28	100	-	11	4	3		Extend to Liman Wilis.

table continues

Table 19.7. (Continued.) Gazetted Reserves, East Java. CA = Cagar Alam or Nature Reserve, SM = Suaka Margasatwa or Wildlife Reserve, TN = Taman Nasional or National Park, TW = Taman Wisata or Nature Recreation Park. The notes are taken from the National Conservation Plan.

Area name	Official area (ha)	Area as mapped (ha)	Genetic value	Socio-economic justification	Management feasibility	Overall priority	Score	Notes
CA. Curah Manis Sempolan	16	100	-	10	7	3		To Raung.
CA. P. Noko P. Nusa	15	-	-	11	9	2		
CA. Manggis Gadungan	12	100	-	8	5	3		Drop.
TW. Tretes	10	-	-	9	5	3		Drop?
CA. Pancur Ijen I/II	9	-	-	9	7	3		To Maelang.
CA. Sungai Kolbu	9	-	-	10	5	3		In Yang, drop.
CA. Besowo Gadungan	7	100	-	10	5	3		Extend to Kawi-Kelud.
CA. Gua Ngilirip	3	-	-	11	4	3		Drop.
TW. Laut Pasir Tengger	3	-	-	-	-	-		Included in Bromo.
Recommended Reserves, East Java.								
CA. P. Saobi extension and marine	2,122	-	-	10	9	2		Extend.
CA. Tambak Rejo	99	100	-	-	-	-		
TW. Arjuno Lalijiwo	20	-	36	12	10	2		Change to CA.
TW. Kawah Ijen Merapi	92	4,300	-	-	-	-		
Proposed Reserves, East Java.								
CA. Mt. Raung	60,000	83,500	294	13	11	2	150	
CA. Mt. Kawi-Kelud	50,000	32,200	541	12	11	1	150	
SM. Dataran Tinggi Yang extension	45,000	44,700	-	-	-	-	-	
SM. Mt. Liman Wilis	45,000	26,200	237	13	11	2	50	
CA. Teluk Lenggosana	16,000	15,700	467	12	8	2	45	Include with Sempu.
SM. Meru Betiri extension	12,000	11,700	-	-	-	-	-	

table continues

Table 19.7. (*Continued.*) Gazetted Reserves, East Java. CA = Cagar Alam or Nature Reserve, SM = Suaka Margasatwa or Wildlife Reserve, TN = Taman Nasional or National Park, TW = Taman Wisata or Nature Recreation Park. The notes are taken from the National Conservation Plan.

Area name	Official area (ha)	Area as mapped (ha)	Genetic value	Socio-economic justification	Management feasibility	Overall priority	Score	Notes
SM. Mt. Beser	4,000	4,400	540	12	10	2	-	
SM. P. Kangean	3,000	-	263	11	9	2	-	
CA. Sepanjang Barat-Timur	2,430	2,900	-	-	-	-	-	
CA. Mt. Ringgit	2,000	4,700	243	12	9	2	-	
SM. Mt. Jaga Tamu	1,860	-	-	10	7	2	-	
SM. Ekoboyo	1,750	1200	-	-	-	-	-	

Sources: MacKinnon and MacKinnon (1986); RePPProT (1989)

Table 19.8. Gazetted Reserves, Bali. * = score when proposed extension is eventually gazetted. CA = Cagar Alam or Nature Reserve, SM = Suaka Margasatwa or Wildlife Reserve, TN = Taman Nasional or National Park, TW = Taman Wisata or Nature Recreation Park.

Area name	Official area (ha)	Area as mapped (ha)	Genetic value	Socio-economic justification	Management feasibility	Overall priority	Score	Notes
SM. Mt. Beser	4,000	4,400	540	12	10	2	-	
TN. Bali Barat	19,560	77,700	517	14	11	1	280*	
CA. Batukau I,II,III	1,763	1,900	-	13	12	2	50*	
TW. Panelokan	540	500	-	-	-	-		
CA. Sangeh	10	-	-	12	9	3		Reassign as TW.
Recommended Reserves, Bali.								
TW. Buyan and Tamblingan	1,000	1,100	-	-	-	-		
Proposed Reserves, Bali.								
TN. Bali Barat extension	58,165	?	?517	14	11	1		
CA. Batukau extension	20,080	14,700	-	13	12	2		
SM. Mt. Abang	3,000	3,800	-	13	12	2		
CA. Trunyan	30	-	-	11	9	3		
SM. P. Kalong, Burung Gadung	-	-	-	-	-	-		
TW. Mt. Batur	-	2,900	-	-	-	-		

Sources: MacKinnon and MacKinnon (1986); RePPProT (1989)

sounds, however, since it is well known that protected areas are often protected on paper only, and sizeable areas have already been converted to other uses (Chap. 20).

It is clear that neither Java nor Bali has a complete system of conservation areas. The situation would be helped on Java, perhaps, if Perhutani could transfer authority to PHPA for those areas currently in their care that have conservation value. It is not that the current inappropriate status makes them inherently unsafe, it simply recognizes that the best care is more likely to come from a body given the national responsibility for conservation, than from a state-owned company given islandwide responsibility for timber production. Similarly, the large forest block on the western mountains of Bali should be transferred from Forest Service authority to the National Park authority. There has, however, been 10 years of wrangling over this issue and it has now been agreed by the parties to subject the issue to a moratorium.

If the remaining, beleaguered conservation areas in Java and Bali are to maintain their integrity, then there has to be a deeper understanding of the surrounding peoples' aspirations, problems, and needs, a will (and a budget) to integrate conservation and development needs, and a working guarding system. Unless protection is ensured, other management activities will be of little use. For example, Baluran National Park would probably benefit from controlled, planned burns. But even contemplating such activities is pointless when fires are lit every year by people living in the coastal enclave, by cattle herders in the north, and by itinerant honey collectors or other gatherers (Watling 1990).

Table 19.9. Summary of gazetted nature reserves in Java and Bali, 1989. National parks at this time were administrative units made up of other gazetted areas and so were not additional to the areas below.

	Nature reserve		Game reserve		Recreation park		Nature marine reserve		Total area
	No.	ha	No.	ha	No.	ha	No.	ha	
Jakarta	4	64	-	-	1	18	1	108,000	108,082
West Java	38	182,233	2	13,528	17	4,366	-	-	212,247
Central Java	23	2,757	-	-	5	254	1	111,625	114,636
Yogyakarta	3	201	-	-	2	132	-	-	333
East Java	19	22,587	5	162,977	6	776	-	-	186,340
Bali	2	1,773	1	19,559	1	540	-	-	21,872
Total area		209,615		196,064		6,086		219,625	13,267,020

Source: RePPProI (1990)

THREATENED SPECIES

Animals

Assessing the degree of threat facing a particular species is fraught with problems but it is attempted regularly by the World Conservation Monitoring Centre, Cambridge, UK, for vertebrates and certain groups of invertebrates. Its most recent *Red List of Threatened Animals* was consulted, and those listed species found on Java and Bali are shown in table 19.10. The degree of threat indicates only their global status, and the status of the species on Java and Bali may be better or worse. Either way, they require and deserve attention. If the species is relatively common on Java and Bali, then the authorities have special responsibility to conserve healthy populations of a globally threatened species. If the species is relatively rare on Java and Bali, then there is a global duty to ensure that it becomes no rarer by losing those populations. Only half of those species listed are protected by Indonesian law, although most of the larger species are included. Indonesian law also protects many other Java and Bali animal species not regarded as globally threatened.

Another listing, of the 12 most critical bird species for conservation on Java, was compiled after extensive surveys, and consideration of their distribution, dependence on forest, breadth of altitudinal distribution, and general rarity (table 19.11). Each of the 12 represents either an endemic species or subspecies (van Balen 1987c). Only four of them are found on the WCMC list above, probably because the criteria were different and because some of the more widespread species do not have a critical status elsewhere in their range.

Plants

If collecting and assessing information on the conservation status of animals is difficult, then doing the same for plants is currently nearly impossible. It is likely that almost all the endemic species listed in Chapter 4 are rare or threatened in some way. As was indicated earlier, however (pgs. 151, 162), extensive surveys over many years looking for Java's orchids allow some measure of threat to be given. Thus, of the 217 endemic orchid species, only 84 appear to have a status which could be regarded as satisfactory. The remaining 133 deserve some form of attention:

- fifty-seven are regarded as rare, confined to restricted but not imminently threatened habitats. Some rare species, such as some of the leafless *Taeniophyllum*, have also been found on tea bushes, so their available 'habitat' has not been reduced as much as other species'. One of the most remarkable of these rare endemic species is the

Table 19.10. Animal species occurring in Java and Bali that are listed in the IUCN *Red List of Threatened Animals* with an indication of their global conservation status. Scientific names brought up-to-date where necessary. Bold indicates species endemic to Java or Bali. K - Insufficiently known, T - threatened, V - vulnerable, E - endangered, Ex - extinct, P - protected under Indonesian law.

MAMMALS

Javan pleated-cheeked fruit bat	*Megaerops kusnotoi*	K	
Javan thick-thumbed bat	*Glischropus javanus*	K	
Bartel's myotis	*Myotis bartelsi*	K	
Javan mastiff bat	*Otomops formosus*	K	
Javan leaf monkey	*Presbytis comata*	E	P
Javan gibbon	*Hylobates moloch*	E	P
Wild dog	*Cuon alpinus*	V	P
Small-clawed otter	*Aonyx cinerea*	K	
Javan ferret-badger	*Melogale orientalis*	K	
Mountain weasel	*Mustela lutreolina*	K	
Binturong	*Arctictis binturong*	K	
Leopard	*Panthera pardus*	T	P
Dugong	*Dugong dugon*	V	P
Javan rhinoceros	*Rhinoceros sondaicus*	E	P
Javan pig	*Sus verrucosus*	V	
Bawean deer	*Axis kuhli*	R	P
Banteng	*Bos javanicus*	V	P

BIRDS

Milky stork	*Mycteria cinerea*	V	P
Lesser adjutant	*Leptoptilos javanicus*	V	P
Javan hawk-eagle	*Spizaetus bartelsi*	V	P
Green peafowl	*Pavo muticus*	V	P
Javan wattled lapwing	*Hoplopterus macropterus*	Ex	P
Javan coucal	*Centropus nigrorufus*	V	
Javan scops owl	*Otus angelinae*	R	
Waterfall swiftlet	*Hydrochuos gigas*	K	
Salvadori's nightjar	*Caprimulgus pulchellus*	K	
Black-banded barbet	*Megalaima javensis*	R	P
Javan cochoa	*Cochoa azurea*	R	
White-breasted babbler	*Stachyris grammiceps*	V	P
Javan white-eye	*Zosterops flavus*	R	
Bali starling	*Leucopsar rothschildi*	E	P

REPTILES

Loggerhead turtle	*Caretta caretta*	V	P
Green turtle	*Chelonia mydas*	E	
Olive ridley turtle	*Lepidochelys olivacea*	E	P
Hawksbill turtle	*Eretmochelys imbricata*	E	P
Leatherback	*Dermochelys coriacea*	E	P
Indian python	*Python molurus*	V	P
Estuarine crocodile	*Crocodylus porosus*	V	P

BUTTERFLIES

Ijen swallowtail	*Atrophaneura luchti*	R	
Milkweed butterfly	*Euploea gamelia*	R	
Milkweed butterfly	*Parantica albata*	R	
Milkweed butterfly	*Parantica pseudomelaneus*	R	

MOLLUSCS

Giant clam	*Tridacna gigas*	V	P

Source: WCMC (1990); Harahap (1991)

monotypic *Silvorchis colorata;*

- one, *Corybas imperatorius,* is regarded as vulnerable because its status could easily slip into 'endangered';
- four are regarded as endangered because, if current trends of habitat loss continue, they are likely to become extinct. Among these are the two beautiful, large-flowered species, the lady's slipper *Paphiopedilum glaucophyllum* and the moon orchid *Phalaenopsis javanica,* which have been collected intensively by commercial traders and by orchid enthusiasts. Unfortunately, the desire to hybridize orchid species and varieties has the effect of making the genuine species threatened, even in captivity. The other two species, *Pseudovanilla affinis, Pteroceras javanicus,* are endangered by the loss of their lowland habitats;
- four are regarded as extinct (*Habenaria giriensis, Liparis lauterbachii, Plocoglottis latifolia,* and *Zeuxine tjiampeana*) because their only known localities have been entirely destroyed; and
- to these must be added 65 species about which even a tentative statement regarding their status is difficult since they are so little known. It is likely, though, that they are not common.

Table 19.11. The 12 bird species with the most critical conservation status on Java.

Javan hawk-eagle	*Spizaetus bartelsi*
Javan scops owl	*Otus angelinae*
Brown hawk owl	*Ninox scutulata*
Salvadori's nightjar	*Caprimulgus puchellus*
Javan frogmouth	*Batrachostomus javensis*
Banded woodpecker	*Picus miniaceus*
Buff-rumped woodpecker	*Meiglyptes tristis*
Scaly-breasted bulbul	*Pycnonotus squamatus*
Asian fairy bluebird	*Irena puella*
White-breasted babbler	*Stachyris grammiceps*
Javan scrub warbler	*Bradypterus montis*
Thick-billed flowerpecker	*Dicaeum agile*

Source: van Balen (1987c)

· SPECIES MANAGEMENT

Determining the most appropriate means of managing Java and Bali's remaining populations of large mammals creates much heated discussion. It is important that correct decisions are made because these animals tend to be flagship species and a major justification for certain protected areas. Although not officially acknowledged, the loss of the last tigers from Meru Betiri means not just a redesign of its logo, but also a reassessment of its conservation purpose. Likewise, if Ujung Kulon ever lost its rhinos, a great proportion of the park's interest would have disappeared, even though very few of its visitors ever set eyes on a rhino. The conservation of the large mammals is also important because many smaller species will benefit directly and indirectly from the management actions taken for the large species.

Pigs

The endemic Javan pig *Sus verrucosus* (p. 215) was feared to be nearly extinct in the late 1970s, but a small population was found in 1981 on the forested slopes of the sacred Mt. Penanggungan near Tretes, East Java (p. 497). A World Wide Fund for Nature survey conducted shortly after this provided basic data on the remaining populations. It appears to favour large areas of scattered teak plantations, secondary growth, or grasslands, below 800 m. The common wild boar *Sus scrofa* is more abundant and adaptable, and occurs at all altitudes.

In some areas with scattered clearings, grasslands, and small villages, crop depredation by pigs is a problem, but the use of poison, by the farmers who suffer, upsets the influential hunting lobby. Some 90% of the population of Java is Muslim and therefore forbidden to eat pig meat; the percentage is not necessarily the same for sport hunters for whom pigs are a popular quarry. These hunters (or at least 600 of the actual total) purchase annual licenses which permit them to take five wild pigs during the nine-month open season without using spotlights at night. As a result of little or no enforcement, these regulations have little influence on hunting practice, and there are at least as many unlicensed hunters. Since hunters are often alerted by villagers when pigs are in the vicinity, there is a mutual feeling that the hunters are performing a beneficial service not provided by the government. Among the villagers themselves, poisoning is commonly employed, and this practice was subsidised by the Government up to 1940. It is now illegal. The continuing use of poisons has caused drastic reductions in pig numbers, with poisoners boasting of having killed 10-40 pigs in a night. The poisons used are many and varied, and apparently very effective, making the consumption of meat by humans very likely. Indeed, pig poisoning is often done expressly to provide meat to specialist meat traders.

Since no one will knowingly buy poisoned animals to eat, the carcasses are stabbed with spears to give the illusion of being a hunted quarry. No data seem to be available on any effects of the meat of poisoned pigs on human consumers.

The prime habitats for the Javan pig are, unfortunately, of little or no value for general conservation, and are anyway already consigned to other forms of land use. The red-maned subspecies *S. v. blouchi* that lives on Bawean is in theory protected within the reserve set up to conserve the Bawean Deer. After considering the problems, three new reserves were proposed: a 13,000 ha area of the Pembarisan Hills between West and Central Java, 16,000 ha around the Lebakharjo limestone forests on the south coast of East Java (p. 802), and Mt. Ringgit and Mt. Beser, east of Surabaya, all of which have values beyond just the existence of the endemic pigs. In addition, Leuweung Sancang and Nusa Kambangan Reserves should both be extended to include good pig habitat while not including human settlements (Blouch 1988). These proposals were first made officially in 1983, but no action has been taken. At least part of the reason is that, with limited budgets, conservation priorities have centred on national parks with touristic (economic) value.

The Javan pig is not a protected species, and it is hard to argue, from existing information, that this status should change. The key to the successful conservation of this unique pig is probably the cadre of sports hunters who have a vested interest in its survival. The control of raiding herds should be the responsibility of licensed hunters. At the same time the government would benefit from better enforcement of hunting regulations because more income would accrue from the sale of licences to all hunters, and local people would benefit from the various services (food, guiding, accommodation) required by the hunters (Blouch 1988).

Banteng

Although vulnerable, and despite claims of a major decline (e.g., Wiriosoepartho 1984), the banteng is not faring badly on Java and Bali. Sizeable and apparently stable populations exist in the national parks of Ujung Kulon, Meru Betiri, Baluran, Alas Purwo, and in Leuweung Sancang and Cikepuh Nature Reserves (fig. 19.2). Some of the smaller populations, such as at Pangandaran and possibly Bali Barat National Park, are of hybrid stock. In discussions of the population sizes of these large forest animals, it should be remembered that there is no easy means of censusing their absolute numbers, although properly conducted surveys can provide reliable information on population trends (S. Hedges pers. comm.).

The populations of banteng at Baluran and Ujung Kulon are the best known, but even so they are not well understood. At Baluran the banteng use only about one-third of the park because of human disturbance, unsuit-

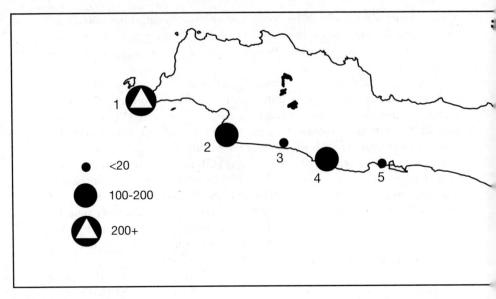

Figure 19.2. Concentrations of banteng in Java and Bali with their approximate numbers. 1 - Ujung Kulon, 2 - Cikepuh, 3 - Bonjonglarang Jayanti, 4 - Leuweung

able vegetation, domestic cattle, and feral buffalo. The buffalo are generally regarded as a major problem because they are assumed to compete with the banteng for drinking water and grazing, and because they are notorious for carrying almost the entire spectrum of parasitic cattle diseases, which may account for the low banteng calf population in Baluran (Wind 1978). One could very easily take the view that all the feral buffalo at Baluran should be removed (caught or culled) as rapidly as possible. In fact, it is not known what the 'natural' fauna of Baluran comprised, and even without the buffalo, the whole reserve is a mosaic of introductions and major human impacts, so to single out the buffalo for eradication is not entirely logical (Watling 1991a). Instead, the appropriate management of the buffalo should be decided simply on the basis of what is good for the Park. The feral buffalo certainly are an attraction for visitors, but the indications are that they compete with banteng for the scarce water resources, either by being more brazen and ousting the banteng, or by wallowing (an activity not indulged in by banteng), which muddies the water. In addition, a surprisingly large number of weak buffalo and calves die in wallows or water holes, and the putrescence of their corpses makes the water undrinkable. In view of this, a reduction of buffalo numbers to about 400 has been suggested (Bismark 1988; Watling 1991a). Exactly how the reduction

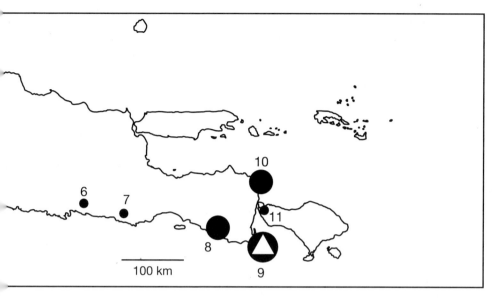

Sancang, 5 - Pangandaran, 6 - south of Blitar, 7 - south of Malang, 8 - Meru Betiri, 9 - Alas Purwo, 10 - Baluran, 11 - Bali Barat.

After Ashby and Santiapillai 1988

is to be achieved is open to discussion. The staff of the park have shown that rounding up buffalo is possible but the work is dangerous, and while most of the Baluran buffaloes adapt to domestication within a month, about one-sixth die (Ahmadi 1985). The alternative of culling agreed numbers using skilled marksmen meets religious objections.

At Ujung Kulon there are no feral buffalo, but there are still management problems. Alikodra (1983) states that forested areas, open grassy areas, freshwater, and access to mineral salts are all basic to the needs of banteng. Although the banteng eat a wide range of plants, there are several common species that are not favoured, such as *Lantana camara* (Verb.), *Melastoma malabaricum* (Mela.), *Chromolaena odoratum* (Comp.), and *Ardisia humilis* (Myrs.). It has long been recognized that very little is known about the nutritional requirements of these animals, the nutritive quality of their food, seasonal or annual cycles in preference, or the optimum habitat composition, and until this is rectified, the management efforts will remain rather random (Sumardja and Kartawinata 1977; Sumardja 1978). It has been suggested that habitat management could involve the deep removal of inedible weeds in the grazing area, and the burning of half the area each year at the end of the dry season (August/September), having first established 15-25 m wide fire breaks (Alikodra 1983). While it may be justifiable to manage the grazing areas to maintain a tourist spectacle, and if it was clear that intervention would genuinely help the situation, there is no need to manage the rest of the area for the benefit of the banteng (or rhino) (Hommel 1987).

Rhino

Around the last days of 1981, five Javan rhino carcasses were found in Ujung Kulon National Park, and some banteng and domestic animals, including 50 buffaloes and over 300 goats, died in villages to the east. Unfortunately, although reports were submitted promptly, official reaction was slow. Eventually a number of teams were sent to the Park, and they concluded that all the rhinos had died of some infectious disease which had caused them to fall over and die quite suddenly, sometimes after a bout of diarrhoea. Poaching and poisoning were ruled out as causes, and the rhinoceros deaths were attributed to them having eaten toxic plants as a result of competition from banteng (Soemandoyo 1984). Had this been the case, further deaths would have been expected. The actual disease could not be identified or isolated from either the wild or the domestic animals, and although the hypothesis that the disease spread from the domestic animals to the wild is unlikely, it is not impossible, given that buffaloes are illegally entered to graze in the Park. Proposals were made by various parties to erect fencing to keep the major wild and domestic species apart as much as possible but the practicability of this was, and is, to be questioned. The most likely disease, or at least a disease whose appearance matches what is known of the rhino deaths, is anthrax, the spores of which can remain inactive in soil for decades. The disease is endemic in Java and elsewhere in Indonesia (Schenkel and Schenkel 1982).

The events of that New Year period worried many people because it demonstrated the vulnerability of the (then) only known population of this large and threatened mammal. Translocation of some animals to neighbouring P. Panaitan or even to Way Kambas National Park (where the last Sumatran 'Javan' Rhinos were shot) were mooted as options, but the difficulties of protecting the animals from poachers in these other locations precluded further consideration (Santiapillai and Suprahman 1986). The possibility of moving some rhinos to another conservation area was discussed by Hoogerwerf (1970), but he foresaw problems having just a few animals in a larger area such as Way Kambas National Park. Since the female oestrous lasts only 24 hours, the animals simply might not meet at the right time for successful breeding. It may, in fact, be very fortunate that Ujung Kulon is not a particularly large conservation area. In addition, certain zoos in the US and elsewhere were very enthusiastic about the possibility of having such rare and little-known large mammals in their collections for captive breeding efforts. A programme began in 1984 to capture 'doomed' Sumatran rhinos from Riau for captive breeding programs in the US, Britain, and Jakarta, but the overall success, either in terms of survival of captured beasts, or of captive breeding results, was far from having been demonstrated. Basically, no one yet really understands rhinoceroses well enough to provide nutritionally balanced diets in captivity to prevent the variety of serious and minor conditions that appear to arise when they are taken from their natural environment. Nine of the 27 Sumatran rhinoceroses caught in recent years in Malaysia and Indonesia died at capture or a few months thereafter. While it would be useful for conservation to understand their molecular genetics, this seems singularly academic when captive animals are generally less than perfectly healthy, do not breed freely, and have low rates of survival.

In 1989 a computer exercise called a 'Population Viability Analysis' was undertaken by a team of those people who knew the most about the animals. This was felt necessary because the Ujung Kulon population had remained at less than 70 individuals for over ten years based on the surveys by PHPA (with assistance from Universitas Nasional, WWF, and IUCN), because erosion of genetic diversity and perhaps extinction were likely, and because removal of one individual every two years by disease or poaching was enough to halt population growth. After the data were analysed, the conclusion was that a captive breeding programme for the Javan rhino was justified in addition to intensified protection at Ujung Kulon (Seal and Foose 1989).

The arguments against a captive breeding programme were not conducive to being quantified but were no less valid. In summary they were that:

- the species is so poorly known that each of the figures entered into the computer model has a high degree of uncertainty. When considered together, the uncertainty becomes compounded and the result therefore highly suspect;
- if foreign zoos are so intent on spending large sums of money capturing, transporting, exhibiting, and maintaining these rare animals, should they not first provide major funding for the pressing conservation priorities of the Park?
- the Sumatran rhino project should demonstrate its success with this relatively common (though highly threatened) species, before a similar project is contemplated for the Javan rhino.

In 1990 and 1991 the debate hotted up, the lobbying of key parties became intense, and nongovernmental organizations, such as the Indonesian Rhino Conservation Forum, comprising students and other interested parties, were established to promote their opinions, to conduct research, and to inform others. Eventually the authorities decided to postpone the option of captive breeding until these three problems could be dealt with. If a small-scale translocation programme were to be initiated, then it should take the form of the establishment of a large enclosure, probably in the east of the park, where the animals could be observed and, hopefully, breed. Much useful information could be gained in this way if captive breeding ever became necessary (MacKinnon et al. 1990). Money from Indonesian and international sources became available to support these efforts.

The first priority was a detailed census of rhino numbers. Surveys had, of course, been conducted for years, but the methodology had some major flaws that may or may not have distorted the real situation. In order to provide better information on the rhino population, 34 automatic cameras triggered by camouflaged pressure mats were set up at likely points around the park. A similar system had been used for photographing large mammals in Sumatra (Griffiths 1990). This has produced a harvest of hundreds of rhino photographs with time and date, and the individual animals are recognizable by features such as body size, horn size and shape, eye wrinkles, neck folds, skin pores, neck plate profile, check profile, ear tears, and skin scars, and in this way the population has become known at a level not dreamt of a few years ago. All those that were positively male had a horn, and those that were positively female did not. Many more adult males have been photographed than adult females, but this apparently skewed sex ratio (noted also by Hoogerwerf (1970)) does not necessarily reflect the actual sex ratio in the entire population. The reason for this is that in such a largely solitary species, it would not be unusual for the males to be more mobile than the females by patrolling their territories and searching for females in oestrous. It is thus more likely for males to be encountered, and preliminary analysis of the photographic results shows that individual males are found over a larger area than are individual females. It may also be that females spend more time off the trails both in search of nutritious forage and to avoid male harassment (Griffiths 1992a, b). The total population of rhinos is now believed to be about 50 animals, including young. The small size of the population rules out a captive breeding programme, but it may allow for a few individuals to be transferred to an enclosure within a suitable forested habitat where they could be studied.

Since Ujung Kulon has clearly defined limits, the rhino population cannot expand indefinitely. It might, however, be possible to encourage the population to grow by increasing the carrying capacity of the peninsula. The most obvious way to do this would be by increasing the food supply. It has been noticed that the rhinos do not visit all parts of the park equally, and that selective cutting of the abundant *Arenga* palms in certain plant communities encouraged the growth of young trees, which were browsed by rhinos some two years later (Schenkel et al. 1978). Such management or manipulation of the vegetation would need to be carried out on a large scale to have any significance for the rhino population, and it is generally

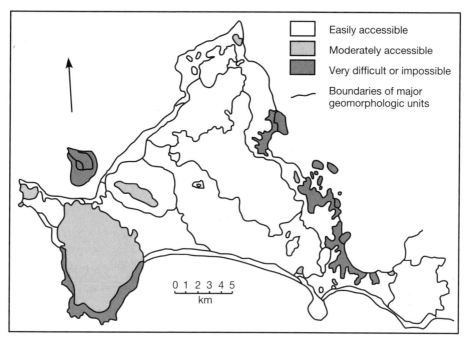

Easily accessible

Moderately accessible

Very difficult or impossible

Boundaries of major
geomorphologic units

0 1 2 3 4 5
km

Figure 19.3. Accessibility for the Javan rhino of the different landscape types in Ujung
Kulon.

After Hommel 1987, 1990a. With permission of the publishers

felt to be difficult to justify such action. Even so, there has been speculation
that the rhino population has reached the carrying capacity of at least the
western part of the park. While not denying the possibility, there are few
data to support such a notion.

One of most detailed ecology field projects ever conducted on Java or
Bali goes some way to elucidating the question of differential habitat use by
the rhinos. Information on the vegetation, soil, and landforms were combined to reveal 26 landscape types which can be understood in terms of climate and history. They are described in detail by Hommel (1987, 1990b).
With these units mapped, they could be compared with what is known of
the Javan rhino (Hoogerwerf 1970; Schenkel and Schenkel-Hulliger 1969;
Schenkel et al. 1978; Djaja et al. 1982; Ammann 1985). It must be remembered, however, that they are very shy animals and much of the area is inaccessible. The topics considered in the analysis were quality and quantity of
forage in the landscape types, the accessibility for rhinos, availability of
water, mud wallows, and salt, the degree of cover, and the presence of pathogenic microorganisms. The first three were the most important, and

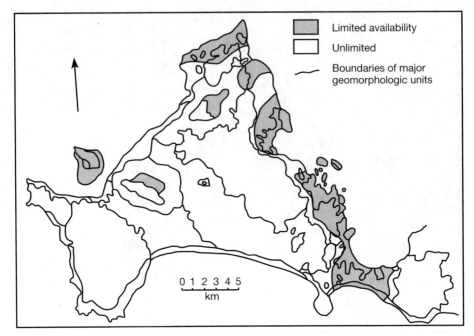

Figure 19.4. Availability of drinking water for the Javan rhino in Ujung Kulon.

After Hommel 1987, 1990a. With permission of the publishers

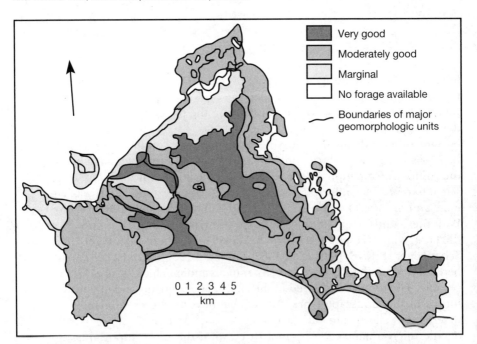

Figure 19.5. Quality and quantity of forage for the Javan rhino in Ujung Kulon.

After Hommel 1987, 1990a. With permission of the publishers

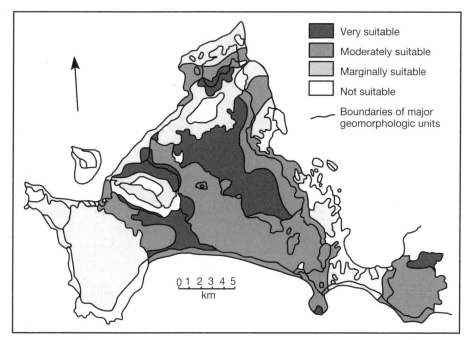

Figure 19.6. Habitat suitability for the Javan rhino in Ujung Kulon.

After Hommel 1987, 1990a. With permission of the publishers

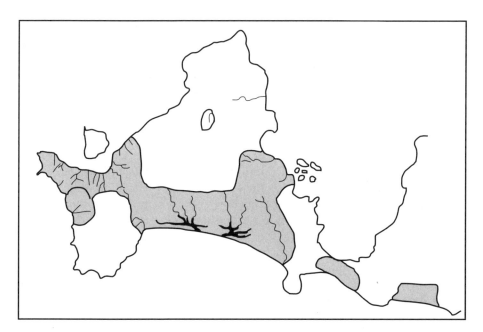

Figure 19.7. Concentrations (shaded) of rhino tracks on Ujung Kulon in 1967-68.

After Schenkel and Schenkel-Hulliger 1969; Hommel 1987, 1990a). With permission of the publishers

their distributions were mapped (figs. 19.3-5). It should be noted that the rhinos are remarkably adept at fleeing from humans, and the rattan scrub which is impassable for people appears to present few problems for the animals.

These three maps were combined to form a map of habitat suitability for the Javan rhino (fig. 19.6), which can be compared with the concentrations of rhino tracks found in the most intense years of study (1967-68) (fig. 19.7). The comparisons are favourable and the apparent disagreements are likely to be due to seasonal differences or to weaknesses of the rhino track censuses, which tend to avoid the dense rattan scrub. It can be concluded that as succession in the rattan scrub areas proceeds, so the supply of rhino food will decrease. However, succession here is progressing very slowly and it will be good few decades before any active management is justified (Hommel 1987, 1990a).

Marine Turtles

Five species of marine turtles are found around Java and Bali, of which four regularly nest here (p. 244). Turtle shells and shell products used to enjoy a lively export market, but greater awareness by customs officials in foreign countries adhering to the regulations of the Washington Convention on Trade in Endangered Species (CITES), and awareness campaigns by pressure groups have greatly reduced this. The primary purchasers of turtle shells are now domestic tourists. Four species of marine turtles are legally protected in Indonesia, the hawksbill only since 1992, but green turtles remain unprotected. Indonesia has tried repeatedly at CITES meetings to convince other parties that trade in Indonesian green turtles would not endanger the large resident population, but the arguments have not been accepted. Indonesia is the only country not to protect this species.

Java's foremost green turtle rookery is at Pangumbahan, just south of Cikepuh Reserve, south of Pelabuhan Ratu, West Java (Sumertha 1976). This has been the site of egg collecting for generations, and in the 1950s some 2.5 million eggs were being taken each year, and this was only a proportion of those laid. Up until the 1970s the collecting was largely for subsistence needs and it was then regulated by the auction of licences. A single company now has the licence, and even though all the eggs are taken, only about 250,000 are found annually – a large number, but a drop of 90%. Unfortunately, no useful data are kept on the numbers harvested, proportion hatched, or numbers released as part of the company's 'conservation' programme. National authorities and NGOs have expressed their disquiet over the enterprise, but it is the provincial authorities which have jurisdiction over it. Profits from the sale of the eggs are not inconsiderable, but little finds its way into the community because rather few local people are employed. The problems are aggravated by the provincial authorities allowing developments, such as intensive tambak, close to the beach which by their accompanying noise, light, and groundwater effects are very disturbing to nesting turtles (Sloan et al. 1994). It seems likely that such blatant mismanagement of an important and valued natural resource will create more bad feeling against Indonesia, and will ensure that pressure is maintained to have green turtles protected both on paper and on the beaches.

Turtle eggs are very popular as food in Java, but the flesh is regarded under Islamic law as *haram* or 'unclean'. In contrast, the demand for the meat to satisfy the traditional consumption at ceremonial and religious feasts on Bali has encouraged the harvesting of turtles in the furthest and remotest parts of the archipelago, often with considerable cruelty and lingering deaths. These turtles are sometimes caught on their nesting beaches, but most are caught in nets in the open sea. Some are taken directly to Bali, but others are held in pens and sold to intermediaries. Bali has been an importer of turtles for many years because, even by 1950, its own turtle populations were said to have been seriously depleted (Sumertha 1974). The Balinese do not eat the eggs, which are instead sold to local Muslims. The former traditional uses of turtles on Bali were probably sustainable, and no criticism could be levelled at the practice when the demand and human population level were low. But the 'slaughter' of green turtles was recently more intensive on Bali than anywhere else in the world. The trade is centred on the village of Tanjung Benoa, just north of the tourist centre of Nusa Dua. The annual import to Bali was probably more than 20,000 turtles during the 1970s (Polunin and Sumertha 1982), reaching a peak of over 30,000 in 1978. By 1988, however, the annual trade was still in the region of at least 25,000.

The impact of killing turtles can be well illustrated as follows. An individual turtle can probably live more than 100 years, although 70-80 would be more likely, and it matures between the ages of 15-50. If we assume conservatively that animals have an average of just two decades of reproductive life, and that a female nests in alternate years, laying 500 eggs per season, then she will lay 5,000 eggs during her life. First-year mortality is very high, probably in the region of over 95%, but adults suffer very low (natural) rates of predation. Thus turtles are very fecund animals and on economic, energetic, and conservation grounds, the killing of adults is very wasteful and certainly more wasteful than taking eggs. If it is assumed that each female killed on Bali would otherwise lay eggs for an average of only five more occasions (probably an underestimate), then an annual killing of 10,000 female turtles means the loss of some 12.5 million eggs. Because of the turtles' long life span, the impacts of intense trade will be felt in the population for decades and, for the same reasons, any serious conservation measures will take at least as long to fully demonstrate their effectiveness. Indeed, if the gross exploitation of adult turtles was as serious as some have surmised, it may already be too late to stop a major population crash when the present breeding adults die (or are killed).

In January 1990, at least in part because of international concern, the Governor of Bali promulgated a decree with the express aim of reducing the number of turtles to 5,000 annually killed, by restricting the use of the turtles for meat to religious occasions, by banning the sale of the meat in public eating places, and by banning the sale of souvenirs made from turtles. Greenpeace, one of the world's major environmental pressure groups, while welcoming the decree, found that in the year following this, almost no changes occurred in the Bali turtle industry. They reported at least 21,000 turtles were killed on Bali in 1990 (twice the official estimate because not all shipments are officially reported, and juveniles are not recorded). Even in the first seven months of 1992, a total of 10,265 turtles had been brought to Bali (E. Sumardja pers. comm.), and over 18,000 were landed in 1993, indicating that little restraint or enforcement were being exercised. Greenpeace felt that for any regulation to be effective, the following would be necessary:

- a ban on the capture of any green turtles for commercial purposes, with only a limited number allowed for use at feasts during religious ceremonies. But since the use of turtle meat is not obligatory at these occasions, this allowed use should have a timetable for its phasing out;
- a ban on the capture or killing of breeding green turtles on or near nesting beaches;
- full protection for all green turtles with a carapace length greater than 80 cm;
- a ban on the sale of turtle meat in public places;
- a licence requirement for all turtle collectors;
- a prohibition on the sale of all stuffed turtles and other sea turtle souvenirs; and
- all turtles to be humanely killed before butchering begins (Greenpeace 1991).

Meanwhile a National Strategy and Action Plan for the Conservation of Marine Turtles was devised by the State Ministry of Population and Environment (KLH 1992d), and it was agreed to by all the relevant national and regional authorities, although no enforceable regulations have been issued. Among its many recommendations is that the annual national quota for green turtles be 3,000.

Many turtle research and management programmes around the world, including a number of sites in Java, have involved the taking of newly laid eggs and hatching them close-by, with some form of protection from predators (e.g., p. 814). Although the results can look impressive, insofar as good numbers of baby turtles hatch, this form of management faces problems. First, it has been known for some time that the sex of hatchling turtles and other reptiles is determined by temperature, with warmer conditions favouring females. Thus, it is possible for human intervention to skew the sex ratio with unforseeable impacts. Second, the hatchlings are able to live for about a week on food reserves within their bodies. Typically the hatchlings are kept for some time (so that visitors can see the fruits of the conservation effort) and then released offshore from a small boat or at the water's edge. The hatchlings will have missed out on adapting to the marine environment and learning to find food, and may even be in a rather weakened state. Third, it may be very important for imprinting the natal beach in the hatchling's mind for it to crawl down the beach. Fourth, despite great worldwide efforts, the number of recoveries of tagged turtles have been minimal.

It has been estimated that even in the absence of human interference the survival rate from hatching to adulthood is about 2.5/1,000 for the green turtle (Hirth and Schaffer 1974). This is an integral part of the life-history strategy of the species and is balanced by high adult survival and longevity. Thus, even if keeping back turtles for a few days on a given beach increased the number of hatchlings reaching 100 m offshore by one thousand, that would only add a couple of adults to the population in a few decades' time. In addition, we are taking out of the ecosystem whatever number of hatchling turtles (ecological meat sandwiches) would die in the first few days (presumably agreed to be a high number, or else the conservation action would be groundless) with unknown effects on other organisms.

On several occasions it has been suggested that a turtle farming facility be established on Sarangan Island, near Bali's international airport, to assist their conservation. The rationale is that such an operation would be, once running, no drain on the wild populations, and traders would prefer to get their supplies of eggs, meat, leather, oil, and shell from a reliable, high-quality source. Even cursory examination reveals how flawed this proposition is. First, turtles are one of the least likely domestic animals: they mature very slowly (taking up to half a century), and holding them in salt-water tanks or coastal impoundments is expensive and highly problematic if high water quality is to be maintained. Second, if one aims to supply about 150 eggs each day, or about 50,000 each year, then about 200-300 adult females (weighing a total of over 30 tons) will be required. This is clearly untenable. In addition, if eggs and bits of turtles are produced commercially, it will merely stimulate an unnecessary market and will further encourage the despoilation of remaining wild populations (Ehrenfeld 1992).

As if our ignorance of the actual effects of current and widespread turtle conservation actions (protected hatcheries and release after a certain period in captivity) were not enough, we also face the problem that such actions 'merely' attempt to increase the number of turtles entering the seas. Admittedly these are easier than enforcing a ban on the threats they will face, such as the collection of their eggs or on the harpooning of adults at sea, but unless they are done in concert, the 'conservation' actions are a waste of time and money (Frazer 1992). Over ten years ago it was stated that further knowledge of turtles was not actually needed for their survival (Ehrenfeld 1982), we simply had to get on and give them a chance. Today that is more true than ever.

BUFFER ZONES AND INTEGRATED CONSERVATION DEVELOPMENT PLANS

Buffer zones are bands of land between a conservation area and the surrounding cultivated areas and settlements, designed to protect the conservation area and its resources against negative influences from outside, and also to protect cultivated areas and settlements or the people and their resources against negative influences originating from the conservation area, such as large pests (figs. 19.8-10) (MacKinnon et al. 1986, 1990; Wind 1991, 1992). Buffer zones are primarily a means to conserve forest, supporting local people as and where possible as a secondary objective so that, when functioning well, there will be no negative influences at the boundary of the conservation area (DHV 1990). It is interesting that a worldwide survey of buffer-zone-type projects concluded that most failed and, for a variety of reasons, the concept could not be translated into reality (Wells et al. 1992). Buffer zones received an Indonesian legal basis in 1990 with the Law No. 5 on Conservation of Natural Resources and Ecosystems. The national parks on Java and Bali have designated buffer zones, generally comprising plantations, but under the current legislation these are outside the authority of PHPA. It is clear that before buffer zones work as it is intended, more thought is required.

The latest addition to the armoury of the conservationist is the concept of integrated conservation development programmes (ICDPs), which grew out of the concept of buffer zones. ICDPs are intended to relieve and divert pressures from human activities which are inappropriate in or around conservation areas (Caldecott 1996). These aims are achieved not just by enforcement of regulations but by positive development projects such as the installation or improvement of irrigation systems, diversification of horticultural crops, rattan plantations, boundary line planting, nature tourism and related facilities, small home industries, and fishponds (Wells and Brandon 1992; but see Angelsen 1995), ideally all under the supervision of the local authorities. Effective programmes will require unprecedented cooperation between different arms of government, an ideal rarely achieved in any country, all of which must have dynamic conservation as

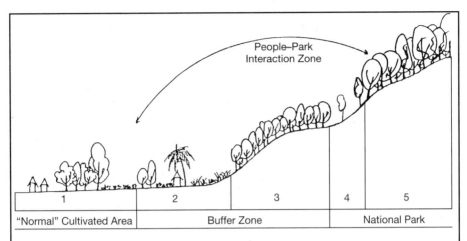

1 - Settlement area/cultivated area.
2 - Nonforest buffer zone. Settlement area/cultivated area.
3 - Forest buffer zone. Natural forest/forest plantations.
4 - Rate of integrity less than 100%.
5 - Rate of integrity about 100%.

Figure 19.8. A hypothetical buffer zone between a cultivated area and a conservation area.

After Wind 1991

Figure 19.9. Tea plantations serve as effective buffer zones between settlements and Mt. Gede-Pangrango/-Telaga Warna conservation areas.

By A.J. Whitten

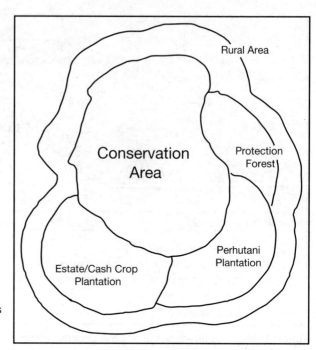

Figure 19.10. Examples
of buffer zones.
After Wind 1991

their main focus for the area concerned. A number of conservation areas
in Indonesia are subject to planning for major ICDPs, but none of these is
on Java or Bali, and none can be pointed to as a model. Efforts to find
models must continue, however, not least because Indonesia has agreed,
under the UN Convention on Biological Diversity, to promote environmentally sound and sustainable development in areas adjacent to conservation areas in order to improve the protection of those areas.

CONSERVATION AREAS AND VISITORS

Among Indonesia's conservation areas, only nature recreation parks, grand
forest parks, and national parks actively encourage visitors (fig. 19.11).
Many of the tourist forests and some other forested (natural and plantation) areas with waterfalls, caves, and small areas of disturbed or planted
forest are managed not by PHPA but by Perhutani to provide recreation
foci which allow visitors to appreciate seminatural conditions (Perhutani
1989). Seminatural areas are those which are derived from natural vegetation but which have been more or less changed by human activities.

Figure 19.11. Tourists travelling slowly up the Kulon River in Ujung Kulon National Park.

By A.J. Whitten

Such areas have very little significance in the conservation of biological diversity. It is interesting that the larger and intrinsically more interesting national parks generally attract far fewer visitors than the small nature recreation parks, and this is illustrated for a selection of contrasting conservation areas in East Java (table 19.12). The assets of these areas are as follows:

- Grojogan Sewu is a 64 ha, privately run nature recreation park 40 km from Solo and 14 km from Madiun with a waterfall, pine forest and other plantation trees, a picnic site, and a swimming pool.
- Sumber Semen is a 17 ha nature recreation park where there is a large freshwater spring, with attractive views, a sacred tomb, with planted trees of *Murraya paniculata* (Ruta.), which exhibits frequent mass flowering like coffee and pigeon orchids. The forest is said to support rhinoceros hornbill *Buceros rhinoceros*, Java sparrow, peafowl, and green jungle fowl (Marjatmo 1987), but the larger of these species cannot be self-sustaining in such a small area.
- Tlogoworno is a 40 ha nature recreation park with one coloured and

one clear lake, views, sacred caves and stones, planted trees, and pond fish. It is 140 km from Semarang, 190 km from Solo, and 60 km from Pekalongan.

- Bromo-Tengger-Semeru National Park, east of Malang, covers 58,000 ha with the famous 'Sand Sea', active volcanoes, a number of lakes, a meditation cave, various religious objects, splendid views, and excellent mountain vegetation (p. 804).

- Baluran National Park is at the northeast tip of Java, 260 km from Surabaya, 160 km from Denpasar, and covers 23,700 ha. It is best known for its savanna woodland with large populations of banteng, deer, and feral buffalo. There is also coral offshore, and an extinct volcano with a forested crater (p. 817).

- Meru Betiri is a 58,000 ha national park, 200 km from Jember, 110 km from Banyuwangi, with a famous turtle beach, wild banteng, fine forests and views (Marjatmo 1987) (p. 812).

The exception to the general trend that natural areas receive relatively few visitors is Bromo-Tengger-Semeru National Park, which is widely promoted internationally (nearly one-third of all visitors are from overseas), where virtually all the visitors come only to see the sunrise from the Bromo crater, after which they leave, and where biological interest is virtually non-existent, at least in those areas seen by most people. It could be debated, in fact, whether isolated nature recreation parks can really be said to contribute much to conservation at all, and whether they would not be better placed under the tourism or provincial authorities. Where they are adjacent to larger, more natural areas, however, they serve a useful role as intensive-use zones.

A major issue here is whether the natural biological diversity of Java and Bali yet generates any significant or genuine interest, inspiration, or con-

Table 19.12. Total visitors to selected East Java conservation areas, 1982-90 (in thousands). D = domestic, F = foreign.

	82		83		84		85		87		88		89		90	
	D	F	D	F	D	F	D	F	D	F	D	F	D	F	D	F
Nature recreation parks																
Grojogan Sewu	296	.6	263	.7	325	.8	327	.4	304	2	316	2	352	2	398	2
Tlogoworno	31	.7	14	.5	10	.7	14	.4	19	1	16	1	30	1	40	4
Sumber Semen	10	0	12	0	26	0	4	0	35	0	24	0	29	0	18	0
National parks																
Bromo-Tengger-Semeru	38	7	71	7	8	4	30	7	42	11	86	17	76	18	55	24
Meru Betiri	5	.3	2	.3	5	.2	8	.2	15	1	8	.4	7	.6	8	.9
Baluran	2	1	3	.5	7	.3	11	.7	15	.8	17	1.1	14	1.3	16	1.8

Source: KSDA Malang

cern among the general populace. It would seem not, and it is relevant, if disappointing, that the nature magazine *Suara Alam* folded in 1988 because of lack of interest. But to dismiss the importance of conservation misses the very noticeable increase in young hikers and campers in Java's wilder areas, the growth in, and increasing activity of, local, national, and Indonesian offices of international NGOs, and the growing understanding of environmental issues among a better-informed public. Unfortunately this group of generally younger people does not yet have a significant voice in the decision-making processes. When its turn comes to bear authority, however, its members will have to dig deep in their hearts to forgive those whose decisions today ultimately affect the quantity and quality of biological diversity tomorrow, unless there is a significant change in attitude and an increase in understanding.

Chapter Twenty

Major Existing and Potential Conservation Areas and Their Problems

The figures for the total number and total extent of the conservation areas on Java and Bali are often quoted as evidence of a commitment to conservation, and as a reassurance that the unique biological resources they contain are safe. This section examines the riches, characteristics, and problems of 27 of the most important areas in Java and Bali (fig. 20.1). The number of conservation areas, or even their size, are clearly not as relevant as the pressures they face, the inappropriateness of their boundaries, the lack of involvement of, and benefits for, the often poor people living around them, their lack of management, their rates of degradation and attrition, and the indifference they face from the police and judiciary.

Ujung Kulon

This westernmost peninsula of Java has had some protection since 1921, and its adjacent islands of Panaitan and Peucang since 1937, but the entire area was not formally gazetted as a conservation area until 1958. Together with the adjacent eastern area of Mt. Honje, it now covers some 76,000 ha. The mainland has been settled in the past, though probably never at high densities. While settlements were recorded on Panaitan in 1771 by Captain James Cook during his circumnavigation of the globe, and two 16th century statues were discovered on the island's Mt. Raksa (MacKinnon 1991), voyagers landing at Peucang Bay in 1816 noted that "the scene was perfectly undisturbed, nor could there be any single trace of inhabitants, or any symptom to tell that the spot had ever been visited since the creation" (Hall 1827). Later, as Java's population grew rapidly (p. 28), the forests of Ujung Kulon used to be exploited for latex from wild *Ficus elastica* (Mora.) trees, and there were a number of villages, most of them around the coast. Food crops, including dry-field rice, were grown inland as part of a shifting cultivation cycle. Thus, in the last century, the vegetation of Ujung Kulon was a mosaic of new fields, recently deserted fields, settlements, and secondary forest. The area covered by primary and even old secondary forest was probably negligible and restricted to the steepest areas, or areas distant

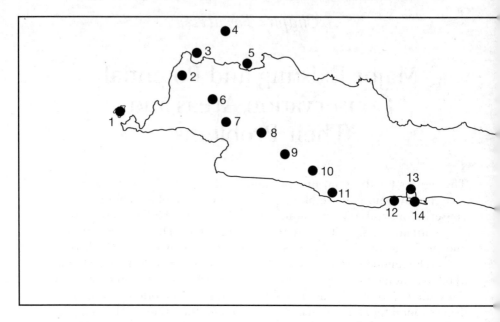

Figure 20.1. Major existing and potential conservation areas discussed in this chapter. 1- Ujung Kulon, 2 - Rawa Danau, 3 - Pulau Dua, 4 - Pulau Seribu, 5 - Pulau Rambut, 6 - Dungus Iwul, 7 - Mt. Halimun, 8 - Mt. Gede-Pangrango, 9 - Mt. Tilu, 10 - Mt. Wayang, 11 - Leuweung Sancang, 12 - Pangandaran, 13 - Segara Anakan, 14

from freshwater (Hommel 1990b).

The most important event in the history of Ujung Kulon was undoubtedly the shattering eruption of Krakatau, only 60 km away, in 1883. The 10-15 m high tidal waves eliminated the villages, crops, and most of the remaining natural vegetation in the coastal areas of the northern shores, although the loss of human life here was not great. It is doubtful that the waves had much effect on the inland vegetation. They did not escape damage, however, because ash falls up to 30 cm thick, still visible as pale levels in the soil, smothered the rice, made the fields unworkable, and brought about a local famine, epidemics of dysentery and malaria, and a large increase in raids from hungry tigers. As a result, the whole peninsula was evacuated. One of the earliest management actions in the area was the maintenance of coastal savannas by cutting and burning in order to provide grazing for the banteng and deer (Hoogerwerf 1970; Hommel 1990b).

Ujung Kulon is famous for its more or less complete assemblage of Javan wildlife including the world's largest population of Javan rhinos, other rare wildlife, such as gibbons, surili leaf monkeys (both of which are less common here than in conservation areas to the east), and banteng, and for its scenic beauty. Although much has been written about the area,

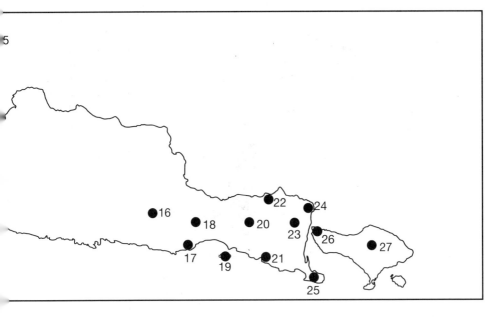

- Nusa Kambangan, 15 - Karimun Jawa, 16 - Mt. Kawi-Kelud, 17 - Lebakharjo and Bantur, 18 - Bromo-Tengger-Semeru, 19 - Nusa Barung, 20 - Yang Plateau, 21 - Meru Betiri, 22 - Mt. Beser and Mt. Ringgit, 23 - Ijen-Maelang-Raung, 24 - Baluran, 25 - Alas Purwo, 26 - Bali Barat, 27 - Batukau.

notably by Hoogerwerf (1970), the vegetation was systematically studied only in the early 1980s (Hommel 1987). In fact, this is the most comprehensive vegetation study undertaken on Java and serves as a model for other areas. Far from being totally focussed on botany, the study also sought to give an estimate of the suitability of the different parts of the park as rhino habitat. Over 300 sample plots chosen after rough analysis of aerial photographs were examined, and within each plot all plant species were recorded, together with their cover and abundance. In addition, observations were made of soils and geomorphology (figs. 20.2 and 20.3). A classification of ecological units and a map of their distribution were produced, and these were then related to available information on the requirements of the Javan rhino (Hommel 1987, 1990a, b) (p. 750).

Sizeable areas of sea around Ujung Kulon were proposed as extensions to the then proposed national park (Halim and Kvalvågnaes 1980) in recognition of the likelihood that they could be as important in conservation and recreation terms as the terrestrial component. Scuba diving is indeed an increasingly popular activity in the park.

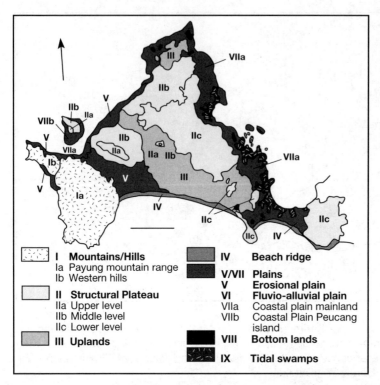

Figure 20.2. Geomorphological units and subunits of Ujung Kulon. Scale bar indicates 5 km.

After Hommel 1990a. With permission of the publishers

I **Mountains/Hills**
Ia Payung mountain range
Ib Western hills

II **Structural Plateau**
IIa Upper level
IIb Middle level
IIc Lower level

III **Uplands**

IV **Beach ridge**

V/VII **Plains**
V **Erosional plain**
VI **Fluvio-alluvial plain**
VIIa Coastal plain mainland
VIIb Coastal Plain Peucang island

VIII **Bottom lands**

IX **Tidal swamps**

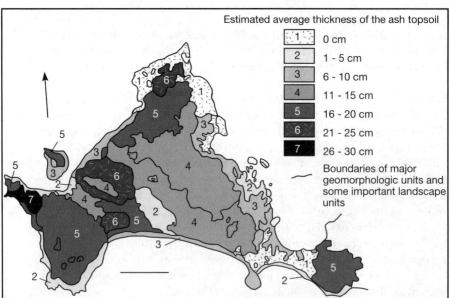

Estimated average thickness of the ash topsoil

1   0 cm
2   1 - 5 cm
3   6 - 10 cm
4   11 - 15 cm
5   16 - 20 cm
6   21 - 25 cm
7   26 - 30 cm

Boundaries of major geomorphologic units and some important landscape units

Figure 20.3. Distribution of thickness classes of Krakatau ash over Ujung Kulon, with boundaries of major geomorphological units and some ecological units. The differences in thickness strongly reflect erosion regimes. Scale bar indicates 5 km.

After Hommel 1990a. With permission of the publishers

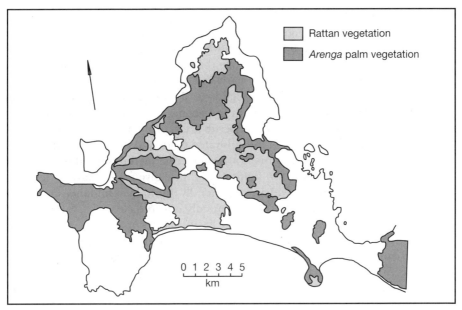

Figure 20.4. Areas dominated by *Arenga* palms (dark shade) and rattans (light shade) on Ujung Kulon. Scale bar indicates 5 km.
After Hommel 1990a. With permission of the publishers

Vegetation

The peninsula is dominated by semi-evergreen rain forest (p. 190). There is a variety of vegetation types, but many are dominated by abundant *Arenga* palm, or dense rattan scrub (fig. 20.4). The palm forests are secondary growth and have an overstorey of trees but a sparse understorey. Such palm-dominated forests are uncommon in Malesia, but have also been observed in Ujung Genteng (M. van Balgooy pers. comm. in Hommel 1987), Nusa Kambangan (Detmer 1907), and Siberut (Whitten 1982). Exactly what factors result in the growth of such forests is not known. They are clearly not a normal successional stage or they would be more common, and their occurrence does not correlate with a particular soil type. The possibility of them arising as a result of the ash falls from Krakatau is negated by a paper written 30 years before the eruption in which the forests described are very similar to those seen today (Anon. 1854). It is also not possible that they are caused by browsing pressure from rhino and banteng because these forests are found on P. Panaitan where rhino and banteng are absent. It has been observed, however, that strong dominance by one tree species in a tropical forest is generally interpreted as a sign of former disturbance (van Steenis and Schippers-Lammerste

1965) (p. 478).

The scrub dominated by rattans and other lianas and vines is almost impenetrable. Some boundary between these scrublands and forests are very clear, but there is no evidence that this is a result of soil differences. Rather, they appear to be an early but persistent successional stage following forest clearance. Such blankets of vines may completely envelop small trees, at best causing them to become stunted, and at worst killing them (van Steenis 1939). These scrub areas probably originated after the Krakatau ash had fallen on the cultivated fields and the human inhabitants were evacuated. Fresh ash is not a conducive medium for the growth of seedlings, and so these ash fields would have given an advantage to light-loving climbing plants which are generally found in abundance at a forest edge and were able to grow horizontally into the clearings. Some of the species would have been browsed by the larger mammals, but not the thorny rattans (Hommel 1987, 1990b). Once a substantial growth has built up, further succession is difficult because very little light penetrates (van Steenis 1939). Thus the distribution of the rattan scrub probably reflects quite closely the pattern of agricultural fields in use around 1883 (Hommel 1987, 1990b).

P. Panaitan was very little known until 1951, when the Bogor Botanical Gardens organized an expedition there. The vegetation was found to comprise many deciduous species such as *Pterocymbium javanicum* (Ster.), *Urandra secundifolia* (Icac.), and *Lagerstroemia speciosa* (Lyth.) (Hoogerwerf 1953). The vegetation of Peucang has also been investigated, and the most prominent species in the central mixed forest were *Parinari corymbosa* (Chry.), *Lagerstroemia speciosa*, *Bombax valetonii* (Bomb.), and *Sterculia macrophylla* (Ster.) (Kartawinata 1965; Kartawinata and Apandi 1977).

Animals

By virtue of its size, low altitude, and relative lack of disturbance, Ujung Kulon has a very rich fauna which is probably the most complete remaining on Java. It is worth noting, however, that not all species of lowland forest animals are abundant throughout. For example, both the Javan gibbon and the surili leaf monkey are rare. The fauna of the peninsula is described in great detail by Hoogerwerf (1969; 1970), and its most famous inhabitants, the rhino and banteng, are discussed elsewhere (pgs. 750, 747).

Social and Conservation Issues

Ujung Kulon National Park is currently inhabited by about 400 families distributed among a number of small settlements mainly in the Mt. Honje region. Most of the families are engaged in dry-field farming, the remainder being primarily fishermen or coconut gatherers. Their presence and their activities are strictly illegal, and a decade ago nearly 100 families were moved to a settlement 50 km to the north. They returned gradually, however, in order to reunite the extended families in their traditional, and in many cases inherited, homes. As a compromise the Park authorities agreed they could have a temporary reprieve, but only on the condition that no forest was cut, no new fields were established, no increase in the number of people occurred, and that they assist in ensuring the security of the Park. Unfortunately, compliance has not been achieved and, what is more, the inhabitants have been implicated in the sale of timber, pumice stone, and even wild animals, including rhino and banteng (Lant and Rusli 1991).

The people themselves find it well nigh impossible to understand what serious damage they can do to the thickly forested and uninhabited Park, and this hinders negotiations. All they want, they claim, is to lead a fulfilling life and provide food, clothing, shelter, and health care for their families from the toil of their hands. It is not clear what options the Government is willing or able to offer (Lant and Rusli 1991); it appears very able and willing to resettle people who have the misfortune to be in the way of development projects, but less willing and able when it comes to unique, irreplaceable, and dwindling natural assets. The declared buffer zone along the western edge of the Mt. Honje forest block is very similar in forest composition to that of the neighbouring wilderness zone and is used by almost all the same species of animals. If its conservation is not to be lost, the human activities within it must be controlled (Wiriosoepartho 1985). It is sometimes said that communities adjacent to important conservation areas are helped if the boundaries are clearly marked on the ground. A few years ago records showed that 196 of 660 border markers on the east of Ujung Kulon National Park had been destroyed (Lant and Rusli 1991). The manner in which the posts were erected probably influenced the peoples' attitude.

Conservation activities over the years have been assisted by the World Wide Fund for Nature and, most recently, by Minnesota Zoo, which has 'adopted' Ujung Kulon and provided a variety of relevant equipment.

RAWA DANAU

The ancient volcanic caldera bottom known as Rawa (or Ranca) Danau was gazetted as a reserve in 1921, and the surrounding Tukung Gede hills to the north and east (caldera rim) followed in 1980 (fig. 20.5). The reserve areas cover 2,500 and 1,700 ha, respectively. Rawa Danau is regarded as a very important conservation area because within it is the last major area of freshwater swamp forest in Java (p. 489). Nowadays, just over half of the

Figure 20.5. View taken early this century over the Rawa Danau from the caldera edge of the Tukang Gede hills.

By G.F.J. Bley, with permission of the Tropen Instituut, Amsterdam

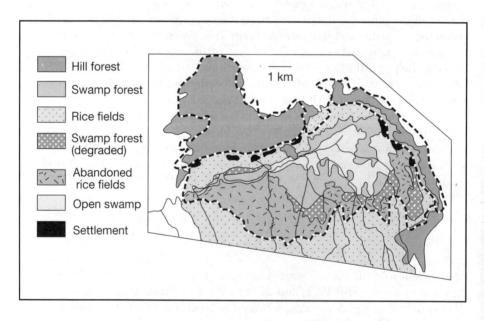

Figure 20.6. Current land use in and around Rawa Danau Reserve. Scale bar indicates 1 km.

After Melisch et al. 1993

gazetted area is in a more or less natural state, and all manner of uncontrolled human activities continue (fig. 20.6) (Melisch et al. 1993). Ten years ago it was found that open water accounted for only 0.4% of the 2,500 ha nature reserve; 45% or 1,110 ha was swamp forest, 25% or 633 ha was reed beds of *Phragmites karka*, and 30% or 714 ha was sawah, scrub, and gardens (Biotrop 1986). This is similar to the current situation.

Vegetation

The swamp vegetation of Rawa Danau contains many species now virtually extinct elsewhere in Java, such as the tree *Elaeocarpus littoralis* (Elae.), wild mango *Mangifera gedebe* (Anac.), *Alstonia spathulata* (Apoc.), *Stemonurus secundiflora* (Icac.), and a huge sedge *Thoracostachyum sumatrana* (Cype.). Others are very rare on Java, such as the sedge *Machaerina rubiginosa* (Cype.), the aroid *Cyrtosperma merkusii* (Arac.) with characteristic large tubers, and floating water plants such as *Hydrocharis dubia* (Hydr.) and water chestnut *Trapa maximoviscii* (Trap.). In addition to the native species there are also many exotic species, the most noxious of which are water hyacinth *Eichhornia crassipes* (Pont.), *Salvinia molesta* (Salv.), and, most recently, *Mimosa pigra* (Legu.) (Endert 1932a; van Steenis 1938; Melisch et al. 1993). The last of these has the potential to become a very serious pest, being favoured by fires and capable of forming impenetrable thickets. It is interesting that the large areas of abandoned rice fields have no woody plants regenerating within them except for this pernicious weed. Every effort should be made now to check its spread. A native water plant, the large lance-leafed *Hanguana malayana* (Hang.) is only locally common but may become a weed. Interestingly, some of the plants found in the Rawa Danau forests are more typical of coastal areas, such as the mangrove fern *Acrostichum aureum* (Adia.), *Barringtonia racemosa* (Lecy.), *Cerbera manghas* (Anac.), and the spiny palm *Oncosperma tigillarium* (Palm.), possibly indicating an earlier link with the coast (Melisch et al. 1993).

Four major vegetation types can be recognized in Rawa Danau: mixed swamp forest, swamp forest dominated by *Ficus retusa*, herbaceous swamp forest, and various types of secondary vegetation. Most of the mixed swamp forest has already been lost and that which remains is generally low (less than 25 m) with only occasional trees reaching 30 m.

Animals

The mammals in the Rawa Danau and adjacent Tukung Gede reserves include the globally threatened surili leaf monkey, possibly the Javan gibbon, and a wide range of other species indicative of relatively undisturbed forest, although some of them must be present in very low numbers. The continued occurrence of the fishing cat *Felis viverrina* is also noteworthy, even if it is not endemic, since this appears to be its only Javan locality outside isolated (and rapidly disappearing) pockets of mangrove forest (p. 387). Among the birds the endemic black-banded barbet *Megalaima javanensis* is quite easily seen and there are at least 23 species present whose status justifies concern (Melisch et al. 1993). In addition to many common fish species, Rawa Danau also supports the Javan endemic *Rasbora aprotaenia*. There also appears to be a rich freshwater mollusc fauna, and the large apple snail *Pila ampullacea* is consumed locally.

Social and Conservation Issues

Interpretation of aerial photographs confirms that nowhere in the entire conservation area is there closed canopy forest, indicating that illegal felling has been, and probably is, common (Melisch et al. 1993). Even 60 years ago illegal encroachment into the reserve was rife (Endert 1932a). Rawa Danau has been the subject of a wide range of hydrological, biological, and conservation studies (listed in Melisch et al. 1993), and it exists under the constant threat from local, illegal farmers and from plans to raise its level to provide a more dependable water supply for the Cilegon industrial estate. Plans to construct a dam at the outlet of the Cidanau have been active for over a decade (BIOTROP 1986). An environmental impact assessment was conducted, but since its report failed to recognize the unique nature of the reserve, the presence of globally threatened species, and the inevitable impacts of raising the water level, plans for the dam were rejected. In addition to direct human influences, the reserve is also threatened by fertilizer runoff and soil erosion from the illegal and legalised agricultural areas within the caldera. Impacts have not been as great as one might expect because some local taboos are still active, such as that against the use of motorboats on the open water; as a result, the only form of transport is dugout canoes made from *Gluta renghas* (Anac.).

If a dam is not constructed and Rawa Danau survives, there is an urgent need for management – not so much of the natural environment but rather of the surrounding area and its population so that negative impacts on the reserve are reduced. Proper boundary demarcation after due consultation, together with active patrolling including local people are needed, and this would allow natural regeneration of the abandoned agricultural land to proceed. Pressure to raise the water level may become intense, however, in order that Cilegon's industrial estate can gain access to more of the reserve's water. If changes were permitted to the water regime of the reserve, it could only be justified if every other conservation need were met in terms of management, rehabilitation, and increasing public awareness. It may be significant that Rawa Danau is a focus of concern for environmental NGOs within and outside Indonesia.

PULAU DUA

This 30 ha reserve in Banten Bay, north of Serang, was gazetted in 1937 to protect the dense roosting and nesting colonies of seabirds. In 1984 the boundary of the original 8 ha nature reserve was redrawn to include all the mangrove forest that has grown between the 'island' and the mainland, a further area to the south which should act as a buffer, neighbouring small P. Satu, and the intervening mudflats to the east (Milton and Marhadi 1984b).

Vegetation

Pulau Dua supports an open forest with mangrove trees primarily around the periphery, a central forest of *Sterculia foetida* (Ster.) and *Erythrina variegata* (Legu.) with *Diospyros maritima* (Eben.) in the north, and *Hibiscus tiliaceus* (Malv.) and *Tamarindus indicus* (Legu.) bushes in the open areas.

Animals

Although about 50 bird species are known from the island (Milton and Marhadi 1986), it is for the colonial-nesting species that the island is famous. A detailed survey of the nesting population of birds on P. Dua was conducted during 1985, and 7,308 nests of the 11 colonial species were recorded. Using data and descriptions of the breeding birds in earlier years (Hoogerwerf 1935, 1948a, 1953b), it was possible to determine which species had increased, decreased, remained stable, or disappeared (Milton and Marhadi 1985, 1986) (table 20.1, fig. 20.7).

Social and Conservation Issues

Human disturbance has been an important factor in the cessation of breeding of the black-headed ibis, milky stork, and oriental darter; their last breeding year (1975) was the first year that overland access to the island became possible. Since then, the 'island' has become increasingly popular as a tourist and study area, particularly during the breeding season. It is unfortunate that even well-formulated management plans can be undermined by undisciplined tourists who get a thrill from watching a colony rise from the trees after being disturbed, and more trained staff are needed to contain the threat these unthinking people pose. The effects of humans are clear because the density of nests is lowest in the areas of greatest human activity, and the main walking trail passes by the large trees that would be favoured by the birds listed above that no longer breed on the island. Another source of disturbance used to be the collecting of tree branches as fuelwood by local people, but improved guarding has now minimized this. The birds may also face the possibly that P. Dua is slowly eroding – between 1-3 ha of the original 8-10 ha reserve have been lost since 1937 (Milton and Marhadi 1985, 1986).

If further degradation of the reserve and its diverse bird population is not to continue, then entry into the reserve for tourists, fishermen, and fuelwood collectors will, unfortunately, have to be completely prohibited during the breeding season (mid-January to mid-July). A PHPA/WWF team proposed in 1986 that visitors should be restricted to a tall watchtower at the south of the reserve outside the *Avicennia* forest. It also proposed that coastal structures would have to be constructed to prevent erosion, and that the treeless part of the central area should be restored by planting it with appropriate species.

Table 20.1. Status of colonial nesting birds which have been recorded as breeding on P. Dua. The 'Cormorants' include both *P. niger* and *P. sulcirostris*. *- but still roosts.

		Approx. no. of nests in 1985	Status
Purple heron	*Ardea purpurea*	0	none*
Oriental darter	*Anhinga melanogaster*	0	none* 1975
Milky stork	*Mycteria cinerea*	0	none* 1975
White spoonbill	*Platalea leucorodia*	0	none
Black-headed ibis	*Threskiornis melanocephalus*	0	none* 1975
Pacific reef egret	*Egretta sacra*	<10	same
Grey heron	*Ardea cinerea*	11	down
Great egret	*Egretta alba*	35	down
Glossy ibis	*Plegadis falcinellus*	40	same
Plumed egret	*Egretta intermedia*	63	down
Black-crowned night heron	*Nycticorax nycticorax*	213	? down
Javan pond heron	*Ardeola speciosa*	978	up
Little egret	*Egretta garzetta*	1,121	up
Cormorants	*Phalacrocorax* spp.	1,286	up
Cattle egret	*Bubulcus ibis*	3,649	up

Source: Milton and Marhadi (1984b, 1985, 1986)

Figure 20.7. Three colonial-nesting bird species which used to breed on P. Dua, but do so no longer: milky stork (above), white spoonbill (centre), and black-headed ibis (below).

By A. Hoogerwerf, with permission of the National Museum of Natural History, Leiden

PULAU SERIBU

The potential of these 110 low islands, scattered in an irregular band up to 80 km north of Jakarta, to become a national park was recognized in 1979 (Kvalvågnaes and Halim 1979b) since very few countries in the world have such excellent marine resources so close to the national capital. There had been, however, a gubernatorial decree in 1970 intended to protect the coral reefs and surrounding waters from excessive fishing, but this received no enforcement. They do not provide the best diving in Indonesia (Muller 1992), but they are one of the best patch-reef systems, and it was felt that it would be possible to ensure valuable recreation for domestic and foreign tourists as well as food and commercial benefits for local residents and visiting fishermen. This could only be achieved, however, by adopting a broad and integrated approach dependent on cooperation between government agencies and the private sector, and a certain amount of compromise on all sides. Only then could long-term benefits for all users be provided (Kvalvågnaes and Halim 1979b). It was further recommended that all the islands should be considered as part of the conservation area but that they should be zoned for different uses (fig. 20.8). It was further suggested that PHPA should administer all existing and future leases of island property, preventing any private vacation uses.

Social and Conservation Issues

Although Pulau Seribu was established as a reserve and proposed as a marine national park back in 1982, it was only in 1992 that it was formally made a marine national park, although the necessary decree has still not yet been issued. The process of establishing the park was made with no significant input from the more than 10,000 people living within its boundaries, and its management staff were a lower echelon than that of other parks, with the result that the head of the park was not able to work at the necessary level within local government (Yates 1993). It is not surprising, then, that it took over a decade to reach agreement on the management of the islands. During and before this period, the reefs have suffered degradation directly from bombing and manual mining, as well as overfishing and the collection of reef fishes and invertebrates. As these species have become rarer, so those seeking them have used increasingly destructive collecting methods. With agreement in place, it is now possible for PHPA to acquire islands that are currently used for commercial purposes in the Protection Zone, and other changes in status must be made and be made known to the range of users and inhabitants. International shipping will be banned from the entire area, and restrictions on certain vessels will be made in the core and protection zones. Enforcement is intended to be a cross-sectoral responsibility. Education and awareness are also high prior-

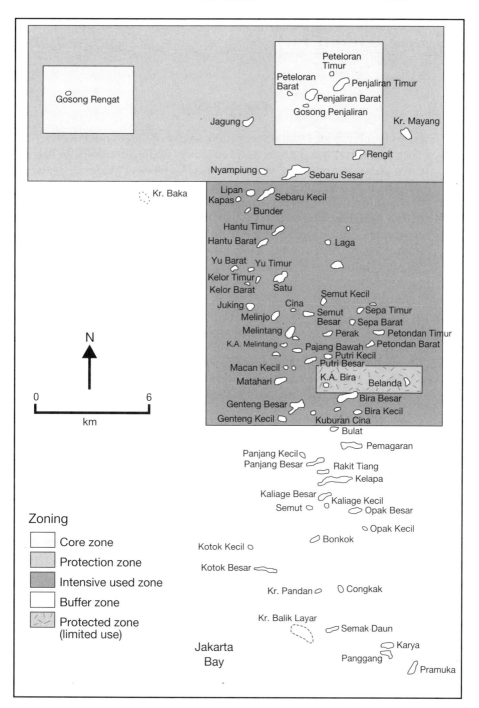

Figure 20.8. Pulau Seribu and its management zones.

After Hutomo et al. 1993

ities, and the World Wide Fund for Nature has been assisting with a visitor centre (Hutomo et al. 1993).

PULAU RAMBUT

Pulau Rambut is an 18 ha reserve at the southern end of the Pulau Seribu, close to Jakarta, it is a nesting and roosting site for thousands of herons, egrets, terns, cormorants, bitterns, ducks, kingfishers, darters, occasional pelicans, ibises, and other wading birds, including the rare milky storks *Mycteria cinerea* (Suwelo 1973; Ganesia 1988) for which this island is the only known breeding site in Java or Bali (Silvius and Verheught 1989), and good views can be had of many of them from the toll road to Soekarno-Hatta airport (which itself claimed about 3,000 ha of wetlands), west of Jakarta (Holmes 1988; Whitten 1989).

Vegetation

More than half the island comprises *Rhizophora* (Rhiz.) mangrove trees around a core of dryland forest mainly comprising *Sterculia foetida* and *Chisocheton pentandrus* (Meli.). In the southeast there is a sandy beach with a belt of *Casuarina equisetifolia* (Casu.) trees (fig. 20.9) (Kartawinata and Walujo 1977; Lambert and Erftemeijer 1989).

Animals

The island has five colonies of nesting birds, each with its own distinct habitat structure and bird composition. Breeding occurs year-round, but most species breed during the rainy season (Pakpahan 1992). In addition to the birds, up to 20,000 flying foxes *Pteropus vampyrus* roost on the island, and there are said to be a few fishing cats *Felis viverrina*.

Social and Conservation Issues

Pulau Rambut is a popular and rewarding destination for birders, but it faces a variety of threats – the ponds and fields that were rich feeding grounds for the birds to the south are disappearing under the asphalt and concrete of industrial and housing estates, and the cossetted greens of golf courses. A good breeding site has no importance if its animals cannot feed. The island also suffers from visitor pressure, and the beaches are decorated with assorted jetsam from passing ships, Jakarta, and surrounding areas – plastic bottles, plastic bags, plastic sandals, and tins. Groups of nature lovers occasionally clear up the mess, but such efforts have no real

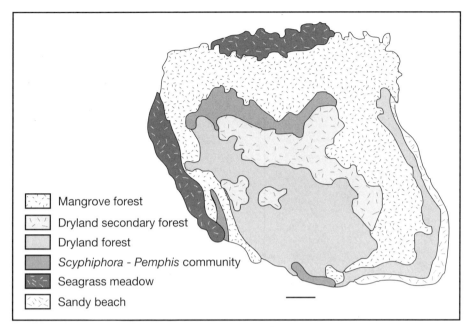

Figure 20.9. Vegetation map of Pulau Rambut. Scale bar indicates 100 m.
After Kartawinata and Walujo 1977

meaning since the next storm or high tide will bring a new load of objects (Sukardjo and Santoso 1982). It is also disturbed by fishermen gathering the large conical shells *Telescopium telescopium* for bait (Lambert and Erfte-meijer 1989).

The integrity of Pulau Rambut depends primarily on the uses to which the feeding grounds of the birds along the northern shore are put. If they are to be converted entirely to real estate and golf courses, then Pulau Rambut has no future.

DUNGUS IWUL

Dungus Iwul is a minute, 9 ha reserve which lies 9 km beyond Jasinga, just off the Bogor-Rangkasbitung road. Although insignificant in area, it has major significance in being one of the very few remnants of low-altitude lowland tropical rain forest left on Java. 'Dungus' is Sundanese for a forest remnant, and 'iwul' for the beautiful, grey-trunked *Orania sylvicola* palms

Figure 20.10. *Orania sylvicola*, the most common tree at Dungus Iwul Reserve. Scale bar indicates 2 m.

After Koorders 1913

Figure 20.11. A view inside Dungus Iwul showing the grey trunks of *Orania sylvicola* palms.

By A.J. Whitten

(figs. 20.10-11) which are extremely abundant (Riswan 1975). The large size (higher than 20 m) and hard trunk of this tree generally mean that woodcutters leave the palm alone when taking wood, opening up spaces in which it can grow. It is found in Malaya and Java on undulating lowlands up to about 300 m. It has a very poisonous fruit and cabbage, with just one fruit being potent enough to kill an elephant (Whitmore 1977). It is now rare in Java because so very little lowland rain forest remains.

Vegetation

The main trees other than those mentioned above are *Aporusa arborea* (Euph.), *Blumeodendron tokbrai* (Euph.), *Bouea macrophylla* (Anac.), *Croton argyratus* (Euph.), *Dacryodes rugosa* (Burs.), *Dialium indum* (Legu.), *Dysoxylon alliaceum* (Meli.), *Pentace polyantha* (Tili.), *Gironniera subaequalis* (Ulma.), *Nephelium lappaceum* (Sapi.), and *Pometia pinnata* (Sapi.) (Riswan 1975). Dungus Iwul is the only place in Java where plants characteristic of Sumatran forests, such as the trees *Baccaurea sumatrana* (Euph.), *Tetradium alba* (Ruta.), *Xylopia malayana* (Anno.), and the herb *Dissochaeta gracilis* (Mela.), appear to grow (Riswan 1975).

Animals

The forest has some of the more common lowland forest birds, but most remarkable is the group of endemic lutung monkeys living within the forest with occasional forays into the surrounding rubber plantation.

Social and Conservation Issues

By comparing recent data with those collected by Hildebrand (1939), it can be concluded that, while *Orania sylvicola* and *Nephelium lappaceum* are still the most common trees in the reserve, the other species have become less abundant. The most likely reason for this is tree felling by local people. Interestingly, early reports do not mention forest damage (Endert 1932b; Hildebrand 1939; van Steenis 1939), whereas now some form of damage can be seen wherever one looks. A major source of this disturbance is the collection of *Dialium indum* fruit which are produced every few years. Since monkeys are very partial to these fruit too, people fell all the trees around each *D. indum* so that the animals cannot reach the tree without going across the ground, something which the group of leaf monkeys still present in the forest (p. 702) at least, are loathe to do. Even though human population density around the reserve is rather low by virtue of the surrounding rubber plantation, it seems as though the forest is slowly degrading. A period of monitoring is required to assess the current extent and causes of the damage and, based on that, proposals should be made to stem the problem.

MT. HALIMUN

Mt. Halimun National Park, 36,000 ha, embraces the largest area of ever-green rain forest in Java (Wind and Soesilo 1978). Only about 20% of the park comprises lowland forest, and this exists in small, discontinuous patches. More lowland forest is found around the park than within it (fig. 20.12). The area has hydrological importance since rivers such as the Cisadane, arising here, feed important industrial areas west of Jakarta. In the centre of the eastern part of the Park is the enclave of the Nirmala tea plantation established during the Pacific War but abandoned soon after-wards. It was opened again in 1973 and has a lease valid until 2002. To the west of the park are found the interesting Baduy (p. 340) and Kasepuhan people (p. 338). Two of the park's peaks are called *Halimun,* a word which in Sundanese means 'cloudy' or 'misty', and indeed much of the forest is enveloped in cloud for most of the year. The fact that the forests of Hal-imun descend to about 1,000 m, whereas the forests around the young and isolated cones of Mt. Gede Pangrango and Salak do not, is due to the rugged terrain of steep and broken ridges, a relatively sparse surrounding human population on rather old volcanic soils, and local superstitions about the dark and cloudy forests.

Mt. Halimun has been very much a poor cousin to West Java's other large forest areas such as Ujung Kulon and Mt. Gede-Pangrango, despite its proximity to Indonesia's biological capital Bogor (40 km) and political cap-ital Jakarta (100 km), and as a result, the literature base about the area is poor. Indeed, only 30% of biology students questioned at the University of Indonesia in 1992 knew where Mt. Halimun was located (Supriatna 1992). Nevertheless, the area has remarkable potential for scientific and social studies as well as various forms of recreation (Amir 1992; Supriatna 1992). It was afforded the status of a national park in 1992, and a permanent field station may be erected there in due course.

Vegetation

Early surveys of the lowland forests found large trees of *Dipterocarpus hasseltii* (Dipt.) and *D. gracilis* at 600-700 m, but these species have disappeared as the lower forests have been converted and faced degradation. *Toona sinensis* (Meli.) was also common at lower altitudes but is now very rare (Wind and Soesilo 1978). The montane forests are very rich and are typical of mon-tane forests as described elsewhere (p. 511), although some species are found only in restricted areas (Dransfield 1971).

Animals

Because of their relatively large area, the Halimun forests are probably of

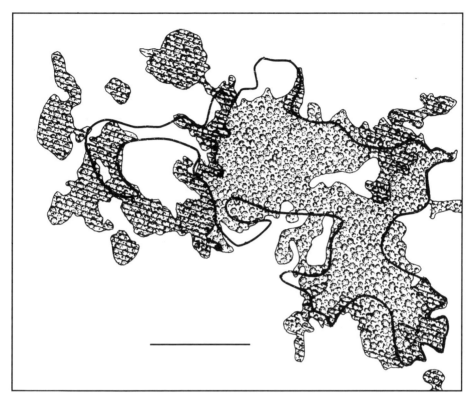

Figure 20.12. Mt. Halimun National Park showing areas of forest within and outside the boundaries. Lowland forest is shown shaded. Scale bar indicates 10 km.
After RePPProT 1989

major importance to many of Java's endemic species – for example, 25 of Java's 30 endemic birds are found there. Unfortunately, because the patches of strictly *lowland* forest are discontinuous, larger animals such as the endemic gibbon and surili leaf monkey, which are dependent on primary forest below 1,250-1,350 m, are not as secure as it may seem. The last reliable reports of tiger were in 1972 (Wind and Soesilo 1978).

Social and Conservation Issues

The people living in small villages around and in the reserve have strong traditions related to forest use, characterized by a long-term perspective in shifting cultivation, and beliefs in forest spirits which can cause sickness in

Figure 20.13. Rice fields adjacent to Mt. Halimun National Park.

By Gerald Cubitt, with permission

people whose behaviour does not conform to the cultural norms (fig. 20.13) (Saleh and Munir 1990; BScC 1992). To the west of the Halimun forests is the Citorek Valley where Kasepuhan people adopted wet-rice cultivation only in about 1900 because their religion forbade the use of agricultural tools (Wind and Soesilo 1978; Adimihardja 1992; Adimihardja and Iskandar 1993). As a result of human pressure from local people and illegal commercial enterprises, the fringes of the park have suffered from widespread illegal cutting of economic timber of *Magnolia blumei* (Magn.), *Cinnamomum sintok* (Laur.), *Dacrycarpus imbricatus* (Podo.), *Schima wallichii* (Thea.), and *Altingia excelsa* (Hama.) (Sukardjo 1982a).

It would seem that the boundaries of the park should be redefined to include as much lowland forest as possible, perhaps by absorbing some adjacent Perhutani land, and to exclude most intensively cultivated land where reclamation is unrealistic. There should be inviolate corridors at the lowest possible altitudes between sections of the park. Part of the northern boundary of the park appears to be threatened by the planned Aneka Tambang gold mine (Marr 1993), and disturbance to the park has long been caused by people around the Cikotok mine to the southeast which was opened in 1954 and is expanding up the Ciusul-Ciawitali Rivers (Munir

1987). The mine operations themselves may cause relatively little direct impact; it is rather the focus of apparently uncontrollable activity they represent, and the lack of any serious management, participation, and integration which threaten the integrity of the park. An environmental impact assessment had been commissioned from the Indonesian Institute of Sciences' Biology Research and Development Centre, but the field team was given just one day at the location to conduct their studies when, unfortunately, it happened to rain.

MT. GEDE PANGRANGO

Twin volcanoes form the Mt. Gede-Pangrango National Park (2,958 and 3,019 m, respectively) which extends over 14,000 ha, is aesthetically exceptional, and includes some excellent montane forest and subalpine vegetation. There is a small area of lowland forest in the southwest but this is much disturbed. Having been the site of many zoological and botanical studies, it is of exceptional scientific interest in tracing long-term changes. Some animals and plants are known only from these mountains, but this probably does not reflect their true distributions (p. 204). The park is very heavily visited, being so close to Jakarta, Bogor, and Bandung, and it is the only conservation area in Indonesia which tries to limit the number of visitors allowed to enter.

The Mt. Gede-Pangrango area has a very special place in the history of ecology and conservation in Indonesia. The Cibodas Gardens at the north east, established in 1830, played an important role in the introduction of cinchona to Indonesia for malaria control (Taylor 1945), and the forest above that, part of which was established as Indonesia's first nature reserve in 1889, has been the site of more ecological research than probably anywhere else in Indonesia. This makes it a 'treasure for international science' (van Steenis 1972), as was confirmed when it was made an UNESCO-MAB Biosphere Reserve in 1977.

There is now a guidebook to the Park (Harris 1994), and a canopy walkway should be open in 1996.

Vegetation

The vegetation of the park is described in detail elsewhere (p. 511), although it is interesting that the vegetation of active Gede and extinct Pangrango volcanoes are quite distinct, with the latter having deeper and better-developed soils with few pioneer plants. The grassy plains or *alun-alun* in the subalpine zone have no forest, possibly as a result of waterlogged soils, fires, and frosts that can occur in such depressions.

Animals

The bird community of Mt. Gede-Pangrango National Park is very well known (Hoogerwerf 1949a; Andrew 1985), with 245 species now recorded, including 25 of Java's 29 endemic species, and interesting informal observations are being made by visitors almost continuously. Although a comprehensive ecological analysis would now be possible, it has not yet been attempted. The loss of bird species from the park is discussed elsewhere (p. 712).

One hundred and fifty years ago there were still rhinoceros, tiger, banteng, and Javan deer in the area of the park (Junghuhn 1854), but these have long since gone. Leopards are now extremely rare, and the binturong or bear cat *Arctictis binturong* and wild dog *Cuon alpinus* are probably no longer present (table 20.2). The smaller carnivores, such as the stink badger *Mydaus javanensis*, leopard cat *Prionailurus bengalensis*, and yellow-throated marten *Martes flavigula*, are probably still quite common, if rarely seen. Perhaps the most threatened of the animals are the endemic surili monkey and Javan gibbon, both of which depend on the lower-altitude forests, such as towards Cibadak in the southwest. A gibbon was once reported from 2,400 m (Docters van Leeuwen 1933), but this is exceptional.

Social and Conservation Issues

It is likely that no other mountain in Indonesia is climbed as much as Mt. Gede even if not all the 30,000 annual visitors actually reach the summit, and as such there is considerable scope for education, although this is not being completely exploited at the moment. This interest in the mountain causes problems of trail wear and graffiti, but the authorities take action by changing the route up the mountain to allow sections of the trail to recover.

Almost everyone who climbs the mountain does so from the Cibodas side, but there are many access routes, and it is in these other areas that the really serious problems of illegal cultivation occur. Many local people enter the forest, and the trail system, even immediately above the Botanical Gardens, is enormously complex as a result. Apart from forest fruits, popular quarries are tree ferns and bird-nest ferns which are uprooted and sold to visitors in the Puncak area. Conservation of this park suffers, perhaps, from being overly focussed on the Cibodas area while damage is being wrought in other parts virtually unnoticed. Those who have studied the problem suggest that in order to make an effective buffer zone around Mt. Gede-Pangrango, different forms of agroforestry systems need to be introduced to induce the villagers not to enter the park to fell trees and gather produce otherwise unavailable to them (Mukhtar and Pratiwi 1986).

Table 20.2. Mammals (excluding bats) recorded from Mt. Gede-Pangrango. Bold type indicates species lost before 1929 (*), and since 1929 (•), '#' indicates species whose loss in the future seems likely.

Common tree shrew	*Tupaia glis*
Javan tree shrew	*T. javanica*
Short-tailed shrew	*Hylomys suillus*
House shrew	*Suncus murinus*
White-toothed shrew	*Crocidura fuliginosa*
Sunda shrew	*C. monticola*
Javan mountain shrew	*C. orientalis*
Flying lemur	*Cynocephalus variegatus*
Surili leaf monkey	*Presbytis comata*
Lutung leaf monkey	*Trachypithecus auratus*
Javan gibbon	*Hylobates moloch* #
Black giant squirrel	*Ratufa bicolor*
Black-banded squirrel	*Callosciurus nigrovittatus*
Three-striped ground squirrel	*Lariscus insignis*
Red giant-flying squirrel	*Petaurista petaurista*
Pencil-tailed tree mouse	*Chiropodamys gliroides*
Javan flat-nailed rat	*Kadarsanomys sodyi*
White-bellied rat	*Leopoldamys sabanus*
Javan mountain spiny rat	*Maxomys bartelsii*
House mouse	*Mus musculus*
Javan volcano mouse	*M. vulcani*
Wooly mountain rat	*Niviventer lepturus*
Polynesian rat	*N. bukit*
Javan mountain rat	*N. cremoriventer*
Javan monkey rat	*Pithecheir melanurus*
Java tree rat	*Rattus exulans*
House rat	*R. rattus*
Javan giant rat	*Sundamys maxi*
Wild dog	***Cuon alpinus*•**
Javan weasel	*Mustela lutreolina*
Yellow-throated marten	*Martes flavigula*
Stinkbadger	*Mydaus javanensis*
Common palm civet	*Paradoxurus hermaphroditus*
Binturong	***Arctictis binturong*•**
Small Indian mangoose	*Herpestes javanicus*
Tiger	***Panthera tigris*•**
Leopard	*P. pardus* #
Leopard cat	*Prionailurus bengalensis*
Javan rhino	***Rhinoceros sondaicus*** *
Wild boar	*Sus scrofa*
Javan pig	***Sus verrucosus*•**
Barking deer	*Muntiacus muntjak*
Javan deer	***Cervus timorensis*•**

Sources: updated from Dammerman (1929b); Boeadi (pers. comm.); R. Beckwith (pers. comm.)

MT. TILU

This is an 8,000 ha forested reserve covering old Quaternary volcanoes, and lies between 900 and 2,434 m, 25 km south of Bandung. It was originally protection forest, but was gazetted in 1978. The reserve, is surrounded by forestry and tea plantations, and villages with gardens and rice fields. Tourism is not developed in the reserve, although youngsters do climb to the craters of Kawah Putih and Kawah Ciwidey where, unfortunately, vandalous depredations can be seen (Brotoisworo et al. 1980).

Vegetation

Mt. Tilu forests are similar to those described elsewhere (p. 511) (Harun and Tantra 1984). Botanical surveys have not been detailed, but a little-known relative of parasitic *Rafflesia* (p. 172), *Rhizanthes zippellii*, once thought to occur only on Mt. Salak (Backer and Bakhuizen van den Brink 1963), was found growing in the lower montane forests by a team from the Institute of Ecology, Pajajaran University, Bandung (Lembaga Ekologi 1980). The endemic bamboo *Nastus elegantissimus* is apparently confined to the forests near Pengelengan.

Animals

The most notable of the animals are the endangered Javan gibbons and the surili leaf monkeys. Leopard and wild dog are also said to occur (Rosanto and Priatna 1982). The endemic snail *Pupisoma tiluanum* is apparently confined to the reserve.

Social and Conservation Issues

Two tea estates form enclaves within the reserve and, at least in the past, the labourers have been poorly paid and had little choice but to take timber, firewood, and game animals from the forest. The reserve faces problems of illegal encroachment, but there are inconsistencies between the formal description of the boundaries and the official map of the reserve. At issue is the status of areas of forestry plantations which connect two major blocks of natural forest. Unclear boundaries, if not resolved, will be a source of endless conflict in the future (Brotoisworo et al. 1980; Lembaga Ekologi 1980).

The northern part of the reserve has suffered from the felling of *Castanopsis* trees to make charcoal for the many small blacksmith industries in nearby Pasirjambu (Lembaga Ekologi 1980). It was estimated that these blacksmiths consumed charcoal from 14 large trees each day and that the damaged Tilu forests had less than one-sixth the basal area of the undam-

aged forests. The activities of the human population around the forests of
Mt. Tilu were also noted by Harun and Tantra (1984), and it was felt that
strict measures would be needed to protect the theft of trees used in the production of charcoal, as well as management of the surrounding forests to
supply fuelwood needs. Alternatives might include encouraging the blacksmiths to use coal rather than charcoal, and establishing fuelwood plantations in village, estate, and Perhutani areas (Rosanto and Priatna 1982).

Mt. Wayang (Mt. Limbang)

This ungazetted and little-known 20,000 ha of forest southwest of Cikajang
(south of Garut) is one of the three most important forested areas in West
Java, together with Ujung Kulon and Mt. Halimun (MacKinnon et al.
1982–where it is referred to as Mt. Limbang). A large area of forest
between the Cikajang-Pameungpeuk and Cikajang-Bungbulang roads, it
supports many species including the critically endangered gibbon and
surili leaf monkey (Asquith et al. 1995). Unless it is gazetted and properly
managed, the expansion of human activities along the southern coast are
likely to be detrimental to it. There are also rumours that at least parts of
it will be cleared for plantation forestry.

Leuweung Sancang

Although only 2,157 ha in extent, this reserve is valuable because it con-
serves a range of habitats from lowland forest (some of it on limestone), to
mangroves and seagrass meadows. About 300 ha of the reserve, between
the Cikalomberan and Cipalawah Rivers, is a rare and arguably the best
example of Javan dipterocarp forest dominated by *Dipterocarpus hasseltii*
(possibly its only remaining site in Java), with *D. gracilis* and *Dillenia obovata*
(Sukardjo 1979).

Vegetation

Dipterocarpaceae is the most common family in the lowland forest, followed by Euphorbiaceae (Sidiyasa et al. 1985). There are six species of
dipterocarp in the reserve – probably the most of any site in Java or Bali:
*Dipterocarpus hasseltii, D. gracilis, Shorea javanica, Anisoptera costata, Hopea
sangal,* and an unnamed species found by Dr. A. Kostermans from which
flowers and fruit have not yet been collected (fig. 20.14). Initially only a
single tree was known but a survey for this book in late 1992 confirmed that

Figure 20.14. The unnamed species of *Shorea* found in Leuweung Sancang.

By A.J. Whitten

it had fruited and the forest floor around it had numerous 20 cm high seedlings. Ashton (1982) in the revision of Dipterocarpaceae mentions only one Javan locality for *Shorea javanica* (Subah, near Pekalongan, where only some seedlings persist), but it has also been found in Leuweung Sancang Reserve (Sidiyasa et al. 1985). The reserve is also famous for its populations of *Rafflesia patma* (p. 172). In Leuweung Sancang *Rafflesia* is found between the beach and inland forest, where the major trees are *Tabernaemontana sphaerocarpa* (Apoc.), *Neesia altissima* (Bomb.), *Millettia pinnata* (Legu.), *Dracontomelon mangifera* (Anac.), *Barringtonia acutangula* (Lecy.), *Terminalia catappa* (Comb.).

Animals

The larger animals include a dozen or so banteng (Setiawati and Mukhtar 1990), a few leopard, hornbills, and peafowl (Hariyadi 1975), but none is common or easy to see.

Social and Conservation Issues

Most of the reserve is to some extent protected by the surrounding rubber plantation, but the coastal areas are damaged by fishermen who live in temporary and not-so-temporary shelters on the south coast. They fell timber, clear small gardens, collect turtle eggs, and collect coral fish. In late 1992 piles of discarded plastic bags were found littering a section of beach forest where fishermen had bagged up their living quarry, caught for the ornamental fish trade.

PANGANDARAN

This small, 457 ha reserve, and adjacent tourist park, is an uplifted limestone peninsula at the southeast corner of West Java and would be rather insignificant were it not for its position at the end of a beach-lined isthmus fringed with hotels and homestays. The tourist park in fact receives more visitors than any other conservation area in Indonesia – possibly 500,000 each year.

Vegetation

Although the vegetation is sometimes described as largely secondary (Sukardjo 1977, 1983), most of it may be a rather dry evergreen forest with *Vitex pinnata* (Verb.), *Dillenia excelsa* (Dill.), and *Cratoxylon formosum* (Gutt.). The flat area that was felled to build the offices and resthouse near the entrance was true rain forest with *Rafflesia patma* plants, and this represents a further loss of an increasingly rare forest type (Kostermans in Sukardjo 1983). In the north near the tourist park gate, there are also planted trees of teak, mahogany, and rosewood.

Animals

The long-tailed macaque *Macaca fascicularis* is abundant at Pangandaran, and indeed it is one of the more successful wild mammals on Java and Bali:
- it ranges both in the trees and on the ground;
- it is omnivorous and happily supplements a largely fruit diet with crabs, termites, peanuts, popcorn, and bread;
- it has a relatively high reproductive rate in large groups, and
- it is audacious in approaching people (Tilsay-Rusoe et al. 1983).

Some 22 mammal species currently survive in the reserve (Sukardjo 1983), or just one-sixth of those known from Java. The gibbon was still present in the mid-1930s (van D. 1938), but probably became extinct before Independence. Also, since 1922, when the reserve was established, wild boar, Javan pigs, otters, and leopard have all become extinct (Sukardjo 1983).

Social and Conservation Issues

The vast majority of visitors visit only the flat tourist park, with its wide paths, tame monkeys, and caves. Meanwhile, local people enter the reserve itself to collect fallen branches for fuelwood, but for full biological diversity, dead wood really should be left so that decomposer organisms (e.g., fungi and beetles) can survive. Concern has also been expressed over the future of the banteng population (believed to be 'hybrids' with Bali cattle) and the two larger deer species because of the total loss of some grazing areas and the gradual degradation of others. Two of the grazing grounds that once supported the population of larger animals have reverted to scrub since 1970 (Sukardjo 1983).

SEGARA ANAKAN

Segara Anakan is the 21,000 ha shallow bay west of Cilacap on the south coast of Java into which the Citanduy and other rivers flow (fig. 20.15). Passage of water from the bay to the ocean is obstructed by Nusa Kambangan, the long, 30,000 ha prison island (p. 800) across its mouth, such that sediment brought down from intensively cultivated agricultural areas settles in the bay and most of the water flows through the relatively broad western channel. The accretion of sediment in Segara Anakan are of two types, extension of the existing shore and the formation of new islands, both of which are quickly colonized and stabilized by mangrove forest (Hadis-umarto 1979). The term 'forest' is not entirely appropriate, however, because as soon as a tree here becomes half-way tall, it is felled for fuelwood, charcoal, or timber. Although Perhutani ostensibly manages the trees, insufficient numbers reach an economic size to permit commercial exploitation.

From an ecological viewpoint, Segara Anakan consists of a complex of mangrove and tidal swamp forests, intertidal mud flats, and a shallow estuarine lagoon. The area of mangroves covers about 13,500 ha and is thus by far the largest on Java, even if illegal land reclamation and timber theft has wrought great damage. It has most of the functions and benefits of extensive mangrove forests: flood mitigation, prevention of salt intrusion, local and offshore fisheries production, fuelwood and timber production, sediment and nutrient, marine fauna nursery ground, and a reservoir of biological diversity.

Figure 20.15. View of Segara Anakan from the guesthouse on the prison island of Nusa Kambangan.
By A.J. Whitten

Vegetation

On the seaward margin of the mangrove forest, the trees are mainly Avicennia marina (Verb.) and Sonneratia alba (Sonn.), followed inland by *Rhizophora* spp. (Rhiz.), *Bruguiera* spp.(Rhiz.), *Xylocarpus granatum* (Meli.), and *Nypa fruticans* (Palm.) (Erftemeijer et al. 1988).

Animals

Among the 85 species of birds recorded, the most notable are the threatened lesser adjutant *Leptoptilos javanicus,* milky stork *Mycteria cinerea,* woolly-necked stork *Ciconia episcopus,* and the endemic Javan coucal *Centropus nigrorufus,* and the mud flats are of international importance as a staging ground for many migratory and wintering wading birds. Among the mammals present are otters, lutung leaf monkeys, long-tailed macaques, and crested porcupines (Erftemeijer et al. 1988).

Cilacap, at the extreme east of the lagoon, is a major fishing port and it seems certain that Segara Anakan represents a very important nursery area for economic species. The areas around the lagoon might be thought to be suitable for the development of tambak, but the tidal range is not

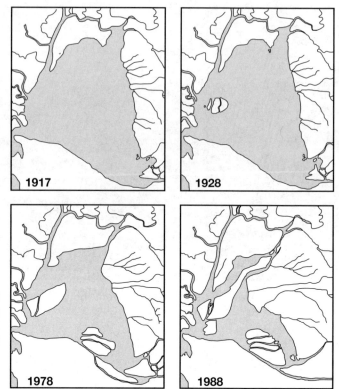

Figure 20.16.
Coastline changes in Segara Anakan between 1917 and 1988. Water is shown shaded.

After various sources in Erftemeijer et al. (1988)

wholly appropriate. In addition, the current productivity is so high, there would be little gain, and also such development would seal the fate of some of the offshore fisheries of species depending on Segara Anakan.

Social and Conservation Issues

Nowhere in Java, perhaps, are the linkages between uplands and the coastal zone more clearly seen than in Segara Anakan since high rates of sedimentation have already caused a decrease in fisheries production. For at least 20 years there has been concern over the high rates of siltation, and it had been predicted that the entire bay could be silted up by 1995 (Kvalvågnaes 1980). Although this will not be the case, it has filled up rapidly (fig. 20.16) and the area deserves special attention for its future management. This has been given by a range of institutions and projects, most recently by the ASEAN/US CRMP project (White et al. 1989; ASEAN/US CRMP 1992), which began in the mid-1980s. After due con-

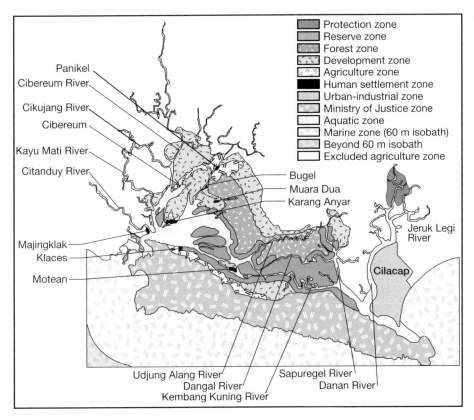

Figure 20.17. Recommended zonation of Segara Anakan.
After ASEAN/US CRMP 1992

sideration of all the uses and potential conflicts, a zonation of the area has been proposed comprising 11 zones (fig. 20.17). These are:

- Protection Zone (480 ha) – a core area of good-quality mangrove forest some kilometres from the main area that will be protected from further exploitation and degradation and that will serve as a sanctuary for fisheries and wildlife;
- Reserve Zone (4,122 ha) – an area designated for conservation to maintain the ecological function of the lagoon as nursery and feeding grounds, especially for shrimps and fishes;
- Forest Zone (2,809 ha) – a buffer between the reserve and agriculture zones to be managed actively by Perhutani, and where harvesting of wood products would be allowed;
- Development Zone (4,779 ha) – designated as a multipurpose economic development zone where local residents can be free to pursue

a variety of livelihood projects such as agriculture, aquaculture, horticulture, agroforestry, livestock, and even light industry and manufacturing;

- Agriculture Zone (1,199 ha) – an area with the ecological features to support agriculture and so has been reserved specifically for the purpose;
- Human Settlement Zone (95 ha) – designated to contain the growth of human settlements in Segara Anakan in centralized areas where government services can be most easily provided;
- Aquatic Zone (6,029 ha) – includes the lagoon and all waterways and aquaculture ponds, and allows the management and planning of fisheries within it and where overfishing and aquaculture promotion will be developed in suitable areas;
- Ministry of Justice Zone (10,258 ha) – although a prison island, Nusa Kambangan has considerable potential for tourism development. The management plan proposes developments that are compatible with the status of the island and to increase the incomes of the local inhabitants, but in early 1994 the Minister of Justice reaffirmed that no change in use or status was being considered;
- Marine Zone (22,000 ha) – an area to allow the regulation of coastal fisheries which are expected to expand in the near future.

At present (1994) no part of the area, except small reserves on Nusa Kambangan (see below), has any legal conservation status, with the result that degradation and destruction of resources continues without hindrance. Appropriate management of the area that would permit and develop nondestructive human activities is beyond the remit of PHPA alone and so a multisectoral team will be required. Not least is the problem of which office has authority over newly formed land. The provincial government should act swiftly. The plans and zones may, in any case, come to nought if approval is given to the vaunted plans of the Department of Public Works to dam off the western outlet of the lagoon in order to increase sedimentation, with a view to converting the whole area, over time, into rice fields and other agricultural land.

NUSA KAMBANGAN

Nusa Kambangan is unique in Indonesia because virtually sole authority on the island is exercised by the Prison Service of the Department of Justice. This is because of the three high-security prisons on the northern plains. As a result, access to the island is very carefully controlled. There are a number of quite large caves on the island with considerable biological and ecological significance (Ko 1983c; Suyanto 1983), and 8,000 ha of largely undisturbed lowland forest in two blocks. Nusa Kambangan currently has three gazetted reserves: a tiny offshore island ostensibly protecting *wijaya kusuma* plants (p. 175), an overgrown Portuguese fort called Karang Bolong, covering 0.5 ha, and about 2,000 ha of the forested southwest corner of the island.

Vegetation

The island supports one of the largest expanses of lowland dipterocarp rain forest in Java, and to wander here is to get some impression of what the original Javan forest might once have been like. Of great interest is the presence of the magnificent timber tree *Dipterocarpus littoralis,* a species known only from Nusa Kambangan (Ashton 1982). Early photographs of the forest have been published (Detmer 1907; Pluygers 1952).

Social and Conservation Issues

Despite the presence of the 2,000 ha reserve and of armed guards patrolling parts of the island, brazen timber thieves have been felling *D. littoralis* trees, sawing them in the forest and carrying them to the western tip of the island before loading them onto boats. There have even been reports of a remote sawmill on the south of the island. Budget cuts in recent years have reduced the area of the island that can be patrolled, and the fact that the situation is no worse is due solely to the personal interest of the current prison chief of the island. The PHPA occasionally visit the forest, but the office is in Cilacap, and budgets have not been given for regular surveys. It is clear that if these wonderful forests are not to become degraded and worthless, protection from well-organized, illegal, commercial exploitation of the timber must be stepped up, and the good relations between the prison and conservation authorities improved further by realistic conservation budgets. The entire 8,000 ha of lowland forest left on the island should become a single conservation area.

KARIMUN JAWA

The Karimun Jawa islands and surrounding seas were declared as a marine park in 1988 on the basis of their coral, seagrass, algal beds, mangrove, lolwland forest, and beach forest, as well as other touristic features such as waterfalls and an old Portuguese fort (Pemda Jateng 1988). Plans for tourist development have been drawn up, but neither involvement of the inhabitants, management activities, nor tourists are much in evidence (Wisudo and Bambang 1992).

MT. KAWI-KELUD

Some 50,000 ha of continuous forest between these two volcanoes has been proposed as a strict nature reserve because of the rare lowland and hill forest on volcanic soils on their slopes, and significant populations of threatened birds, notably the Javan hawk-eagle *Spizaetus bartelsi* (MacKinnon et al. 1982; S. van Balen pers. comm.). Unfortunately, no action has yet been taken and attrition and degradation continue. The northern slopes of the mountains formed part of the Kali Konto project area (e.g., PSL Unibraw 1984a; RINM 1985; Smiet et al. 1989).

LEBAKHARJO AND BANTUR

These ungazetted forest areas on the coast south of Malang are 13,000 and 5,000 ha, respectively, and constitute two of the most important areas of lowland forest in Java (fig. 20.18) (Bekkering and Kucera 1990). They were ignored as potential conservation areas because early maps marked them as designated teak forests, but they had not been converted. Lebakharjo is best known as the location of the world's first International Community Development Camp (COMDECA) in late 1993, organized by the Indonesian Scout Movement.

Vegetation

The vegetation of the area does not seem to have been investigated, but Perhutani reports that the trees around the camping ground include *Anthocephalus* (Rubi.), *Antidesma* (Euph.), *Artocarpus* (Mora.), *Gironniera* (Ulma.), *Lithocarpus* (Faga.), *Mallotus* (Euph.), *Pangium* (Flac.), *Pterospermum* (Ster.), and *Sandoricum* (Meli.) (R.M. Sardjono pers. comm.), indicative of both primary and secondary forest. Perhutani has planted out hundreds of thousands of rattan seedlings, but it is not known how successful this has been.

Animals

The bird fauna of the Lebakharjo forests is as rich as that of highly esteemed areas such as Ujung Kulon and Meru Betiri. The call of an argus pheasant was heard here during a survey in 1989 (van Balen 1989), and it is an intriguing possibility that this might have been the double-barred argus pheasant, known only from part of a single feather (p. 718). Other globally threatened bird species endemic to Java (and Bali), and present in good numbers, are the Javan hawk-eagle, yellow-throated hanging parrot

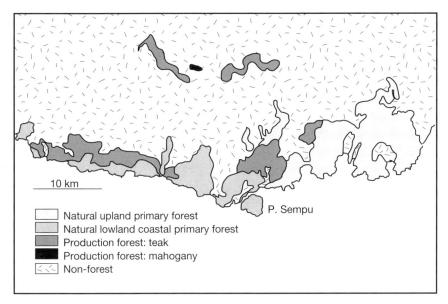

Figure 20.18. The Bantur (left) and Lebakharjo (right) forest areas, south of Malang, showing the existing Sempu Island Nature Reserve. Scale bar indicates 10 km.

After Rijksen 1989

Loriculus pusillus, black-banded barbet *Megalaima javensis,* and the white-breasted babbler *Stachyris grammiceps.*

Social and Conservation Issues

The richer of the two areas, Lebakharjo, is threatened because in 1987 certain parts had their status as Protection Forest changed to Production Forest, paving the way for conversion. Various management plans have been produced by Perhutani, covering conversion to plantations of *Paraserianthes falcataria* and mahogany *Swietenia macrophylla* trees, or development of a rattan estate. The 'protected status' that it had before had not been particularly effective since many coffee gardens were established in the area, and as of early 1994, forest clearance by smallholders was continuing. Both these forests should be given serious and immediate consideration as new conservation areas with adequate management (Bekkering and Kucera 1990). In 1993 the eastern forests suffered intrusion from people whose adjacent land was inundated by a cold lahar (W. Saleh pers. comm.). Perhutani has plans to develop the Pujiharjo bay for tourism, including the making of trails in the fine coastal forest.

BROMO-TENGGER-SEMERU

This 57,000 ha park, declared in 1982, is best known for the spectacular 10 km wide Tengger Caldera, within which squats the active Bromo volcano and the surrounding 'Sand Sea', and this small area was first protected in 1919. The Bromo crater has and still does play an important part in the traditional religion and life of the distinct Tenggerese people and the annual festivals are of great local significance as well as being a major attraction to domestic and foreign visitors. There is, however, much more to this park than that seen by most tourists (fig. 20.19).

Vegetation

The Sand Sea, near where the majority of visitors cross to the Bromo crater, is largely bare of vegetation because the 'soil' is new and extremely porous, although in places a number of sand-binding plant species can be found (Beudels and Hardi 1980). Towering over the Tengger Caldera is the active ash cone of Mt. Semeru (3,676 m), Java's highest and most consistently active volcano, which has some fine undisturbed lower and upper montane forests and *Casuarina* forests on its slopes. Accidental and deliberate fires have created grasslands containing mainly indigenous species on many of the slopes (fig. 20.20). A vegetation survey is required to identify communities of special interest (Watling 1990). Some endemic species of plants are known only from the Tengger area (table 4.4).

Animals

Despite there being extensive grasslands, there are few grazing animals, partly due to hunting, and partly due to lack of access to salt (p. 809). The fauna is poorly known, although a bird list is available (Beudels and Hardi 1980) and additions are being made (van Balen 1992). Some endemic species of snakes, moths, stick insects, damselflies, and snails appear to be confined to the Tengger area (chap. 5). Surveys of invertebrates, reptiles, amphibians, birds, and smaller mammals are almost certain to reveal new records if not new species, although the larger mammals are conspicuous by their absence.

Social and Conservation Issues

The general paucity of the wildlife is due primarily to the long history of human activity in the northern sections of the park, with the enclave onion-growing villages of Ngadas and Ranu Pani having been in existence for at least 80 years. The intensive vegetable cultivation on the dramatically steep outer rim near Ngadisari dates back many years, but little of it can be

Figure 20.19. Lake Kumbolo on the slopes of Mt. Semeru.
By 's Jacob, with permission of the Tropen Instituut, Amsterdam

Figure 20.20. Fire-generated grasslands on slopes in the Tengger Caldera.
By A.J. Whitten

said to be traditional. The major force behind the intensive agriculture is absentee landlords supplying urban populations with cold-loving vegetables. Almost all the attention given to the park is around that very small area visited by tourists. A management plan has been written (Beudels and Hardi 1980) and proposals for the development of buffer zones have been made (Wagito 1986), but little action has been taken.

NUSA BARUNG

This 6,000 ha rugged limestone island (fig. 20.21) lies 10 km off the southeast coast of Java, and was established as a reserve in 1920 to protect its deciduous forest, seabird colonies, turtle beaches, and possibly because its virtual lack of surface freshwater made it unsuitable for any other use. The only permanent river is the Kedokwatu, which flows south in the west of the island.

Vegetation

The only vegetation survey was made nearly 40 years ago (Jacobs 1958), and it was found that no particular genera or species dominated, although the locally abundant species *Diospyros maritima* was common on the flat lowlands and *Croton tiglium* (Euph.) on hills. Wijayakusuma *Pisonia grandis* (Nyct.) and *Kleinhovia hospita* (Ster.) were also noticeable.

Animals

Pig-tailed macaques, lutung monkeys, wild boar, and deer are the largest inhabitants on the island, but the most significant species are the seabirds (p. 381), and turtles. Two clausiliid snails endemic to Java have been recorded from the island, but neither is confined to it (table 5.28).

Social and Conservation Issues

Fishermen have cleared small areas behind some of the accessible bays, but most of the island is still forested. Little management has been afforded to the island, however, with the result that marketable timber has been removed, and this can be seen from the bare earth slides on the steep, vegetated cliffs down which trees are tumbled to the sea. The disturbance is also indicated by the abundance of climbing plants in the forest canopy. It is hard to gauge whether the current intensity of timber theft is causing degradation of the forest, and so it is important to institute a programme of monitoring, using either satellite imagery or air photos. Turtle eggs, bat

Figure 20.21. A bay on Nusa Barung.

Photographer unknown, with permission of the National Museum of Natural History, Leiden

guano, and swiftlet nests are also taken from around the coast. The status of the seabird colonies is not known, but fishermen report the continued existence of at least some nesting seabirds, particularly in June and July. Tourists may be attracted to the island's beaches but the nearest harbour of Puger currently has no accommodation of any sort. The difficulties of walking across the cone karst of the interior are likely to put off all but the most hardy expeditions.

YANG PLATEAU

This 15,000 ha reserve, north of Jember and southwest of Bondowoso, comprises a rolling plateau between 1,700 and 2,400 m above sea level. The peak of Argapura (3,088 m) lies to the west of the reserve and is of particular interest because it has the highest ancient remains on Java comprising temples with adjacent buildings or other shelters. The walls of some of these are tumbling, in part because of antiques thieves, partly because of an earthquake in 1954, and partly because they are covered with aged blueberry *Vaccinium varingiaefolium*, the roots of which push the bricks apart. The reserve also has a number of lakes (fig. 20.22).

Figure 20.22. Taman Hidup, the largest lake on the Yang Plateau.
By J.G. Kooiman, with permission of the Tropen Instituut, Amsterdam

Vegetation

The vegetation of the highly undulating plateau comprises mainly *Casuarina*-covered hills and grassy meadows, on the edges of which grow 5 m tall edelweiss *Anaphalis viscida* (Comp.). In the past it has been famous for its diverse and beautiful meadow flowers, but for reasons described below, these are not so common now. The plateau is encircled by some excellent montane forests with large podocarp trees (p. 515).

Animals

In the middle of last century Junghuhn visited the plateau and reported seeing up to 50,000 deer on a single day, in groups of up to 1,000 (Junghuhn 1854). At the beginning of this century the meadow parts were lent or leased to individuals to develop a health and recreation centre. After nearly ten years of inactivity, the leases were given instead to the two infamous Ledeboer brothers, both of whom were fanatical big game hunters, and one of whom had already leased a small area of the meadows on the condition it was used for dairy farming. It appears, in fact, that their only interest was to develop a private hunting reserve, and to that end they built an airstrip and a lodge for their guests, magnificently situated on the largest meadow, next to the Dolbu spring. They had realized, that if the habitat were managed and the poaching stopped, then the deer numbers would increase again. Thus guards were brought in, and every conceivable deer predator – leopard, jungle cat, wild dog, and pig – was sought and killed. In 1932 alone, 140 wild dogs are reported to have been killed, and between 1915 and 1930, hundreds of leopards were shot (Hoogerwerf 1974). The frequent fires were also controlled. As a result of these extreme efforts, the deer population grew to an estimated 8,000-10,000, and by the time of the Japanese Occupation the barking deer, peafowl, and jungle fowl had also greatly benefited. At about this time the Yang Plateau was described as 'paradise on earth' (Loogen 1940) because of the abundant and visible wildlife and its meadow flowers; every streamside had *Primula prolifera* (Prim.), and between the grazed grass tussocks were abundant colourful herbs such as *Viola pilosa* (Viol.), *Swertia javanica* (Gent.), and *Gentiana quadrifaria* (Gent.). These were obvious because the deer kept the *Casuarina* in check and grazed and fertilized the turf, thereby encouraging a wide range of herbaceous species. There were, however, complaints about the damage done by the large numbers of deer to adjoining plantations and agriculture. As a result, the Ledeboer family was pressured to promise a large-scale shooting programme and the building of a 'corned beef' factory to process the deer carcasses. This was never built (Hoogerwerf 1974).

By 1971 the plateau was a shadow of its former glory, with regular fires having greatly impoverished the once-diverse meadows. There were some deer, but they were extremely wary, taking cover when the observer was still 300 m distant, but there was ample evidence of leopard, which were presumably feeding on deer and pig (fig. 20.23). During June 1993 a four-day survey found no signs of Javan deer, even though this should have been the start of the rut (breeding season). Pig were present, however, and their hair formed the bulk of the commonly encountered leopard faeces. One faecal deposit even had a foot of a wild dog within it, which at least confirms their continued existence there. Some barking deer and peafowl were also encountered.

Hunting was probably the major cause of the decline in the deer population, but a lack of salt, essential to grazing animals, was almost certainly another important factor. In days past, the deer used to journey to the coast to drink sea water, but the dangers of such an expedition today would be too great. During the period that the Ledeboer brothers were in control, they provided salt 'cakes' in shelters to protect them from dissolving in the rain (van Steenis 1972).

The lake of Telaga Tunjung within the reserve is one of the very few localities known in Java for the Australian pochard *Aythya australis*, but there are no recent records (Hoogerwerf 1974; MacKinnon 1988, 1990). On the lake at Taman Hidup, coots *Fulica atra* with young and little grebe *Tachybaptus ruficollis* have been seen (Kooiman and van der Veen 1936). These birds are distributed widely from Europe to Australia, but they are

Figure 20.23. Deer are now scarce on the Yang Plateau, where once 50,000 could be seen on a single day.

By J.G. Kooiman, with permission of the Tropen Instituut, Amsterdam

rare on Java and Bali (MacKinnon 1988, 1990). Peafowl can be seen in the Yang grasslands and heard in the surrounding forests, but they are very wary of people indicating that they are subject to hunting. A new sub-species of rat was found in the *Casuarina* forests in 1993.

Social and Conservation Issues

The Yang Plateau Reserve, together with the excellent surrounding podocarp forests (mostly under the jurisdiction of Perhutani), have great potential as a national park if some management were applied to the higher regions, and this was first proposed in 1978 (van der Zon and Supriadi 1978) (fig. 20.24). This has been repeated since then, but the reason most frequently given against it is that East Java already has Bromo-Tengger-Semeru and Baluran-Ijen-Alas Purwo as composite National Parks, and this is already more than the 'average' (KSDA IV 1985). Such reason fails to appreciate that East Java just happens to have more large areas of important 'natural' habitat with tourism potential than most other areas, and failing to protect areas for fear of being thought unfair is illogical.

Management would be required: deer would have to be encouraged and even translocated to the main meadow area and provided with salt

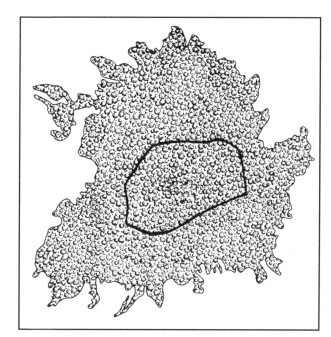

Figure 20.24. Location of the Yang Plateau Reserve within its surrounding forest block. Scale bar indicates 10 km.
After RePPProT 1989

licks. The deer would reduce the amount of tinder on the meadow, and their presence would encourage the growth of the once-abundant colourful herbs. That these are still present is indicated by their presence after fires (Subari pers. comm.), but they are then swamped by the ungrazed grass, mainly the brown-tipped and curly-leaved *Pennisetum alopecuroides*. Accommodation for hikers could be established at the old Ledeboer lodge if it were rebuilt, and appropriate tourism encouraged, since access is already quite easy. Money is available for development, as was shown when the local government drew up plans to drive a road straight through the reserve to the lodge site (rightly vetoed by PHPA) but the problem is to make sure it is spent in the best way.

Negative impacts on local people as a result of conservation actions would be minimal since the villagers rarely enter the plateau area, and the natural forests around it (albeit logged about fifty years ago) are surrounded themselves by Perhutani plantations planted in many areas with *Calliandra* to provide subsistence and marketable fuelwood. Careful planning and thoughtful management could provide significant conservation benefits to surrounding communities.

MERU BETIRI

This national park lies southeast of Jember and 50 km southwest of Banyuwangi and was declared as reserve in 1972 and as a national park in 1982. Meru Betiri comprises mainly tertiary deposits which have created very steep but low mountain ridges (mostly lower than 500 m but reaching 1,223 m) of little agricultural use. Before the Second World War these forests were continuous with those in Alas Purwo National Park to the east, but agricultural developments have now severed this link. Within the park and along the rich alluvial valleys are two plantation enclaves, Bandealit in the west and Sukamade in the east, totalling some 2,500 ha of the entire 50,000 ha.

Vegetation

Most of the forest is of the moist deciduous type (fig. 20.25). The larger trees present include *Bischofia javanica* (Euph.), *Planchonia valida* (Lecy.), *Kleinhovia hospita* (Ster.), *Lagerstroemia flos-reginae* (Lyth.), *Pterospermum javanicum* (Ster.), *Spondias pinnata* (Anac.), *Sterculia* spp.(Ster.), and *Ficus* spp. (Mora.), but only rarely do these form a continuous, closed canopy. As a result, the undergrowth is a dense growth of rattan, gingers, and bamboos (Hoogerwerf 1974). A species of bamboo, and an orchid, appear to be endemic to the park (tables 4.5 and 4.6).

Animals

To many people Meru Betiri is synonymous with the last Javan tigers, and the story of their demise is told elsewhere (p. 706). In 1971 the director of the Sukamade plantation opened up an area dominated by bamboo to provide grazing for deer (which had to be introduced) and banteng, to increase the prey animals available to the tigers (Hoogerwerf 1974). Later, banteng were released into the Sukamade forest but they did not persist (most of the Park's banteng are in the west and north). Several batches of deer were released after 1978, and there are now 'wild' herds providing prey for the leopards and wild dog.

Meru Betiri is justly famous for its turtles, which visitors are allowed to watch hauling themselves up the beach to lay their eggs. While reliable data are not many, it appears that turtles breed on a roughly four-year cycle of peaks and troughs, the magnitudes of which have been decreasing. Unfortunately it is impossible to tell whether this might be due to the existing management actions, to inadequate management, or completely unrelated to either. Management is certainly intensive. Between 1984 and 1990, 802 tags were placed on turtles laying eggs on Sukamade beach, representing 20%-30% of the total. In the 11 years up to the end of 1990, 8,534

Figure 20.25. Moist deciduous forest near Sukamade, Meru Betiri National Park, showing a number of leafless trees.
By A.J. Whitten

nests had been recorded, and about 90% of the 708,000 eggs placed in artificial nests hatched successfully. Of the 645,000 young turtles released, 622,000 were green turtles (KSDA 1991) (fig. 20.26). The value and problems associated with this type of work are discussed elsewhere (p. 756).

Social and Conservation Issues

Since concern for the tigers developed, the role or threat of the plantations in conservation has been much debated. In 1979 it was generally agreed that the two plantation enclaves were inappropriate to the conservation of the remaining tigers, although no steps were taken to remove them. Now that the tigers can be said to be extinct (Thohari 1988), the fate of the enclaves has still not been decided, which means that the owners are reluctant to invest further in the areas. On balance, the plantations are probably a net benefit to the park, providing access into it along maintained roads, employing the majority of people in their areas, and having a management committed to the park's conservation. The companies are using their own money for work and infrastructure which the government would otherwise have to find among its already over-stretched budgets. Meru Betiri's worst problems are not, in any case, from the enclave plantation areas, but from remote villages around the north of the park adjacent to the peripheral plantations.

Figure 20.26. The hatching enclosure at Sukamade, Meru Betiri National Park.

By A.J. Whitten

The major draw of Meru Betiri, as far as local villagers are concerned, is the bamboo. Its exploitation has been likened to parasitism whereby one party (the local gatherer) benefits while the other (the park) loses (Setiawan 1985). There is a clear need not just for a physical buffer, in which desired bamboos could be planted and harvested, but also a social buffer to improve the perception of the park among the surrounding people by educational improvements, provision of interpretative materials, increasing incomes through intensification and diversification, and creating jobs that draw people away from the park (p. 760) (Atmodjo 1987).

Workable systems of buffer zones have been designed (Alikodra 1984), but implementation on the ground has been limited by inadequate budgets. There has, however, been a programme of cultivating some of the medicinal plants (p. 182) sought in the forest (KSDA Jatim 1990), notably the long pepper *Piper retrofractum* (Pipe.) and tailed pepper *Piper cubeba*, both of which have been important items of trade and export for hundreds of years (Burkill 1966). This will make a positive contribution, but in the context of the entire park, it is clear that more needs to be done. The plantations need to be brought in as positive partners in conservation, but this cannot begin until they are sure that their licences will be extended.

The park needs to be promoted as a destination for visitors. It has been calculated that the current facilities of Meru Betiri have a carrying capacity of 114 people/day, but the number of visitors is generally far

Figure 20.27. Mt. Ringgit, and Mt. Beser to the south, have relatively extensive deciduous forest and are worthy of immediate conservation surveys and action.

By A.J. Whitten

below this, except at weekends and holidays (KSDA Jatim 1989).

MT. BESER AND MT. RINGGIT

These neighbouring 4,000 ha and 2,000 ha areas respectively, north of Bondowoso (fig. 20.27), are among the few remaining areas of deciduous forest on Java and have been proposed as a single game reserve because of their interesting and distinct flora and their potential for visitors from the increasingly popular resort of Pasir Putih (MacKinnon et al. 1982). Unfortunately no action has been taken. The opinion has been expressed that if not protected soon, then the value of the area will have been lost (Setiadi and Setiawan 1992).

IJEN-MAELANG-RAUNG

The proposed national park covering the largest and easternmost volcanic massif on Java comprises a large and wild area of over 130,000 ha which, although of lesser conservation value than neighbouring Baluran, Meru

Betiri, and Alas Purwo, is still worthy of immediate and serious management. The 60,000 ha of forest around the spectacular peak and caldera of Mt. Raung is the largest single expanse of forest on Java, although similar in area to the main block of forest that was hoped to become part of Bali Barat National Park (p. 827). The area is best known for the Ijen Crater in which men collect condensed sulphur from the active volcanic vents around the milky green lake. Almost the whole of the central plateau has been converted to coffee plantations, but the outer slopes of the massif are still covered with forest. Inadequate management has permitted the annual burning of the grasslands and *Casuarina* forest to continue unabated, leading to progressive degradation and deforestation.

Vegetation

To the north of the park there is 30 m high monsoon forest with abundant *Schleichera oleosa* and species of *Ficus*, although some of this vegetation has been replaced by teak plantations. Higher up there are montane forests, although much of the original *Casuarina* forests have been felled to make way for coffee plantations. The high peaks do have interesting herbs, however (van Steenis 1940a).

Animals

An decade ago in the less accessible areas, there were said to be good populations of peafowl, banteng, leopard, wild dog, and deer (van der Zon and Supriadi 1978; FAO 1982), but there are no recent data.

Social and Conservation Issues

The most famous users of the proposed conservation area are the sulphur carriers. They cause path erosion at higher altitudes, and toxic washoff of sulphur at resting points is causing the death of vegetation, but both of these problems could be solved easily. Much more widespread and worrying is the burning of high-altitude grassland which is preventing forest growth. Until guards are posted, public education started, and alternatives devised for the miscreants, it is unlikely that any improvement can be expected. An additional serious problem is the planting of the introduced *Acacia decurrens* as a shade tree for the coffee gardens. This tree seems to be spreading out of control away from the main paths at higher altitudes.

BALURAN

The Baluran Game Reserve was established in the northeast tip of Java in 1937. In 1954 a strip of over 3,000 ha along both sides of the western road were withdrawn to become plantation, but this was restored to the conservation area when the National Park was declared in 1982. The park is dominated by the 1,268 m old volcano with a split side known as Mt. Baluran. The marine resources of Baluran are considered valuable, and the boundary of the park extends into the sea. However, very few of the recommendations made in 1979 to safeguard the resources (Halim et al. 1979) have been adopted.

The park is one of the driest parts of Java, receiving an average of less than 1,000 mm of rain each year, concentrated between December and February. The average figure disguises, however, considerable year-to-year variation. The long dry period and the desiccating northeast monsoon winds contribute to the savanna-like landscape, although fire has also been a major factor (Wind and Amir 1977).

Vegetation

The species composition of the vegetation is described elsewhere (p. 477). *Acacia nilotica* now dominates the savanna vegetation around Bekol (Suharti et al. 1987) (figs. 20.28-29), and there is little reason to believe that it will not continue to spread throughout the open woodland areas. *Acacia nilotica* was first planted as a living firebreak at the edge of the Bekol grass savanna in 1969 (Nazif 1988). The intention was that the tree would prevent fires lit on the savanna to improve grazing from moving into the wooded areas. This would be achieved because the shading of the tree would prevent the growth of grasses and other herbs that die and form tinder during the dry season. What was not considered, however, was how the tree was to be contained or even if it needed to be. It grew and its seeds spread rapidly on the feet and in the guts of the grazing animals, causing a serious decrease in the tourism value of that area, although its long-term effect on the wildlife is not entirely clear.

Animals

Nothing appears to be known about the Baluran fauna from before 1932, when a Dutch hunter reported finding tracks of banteng, deer, and wild dog but no leopard or tiger. By 1936 there were already an estimated 100-150 feral water buffaloes, and reports indicated that leopards were common and tigers present, albeit in very low numbers. By the late 1960s, however, the tiger had probably become extinct (Hoogerwerf 1974). The

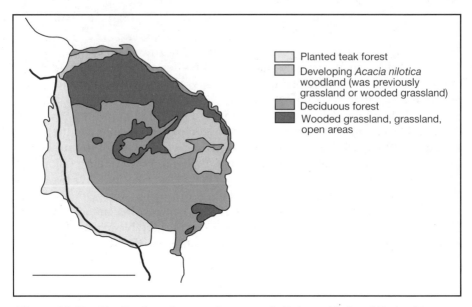

Figure 20.28. Distribution of vegetation types in Baluran National Park with areas. Broad line indicates road, scale bar indicates 10 km.

After Wind and Amir 1977; S. Hedges pers. comm.

Figure 20.29. The Bekol 'savanna' in Baluran National Park, covered with a growth of young *Acacia nilotica*.

By A.J. Whitten

large herds of deer and banteng are now the most impressive sight, although they are increasingly difficult to see because of the growth of *A. nilotica*.

About 150 species of birds have been recorded, and the most striking at Baluran is the peafowl *Pavo muticus*. The males move around alone, whereas the females are found in groups of two to nine individuals with their young. They appear to breed from the onset of the rains. They roost in tall, open trees such as *Acacia leucophloea* and *Corypha utan* (Pattaratuna 1976). The choice of such trees presumably gives the maximum protection from predators.

Baluran has about 16,500 ha of habitat which seems potentially suitable for the critically rare Bali starling, now restricted to the northern part of Bali Barat National Park. The floral and faunal communities are very similar, and the starlings' main competitor, the black-winged starling *Sturnus melanopterus,* is less common than in northwest Bali, though perhaps only because it has been trapped more intensively (S. van Balen pers. comm.). If it were not for the issue of 'faunal falsification' (the fact that there is no record of the species from Java), it would be an ideal location for the establishment of a second population of Bali starlings (van Helvoort et al. 1986). The increased management that would be necessary would, however, also favour the black-winged starling which could potentially compete successfully against the Bali starling (S. van Balen pers. comm.).

Social and Conservation Issues

Local people enter Baluran National Park to collect wood for timber and fuel, gather grass and other leaves for livestock, collect honey, trap birds, pick fruit of tamarind *Tamarindus indica* (Legu.), seeds of *Aleurites moluccana* for candlenut, and acacia seeds for propagation or to bulk out coffee. They also dig tubers of the large wild yam *Dioscorea hispida* (Dios.), leaves and other parts for traditional medicine, graze their cattle and other livestock, hunt deer, banteng, and buffalo, and set up virtually permanent encampments for fishing and collecting milkfish fry (Fahut IPB 1986). In the northern part of the park there are now some very secure-looking settlements and associated cultivated areas. The numbers of people living in or entering the park are a serious problem, not least because they increase the risks of accidental and deliberate fires being started.

One of the main priorities for the park is to conserve the banteng, and in 1985 a programme began to remove the feral buffalo which were numerically dominant and likely competitors of the banteng (p. 747). Nearly 300 buffalo were caught, using special Presidential funds, and sent to transmigrants and other farmers. This had little impact and it was agreed that further funds would be made available to remove 800 more buffalo over four years from 1989 (Sumardja 1989). The programme has met with numerous problems, not the least of which is the fact that the feral buffalo are far from easy to catch. In 1990 about 200 buffalo were caught, but none have been caught since then.

A major problem with establishing management regimes for the large mammals is that surveying them at Baluran is extremely difficult (Watling 1990; S. Hodges pers. comm.). The very hard and dry ground that exists for most of the year, the density of much of the vegetation, the similarity between the faecal deposits of buffalo and banteng, the scattered water holes, and the nocturnal habits of most of the animals all confound the best intentioned efforts, and this explains the wide variation among the estimates of population sizes over the years. Unless large numbers of trained people are available over a period, only general statements such as 'the population of buffalo exceeds that of banteng' (Sugardjito 1984) can properly be made.

Another major priority is to decide a strategy and methodology for control of the *A. nilotica*. Various methods have been suggested or tried: herbicides (Nazif 1988), uprooting using tractors, manual uprooting, utilization for fuelwood and other needs (Alikodra 1986), and there was even a plan for establishing a village on the Bekol savanna to control the spread and growth of the tree! Its thorns make it an extremely unpleasant tree to handle, and it is unlikely to be used while other trees are available. A comparison of manual uprooting of the trees and impregnation of the tree stems with herbicides found that the mechanical method was more cost effective (Suharti and Santoso 1989). The most recent 'control' method (early 1994) has been bulldozing the *A. nilotica* (and all other trees in the bulldozer's path) into ridges around Bekol. The controlled use of fire, as in the past, could be important in the management of the grazing ground, but should only be considered if the staff are fully trained and able to deal with wild fires (Alikodra 1986). Fire may stimulate the germination of the tree seeds, but it also weakens the saplings (S. Hedges pers. comm.). It is of interest that there is no *Acacia* problem in the north of the park where inhabitants cause uncontrolled fires.

The control methods that have been, are, and are likely to be used in the near future are only capable of cosmetic improvements in the situation, and it seems inevitable that the tree will remain a permanent feature of Baluran. To a certain extent this is a disaster as the potential for viewing the large animals will be severely reduced and the special landscape will be changed in as yet unknown ways. From an ecological point of view, however, it may not be particularly significant, at least not in the context of all the other introduced plants in the reserve. It may even come to be regarded as beneficial once it has grown up to form open woodland, as has already happened in areas away from the road. This is favoured by banteng which prefer to live under cover or at least with easy access to cover. Even when the tree is young and growing in dense thickets, it is not avoided by banteng, buffalo, or deer which rest in its shade during the day. It is possible that its shading may significantly reduce the amount of forage available to the banteng but, even if it does, it remains to be seen whether the reduction will be any greater than that caused by the arrival of the toxic *Lantana camara* (Verb.) (Watling 1991a). In addition, it would be wrong to extrapolate its advancement through the entire park from its gallop across the open grazing ground because its seedlings are not tolerant of shade.

That being said, if many of the values for which Baluran was made a national park are not to be severely damaged, an efficient and cost-effective means of controlling the tree in at least some of the area around Bekol must be found. The slashing of the young trees at ground level has been going on more or less continuously for over ten years, and there appears to be an embarrassingly positive correlation between frequency of slashing and density of *A. nilotica* growth. This is the opposite of what was intended but was foreseen long ago (Robinson et al. 1982). The result of the slashing is that if or when the stand does grow up, it will be particularly thick and impenetrable (Watling 1991a). This can now be seen close to Bekol. A much preferable method is the uprooting of trees using manual labour over a relatively limited area, perhaps about 40 ha, to form an open area with many edges, favoured by most grazing animals (figs. 20.30-31). Uprooting was tried in 1991-92 but this was abandoned, perhaps because it appeared to be much slower.

Plans for buffer zones to contain and control the impacts of local people have been made, but they have not been implemented and there is confusion among the parties over the issues of traditional rights (p. 68) (Wind and Amir 1977; Robinson et al. 1982; Fahut IPB 1986; Wustamidin 1988; Nugroho et al. 1991).

ALAS PURWO

In terms of its wildlife, Alas Purwo easily surpasses Baluran in importance (fig. 20.32). This area in the extreme southeast peninsula of Java was established as a 42,000 ha nature monument in 1913 known as Purwo-Jati Ikan. In 1939 an extension of 20,000 ha of grassy savanna lowlands known at that time as the richest wildlife area in Java was added to the west of the peninsula to ensure that they were not planted with teak. The whole area was declared as the South Banyuwangi Game Reserve. That crucial extension was cancelled in 1954 and was cleared and planted with mahogany and teak (Hoogerwerf 1974). Much of the extension was returned to a conservation function in 1968, but clearing and planting continued. This ceased only in 1990 when the whole area was declared a national park. Under the Conservation Act, buffer zones are understood to lie outside conservation

Figure 20.30. Manual uprooting of *Acacia nilotica* in Baluran National Park is by far the preferred means of controlling it.

By A.J. Whitten

Figure 20.31. Suggested configuration of the grazing ground of the area around Bekol, Baluran National Park, as part of *Acacia nilotica* management.

After Watling 1991b

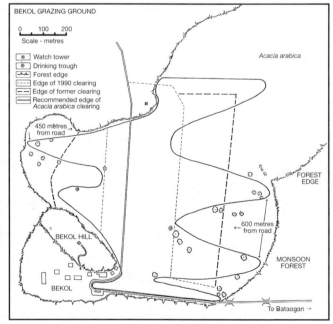

BEKOL GRAZING GROUND

0   100   200
Scale - metres

☒ Watch tower
⊕ Drinking trough
Forest edge
Edge of 1990 clearing
Edge of former clearing
Recommended edge of *Acacia arabica* clearing

450 metres from road

Acacia arabica

FOREST EDGE

600 metres from road

BEKOL HILL

MONSOON FOREST

BEKOL

To Bataogan →

Figure 20.32. The herds of banteng and deer, and the occasional peacock and wild dog, make the view from Sadengan in Alas Purwo National Park one of the greatest terrestrial wildlife spectacles in Java and Bali.

By A.J. Whitten

areas, and so responsibility for the extension has passed away from the conservation authorities again.

The 30 km long peninsula itself comprises an uplifted limestone massif which has eroded into very sharp and steep ridges. The slopes are covered with only shallow soil. Rainfall is quite low and the whole area has only one permanent stream. It is in fact a very inhospitable region with no footpaths in the central rugged section, and coast walking is hampered by the lack of fresh water, and the sea which covers almost all the beach at high tide. There is no direct evidence (dwellings, wells, etc.) of former habitation (mainly because of the lack of fresh water), but the abundance of bamboo, and once teak, throughout the more accessible areas is a good indication of former human disturbance, if not clearance (Jacobs 1958; Beudels and Kurnianto 1982).

To the west of the park is the Segoro Anak lagoon and mangrove forest. This is a very rich wetland area subject to a wide variety of traditional fishing practices exploiting oysters, crabs, milkfish larvae, cockles, prawns, and fish. Despite the intensity of use, the activities do not yet seem to be actually damaging (Ritchie 1991). The park also has a 'surf camp' at Plengkung, in the south, where, in about 1977, surfers discovered "a unique freak of nature that is without doubt the home of the world's most awe-inspiring lefthand reefbreak" (Neely 1992), where waves run for nearly half a kilometre, particularly between April and October. The attention of this group of visitors is almost exclusively seaward and there is virtually no impact on the park.

Vegetation

Alas Purwo comprises moist deciduous forest, with more than 50% of the trees being deciduous and with the usual secondary trees, such as *Macaranga* (Euph.) and *Trema* (Ulma.), being scarce (fig. 20.33). The large trees found in Alas Purwo are *Ficus* spp. (Mora.), *Sterculia foetida* (Ster.), *Kleinhovia hospita* (Ster.), *Terminalia catappa* (Comb.), *Lagerstroemia flos-reginae* (Lyth.), *Tetrameles nudiflora* (Dati.), and *Eugenia* spp. (Myrt.), none of which exceed 30 m in height. The forest used to contain many fine specimens of valuable *Vitex pubescens* (Verb.) and *Manilkara kauki* (Sapo.), but all those that could be felled and dragged away have gone. Jacobs (1958) noted that the large *Corypha utan* was "rather frequently scattered, from the shore to the hills, with many youngsters." In 1992, however, *only* youngsters could be found around Trianggulasi, Kali Pancur, the grazing ground and the hills visible from them. Adults are generally very striking and are easily seen against the skyline. Since these palms are monocarpic, flowering once when mature and then dying, it may be that the phenological cycle is somewhat synchronous.

After recommendations that a grazing ground be established (Hoogerwerf 1974), a 100 ha pasture was opened between 1975-79 at Sadengan, near the guesthouses at Trianggulasi. This was done using a tumpangsari system, and before the farmers left, they had to plant the indigenous grasses *Arudinella setosa, Dischantium caricosum, Polytrias amaura,* and *Heteropogon contortus* (brought from Baluran), as well as the introduced elephant grass *Pennisetum purpureum* (Beudels and Kurnianto 1982; Watling 1991b).

Animals

The Sadengan grazing ground provides a wonderful spectacle with the largest herd of banteng in Java, many Javan deer, jungle fowl, peafowl, and other animals easily seen. In the surrounding forests monkeys and other animals are not difficult to watch (Saim 1986). No one has attempted to describe the distribution of the wildlife in the national park, but it is quite

Figure 20.33. Moist deciduous forest in Alas Purwo National Park.
By A.J. Whitten

likely that the dry, rugged hills do not support a rich or diverse fauna, because many of the larger species are dependent on drinking sites. It is for this reason that the temporary encampments of fishermen and others near the limited number of water holes is very disturbing to the·animals (van Assendelf 1991). Notable among the birds are three species of hornbill (Beudels and Kurnianto 1982). Green, hawksbill, olive ridley, and leatherback turtles nest along at least the southwest beaches, with green turtles being the least frequent species. There is an active hatching and release programme (p. 754) (fig. 20.34).

The Segoro Anak or Grajagan Estuary was the focus of attention of the Living Resources in Coastal Areas part of the ASEAN-Australia Cooperative Programme on Marine Science. Studies made included the diversity, seasonal changes, and habitat requirements of the soft-bottom benthic community. Other studies investigated the composition of the fish fauna in the mangrove areas, and the mangrove forest itself, in terms of structure, species frequency, density, dominance, and net potential primary productivity.

Social and Conservation Issues

Segoro Anak is used quite intensively for the gathering of mangrove ani-

Figure 20.34. Young hawksbill turtle in the rearing facilities at Alas Purwo National Park.
By A.J. Whitten

mals and plants, but its management should attempt to keep these locally important economic activities to their present levels until it can be shown that increases in use would be sustainable (Ritchie 1991). The sandy coasts include the only protected beach in Java or Bali where four turtle species breed annually. Eggs are collected by the guards, and the hatchlings released after a short period of captivity, and the wisdom of this management needs to be reappraised (p. 754).

Monitoring of the Sadengan feeding ground needs to be instituted immediately to determine whether the pasture is spreading or being encroached upon, whether its composition and quality are being maintained, and whether sheltering rows of trees would benefit the wildlife. Most important of all, however, is the need to make formal and urgent representation to Perhutani to remove all adult, young, and seedling *Acacia nilotica* trees which were inadvizedly planted along the road to Alas Purwo and in the tumpangsari areas around the feeding ground. The presence of seedlings on the feeding ground should also be monitored.

The presence of the extensive plantation areas means that human pressure on many parts of the inhospitable reserve is not great, but the Sembulungan peninsula, which juts north along the east coast of Java, is highly disturbed and needs management to maximize its benefits to the wildlife (Watling 1990).

Figure 20.35. Rain forest in the proposed extension to Bali Barat National Park.
By A.J. Whitten

BALI BARAT

The Bali Barat National Park as proposed would have covered 77,000 ha. The significance of the proposed park would have been the generally undisturbed nature of the 60,000 ha of rain forest covering the hills (fig. 20.35), the high-quality marine habitats, critically endangered wildlife, and great tourist potential (FAO 1982; Rinjin and Sarna 1990). When the park was declared in 1982, it extended over only the 19,000 ha which had previously been the Game Reserve agreed by the Council of Kings just after the end of the Independence struggle. National parks should be large, of a relatively natural character, and have tourism potential, and in this context 19,000 ha is really rather small. In addition, the extensive human disturbance of the land and coasts had earlier been felt to be so 'overwhelming' that it was disqualified as a potential national park (Halim et al. 1979). Efforts have been made to extend the park to its full size, but the negotiations between the provincial Forest Service (which currently has jurisdiction over the area), and the National Park authorities have broken down and are now subject to a moratorium.

The exceptional nature of the marine environment around P. Menjangan became known only in the mid-1970s, and a marine extension to the proposed national park was suggested (Halim et al. 1979; Polunin et al. 1980). The reefs off Menjangan Island are the very best Bali has to offer in terms of marine tourism potential, given their closeness to the shore, freedom from dangerous currents, shallow and deep areas, and healthy and diverse coral growth in undeveloped surroundings (Polunin et al. 1980; Dono 1987).

Vegetation

In the savanna of the Prapat Agung peninsula, *Corypha* and *Borassus* palms are common, as are *Ziziphus jujuba*, *Schleichera oleosa*, *Acacia leucophloea*, *Albizia lebbekoides*, and the neem tree *Azadirachta indica*, which stays green all year and produces fruit that are much favoured by birds (de Voogd 1937). Across the road to the east, the dominant trees are *Vitex pubescens* (Verb.), *Toona sureni* (Meli.), and *Pterospermum javanicum* (Ster.) (Kanwil Kehutanan Bali 1993). The vegetation in the planned extension to the park is variable, but on parts of the southern slopes the dominant large tree is *Planchonia valida* (Lecy.), the leaves of which turn a deep red before falling. Other common trees include *Palaquium javense* (Sapo.), *Duabanga moluccana* (Sonn.), *Pterospermum javanicum* (Ster.), and *Meliosma ferruginosa* (Sabi.) (Soepadmo 1961).

The mangroves in the bay behind Gilimanuk in the Bali Barat National Park are very similar to those on Java in terms of composition, although the presence of *Osbornia octodonta* (Myrt.) is noteworthy (Polunin et al. 1983). This is typically a mangrove tree of eastern Indonesia and northeast Australia (van Steenis 1936).

Animals

The park's best-known animal is the very rare Bali starling, which is described in detail elsewhere (p. 227). It is worth noting, however, that there is not, and nor has there ever been, enough suitable habitat within the park boundaries to support more than 150-200 starlings. Even if these numbers were attained, they would not represent a viable population, and so a new population is being started on Menjangan Island, and the restoration of the enclave areas within the park is being planned (van Balen and Gepak 1994).

The largest animal in the park is the banteng (though it may not be native to Bali), but the remaining herd is probably less than 20 animals and mixed with domesticated Bali cattle (pgs. 220, 295). Deer (and perhaps banteng) used to be very common in the savanna area, but hunters from Java seriously reduced their numbers more than sixty years ago (de Voogd 1937). The park may have the last viable populations of the Bali subspecies of the lutung monkey (Wheatley et al. 1993).

Social and Conservation Issues

Javanese people were brought over to Gilimanuk and surrounding areas in the 1930s to work in forestry and plantation operations which the Balinese themselves were not prepared to do (Halim et al. 1979), and they have represented a major threat by gathering wood from the forest, and possibly trapping Bali starlings. The first lease of the enclave coconut and kapok plantation expired in 1993, and a few dozen families have already transmigrated to Sulawesi. The final licence expires in 2005.

The marine environment suffers continuously from human activities, but fishing using explosives has not been a major problem around Bali because the ashes of cremated corpses are cast into the sea, and the dead should not be disturbed. The bombing that occurs is generally the work of Madurese and Javanese people. The importance of Menjangan can be judged if conditions there are compared with those around Tabuan, a small island just to the north which has been mined intensively and where conspicuous fish species are rare (Polunin et al. 1983). Also from the west comes rubbish, especially plastic bags, which are extremely unsightly and appear in vast quantities when heavy rains in East Java combine with certain sea currents.

The size of the park and the status of the excellent hill forests will not be discussed formally again for some years. Given that situation, it is important that the Bali Forestry Service be given meaningful support for the effective and acceptable management of the forests, particularly those above the north coasts which are relatively dry and easily damaged by fire, some of it deliberately set.

BATUKAU

Batukau is a massif of relatively recent volcanic origin including five of Bali's highest mountains, three of them over 2,000 m. Most of it is covered with forest, it has three crater lakes, and a dozen hot springs. The peak of the tallest mountain, Batukau (2,276 m), is cleared of trees and a small temple and stage have been built there. At the present there are three small and separate reserves around three of the three smaller peaks, totalling only about 1,700 ha.

Vegetation

The vegetation is dominated by *Dacrycarpus imbricatus, Ficus* spp., *Crypteronia paniculata* (Cryp.), *Eugenia* spp. (Myrt.), *Dysoxylum* spp. (Meli.), and *Magnolia blumei.* Some of the clearings and land slips are natural, having been formed when trees were knocked over by earthquakes in the 1960s and 1970s. Around the Pura Luhur temple, on the southern slopes of Mt. Batukahu, the most common tree is a species of *Tabernaemontana* (Apoc.).

Animals

The forests appear to have a full complement of Bali's montane forest birds and mammals.

Social and Conservation Issues

To the south the forest abuts the beautiful Pura Luhur temple and the forest is scarcely disturbed. Elsewhere, however, pressures to increase the land area devoted to market gardening or coffee is resulting in illegal clearing, such as around Lake Tamblingan and southeast of Mundok. On the slopes of Mt. Lesung, 50 ha of Protection Forest has recently been converted to coffee plantations, and farmers are making charcoal from the forest trees, netting barking deer, and snaring the green jungle fowl (Boeadi pers. comm.).

It is clear that the three small existing reserves around the lower peaks of Mts. Lesung, Pohen, and Tapak really should be combined and extended to include forest adjacent to the Eka Karya Botanical Gardens and Pura Luhur temple in order to form a single and much more useful reserve totalling nearly 20,000 ha. To do this, it would first be necessary to bring the surrounding communities, particularly those to the north, into the planning process. The area has immense potential for nature tourism, which is only just starting to be exploited, and this should be brought into the discussions.

Part E

Finding a Path for the Future

All the problems related in this book, such as the failure of economic arguments in conserving biological diversity, the increasingly consumptive lifestyles of the more prosperous members of society, the increasing reliance on materialism as a means of finding happiness, and the serious threats to the remaining areas of natural ecosystems, all seem to show that something is intangibly wrong with Java and Bali, and that the ultimate result could be irreversible environmental loss or degradation. To deal with this effectively demands focussed attention and broad, unified participation driven by people's whole body, mind, and spirit.

The intangibility can be approached with a philosophical mind influenced by ecological principles. Thus, the first chapter in this part examines the background to, and different forms of, environmental ethics. We conclude the book by examining the challenges facing people living on, and visiting, Java and Bali, hoping to provoke thought and actions which will make it possible for Java and Bali to proceed in a deep groove towards the goal of sustainable development.

Chapter Twenty-One

Environmental Ethics

We do not pretend to be philosophers but we have attempted to present below a very basic outline of environmental ethics, a field which is developing rapidly due to the increasing pressure on all environmental systems, the progressive failure of economic arguments in conserving natural areas, and the increasing conviction among concerned parties that there is something intangibly wrong with the irreversible loss or degradation of nature. That intangibility can be approached with a philosophical mind.

PHILOSOPHICAL BASIS

A clarification of terms is needed, notwithstanding the fact that not all their users agree on their scope of meaning. 'Morality' refers to human behavioural requirements; for example, it is generally regarded as immoral to kill other human beings. Indeed, those things that are considered immoral tend to be actions that cause people harm. The term 'ethics' refers to an agreed code of behaviour among a particular group of people. For example, doctors have a code of ethics and their governing bodies can take action or impose sanctions against those who do not adhere to the code. Ethics thus stands between morality and the law in guiding human behaviour. The law prohibits certain types of behaviour within a given country, and punishment can be meted out by judges against any individual violating those laws within the area of jurisdiction. Morality, ethics and law frequently coincide; for example, it is morally wrong in general, legally wrong throughout, and ethically wrong for a doctor to kill another person. This does not, however, mean that they are the same thing:

- the legal system is established so that society can run smoothly, but in the event that the law allowing killing is changed (in the case of euthanasia or abortion), it could be argued that the morality of the situation had not changed;
- whereas laws are passed by bodies having authority to do so, morals are standards which remain whether or not they are codified, and in the absence of applicable ethical codes or laws, are left up to us as

individuals to adopt or abandon;

- law concerns actions rather than inward thoughts or feelings. Thus morality requires not just adherence to a code of action (e.g., prevention of polluting effluent entering a river) but also a belief or concern related to emotions. In contrast, the law cares not a bit if we obey a regulation with reluctance, wishing we had done otherwise; and

- since certain laws may be thought to be morally defective (as is sometimes the case in certain aspects of human rights), then morality and the law must be distinct.

In a similar vein, just because certain behaviour is immoral (e.g., causing the extinction of a species), it does not follow that such behaviour is illegal, if only for practical reasons. In any case, the law exists to allow maximum freedom to live without undue interference from others, and it is generally agreed that if certain immoral behaviour does not affect people, there is no need to make it illegal. Thus, although we are wont to talk of environmental ethics – a code of acceptable and required behaviour to benefit the environment – we really need to first explore consensual morality as it relates to the environment so that those of our actions that are neither regulated by, nor the concern of, ethics or the law are still worthy of aware, global citizens.

In order to ascertain what behaviour affecting the environment falls within the bounds of morality, one has to decide between two moral theories:

- that certain actions are morally wrong because of their consequences, or

- that certain actions are simply morally wrong in themselves irrespective of their consequences.

If the former theory is adopted, then its outcome is that a certain action is morally required if it has better or more valuable consequences than all of the alternative actions. Determining which possible outcome is better can be problematic because an action may affect values attached to life, beauty, welfare, liberty, equality, and so forth in different ways, and one, such as landscape beauty, may have to be sacrificed to provide marked improvements in another, such as human welfare. In such a situation, when one consequence is positive and another is negative but they are not strictly comparable, then it is extremely difficult to judge which action is correct.

An alternative way to reach a decision is to deny a multiplicity of beneficial consequences and to hold that one thing, and one alone, is intrinsically valuable. Whereas instrumental value accrues by virtue of us being able to do something, intrinsic value applies to something that does not necessarily have any utility. There has been long debate over what, if anything, is intrinsically good, but the classic view is that only happiness is intrinsically good. Thus an action is correct if, on balance, it creates greater happiness. Sometimes it is said to be the greatest happiness for the greatest number, but by adding the breadth of the happiness, this goes against the thought that only one thing is intrinsically valuable. Another view is that an action is right if there is no alternative action that would produce a greater satisfaction of desires. In this case the satisfaction is not necessarily a mental state and can occur even if the desires are satisfied after ones' death. Both views require than only creatures with sense and feelings are considered, and no other aspect of nature can have any intrinsic value. This does not necessarily lead to destruction of ecological systems, for as long as they have some instrumental value that leads to great happiness, then they are safe. If, on the other hand, it is judged that great well-being will be produced by turning a national park into an open cast mine, then the development will go ahead. It could be held that two (or more) factors are intrinsically valuable (e.g., happiness and the existence of natural beauty), but this ties itself in knots when one gains and another loses.

Believing that happiness is intrinsically good has problems because it concludes that collective happiness is more important that individual happiness, virtually denying that an individual has any rights. And who may be said to have rights: adult humans and children, yes; babies and foetuses, probably; animals, possibly; trees, doubtful. Philosophers, politicians, and others have for many reasons resisted the idea that inanimate things have rights, and have argued that, to have rights, something must have sense and feelings as well as interests that can be served. Plants still fall into a grey area.

Thus, neither the beliefs that consequences define morality nor the rights-based theories offer any great basis for an environmental morality. However, it can also be held that rational agents are owed respect and should be treated as ends and not as means to some other goal. This idea can, in fact, be extended and freed of the requirement of rationality, if religious respect for the Gaian world (p. 75) were required. Ethical philosophers can take Gaia (the name of the Greek Earth goddess) to be analogous to a 'person' with vital organs (wetlands, tropical forests, continental shelves), the same values, and requiring the same respect as *Homo sapiens* (e.g., Devereux et al. 1989). It should be stressed, however, that Gaia alone does not form the foundation of a complete environmental ethic since all value is set in the homeostasis, and little or no value is given to species or individuals (Weston 1987). The respect for Gaia or biosphere is due because it is not ours to use and dispose of, but rather something over which we have been awarded stewardship because we cannot own, either individually or collectively, the wider environment. If we are stewards, then we should talk not of rights but of duties. There are some who would like to remove such ideas from what they perceive as a threatening religious context, but in Indonesia this is unnecessary and probably inadvisable since belief in a Godhead is the first of the five state principles (Pancasila, p. 842).

RELIGIOUS BASIS

It is fortunate, perhaps, that no major religion can claim a monopoly on practical, wholesale, and active concern for the environment. All religions are able to join, or even lead, the calls for ecological and economic justice, peace, and the integrity of creation, and all the major religions signed a variety of global declarations prior to the 1992 Earth Summit in Rio de Janeiro, Brazil. This ability to call with a common voice is the result of re-reading scriptures in the light of the global environmental crisis and the realization by many that the Earth is not for us to impoverish and pollute, but rather it is a spiritual inheritance demanding a global acceptance of responsibility for its stewardship (Küng 1991). Some major international NGOs have lent support to the exploration of the religious aspects of environmental care, notably the World Wide Fund for Nature.

The level to which this spirituality can be taken depends on the individual and his or her response to experience and available information. Some excellent texts are available, some for specific religions, such as for Christianity (Wilkinson 1980; Stott 1984; McDonagh 1986; Serrini 1986, Duke of Edinburgh and Mann 1989; Cooper 1990; Breuilly and Palmer 1992; KLH-DA 1993), Islam (Zaidi 1981; Nasseef 1986; Khalid 1992; KLH 1992a), Hinduism (Singh 1986; Prime 1992), and Buddhism (Rinpoche 1986; Batchelor and Brown 1992), and some dealing with all these (Chee 1987; Küng 1991; Marshall 1992). Whatever the religion, religious leaders have a unique responsibility to meet the environmental challenges. To the affluent, the need for thrift, prudence, sufficiency, and redistribution can be preached, and the sins of overconsumption, waste, and greed preached against (Goodland 1991).

The essence of the general positions held by the major faiths in relation to care for the environment are as follows:

Islam

The entire universe is God's creation and thus belongs to Him. He created people to be unique, giving them the power to reason, and even the option of rebelling against Him. People have the potential to acquire a status higher than that of angels or to sink lower than the lowest animals. Even given our special position, or perhaps because of it, we can properly understand ourselves only when we recognize that submission to the Creator God is of paramount importance. Only then will there be peace – between one person and another, and between people and nature. That submission casts us in the mould of servants, yet able to exercise our powers, potentials, skills, and knowledge, appreciating that all our achievements derive from the mercy of God. When we return proper thanks and worship to God for our nature and creation, then we become free, aware, responsible trustees, or *khalifa* of God's gifts and bounty. Trustees are not to be masters, but are answerable to God, the Master, because the safekeeping of the earth has been entrusted to us by Him. We will be judged at the Day of Reckoning, or *akhrah*, for things done and not done, and the practice of Islam provides the means of preparing for and facing the judgement.

The cornerstone of Islamic teaching on the environment is maintaining the unity of God, the *tauhid*, which should be reflected in the unity of all peoples, and of people with nature, to ensure the integrity of the physical and living elements of the Earth. Unity will be achieved by walking the middle path of balance and harmony, not allowing one specific need to dominate over the mass of others. The concepts of *khalifa, akhrah,* and *tauhid* are central to Islam in general and to Islamic environmental ethics in particular. We only need to adopt the values and act them out in our lives. All followers of Islam are *khalifa,* and they must be adequately informed in order for them to make decisions and choices for which they will be held accountable to God (Zaidi 1981).

The practicality of seemingly abstract and pious metaphysical discussions was demonstrated long ago in the *Shariah* book of Law, within which provision is made for *haram* land within which no development is allowed that would damage natural resources, and *hima* areas intended purely to conserve wildlife and forests. The *Shariah* has also been used to formulate an animals' bill of rights, prevent overgrazing, safeguard water resources, conserve forests, limit city growth, and protect cultural property. The overriding matter of importance now is imbibing the Islamic ethic to counter the environmental destructive thoughts and actions that dominate today's world (Nasseef 1986).

Christianity

An ancient Psalm (No. 148) exhorts God's people to praise Him for all aspects of the world about us, for to do so is to confess that God is the Creator of all things visible and invisible and to thank him for the gifts he bestows. God has declared everything that he created to be good, indeed very good, and nothing was created unnecessarily, and nothing necessary was not created. The created universe thus exhibits a harmony that should, by the interdependence of its parts, manifest and reflect God's truth, beauty, love, goodness, wisdom, majesty, glory, and power. This harmony exists irrespective of whether the elements are useful or attractive or convenient from the viewpoint of humans. We humans are unique, made in the image and likeness of God and, amazingly, entrusted to be stewards over nature, and to manifest God's goodness. People have the responsibilities of rulers, yet are subject to heavenly authority. There is no way that the authority awarded to us should be used to despoil, abuse, squander, or destroy what God has set in place to reveal his glory.

Christians believe that when Adam and Eve, the first human couple, rebelled against God's authority in Eden, disharmony was introduced into the relationships between God and people, and people and nature. This sin has continued through history, exhibited in disharmony, injustice, aggression, and unsustainable exploitation. But God, as promised to Abraham and Noah, has provided a way out of the morass by allowing that essence of Him known as Jesus to live on Earth and by his suffering, death, and resurrection to bring redemption and healing to all people and all things. Through him and his life-giving Spirit, God reconciles all things, visible and invisible.

While much environmental damage has been wrought by those appropriating the name of Christianity (White 1967; see also Moncrief 1978), the gospel, or Good News of Jesus, has also inspired people to good works based on love for fellow men, creatures, and the environment. The man held up as a patron to ecologists by Pope John Paul in 1979 is Francis of Assisi (1182-1226), a pious Italian wedded to 'Lady Poverty', and who regarded all of creation as God's gifts and signs of his reconciling love. He adopted a simple lifestyle, and saw work as a daily grace given by God to be exercised in such a way that would lead to mutual enrichment of people and their environment. The corrupt and wealthy church of the time found this man problematic, and he was nearly expelled because of his views. Indeed, after his death his teachings on stewardship were watered down or completely ignored. Today, however, the interpretation of scriptures by this man are finding new and urgent relevance (Hooper and Palmer 1992).

In summary, Christians reflect Jesus and repudiate actions, such as aggression, discrimination, and cultural destruction, that do not respect peoples' authentic interests, and do not allow them to pursue and fulfil their total vocation within a universal harmony. They also repudiate exploitation of nature that threatens to destroy it and thereby make people the victim of degradation. There is a gulf between Christian scripture and much modern Christian practice which can only be bridged by a turning to God (Serrini 1986).

Buddhism

The Buddha recognized that causes, or *karma*, and effects are linked, and that happiness and suffering do not occur simply by chance. Thus a human action motivated by healthy and positive attitudes constitutes a major cause of happiness, whereas actions based on ignorance, greed, and negative attitudes bring about suffering and misery. Positive human attitudes are rooted in selflessness and love, and Buddhism is a nonviolent religion of love, understanding, and compassion. Since all beings depend on the environment for their survival, Buddhism, by extension, attaches great importance to environmental protection.

All creatures other than humans must also be taken into account because they can experience suffering and seek happiness even if they cannot necessarily communicate their feelings. Unfortunately, history shows that this view has rarely prevailed. Instead, utility has been used as the sole criterion of concern, and this general view has caused great cruelty and violence. There are many parallels between the wilful termination of the lives of animals and of people, and utility is not a valid criterion.

Buddhism holds that there is rebirth and life after death of all sentient creatures not just on Earth but in the entire universe. Within this belief all beings are related, just as we and our families are related in this life. Extending this, the dependence on, and respect for, our parents must be expressed to other creatures, and the fact that we are unaware of these cosmic relationships does not undermine their importance. Most humans regard their own survival as a right which cannot be compromised, and now it is necessary to extend this to other species. Humans and non-humans depend on a healthy environment for their survival and thus, restoring, protecting, and appreciation of the world's environment must be wholeheartedly pursued.

His Holiness the Dalai Lama has said:

> Our ancestors left us a world rich in its natural resources and capable of fulfilling our needs. This is a fact. It was believed in the past that the natural resources of the earth were unlimited, no matter how much they were exploited. But we know today that without understanding and care these resources are not inexhaustible. It is not difficult to understand and bear the exploitation done in the past out of ignorance, but now that we are aware of the dangerous factors, it is very important that we examine our responsibilities and our commitment to values, and think of the kind of world we are to bequeath to future generations. We are the generation with the awareness of a great danger. We are the ones with the responsibility and the ability to take steps of concrete action, before it is too late (Rinpoche 1986).

Hinduism

In the roots of Hinduism, humans are regarded as part of nature linked spiritually and psychologically with all the other physical and biological elements of the environment, all of which share and are pervaded by the same spiritual power. While humans are, at present, the peak of the evolutionary process, they arose from and remain part of nature. Hindu belief is permeated by a respect and reverence for life and by the awareness that all creatures, plants, and natural forces are inextricably linked. The divine being is not exterior to nature, but rather, nature is the expression of the divinity (Prime 1992).

Numerous Hindu texts advise that all animal species should be regarded as children, and in the rich mythology the various deities (all of the same divine power) ride on, and are associated with, particular animals. This should lead to a reverence for animal life which is reflected in widespread vegetarianism and the philosophies of simplicity and non-violence. In addition the plant world represented by individual or groups of trees associated with the Hindu pantheon, is held as sacred and is to be protected. The reverence for nature has not been acted out in daily life but is a powerful tradition worthy of renurturing and reapplication. For example, animals do not arouse aesthetic or intrinsic interest among the Balinese, with the exception of certain birds, and species that shed their skins (e.g., snakes), which are used symbolically in rituals. Animals that occupy sacred space, such as the monkeys at Sangeh and the bats in Goa Lawah, are treated with some respect, but the calls and activities of animals that share human space generally symbolize spiritual disharmony or are warnings of future events (Lovric 1990). The effects of centuries of rapacious exploitation are catching up with the world, and a new attitude is required. Hinduism pictures the Earth as the Universal Mother that has cosseted her children from the beginning. Despoilation of nature is tantamount to abusing her, and people must rediscover their earlier reverence (Singh 1986).

The fundamental Balinese-Hindu principle of *tri hita karana* (three-happiness-sources) undergirds all understanding of the natural and socio-cultural worlds of Bali. It regards material and spiritual happiness as being dependent on there being harmony between Sang Hyang Widhi Wasa (God), humans, and the environment. This harmony is sought at all times and in all places, in lifestyle, architecture, and relationships. 'Nature' or 'natural world' are not Balinese concepts since nonhuman and human species are all viewed as natural, all of which coexist in different ways, times, and places (Hobart 1990). The clearest demonstration of this is the strong belief in enchanted places, such as the enormous fig trees found dotted around the Bali landscape.

DEEP ECOLOGY

In reaction to the failure of classical philosophy to come to grips with environmental problems, the Norwegian philosopher Arne Naess coined the term 'deep ecology' in 1972. Deep ecology aims to be an articulated wisdom, a synthesis of theory and practice, that enables people to find their bearings and to ground themselves through fuller experience of our connection to the Earth. Deep ecology has been defined as 'ecological realism' (Devall 1990), and identifies the common threads among religious and philosophical systems. Deep ecology has been developed and extended in the last 20 years such that, while not a religion in itself, some adherents to the concepts have evolved 'pantheistic' beliefs akin to Hinduism, coupling them to the concept of Gaia (Devereux et al. 1989). It should be stressed, though, that deep ecology is not a religion, and can be grasped and acted out within the beliefs of the major religions of Indonesia.

General statements elaborating the meaning of deep ecology were formulated by Naess and Sessions (Devall and Sessions 1985) and have been much discussed. They are summarized below with minor changes or elaborations (Devall 1990) to reduce possible misunderstandings:

It is interesting how close the last two points are to Indonesia's Law No. 4 of 1982 on Basic Provisions on Environmental Management. In this, the contrast between the existing and desired environmental conditions are clear, and each citizen is given both a right to live in, and the duty to work for, a healthy environment.

Deep ecology was established to be distinct from the relatively comfortable shallow (or reform) ecology or environmentalism because it rejected the sole promotion of the health and affluence of people in developed countries that was espoused, perhaps inadvertently, by environmental bodies. The difference between the conventional reform or shallow ecology and the newer deep ecology can be illustrated by comparing typical slogans used by the respective proponents (table 21.1).

Many of the views underlying deep ecology are not, in fact, particularly radical and were even formalized in the World Charter for Nature passed by the U.N. General Assembly in 1982 (and then catapulted into obscurity),

Table 21.1. Slogans typical of reform (shallow) ecology and deep ecology.

Reform ecology	Deep ecology
Plant species should be saved because of their value as genetic reserves for agriculture and medicine.	Plant species should be saved because of their intrinsic value.
It is nonsense to talk about value except as a value for people.	Equating value with value for humans reveals a racial prejudice.
Pollution should be decreased if it threatens economic growth.	Decrease of pollution has priority over economic growth.
People will not tolerate a broad decrease in their standard of living.	People should not tolerate a broad decrease in the quality of life, but tolerate a broad decrease in the standard of living in overdeveloped countries.
Developing nations' population growth threatens ecological equilibrium.	World population at the present level threatens ecosystems, but the population and behaviour of industrial states more than any others.

Source: Devall (1990)

which states that the signatories are:

- Aware that
    - mankind is a part of nature, and life depends on the uninterrupted functioning of natural systems which ensure the supply of energy and nutrients;
- Convinced that
    - every form of life is unique, warranting respect regardless of its worth to man, and to accord other organisms such recognition, man must be guided by a moral code of action.

It has been argued that the relevance of deep ecology to the Third World has been perverted because its western exponents equate environmental protection with wilderness preservation, with the result that the root causes of environmental degradation are obscured (Guha 1989). The undeniable emphasis on wilderness preservation in many papers on deep ecology merely reflects the culture of those writing on the topic, and does not undermine the profound principle that humans have their place in nature (Johns 1990). Deep ecology writings should stress the distinction between the prevailing people-centred approaches and the necessary nature-centred approaches. Where the latter incorporates human concerns, acknowledging that environmental degradation occurs as a consequence of overconsumption and militarism, then meaningful and sustainable progress may be made (Johns 1990). This is as true in the Third World as elsewhere.

PANCASILA

The endogenous philosophy of most Indonesians regarding nationhood and life is found in the five philosophical principles, or Pancasila, of the Indonesian State. They are:

- **Belief in God**, the practical expression of which must be to maintain peaceful relationships among and between the various groups of religious believers, and to keep and practice basic spiritual, moral, ethical disciplines in order to sustain national development. Each religion provides guiding principles for achieving compatible and harmonious relationships with the environment.
- **Humanity which is fair, and a civilized society,** which includes the participation of Indonesian individuals in combating poverty, developing solidarity, and eliminating injustice. In every aspect of these, the subject of human interaction with, and interdependence on, the environment and natural resources are implicitly and inseparably related. Indeed, issues such as poverty and injustice are basic concerns of environmentally sound and sustainable development at

both the national and global levels.

- **Indonesian unity** enhances the development of Indonesian people not just as responsible individual citizens but also into living communities, a nation, and a state whereby solidarity in creed and action is strengthened. In this context, concern for, and sharing with, others in the culturally and biologically diverse archipelago is the ultimate goal.

- **Democracy led by wisdom in consensus representation,** sometimes known as the Pancasila Democracy, covers ideas for initiating and developing a democratic political system with dynamic national stability. The ultimate objective is to let every citizen develop active political awareness and responsibility in order to fully participate in the country's decision-making process. Political awareness and responsibility must be generated not only through people's socio-economic and socio-cultural background, but also through an awareness of their compatibility and harmony with the environment.

- **Social justice for all citizens** covers the development of economic growth at the highest, but environmentally constrained, level with equitable welfare distribution. To achieve this requires that the economic system must be developed with equal community participation in deciding which ecological and economic considerations should be integrated. This part of the Pancasila urges fairness in welfare distribution which, in its broad sense, includes the quality of the environment and the sustainability of its natural resources.

Based on these five pillars of Pancasila, the spirit and ultimate goal of national development is to develop an Indonesian individual in his totality, meaning a person who is in harmony with his God the Creator, with his fellow citizens, and with his environment. The pillars of Pancasila provide every Indonesian, both as an individual and as a member of her or his community or nation, with all the guidance necessary for living in harmony with this environment. This is in line with the national commitment to pursue environmentally sound and sustainable development as it stated in the objectives and target of the second long-term development plan (1993-2018). In addition, the General State Guidelines of 1993 (as promulgated by the People's General Assembly) provides the following guidance as far as the uses of natural resources and environment are concerned:

> the use of natural resources as a source of people's welfare must be carried out according to rational, optimal and responsible planning, and be adapted to the carrying capacity of the environment, aiming at achieving the greatest possible gain for the welfare of the people while paying attention to the conservation of ecological functions and the balance of the environment for pursuing sustainable development (Republik Indonesia 1993).

Chapter Twenty -Two

Challenges

The chapters in this book present a wide range of challenges which must be met for the sustainable development of Java and Bali. This requires not only our consideration, not only our assent, but also our response and action. Some of these challenges are discussed briefly below.

INSTITUTIONAL REFORM

While it may sometimes seem as if government institutions are immovable and immutable, it behoves us well to stand back and consider whether the current structures and institutions really serve the environment well. Might there be other structures within which sustainable development would have a better chance of succeeding?

For example, we believe that the country's official conservation agency (PHPA) will not become truly effective until it is removed from the Department of Forestry. This problem is not unique to Indonesia, but a number of other countries have recognized that the inherent conflicts between conservation and exploitation are not best served when one is under the thumb of the other. It could be argued that PHPA has the same status (a directorate general) within the department as forest exploitation (PH), but the dominant flavour of the department is exploitative, even if this has mellowed over the last few years. In addition, PHPA is increasingly committed to marine conservation, and this sits uncomfortably within a department which is otherwise entirely devoted to terrestrial matters. Other countries have coped with this problem in different ways. Some have elected to have wildlife and conservation affairs handled regionally (certainly an option, at least, for Grand Forest Parks and Tourism Parks), whereas others have taken wildlife and conservation out of Forestry and into other departments, such as Tourism, Environment, Land or Internal Affairs (IUCN 1992). Urgent and open discussion is needed on the most appropriate placing in the Indonesian context. Opposition is likely to come from those who would see this as removing power and influence from certain quarters, but this is precisely why a broad debate is required.

Another possible, and even more radical, change would be to redraw the kabupaten boundaries so that they coincide with watershed or sub-watershed boundaries.

SUSTAINABLE DEVELOPMENT AT LOCAL AND HIGHER LEVELS

It must be clearly understood that sustainable development is not the exclusive domain of national governments or even of groups of environmental enthusiasts. It can be practiced at all levels – province, community, watershed, or production unit. In view of this, it is worthwhile to spell out some of the necessary elements of sustainable development practice, in order to challenge all those with even minor influence on, or interest in, the development process.

The main characteristics of ecologically sound sustainable development require policies and nonconflicting regulations to:
- improve and maintain the integrity of ecosystems by ensuring that modifications to them will be introduced only after careful evaluation of likely consequences;
- minimize environmental degradation and resource depletion, and find ways to mitigate impacts when and where necessary;
- minimize waste and maximize recycling and reuse of resources;
- determine acceptable and equitable access, control, and allocation;
- tackle the difficulties of integrated ecosystem improvement and socio-economic development;
- value biological and human diversity.

Thus, sustainable development must:
- involve all those who have the bureaucratic responsibility to make decisions, those who will be affected, and those who will effect the development;
- be compatible with environmental, spatial, temporal, sociocultural, and socioeconomic conditions;
- aim at resolving future, as well as immediate, problems;
- be acceptable not just on the grounds of economic viability but also on the grounds of sustainability and the involvement of the affected communities;
- be subject to evaluation and monitoring, allowing for response, mitigation, and feedback.

A current challenge is that presented by government efforts to combat poverty, such as through the Presidential Instruction No. 5 1993 on 'disadvantaged villages': those which the benefits of economic development seem to have passed by, or those which had been objects of development rather than its subjects. Under this initiative, each of the 20,000 named villages will receive Rp.20 million (US$10,000) to spend more or less as they wish, but with the aim of increasing their self-reliance and reducing their dependence. Before now these villages almost certainly had no participatory role in the largely top-down process of most decision making. As a result they have followed their own agendas for survival, often with serious ecological repercussions. It would be quite easy to fritter the money away, and they face a real challenge in making it contribute in some sustainable way. Money is only one part of their problem, of course, and some means should be devised to devolve a measure of power and authority to local institutions so that they can control the management of some of their surrounding natural resources.

CONFLICTS AND DECISION MAKING IN NATURAL RESOURCE MANAGEMENT

Conflicts often arise among the different groups of existing and potential future users of natural resources. When complaints are raised against some proposed exploitation, it does not automatically mean the protagonists are anti-development, and decision makers must become more aware that every project has its winners and losers, and that conflicts need to be managed in order to reach resolution rather than snuffed out.

Decision makers, especially where there is unofficial collusion with the project proponents, often do not realize the full impacts of their decisions, and there may be a reluctance to delve deep into the issues because even the superficial consideration of environmental aspects is still sometimes regarded as an obstacle to development. The suggestion may be made that people should be prepared to make voluntary sacrifices for the sake of development. But the challenge is to answer honestly the question of whether it is right, reasonable, and humane to ask for sacrifices from the poor and weak in society. Is it not more reasonable for the sacrifice to be made by the people who have a greater range of options open to them? Would it not be reasonable to make every effort to inform all levels of the community of the radical provisions and protection afforded under the Act No. 10 1992 concerning population development and prosperous families, even though they still have no implementing regulations?

In consideration of this, decisions on natural resource use require a process which fulfils the principles of both conservation and social justice. A mechanism is needed by which each group concerned has the same access to the information and to the decision-making process which will determine allocation of the resource. Unless this is done, it cannot be said that the eventual decision has taken all factors into consideration, or that there will be mitigation of unavoidable impacts. Opportunities must be given for each of the opposing parties to try to understand the views of the other. Resolution can be sought through the courts, but before this stage is reached, attempts should be made to find solutions through conciliation, facilitation, negotiation and fact finding, and mediation. Third parties should be mutually trusted individuals or groups, rather than government representatives. These steps are all essential for a demonstrably clean and transparent government. The process would be helped if those passages in the Act No. 4 1982 on environmental management which concern public participation were brought into effect.

Also relevant to decision making is the problem of law enforcement which cannot yet be regarded as widely effective. It has been said that the law is like a spider's web: large birds can fly straight through, barely aware of its existence, whereas small flies need get only one leg caught on a sticky strand to be irretrievably caught. It is clear that an upright and clean legal system and a real understanding of environmental issues are required by the judiciary. Transparency through the whole process will encourage decision makers to be consistent in their carrying through of policies. Informed and balanced public pressure can encourage clean government and environmental management based on conservation and social justice.

CHANGING LIFESTYLES

The last, and perhaps greatest, challenge is that which concerns each one of us and our personal actions. While conservation and environmental protection are ultimately the government's responsibility, the Act No. 4 1982 on environmental management gives all citizens the responsibility of being involved. It must be obvious by now that, while appropriate government policies will achieve much, living within a society committed to sustainable development depends at least as much on individual and community involvement. Everyone must start to apply ecology in its simplest forms – ensuring that we do not extend our activities beyond the capability of the ecosystems within which we live, and showing that this applies to all of us. The relentless pursuit of growth in a finite world will bring inevitable environmental degradation, and we might as well venture as far as we can in terms of living an ecologically appropriate lifestyle before prohibitions force the issue. If policies are put in place which will encourage or facilitate the necessary changes, so much the better, but it is a hollow excuse to delay action until every other constraining factor has been dealt with.

The questions most often posed in consideration of lifestyles is whether individual or small-scale changes are of any use, and whether even the cumulative impacts of reducing consumption will have any significant effects, given both the severity and scale of the problems, and the fact that many of the problems are distant from our homes, work places, or schools. The answer to both must be 'yes', although it must be acknowledged that change will not occur overnight. Adopting more sustainable lifestyles will, in any case, probably lead to greater health, greater happiness, less 'wasted' money, greater ability to help others, greater environmental safety, and hence a richer quality of life. Most of the people reading this book will be among the economic groups most able to make choices and decisions about individual and corporate lifestyle. They have the luxury of examining valid options and the responsibility of knowing that those choices will make a difference, down the line, to other people and to the natural resources. We recognize that not all people are in that position.

Despite the many contrasts, writing about lifestyle changes simultaneously for a western and an Indonesian audience is possible. There are many Indonesians who have a material standard of living and quality of life, far in excess of that of many people from western countries, and suggesting that the situation in Indonesia is different from that in the West is missing the point. So, for the purposes of this book, a simple list of guidelines or thought-points is provided for honest consideration (Elgin 1981):

- does what I own or buy promote activity, self reliance, and involvement, or does it induce passivity and dependence?
- are my consumption patterns basically satisfying, or do I buy much that serves no real need?
- how tied is my work and lifestyle to credit repayments, maintenance and repair costs, or to the expectations of others?
- do I consider the impact of my consumption patterns on other people and on the Earth?

For example:

- do I waste petrol, electricity, paper, and other common resources?
- are my consumption patterns at work and at home the same?
- do I discriminate against overwrapped products?
- do I favour local produce?
- do I refuse unnecessary plastic bags?
- do I favour or ask for food produced using organic methods – with no or very limited use of pesticides or inorganic fertilizers (Fukuoka 1978, 1991)? One can be certain they will not be available until there is a demand.
- do I consider options for recycling and reusing the products which I purchase?
- do I minimize the consumption of meat? Remember that 80% of our planet's agricultural land is used to feed animals, and only 20% to feed humans directly.

The aim of these lists is not to promote self-denial leading to suffering, pain, deprivation, and new forms of dependence, but rather to find a balance of simplicity, the 'middle way', producing richness without compromising other people or the support mechanisms of the planet. While the Pancasila encourages the maintenance of spirit and matter, increasing influence from outside is tearing the two apart and laying more emphasis on the second, creating more desires for more material goods which often do not serve to reinforce our relationship with nature and the global human community. Maybe the most important challenge is questioning what is actually important or meaningful in our lives, and to question whether we have to go along with the economists and commercial giants who require us all to consume more to keep the economy growing and disappearing into irretrievable unsustainability.

Finally, it is action that matters. There is little point being convinced if one does not let others know and encourage them to join your ecologically appropriate way of life. This is particularly the case when the 'others' are in a position to make significant change or to have disproportionate influence. Some of this can be done verbally, but the power of the pen is much greater. For example, one could write to:

- local and national newspapers and magazines to generate discussion and responsible coverage of environmental issues;
- your bank to enquire how your money is invested;
- manufacturers of goods you feel may be environmentally damaging, persisting until you get an acceptable and useful reply;
- environmental NGOs and join them to give support to their activities;
- ministers and governors to encourage them to initiate environmental programmes such as recycling;
- shop managers to complain, if necessary, about excessive packaging (which you, the consumer, pay for);
- local government offices to inform them of environmental impacts that concern you; and
- managers in the private sector to let them know how good (or bad) you feel their products are from an environmental point of view, and let your purchasing power reflect your understanding of the issues.

We (we who write, we who read, we who care) cannot just sit back and allow the world to continue its possible descent into the vortex. We are the last generation to be able to do anything to reverse environmental deterioration.

Appendices

Appendix 1. Key to plant family abbreviations.

Acan.	Acanthaceae	Cype.	Cyperaceae
Acer.	Aceraceae	Dati.	Datiscaceae
Acti.	Actinidiaceae	Denn.	Dennstaedtiaceae
Adia	Adiantaceae	Dict.	Dictyotaceae
Agav.	Agavaceae	Dill.	Dilleniaceae
Aizo.	Aizoaceae	Dios.	Dioscoreaceae
Alis.	Alismataceae	Dipt.	Dipterocarpaceae
Amar.	Amaranthaceae	Diptd.	Dipteridaceae
Anac.	Anacardiaceae	Eben.	Ebenaceae
Anno.	Annonaceae	Elae.	Elaeocarpaceae
Apoc.	Apocynaceae	Epac.	Epacridaceae
Arac.	Araceae	Eric.	Ericaceae
Arau.	Araucariaceae	Erio.	Eriocaulaceae
Ascl.	Asclepiadaceae	Euph.	Euphorbiaceae
Aspl.	Aspleniaceae	Faga.	Fagaceae
Bals.	Balsaminaceae	Flac.	Flacourtiaceae
Bego.	Begoniaceae	Fuca.	Fucaceae
Berb.	Berberidaceae	Gent.	Gentianaceae
Blec.	Blechnaceae	Gesn.	Gesneriaceae
Bomb.	Bombacaceae	Glei.	Gleicheniaceae
Bora.	Boraginaceae	Gnet.	Gnetaceae
Brom.	Bromeliaceae	Good.	Goodeniaceae
Burs.	Burseraceae	Grac.	Gracilariaceae
Cact.	Cactaceae	Gram.	Gramineae
Cann.	Cannaceae	Gramt.	Grammitidaceae
Capp.	Capparidaceae	Gutt.	Guttiferae
Capr.	Caprifoliaceae	Hama.	Hamamelidaceae
Cari.	Caricaceae	Hang.	Hanguanaceae
Casu.	Casuarinaceae	Hydrc.	Hydrocaryaceae
Caul.	Caulerpaceae	Hydr.	Hydrocharitaceae
Cecr.	Cecropiaceae	Hype.	Hyperiacaceae
Cera.	Ceratophyllaceae	Jugl.	Juglandaceae
Chen.	Chenopodiaceae	Junc.	Juncaceae
Chry.	Chrysobalanaceae	Labi.	Labiatae
Comb.	Combretaceae	Laur.	Lauraceae
Comm.	Commelinaceae	Lecy.	Lecythidaceae
Comp.	Compositae	Legu.	Leguminosae
Conv.	Convolvulaceae	Lemn.	Lemnaceae
Cucu.	Cucurbitaceae	Lent.	Lentibulariaceae
Cuno.	Cunoniaceae	Lili.	Liliaceae
Cyat.	Cyatheaceae	Limn.	Limnocharitaceae
Cyca.	Cycadaceae	Loga.	Loganiaceae

Lyth.	Lythraceae	Rham.	Rhamnaceae
Magn.	Magnoliaceae	Rhiz.	Rhizophoraceae
Malv.	Maranthaceae	Rosa.	Rosaceae
Mara.	Marattiaceae	Rubi.	Rubiaceae
Mars.	Marsiliaceae	Ruta.	Rutaceae
Mela.	Melastomaceae	Sabi.	Sabiaceae
Meli.	Menispermaceae	Sali.	Salicaceae
Moll.	Molluginaceae	Salv.	Salvadoraceae
Mora.	Moraceae	Sapi.	Sapindaceae
Musa.	Musaceae	Sapo.	Sapotaceae
Myri.	Myristicaceae	Sarg.	Sargassaceae
Myric.	Myricaceae	Saur.	Saururaceae
Myrs.	Mysinaceae	Scro.	Scrophulariaceae
Myrt.	Myrtaceae	Sima.	Simaroubaceae
Nyct.	Nyctaginaceae	Sola.	Solanaceae
Nymp.	Nymphaeaceae	Sonn.	Sonneratiaceae
Ochn.	Ochnaceae	Ster.	Sterculiaceae
Palm.	Palmae	Symp.	Symplocaceae
Pand.	Pandanaceae	Thea.	Theaceae
Pass.	Passifloraceae	Tili.	Tiliaceae
Pipe.	Piperaceae	Trap.	Trapaceae
Podo.	Podocarpaceae	Ulma.	Ulmaceae
Podos.	Podostemaceae	Umbe.	Umbelliferae
Poly.	Polygonaceae	Urti.	Urticaceae
Polyp.	Polypodiaceae	Verb.	Verbenaceae
Pont.	Pontederiaceae	Viol.	Violaceae
Prim.	Primulaceae	Vita.	Vitaceae
Raff.	Rafflesiaceae	Xyri.	Xyridaceae
Ranu.	Ranunculaceae	Zing.	Zingiberaceae

Appendix 2. Selected vegetation studies and surveys.

1. Ujung Kulon NP
Hartoto 1986
Hommel 1987
Hoogerwerf 1952, 1970
Kartawinata and Apandi 1977
Partomihardjo and Tagawa 1987
Soenarko 1971
Soerianegara 1970

2. Pulau Dua
Partomihardjo 1986

3. Dungus Iwul
Riswan 1975

4. Mt. Haliman NP
UNDP/FAO 1978

5. Ciampea
van Steenis 1931

6. Dapur Island
van Steenis 1935

7. Mt. Salak
Hartman 1939

8. Jakarta
Visser Smits 1914

9. Lengkong
Kostermans 1970
Wiriadinata 1977

10. Cibinong
van Steenis 1934

11. Mt. Gede-Pangrango NP
Bruggeman 1927a, b
Docters van Leeuwen 1924
Kramer 1933
Meijer 1954a, b, 1955, 1959
van Steenis 1928, 1937, 1941
Yamada 1975, 1976a, b, 1977

12. Telaga Warna
Wulijarni-Soetjipto and Hambali
1974

13. Cibitung
Den Hoed and van der Meer 1941

14. Cariu
Erythrina 1915

15. Mt. Sanggabuana
van Steenis 1934

16. Mt. Cadaspanjang and Tikukur
Kartawinata 1977

17. Mt. Patuha
Lörzing 1921

18. Padalarang
van der Pijl 1933

19. Burangrang
Danimihardja and Anggarwulan 1976

20. Mt. Tilu Reserve
Harun and Tantra 1984
Lembaga Ekologi 1980
Sukardjo 1978

21. Mt. Malabar Reserve
van Welsem 1919

22. Mt. Papandayan
Docters van Leeuwen 1930
Riswan 1976
van Steenis 1920, 1935, 1937

23. Mt. Guntur
van der Pijl 1938

24. Mt. Tampomas
van der Pijl 1929

25. Leuweung Samcang Reserve
Sidiyasa 1985, 1986

26. Indramaju
van Steenis 1936

27. Mt. Ciremai
Lam 1925

28. Pangandaran Reserve
Sukardjo 1977, 1983
Kartawinata and Sumardja 1977

29. Cilacap
Iskandar 1984

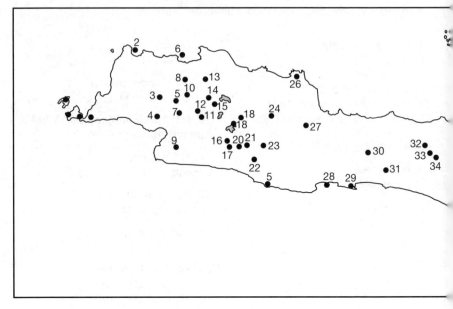

Figure A2.1. Sites of vegetation studies and surveys.

30. Mt. Slamet
Backer 1914
Lam 1924
Abdulhadi 1978

31. Serayu
Bruijnzeel 1984

32. Dieng Plateau
Bunnemeyer 1918
Irawati 1982
Loogen 1941
Wiljes-Hessink 1953

33. Mt. Sindoro
Docters van Leeuwen 1929
Gunadi and Notosoedarmo 1988

34. Mt. Sumbing
Docters van Leeuwen 1929

35. Mts. Merbabu-Merapi
Gunadi and Notosoedarmo 1988

36. Parangtritis
Lutjeharms 1937

37. Mt. Lawu
Docters van Leeuwen 1925

38. Rembang
Beumee 1921

39. Mt. Wilis
Lorzing 1914

40. Rawah Bening
Coert 1933, 1934

41. Jombang
de Voogd 1927, 1928

42. Mt. Kawi-Kelud Reserve
Clason 1935
Docters van Leeuwen 1935

43. Arjuno Reserve
van Steenis 1926

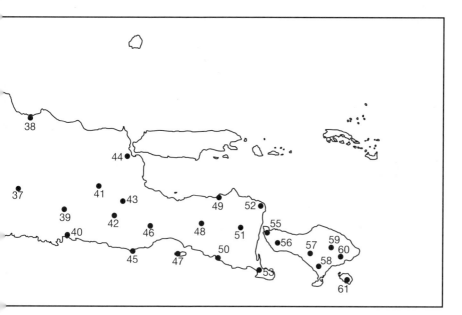

44. Suci
Geerts-Ronner 1924, 1940

45. Sempu Reserve
Appelman 1940

46. Bromo-Tengger-Semeru NP
Leefmans 1914
Postma 1960

47. Nusa Barung Reserve
Jacobs 1958

48. Yang Plateau Reserve
Kooiman and van der Veen 1936
van Steenis 1937
Loogen 1940

49. Mt. Ringgit
Clason and Laarman 1932

50. Meru Betiri NP
Dransfield 1973

51. Ijen-Merapi Reserve
van Steenis 1940

52. Baluran NP
Appelman 1937
Clason 1934
Partomihardjo and Mirmanto 1986
Wind and Amir 1977
Wiriosoepartho 1984

53. Alas Purwo NP
Jacobs 1958
Saim 1986

54. BALI - General
Beumee1929
Meijer 1985
Sukardjo1982
van Steenis 1937

55. Prapat Agung
Daryono 1985

Soepadmo 1961

56. Bali Barat NP
Soepadmo 1985
de Voogd 1937

57. Batukau
Soepadmo 1961
Tantra and Anggana 1977

58. Sangeh
de Voogd 1937
Meijer 1985
Soepadmo 1961

Tantra 1982

59. Mt. Batur
de Voogd 1937
Lieftinck and van Steenis 1940
Soepadmo 1961

60. Mt. Agung
Dilmy 1965

61. Nusa Penida
de Voogd 1937

Appendix 3. Sites of selected bird surveys and observations on Java and Bali.

WEST JAVA

1. Ujung Kulon
Compost and Milton 1986
Hoogerwerf 1948, 1969, 1970

2. Panaitan
Hoogerwerf 1953

3. Cimungkat
Hoogerwerf 1948

4. Carita
van Balen and Lewis 1991

5. P. Deli
Holmes and van Balen 1990

6. P. Tinjil
Holmes and van Balen 1990

7. Labuan
Hoogerwerf 1948

8. Ranca Danau
Milton 1984

9. Pulau Rambut
Wiriosoepartho 1986
Sukardjo and Santosa 1982
van Strien 1981
Allport and Milton 1988
Lambert and Erftemeijer 1989
Pakpahan et al. 1991

10. Pulau Dua/Sawah Luhur
Allport and Milton 1988
Milton and Marhadi 1984, 1985, 1986
van Strien 1981

11. Cengkareng
Holmes 1988
Whitten 1989

12. Muara Angke
Allport and Milton1988
Milton and Marhadi 1984

13. Tangkuban Perahu (Pelabuhan Ratu)
Erftemeijer 1987

14. Pelabuhan Ratu
Hoogerwerf 1948

15. Jakarta
Vorderman 1882-85
Hoogerwerf and Siccama 1938

16. Mt. Pancar
Erftemeijer 1987

17. Bogor
van Balen et al. 1988
Farimansyah 1981
Diamond et al. 1987
Hoogerwerf 1948, 1949a

18. Mt. Salak
Vorderman 1885
Erftemeijer 1987
Hoogerwerf 1948

19. Mt. Gede Pangrango
Andrew and Milton 1988
Erftemeijer 1987
Andrew 1985
Hoogerwerf 1949b

20. Cibodas
Spennemann 1923
Dammerman 1929
Hoogerwerf 1948, 1949

21. Megamendung
Erftemeijer 1987

22. Telaga Warna
van Balen 1987
Erftemeijer 1987

23. Halimun
Erftemeijer 1987

24. Ciomas
Hoogerwerf 1948

25. Mt. Walet
Erftemeijer 1987
Hernowo 1989

26. Cikidang
van Helvoort 1984

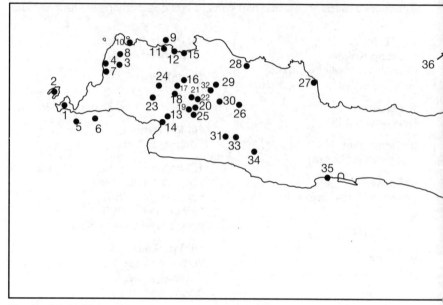

Figure A3.1. Sites of bird surveys on Java and Bali.

Iskandar 1984

27. Indramayu, Cirebon, and Karawang
Allport and Milton 1988
Sudarmana 1981

28. Ciasem Bay
Spennemann 1924
van Balen and Noske 1991
van Helvoort 1985

29. Cilangkap
van Helvoort 1987
Iskandar 1984

30. Pasir Jatiluhur
van Helvoort 1981
Iskandar 1984

31. Telaga Patenggang
Erftemeijer 1987

32. Mt. Sanggabuana

Erftemeijer 1987

33. Mt. Manglayang S. Bandung
Erftemeijer 1987

34. Papandayan
Stresemann 1930
Hoogerwerf 1948

35. Pangandaran
Sanguila 1976

CENTRAL JAVA

36. Bawean
Oberholser 1917

37. Rawa Pening
Quirk et al. 1979

38. Manggir S. Yogyakarta
Buil 1984

39. Gajah Mungkur
Kuslestari 1983

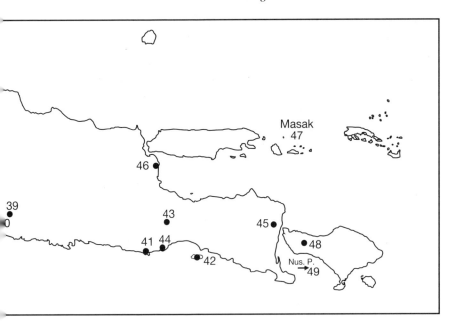

40. Rongkop
Bartels and Bouma 1937
de Korte and Silvius 1990

EAST JAVA

41. Sempu
Appelman 1940

42. Nusa Barung
Kooiman and van der Veen 1940
Kooiman 1940
Hoogerwerf 1947

43. Tengger
Hoogerwerf 1948
Bemmel-Lieneman and van Bemmel
　1940

44. Lebakharjo
van Balen 1989

45. Ijen

Hoogerwerf 1948

46. Brantas and Solo
Erftemeijer and Djuharsa 1988
Hoogerwerf 1935

47. Masakambing
Atmosoedirdjo 1985

48. BALI
Alikodra and Amzu 1984
Ash 1984
Bruce 1982
Green 1991
Klapste 1984
Mason 1988, 1990
Mason and Jarvis 1989
Rensch 1930
Stresemann 1913
van Balen 1991
van Helvoort 1987

van Helvoort and Soetawijaya 1987
Wiegant and van Helvoort 1987

49. Nusa Penida
van Helvoort 1989

Note: Other papers are listed in Hoogerwerf (1953a, b, c, d) and an important summary is due to be published (van Balen in prep.).

Java general
Hardjowigeno 1989
Ongkosongo 1989
Soerianegara 1971
Sukardjo 1989b
Sukardjo 1989c
Sukardjo and Akhmad 1982
Usman 1984

Northwest Java
Alrasjid 1986
Effendi 1989
Harminto and Yusuf 1978
Jafarsidik and Anwar 1987
Jhamtani and Djaja 1985
Kusmana 1983
Sukardjo 1982
Sukardjo 1986
Sukardjo 1987
Sukardjo 1989
Widiarti and Effendi 1989

Ujung Kulon
Moro et al. 1986
Reksodihardjo et al 1986

North Java
Hendrato and Hutabarat 1981

Appendix 4. Selected coral studies and surveys.

Karimun Jawa
Indriati 1984
Kusyanto 1984
PSL-UNDIP 1982

P. Air
Romimohtarto and Moosa 1977
Sukarno 1975

P. Panjang
Aryono 1987
Foster-Smith 1977

P. Pari
Kaniawati 1985
Brown et al. 1983
Aziz 1981
Darsono et al. 1978
Heryanto 1984

P. Rambut
Prahoro 1986

Nusa Barung
Usher 1984

Meru Betiri
Usher 1984

Bandengan Bay
Supriharyono 1985, 1986

Kangean
Beudels et al. 1981

Sapudi
Usher 1983

Thousand Islands
Verwey 1930, 1931a, b
Brown et al. 1983
Verstappen 1975
Zanefeld and Verstappen 1952a, b
Scrutton 1977

Baluran
Ritchie 1991

P. Rakit
Zanefeld and Montagne 1958

P. Menjangan
Halim et al. 1983
Ritchie 1991

Appendix 5. Selected mangrove studies and surveys.

Java general
Hardjowigeno 1989
Ongkosongo 1989
Soerianegara 1971
Sukardjo 1989b
Sukardjo 1989c
Sukardjo and Akhmad 1982
Usman 1984

Northwest Java
Alrasjid 1986
Effendi 1989
Harminto and Yusuf 1978
Jafarsidik and Anwar 1987
Jhamtani and Djaja 1985
Kusmana 1983
Sukardjo 1982
Sukardjo 1986
Sukardjo 1987
Sukardjo 1989
Widiarti and Effendi 1989

Ujung Kulon
Moro et al. 1986
Reksodihardjo et al 1986

North Java
Hendrato and Hutabarat 1981
PSL-UNDIP 1982
Sukardjo 1986

P. Dua
Munaf 1978
Boeadi 1978

P. Rambut
Kartawinata and Waluyo 1977
Brotonegoro and Abdulkadir 1978
Dwisasanti 1987

North Java - Central
Martodigdo et al. 1985
Hendrarto 1980

Segara Anakan/Cilacap
Anwar 1987
Anwar and Sumarna 1987
Azkab et al. 1986
De Haan 1931

Djohan 1982
Hardjosuwarno 1989
Hutomo and Djamali 1978
Marsono 1989
Martodigdo et al. 1986
Martodigdo et al. 1987
Pratiwi et al. 1986
Soeroyo 1956
Toro 1978
Toro and Sukardjo 1989
Wirjodarmodjo et al. 1978

Solo-Brantas
Schuster 1948
Erftemeer and Djuharsa 1988
Enex 1994

Segoro Anak/Grajagan
Ritchie 1991
Soerojo and Sukardjo 1991
Sulistiadi 1986

Baluran
Indiarto et al. 1986

Bali
Ritchie 1991
Sukardjo and Akhmad 1982
Sukarme and Nuitja 1986
Sulistijo and Atmadja 1980

Bibliography

Aak. 1993. *Teknik bercocoktanam jagung.* Yogyakarta: Kanasius.

Abdoellah, O. S. 1990. Homegardens in Java and their future development. In *Tropical home gardens*, ed. K. Landauer, and M. Brazil, 69-79. Tokyo: United Nations Univ. Press.

Abdulhadi, R. H. 1978. Fitososiologi hutan pegunungan di lereng selatan Gunung Slamet, dekat Purwokerto, Jawa Tengah. UNSOED, Purwokerto.

Abdullah, A. 1991. Rasionalisasi pemanfaatan ekosistem mangrove dipandang dari sudut konservasi. In Prosidings Seminar IV Ekosistem Mangrove, ed. S. Soemodihardjo et al., 65-68. Bandar Lampung, 7 August 1990. Jakarta: MAB-LIPI.

Abdullah, A. R. 1991. Persistent organochloride pesticides in the tropical environment. *Wallaceana* 64: 1-6.

Abdullah, O. S., and G. Marten. 1983. Production of human nutrients from pekarangan, kebun and sawah agricultural systems in Jatigede area (West Java). EAPI-SUAN Symposium on research on impact of development in human activity systems in southeast Asia, Bandung: Institute of Ecology UNPAD.

Abdurrahman, F. 1980. *Studi jenis-jenis siput dan penyebarannya di DKI Jakarta.* FIPIA UI, Jakarta.

Ackery, P. R., and R. I. Vane-Wright. 1984. *Milkweed butterflies: Their cladistics and biology.* London: British Museum (Natural History).

Acosta, and Rullin. 1991. *Environmental impact of the golden snail (*Pomacea *sp.) on rice farming systems in the Philippines.* Manila: ICLARM.

Adams, W. M. 1990. *Green development: Environment and sustainability in the Third World.* London: Routledge.

Adianto, and S. Sastrodihardjo. 1982. Soil arthropods from three different communities in the surrounding area of Mt. Tangkuban Perahu (West Java). In BIOTROP 10 months Training Course on Forest Ecology, Bogor, 1 June 1982. Bogor: BIOTROP.

Adimihardja, K. 1992. *Kasepuhan yang tumbuh di atas yang luruh: Pengelolaan lingkungan secara tradisional di kawasan Gunung Halimun, Jawa Barat.* Bandung: Tarsito.

Adimihardja, K., and J. Iskandar. 1993. Kasepuhan: ecological influences on traditional agriculture and social organization in Mt. Halimun of West Java, Indonesia. Paper given at Pithecanthropus Centennial 1893-1993, Leiden, 26 June.

Adisewojo, S. S., S. Tjokronegoro, and R. Tjokronegoro. 1984. Natural biological compounds traditionally used as pesticides and medicines. *Environmentalist* 4 (Suppl. 7): 11-14.

Adisoemarto, S. 1977. Mosquitoes of the Bogor Botanical Garden. *Berita Biologi* 2: 25-30.

Adisusmianto. 1984. *Telaah kemungkinan pemanfaatan daerah penyangga Taman Nasional Baluran ditinjau dari kebutuhan masyarakat terhadap kayu bakar.* IPB, Bogor.

Adiwibowo, S. 1983. Sistem sosial ekologi tambak dan sawah di wilayah pesisir Kabupaten Karawang. Ph.D. diss., Fakultas Pasca Sarjana IPB, Bogor.

Agung, A. A. G. 1991. *Bali in the 19th century.* Jakarta: Obor.

Ahmad, H. 1987. *Tanggapan masyarakat nelayan terhadap pelestarian laut utara di Kabupaten Situbondo.* Pusat Penelitian UNEJ, Jember.

Ahmadi, Y. 1985. *Studi perilaku kerbau liar (*Bubalus bubalis *L.) dalam proses domestikasi di Taman Nasional Baluran Jawa Timur.* Fakultas Kehutanan IPB, Bogor.

Aimi, M. 1981. Fossil *Macaca nemestrina* (Linnaeus 1766) from Java, Indonesia. *Primates* 22: 409-413.

—-. 1989. A mandible of *Sus stremmi* Koenigswald 1933, from Cisaat, Central

Java, Indonesia. *Publ. Geol. Res. Dev. Cen. Palaeontol. Ser.* 6: 4-10.

Akhtar, M. S., and M. Achmad. 1985. A new nasute termite from Java, Indonesia (Isoptera: Termitidae: Nasutitermitinae). *Pakistan J. Zool.* 17: 215-218.

Akmad, M. 1986. *Budidaya ikan di sawah tambak.* Jakarta: Simplex.

Akratanakul, P. 1986. Beekeeping in Asia. FAO, Rome.

—. 1987. Honeybee diseases and enemies in Asia: a practical guide. FAO, Rome.

Ali, F. 1990. Kehidupan ikan baung (*Macrones nemurus*) di Lubuk Senggotan, Sungai Cisadane. In *Biologi perairan sekitar Bogor.* F. Sabar, 7-14. Bogor: Balitbang Biologi Perairan - LIPI.

Aliadi, A., E. A. M. Zuhud, and E. Djamhuri. 1990. Possibilities of cultivating edelweiss with stem cuttings. *Media Konservasi* 3: 37-45.

Alibasah, M. M. 1990. *Folk tales from Bali and Lombok.* Jakarta: Djambatan.

Alikodra, H. S. 1978. *Pola pembinaan dan pengembangan Suaka Margasatwa Bali Barat.* Sekola Pasca Sarjana IPB, Bogor.

—. 1983. Ekologi banteng (*Bos javanicus* d'Alton) di Taman Nasional Ujung Kulon. Ph.D. diss., Fakultas Pasca Sarjana IPB, Bogor.

—. 1984. *Penelitian pengembangan buffer zone pelestarian alam: Taman Nasional Meru Beteri.* Kantor Menteri Negara KLH, Jakarta.

—. 1986. *Tanaman eksotik akasia* (Acacia nilotica) *dan masalahnya bagi ekosistem savana di Taman Nasional Baluran.* Fakultas Kehutanan IPB, Bogor.

—. 1987. Masalah pelestarian Jalak Bali. *Media Konservasi* 1: 21-27.

Alikodra, H. S., and E. Amzu. 1984. *Studi tentang pengaruh tanaman pekarangan terhadap kelestarian burung di wilayah Dati II Kabupaten Bogor.* Fakultas Kehutanan IPB, Bogor.

Alkantan, C., and Lambertus. 1986. *Map of Luweng Ngepoh.* HIKESPI, Bogor.

Alldredge A.L., and J. M. King. 1977. Distribution, abundance and substrate preferences of demersal reef zooplankton at Lizard Island Lagoon, Great Barrier Reef. *Mar. Biol.* 41: 317-333.

Allport, G., and G. R. Milton. 1988. A note on the recent sighting of *Zosterops flava,* Javan White-eye. *Kukila* 3: 142 -149.

Alrasyid, H. 1977. Kelestarian hutan mangrove. Prosiding Seminar II Perikanan Udang, 382-392. Jakarta, 15 March 1977.

—. 1986. Jalur hijau untuk pengelolaan hutan mangrove Pamanukan, Jawa Barat. *Bull. Pen. Hutan* 475: 29-74.

Altieri, M. A. 1991. Increasing biodiversity to improve insect pest management in agroecosystems. In *The biodiversity of microorganisms and invertebrates: Its role in sustainable agriculture,* ed. D. L. Hawksworth, 165-182. Wallingford: CAB International.

Altona, T. 1912. Over de waarde van inlandsche namen van Javaansche woudboomen. *Tectona* 5: 417-419.

—. 1913. *Rapport ropens het voorloopig hydrologisch onderzoek van het Brantas gebied.* Dep v. L.N. en H.

—. 1924. Djati en Hindoes: De invoer van djati op Java. *Tectona* 17: 865-906.

—. 1926. Djati en Hindoes. *Tectona* 19: 939-1011.

Alvise, J. C. 1989. A role for molecular genetics in the recognition and conservation of endangered species. *Trends Ecol. Evol.* 4: 279-281.

Amann, V. et al. 1989. Remote sensing of water parameters in Madura Bay, Java, Indonesia. *Neth. J. Sea Res.* 23: 473-482.

Amas, K., and I. Purwanto. 1992. Run-off and soil erosion under various land cover in Kadipaten, West Java. *Bull. Pen. Hutan* 547: 1-11.

Ami, M. 16 June 1993. Hutan bakau Losarang lindungi tambak dari serangan virus. *Kompas.*

Aminudin, S., A. Gozali, T. P. Sinaga, and P. Brahmana. 1985. *Penelitian tentang ekosistem pantai selatan Cilacap dan Kebumen.* PSL UNSOED, Purwokerto.

Amir, M. 1992. Pengembangan penelitian ilmiah di Cagar Alam Gunung Halimun, Jawa Barat. Workshop Cagar Alam Gunung Halimun, Jakarta, January 1992.

Amir, M., and A. H. S. Atmowidjojo. 1983. *Chironomus dan invertebrata air tawar pakan ikan lainnya di kolam Kebun Raya Bogor, kolam ikan Cisaat dan Gunung Jaya, Sukabumi.* Paper presented at Kongres Entomologi II, Jakarta, 24 January 1983.

Ammann, H. 1985. Contributions to the ecology and sociology of the Javan rhinoceros (*Rhinoceros sondaicus* Desm.). Ph.D. diss., Univ. of Basel, Basel.

Anderson, B. 1993. The Philippine golden snail disaster. *Ecologist* 23: 70-72.

Anderson, I. October 1993. Keep taking the frog skin. *New Scient.,* 10.

Andrew, P. 1985. An annotated checklist of the birds of the Cibodas-Gunung Gede Reserve. *Kukila* 2: 10-28,

—. 1992. *The birds of Indonesia: A checklist*

(Peter's sequence). Jakarta: Indonesian Ornithological Society.

Andrew, P., and G. R. Milton. 1988. A note on the Javan scops-owl *Otus angelienae* Finsch. *Kukila* 3: 79-81.

Andrews, T. 1993. The first record of the Chinese egret on Java. *Kukila* 6: 133.

Annis, B. et al. 1990. *Toxorhychites amboinensis* larvae released in domestic containers fail to control dengue vectors in a rural village in Central Java, Indonesia. *J. Am. Mosquito Control. Assoc.* 6: 75-78.

Anon. 1854. Verslagen van de natuurkundige vereniging over bezoeken aan Bantam. *Natuurk. Tijds. Ned.-Ind.* 6: 15.

——. 1981. *Pra ekspedisi Lueweng Ombo, Pacitan (Jawa Timur)*. Specavina Jakarta - Mapala UI, Jakarta.

——. 1983. *Java caves 1983*. The Federation of Indonesian Speleological Activities, Bogor.

——. 1984. *Laporan hasil karya wisata di Cagar Alam Pananjung Pangandaran*. FMIPA IPB, Bogor.

——. 1984. *Penelitian pengembangan buffer zone pelestarian alam Taman Nasional Meru Betiri*. Kantor Menteri Negara KLH, Jakarta.

——. 1984. Upaya pelestarian hutan di daerah batuan kapur Gombong Selatan. Paper presented at Simposium Upaya Pengelolaan Sumber Daya Alam dan Pelestarian Kemampuan Daya Dukung Lingkungan Hidup di Daerah Gombong Selatan, Kabupaten Kebumen, Semarang, 26 November 1984.

——. 1988. *Penangkaran rusa Bawean (Axis kuhlii) di Desa Sendang Biru, Kecamatan Sumber Manjing Wetan, Kabupatan Malang*. PSL UNIBRAW, Malang.

——. 1990. *The Oxford Univ. Herpetological Expedition to West Java 1990*. Unpublished.

Anwar, A. 1982. Analisis produksi usaha tani dalam rangka melihat kemungkinan pengingkatan daya dukung sumber daya alam dan lingkungan hidup di Wadas Cimanuk. *Lingkungan dan Pembangunan* 2: 143-159.

Anwar, C. 1987. Pengaruh relung terhadap laju pelapukan serasah daun mangrove di Rawa Timur, Cilacap. *Bull. Pen. Hutan* 492: 11-19.

Anwar, J., S. J. Damanik, N. Hisyam, and A. J. Whitten. 1984. *Ekologi ekosistem Sumatera*. Yogyakarta: Gadjah Mada Univ. Press.

Aoki, J. I., G. Takaku, and F. Ito. 1994. Aribatidae, a new myrmecophilous oribatid mite family from Java. *Int. J. Acarol.* 20: 3-10.

Appelman, F. J. 1930. Wildbescherming. *Tectona* 23: 582-606.

——. 1937. De Baloeran. In *Album van Natuurmonumenten in Nederlandsch-Indie*, ed. C. G. G. J. van Steenis, 50-56.

——. 1939. Het schiereiland Poerwo: Bosch en wild in Java's Zuidoost-hoek. In *3 jaren indisch natuurleven*, ed. C. G. G. J. van Steenis, 293-298. Batavia: Nederlandsch-Indische Vereeniging tot Natuurbescherming.

——. 1940. Poeloe Sempoe. *Trop. Nat.* 29: 164-168.

Appelman, F. J., and G. F. H. Siccama. 1939. Noesa Baroeng: Natuurmonument tegenover Besoeki's zuidkust. In *3 jaren indisch natuurleven*, ed. C. G. G. J. van Steenis,, 289-292. Batavia: Nederlandsch-Indische Vereeniging tot Natuurbescherming.

Arief, A. J. 1987. The composition and distribution of coral-reef molluscs community in Leuweung Sancang Reserve, West Java, Indonesia. *Berita Biologi* 3: 310-316.

Arthington, A. H., and L. N. Lloyd. 1990. Introduced poeciliids in Australia and New Zealand. In *Ecology and evolution of live-bearing fishes (Poeciliidae)*, ed. G. K. Meffe, and F. F. Snelson, 333-348. Englewood Cliffs: Prentice Hall.

Aryono, B. 1987. Penelitian tentang kepadatan teripang (Holothuroidea) di perairan karang pantai Bandengan dan perairan karang Pulau Panjang Timur, Jepara. Fakultas Peternakan UNDIP, Semarang.

ASEAN/US CRMP. 1992. *The integrated management plan for Segara Anakan-Cilacap, Central Java, Indonesia*. ICLARM, Manila.

Ash, J. S. 1984. Bird observations on Bali. *Bull. Brit. Orn. Cl.* 104: 24-35.

——. 1993. Raptor migration on Bali, Indonesia. *Forktail* 9: 3-11.

Ashby, K., and C. Santiapillai. 1988. The status of the banteng *(Bos javanicus)* in Java and Bali. *Tigerpaper* 15: 16-25.

Ashton, P. 1989. Sundaland. In *Floristic inventory of tropical countries*, ed. D. J. Campbell, and H. D. Hammond, 91-99. New York: New York Botanical Garden.

Ashton, P. S. 1982. Dipterocarpaceae. *Fl. Mal. I* * 9: 237-552.

Askew, A. 1991. Learning to live with floods. *Nat. Res.* 27: 4-8.

Asmoro, P. B. 1993. *Survey ekologis Rawa Lakbok Kabupaten Ciamis (Jawa Barat)*. Asian Wetland Bureau, Bogor.

Asnawati. 1990. Macrofauna bentik di beberapa anak sungai Cisadane - Bogor. In *Biologi perairan sekitar Bogor*, ed. F. Sabar, 32-36. Bogor: Balitbang Biologi Perairan - LIPI.

Asquith, N. M. 1993. The status of the silvery gibbon (*Hylobates moloch*) in the Ujung Kulon National Park. *Trop. Biodiv.* 1: 185-188.

Asquith, N. M. 1995. Javan gibbon conservation: why habitat protection is crucial. *Trop. Biodiv.* 3: 63-65.

Asquith, N. M., Martarinza, and R. M. Sinaga. 1995. The Javan gibbon; status and conservation recommendations. *Trop. Biodiv.* 3: 1-14.

Astuti, S. R. 1983. *Monitoring bencana alam Galunggung.* PSL ITB, Bandung.

Atmadja, W. S. 1977. Notes on the distribution of red algae (Rhodophyta) on the coral reef of Pari Islands, Seribu Islands. *Mar. Res. Indon.* 17: 15-27.

Atmadja, W. S., and Sulistijo. ????. Beberapa aspek vegetasi dan habitat tumbuhan laut bentik di pulau-pulau Seribu. In *Teluk Jakarta: Biologi, budidaya, oseanografi, geologi dan kondisi perairan*, ed. M. K. Moosa, D. P. Praseno, and Sukarno. LON - LIPI, Jakarta.

——. 1980. Komunitas rumput laut di pantai Pananjung Pangandaran, pantai selatan Jawa Barat. In *Sumberdaya hayati bahari*, ed. Burhanuddin, M. K. Moosa, and H. Razak. LON - LIPI, Jakarta.

——. 1983. The distribution and some ecological aspects of marine algal genus *Euchema* in the Indonesian waters. Paper presented at the International Conference on Development and Management of Tropical Living Aquatic Resources, Serdang, Malaysia, 2 August 1983.

Atmodjo. 1987. *Wilayah pedesaan sebagai penunjang pelestarian Taman Nasional Meru Betiri di Bandialit, Kabupaten Jember.* Pusat Penelitian UNEJ, Jember.

Atmodjo, J. S., and N. Edwin. 1988. Mount Pananggungan: home of the gods. *Voice of Nature* 58: 37-46.

Atmosoedardjo, S., and S. G. Banyard. 1978. The prosperity approach to forest community development in Java. *Commonwealth For. Abs. Rev.* 57: 89-96.

Atmosoedardjo, S., P. Soedibjo, and H. Soewondo. 1974. *Beberapa segi kesehatan masyarakat dalam hubungannya dengan perubahan ekologi di sekitar Waduk Bendungan Karangkates, 1972-1974.* Fakultas Kedokteran UNAIR, Surabaya.

Atmosoedirdjo, S. 1986. Kakatua masakambing. *Suara Alam* 35: 41-43.

Atmosoedjono, S. 1990. Malaria control in Indonesia since World War II. In *Environmental measures for malaria control in Indonesia - An historical review on species sani-*tation, ed. W. Takken et al., 141-154. Wageningan: Agricultural Univ..

Atmowidjojo, A. H. S., and Boeadi. 1986. Food prey in the stomach contents of frogs. *BIOTROP Spec. Publ.* 32: 77-82.

Audley-Charles, M. G. 1981. Geological history of the region of Wallace's line. *Wallace's line and plate tectonics.* T. C. Whitmore, 24-35. Oxford: Clarendon.

——. 1987. Dispersal of Gondwanaland: relevance to evolution of the angiosperms. In *Biogeographical evolution of the Malay archipelago.* T. C. Whitmore, 5-25. Oxford: Clarendon.

Audley-Charles, M. G., A. M. Hurley, and A. G. Smith. 1981. Continental movements in the Mesozoic and Cenozoic. In *Wallace's line and plate tectonics*, ed. T. C. Whitmore, 9-23. Oxford: Clarendon.

Aziz, A. 1981. Fauna echinodermata dari terumbu karang Pulau Pari, Kepulauan Seribu. *Oseanol. Indo.* 14: 41-50.

Aziz, F. 1989. *Macaca fasiculata* (Raffles) from Ngandong, East Java. *Publ. Geol. Res. Dev. Cen. Palaeontol. Ser.* 5: 50-56.

Aziz, F., and J. de Vos. 1989. Rediscovery of the Wadjak site Java, Indonesia. *J. Anthro. Soc. Nippon* 97: 133-144.

Azkab, M. H., and S. Sukardjo. 1986. Komunitas semai mangrove di Pulau Pari, Pulau Seribu. Paper presented at Seminar III Ekosistem Mangrove, Denpasar, 5 August 1986.

Baas Becking, L. G. M. 1948. Notes on jungle trees. *Chron. Nat.* 104: 271-277.

Bachri, S. 1978. *Goa Jatijajar dan obyek-obyek wisata.* Privately published.

Backer, C. A. 1912. Sawah planten. *Trop. Nat.* 1: 129-135.

——. 1913. Sawah planten. *Trop. Nat.* 2: 74-76, 81-85, 118-122, 132-133.

——. 1914. Sawah planten. *Trop. Nat.* 3: 55-62.

——. 1917-1922. Indische duinplanten. *Trop. Nat.* 6: 73-78, 89-92, 97-100, 145-147; 7: 5-11, 55-59; 8: 6-10; 9: 173-191; 10: 12-17; 11: 131-140;. 12: 17-22.

——. 1934. *Onkruidflora der Javasche suikerrietgronden.* Pasuruan: Proefstation voor de Java-Suikerindustrie.

——. 1936. Verwilderingscentra op Java van uitheemsche planten. *Trop. Nat.* 25: 51-60.

——. 1973. *Atlas of 220 weeds of sugar cane fields in Java.* Ysel: Deventer.

Backer, C.A., and C. G. G. J. van Steenis. 1951. Sonneratiaceae. *Fl. Mal.* I 4: 280-289.

Backer, C.A. and D.F. van Slooten. 1924. *Geillustreerd handbook der Javaansche theeonkruiden.* Batavia: Ruygrok.

Backer, C. A., and R. C. Bakhuizen van den

Brink. 1963. *Flora of Java Volumes I-III.* Groningen: Noordhoff.

Backer, C. A., and O. Posthumus. 1939. *De varenflora voor Java.* Buitenzorg: 's Lands Plantentuin.

Backer, K. 1914. Ranoe Bedali. *Trop. Nat.* 3: 155-158.

Baehr, M. 1990. *Ochterus noualhieri* n. sp. a new ochterid bug from Java, Indonesia (Heteroptera: Ochteridae). *Rev. Fr. Entomol (Nouv. ser)* 12: 91-93.

Bahnemann, H. G. 1985. Rabies in southeast Asia. In *Rabies in the tropics,* ed. E. Kuwert et al., 541-544. Berlin: Springer Verlag.

Bakels, J. 1988. The Baduy of Baten: the hidden people. In *Indonesia in focus.* ed R. Schefold, 38-45. Meppel: Edu'actief.

Bakhuizen van den Brink, R.C. 1922. Lalab. *Trop. Nat.* 11: 115-124, 150-155.

Bakornas PBA. 1982. *National workshop on Mt. Galunggung volcanic risk management.* Badan Koordinasi Nasional Penanggulangan Bencana Alam, Jakarta.

Baldia, J. P., and J. B. Pantastico. 1991. An overview of golden snail research in the Philippines. In *Environmental impact of the golden snail (Pomacea sp.) on rice farming systems in the Philippines.* B. O. Acosta, and R. S. V. Pullin (Eds.). Manila: ICLARM.

Balitbang PU. 1991. *Kualitas air untuk menunjang Program Air Bersih (Prokasih).* Badan Penelitian dan Pengembangan Pekerjaan Umum, Bandung.

Bambang, A. M. 1989. *Tambak air payau: Budidaya udang dan bandeng.* Jakarta: Kanisius.

Bang, Y. H., S. Arwati, and S. Gandahusada. 1982. A review of insecticide use for malaria control in Central Java, Indonesia. *Malay. Appl. Biol.* 11: 85-96.

Bansgrove, A. J. 1991. Sources of water pollution in the Ciliwung river, Jakarta: an appliction of the rapid assessment approach. Department of Geography, Univ. of Waterloo, Waterloo.

Bänziger, H. 1988. The heaviest tear drinkers: ecology and systematics of new and unusual notodontid moths. *Nat. Hist. Bull. Siam Soc.* 36: 17-53.

—-. 1989a. A persistent tear drinker: notodontid moth *Poncetia lacrimisaddicta* sp. n., with notes on its significance to conservation. *Nat. Hist. Bull. Siam Soc.* 37: 31-46.

—-. 1989b. Skin-piercing blood sucking moths V: attacks on man by *Calyptra* spp. (Lepidoptera, Noctuidae) in S. and S.E. Asia. *Mitt. Schweiz, Entomol. Ges.* 62: 215-233.

—. 1991. Stench and fragrance: unique pollination lure of Thailand's largest flower, Rafflesia kerri Meijer. *Nat. Hist. Bull. Siam Soc.* 39: 19-52.

Baran, G. W., and M. Frith. 1986. *Urban dog ecology in Guayaquil, Ecuador: A report to the World Health Organisation.* Dept. Veterinary, Microbiology and Preventative Medicine, Iowa State Univ.,

Barbier, C.; and C. Courvoisier. 1980. *Approach to the traditional uses of medicinal plants in Java.* Université des Sciences et Techniques du Languedoc, Montpellier.

Barbier, E. B. 1988. *The economics of farm-level adoption of soil conservation measures in the uplands of Java.* Environment Dept, World Bank, Washington D.C.

Barlow, C. 1991. Developments in plantation agriculture and small holder cash-crop production. In *Indonesia: resources, ecology and environment,* ed. J. Hardjono, 85-103. Singapore: Oxford Univ. Press.

Barlow, H. S. 1982. *An introduction to the moths of south east Asia.* Kuala Lumpur: Malayan Nature Society.

Barnes, R. S. K. 1984. *A synoptic classification of living organisms.* Oxford: Blackwell.

Barnes, R. S. K., and R. N. Hughes. 1982. *An introduction to marine ecology.* Oxford: Blackwell.

Barodji. 1980. *Jenis dan populasi nyamuk di daerah persawahan desa Darmaga dan Cidereum, Kabupaten Bogor.* Sekolah Pasca Sarjana IPB, Bogor.

Barrau, E. M., and K. Djati. 1985. The Citanduy project in Java: Toward a new approach to watershed stabilization and development (soil erosion and conservation). *Soil Cons. Soc. Am. Rep.* 1985: 729-740.

Barrau, E. M., K. Djati, S. A. El-Swaify, and W. C. Moldenhauer. 1983. The Citanduy project in Java: Toward a new approach to watershed stabilization and development. Paper presented at the International Conference on Soil Erosion and Conservation, Jakarta, 16 January 1983.

Bartels, M. 1906. Systematische übersicht einer Java vogel. *J. Ornithol.* 54: 383-407, 497-519.

—. 1929. Kiara-gasten. *Trop. Nat.* 18: 37-42.

—. 1937a. A new rat from Java. *Treubia* 16: 45-47.

—. 1937b. Zur kenntnis der verbreitung und der lebensweise Javanischer saugetiere. *Treubia* 16: 149-169.

—. 1938. A new species of *Rattus bartelsii* (Jentink) from Central Java. *Treubia* 16: 323-329.

—-. 1942. Nogmaals: kenmaekende

*
- Note that new and back issues of *Flora Malesiana* can be ordered from the Rijksherbarium/Hortus Botanicus, Leiden University, Leiden.

verrschillen tusschen zeugen van wratten-en streepenzwijnen. *Ned. Ind. Jager* 12: 6.

Bartels, M., and P. J. Bouma. 1937. Keerkringsvogels en slechtvalken aan Java's zuidkust. *Trop. Nat.* 26: 108-111.

Bartels, M., and E. Stresemann. 1929. Systematische uebersicht der bisher van Java nachgewiesenen vogel. *Treubia* 11: 89-146.

Bartstra, G. J. 1983. Some remarks upon fossil man from Java, his age and his tools. *Bijdragen tot de Taal-, Land- en Volkenkunde* 139: 421-434.

Bartstra, G.J., S. Soegondho, and A. van der Wijk. 1988. Ngandong Man age and artifacts. *J. Human Evol.* 17: 325-338.

Bartstra, G. J., and Basoeki. 1989. Recent work on the Pleistocene and the Paleolithic of Java. *Current Anthrop.* 30: 241-244.

Basmi, J. 1986. Spatial and seasonal distribution of zooplankton in Lake Rawa Pening, Central Java, Indonesia. *BIOTROP Spec. Publ.* 27: 47-67.

Basuni, S., and D. G. Setiyani. 1989. Bird trading at the Pasar Pramuka, Jakarta and the bird catching techniques. *Media Konservasi* 2: 9-18.

Batchelor, M., and K. Brown. 1992. *Buddhism and ecology.* London: Cassel.

Bater, J. H., W. C. Found, S. Martopo, and M. Soerjani. 1995 Implications of the Bali Sustainable Development Project for sustainable development. In *Bali: Balancing environment, economy and culture,* ed. S. Martopo and B. Mitchel. 613-631. Waterloo: Dept. Geography, University of Waterloo.

Bauchop, T. 1978. Digestion of leaves in vertebrate arboreal folivores. In *The ecology of arboreal folivores,* ed. G. G. Montgomery, 193-204. Washington D.C.: Smithsonian Institution Press.

Baum, V. 1937. *A tale from Bali.* Singapore: OUP(reprinted 1973).

Bawa, W. 1987. Studi tentang pelestarian lembu putih *(Bos sondaicus)* di Bali. In Kumpulan Abstrak Kongres Nasional Biologi VIII, Purwokerto, 8 October 1987.

——. 1988. *Studi tentang ekologi burung belekok (Ardeola speciosa) di Petulu (Bali) dan sekitarnya.* FKIP UNUD, Denpasar.

Beamann, R. S., P. J. Decker, and J. N. Beaman. 1988. Pollination of *Rafflesia* (Rafflesiaceae). *Am. J. Bot.* 75: 1148-1162.

Beattie, A. J. 1985. *The evolutionary ecology of ant-plant mutualisms.* Cambridge: Cambridge Univ. Press.

Becker, P., M. Leighton, and J. B. Payne. 1985. Why tropical squirrels carry seeds out of source crowns. *J. Trop. Ecol.* 1: 183-186.

Beckerman, W. 1992. Global warming and international action: an economic perspective. In *The international politics of the environment,* ed. A. Hurrell, and B. Kingsbury, 253-289. Oxford: Oxford Univ. Press.

Becking, J. H. 1935. De ontwikkeling van de Dienst der wildhoutbosschen op Java gedurende de laatsche vijf jaren. *Tectona* 28: 343-433.

——. 1976. Feeding range of Abbott's booby *Sula abbotii* at the coast of Java. *Ibis* 118: 589-590.

——. 1989. *Henri Jacob Victor Sody (1892-1959): His life and work.* Leiden: Brill.

Bedford, G. O. 1978. Biology and ecology of the Phasmatodea. *Ann. Rev. Entomol.* 23: 125-149.

Beeby, A. 1993. *Applying ecology.* London: Chapman and Hall.

Beehler, B. 1982. Ecological structuring of forest bird communities in New Guinea. *Ecol. Monogr.* 42: 837-861.

Begon, M., J. L. Harper, and C. R. Townsend. 1990. *Ecology: Individuals, populations and communities.* Oxford: Blackwell.

Bekkering, T. D., and K. P. Kucera. 1990. *The Lebakharjo forest area: natural resources and human interference.* Konto River Project Phase III, Malang.

Bell, T. I. W. 1973. Erosion in the Trinidad teak plantations. *Commonw. For. Rev.* 52: 223-233.

Bellwood, P. S. 1980. The peopling of the Pacific. *Scient. Amer.* 243: 174-185.

——. 1984. The great Pacific migration. In *1984 Yearbook of science and the future.* pp.80-93. Chicago: Encyclopaedia Britannica.

——. 1985. *Prehistory of the Indo-Malayan archipelago.* London: Academic Press.

Belo, J. 1966. *Bali: Temple festival.* Seattle: Univ. of Washington.

Bennett, D. H. 1986. Triage as a species preservation stratergy. *Env. Ethics* 8: 47-58.

Bennett, R., and S. Bennett. 1980. *Bandung and beyond.* Bandung: Aneka Karya.

Bergmans, W., and P.J.H. van Bree. 1986. On a collection of bats and rats from the Kangean island, Indonesia (Mammalia: Chiroptera and Rodentia). *Zeits. Saügetierk.* 51: 329-344.

Bernsten, R. J., B. H. Siwi, and H. M. Beachell. 1982. *The development and diffusion of rice varieties in Indonesia.* Los Baños: IRRI.

Berry, A. J. 1972. The natural history of West Malaysian mangrove faunas. *Malay. Nat. J.* 25: 135-162.

Beudels, R. C., and H. B. Hardi. 1980.

Bromo-Tengger-Gunung Semeru proposed national park management plan 1981-5. *Field report No. 12.* FAO, Bogor.

Beudels, R. C., and Kurnianto. 1982. Blambangan Nature Reserve management plan. *Field report No. 40.* FAO, Bogor.

Beumée, J. G. B. 1921a. Een auto-botaniseertocht in Rembang. *Trop. Nat.* 10: 81-91.

—. 1921b. Nog enkele bewijzen voor de alluviale rijzing van het eiland Java. *Trop. Nat.* 10: 141-144.

—. 1929. Djokja-Tengger-Soerabaja (botanical trip). Proceedings of the Fourth Pacific Science Congress. Vol. III: Biological Papers.

Bibby, C. J., N. J. Collar, M. J. Crosby, M. F. Heath, and C. Imboden. 1992. *Putting biodiversity on the map: priority areas for global conservation.* Cambridge: ICBP.

Bingham, A. 1982. Cave water: Java's hidden resource. *World Water* 5: 49-51.

BIOTROP. 1980. *Studi ekologi Waduk Wonogiri tahap pra-inundasi.* BIOTROP, Bogor.

—. 1986. *Penelitian usaha pelestarian sumber air Rawa Danau, Kabupaten Serang.* BIOTROP, Bogor.

Bird, E. C. F., and O. S. R. Ongkosongo. 1980. *Environmental changes on the coasts of Indonesia.* Tokyo: United Nations Univ.

Birkeland, C. 1984. Sources and destinations of nutrients in coral reef. *UNESCO Rep. Mar. Sci.* 21: 3-20.

Birkes, F. 1993. Application of ecological economics to development: the institutional dimension. In *Ecological economics: emergence of a new development paradigm.* pp.61-74. Ottawa: Univ. of Ottawa.

Birley, M. H. 1991. *Guidelines for forecasting the vector-borne disease implications of water resources development, 2nd edn.* Geneva: WHO.

Bishop, K. D. 1988. The long-legged pratincole *Stiltia isabella* in Bali. *Kukila* 3: 141.

Bishop, M. J. 1977. Anatomical notes on some Javanese *Amphidromas* (Pulmonata: Camaenidae). *J. Conchol.* 29: 199-205.

Bismark, M. 1988. Prospek penangkaran kerbau liar (*Bubalus bubalis*) di Taman Nasional Baluran, Jawa Timur. *Bull. Pen. Hutan* 495: 21-30.

Bismark, M., and A. S. Wiriosoepartho. 1980. *Beberapa aspek ekologi lutung (Presbytis cristata Raffles) di Suaka Margasatwa Meru Betiri, Jawa Timur.* Lembaga Penelitian Hutan, Bogor.

Biyono, B., and K. D. Carlander. 1979. Estimasi jumlah dan komposisi ikan yang ditangkap di Danau Rawa Pening. In Proceedings of the 2nd Seminar in Aquatic Biology and Aquatic Management of the Rawa Pening Lake, Salatiga, 29 October 1979. Salatiga: Satya Wacana Christian Univ.

Blair, K. G. 1936. A new cave-dwelling tenebrionid from Java. *Treubia* 15: 321.

Blanke, W. 1939. Vulkaan-ruines bij Plered. *Trop. Nat.* 28: 114-117.

Blaustein, A. R., and D. B. Wake. 1990. Declining amphibian populations: A global phenomenon? *Trends Ecol. Evol.* 5: 203-204.

Blouch, R. A. 1982. *Proposed Meru Betiri National Park management plan; 1983-1988.* WWF Indonesia, Bogor.

—. 1988. Ecology and conservation of the Javan warty pig, *Sus verrucosus* Muller 1840. *Biol. Conserv.* 43: 295-307.

Blouch, R. A., and S. Atmosoedirdjo. 1980. *Kuhl's Deer, Bawean Island, Indonesia.* IUCN-WWF, Bogor.

—. 1987. Biology of the Bawean deer and prospects for its management. In *The Biology and Management of the Cervidae,* C. M. Wemmer, 320-327. Washington, D.C.: Smithsonian Inst. Press.

Blouch, R. A., and C. P. Groves. 1990. Naturally occurring suid hybrid in Java. *Zeits. Saügetierk.* 55: 270-275.

Blower, J. H. 1975. *Report on a visit to Pulau Bawean.* Nature Conservation and Wildlife Management Project, Bogor.

Boeadi. 1978. Hutan bakau di Pulau Dua, Teluk Banten, Jawa Barat. In Prosiding Seminar Ekosistem Hutan Mangrove. Jilid I, Jakarta, 27 February 1978. Jakarta: LON - LIPI.

—. 1980. *Catatan tentang kelelawar yang ditemukan di dalam gua-gua di Cidolog, Segaranten, Sukabumi.* HIKESPI, Bogor.

—. 1983. *Studi reproduksi tikus sawah (Rattus argentiventer) di Pamanukan (Jawa Barat) dan Randudongkal (Jawa Tengah).* FIPIA UI, Jakarta.

—. 1990. On two new specimens of the insectivorous bat *Otomops formosus* Chasen, 1939. *Mammalia* 50: 263-266.

Boeadi, M. Amir, and Suyanto. 1983. An insectivorous bat *Tadarida plicata* (Buchanan) (Chiroptera: Molossidae) as a possible component in biological control of insect pests. *BIOTROP Spec. Publ.* 18: 243-247.

Boeadi, and J. E. Hill. 1986. A new subspecies of *Aethalops alecto* Thomas 1923, Chiroptera: Pteropoda from Java. *Mammalia* 50: 263-266.

Boedhihartono. 1933. Tiger mandible fossil:

a preliminary fossil. International Conference on Human Palaeoecology, Jakarta, 13 October. Jakarta: LIPI.

Boehmer, K. 1992. Environmental behaviour in Petulu, Bali: what traditional planning can contribute to sustaibable development. School of Urban and Regional Planning, University of Waterloo: Waterloo.

Bohmer, K. and T. Wickham, T. 1995. Linking Bali's past with a sustainable future. In *Bali: Balancing environment, economy and culture*, ed. S. Martopo and B. Mitchell. 437-464. Waterloo: Dept. Geography, University of Waterloo.

Boelman-Caspare, E. 1926. Aanteekeningen over Noesa Kambangan (*Rafflesia patma* Bl.). *Trop. Nat.* 15: 150-153.

Boeseman, M. 1949. On Pleistocene remains of *Ophicephalus* from Java, in the "Collection Dubois". *Zool. Mededel.* 30: 83-94.

Bohnsack, J. A., and D. L. Sutherland. 1985. Artificial reef research: A review with recommendations for future priorities. *Bull. Mar. Sci.* 37: 11-39.

Bompard, J. M., P. Hecketsweiler, and G. Michon. 1980. *A traditional agricultural system: Village-forest garden in West Java*. Montpellier: Acad. Montpellier.

Bond, W. J. 1993. Keystone species. *Biodiversity and ecosystem function*, ed. E. D. Schulze, and H. A. Mooney, 237-254. Berlin: Springer.

Bonne-Wepster, J. 1954. *Synopsis of one hundred common non-anopheline mosquitoes of the Greater and Sunda Sundas, the Moluccas and New Guinea*. Amsterdam: Elsevier.

Bonne-Wepster, J., and N. H. Swellengrebel. 1953. *The anopheline mosquitoes of the Indo-Australian region*. Amsterdam: de Bussy.

Booberg, G. 1928. Een planten sociologische ondercoek van de duinen bij Poeger. 5th Nederlandsch-Indisch Natuurwetenschappelijk Congres, 366-377.

—. 1929. Een planten sociologisch onderzoek van de duinen bij Poeger. *Hand. Ned. Ind. Natuurwet. Congress* 1928 pp.366-378.

Boomgaard, P. 1988. Forests and forestry in colonial Java, 1677-1942. *Changing tropical forests: historical perspectives on today's challanges in Asia, Australasia and Oceania.* eds J. Dargavel, K. Dixon, and N. Semple, 59-88. Canberra: Australian National Univ..

—. 1991. Sacred trees and haunted forests. *Asian perceptions of nature*, ed. O. Brunn, and A. Kaland, 39-53. Copenhagen: NIAS.

—. 1994. Death to the tiger! The development of tiger and leopard rituals in Java, 1605-1906. *S.E. Asia Res.* 2: 141-175.

Boon, J. A. 1977. *The anthropologicial romance of Bali: 1597-1972.* Cambridge: Cambridge Univ. Press.

Boon, P. J. 1977. *A survey of freshwater habitats in Central Java.* Faculty of Fisheries and Animal Sciences, Newcastle upon Tyne, Newcastle upon Tyne.

—. 1979. Adaptive strategies of Amphipsyche larvae (Trichoptera: Hydropsychidae) downstream of a tropical impoundment. In *The ecology of regulated streams.* J. V. Ward, and J. A. Stanford, 237-255. New York: Plenum.

—. 1984. Habitat exploitation by larvae of *Amphipsyche meridiana* (Trichoptera: Hydropsychidae) in a Javanese lake outlet. *Freshw. Biol.* 14: 1-12.

Booth, A. 1985. Accommodating a growing population in Javanese agriculture. *Bull. Indo. Econ. Studies* 21: 115-145.

—. 1988. *Agricultural development in Indonesia.* Sydney: Allen and Unwin.

Booth, A., and A. Damanik. 1989. Central Java and Yogyakarta: Malthus overcome? In *Unity and diversity: Regional economic development in Indonesia*, ed. H. Hill, 283-306. Singapore: Oxford Univ. Press.

Bor, N. L. 1970. Graminae. *Flora Iranica*, ed. K. H. Rechinger, Graz: Akademische Drukund Verlagsanstalt.

Bordet, M. 1979. *Some aspects of fish population dynamics in Jatiluhur Reservoir.* BIOTROP, Bogor.

Bornkamm, R., J. A. Lee, and M. R. D. Seaward. 1982. *Urban ecology.* Oxford: Blackwell.

Bosmi, J. Spatial and seasonal distribution of zooplankton in Lake Rawa Pening, Central Java, Indonesia. Paper presented at Seminar on Production and Exploitation of Open Waters (with special emphasis on reservoirs), Bogor, 15 June 1982.

BOSTID. 1993. *Vetiver grass: A thin green line against erosion.* Washington D.C.: National Academy Press.

Boto, K. G., J. S. Bunt, and J. T. Wellington. 1984. Variations in mangrove forest productivity in Northern Australia and Papua New Guinea. *Estuar. Coast. Shelf Sci.* 19: 321-329.

Bouharmont, P. 1992. Selection of the Java variety and its use in Arabica coffee regeneration in Cameroon. *Café Cacao Thé* 36: 247-262.

Boulière, F., and M. Hadley. 1983. Present day savannas. In *Tropical savannas.* Ed F. Boulière, New York: Elsevier.

BP3U. 1991. Kualitas air untuk menunjang

Program Kali Bersih (Prokasih). Badan Penelitian dan Pengembangan Pekerjaan Umum, Bandung.

BPS. 1985. *Buku saku statistik Indonesia 1984*. Jakarta: Biro Pusat Statistik.

———. 1987. *Survei pertanian luas lahan menurut penggunaannya di Jawa: 1986*. Jakarta: Biro Pusat Statistik, Bagian Statistik Pertanian Tanaman Pangan.

———. 1989. *Kecamatan Grogol Petamburan dalam angka: 1989*. Kantor Statistik Jakarta Barat, Jakarta.

———. 1991. *Vehicles and length of road statistics*. Jakarta: Biro Pusat Statistik.

Braatz, S. 1992. *Conserving biological diversity: A strategy for protected areas in the Asia-Pacific region*. World Bank, Washington D.C.

Brafield, A. E. 1978. *Life in sandy shores*. London: Edward Arnold.

Bragg, P. E. 1992. The phasmid database. *Phasmid Stud.* 1: 38-46.

Brandon-Jones, D. 1995. A revision of the Asain pied leaf-monkeys with a description of a new subspecies. *Raffles Bull. Zool.* 43:3-44.

Bratamihardja, M. 1988. "Social forestry" pada tanah hutan negara. *Duta Rimba* 101: 17-32.

———. 1990. Agroforestry on forest land in Java. *BIOTROP Spec. Publ.* 39: 141-146.

———. 1991. Pengelolaan hutan payau di pantai utara Pulau Jawa. Prosidings Seminar IV Ekosistem Mangrove, S. Soemodihardjo et al., 59-64. Bandar Lampung, 7 August 1990. Jakarta: MAB-LIPI.

Bray, F. 1986. *The rice economies: Technology and development in Asian societies*. Oxford: Basil Blackwell.

Breiully, E., and M. Palmer. 1992. *Christianity and ecology*. London: Cassel.

Brinck, P., L. M. Nilsson, and U. Svedin. 1988. Ecosystem redevelopment. *Ambio* 17: 84-89.

Brock, P. 1992. *Rearing and studying stick and leaf-insects*. Colchester: Amateur Entomologists' Society.

———. in press. *The stick and leaf-insects of West Malaysia and Singapore*.

Brongersma, L. D. 1930. Notes on the list of reptiles of Java. *Treubia* 12: 299-303.

———. 1937. On fossil remains of a hyalind from Java. *Zool. Mededel.* 20: 186-201.

———. 1940. Note on *Mustela lutreolina* Rob and Thos. *Temminckia* 5: 257-263.

———. 1950. Notes on *Pseudoxenodon inornatus* (Boie) and *Pseudoxenodon jacobsonii* (Linth). *Proc. K. Ned. Akad. Wet.* 53: 1498-1505.

———. 1958. Note on *Vipera russelii* (Shaw). *Zool.*

Mededel. 36: 55-79.

Brookfield, N. C. 1990. An approach to islands. In *Sustainable development and environmental management of small islands*. W. Beller, P. d'Ayala, and P. Hein, 23-33. Paris: MAB-UNESCO.

Brotoisworo, E. 1979. The lutung *(Presbytis cristata)* in Penanjung Pangandaran Nature Reserve: social adaptation to species. Ph.D. diss., Kyoto.

———. 1983. Population dynamics of lutung *(Presbytis cristata)* in Pananjung Pangandaran Nature Reserve, West Java. In Proceedings on Training Course on Wildlife Ecology, Bogor. Indonesia, Vol. IIa. Bogor: BIOTROP.

Brotoisworo, E., and J. Iskandar. 1984. Problems of bird protection in Indonesia: a case study on Java. 10th Asian Continental Conference, Sri Lanka.

Brotisworo, E., and I. W. A. Dirgayusa. 1991. Ranging and feeding behaviour of *Presbytis cristata* in the Pangandaran Nature Reserve West Java Indonesia. In *Primatology today*, 115-118. Amsterdam: Elsevier.

Brotoisworo, E., K. Gurmaya, and O. Soemarwoto. 1980. Problems of Gunung Tilu. Ciawi Conservation School, Bogor.

Brotonegoro, S., and S. Abdulkadir. 1978. Penelitian pendahuluan tentang kecepatan gugur daun dan penguraiannya dalam hutan bakau Pulau Rambut. In Prosiding Seminar Ekosistem Hutan Mangrove. Jilid I, Jakarta.

Brouen, B. E. et al. 1983. Coral assemblages of reef flats around Pulau Pari, Thousand Islands, Indonesia. *Atoll Res. Bull.* 281: 1-14.

Brown, B. E., and Suharsono. 1990. Damage and recovery of the coral reefs affected by El Niño related sea warming in the Thousand Islands, Indonesia. *Coral Reefs* 8: 163-170.

Brown, B. E., L. Sya'rani, and M. Le Tissier. 1985. Skeletal form and growth in *Acropora aspera*. *J. Exp. Mar. Biol. Ecol.* 86: 139-150.

Brown, J. H., and M. R. Muskanofola. 1985. An investigation of stocks of giant clams (family Tridacnidae) in Java and of their utilization and potential. *Aquacul. Fish. Mgmt.* 16: 25-40.

Brown, J. H., and B. Prayitno. 1987. Backyard fish farming in Java, Indonesia. *Community Dev. J.* 22: 237-241.

Brown, L. R., S. Postel, and C. Flavin. 1992. From growth to sustainable development. In *Population, technology and lifestyle*. eds R. Goodland, H. E. Daly, and S. El-

Serafy,Washington D.C.: Island Press.

Bruce, M. D. 1982. Occurrence of the lesser adjutant stork *Leptoptilos javanicus* on Bali, Indonesia. *Bull. Brit. Orn. Cl.* 102: 39-41.

Bruce-Chwatt, L. J. 1980. *Essential malariology.* London: Heinemann.

Bruggeman, M. L. A. 1927. *Gids voor den bergtuin te Tjibodas.* s'Lands Plantentuin, Buitenzorg.

—. 1927. The numbered trees, shrubs and lianas in the forest of Mount Gede near Tjibodas, West Java. *Bull. Jard. Bot. Buitenz.* 9: 196-219.

Bruijnzeel, L. A. 1982. Hydrological and biogeochemical aspects of man-made forests in south-central Java, Indonesia. Ph.D. diss., Free Univ., Amsterdam.

—. 1984. Elemental content of litterfall in a lower montane rain forest in Central Java, Indonesia. *Malay. Nat. J.* 37: 199-208.

—. 1985. Nutrient content of litterfall in coniferous forest plantations in Central Java, Indonesia. *J. Trop. Ecol.* 1: 353-372.

—. 1988. Biochemical aspects of coniferous forest plantations in Central Java, Indonesia. I: Composition of bulk precipitation. *Malay. For.* 48: 206-222.

—. 1989. Nutrient content of bulk precipitation in south-central Java, Indonesia. *J. Trop. Ecol.* 5: 187-202.

—. 1990. *Hydrology of moist tropical forests and effects of conversion: a state of knowlege review.* Amsterdam: Free Univ..

—. 1991. Hydrology of moist tropical forests and effects on conversion: A state of knowledge review. *Wallaceana* 64: 10-14.

—. 1993a. Hydrological observations in montane rain forests in Gunung Silam, Sabah, Malaysia, with special reference to the '*Massenerhebung*' effect. *J. Ecol.* 81: 145-167.

—. 1993b. Land-use and hydrology in warm humid regions: where do we stand. In *Hydrology of warm humid regions,* ed. J. S. Gladwell, 3-34. Yokohama, 13 July 1933. IAHS.

Bruijnzeel, L.A. et al. 1993. Hydrological observations in montane rain forests on Gunung Silam, Sabah, Malaysia, with special reference to the 'Massenerhebung' effect. *J. Ecol.* 81: 145-167.

Bruijnzeel, L. A., and J. Proctor. 1993. Hydrology and biogeochemistry of tropical montane "Cloud Forests": What do we really know? Tropical Montane Cloud Forests Symposium and Workshop, San Juan, Puerto Rico, May 1931.

Bryant, N. A. 1973. *Population pressure and agricultural resources in Central Java.* Ph.D. diss. Michigan State Univ..

Brygoo, E. R. 1988. Les types d'Agamidés (Reptiles, Sauriens) du Museum national d'Histoire naturelle: Catalogue critique. *Bull. Mus. nat. d'Hist. nat.* 10: 1-56.

BScC. 1992. Keanekaragaman hayati Cagar Alam Gunung Halimun, Jawa Barat. Workshop Cagar Alam Gunung Halimun, Jakarta, 23 January 1992.

BSDP. 1992. *Sustainable development strategy for Bali.* Bali Sustainable Development Project, Gadjah Mada Univ., Yogyakarta.

Budi-Asmoro, P. 1992. *Survei ekologis Rawa Lakbok dan sekitarnya di Kabupaten Ciamas (Java Barat).* Asian Wetland Bureau, Bogor.

—. 1993. *Laporan kegiatan singkat pengenalan dan teknik penangkapan amphibia dan reptilia.* PHPA-AWB, Bogor.

Budiman, A. 1988. Some aspects on the ecology of mangrove whelk, *Telescopium telescopium* Linne 1758 (Mollusca: Gastropoda: Potamididae). *Treubia* 29: 237-246.

Budiman, A., M. Djajasasmita, C. S. Kaswadji, and F. Sabar. 1979. Beberapa telaga di Jawa Tengah. *Alam Kita* 4: 18-21.

Budiman, A., and M. Djajasasmita. 1981. Berburu Tambra di beberapa daerah aliran sungai di Jawa Tengah dan Jawa Barat. *Alam Kita* 6: 9-15.

Buil, M. A. M. 1984. Birds of Manggiri: A preliminary inventory of birds in the Gajah Mada Univ.'s research area Manggiri, south of Yogyakarta. *Report No. 751.* Nature Conservation Dept., Agricultural Univ., Wageningen.

Bullock, J. A. 1966. The ecology of Malaysian caves. *Malay. Nat. J.* 19: 57-63.

Bulte, E. H., and L. T. C. van Wee. 1990. *Nature conservation on the slopes of the Merapi.* Fakultas Kehutanan UGM, Yogyakarta.

Bunnemeijer, H. A. B. 1918. Een tocht naar het Dieng plateau (20-29 January 1917). *Trop. Nat.* 7: 43-48, 67-74, 101-104, 122-124, 135-138.

Burbridge, P. R., and J. E. Maragos. 1985. *Coastal resources management and environmental assessment needs for aquatic resources development in Indonesia.* Washington: International Institute for Environment and Development.

Bureau of Statistics Office. 1983. *Central Java handbook.* Jakarta: Provincial Govt. Central Java.

Burger, D. 1930. Brand in gebergtebosch. *Tectona* 23: 392-407.

Burgers, J. G. P. 1988. *Ngalau Lida Ayer (Sumatra) en Gvea Djimbe (Java): Palaeoecolo-*

gische onderzoek van twee grotten fauna's van Indonesia. Unpublished report, Burhanuddin, and S. Martosewojo. 1978. Pengamatan terhadap ikan gelodok *Periopthalmus koelenteri* (Pallas) Pulau Pari. Paper presented at Seminar Ekosistem Hutan Mangrove, Jakarta, 27 February 1978.

Burkill, I. H. 1917. The flowering of the pigeon orchid, *Dendrobium crumenatum*, Lindl. *Gdns' Bull. Straits Settl.* 1: 400-405.

—. 1966. *A dictionary of the economic products of the Malay Peninsula.* Kuala Lumpur: Ministry of Agriculture.

Burkill, L. H. 1917. The flowering of the pigeon orchid, *Dendrobium crumenatum* Lindl. *Gdns' Bull. Straits Settl.* 1: 400-405.

Burnham, C. P. 1984. The forest environment: soils. *Tropical rain forests of the Far East,* ed. T. C. Whitmore, 137-154. Oxford: Clarendon.

Burr, M. 1912. A new species of *Arixenia. Entomol. Month. Mag.* 23: 105-106.

Burr, M., and K. Jordan. 1913. On Arixenima Burr, a suborder of Dermaptera. *Trans. 2nd int. Congr. Ent.* 398-421.

Burrett. C. et al. 1991. Asian and southwestern Pacific continental terranes derived from Gondwana, and their biogeographic significance. *Aust. Syst. Bot.* 4: 13-24.

Butot, L. J. M. 1953. Molluscs from Ujung Kulon Peninsula, Banten. *Trop. Nat.* 33: 28-29.

—. 1955. The mollusc fauna of Pulau Panaitan (Prinsen eiland), land and freshwater molluscs. *Treubia* 23: 69-135.

Cabaton, A. 1911. *Java, Sumatra, and the other islands of the Dutch East Indies.* London: Fisher Unwin.

Cagauan, A. G. 1991. Fish as biological tool for aquatic weed management in integrated rice-fish culture system. *BIOTROP Spec. Publ.* 40: 217-230.

Cahoun. 1984. *Aust. J. Bot.* 32: 37-373.

Caldecott, J. O. 1980. Habitat quality and populations of two sympatric gibbons (Hylobatidae) on a mountain in Malaya. *Folia primatol.* 33: 291-309.

—. 1992. Designing protected area projects to reduce conflicts with local people. IVth World Congress on National Parks and Protected Areas, Caracas, Venezuela, 10 February 1992.

—. 1996. *Designing conservation projects.* Cambridge: Cambridge University Press.

Calder, I. 1991. Water use of eucalypts - a review. *Growth and water use of forest plantations,* ed. I. Calder, R. L. Hall, and P. G.

Adlard, 167-179. Wallingford: Institute of Hydrology.

Calder, I. R., I. R. Wright, and D. Murdiyarso. 1986. A study of evaporation from tropical rain forest: West Java. *J. Hydrol.* 89: 13-31.

Caldwell, M. M., A. H. Teramura, and M. Tevini. 1989. The changing solar ultraviolet climate and the ecological consequences for higher plants. *Trends Ecol. Evol.* 4: 363-367.

Campbell, A. 1989. *The sense of well being in America: Recent patterns and trends.* New York: McGraw-Hill.

Cann, R. L., M. Stoneking, and A. C. Wilson. 1987. Mitochondrial DNA and human evolution. *Nature* 325: 31-36.

Capocaccia, L. 1976. Contributo allo studio dei serpenti dell'isola di Giava. *Annal. Mus. Civ. Stor. Nat. Genova* 81: 51-95.

Carlander, K. D. 1979. Further thoughts on overfishing of Rawa Pening Lake. *J. Satya Wacana Res.* 2: 19-25.

—. 1980. Water hyacinth, *Eichhornia crassipes,* and over-fishing problems in Indonesian lakes. *Proc. Iowa Acad. Sci.* 87: 20-22.

Carlander, K. D., and H. Kastowo. 1978. Amounts and species composition of fish harvested from Rawa Pening 1978. *J. Satya Wacana Res.* 1: 474-492.

Carson, B., and W. H. Utomo. 1986. *Erosion and sedimentation processes in Java.* Badan Penelitian dan Pengembangan Pertanian, Jakarta.

Carson, C. 1962. *Silent spring.* New York: Fawcett Crest.

—. 1990. *Musim bunga yang bisu.* Jakarta: Obor.

Carson, H. L., and R. G. Wisotzkey. 1989. Increase in genetic variance following a population bottleneck. *Amer. Nat.* 134: 668-673.

Castles, L. The growing centre. In *Unity and diversity: Regional economic development in Indonesia since 1970,* ed. H. Hill, 233-254. Singapore: Oxford Univ. Press.

Caughley, G. 1994. Directions in conservation biology. *J. Anim. Ecol.* 63: 215-244.

Chambers, R. 1983. *Rural development: putting the last first.* London: Longman.

Champion, H. G., and S. K. Seth. 1968. *A revised survey of the forest types of India.* Delhi: Govt of India Press.

Chapman, J. L., and M. J. Reiss. 1992. *Ecology: principles and applications.* Cambridge: Cambridge Univ. Press.

Chasen, F. N. 1940. Four new mammals from Java. *Treubia* 17: 185-188.

Chasen, F. N., and C. B. Kloss. 1932. On a

small collection of birds from the Karimoendjawa islands. *Treubia* 14: 165-171.

Chee, Y. L. 1987. *Humanity must protect nature.* Penang: Third World Science Movement - Consumers' Association of Penang.

Cherrett, J. M. 1989. Key concepts: the results of a survey of our members' opinions. In *Ecological concepts: The contribution of ecology to an understanding of the natural world,* ed. J. M. Cherrett, 1-16. Oxford: Blackwell.

Chester, R. H. 1969. Destruction of Pacific corals by the sea star *Acanthaster planci. Science* 165: 280.

Chin, S. C. 1977. The limestone hill flora of Malaya I. *Gdns' Bull. S'pore* 30: 165-219.

—. 1983. The limestone hill flora of Malaya IV. *Gdns' Bull. S'pore* 36: 31-91.

Chittister, J. 1992. Monasticism: an ancient answer to modern problems. *Christianity and ecology.* E. Breuilly, and M. Palmer, 65-75. London: Cassell.

Christanty, L. 1982. Traditional agroforestry in Java, Indonesia. *EAPI East West Center report.* East West Center, Honolulu.

—. 1990. Homegardens in tropical Asia, with special reference to Indonesia. In *Tropical home gardens,* ed. K. Landauer, and M. Brazil, 9-20. Tokyo: United Nations Univ. Press.

Christanty, L., and J. Iskandar. 1985. Community forestry: Socio-economic aspects. In *Development of decision-making and management skills in traditional agro-forestry: examples in West Java,* ed. Y. R. Rao, N. T. Vergara, and G. W. Lovelace, 198-214. Bangkok: FAO.

Christie, G., and H. H. Regier 1988. *Can. J. Fish. Aquat. Sci.* 45: 301-314.

Christie, J. W. 1983. Raja and Rama: The classical state in early Java. In *Centers, symbols and hierarchies: Essays on the classical states of Southeast Asia,* ed. L. Gesick. New Haven: Yale Univ. Press.

Church, G. 1959. Size variation in *Bufo melanostictus* from Java and Bali (Amphibia). *Treubia* 25: 113-126.

—. 1960a. Annual and lunar periodicity in the sexual cycle of the Javanese toad *Bufo melanostictus* Schneider. *Zoologica* 44: 181-188.

—. 1960b. The invasion of Bali by *Bufo melanostictus. Herpetologica* 16: 15-21.

—. 1960c. The effects of seasonal and lunar changes on the breeding pattern of the edible Javanese frog, *Rana cancrivora* Gravenhorst. *Treubia* 25: 215-233.

—. 1961. Seasonal and lunar variation in the numbers of mating toads in Bandung. *Herpetologica* 17: 122-226.

—. 1962. The reproductive cycles of the Javanese house geckos, *Cosymbotus platyurus, Hemidactylus frenatus* and *[Peropus mutilatus]. Copeia* 2: 262-269.

Church, G., and L. C. Lim. 1961. The distribution of three species of home geckos in Bandung (Java). *Herpetologica* 17: 199-201.

Chye, H. S., and J. I. Furtado. 1982. The limnology of lowland streams in West Malaysia. *Trop. Ecol.* 23: 84-97.

Clark, L. R., R. L. Kitching, and P. W. Geier. 1979. On the scope and value of ecology. *Protect. Ecol.* 1: 223-243.

Clason, E. W. 1934. Het Noord-Westelijk Baloerangebied. *Trop. Nat.* 23: 121-128.

—. 1935. The vegetation of the upper-Badak region of Mount Kelud (East Java). *Bull. Jard. Bot. Buitenz.* 13: 509-518.

Clason E.W., and E. H. H. Clason-Laarman. 1932. De Goenoeng Ringgit in Oost-Java. *Trop. Nat.* 21: 1-7.

Cleveland, C. J. 1993. Basic principles and evolution of ecological economics. In *Ecological economics: emergence of a new development paradigm.* pp.25-37. Ottawa: Univ. of Ottawa.

Cockrill, W. R. 1984. Water buffalo. *Evolution of domesticated animals,* ed. I.L. Mason, 52-62. London: Longman.

Coert, J. H. 1933. Excursies in Oost-Java. *Trop. Nat.* 22: 62-68, 80-86.

—. 1934. Excursies in Oost-Java (De Rawah Pening, Kediri). *Trop. Nat.* 23: 6-31.

Cohen, J. E., S. L. Pimm, P. Yodzis, and J. Saldaña. 1993. Body sizes of animal predators and animal prey in food webs. *J. Anim. Ecol.* 62: 67-78.

Cole, G. A. 1983. *Textbook of limnology.* St. Louis: Masby.

Cole, M. M. 1986. *The savannas: Biogeography and geobotany.* New York: Academic.

Colinvaux, P. 1993. *Ecology 2.* London: Wiley.

Collier, W. L. 1978a. Food problems, unemployment and the green revolution in rural Java. *Prisma* Febr.: 1-24.

—. 1978b. Rural development and the decline of traditional village welfare-institutions in Java. Paper presented at the Western Economics Association 1978 Conference, Honolulu, .

—. 1981. Agricultural evolution in Java: Decline in shared property and involution. In *Agricultural and rural development in Indonesia,* ed. G.E.Hansen, 147-175. Boulder, Co.: Westview.

Collier, W. L., and Soentoro. 1978. Rural development and the decline of traditional village welfare institutions in Java. Conference of Western Economics Association, Honolulu, June 1978.

Collier, W. L., Soentoro, G. Wiradi, and Makali. 1974. Agricultural technology and institutional change in Java. *Food Res. Inst. Studies* 13: 169-193.

Collier, W. L., Soentoro, G. Wiradi, E. Pasandaran, and K. Santoso. 1982. Acceleration of rural development of Java. *Bull. Indo. Econ. Studies* Nov. 82: 84-101.

Collier, W. et al. 1993. *A new approach to rural development in Java: twentyfive years of village studies.* Jakarta: International Labour Organisation.

Collins, M. 1984. *Urban ecology: A teachers' resource book.* Cambridge: Cambridge Univ. Press.

Collins, N. M. 1980. The distribution of soil macrofauna on the west ridge of Gunung (Mount) Mulu, Sarawak. *Oecologia* 44: 263-275.

Collins, N. M., and M. G. Morris. 1985. *Threatened swallowtail butterflies of the world: The IUCN Red Data Book.* Gland: IUCN.

Comber, J. B. 1990. *Orchids of Java.* Kew: Bentham-Moxon Trust (Royal Botanical Gardens).

Compost, A. R., and G. R. Milton. 1986. An early arrival of the Malayan night-heron *Gorsachius melanolophus* in Java? *Kukila* 2: 88-89.

Conway, G. R. in press. Sustainability in agricultural development: trade-offs with productivity, stability and equitability. *J. Farm. Syst. Res. Ext.*

—. 1985a. Agroecosystem analysis. *Agric. Admin.* 20: 31-55.

—. 1985b. Agricultural ecology and farming systems research. *Agricultural systems research for developing countries.* J. Remenyi, 43-59. Canberra: ACIAR.

Coomans de Ruiter, L. 1953. *De wonderen der tropische natuur: lets over levensgemeenschappen en biotopen.* Groningen: Wolters.

Cooper, T. 1990. *Green Christianity: Caring for the whole creation.* London: Spire.

Corbet, G. B., and J. E. Hill. 1992. *The mammals of the Indomalayan region.* London: Natural History Museum.

Corbet, G. B., and H. N. Southern. 1977. *The handbook of British mammals.* Oxford: Blackwells.

Corbett, A. S., and H. M. Pendlebury. 1978. *The butterflies of the Malay Peninsula.* Kuala Lumpur: Malayan Nature Society.

Cordes, J. W. H. 1881. *De Djatibosschen op Java: Hunne natuur, verspreiding, geschiedenis en exploitatie.* Batavia: Ogilvie.

Cork, V. 1979. *Stories from the morning of the world.* London: Macmillan.

Corlett, R. T. 1986. The mangrove understorey: some additional observations. *J. Trop. Ecol.* 2: 93-94.

—. 1988. The naturalised flora of Singapore. *J. Biogeog.* 15: 657-663.

Corlett, R. T., and P. W. Lucas. 1990. Alternative seed handling strategies in primates: seed spitting by long-tailed macaques. *Oecologia* 82: 166-171.

Cornell, H. V., and J. H. Lawton. 1992. Species interactions, local and regional processes, and limits to the richness of ecological communities: a theoretical persective. *J. Anim. Ecol.* 61: 1-12.

Corner, E. J. H. 1952. *Wayside trees of Malaya.* Kuala Lumpur: Malayan Nature Society.

—. 1978. The freshwater swamp forest of south Johore and Singapore. *Gdns' Bull. S'pore Suppl.* 1.

Cornwell, P. B. 1989. *The cockroach. Vol. 2.* East Grinstead: Rentokill.

Costanza, R. 1992. The ecological economics of sustainability: investing in natural capital. In *Population, technology and lifestyle,* ed. R. Goodland, H. E. Daly, and S. El-Serafy, ??????. Washington D.C.: Island Press.

Costanza, R., and H. E. Daly. 1992. Natural capital and sustainable development. *Conserv. Biol.* 6: 37-46.

Costa-Pierce. B.A., and C. M. Roem. 1990. Waste production and efficiency of feed use in floating net cages in a eutrophic tropical reservoir. In *Reservoir fisheries and aquaculture development for resettlement in Indonesia,* ed. B. A. Costa-Pierce, and O. Soemarwoto, 257-271. Manila: ICLARM.

Costa-Pierce, B., and O. Soemarwoto. 1990a. Biotechnical feasibility stuides on the importation of *Clupeichthys aesarnensis* from northeastern Thailand to the Saguling reservoir, West Java, Indonesia. In *Reservoir fisheries and aquaculture development for resettlement in Indonesia,* ed. B. Costa-Pierce, and O. Soemarwoto, 329-363. Manila: ICLARM.

—, (eds.). 1990b. *Reservoir fisheries and aquaculture development for resettlement in Indonesia.* Manila: ICLARM.

Costa-Pierce, B. A., H. Y. Hadikusumah, and Y. Dhahiyat. 1989. Tilapia (*Oreochromis* sp.) and carp (*Cyprinus carpio*) production in cage systems in West Java, Indonesia. In *Aquaculture research in Asia: Management techniques and nutrition,* ed. E. A. Huisman, N. Zonneveld, and A. H. M. Bouwmans, 84-96 Wageningen: PUDOC.

Costa-Pierce, B. A., G. W. Atmadja, P. Effendi, and S. Zainal. 1988. Integrated aquaculture systems in the Saguling Reservoir, West Java, Indonesia. In *Reservoir fishery management and development in Asia*, ed. S. S. da Silva, 224-233. Ottawa: IDRC.

Coster, C. 1926. Periodische Blutterscheinungen in den Tropen. *Ann. Jard. Bot. Buitenz.* 35: 125-162.

——. 1938. Bovengrondsche afstrooming en erosie op Java. *Landbouw* 16: 3-118.

Courtenay, W. R., and G. K. Meffe. 1990. Small fishes in strange places: a review of introduced Poeciliids. In *Ecology and evolution of livebearing fishes (Poeciliidae)*, ed. G. K. Meffe, and F. F. Snelson, 319-348. Englewood Cliffs: Prentice Hall.

Cousins, S. H. 1991. Species diversity measurement: choosing the right index. *Trends Ecol. Evol.* 6: 190-192.

Covarrubias, M. 1937. *Island of Bali.* Singapore: OUP (reprinted 1972).

——. 1972. *Bali.* Oxford: Oxford Univ. Press.

Cox, C. B., and P. D. Moore. 1985. *Biogeography: an ecological and evolutionary approach. 4th Ed.* Oxford: Blackwell.

Crawford, P. 1985. *The living isles: a natural history of Britain and Ireland.* London: BBC.

Crawfurd, J. 1856. *History of the Indian archipelago; containing an account of the manners, arts, languages, religions, institutions, and commerce of its inhabitants.* London: Bradbury and Evans.

Critchfield, R. 1980. Java: bursting at the seams. *Int. Wildlife* 10 (March-April): 36-42.

Crowther, J. 1982. Ecological observations in a tropical karst terrain, West Malaysia. I. Variations in topography, soils and vegetation. *J. Biogeog.* 11: 65-78.

da Silva, S. S. 1989. *Exotic aquatic organisms in Asia*, ed. S. S. da Silva. Asian Fisheries Society-IDRC-AIDAB, Ottawa.

D'Abrera, B. 1982. *Butterflies of the Oriental region: I Papilionidae, Pieridae and Danaidae.* Ferncy Creek: Hill House.

——. 1985. *Butterflies of the Oriental region: II Nymphalidae, Satyridae and Amuthusidae.* Melbourne: Hill House.

——. 1986. *Butterflies of the Oriental region: III Lycaenidae and Riodinidae.* Melbourne: Hill House.

Daerah Tk. I Bali. 1992. *Program pengembangan ekspor komoditas di luar migas.* Panitia Kerja Tetap Pengembangan Ekspor Daerah Tk. I Bali, Denpasar.

Dahler. 1940. Rallenvangst in de sawah. *Trop. Nat.* 29: 118-120.

Daily, G. C., and P. R. Ehrlich. 1992. Population, sustainability, and Earth's carrying capacity. *BioScience* 42: 761-771.

Daldjoeni, N. 1984. Pranatamangsa, the Javanese agricultural calendar: Its bioclimatological and sociocultural function in developing rural life. *Environmentalist* 4: 15-19.

d'Almeida, W. B. 1864. *Life in Java with sketches of the Javanese.* London: Hurst and Blackett.

Daly, H. 1989. *Sustainable development: towards an operational definition.* Washington D.C.: World Bank.

Daly, H. E. 1992a. Sustainable development: from religious insight to ethical principle to economic policy. World Council of Churches Conference, Rio de Janeiro, June 1992.

——. 1992b. *Steady state economics.* London: Earthscan.

Daly, H. E., and J. B. Cobb. 1990. *For the common good: Redirecting the economy toward community, the environment, and a sustainable future.* London: Green Print.

Daly, H., and R. Goodland. 1994. An ecological-economic assessment of deregulation of international commerce under GATT. *Ecol. Econ.* 9: 73-92.

Dammerman, K. W. 1916. Gegevens over rattenplag in de afdeling Malang. *Mededel. Lab. Plantenziekten* 24: 1-45.

——. 1918. Gegevens over de veldratenplag op Java. *Mededel. Lab. Plantenziekten* 31: 5-17.

——. 1929a. On the zoogeography of Java. *Treubia* 11: 1-88.

——. 1929b. The fauna of the nature reserve Tjibodas-Gunung Gede. *Proc. 4th Pacific Sci. Cong. Excursion C3.*

——. 1929c. *The agricultural zoology of the Malay archipelago.* Amsterdam: Bussey.

——. 1934. On prehistoric mammals from the Sampoeng Cave, Central Java. *Treubia* 14: 477-486.

——. 1937. Het javaansche schubdier. In *Album van Natuurmonumenten in Nederlandsche-Indies*, ed. C. G. G. J. van Steenis, 46-48.

Danaatmadja, O., and H. Natawidjaja. 1990. Menimang-nimang Gunung Pinang. *Duta Rimba* 16: 40-50.

Danaatmadja, O., and R. Tirtakusumah. 1990. Pola perhutanan sosial di hutan payan Cikiong. *Duta Rimba* 16: 33-39.

Darjamuni. 1988. The development of seaweed culture in Kepulauan Seribu, Jakarta. In Report on the Training Course on Seaweed Farming, p.99. Manila, 2 May 1988.

Darmatin, V. 1978. *Permudaan alam jamuju (Podocarpus imbricata Bl.) di hutan lindung*

Komplek Gunung Ciremai, Jawa Barat. Akademi Ilmu Kehutanan, Bandung.

Darmawiredja, M. R. 1979. *Inventarisasi dan kelimpahan ikan-ikan di wilayah Lahor, Jawa Timur.* Fakultas Perikanan IPB, Bogor.

Darsidi, A., S. Hadisepoetro, and H. L. Djwa. 1985. Pemugaran dan perluasan kawasan hutan di Pulau Jawa dalam kerangka kebijaksanaaan dan strategi pengukuran dan penatagunaan hutan nasional. In Prosiding Lokakarya Pemugaran Kawasan Hutan di Pulau Jawa, 79-96. Semarang, 12 September 1985.

Darsono, P., A. Aziz, and A. Djamali. 1978. Pengamatan terhadap populasi bintang laut *Archaster typicus* di daerah rataan gugus Pulau Pari. *Oseanol. Indo.* 10: 33-41.

Darsoprajitno, S. 1984. Pencagaran endapan batu gamping dan pemanfaatannya untuk menunjang pembangunan negara. In *Kumpulan Makalah Simposium Upaya Pengelolaan Sumber Daya Alam dan Pelestarian Kemampuan Daya Dukung Lingkungan Hidup di Daerah Gombong Selatan, Kabupaten Kebumen, Jawa Tengah.* Semarang, 26 November 1984.

—. 1985. *Tangki air alamiah di Karangbolong, Jawa Tengah.* HIKESPI, Bogor.

—. 1990. *Buku panduan Museum Geologi.* Bandung: Pusat Penelitian dan Pengembangan Geologi.

Darsoprajitno, S., and I. N. Astawa. 1982. *Gua Kisendo, Kendal, Jawa Tengah.* Puslitbang Geologi, Bandung.

Darsoprajitno, S., and Y. Ermadi. 1979. *Pengamatan geologi umum Guwa Lawa dan sekitarnya.* Direktorat Geologi, Laboratorium Palaeontologi, Bandung.

Darsoprajitno, S., and K. Handokotresno. 1981. *Gua Seplawan Purworejo, Jawa Tengah.* Puslitbang Geologi, Bandung.

Darwin, E. 1789. *Loves of the plants.* London: Johnson.

Daryono, H. 1985a. Effects of age on the composition and development of teak *Tectona grandis* undergrowth flora. *Bull. Pen. Hutan* 469: 67-93.

—. 1985b. Studi ekologi hutan alam sawo kecik (*Manilkara kauki* Dubard) di Prapat Agung, Bali Barat. *Bull. Pen. Hutan* 470: 46-55.

Davenport, J. 1988. The turtle industry of Bali. *Brit. Herpetol. Soc. Bull.* 25: 16-24.

Davies, A. G., E. L. Bennett, and P. G. Waterman. 1988. Food selection by two Southeast Asian colobine monkeys (*Presbytis rubicunda* and *Presbytis melalophos*) in relation to plant chemistry. *Biol. J. Linn.*

Soc. 34: 33-56.

Davis, C. C. 1955. *The marine and freshwater plankton.* Michigan Univ. Press: Michigan.

Davis, T. A., H. Sudasrip, and S. N. Darwis. 1985. *Coconut Research Institute, Manado.* Manado: Coconut Research Institute.

Davison, G. W. H. 1981. Diet and dispersion of the great argus pheasant *Argusianus argus. Ibis* 123: 485-494.

—. 1983. Notes on the extinct *Argusianus bipunctatus. Bull. Brit. Orn. Cl.* 103: 86-88.

Dawson, M. R. 1971. Fossil mammals of Java I: Notes on Quaternary Leporidae (Mammalia: Lagomorpha) from Central Java. *Proc. K. Ned. Akad. Wet., Ser. B* 74: 27-32.

Day, R. W. 1983. Effects of benthic algae on sessile animals: observational evidence from coral reef habitats. *Bull. Mar. Sci.* 33: 53-58.

de Beaufort, L. F., and J. C. Briggs. 1962. *The fishes of the Indo-Australian archipelago XI. Scleroparei, Hypostomides, Pediculati, Plectognathi, Opisthomi, Discocephali, Xenopterygii.* Leiden: Brill.

de Graaf, H. J., and T. G. T. Pigeaud. 1974. *De eerste Moslimse vorstendommen of Java.* The Hague: Nijhoff.

de Graaff, J., and K. F. Wiersum. 1992. Rethinking erosion on Java: a reaction. *Neth. J. Agric. Sci.* 40: 373-379.

de Haan, J. H. 1931. De Tjilatjapsche vloedbosschen. *Tectona* 24: 39-76.

—. 1952. Silt transports of rivers in Java. *Flood Control. Ser.* 3: 179-184.

—. 1953. Reis rapport No. 17: Naar het panglongegebied. *Rimba Indonesia* 2: 365-369.

de Haas, C. P. 1941. Some notes on the biology of snakes and their distribution in two districts of West Java. *Treubia* 18: 327-375.

—. 1950. Checklist of the snakes of the Indo-Australian archipelago. *Treubia* 20: 511-625.

de Iongh, H. 1980. De Balinese tijger natuurbescherming op Bali. *Panda* 16: 88-89.

—. 1982. A survey of the Bali mynah *Leucopsar rothschildi* Stresemann 1912. *Biol. Conserv.* 23: 291-295.

—. 1983. Is there still hope for the Bali Mynah? *Tigerpaper* 10.

de Iongh, H., and Y. Ruhyat. 1974. *An ecological survey in the Citarum and the carp ponds of the area near Gunung Wayang Windu at West Java during the period of February - April 1973.* Lembaga Ecologi UNPAD, Bandung.

de Iongh, H. et al. 1982. *An ecological survey of the Kangean Island archipelago in Indonesia.* WWF-Ijsselstein, Bogor.

de Klerck, E.S. 1908. *De Java-oorlog van 1825-*

'30. 'shage: M. Nijhoff.

de Klerck, L. G. 1983. Zeespiegels riffen en kustvlaken in. Ph.D. diss., Univ. of Utrecht, Utrecht.

de Korte, J. 1984. Status and conservation of seabird colonies in Indonesia. *ICBP Tech. Publ.* 2: 527-545.

de Korte, J., and M. J. Silvius. 1990. Pelicaniformes in Indonesia: status, recent changes and management. Paper for Seabird Specialists' Group Pre-Conference Workshop on Seabirds on islands: threats, case studies and action plans at the ICBP World Conference: Seabird Workshop, Univ. of Waikato, Hamilton, NZ, 19 November 1990.

de Laubenfels, D. J. 1988. Coniferales. *Fl. Mal. I* 10: 351-453.

de Man, J. G. 1892. Decapoden des Indischen Archipels. In *Zoologische Ergebnisse einer Reise in Niederländisch Ost-Indien, Vol. 2*, ed. M. Weber, 265-524. Leiden: Brill.

de Rooij, N. 1915. *The reptiles of the Indo-Australian archipelago. I. Lacertilia, Chelonia, Emydosauria.* Leiden: Brill.

——. 1917. *The reptiles of the Indo-Australian archipelago. II. Ophidia.* Leiden: Brill.

de Visser Smits, D. 1914. Excursie in de omgeving van Batavia. *Trop. Nat.* 3: 88-91, 107-111.

de Voogd, C. N. A. 1927. Voorjaaar in het djatibosch. *Trop. Nat.* 16: 165-168.

——. 1937a. Het heilige bosch van Sangeh (Zuid-Bali). In *Album van Natuurmonumenten in Nederlandsch-Indie*, ed. C. G. G. J. van Steenis,, 43-45.

——. 1937b. Een tocht naar Noesa Penida. *Trop. Nat.* 26: 161-165.

——. 1937c. Botanische aanteekeningen van de Kleine Soende eilanden (II): Bali zoals een toerist het niet ziet. *Trop. Nat.* 26 (1-2): 1-9.

——. 1937d. Botanische aanteekenigen van de Kleine Soenda Eilanden (III): Bali zoals een toerist het niet ziet. *Trop. Nat.* 26: 37-40.

——. 1940. Botanische aantekeningen van de Kleine Soenda Eilanden (VIII): De Batoer op Bali. *Trop. Nat.* 29: 38-53.

——. 1950. *Ken je die plant? Beknopte plantenatlas voor Indonesië.* 's-Gravenhage: van Hoeve.

de Voogd, C. N. A., and G. F. H. Siccama. 1938. Java en Madoera. In *3 jaren indisch natuurleven*, ed. C. G. G. J. van Steenis, 50-106.

de Vos, J. 1987. Note on two upper canines of *Megantereon* sp. (Mammalia: Felidae) from the Pleistocene of Java, Indonesia.

Proc. K. Ned. Akad. Wet., Ser. B 90: 57-64.

de Vos, J., and F. Aziz. 1989. The excavations by Dubois 1891-1900, Selenka 1906-1908 and the geological survey by the Indonesian-Japanese team 1976-1977 at Trinil, Java, Indonesia. *J. Anthro. Soc. Nippon* 97: 407-420.

de Vos, J., and P. Sondaar. 1994. Dating hominid sites in Indonesia. *Science* 266: 1726-1727.

de Vos, J., P.Y. Sondaar, G.D. van der Berg, and F. Aziz. in press. *The hominid bearing deposits of Java and its ecological context.* Schekenberg: Cour. Forsch. Inst.

de Vries, E. 1985. *Pertanian dan kemiskinan di Jawa.* Jakarta: Yayasan Obor.

de Wilde, A. 1830. *De Preanger Regentschappen op Java gelegen.* Amsterdam: Westerman.

de Wiljes-Hissink, E. A. 1953. Twee excursies naar de Tangkuban. *Trop. Nat.* 33: 115-119.

De Wit, A. 1984. *Java: Facts and fancies.* Singapore: Oxford Univ. Press.

dela Cruz, C. R. 1990. The pond refuge in rice-fish systems. *Aquabyte* 3: 6-7.

Delacour, J. 1951. *The pheasants of the world.* London: Country Life.

Delsman, H. C. 1923. Een stranding in straat Madoera. *Trop. Nat.* 12: 33-39.

——. 1925. On the "radjungans" of the bay of Batavia. *Treubia* 6: 317-321.

——. 1926a. On the distribution of the freshwater eels on Java. *Treubia* 9: 317-337.

——. 1926b. Over zoetwaterpalingen op Java. *Trop. Nat.* 15: 163-169.

——. 1926c. Vogelleven in het oerbosch. *Trop. Nat.* 15/16/17: 193-197, 82-89, 26-29.

——. 1951. *Dierenleven In Indonesie.* s'Gravenhage: W. van Hoeve.

den Berger, L. G. 1927. Unterscheidungsmerkmale von recenten und fossilen Dipterocarpaceae engattungen. *Bull. Jard. Bot. Buitenz.* 8: 495-498.

den Hartog, C. 1957. Hydrocharitaceae. *Fl. Mal. I* 5: 381-413.

——. 1970. *Sea grasses of the world.* Amsterdam: North Holland.

Dent, F. J., J. R. Desaunettes, and J. P. Malingreau. 1977. Detailed reconnaissance land resources survey of the Cimanuk watershed area (West Java): A case study of land resource survey and land evaluation procedures designed for Indonesian conditions. *Land capability appraisal project.* Soil Research Institute, Bogor.

Departemen Kehutanan. 1986. *Sejarah Departemen Kehutanan.* Jakarta: Departemen Kehutanan.

Departmen Kesehatan. 1980. *Pemanfaatan*

tanaman obat. Jakarta: Departmen Kesehatan.

der Paardt, V. 1926. Manoek poetih, *Leucopsar rothschildi. Trop. Nat.* 15: 169-173.

Dermawan, A. 1987. Mencegah punahnya si Tukik. *Suara Alam* 50: 57-58.

Desaunettes, J. R. 1977. *Detailed reconnaissance land-resources survey climax, Cimanuk Watershed area, West Java.* Soil Research Institute, Bogor.

Deshmukh, I. 1986. *Ecology and tropical biology.* Blackwell: Oxford.

——. 1992. *Ekologi dan biologi tropika.* Jakarta: Yayasan Obor.

Detmer, W. 1907. *Botanische und landwirtschaftliche studien auf Java.* Jena: Gustav Fischer.

Devall, B. 1990. *Simple in means, rich in ends: Practicing deep ecology.* London: Green Print.

Devall, B., and G. Sessions. 1985. *Deep ecology: Living as if nature mattered.* Salt Lake City: Gibbs Smith.

Devereux, P., J. Steele, and D. Kubrin. 1989. *Earthmind: Communicating with the living world of Gaia.* Rochester: Destiny.

Dhalar, M. A., and D. K. Kalsim. 1982. Jalur hijau di kawasan pesisir, fungsi serta lebar optimum ditinjau segi ekologi dan sosial ekonomi. In Prosiding Pertemuan Teknik Evaluasi Hasil Survei Hutan Bakau, 77-81. Jakarta, 1 June 1982.

Dharma, B. 1988. *Siput dan kerang Indonesia (Indonesian shells) I.* Jakarta: Sarana Graha.

——. 1992. *Siput dan kerang Indonesia (Indonesian shells) II.* Wiesbaden: Christa Hemmen.

DHV, Research Institute for Nature Management, and P.T. Indra Development Consultants. 1990. *Action plan for buffer zone development.* DHV, Jakarta.

Diakonoff, A. D. 1973. The south Asiatic Olethreutini. *Zool. Monogr. Rijksmus. Nat. Hist. Leiden* 1: 1-700.

Diamond, J. M., K. D. Bishop, S. and van Balen. 1987. Bird survival in an isolated Javan woodland: island or mirror. *Conserv. Biol.* 1: 132-142.

Diamond, J. M., and R. M. May. 1976. Island biogeography and the design of nature resserves. In *Theoretical ecology: principles and applications,* ed. R. M. May, 163-186. Oxford: Blackwell.

Diemont, W. H., A. C. Smiet, and Nurdin. 1991. Re-thinking soil erosion on Java. *Neth. J. Agric. Sci.* 39: 213-224.

Dilmy, A. 1965. Pioneer plants found one year after the 1963 eruption of Mt. Agung in Bali. *Pacific Sci.* 19 (498-501):

Ding Hou. 1958. Rhizophoraceae. *Fl. Mal. I*

5: 429-493.

——. 1978. Anacardiaceae. *Fl. Mal. I* 8: 395-548.

Diponegoro, S. L. 1981. The pattern of land use on the north-west coasts of Java (Labuan to Anyer). In Proceedings of the Workshop on Coastal Resources Management of Krakatau and the Sunda Strait Region, Jakarta, 19 August 1981.

Direktorat PPA. 1982. *Kematian badak Jawa di Taman Nasional Ujung Kulon tahun 1981-1982.* Direktorat Perlindungan dan Pengawetan Alam Dep. Kehutanan, Bogor.

Dirjen Perkebunan. 1993. *Statistik Perkebunan.* Jakarta: Dirjen Perkebunan.

Dirsh, V. M. 1975. *Classification of the acridomorphoid insects.* Faringdon: Classey.

Dit. Jen. Pencegahan dan Pemberantasan Penyakit Menular dan Penyehatan Lingkungan Pemukiman. 1990. *Malaria 1: Epidemiologi.* Jakarta: Dit. Jen PPPMPLP.

Dit. Jen. Perikanan. 1992. Statistik perikakan Indonesia 1990. Dit.-Jen. Perikanan, Jakarta.

Dixon, J. A., and K. W. Easter. 1991. Integrated watershed management: an approach to watershed management. In *Watershed resources management: studies from Asia and the Pacific,* ed. K. W. Easter, J. A. Dixon, and M. M. Hufschmidt, 3-16. Singapore: Institute of Southeast Asian Studies.

Dixon, J. A., and L. A. Fallon. 1989. The concept of sustainability: origins, extensions, and usefulness for policy. *Soc. Nat. Res.* 2: 73-84.

Dixon, J. A., and P. B. Sherman. 1990. *Economics of protected areas: A new look at benefits and costs.* Washington D.C.: Island Press.

Djaja, B., and D. S. Hadi. 1979. *Inventarisasi keadaan primata di RPH Muara Gembong dan RPH Singkil, BKPH Ujung Karawang, Jawa Barat, Desember 1979.* Biological Science Club, Jakarta.

Djaja, B., H. R. Sajudin, and L. Y. Khain. 1982. *Studi vegetasi untuk keperluan makanan bagi badak Jawa.* Fakultas Biologi UNAS, Jakarta.

Djaja, B., G. Sudarso, and B. Indrasuseno. 1984. Hutan mangrove di Tanjung Karawang, Bekasi, Jawa Barat. In Prosiding Seminar II Ekosistem Mangrove, 156-161. Jakarta: LON - LIPI.

Djajadiningrat, S. T., and H. H. Amir. 1992. *Kualitas lingkungan hidup 1992: 20 tahun setelah Stockholm.* Jakarta: Kantor Menteri KLH.

Djajadiredja, R. et al. 1983. Indonesia. *Fish*

quarantine and fish diseases in southeast Asia. International Development Research Centre, Ottawa.

Djajasasmita, M. 1977. A new species of freshwater clam from Java, Indonesia. *Veliger* 19: 425-426.

—. 1982. The occurrence of *Anodonta woodiana*, new record (Pelecypoda: Unioidae). *Veliger* 25: 175.

—. 1987. The apple snail *Pila ampullacea*, its food and reproduction (Gastropoda: Ampullariidae). *Berita Biologi* 3: 342-346.

Djamal, R., Y. Soselisa, and S. Marzuki. 1985. Survey on culture location of macroscopic marine algae in the Karimun Jawa Island. *J. Pen. Perikanan Laut* 33: 27-34.

Djatmiko, W. A., and A. T. Yusuf. 1988. Seeing the remains of Jakarta's forest. *Voice of Nature* 59: 38-44.

Djelantik, A. A. M. 1990. *Balinese paintings.* Singapore: Oxford Univ. Press.

Djenal, S., K. S. Kusmahadi, R. K. Rinanti, and I. Krisnamurti. 1983. Sebaran dan keanekaragaman jenis nyamuk di bagian barat Pulau Panaitan. Paper presented at Kongres Entomologi III, Jakarta, 30 September 1983.

Djoewisno. 1987. *Potret kehidupan masyarakat Baduy.* Jakarta: Cipta Pratama.

Djohan, T. S. 1980. *Species diversity of mangrove forest floor fauna in Segara Anakan and the Donan River.* Faculty of Biology, Gadjah Mada Univ., Yogyakarta.

Djuangsih, N., and O. Soemarwoto. 1985. Environmental lead in the rice field in Padalarang, Bandung. In *Health ecology in Indonesia,* ed. S. Suzuki, 197-201. Tokyo: Gyosei Corporation.

Djuwarso, M. S. Ular tanah bahaya laten masyarakat Banten. *Variasi* 10: 101-103.

Docters van Leeuwen, W. M. 1925. De alpine vegetatie van de Lawoevulkan in Midden Java. *Natuurk. Tijds. Ned.-Ind.* 85: 23-48.

—. 1927. Uit het leven van planten en dieren op de top van de Pangrango. *Trop. Nat.* 13/15/16: 57-65, 97-103, 111-119, 141-146, 152-159, 185-194.

—. 1929. Beitrag zur kenntnis der avifauna der mittel-Javanischen Vulkane Soembing und Sindoro. *Treubia* 10: 439-446.

—. 1930a. De Krater Tegal Primula van de Goenoeng Ipis (bij de Papandajan). *Trop. Nat.* 19: 121-123.

—. 1930b. Beitrag zur Kenntnis der gipfelvegetation der in Mittel-Java gelegenen Vulkane Soembing und Sindoro. *Bull. Jard. Bot. Buitenz.* 11: 28-56.

—. 1933. Biology of the plants and animals occuring in the higher parts of Mount Pangrango Gedeh in West Java. *Verh. K. Ned. Akad. Wet., Afdeling Natuurk.* 31: 1-278.

—. 1935. Op de toppen van den Goenoeng Kawi in Oost Java. *Levende Nat.* 5: 57-62.

—. 1938. Observations about the biology of tropical flowers. *Ann. Jard. Bot. Buitenz.* 48: 27-68.

Donner, W. 1987. *Land use and environment in Indonesia.* Honolulu: Univ. Hawaii.

Dono, T. 1987. Taman Nasional Bali Barat. *Suara Alam* 49: 20-25.

Douglas, I. 1977. *Humid landforms.* Cambridge Mass.: MIT Press.

Douglas, S. 1990. Bali high-tech (Agricultural technology in Bali). *Omni* 12 (June): 22-24.

Douthwaite, R. 1992. *The growth illusion.* Bideford: Green.

Dove, M. R. 1988. Traditional culture and development in contemporary Indonesia. In *The real and imagined role of culture in development,* ed. M. R. Dove, 1-40. Honolulu: Univ. of Hawaii Press.

—. 1985. The agroecological mythology of the Javanese and the political economy of Indonesia. *Indonesia* 39: 1-36.

—. 1986. The practical reason of weeds in Indonesia: peasant vs State views of *Imperata* and *Chromolaena. Human Ecol.* 14: 163-190.

—. 1990. Review article: Socio-political aspects of home garden in Java. *J. SEA Studies* 21: 155-163.

—. 1991. Review of "Agricultural Development in Indonesia" by Anne Booth. *J. SEA Studies* 22: 177-180.

—. 1993. A revisionist view of tropical deforestation and development. *Environ. Conserv.* 20: 17-24.

—. 1993. Review of *The political economy of mountain Java. Amer. ethnol.* 20: 207-208.

Dover, M., and L. M. Talbot. 1987. *To feed the earth: agro-ecology for sustainable development.* Washington D.C.: World Resources Institute.

Dozy, F., and J. H. Molkenboer. 1970. *Bryologia Javanica seu descriptio muscorum frondosorum Archipelagici Indici.* Leiden: Brill.

Dransfield, J. 1971. The Javanese palm flora, first impression. *Berita Biologi* 1: 221-25.

—. 1973. *Expedition to East Java and Bali.* BIOTROP, Bogor.

Dransfield, J., and N. W. Uhl. 1986. An outline of a classification of palms. *Principes* 30: 3-11.

Dransfield, S. 1980. Bamboo taxonomy in the Indo-Malesian region. In *Bamboo research in Asia.* G. Lessard, and A.

Chouinard, 121-130. Singapore: IDRC.

Dressen, W. 1937. *Hunderd tag auf Bali*. Hamburg: Broschek.

du Bus de Giesignies, L. P. J. 1827. *Rapport vande Commisaris-general Dubus over het stelsel van kolonisatie*. Batavia:

Du, N. Z. 1988. On some silicified woods from the Quaternary of Indonesia. *Palaeobot.* 91: 339.

Dubois, E. 1894. *Pithecanthropus erectus: Eine menschenachliche uebergangsform aus Java*. Landesdruckerei, Batavia.

—-. 1908. Das geologische Alter der Kendeng-order Trinil Fauna. *Tijdschr. Kron. Ned. Aardr. Gen., Ser 2* 25: 1235-1270.

Duchin, D. 1986. The Subak agriculture of Bali. *Garden* 10: 6-11.

Dudgeon, D. 1983. The importance of streams in tropical rain forest systems. In *Tropical rain forest: Ecology and management*, ed. S. L. Sutton, A. C. Chadwick, and T. C. Whitmore, 71-82. Oxford: Blackwell.

—. 1986. The life cycle, population dynamics and productivity of *Melanoides tuberculata* (Muller, 1774) (Gastropoda: Prosobranchia: Thiaridae) in Hong Kong. *J. Zool.* 208: 37-53.

Dudgeon, D., and M. W. Yipp. 1983. The diets of Hong Kong freshwater gastropods. In Proceedings of the Second International Workshop on the Macro-fauna of Hong Kong and Southern China, ed. B. Morton, and D. Dudgeon, 491-509. Hong Kong: Hong Kong Univ. Press.

Dudley, R. G., and K. C. Harris. 1987. The fisheries statistics system of Java, Indonesia: Operational realities in a developing country. *Aquacul. Fish. Mgmt.* l8: 365-76.

Dudley, R. G., and G. Tampubolon. 1986. The artisanal seine- and lift-net fisheries of the north coast of Java. *Aquacul. Fish. Mgmt.* 17: 167-175.

Duke of Edinburgh, and M. Mann. 1989. *Survival or extinction: A Christian attitude to the environment*. Salisbury: Michael Russell.

Dumarcay, J. 1986. *The temples of Java*. Singapore: Oxford Univ. Press.

Dunn, B. 1979. *The bamboo express*. Chicago: Adams Press.

Dunn, F. L., and D. F. Dunn. 1977. Maritime adaptations and exploitation of marine resources in Sundaic southeast Asian prehistory. *Mod. Quat. Res. SE Asia* 3: 1-28.

Dupont, F. 1935. Uit het leven van Javaansche onrustlinders. *Trop. Nat.* 24: 57-64.

Durand, F. 1989. *L'evolution du couvert foresteier en Indonesie: histoire et cartographie des grands sous-ensembles regionaux*. Jussieu:

Universite Paris VII.

Dwisasanti, V. I. 1987. Struktur komunitas mollusca di hutan bakau Pulau Rambut, khususnya mengenai suku Potamididae. FMIPA UI, Jakarta.

Dyer, M. I. 1990. Ecosystem redevelopment: prospects for the future. In *Environmental rehabilitation: preamble to sustainable development*. M. K. Wali. The Hague: SPB Academic.

Eavis, A. J., and A. C. Waltham. 1983. Deep cave waters for Gunung Sewu. *Geog. Mag.* 55: 460-469.

Eddy, A. 1988. *A handbook of Malesian mosses. Volume I: Sphagnales to Dicranales*. London: British Natural History Museum.

—. 1990. *A handbook of Malesian mosses. Volume 2: Leubryaceae to Buxbaumiaceae*. London: British Natural History Museum.

Edeling, A. C. J. 1870. Botanische wandeling in den ontrek van Bidara Tjina. *Natuurk. Tijds. Ned.-Ind.* 31: 170-287.

Edwards, P. J. 1982. Studies of mineral cycling in a montane rain forest in New Guinea. V. Rates of cycling in throughfall and litter fall. *J. Ecol.* 70: 649-666.

Edwin, N. 1988a. Gajahmungkur, waduk yang makin tersungkur. *Suara Alam* 54: 28-33.

—. 1988b. Wonogiri's Gajahumungkur dam. *Voice of Nature* 54: 28-33.

—. 1988c. In the belly of the earth. *Voice of Nature* 62: 28-35.

—. 1988d. The *reog ponorogo:* conflict between nature and tradition. *Voice of Nature* 58: 28-35.

—. 1988e. How Mt. Semeru came to Java. *Voice of Nature* 57: 46.

Edwin, N., and K. Arifin. 1988. Trowulan: human greed and the wrath of nature. *Voice of Nature* 57: 13-18.

Edy, Y. 1986. Bertanam lamtoro. *Trubus* 17: 24.

Effendi, R. 1989. The growth of *Rhizopora mucronata* in the tambak forest system at the mangrove forest complex of Indramayu, West Java. *BIOTROP Spec. Publ.* 37: 271-274.

—. 1990. An evaluation of a mangrove plantation established by the tambak-taungya system in Cangkring Indramayu forest district, West Java, Indonesia. *Bull. Pen. Hutan* 523: 31-36.

Effendie, I. 1972. *Rawa-rawa Bekasi perlu di up-grade*. IPB, Bogor.

Ehrenfeld, D. 1982. Options and limitations in the conservation of sea turtles. In *Biology and conservation of sea turtles*. K. A. Bjorndal,

457-464. Washington D.C.: Smithsonian Institution.

—. 1992. The business of conservation. *Conserv. Biol.* 6: 1-3.

Ehrlich, P. R. 1993. Biodiversity and ecosystem function: need we know more? *Biodiversity and ecosystem function,* ed. E. D. Schulze, and H. A. Mooney, VII-XI. Berlin: Springer.

Ehrlich, P. R., and A. H. Ehrlich. 1982. *Extinction: The causes and consequences of the disappearance of species.* London: Gollancz.

—. 1992. The value of biodiversity. *Ambio* 21: 219-226.

Eiseman, F. B. 1989. *Bali sekala and niskala. Vol. 1: Essays on religion, ritual and art.* Berkeley: Periplus Editions.

—. 1990. *Bali sekala and niskala. Vol. 2: Essays on society, tradition and craft.* Berkeley: Periplus Editions.

Eiseman, F., and M. Eiseman. 1988a. *Flowers of Bali.* Singapore: Periplus Editions.

—. 1988b. *Fruits of Bali.* Berkeley: Periplus Editions.

Ekha, I. 1991. *Dilema pesticida: tragedi revolusi hijau.* Jakarta: Kanisius.

Ekins, P., C. Folke, and R. Costanza. 1994. Trade, environment and development: the issues in perspective. *Ecol. Econ.* 9: 1-12.

Ekowati, M. 1991a. Bunga wijayakusuma dalam mitos raja Jawa. *Trubus* 263: 148-150.

—. 1991b. Mitos keng hwah alias wijayakusuma. *Trubus* 263: 152.

El Serafy, S. 1988. The proper calculation of income from depletable natural resources. In *Environmental and resource accounting and their relevance to the measurement of sustainable income,* ed. E. Lutz, and S. El Serafy. World Bank, Washington, D.C.

Elgin, D. 1981. *Voluntary simplicity: Toward a way of life that is outwardly simple, inwardly rich.* New York: William Morrow.

Elisofan, E. 1969. *Java diary.* New York: Macmillan.

Elton, C. S. 1958. *The ecology of invasions by animals and plants.* London: Chapman and Hall.

Emmiuarti, Y., and O. Rosjidin. 1983. *Laporan perjalanan ke Gunung Muria dan sekitarnya, 15 Februari - 1 Maret 1983.* LBN - LIPI, Bogor.

Emmons, L. H., J. Nias, and A. Briun. 1991. The fruit and consumers of *Rafflesia keithii. Biotropica* 23: 197-199.

Endert, F. H. 1932a. Het natuurmonument Danau in Bantam. *Tectona* 25: 963-987.

—. 1932b. Het natuurmonument Doengoes Iwul. In *Verslag over 1932.* pp.19-20. Jakarta:

Nederlandsch-Indische Vereeniging bot Natuurbescherming.

—. 1935. *Shorea javanica* K.V., een belangrijke harsproducent, in het Natuurmonument Soebah. *Tectona* 28: 488-491.

ENEX. 1994. *The Solo-Brantas region: integrated mangrove conservation and land use management plan.* Jakarta: ENEX.

Epton, N. 1956. *The palace and the jungle.* London: Oldbourne.

Erawan, T. S. 1984. *Penelitian perkembangbiakan penyu di Pantai Sukamade, Pantai Ciramea dan Pantai Citirem.* FMIPA UNPAD, Bandung.

Erftemeijer, P. 1987. *Forest avifauna on West Java: An analysis of mixed flocks of insectivorous birds in lowland and mountain forest on West Java, Indonesia.* Landbouw Universiteit, Wageningen.

—. 1989. The occurence of nests of nankeen night heron *Nycticorax caldonicus* in East Java. *Kukila* 4: 146-147.

Erftemeijer, P., and Boeadi. 1991. The diet of *Microhyla heymonsi* (Microhylidae) and Rana chalconata Schlegel (Ranidae) in a pond in West Java. *Raffles Bull. Zoology* 39: 279-282.

Erftemeijer, P., and E. Djuharsa. 1988. *Survey of coastal wetlands and waterbirds in the Brantas and Solo deltas, East Java, Indonesia.* Asian Wetlands Bureau/Interwader, Bogor.

Erftemeijer, P., and C. Swennen. 1990. Densities and biomass of macrobenthic fauna of some intertidal areas in Java, Indonesia. *Wallaceana* 59/60: 1-6.

Erftemeijer, P., S. van Balen, and E. Djuharsa. 1988. The importance of Segara Anakan for nature conservation, with special reference to its avifauna. *Report No. 5.* Asian Wetland Bureau/Interwader - PHPA, Bogor.

Erkert, H. G. 1982. Ecological aspects of bat activity rhythms. In *The ecology of bats,* ed. T. H. Kunz, 201-242. New York: Plenum.

Erythrina. 1915. Op' Pad. *Trop. Nat.* 4/5: 165-170, 180-182; 9-14.

Eudey, A. 1987. *Action plan for Asian primate conservation: 1987-1991.* Gland: IUCN.

Eussen, J. H. H. 1981. *Weeds in upland rice.* Bogor: BIOTROP.

Evans, F. C. 1956. Ecosystems as the basic unit in ecology. *Science* 123: 1127-1128.

Evans, J. 1992. *Plantation forestry in the tropics.* Oxford: Clarendon.

Evelyn, J. 1664. *Sylva, or a discourse on forest trees.* London.

Everaarts, A. P. 1981. *Weeds of vegetables in the highlands of Java.* Lembaga Penelitian Hortikultura, Jakarta.

Ewell, J. J. 1987. Restoration is the ultimate test of ecological theory. In *Restoration ecology: a synthetic approach to ecological research*, ed. W. R. Jordan, M. E. Gilpin, and J. D. Aber, 31-33. Cambridge: CUP

Fahut IPB. 1986. *Penelitian pengembangan wilayah penyangga kawasan hutan konservasi: Taman Nasional Baluran.* Fakultas Kehutanan IPB, Bogor.

Falcon, W. P. et al. 1984. *The cassava economy of Java.* California: Stanford Univ. Press.

FAO. 1977. *Proposed Bromo Tengger - Gunung Semeru Mountain National Park, East Java.* FAO, Bogor.

—. 1978a. *Field survey to the Kangean islands feasibility study.* FAO, Bogor.

—. 1978b. *Ijen management plan.* FAO, Bogor.

—. 1978c. *Karimun Jawa marine survey.* FAO, Bogor.

—. 1978d. *Marine resources of the proposed Baluran National Park.* FAO, Bogor.

—. 1978e. *Marine resources of the proposed Ujung Kulon National Park.* FAO, Bogor.

—. 1978f. *On the possibilities of making a marine national park in Pulau Seribu, Jakarta Bay: A preliminary survey.* FAO, Bogor.

—. 1978g. *Pananjung Pangandaran marine survey.* FAO, Bogor.

—. 1978h. *Proposed Pulau Seribu Marine National Park management plan; 1982-1983.* FAO, Bogor.

—. 1978. *The Segara Anakan-Nusa Kambangan area as a National Park.* FAO, Bogor.

—. 1978. *Proposed Gunung Gede Pangrango National Park management plan 1979/80-1983/4.* FAO, Bogor.

—. 1980. *Bromo-Tengger/Gunung Semeru proposed national park management plan 1981-1985.* FAO, Bogor.

—. 1982. *Maelang-Ijen-Raung preliminary management plan.* FAO, Bogor.

—. 1982. *Proposed Bali Barat National Park.* FAO, Bogor.

—. 1986. *Tropical and subtropical agriculture.* FAO, Rome.

—. 1990. *Situation and outlook of the forestry sector in Indonesia.* FAO, Jakarta.

Farimansyah. 1981. *Keragaman jenis burung pada berbagai lingkungan di Bogor dan sekitarnya.* Fakultas Kehutanan IPB, Bogor.

Fenton, M. B. et al. 1979. Activity patterns, habitat use, and prey selection by some African insectivorous bats. *Biotropica* 9: 73-85.

—. 1983. *Just bats.* Toronto: Univ. of Toronto Press.

Fenton, M. B., and J. H. Fullard. 1981. Moth hearing and the feeding strategies of bats.

Am. Scient. 69: 266-275.

Fernandes, E., and P. Nair. 1986. An evaluation of the structure and function of tropical homegardens. *Agric. Sys.* 21: 1-14.

Fernando, C. H., and J. Holcik. 1991. Some impacts of fish introductions into tropical freshwaters. In *Ecology of biological invasion in the tropics.* P. S. Ramakrishnan, 103-129. New Dehli: International Scientific Publications.

Fidhiany, L. 1983. Distribusi dan struktur populasi *Macrobrachium pilimanus* (de Man) di Sungai Cisadane, pada anak sungai Cigombong dan anak sungai Cisalopa, Kecamatan Cigombong, Kabupaten Bogor. Fakultas Perikanan, Bogor.

Finlayson, M. C., T. P. Farrell, and D. J. Griffiths. 1984. Studies of the hydrobiology of a tropical lake in North-western Queensland. III. Growth, chemical composition and potential for harvesting of the aquatic vegetation. *Aust. J. Inter. Freshw. Res.* 31: 584-96.

Firdausy, C., and C. Tisdell. 1991. Economic returns from seaweed *(Eucheuma cottonii)* farming in Bali, Indonesia. *Asian Fish. Sci.* 4: 61-73.

Firman, T. 1991. Population mobility in Java: in search of theoretical explanation. *Sojourn* 6: 71-105.

Fleischer, M. 1923. *Die Musci der Flora von Buitenzorg.* Leiden: Brill.

Flenley, J. 1985. Man's impact on the vegetation of southeast Asia: the pollen evidence. In *Recent advances in Indo-Pacific prehistory.* V. Misra, and P. Bellwood, 297-305. Leiden: Brill.

Flint, M. L., and R. van den Bosch. 1981. *Introduction to integrated pest management.* New York: Plenum.

—. 1990. *Pengendalian hama terpadu.* Yogyakarta: Kanisius.

Foley, A. S. 1984. *Rice cultivation in Bali: an energy analysis.* Population and World Resources Programme, Murdoch Univ..

—. 1987. Ecological transition in Bali. Ph.D. diss., Australian National Univ., Canberra.

Fontanel, J., and A. Chantefort. 1978. Bioclimats du monde Indonésien. *Trav. sec. sci. techn. Inst. fr. Pondichéry* 16: 1-104.

Fooden, J. 1982. Ecogeographic segregation of macaque species. *Primates* 23: 574-579.

Forbes, H. O. 1885. *A naturalist's wanderings in the eastern Archipelago.* New York: Harper.

—. 1989. *A naturalist's wanderings in the eastern Archipelago.* Singapore: Oxford Univ. Press.

Forman, W. 1983. *Bali: The split gate to heaven.* London: Orbis.

Forshaw, J. M. 1973. *Parrots of the world.* Melbourne: Lansdowne.

Fortune Hopkins, H. in press. The Indo-Pacific species of *Parkia* (Leguminosea: Mimusoideae). *Kew Bull.*

Fox, A. 1986. *Common Malaysian moths.* Kuala Lumpur: Longman.

Fox, J. J. 1991. Managing the ecology of rice production in Indonesia. In *Indonesia: resources, ecology and environment,* ed. J. Hardjono, 61-84. Singapore: Oxford Univ. Press.

Fox, M. W. 1984. *The whistling hunters.* Albany: Univ. of New York Press.

Francillon, G. 1979. *Bali: Tourism, culture, environment.* Paris: UNESCO.

——. 1990. The dilemma of tourism in Bali. In *Sustainable development and environmental management of small islands,* ed. W. Beller, P. d'Ayala, and P. Hein, 267-272. Paris: MAB-UNESCO.

Franck, P. F. 1937. Het Hiang-plateau als natuurreservaat. In *Album van Natuurmonumenten in Nederlandsch-Indie,* ed. C. G. G. J. van Steenis, Batavia.

Frankel, O. H., and M. E. Soulé. 1981. *Conservation biology.* Sunderland, MA: Sinauer.

Frankie, R. W. 974. Miracle seeds and shattered dreams in Java. *Nat. Hist.* 83: 10-19.

Franssen, C. J. H. 1935. Insecten, schadelijkaan het batatengewas op Java. *Landbouw* 10: 205-225.

——. 1987. *Insect pests of sweet potato in Java* (translation of *Insecten, schadelijk aan het batatengewas op Java).* Tainan: Asian Vegetable Research and Development Centre.

Frazer, J. G. 1939. *The native races of Australasia.* London: Percy Lund Humphries.

Frazer, N. B. 1992. Sea turtle conservation and halfway technology. *Conserv. Biol.* 6: 179-184.

Freeman, P., and T. Fricke. 1984. The success of Javanese multi-storied gardens. *The Ecologist* 14: 150-152.

Freeze, R. A., and J. A. Cherry. 1979. *Groundwater.* New Jersey: Prentice Hall.

Friend, A. M. 1993. Environmental and resource accounting in developing countries. In *Ecological economics: emergence of a new development paradigm.* pp.75-84. Ottawa: Univ. of Ottawa.

Frodin, D. G. 1988. Studies in *Schefflera.* III: A new species from Java, Indonesia. *Schefflera reiniano* new status, new name, Araliales: Araliaceae. In Proceedings of the Academy of Natural Sciences of Philadelphia, 336-338.

Fryer, G. 1991. Biological invasions in the tropics: hypotheses versus reality. In *Ecology of biological invasion in the tropics.* P. S. Ramakrishnan, 87-101. New Dehli: International Scientific Publications.

Fujita, M. 1991. *Flying fox (Chiroptera: Pteropodidae) pollination, seed dispersal, and economic importance: A tabular summary of current knowlege.* Bat Conservation International, Austin.

Fujita, M., and M. D. Tuttle. 1991. Flying foxes (Chiroptera: Pteropodidae): Threatened animals of key ecological and economic importance. *Conserv. Biol.* 5: 453-463.

Fukuoka, M. 1978. *The onestraw revolution: an introduction to natural farming.* New York: Rodale.

——. 1991. *Revolusi sebatang jerami: sebuah pengantar menuju pertanian alami.* jakarta: Yayasan Obor.

Gadagkar, R., K. Chandrashekara, and P. Nair. 1990. Insect species diversity in the tropics: sampling methods and a case study. *J. Bombay Nat. Hist. Soc.* 87: 337-353.

Gaisler, J. 1979. The ecology of bats. In *Ecology of small mammals.* Ed D. R. Stoddart, 281-342. London: Chapman and Hall.

Gallagher, K. 1984. Effects of host plant resistance on the microevolution of the rice brown planthopper *Nilaparvata lugens.* Ph.D. diss., Univ. of California, Berkeley.

Ganesia, O. 1988. Rambut Island: bird kingdom in Jakarta Bay. *Voice of Nature* 59: 6-8.

Garna, J. 1987. *Orang Baduy.* Bangi: Penerbit Universiti Kebangsaan Malaysia.

Geerts-Ronner, S. J. 1924. Uit Soerabaia's omstreken. I. Soetji. *Trop. Nat.* 13: 161-173.

——. 1940. Uit Soerabaia's omstreken. II. Soetji in den regentijd. *Trop. Nat.* 29: 101-106.

Geertz, C. 1960. *The religion of Java.* Chicago: Univ. of Chicago Press.

——. 1963. *Agricultural involution: The process of ecological change in Indonesia.* Chicago: Univ. of Chicago Press.

——. 1968. Tihingan: A Balinese vilage. In *Villages in Indonesia,* ed. Koentjaraningrat, 210-225. Ithaca: Cornell Univ. Press.

——. 1972. The wet and the dry: Traditional irrigation in Bali and Morocco. *Human Ecol.* 1: 23-39.

——. 1973. Comments on Benjamin White's "Demand for labor and population growth in colonial Java". *Human Ecol.* 1: 237-239.

——. 1980. *Negara: The theatre state in nineteenth century Bali.* Princeton, N.J.: Princeton Univ. Press.

——. 1983. *Involusi pertanian.* Jakarta: Bhratara Karya Aksara.

Geertz, H. 1991. *State and society in Bali.* Leiden: KITLV.

Geesink, R. 1990. The general progress of the Flora Malesiana. In *The plant diversity of Malesia,* ed. P. Baas, K. Kalkman, and R. Geesink, 11-16. Dordrecht: Kluwer.

Geisen, W. 1991. *Checklist of Indonesian freshwater aquatic herbs including an introduduction to freshwater aquatic vegetation.* PHPA/AWB, Bogor.

Gennardus, B. 1984. Verdere waarnemingen aan vlinders na een vulkanische aregen (Lepidoptera). *Entomol. Berichte* 44: 2-4.

George, S. 1992. *The debt boomerang.* London: Pluto.

Ghofir, A. 1989. *Peranan vegetasi bagi kehidupan owa di Cagar Alam Gunung Halimun, Jawa Barat.* FMIPA UI, Jakarta.

Gibbons, L. M., M. T. Crawshaw, and A. E. Rumpus. 1992. *Molinacuaria indonesiensis* n. sp. (Nematoda: Acuarioidea) from *Rattus argentiventer* in Indonesia. *Syst. Parisitol* 23: 175-181.

Gibson, I. A. S. 1980. Control of disease caused by biotic agencies in the forest. Paper given at Commonwealth Forestry Conference, Trinidad,

Gibson, I. A. S., and T. Jones. 1977. Monoculture as the origin of major forest pests and diseases. In *Origins of pests, parasites, diseases, and weed problems,* ed. J. M. Cherrett, and G. R. Sagar, 139-161. Oxford: Blackwell.

Giesen, W. 1991. Cheklist of Indonesia freshwater aquatic herbs (including an introduction to freshwater aquatic vegetation). *PHPA - AWB Sumatra wetland project report No. 27.* Asian Wetland Bureau, Bogor.

Gilert, E., and M. Neuendorf. 1991. A new species of *Lamproderma* (Myxomycetes) found in Java, Indonesia. *Nord. J. Bot.* 10: 661-664.

Gillis, M. 1988. Indonesia: public policies, resource management, and the tropical forest. *Public policies and the misuse of forest resources.* R. Repetto, and M. Gillis, 43-113. Cambridge: Cambridge Univ. Press.

Ginting, N., Yuningsih, and Indraningsih. 1981. Tanam-tanaman beracun di Jawa Barat. *Bull. Pen. Hutan* 13: 63-74.

Gisius, A. 1930. Het sterven der dieren in de wildernis. *Trop. Nat.* 19: 197-198.

Glover, I. C. 1979. Prehistoric plant remains from southeast Asia, with special reference to rice. *South Asian archaeology,* ed. M. Taddie, 7-37. Naples: Instatato Universitario Orientales.

——. 1985. Some problems relating to the domestication of rice in Asia. In *Recent advances in Indo-Pacific prehistory,* ed. V. Misra, and P. Bellwood, 265-274. Leiden: Brill.

Glynn, P. 1991. Coral reef bleaching in the 1980s and possible connections with global warming. *Trends Ecol. Evol.* 6: 175-179.

Goff, M. L. 1983. A new species of *Gahrliepa* (Acari: Trombiculidae) from Java. *Int. J. Entomol.* 25: 281-284.

Göltenboth, F. 1978. The Rawa Pening Lake: Background information, facts and problem. In Proceedings of the 1st Seminar on Aquatic Biology and Aquatic Management of the Rawa Pening Lake, 4-19. Salatiga, 20 November 1978. Salatiga: Satya Wacana Christian Univ.

——. 1979. Hydrobiological studies on the Muncul River system in the Rawa Pening Lake area. In Proceedings of the Second Seminar in Aquatic Biology and Aquatic Management of the Rawa Pening Lake, Salatiga, 29 October 1979. Salatiga: Satya Wacana Christian Univ.

——. 1992. *Bibliography of Rawa Pening.* Salatiga, Satya Wacana Christian Univ..

Göltenboth, F., and D. Bolsehner. 1978. The ecology of the Rawa Pening Lake: Contribution to the knowledge of some physical properties. In Proceedings of the 1st Seminar on Aquatic Biology and Aquatic Management of the Rawa Pening Lake, Salatiga, Central Java, Salatiga, 20 November 1978. Salatiga: Satya Wacana Christian Univ.

Göltenboth, F., and A. I. A. Kristyanto. 1979a. The Desmidiaceae as revealed by a quantitative and systematic investigation in the Rawa Pening Lake (Central Java). *J. Asian Ecol.* 1: 42-50.

——. 1979b. The cyclic variation of some abiotic parameters and the appearance of volcanic influences on the Rawa Pening Lake. In Proceedings of the 2nd Seminar on Aquatic Biology and Aquatic Management of the Rawa Pening Lake, Salatiga, Central Java, Salatiga, 29 October 1979. Salatiga: Satya Wacana Christian Univ.

——. 1987. Ecological studies on the Rawa Pening Lake (Central Java): Some biotic parameters of the semi-natural lake. *Trop. Ecol.* 28: 101-116.

Göltenboth, F., and A. J. A. Kristyanto. 1993. *Fisheries in Lake Rawa Pening - facts, problems, prospects.* FIPIA, Univ. Kristen Satya Wacana, Salatiga.

Göltenboth, F., and K. H. Timotius. 1992. *The Rawa Pening Lake/ Danau Rawa Pening.* Fakultas Sains dan Matematika Universitas

Kristen Satya Wacana, Salatiga.

Gomez, E. d., and L. C. Alcala. 1982. Growth of some corals in an artificial reef off Dumaguete, Central Visayas, Philipines. *Phil. J. Biol.* 11: 148-157.

Goodland, R. 1991. *Tropical deforestation: solutions, ethics and religions.* World Bank, Washington D.C.

—. 1992a. The case that the world has reached limits: more precisely that current throughput growth in the global economy cannot be sustained. *Pop. Env.* 13: 167-182.

—. 1992b. The case that the world has reached limits. In *Population, technology and lifestyle.* eds R. Goodland, H. E. Daly, and S. El-Serafy, 3-22. Washington D.C.: Island Press.

—. 1995. The concept of environmental sustainability. *Ann. Rev. Ecol. Syst.* 26: 1-24.

Goodland, R., and H. Daly. 1993. *Poverty alleviation is essential for environmental sustainability.* Environment Department, World Bank, Washington D.C.

Goodman, D., and M. Redclift. 1991. *Refashioning nature: Food, ecology and culture.* London: Routledge.

Göppert, H. R. 1854. *Die Tertiärflora auf der Insel Java, nach den Entdeckungen des Herrn Fr. Junghuhn.* s'Gravenhage.

Gould, E. 1978. Foraging behaviour of Malaysian nectar-feeding bats. *Biotropica* 10: 184-193.

Gradwohl, J., and R. Greenberg. 1991. *Menyelamatkan hutan tropika.* Jakarta: Yayasan Obor.

Grant, J. 1984. Sediment microtopography and shorebird foraging. *Mar. Ecol. Progr. Ser.* 19: 293-296.

Green, A. J. 1992. *The status and conservation of the white-winged wood duck Carrina scutulata.* IWRB, Slimbridge.

Green, C. A. 1991. First sight record of yellow-vented flowerpecker for Bali. *Kukila* 5: 143.

Green, J., S. A. Corbet, E. Watts, and B. L. Oey. 1976. Ecological studies on Indonesian lakes, overturn and restratification of Ranu Lamongan. *J. Zool.* 180: 315-354.

—. 1978. Ecological studies on Indonesian lakes: The montane lakes of Bali. *J. Zool.* 186: 315-354.

Greenpeace. 1991. *Slaughter in paradise: The exploitation of sea turtles in Indonesia.* Amsterdam: Greenpeace International.

Grey-Wilson, C. 1989. *Semeiocardium* Zoll.; is it a good genus? *Kew Bull.* 44: 107-113.

Griffiths, M. 1990. *Indonesian eden.* Baton Rouge: Louisiana State Univ. Press.

—. 1992a. *Javan rhino census. July 16 1992.* WWF, Jakarta.

—. 1992b. *Javan rhino census, Jan. 91 through Aug. 92.* WWF, Jakarta.

Grist, D. H. 1978. *Rice.* London: Longman.

Groves, C. P. 1981. *Ancestors for the pigs: taxonomy and phylogeny of the genus Sus.* Department of Prehistory, Australian National Univ., Canberra.

—. 1985. Plio-Pleistocene mammals in island Southeast Asia. *Mod. Quat. Res. SE Asia* 9: 43-54.

—. 1989. *A theory of human and primate evolution.* Oxford: Clarendon.

Groves, C. P., and P. Grubb. 1987. Relationships in living Cervidae. In *Biology and management of the Cervidae,* ed. C. M. Wemmer, 21-59. Washington D.C.: Smithsonian Institution.

Grubb, P. J. 1974. Factors controlling the distribution of forest types on tropical mountains: New factors and a new perspective. In *Altitudinal zonation in Malesia,* ed. J. R. Flenley, 13-46. Hull: Univ. of Hull.

—. 1977. Control of forest growth and distribution on wet tropical mountains with special reference to mineral nutrition. *Ann. Rev. Ecol. Syst.* 8: 83-107.

Grubb, P. J., and T. C. Whitmore. 1966. A comparison of montane and lowland rain forest in Ecuador. II. The climate and its effects on the distribution and physiognomy of the forests. *J. Ecol.* 54: 303-333.

Guha, R. 1989. Radical American environmentalism and wilderness preservation: a Third World Perspective. *Environ. Ethics*

Guillot, C. 1990. *The Sultanate of Banten.* Jakarta: Gramedia.

Guinness, P. 1986. *Harmony and hierarchy in a Javanese kampung.* Singapore: Oxford Univ. Press.

Gunadi, B. 1993. Decomposition and nutrient flow in a pine forest plantation in Central Java. Ph.D. diss., Free Univ., Amsterdam.

Gunadi, B., and S. Notosoedarmo. 1988. *Struktur arthropoda tanah dan vegetasi lantai hutan pinus di Jawa Tengah.* Fakultas Biologi, Satya Wacana Christian Univ., Salatiga.

Gunadi, S. Notosoedarmo, and E. N. G. Joosse. 1991. Effects of disturbances on the guild structure of soil arthropods in pine forests in central Java - a preliminary survey. In *Advances in management and conservation of soil fauna,* ed. A. Veeresh, 391-398. New Delhi: Oxford Univ. Press.

Guppy, N. 1990. Conflicts between sustainable development and tourism. In *International Seminar on Human Ecology, Tourism, and Sustainable Development.* 108-114. Denpasar: Human Ecology Study Group.

Haak, J. 1889. *Observations sur les Rafflesias* (Rafflesia patma *Blume).* Amsterdam: Scheltema en Holkema.

Hackert, E. C., and S. Hastenrath. 1986. Mechanism of Java rainfall anomalies. *Mo. Weather Rev.* 114: 745-757.

Hadipoernomo. 1981. *Acacia arabica* as a hedge crop (Kalikonto Watershed Project, Malang, Indonesia). *Duta Rimba* 7: 13-15.

Hadisubroto, I. 1988. A trial improvement on coral reef in Jepara. In Report of the Workshop on Artificial Reefs Development and Management, 93-96. Penang, 13 September 1988.

Hadisumarto, S. 1979. Coastline accretion in Segara Anakan, Central Java, Indonesia. *Indo. J. Geog.* 9: 45-52.

Hadiwinoto, S., and G. Clarke. 1990. *The environmental profile of Jakarta.* Metropolitan Environmental Improvement Program, Jakarta.

Haeruman, H. et al. 1977. *Studi konsumsi sumber daya enersi pedesaan, terutama kayu bakar di Jawa Barat.* Fakultas Kehutanan IPB, Bogor.

———. 1979. Energy consumption in rural areas of West Java. Paper presented at the 7th General Assembly of the World Federation of Engineering Organizations, Jakarta, 14 November 1979.

Haile, N. S. 1958. The snakes of Borneo with a key to species. *Sarawak Mus. J.* 8: 743-771.

Hails, C. J., and A. Amiruddin. 1981. Food samples and selectivity of white-bellies swiftlets *Collocalia esculenta. Ibis* 123: 328-333.

Haines, C. P., and S. M. T. Lynch. 1987. A new species of *Madaglyphus* (Acarina: Acaridae) from a rice mill in Java, Indonesia. *Acarologia* 28: 255-264.

Haitlinger, R. 1990. Four new species of *Leptus* Latrille 1796 (Acari: Prostigmata: Erythraeidae) from insects of Australia, New Guinea and Asia. *Wiad. Parazytol* 36: 47-54.

Halder, V. 1975. *Oekologie und verhalten des Banteng* (Bos javanicus) *in Java.* Hamburg: Paul Parey.

Halim, M. H., K. Kvalvågnaes, A. H. Robinson, and N. V. C. Polunin. 1979. Preliminary report on marine conservation in Bali. FAO, Bogor.

Halim, M. H., and K. Kvalvågnaes. 1980. Marine resources of the proposed Ujung Kulon National Park. FAO, Bogor.

Halim, M. H., K. Kvälvagnaes, and N. V. C. Polunin. 1980. Coral reefs of the Bali Barat Reserve, Indonesia. Paper presented at the Symposium on Recent Research Activities on Coral Reefs in South-East Asia, Bogor, Bogor, 6 May 1980.

———. 1983. Bali Barat: An Indonesian marine protected area and its resources. *Biol. Conserv.* 25: 171-180

Hall, B. 1827. *Voyage to the Eastern Seas in the year 1816; including an account of Captain Maxwell's attack on the batteries at Canton and notes of an interview with Bonaparte at St. Helena in August 1817.* New York: Carvill.

Hamidy, R. 1984. Keragaman dan kesamaan hewan benthos dua buah sungai kecil di kawasan Balai Penelitian Teh dan Kina, Gambung, Jawa Barat. *Terubuk* 10: 3-14.

Hamilton, L. S., and P. N. King. 1983. *Tropical forested watersheds: hydrologic and soils response to major uses or conversions.* Boulder, Colorado: Westview.

Hamilton, L. S., and A. J. Pearce. 1991. Biophysical aspects in watershed management. In *Watershed resources management: studies from Asia and the Pacific,* ed. K. W. Easter, J. A. Dixon, and M. M. Hufschmidt, 33-52. Singapore: Institute of Southeast Asian Studies.

Hamilton, W. 1979. Tectonics of the Indonesian region. *Geological survey professional paper No. 1078.* Washington, D.C.: US Printing Office.

Hammer, M. S. J. 1979. *Investigations on the oribatid fauna of Java.* Kobenhavn: Kommissionor, Munksgaard.

Hammond, R., and M. King. 1984. *Nature by design: a teacher's guide to practical nature conservation.* Birmingham: Urban Wildlife Group.

Hamzah, Z. 1976. Laporan survey ekologi cendana di pulau Bali. *Laporan No. 241.* Lembaga Penelitian Hutan, Bogor.

———. 1977. Survey ekologi sawo kecik di ujung timur Pulau Jawa. *Laporan No. 252.* Lembaga Penelitian Hutan, Bogor.

Hanapi, D. K. ????//. *Suatu studi tentang kehidupan penyu* (Chelonia mydas) *di Pantai Pangumbahan, Sukabumi, Jawa Barat.* Fakultas Kehutanan IPB, Bogor.

Handschin, E. 1926. Ost-Indische Collembolen III: Beitrag zur Collembolen fauna von Java und Sumatra. *Treubia* 8: 446-461.

———. 1928. Collembolen aus java, nebts einem beitrag zur einer monographie der gattung *Gremato cephalus. Treubia* 10: 245-270.

Hannig, W. 1988. *Towards a blue revolution.*

Yogyakarta: Gadjah Mada Univ. Press.

Hanski, I. 1985. What does a shrew do in an energy crisis? In *Behavioural ecology: Ecological consequences of adaptive behaviour*, ed. R. M. Sibly, and R. H. Smith, 247-252. Oxford: Blackwell.

Hanski, I., and J. Krikken. 1990. Dung beetles in tropical forests in south-east Asia. In *Dung beetle ecology*, ed. I. Hanski, and Y. Cambefort, 179-197. Princeton N.J.: Princeton Univ. Press.

Harahap, H. 1991. *Keputusan Menteri Kehutanan 301/Kpts-II/1991 tentang inventarisas: satwa yang dilindungi undang-undang dan atau bagian-bagiannya yang dipelihara oleh perorangan*. Departmen Kehutanan, Jakarta.

Hardenberg, J. D. F. 1938. De Koralleilanden in de Baai van Batavia. In *3 jaren indisch natuurleven*, ed. C. G. G. J. van Steenis, Batavia: Nederlandsch-Indische Vereeniging tot Natuurbescherming.

Hardi, T., and S. Santosa. 1992. The growth and the attack of psyllid on some leucaena varieties. *Bull. Pen. Hutan* 549: 19-26.

Hardjamulia, A., and P. Suwignyo. 1988. The present status of the reservoir in Indonesia. In *Reservoir fishery management and development in Asia*, ed. S. S. da Silva, 8-12. Ottawa: IDRC.

Hardjamulia, K., E. Setiadi, and N. S. Rabegnatar. 1988. Some biological aspects of the predominant fish species in the Jatiluhur Reservoir, West Java, Indonesia. In *Reservoir fishery management and development in Asia*, ed. S. S. da Silva, 98-104. Ottawa: IDRC.

Hardjasasmita, S. 1987. Suidae in Indonesia. *Scripta Geol.* 85: 1-30.

Hardjono, J. 1986. Environmental crisis in Java. *Prisma* 39: 3-13.

——. 1991a. The dimensions of Indonesia's environmental problems. In *Indonesia: resources, ecology and environment*, ed. J. Hardjono, 1-16. Singapore: Oxford Univ. Press.

——. 1991b. Environment or employment: vegetable cultivation in West Java. In *Indonesia: resources, ecology and environment*, ed. J. Hardjono, 133-153. Singapore: Oxford Univ. Press.

Hardjono, J., and H. Hill. 1989. West Java: Population pressure and regional diversity. In *Unity and diversity: Regional economic development in Indonesia since 1970*, ed. H. Hill, 255-282. Singapore: Oxford Univ. Press.

Hardjosuwarno, S. 1980. Biological monitoring in the Cilacap refinery area. In Asian Symposium on Mangrove Environmental Research and Management, Kuala Lumpur, 25 August 1980.

——. 1989. The impact of oil refinery on the mangrove vegetation. *BIOTROP Spec. Publ.* 37: 187-192.

Hardjosuwarno, S., A. Pudjoarinto, and M. Nasir. 1974. *Penelitian studi ecologi Rawa Pening*. Universitas Gadja Mada, Yogyakarta.

Harger, J. R. E. 1986. Community structure as a response to nature and man-made environmental variables in the Pulau Seribu island chain. In Proceedings of the Regional Seminar on Coral reef Ecosystem: Their Management Practice and Corresponding Research/Training Needs, 1-121. Ciloto, 4 March 1986.

Hariyadi. 1975. Cagar Alam/Suaka Margasatwa Leuweung Sancang, Garut Selatan. *Biologica* 2: 5-9.

Harminto, S., and Y. Yusuf. 1978. Keadaan vegetasi hutan bakau di pesisir sebelah barat Teluk Jakarta. Paper presented at Seminar Ekosistem Hutan Mangrove. Jilid I, Jakarta, 1 February 1927.

Harrington, B. J. 1988. Comments on the blood-feeding tribe Cleradini (Hemiptera: Lygaeidae: Rhyparochrominae) and description of a new genus and a new species with the legs modified for grasping. *Ann. Entomol. Soc. Am.* 81: 577-580.

Harris, K. M. 1994. Cibodas to Cibeureum: Mt. Gede-Pangrango National Park. National Parks Directorate, Information Book Series No. 1.

Harrison, P. L. et al. 1984. Mass spawning in tropical reef corals. *Science* 223: 1186-1189.

Hart, T. B. 1990. Monospecific dominance in tropical rain forests. *Trends Ecol. Evol.* 5: 6-11.

Hartadi, R. 1984. Pembangunan pariwisata yang memperhatikan lingkungan. Paper presented at Simposium Upaya Pengelolaan Sumber Daya Alam dan Pelestarian Kemampuan Daya Dukung Lingkungan Hidup di Daerah Gombong Selatan, Kabupaten Kebumen, Jawa Tengah, Semarang, Semarang, November 1984.

Harte, J. et al. 1991. *Toxics A to Z: a guide to everyday pollution hazards*. Berkeley: Univ. of California Press.

Hartmann, M. A. 1939. De Goenoeng Salak in West Java. *Trop. Nat.* 28: 177-188.

Hartojo, P., and I. S. Suwelo. 1987. Upaya pelestarian jalak Bali *Leucopsar rothschildi*. *Media Konservasi* 1: 29-41.

Hartono, P. 1979. *Perjalanan ke Gua Lawa di Jawa Tengah*. HIKESPI, Bogor.

Hartoto, D. I. 1986. Distribusi lokal dan

spasial *Puntius binotatus* dan *Rasbora lasteri-triata* di Ci Taman Jaya dan Ci Binua, Ujung Kulon. *Berita Biologi* 3: 261-267.

Hartoto, D. I., and R. M. Marwoto. 1987. Community structure and spatial distribution of *Melanoides placaria* at Cibanua, Ujung Kulon. *Berita Biologi* 3: 155-158.

Harun, W. K., and I. G. M. Tantra. 1984. Flora Cagar Alam Gunung Tilu, Jawa Barat. *Laporan No. 425.* Lembaga Penelitian Hutan, Bogor.

Haryadi. 1975. Cagar Alam Leuweung Sancang Garut Selatan. *Biologica* 2: 5-9.

Hasanuddin, L. 1988a. Jakarta needs its green belt. *Voice of Nature* 59: 28-30.

—. 1988b. The spotted deer of Bogor Palace. *Voice of Nature* 61: 18-21, 65.

Hasegawa, H., S. Shiraishi, and Rochman. 1992. *Tikusnema javaense* (Nematoda: Acuarioidae) and other nematodes from *Rattus argentiventer* collected in West Java, Indonesia. *J. Parisitol.* 78: 800-804.

Haskoning. 1988a. *Environmental profile, West Java.* Haskoning, Nijmegen.

—. 1988b. *Profil lingkungan, Jawa Barat.* Haskoning, Nijmegen.

Hassall and Associates. 1992. Comprehensive tourism development plan for Bali. Hassall and Associates, Denpasar.

Hastenrath, S. 1987. Predictability of Java monsoon rainfall anomalies: A case study. *J. Climatol. Appl. Meteorol.* 26: 133-141.

Hatcher, B. G. 1981. The interaction between grazing organisms and the epilithic algal community of a coral reef; a quantitative assessment. Proc. 4th Int. Coral Reef Symp., 515-524. Manila.

—. 1983. Grazing in coral reef ecosystems. In *Perspectives on coral reefs,* ed. D. J. Barnes, 164-179. Manuka: Cloustan.

Hatcher, B. G., and A. W. D. Larkum. 1983. An experimental analysis of factors controlling the standing crop of the epilithic algal community on a coral reef. *J. Exp. Mar. Biol. Ecol.* 69: 61-84.

Hawksworth, D. L., and L. A. Mound. 1991. Biodiversity databases: the crucial significance of collections. In *The biodiversity of microorganisms and invertebrates: its role in sustainable agriculture,* ed. D. L. Hawksworth, 17-29. Wallingford: CAB International.

Hayashi, Y. 1984. A radio-telemetry of the red junglefowl and the green junglefowl in Indonesia. *Jap. J. Zootech. Sci.* 55: 439-443.

Heaney, L. R. 1984. Mammalian species richness on islands on the Sunda Shelf, Southeast Asia. *Oecologia* 61: 11-17.

Hecht, J. 1990. A climate of change sweeps the tropics. *New Scient.* 22 Dec: 13.

Heddy, S., S. B. Soemitro, and S. Soekartomo. 1986. *Pengantar ekologi.* Jakarta: Rajawali.

Heekeren, H. R. 1941. Naar de jongste uitbarsting van den Semeroe. *Trop. Nat.* 30: 165-170.

Hefner, R. W. 1985. *Hindu Javanese: Tengger tradition and Islam.* Princeton N.J.: Princeton Univ. Press.

—. 1990. *The political economy of mountain Java: An interpretive history.* Berkeley: Univ. of California Press.

—. 1991. Java's five regional cultures. In *Java,* ed. E. Oey, 66-69. Singapore: Periplus Editions.

Hegewald, E., and B. O. Zanten. 1986. A list of bryophytes from Bali, Indonesia, collected by E. and P. Hegwald in 1981. *J. Hattori Bot. Lab.* 60: 263-270.

Hehanusa, P. E. 1987. Groundwater potential in the coastal plain of West Java, Indonesia; sustainable clean water. *Ergeb. Limnol.* 28: 35-38.

Heilbronner, R. 1980. *An enquiry into the human prospect: updated and reconsidered for the 1980s.* New York: Norton.

Hellebrekers, W. P. J., and A. Hoogerwerf. 1967. A further contribution to our zoological knowledge of the island of Java (Indonesia). *Zool. Verhand.* 88: 1-164.

Hellier, C. 1988. The mangrove wastelands. *The Ecologist* 18: 77-79.

Hemmer, H. 1971. Fossil mammals of Java. III: Zur kenntnis der evolution Javanishcher Kleinkatzen *Prionailurus bengalensis koenigswaldi* ssp. n. und *Felis chaus* ssp. aus dem Neolithikum von Sampung, Mittel-Java. *Proc. K. Ned. Akad. Wet., Ser. B* 74: 363-375.

—. 1987. The phylogeny of the tiger. In *Tigers of the world: The biology, biopolitics, management, and conservation of an endangered species,* ed. R. L. Tilson, and U. S. Seal, 28-35. New Jersey: Noyes Publications.

Henderson, H. 1978. *Creating alternative futures.* New York: Berkley.

Hendrarto, B. 1979. The ecology of biological resources of the north coasts of Central Java, Indonesia. In Proceeding of the 5th International Symposium on Tropical Ecology. Part 1, ed. J. I. Furtado, Kuala Lumpur, 16 April 1979.

Hendrarto, B., and L. Hutabarat. 1981. Aspek sumberdaya pantai di pantai utara Jawa Tengah. Paper presented at Kongres Ilmu Pengetahuan Nasional III, Jakarta, 15 September 1981.

Henrey, L. 1982. *Coral reefs of Malaysia and Singapore.* Kuala Lumpur: Longman.

Hernowo, J. B. 1989. Ornithological news. *Media Konservasi* 2: 57-8.

Heryanto. 1984. *Suatu studi tentang kepadatan dan penyebaran berbagai jenis tripang (Echinodermata: Holothuroidea) di perairan gugus Pulau Pari, Teluk Jakarta.* Fakultas Perikanan IPB, Bogor.

Heryanto, I. 1976. The seasonal variation of plankton organic and mineral contents in the Rawa Pening lake sediment. *J. Satya Wacana Res.* 1: 428-446.

Heyne, K. 1913. *Nuttige planten van Nederlandsche Indie (1913-1917).* Wageningen:

—. 1917. *Nuttige planten van Nederlandsch Indie.* Wageningen:

—. 1987. *Tumbuhan berguna Indonesia.* Jakarta: Yayasan Sarana Wana Jaya.

Heywood, V. H., and S. N. Stuart. 1992. Species extinctions in tropical forests. In *Tropical deforestation and species extinction,* ed. T. C. Whitmore, and J. A. Sayer, 91-117. London: Chapman and hall.

Hickson, S. J. 1889. *A naturalist in Celebes.* London: Murray.

Hidayat, M. et al. 1982. *Laporan expedisi II ke Gua Ngerong.* Persatuan Caving dan Speleologi Indonesia, Malang.

HIKESPI. 1983. *Laporan survei daerah karst Cibodas di Ciampea.* HIKESPI, Bogor.

Hildebrand, F. H. 1939. Doengoes Iwul. In *3 jaren indisch natuurleven.* pp.310-313. Jakarta: Nederlandsch-Indische Vereeniging bot Natuurbescherming.

—. 1951. Daftar nama pohon-pohonan Jawa-Madura, dengan keterangan-keterangan tentang penjebaran dan ukurannja (Telah diperbaiki). *Laporan No. 50.* Lembaga Penelitian Hutan, Bogor.

—. 1954. Aantekeningen over Javaanse bambu-soorten. *Lap. Balai Penjelidikan Kehutanan* 66: 1-52.

Hill, J. E., and Boeadi. 1978. A new species of *Megaerops* from Java (Chiroptera: Pteropodidae). *Mammalia* 42: 427-434.

Hill, J. E., and J. D. Smith. 1984. *Bats: a natural history.* London: British Museum (Natural History).

Hirayama, N. 1990. Immune state of dogs injected with rabies in West Java Indonesia. *Jap. J. Vet. Sci.* 52: 1099-1102.

Hirth, H. F., and W. Schaffer. 1974. Survival rates of the green turtle *Chelonia mydas,* necessary to maintain stable populations. *Copeia* 1974: 544-546.

Hobart, M. 1990. The patience of plants: a note on agency in Bali. *Rev. Indon. Malay.*

Aff. 24: 990-135.

Hodges, R. 1993. Snakes of Java with special reference to East Java Province. *Brit. Herpetol. Soc. Bull.* 43: 15-32.

Hoeksema, B. W. 1991. Control of bleaching in mushroom coral populations Scleractinia, Fungiidae in the Java Sea: stress tolerance and interference by life history strategy. *Mar. Ecol. Progr. Ser.* 74: 225-237.

Hoekstra, P. 1987. The development of two major Indonesian river deltas: Morphology and sedimentary environments of the Solo and Porong deltas, East Java. In Proceedings of the Symposium on coastal lowlands: Geology and geomorphology, The Hague.

Holling, C. S. 1973. Resilience and stability of ecological systems. *Ann. Rev. Ecol. Syst.* 4: 1-23.

Hollis, D. 1975. A review of the subfamily Oxynae (Orthoptera: Acridoidae). *Bull. Brit. Mus. (N.H.), Entomology* 31: 191-234.

Holloway, J. D. 1986. Origin of lepidopteran faunas in high mountains of the Indo-Australian tropics. In *High altitude tropical biogeography,* ed. F. Vuilleumier, and M. Monasterio, 533-556. Oxford: Oxford Univ. Press.

—. 1990. Patterns of moth speciation in the Indo-Australian archipelago. In *The unity of evolutionary biology,* ed. E. C. Dudley, 340-372. Portland: Dioscorides.

Holloway, J. D., and N. E. Stork. 1991. The dimensions of biodiversity: the use of invertebrates as indicators of human impact. In *The biodiversity of microorganisms and invertebrates: Its role in sustainable agriculture,* ed. D. L. Hawksworth, 37-62. Wallingford: CAB International.

Holm, L. G. et al. 1977. *The world's worst weeds: Distribution and biology.* Honolulu: Hawaii Univ. Press.

Holmes, D. A. 1988. Berwisata burung ke Cengkareng. *Suara Alam* 56: 61-64.

Holmes, D. A., and S. van Balen. 1990. The birds of Pulau Tinjil. *Media Konservasi* 3: 53-54.

Holmes, D. A., and K. Burton. 1987. Recent notes on the avifauna of Kalimantan. *Kukila* 3: 19.

Holmes, D. A., and S. Nash. 1989. *The birds of Java and Bali.* Kuala Lumpur: Oxford Univ. Press.

Holthuis, L. B. 1984. Freshwater prawns (Crustacea: Decapoda: Natantia) from subterranean waters of the Gunung Sewu area, Central Java, Indonesia. *Zool. Mededel.* 58: 141-148.

Holttum, R. E. 1963. Cyathaceae. *Fl. Mal. II* I:

65-176.

—. 1974. Additions to the fern flora of Java. *Reinwardtia* 8: 499-501.

Holttum, R. E., and I. Enoch. 1991. *Gardening in the tropics.* Singapore: Times Editions.

Holway, P. 10 October 1992. Can we afford to be affluent? *New Scient.*, 47.

Hommel, P. W. F. 1987. Landscape ecology of Ujung Kulon, West Java, Indonesia. Ph.D. diss., Soil Survey Institute, Wageningen.

—. 1990a. Ujung Kulon: landscape survey and land evaluation as a habitat for the Javan rhinoceros. *ITC J.* 1990: 1-15.

—. 1990b. A phytosociological study of a forest area in the humid tropics (Ujung Kulon, West Java, Indonesia). *Vegetatio* 89: 39-54.

Hoogerwerf, A. 1935. Broedende witte ibissen in West Java. *Trop. Nat.* 24: 161-171.

—. 1935. Ornithologische merk-waardigheden in de Brantas-delta. *Trop. Nat.* 24: 89-96.

—. 1936. Vogels, die van verse homen. *Trop. Nat.* 25: 11-18.

—. 1937. Over den massamoord op enkele reigerachtigen en sterns en de becherming der vogels. In *Album van Natuurmonumenten in Nederlandsch-Indie*, ed. C. G. G. J. van Steenis. Batavia.

—. 1938a. Honderd uren de "ogneziene" gast der bantengs in Oedjoeng Koelon. *Trop. Nat.* 27: 25-35, 39-47, 64-72, 77-86.

—. 1938b. Bij de badaks en bantengs in het wilderservaat Oedjoeng Koelon. In *Album van Natuurmonumenten in Nederlandsch-Indie*, ed. C. G. G. J. van Steenis. Batavia.

—. 1947a. Iets over ontmoetingen tussen een tijger en een bantengstier. *Weer en Wind* 9: 201-219.

—. 1947b. Mededelingen over sterns bij Java waargenomen. *Limosa* 20: 197-200.

—. 1948a. Contribution to the knowledge of the distribution of the birds on the island of Java, with remarks on some new birds. *Treubia* 19: 83-137.

—. 1948b. Enkele waarnemingen in het Natuurmonument Oedjoeng Koelon in de Bezettingtijd. *Tectona* 38: 205-214.

—. 1948c. Het wildreservaat Baloeran. *Tectona* 38: 33-50.

—. 1949a. *De avifauna van Tjibodas en omgeving.* Buitenzorg: Koninklijk Planten-tuin van Buitenzorg.

—. 1949b. *De avifauna van de plantentuin to Buitenzorg.* Buitenzorg: 'sLands Plantentuin.

—. 1949c. Een bijdrage tot de oologie. *Zool. Verhand.* 88: 1-320.

—. 1953a. Some notes about the Nature Reserve Pulau Panaitan (Prinsen Eiland) in Sunda Strait, with special reference to the avifauna. *Treubia* 21: 481-505.

—. 1953b. Zwarte ibissen, *Plegadis falcinellus peregrinus* (BP), in het Vogelreservat Pulau Dua in de jaren 1951 en 1952. *Limosa* 26: 20-30.

—. 1953c. An ornithological bibliography having particular reference to the study of the birds of Java (I). *Org. Sci. Res. Bull.* 13: 1-28.

—. 1953d. An ornithological bibliography having particular reference to the study of the birds of Java (II). *Org. Sci. Res. Bull.* 14: 29-72.

—. 1953e. An ornithological bibliography having particular reference to the study of the birds of Java (III). *Org. Sci. Res. Bull.* 15: 73-120.

—. 1953f. An ornithological biobliography having particular reference to the study of the birds of Java (IV). *Org. Sci. Res. Bull.* 16: 121-169.

—. 1962a. Notes on Indonesian birds with special reference to the avifauna of Java and the surrounding small islands (I). *Treubia* 26: 11-38.

—. 1962b. Notes on Indonesian birds with special reference to the avifauna of Java and the surrounding small islands (II). *Treubia* 26: 39-210.

—. 1965. Notes on Indonesian birds with special reference to the avifauna of Java and the surrounding small islands (III). *Treubia* 29: 211-291.

—. 1966. Notes on the island of Bawean (Java Sea) with special reference to the birds. *Nat. Hist. Bull. Siam Soc.* 21: 313-340.

—. 1967. Notes on the island of Bawean (Java Sea) with special reference to the birds. *Nat. Hist. Bull. Siam Soc.* 22: 15-103.

—. 1969. On the ornithology of the Rhino sanctuary Udjung Kulon in West Java Indonesia. *Nat. Hist. Bull. Siam Soc.* 23: 9-135.

—. 1970. *Udjung Kulon, the land of the last Javan Rhinoceros.* Leiden: E.J. Brill.

—. 1974. *Report on a visit to wildlife reserves in East Java, Indonesia (August to November 1971).* Mededelingen Nederlandsche Commissie voor Internationale Natuurbescherming No. 21,

Hoogerwerf, A., and G. F. H. Siccama. 1938. *De avifauna van Batavia en omstreken.* Leiden: E.J. Brill.

Hoogland, R. D. 1978. Saurauiae gerontogea. II Notes on some species of Java. *Gdns' Bull.*

S'pore 31: 67-72.

Hoogstraal, H., S. Kadarsan, M. N. Kaiser, and F. V. D. Peenen. 1974. *Ornithodoros (Alectorobius) collocalinae*, new species (Ixodoidae: Argasidae) parasitizing cave swiftlets (Aves: Apodidae) in Java. *Ann. Entomol. Soc. Am.* 67: 224-230.

Hooijer, D. A. 1946. The evolution of the skeleton of *Rhinoceros sondaicus* Desmarests. *Proc. K. Ned. Akad. Wet.* 49: 671-676.

——. 1947. On fossil and prehistoric remains of *Tapirus* from Java, Sumatra and China. *Zool. Mededel.* 27: 253-299.

——. 1952. Fossil mammals and the Plio-Pleistocene boundary in Java. *Proc. K. Ned. Akad. Wet.* 55: 436-447.

——. 1954. A pygmy *Stegodon* from the middle Pleistocene of Eastern Java. *Zool. Mededel.* 33: 91-104.

——. 1955. Fossil Probiscidae from the Malay Archipelago and the Punjab. *Zool. Verhand.* 28: 1-146.

——. 1960. Quarternary gibbons from the Malay Archipelago. *Zool. Verhand.* 46: 1-41.

——. 1962a. The middle Pleistocene fauna of Java. In *Evolution and hominisation*, ed. G. Kurth,

——. 1962b. Quaternary langurs and macaques from the Malay Archipelago. *Zool. Verhand.* 55: 1-64.

——. 1964. New records of mammals from the middle Pleistocene of Sangiran, Central Java. *Zool. Mededel.* 40: 73-88.

——. 1973. Tenzenschildpadden en dwergolifanten. *Museologia* 1: 9-14.

——. 1974a. *Manis palaeojavanica* Dubois from the Pleistocene of Gunung Butok, Java. *Proc. K. Ned. Akad. Wet., Ser. B* 77: 198-200.

——. 1974b. *Elephas celebensis* (Hooijer) from the Pleistocene of Java. *Zool. Mededel.* 48: 85-93.

——. 1975. Quaternary mammals west and east of Wallace's Line. *Mod. Quat. Res. SE Asia* 1: 37-46.

——. 1982. The extinct giant land tortoise and the pygmy stegodont of Indonesia. *Mod. Quat. Res. SE Asia* 7: 171-176.

Hooijer, D. A., and B. Kurten. 1984. Trinil and Kedungbrubus: The Pithecanthropus bearing fossil faunas of Java and their relative age. *Ann. Zool. Fennici* 21: 135-142.

Hooper, P., and M. Palmer. 1992. St Francis and ecology. *Christianity and ecology*. E. Breuilly, and M. Palmer, 76-85. London: Cassell.

Hopkins, H. C. F. 1992. *Parkia. Fl. Mal. I* 11: 193-204.

Horsfield, T. 1821. Systematic arrangement and description of birds from the island of Java. *Trans. Linn. Soc. London* 13: 133-200.

——. 1824. *Zoological researches in Java and neighbouring island*. Kuala Lumpur (reprinted 1991): Oxford Univ. Press.

Horstkotte, G., C. Lightfoot, H. Waibel, and P. Kenmore. July 1992. Integrated pest management and aquatic life management: a natural partnership for rice farmers? *Naga*, 15-16.

Houghton, J. T., G. J. Jenkins, and J. J. Ephraums. 1990. *Climate change: the IPCC assessment*. Cambridge: Cambridge Univ. Press.

Howarth, F. G. 1983. Ecology of cave arthropods. *Ann. Rev. Entomol.* 28: 365-389.

——. 1991. Environmental impacts of classical biological control. *Ann. Rev. Entomol* 36: 485-509.

Howarth, F. G., and G. W. Ramsay. 1991. The conservation of island insects and their habitats. In *The conservation of insects and their habitats*, ed. N. M. Collins, and J. A.Thomas, 71-107. London: Academic.

Hugo, G. J. 1978. *Population mobility in West Java*. Yogyakarta: Gadjah Mada Univ. Press.

Hugo, J. H. et al. 1987. *The demographic dimension in Indonesian development*. Singapore: Oxford Univ. Press.

Hull, R. B., and G. R. B. Revell. 1989. Cross-cultural comparison of landscape scenic beauty evaluations. *J. Env. Psy.* 9: 177-192.

Hull, V. J. 1986. Dietary taboos in Java: Myths, mysteries, and methodology. In *Shared wealth and symbol: Food, culture, and society in Oceania and Southeast Asia*, ed. L. Manderson, Cambridge: Cambridge Univ. Press.

Hunink, R. B. M., and J. W. Stoffers. 1984. *Mixed and forest gardens on Central Java: An analysis of socio-economic factors influencing the choice between different types of land use*. Netherlands: Geografisch Instituut Rijks Universiteit Utrecht.

Hurley, B. 1992. Tropical ozone loss. *New Scient.* (14 March): 55.

Hutabarat, H. P. 1992. *Studi awal keberhasilan penetasan dan masa inkubasi telur penyu sisik (Eretmochelys imbicata) di empat kedalaman sarang semi-alam di Taman Nasional Laut Kepulauan Seribu*. Fakultas Biologi UNAS, Jakarta.

Hutabarat, J., L. Sya'rani, and M. A. K. Smith. 1986. The use of freshwater hyacinth *Eichhornia crassipes* in cage culture in Lake Rawa, Pening, Central Java. In Proceedings of the 1st Asian Fisheries Forum, 577-580. Manila, 26 May 1986.

Hutabarat, S. 1991. Benefit cost analysis of agroforsestry practices: tumpangsari and INMAS tumpangsari in Cepu Forest District, Java, Indonesia. Ph.D. diss., Michigan State Univ., Michigan.

Hutomo, K., and A. Djamali. 1980. Komunitas ikan pada padang "seagrass" di pantai selatan Pulau Tengah, Pulau-pulau Seribu. In *Sumberdaya hayati bahari*, ed. Burhanuddin, M. K. Moosa, and H. Razak. Jakarta: LON - LIPI.

Hutomo, M. 1985. *Telaah ekologik komunitas ikan pada padang rumput laut (seagrass, Anthrophyta) di perairan teluk Banten.* Fakultas Pasca Sarjana IPB, Bogor.

Hutomo, M., and A. Djamali. Penelaahan pendahuluan tentang komunitas ikan di daerah mangrove pulau Pari, Kepulauan Seribu. Paper presented at Seminar Ekosistem Hutan Mangrove. Jilid I, Jakarta, 1 February 1927.

Hutomo, M., and S. Martosewojo. 1977. The fishes of seagrass community on the west side of Burung island (Pari island, Seribu islands) and their variations in abundance. *Mar. Res. Indon.* 17: 147-172.

Hutomo, M., et al. 1993. Marine conservation areas in Indonesia: two case studies of Kepulauan Seribu, Java , and Bunaken, Sulawesi. UNEP-COBSEA/NOSTE Workshop EAS 25: Case studies in planning and management of marine protected areas, Penang, Malaysia, February 1993. Penang:

Hutterer, K. L. 1985. The Pleisstocene archaeology of Southeast Asia in regional context. *Mod. Quat. Res. SE Asia* 9: 1-23.

Ibkar-Kramadibrata, H., and B. L. Oey. 1987. Some ecological exploitation and managerial aspects of Telaga Pasir reservoir, East Java: a reconnaisance. *BIOTROP Spec. Publ.* 27: 13-25.

Ihle, J. E. W. 1912. Ueber eine kleine Brachyuren-Sammlung aus unterirdischen Flüssen von Java. *Notes Leyden Mus.* 34: 177-182.

Indarto, Y., Suhardjono, and Mulyadi. 1991. Pola variasi produksi serasah hutan mangrove Pulau Dua, Jawa Barat. Prosidings Seminar IV Ekosistem Mangrove, S. Soemodihardjo et al., 169-174. Bandar Lampung, 7 August 1990. Jakarta: MAB-LIPI.

Indiarto, Y. et al. 1986. Analisis vegetasi hutan mangrove Baluran, Jawa Timur. Paper presented at Seminar III Ekosistem Mangrove. Denpasar, 5 August 1986. Jakarta: MAB-LIPI.

Indrawan, D. 1991. A winter flock of pheasant tailed jacanas at Ciamis, West Java. *Kukila* 5: 138-139.

Indrawan, M. 1990. A survey of Rawa Lakbok: The largest wetland damaged in Java? *Voice of Nature* 86: 12-13.

Indrawan, M., S. van Balen, and Y. R. Noor. 1993. *Coastal wetlands and wildlife of urban and rural Jakarta: a status review.* AWB, Bogor.

Indrawan, M., W. Lawler, W. Widodo, and Sutandi. 1993. Notes on the feeding behaviour of milky storks *mycteria cinerea* at the coast of Indramayu, west Java. *Forktail* 8: 143-144.

Indrawan, M., and Sunarto. 17 September 1992. Yang memudar dari Muaraangke. *Suara Pembaruan,*

Indriati, S. E. 1984. Keanekaragaman dan kepadatan Echinodermata di perairan pantai Pulau Menjangan Kecil, Kepulauan Karimun Jawa. *Laporan No. 311.* Lembaga Penelitian Hutan, Semarang.

Inger, R. F., and R. B. Stuebing. 1990. *Frogs of Sabah.* Kota Kinabalu: Sabah Parks Department.

Iniguez, L. 1991. Towards the conservation and improvement of the Javanese Fat-tail sheep. In *Production aspects of Javanese Fat-tail sheep in Indonesia*, ed. I. K. Sutama, and L. Iniguez, 79-84 Bogor: Indonesian Small Ruminant Network.

Institute of Ecology. 1980. *Environmental impact analysis of the Saguling Dam: mitigation of impact.* Institute of Ecology UNPAD, Bandung.

Institute of Hydraulic Engineering. 1984. *Ecological aspects of Segara Anakan in relation to its future management.* Ministry of Public Works - IPB, Bogor.

Institute of Hydrology. *Eucalyptus: forest friend or foe?* Institute of Hydrology, Wallingford.

Intari, S. E. 1986. Efficacy of some insecticides on *Chaetocnema* sp. beetles attacking leaves of *Bruguiera* spp. in Tritih, Cilacap, Central Java. *Bull. Pen. Hutan* 477: 1-68.

——. 1991. Effect of *Neotermes tectonae* Damm. attack on the quality and quantity of teak timber in forest district Kebonharjo, Central Java, Indonesia. *Bull. Pen. Hutan* 530: 25-36.

IPB. 1984. *Ecological aspects of Segara Anakan in relation to its future management.* Faculty of Fisheries IPB, Bogor.

——. 1986a. *Penelitian pengembangan wilayah penyangga kawasan hutan konservasi: Taman Nasional Baluran.* Fakultas Kehutanan IPB, Bogor.

——. 1986b. *Analisa dampak lingkungan pada proyek Tulungagung.* Lembaga Penelitian,

Institut Pertanian Bogor, Bogor.

IPCC. 1990. *Climate change: the IPCC scientific assessment.* Cambridge: Cambridge Univ. Press.

Irwan, Z. D. 1992. *Prinsip-prinsip ekologi dan organisasi ekosistem komunitas dan lingkungan.* Jakarta: Bumi Aksara.

Ischak, T. M. 1976. Daftar jenis-jenis burung di cagar alam Pulau Rambut, Teluk Jakarta. *Biologica* 3: 15-19.

Iskandar, D. T. 1987. The occurrence of *Enhydris alternans* new record in Java, Indonesia. *Snake* 19: 72-73.

Iskandar, J. 1980. Penelitian ekologi burung di beberapa pedesaan di daerah aliran sungai Citarum. Jurusan Biology UNPAD, Bandung.

—. 1984. Development and bird population: Socio-economic effects in rural Java. *Voice of Nature* 22: 17-18.

—. 1989. *Jenis burung yang umum di Indonesia.* Jakarta: Djambatan.

—. 1992. *Ekologi perladangan di Indonesia: Studi kasus dari daerah Baduy, Banten Selatan, Jawa Barat.* Jakarta: Djambatan.

Iskander, D. K. 1984. Nipa di Sungai Adiraja. *Suara Alam* 7: 41-42.

Ismail, R. I. et al. 1981. *Studi penyebaran, kepadatan populasi dan komposisi nutrient gonad bulu babi (Diadema setosum Leske) di terumbu karang Pulau Pari, Kepulauan Seribu.* BPP Teknologi, Jakarta.

Ismail, W. 1985. Productivity and salinity tolerance of guppy (*Lebistes reticulatus* P.) and its possibility as live bait in skipjack pole and line fishing. *Lap. Pen. Perikanan Laut* 32: 93-102.

Itino, T., M. Kato, and M. Hotta. 1991. Pollination ecology of the two wild bananas *Musa acuminata* and *M. salaccensis*: chiropterophily and ornithophily. *Biotropica* 23: 151-158.

IUCN. 1977. *Red data book: Birds.* Gland: IUCN.

IUCN/UNEP/WWF. 1993. *Bumi Wahana: Strategi menuju kehidupan yang berkelanjutan.* Jakarta: WALHI.

Jaag. 1943. *Mitt. Naturf. Ges.* 12: 211-217.

Jacobs, M. 1958. Botanical reconnaissance of Nusa Barung and Blambangan, Southeast Java. *Blumea* 4 (Suppl.): 68-86.

Jacobs, M. 1992. *The green economy.* London: Pluto.

Jacobson, E. 1912. Iets over blinde visschen. *Levende Nat.* 16: 513-516.

—. 1933. Über javanische Tigerkatzen. *Zool. Garten* 6: 238-244.

Jafarsidik, J. S., and C. Anwar. 1987. Kompo-sisi hutan mangrove Komplek Kali Bedakan, Pamanukan, Jawa Barat dikaitkan dengan proses penambakan lahan hutan pantai. *Bull. Pen. Hutan* 491: 34-41.

Jafarsidik, J. S., and W. Meijer. 1983. Rafflesiaceae di Jawa. *Bull. Kebun Raya* 6: 73-76.

Jafarsidik, J. S., and A. P. Soewanda. 1985. Jenis-jenis *Erythrina* di Jawa. *Bull. Pen. Hutan* 469: 15-38.

Jafarsidik, J. S., and M. Sutarto. 1980. Jenis tumbuh-tumbuhan obat di beberapa hutan Jawa Timur dan Bali: Aspek-aspek pemanfaatan dan pengembangannya. *Laporan No. 346.* Lembaga Penelitian Hutan, Bogor.

Jafarsidik, J. S., and S. Sutomo. 1983. Jenis-jenis tumbuhan obat di bawah tegakan hutan jati dan beberapa macam obat tradisional penduduk Jatisari (Subah, Jawa Tengah). *Laporan No. 411.* Lembaga Penelitian Hutan, Bogor.

Jain, S. K. 1990. Conservation of aquatic plants. In *Ecology and management of aquatic vegetation in the Indian subcontinent,* ed. B. Gopal, 237-241. Dordrecht: Kluwer.

James, I. 1986. *Army caving expedition to southeast Java, Exercise Phreatic Diamond.* Army Caving Association, London.

Jameson, S. C. 1976. Early life history of the giant clams *Tridacna gigas, T. maxima* and *Hippopus hippopus. Pacific Sci.* 30: 219-233.

Jani, C., and S. Sukardjo. 1982. Gunung Halimun, cagar alam yang mengagumkan. *Suara Alam* 15: 38-42.

Jansen, P. 1953. *Unpublished draft of revision of Graminae for Flora Malesiana.*

Jansen, P. C. M. et al. 1993. *Basic list of species and commodity groupings.* Bogor: PROSEA.

Janzen, D. H. 1973. Dissolution of mutualism between *Cecropia* and its *Azteca* ants. *Biotropica* 5: 15-28.

—. 1983. No park is an island: increase in interference from outside as park size decreases. *Oikos* 41: 402-410.

—. 1985. Mangroves: where's the understorey? *J. Trop. Ecol.* 1: 89-92.

—. 1986a. The eternal external threat. In *Conservation biology,* ed. M. E. Soulé, 286-303. Sunderland, MA: Sinauer.

—. 1986b. The future of tropical ecology. *Ann. Rev. Ecol. Syst.* 17: 305-324.

—. 1987a. Oh, I forgot about zoos. *Oikos* 48: 241-242.

—. 1987b. Insect diversity of a Costa Rica dry forest: why keep it, and how? *Biol. J. Linn. Soc.* 30: 343-356.

—. 1988. Tropical ecological and biocultural restoration. *Science* 239: 243-244.

—. 1990. Sustainable society through

applied ecology: the reinvention of the village. In *Race to save the tropics*. R. Goodland, xi-xiv. Washington D.C.: Island Press.

——. 1991. Masa depan ekologi tropik. In *Krisis biologi: hilangnya keanekaragaman biologi*, ed. K. Kartawinata, and A. J. Whitten, 234-261. Jakarta: Yayasan Obor.

Jarvis, F. 1984. Birds of Bali Barat. *Voice of Nature* 22: 11-14.

Jay, R. 1969. *Javanese villagers: Social relations in rural Modjokerto.* Cambridge: MIT Press.

Jayasuriya, S., and I. K. Nehen. 1989. Bali: economic growth and tourism. In *Unity and diversity: Regional economic development in Indonesia since 1970*, ed. H. Hill, 331-363. Singapore: Oxford Univ. Press.

Jellink, C. 1991. *The wheel of fortune: The history of a poor community in Jakarta.* Sydney: Allen and Unwin.

Jennings, J. N. 1971. *Karst.* Cambridge, Mass: MIT Press.

Jensen, G. D., and L. K. Suryani. 1992. *The Balinese people: a reinvestigation of character.* Singapore: Oxford Univ. Press.

Jentink, F. A. 1897. The dog of the Tengger. *Notes Leyden Mus.* 18: 217-220.

Jhamtani, H. P., and Djaya. 1985. The case of mangrove destruction in Muara Gembong, mismanagement or poverty? *Tigerpaper* 12: 18-24.

Jimenez, J. A., A. E. Lugo, and G. Cintron. 1985. Tree mortality in mangrove forests. *Biotropica* 17: 177-185.

Johannes, R. E., and D. W. Rimmer. 1984. Some distinguishing characteristics of nesting beaches of the green turtle *Chelonia mydas. Mar. Biol.* 83: 149-154.

Johns, A. D. 1989. Recovery of a Peninsular Malaysian rain forest avifauna following selective timber logging: the first twelve years. *Forktail* 4: 89-105.

——. 1991. Responses of Amazonian rain forest birds to habitat modification. *J. Trop. Ecol.* 7: 417-437.

Johns, D. M. 1990. The relevance of Deep Ecology to the Third World: Some preliminary comments. *Env. Ethics* 12: 233-252.

Johnson, A. 1980. *Mosses of Singapore and Malaysia.* Singapore: S'pore Univ. Press.

Johnson, R., W. Lawler, Y. R. Noor, and M. Barter. 1993. *Migratory waterbird survey and bird banding training project in the Indramaya-Cirebon region, West Java: the oriental pratincole Glareola maldivarum as a case study.* PHPA/Asian Wetland Bureau/Australian Wader Studies Group, Bogor.

Johnson, W. 1978. *Muddling toward frugality.* San Fransisco: Sierra Club Books.

Johnstone, I. M. 1981. Consumption of leaves by herbivores in mixed mangrove stands. *Biotropica* 13: 252-259.

Joongjai, P., and N. Rosengren. 1983. Coastal geomorphology between Labuan and Merak, West Java. In Proceeding of the Workshop on Coastal Resources Management of Krakatau and the Sunda Strait Region, Indonesia, ed. E. C. F. Bird, A. Soegiarto, and K. A. Soegiarto, 242-249. Jakarta, 19 August 1983.

Jordan, D. S., and A. Scale. 1907. List of fishes collected in the river at Buitenzorg, Java, by Dr. Houghton Campbell. *Proc. US Nat. Mus.* 33: 525-543.

Joseph, S. W., P. F. D. van Peenen, A. E. New, and D. W. Eggena. 1978. Attempts to identify reservoirs of rabies in Indonesia. *Bull. Penelitian Kesehatan Health Stud. Indones.* 6: 1-8.

Junghuhn, F. 1845. *Topographische und naturwissenschaftliche Reisen durch Java.* Magdeburg: Baensch.

Junghuhn, F. W. 1854. *Java, zijne gedaante, zijn plantetooi en inwendije bouw.* n.p.

Kabata, Z. 1985. *Parasites and diseases of fish cultured in the tropics.* London: Taylor and Francis.

Kadar. 1985. *Inventarisasi jenis tikus dan kepadatan populasinya di resort hutan KPA Cibodas, Taman Nasional Gunung Gede Pangrango, Jawa Barat.* Fakultas Kehutanan IPB, Bogor.

Kader, A. B. A. et al. 1994. *Environmental protection in Islam.* Cambridge: IUCN.

Kahlke, H. D. 1972. A review of the Pleistocene history of the Orang Utan. *Asian Persp.* 15: 5-14.

Kahn, M. H. 1963. Gunung Kidul: An introduction to a problem area in Java. *Geografi* 3: 4-6.

Kakisina, S., and P. Yuwono. 1984. *Deskripsi kondisi dan permasalahan di sektor pertanian di Kabupaten Boyolali.* Fakultas Ekonomi, Satya Wacana Christian Univ., Salatiga.

Kalima, T. 1989. *Gigantochloa* in Java, Indonesia. *Bull. Pen. Hutan* 506: 29-49.

——. 1992. The composition and distribution of ground vegetation under four tree stands at Yanlapa Experimantal Garden, West Java. *Bull. Pen. Hutan* 550: 13-26.

Kalima, T., U. Sutisna, H. C. Soeyatman, and Pratiwi. 1988. Analisis komposisi vegetasi di Cagar Alam Leuweung Sancang, Jawa Barat. *Bull. Pen. Hutan* 498: 45-55.

Kalshoven, L. G. E. 1953. Important outbreaks of insect pests in the forests of Indonesia. *Trans. Ninth Int. Congr. Entomol.*

2: 229-234.

———. 1954. *Survival of Neotermes colonies in infested teak trunks after girdling or felling of the trees.* Bogor: Balai Penjelidikan Kehutanan, Kementerian Pertanian.

———. 1981. *Pests of crops in Indonesia.* Jakarta: Ichtiar baru.

Kaniawati, S. C. 1985. *Studi komunitas karang batu di perairan pantai Pulau Pari, Kepulauan Seribu.* Fakultas Kehutanan IPB, Bogor.

Kanwil Kehutanan Bali. 1991. *Monografi Kehutanan Propinsi Bali.* Kantor Wilayah Departemen Kehutanan, Denpasar.

———. 1993. Taman Nasional Bali Barat sebagai bank plasma nutfah khas Bali. Dep. Kehutanan, Kantor Wilayah Bali, Denpasar.

Kappeler, M. 1984a. The gibbon in Java. In *The Lesser apes: Evolutionary and behavioural biology,* ed. H. Preuschoft, D. J. Chivers, W. Y. Brockelman, and N. Creel, 19-31. Edinburgh: Edinburgh Univ. Press.

———. 1984b. Diet and feeding behaviour of the moloch gibbon. *The Lesser Apes: Evolutionary and Behavioural Biology,* ed. L. Preuschoft, D. J. Chivers, W. Y. Brockelman, and N. Creel, 228-241. Edinburgh: Edinburgh Univ. Press.

Karny, H. H. 1925. Malaysian Gryllacridae. *J. Fed. Malay St. Mus.* 13: 1-67.

Kartamihardja, E. S. 1977. *Berberapa aspek biologis ikan jambal Pangasius pangasius (Hamilton and Buchanan) di Wasuk Jatiluhur, Jawa Barat.* Fakultas Perikanan, Institut Pertanian Bogor, Bogor.

Kartawinata, K. 1965. Notes on the vegetation of Peutjang Island (SW Java). Symposium in ecological research in humid tropics vegetation, 26-30. Jakarta: UNESCO.

———. 1975. Structure and composition of forests in some nature reserves in West Java, Indonesia. *13th Pacific Sci. Cong.*

———. 1977. The forest on Gunung Cadaspanjang and Gunung Tikukur, West Java. *Bull. Kebun Raya* 3: 35-38.

———. 1987. Kota sebagai ekosistem. Lecture given at Universitas Atmajaya, Yogyakarta, 19 September 1987.

——— 1990a. A potential application of some non-timber forest plants in agroforestry. *BIOTROP Spec. Publ.* 39: 93-97.

———. 1990b. A review of natural vegetation studies in Malesia. In *The plant diversity of Malesia,* ed. P. Baas, K. Kalkman, and R. Geesink, 177-191. Dordrecht: Kluwer.

Kartawinata, K., and A. Apandi. 1977. Checklist of plant species on the Peucang Island (Ujung Kulon Nature Reserve, West Java).

Berita Biologi 2: 13-18.

Kartawinata, K., A. Apandi, and T. B. Suselo. 1985. The forest of Peucang Island, Ujung Kulon National Park, West Java. In Proceedings of the Symposium on 100 year development of Krakatau and its surroundings. Jakarta 23-27 Aug. 1983. LIPI, 448-451. Jakarta: LIPI.

Kartawinata, K., and E. B. Walujo. 1977. A preliminary study of the mangrove forest on Pulau Rambut, Jakarta Bay. *Mar. Res. Indon.* 18: 119-129.

Kartawinata, K., and A. J. Whitten. 1990. *Krisis biologi: Hilangnya keanekaragaman hayati.* Jakarta: Obor.

Kartikasari, S. N. 1986. *Studi populasi dan perilaku lutung (Presbytis cristata, Raffles) di Taman Nasional Baluran, Jawa Timur.* Fakultas Kehutanan IPB, Bogor.

Karyono. 1990. Homegardens in Java: Their structure and function. In *Tropical home gardens,* ed. K. Landauer, and M. Brazil, 138-146. Tokyo: United Nations Univ. Press.

Kasno. 1990. Possible inclusion of beekeeping in agroforestry systems in Indonesia. *BIOTROP Spec. Publ.* 39: 135-139.

Katili, J. A., and A. Sudradjat. 1984. *Galunggung: the 1982-1983 eruption.* Bandung: Volcanological Survey of Indonesia.

Kattan, G. H. 1992. Rarity and vulnerability: The birds of the Cordillera, Central of Columbia. *Conserv. Biol.* 6: 64-70.

Kempers, A. J. B. 1988. The kettledrums of southeast Asia. *Mod. Quat. Res. SE Asia* 10: 1-471.

Kenmore, P. 1980. Ecology and outbreaks of a tropical insect pest of the green Revolution, the rice brown planthopper *Nilaparvata lugens.* Ph.D. diss., Univ. of California, Berkeley.

Kern, J. H. 1974. Cyperaceae. *Fl. Mal. I* 7: 435-753.

Kertonegoro. 1987. *Man from behind the mist.* Bali: Harkat Foundation.

Khairijon. 1991. Produksi dan laju dekomposisi serasah di hutan bakau hasil reboisasi yang berbeda kelas umurnya. Prosidings Seminar IV Ekosistem Mangrove, S. Soemodihardjo et al., 145-154. Bandar lampung, 7 August 1990. jakarta: MAB-LIPI.

Khalid, F. 1992. *Islam and ecology.* London: Cassel.

Khan, M. H. 1963. Gunung Kidul. *Maj. Geog. Indon.* 3: 47-60.

Kijne, A. 1990. Changes in composition of homegardens of Gunung Kidul, Java. *FONC project communication No. 1990 - 3.* FONC - Fakultas Kehutanan UGM, Yogyakarta.

Kikkawa, J., and P. D. Dwyer. 1992. Use of scattered resources in rain forest of humid tropical lowlands. *Biotropica* 24: 293-308.

Killick, L. M., M. J. Kennedy, and M. Beverley-Burton. 1982. *Oochoristica javaensis* (Eucestoda: Linstowiidae) from *Gehyra mutilata* and other gekkonid lizards (Lacertilia: Gekkonidae) from Java, Indonesia. *Can. J. Zool.* 60: 2459-2464.

King, B. F., M. W. Woodcock, and E. C. Dickinson. 1975. *A field guide to the birds of southeast Asia.* London: Collins.

Kirnowardoyo, S., and Supalin. 1986. Zooprophylaxis as a useful tool for control of *Anopheles aconitus* transmitted malaria in Central Java, Indonesia. *J. Comm. Dis.* 18: 90-94.

Kirnowardoyo, S., and G. P. Yoga. 1987. Entomological investigations of an outbreak of malaria in Cilacap on south coast of central Java, Indonesia during 1985. *J. Comm. Dis.* 19: 121-127.

Kiswara, W. et al. 1984. Preliminary study on artificial regeneration of *Rhizophora* spp. in the Pulau Pari Group, Seribu Islands. In Prosiding Seminar II Ekosistem Mangrove, ed. S. Soemodihardjo et al., 321-328. Jakarta, 1984. Jakarta: LIPI.

Kitchener, D. J., and S. Foley. 1985. Notes on a collection of bats (Mammalia: Chiroptera) from Bali Is., Indonesia. *Rec. W. Aust. Mus.* 12: 223-232.

Kitchener, D. J., S. Hisheh, L. H. Schmitt, and I. Maryanto. 1993. Morphological and genetic variation in *Aethalops alecto* from Java, Bali and Lombok Islands, Indonesia. *Mammalia* 57: 255-272.

Kitchener, D. J., and I. Maryanto. 1993. Taxonomic reappraisal of the *Hipposideros larvatus* species complex in the Greater and Lesser Sunda islands, Indonesia. *Rec. W. Aust. Mus.* 16: 119-173.

Kiteartika, K. 1988. Report on *Eucheuma* culture production in Bali Province, Indonesia, 1987. Report on the Training Course on Seaweed Farming, 100-105. Manila, 2 May 1988.

Kitting, C. L., B. Fry, and M. D. Morgan. 1984. Detection of inconspicuous epiphytic algae supporting food webs in seagrass meadows. *Oecologia* 62: 145-149.

Klapste, J. 1984. Occurrence of the long-billed dowitcher on Bali, Indonesia and other observations. *Aust. Bird Watcher* 10: 186-195.

KLH. 1984. *Penelitian pengembangan buffer zone pelestarian alam: Taman Nasional Meru Betiri.* Kantor Menteri KLH, Jakarta.

———. 1992a. *Pandangan Islam tentang kependudukan dan lingkungan hidup pegangan bagi para dai.* Jakarta: Kantor Menteri Negara KLH.

———. 1992b. *Indonesian country study on biological diversity.* Kantor Mentri Negara KLH, Jakarta.

———. 1992c. *National strategy and action plan on coral reef ecosystem conservation and management.* KLH, Jakarta.

———. 1992d. *Conservation of marine turtles: national strategy and action plan.* KLH, Jakarta.

KLH-DA. 1993. *Kependudukan dan lingkungan hidup melalui jalur agama Katolik.* Jakarta: Kantor Menteri Negara KLH and Departemen Agama.

Klop, A., and F. van Ogtrop. 1985. *Forestry on Java: Report of a field practice with Perum Perhutani from October 1981 till April 1982.* Dept. Sylviculture, Agri. Univ. Wageningen, Wageningen.

KNAG. 1938. *Atlas van tropisch Nederland.* Amsterdam: Koninklijk Nederlandsch Aardrijkskundig Genootschap.

Ko, R. K. T. 1983a. *The karst formation near Pangkalan and the potholes.* HIKESPI, Bogor.

———. 1983b. *Pengunduhan sarang burung walet.* HIKESPI, Bogor.

———. 1983c. *Laporan survei kota Cilacap, pulau Nusakambangan, Bukit Selok, Bukit Srandil dan Gua Petruk.* HIKESPI, Bogor.

———. 1984d. Urgensi konservasi karst Gombong Selatan ditinjau dari segi speleologi. Paper presented at Simposium Upaya Pengelolaan Sumber Daya Alam dan Pelestarian Kemampuan Daya Dukung Lingkungan Hidup di Daerah Gombong Selatan, Kabupaten Kebumen, Jawa Tengah, Semarang, 26 November 1984.

———. 1986. Conservation and environmental management of subterranean biota. *BIOTROP Spec. Publ.* 30:

Kochummen, K. M., J. V. Lafrankie, and N. Manokaran. 1990. Diversity of trees and shrubs in Malaya at regional and local levels. International Conference of Tropical Biodiversity, Kuala Lumpur, 12 June 1990.

Koeniger, N. 1987. The eastern honeybee *(Apis cerana)* and its mite *Varroa jacobsoni. Imkerfreund* 42: 303-306.

Koentjaraningrat. 1985. *Javanese culture.* Singapore: Oxford Univ. Press.

Koesharyono, C., R. J. Theos, and G. Simanjuntak. 1985. The epidemiology of rabies in Indonesia. In *Rabies in the tropics*, ed. E. Kuwert et al., 545-555. Berlin: Springer.

Koesoemadinata, S., and B. Costa-Pierce. 1988. Development of rice-fish farming in Indonesia: past, present and future. Paper at 1st International Workshop on Rice-Fish Culture, Ubon, Thailand, 21 March 1988. Manila: ICLARM.

Koeswara, W. 1985. *Penyebaran dan kelimpahan jenis keong gondang* (Pila *spp.*) *di perairan Rawa Pening, Jawa Tengah.* Fakultas Perikanan IPB, Bogor.

Koh, J. K. H. 1989. *A guide to common Singapore spiders.* Singapore: Science Centre.

Kohlbrugge, J. H. F. 1896. Zoogdieren van der Tengger. *Natuurk. Tijds. Ned.-Ind.* 55: 261-298.

Komara, A. 1981. *Tinjauan ekologis penyu hijau* (Chelonia mydas) *di Ujung Kulon.* WWF Indonesia, Bogor.

Komaruddin, M. 1977. Penelitian konsumsi kayu bakar di Desa Babakan, Bogor. Fakultas Kehutanan IPB, Bogor.

Koningsberger, J. C. 1909. *De vogels van Java.* Batavia: Kloff.

—. 1915. *Java: Zoologisch en biologisch.* Buitenzorg: Department van Landbouw, Nijverheid.

Kooiman, J. G. 1938. Excursie-verslag van de Afd. Besoeki naar de kuststrook ten Z van Loemadjang. *Trop. Nat.* 27: 126-130.

—. 1940. Walrissen, stormvogeltjes en sterns. *Trop. Nat.* 29: 195-197.

Kooiman, J. G., and R. van der Veen. 1936. Een excursie naar het Jang-plateau. *Trop. Nat.* 25: 161-167.

—. 1940. Een tocht naar het Natuurmonument Noesa Baroeng. *Trop. Nat.* 29: 61-64.

Kool, K. M. 1992a. Food selection by the silver leaf monkey, *Trachypithecus auratus sondaicus*, in relation to plant chemistry. *Oecologia* 90: 527-533.

—. 1992b. The status of endangered primates in Gunung Halimun Reserve, Indonesia. *Oryx* 26: 29-33.

—. 1993. The diet and feeding behaviour of the silver leaf monkey *Trachypithecus auratus sondaicus* in Indonesia. *Int. J. Primatol.* 14:

Kool, K. M., and D. B. Croft. 1992. Estimates for home range areas of arborial colobine monkeys. *Folia Primatol.* 58: 210-214.

Koop, H. 1992. *Bufferzone management: Tools to diagnose forest integrity.* DHV Consultants, Bogor.

Kooper, W. J. C. 1927. Sociological and ecological studies on the tropical weed vegetation of Pasuruan. *Res. Trav. Bot. Neer.* 24: 1-246.

Koorders, S. H. 1911. *Exkursion flora von Java. Vol. I.* Jena: Fischer.

—. 1912a. *Exkursion flora von Java. Vol. II.* Jena: Fischer.

—. 1912b. *Exkursion flora von Java. Vol. III.* Jena: Fischer.

—. 1913. *Exkursion flora von Java. Vol. IV.* Jena: Fischer.

Koorders, S. H., and T. Valeton. 1913. *Atlas der baumarten von Java im Anschluss an die "Bijdragen tot de kennis der boomssoorten van Java". Vol. I.* Leiden: P.W.P. Trap.

—. 1914. *Atlas der baumarten von Java im Anschluss an die "Bijdragen tot de kennis der boomssoorten van Java". Vol. II.* Leiden: P.W.M. Trap.

—. 1915. *Atlas der baumarten von Java im Anschluss an die "Bijdragen tot de kennis der boomssoorten van Java". Vol. III.* Leiden: P.W.P. Trap.

—. 1918. *Atlas der baumarten von Java im Anschluss an die "Bijdragen tot de kennis der boomssoorten van Java". Vol. IV.* Leiden: P.W.P. Trap.

Kopstein, F. 1926. De schorpioenen van Java. *Trop. Nat.* 15: 109-118.

—. 1928. De Javaansche brilslang: *Naja tripudians sputatrix. Trop. Nat.* 17: 191-199.

—. 1929. Herpetologische notizen 1: Ein neuerfall von Termitophilie. *Treubia* 10: 467-9.

—. 1930a. *De Javaansche gifslangen en haar beteekenis voor den mensch.* Nederlandsch-Indische Natuurhistorische Vereeniging.

—. 1930b. Herpetologische notizen III: Reptilien des ostlichen Preanger (West Java). *Treubia* 12: 273-276.

—. 1931. Die okologie der Javanischen Ratten und ihre Bedeutung fur die epidemi. *Zeits. Morphol. Oekol. Tiere* 22: 774-807.

—. 1932a. Herpetologische notizen V: *Bungarus javanicus*, eine neue giftschlange von Java. *Treubia* 14: 73-77.

—. 1932b. Herpetologische notizen VI: Weitere beobachtungen ueber die fortpfalnzung west-Javanischer reptilien. *Treubia* 14: 78-84.

—. 1935. Herpetologische notizen X: Weitere beobachtungen uber die fortplanzung west-Javanischer Reptilien. *Treubia* 15: 55-56.

—. 1936. Herpetologische notizen XIII: Ueber *Vipera russellii* von Java. *Treubia* 15: 259-264.

—. 1938. Ein beitrag zur morphologie, biologie und oekologie von *Xenodermus javanicus* Reinhardt. *Bull. Raffles Mus.* 14: 168-174.

Koschier, E., M. Meredith, and W. Waagner.

1983. *Java caves*. HIKESPI, Bogor.

Kostermans, A. J. G. 1970. *Laporan perjalanan koleksi ke suatu hutan di Lengkong (Jampang Tengah), September 26 - 28, 1970*. Herbarium Bogoriense, Bogor.

Kostermans, A. J. G., and J. M. Bompard. 1993. *The mangoes (Anacardiaceae)*. London: Academic.

Kostermans, A. J. G., S. Wirjahardja, and R. J. Dekker. 1987. The weeds: description, ecology and control. In *Weeds of rice in Indonesia*, ed. M. Soerjani, A. J. G. Kostermans, and G. Tjitrosoepomo, 24-565. Jakarta: Balai Pustaka.

Kottelat, M. 1992. *Fishes obtained and observed in the Kali Brantas (Jawa Timur) in July 1992*. EMDI Project, Kantor Menteri KLH, Jakarta.

Kottelat, M., A. J. Whitten, S. N. Kartikasari, and S. Wiryoatmodjo. 1993. *Freshwater fishes of western Indonesia and Sulawesi*. Jakarta: Periplus.

———. 1996. *Asia-wide assessment of freshwater biodiversity*. Washington, D.C.: The World Bank.

Koumans, F. P. 1949. On some fossil fish remains from Java. *Zool. Mededel.* 30: 77-81.

Kramadibrata, K., and J. N. Hedger. 1990. A new species of *Acaulospora* associated with cocoa in Java and Bali, Indonesia. *Mycotaxon* 37: 73-78.

Kramer, A. 1990. The ancestry of modern Australasians evidence from the early Pleistocene of Java. *Am. J. Phys. Anthro.* 81: 253-258

Kramer, F. 1926. Onderzoek maar de natuurlijke verjoning in den uitkap in Preanger gebergtebosch. *Med. Proefs. Boschw. Bogor* 14: 1-182.

———. 1933. De natuurlijke verjonging in het Goenoeng-Gedeh complex. *Tectona* 26: 155-185.

Kramer, K. 1974a. Die tertiären Hölzer Südost-Asiens (unter ausshluss der Dipterocarpaceae). Teil 1. *Palaeontographica* 144: 45-181.

———. 1974b. Die tertiären Hölzer Südost-Asiens (unter ausshluss der Dipterocarpaceae). Teil 2. *Palaeontographica* 145: 1-150.

Krause, G., and K. With. 1922. *Bali*. Hagen: Folkwang.

Krebs, C. J. 1994. *Ecology: The experimental analysis of distribution and abundance, 4th Ed.* London: Harper and Row.

Krikken, J., and J. Huijbregts. 1988. A new relative of *Onthophagus palatus* from Java, Indonesia (Coleoptera: Scarabidae).

Entomol. Berichte 48: 13-15.

Krismono, A., and H. Atmadja. 1983. Penelitian populasi ikan di Waduk Jatiluhur, Jawa Barat. *Bull. Pen. Perikanan Darat* 4: 50-53.

———. 1986a. Distribusi vertikal oksigen terlarut, suhu air dan kandungan bahan organik di Waduk Jatiluhur, Jawa Barat. *Bull. Pen. Perikanan Darat* 5: 83-89.

———. 1986b. Kondisi fisika dan kimiawi air di Waduk Jatiluhur, Jawa Barat dalam tahun 1983. *Bull. Pen. Perikanan Darat* 5: 100-110.

Kristyanto, A. I. A. 1979. Tumbuhan air dan populasi gulma di daerah Rawa Pening. In Proceeding of the 2nd Seminar in Aquatic Biology and Aquatic Management of the Rawa Pening Lake, Salatiga, Central Java, 115-131. Salatiga, 29 October 1979. Salatiga: Satya Wacana Christian Univ.

———. 1991. Menjaga kelestarian danau Rawa Pening. Seminar tentang lingkungan hidup: Tantangan terhadap Pembangunan Berkelanjutan, Salatiga, 26 November 1991.

Kristyanto, A. I. A., and F. Göltenboth. 1979. The ecology of the Rawa Pening Lake. 2: Contribution to the knowledge of some chemical properties. *J. Satya Wacana Res.* 1: 455-473.

Kristyanto, A. I. A., and E. Widiyastuti. 1986. *Potensi Rawa Kele untuk pengembangan tambak*. PSL UNDIP, Purwokerto.

KSDA. 1991. *Laporan usaha pelestarian dan pembinaan populasi penyu di pantai Sukamade, Taman Nasional Meru Beteri*. Sub Balai KSDA Jawa Timur II, Jember.

KSDA IV. 1985. *Laporan hasil penilaian potensi Suaka Margasatwa Dataran Tinggi Yang*. Balai KSDA IV, Malang.

KSDA Jatim. 1989a. *Penghitungan daya dukung objek wisata di Taman Nasional Meru Beteri*. Sub Balai KSDA Jawa Timur II, Jember.

———. 1989b. *Usaha pelestarian dan pembinaan populasi penyu di pantai Sukamade, Taman Nasional Meru Beteri*. Sub Balai KSDA Jawa Timur II, Jember.

———. 1990. *Pembinaan daerah penyangga Taman Nasional Meru Beteri*. Sub-Balai KSDA Jatim II, Jember.

———. 1991. *Usaha pelestarian dan pembinaan populasi penyu di pantai Sukamade, Taman Nasional Meru Beteri*. Sub-Balai KSDA Jawa Timur II, Jember.

Kuiter, R. H. 1992. *Tropical reef-fishes of the Western Pacific, Indonesia and adjacent waters*. Jakarta: Gramedia.

Kulczynski. 1908. Spiders from Sumatra and Java. *Bull. Acad. Cracovie* 37: 527-581.

900 BIBLIOGRAPHY

Kullberg, R. G. 1982. Algal succession in a hot spring community. *Am. Midl. Nat.* 108: 224-244.

Küng, H. 1991. *Global responsibility: in search of a new world ethic.* New York: Crossroad.

Kuroda, N. 1933. *Non-passeres: Birds of the island of Java, Vol 1.* Tokyo: Kuroda.

——. 1936. *Passeres: Birds of the island of Java.* Tokyo: Kuroda.

Kusch, H. 1978. Die Höhlen der Insel Bali (Indonesien). *Höhle* 30: 1-14.

——. 1980. Die Gua Gede und der Kuppenkarst auf der Insel Nusa Penida (Indonesien). *Höhle* 31: 103-111.

Kuslestari. 1983. *Beberapa species burung diurnal yang hidup di daerah Waduk Serbaguna Gajah Mungkur dan sekitarnya.* Fakultas Biologi UGM, Yogyakarta.

Kusmana. 1900. *Analisa vegetasi hutan mangrove di Muara Angke, Jakarta.* Fakultas Kehutanan IPB, Bogor.

Kusumadinata, K, ed.. 1979. *Data dasar gunung api Indonesia.* Bandung: Dir. Jen Pertambangan Umum.

Kusyanto, D. Inventarisasi/identifikasi karang pantai di Jepara-Karimun Jawa. Paper given at Widyakarya pengembangan sistem pengelolaan data oseanologi, Jakarta: LON-LIPI.

Kvalvågnaes, K. 1980. *Marine survey report of the Segara Anakan/Nusa Kambangan area.* FAO, Bogor.

Kvalvågnaes, K., and M. H. Halim. 1979a. *Report of marine survey Pananjung Pangandaran.* Direktorat Perlindungan dan Pengawetan Alam, Bogor.

——. 1979b. *Survey pendahuluan kemungkinan Taman Nasional Laut Pulau Seribu.* FAO, Bogor.

Laksono. 1985. *Masalah penggalian pasir di Tangerang.* Fakultas Pasca Sarjana UI, Jakarta.

Laksono, P. M. 1988. Perception of volcanic hazards: Villagers versus government officials in Central Java. *The real and imagined role of culture in development,* ed. M. R. Dove, 183-200. Honolulu: Univ. of Hawaii Press.

Lal, R. 1991. Conservation and biodiversity. In *The biodiversity of microorganisms and invertebrates: Its role in sustainable agriculture,* ed. D. L. Hawksworth, 89-104. Wallingford: CAB International.

Lam, H. J. 1924. Aanteekeningen betreffende een bestijging van den G. Slamet. *Trop. Nat.* 13: 17-25.

——. 1925. Een bestijging van den G. Tjareme. *Trop. Nat.* 14: 2-10.

Lamb, D. 1988. *IUCN guidelines for restoration of degraded ecosystems.* Gland: IUCN.

Lambert, F. R., and P. Erftemeijer. 1989. The waterbirds of Pulau Rambut, Java. *Kukila* 4: 3-4.

Lambert, F. R., and A. G. Marshall. 1991. Keystone characteristics of bird-dispersed *Ficus* in a Malaysian lowland rain forest. *J. Ecol.* 79: 793-809.

Lande, R. 1988. Genetics and demography in biological conservation. *Science* 241: 1455-1460.

Lansing, J. S. 1987. Balinese "water temples" and the management of irrigation. *Am. Anthro.* 89: 326-341.

——. 1991. *Priests and programmers: Technologies of power in the engineered landscape of Bali.* Princeton N.J.: Princeton Univ. Press.

Lant, D., and I. Rusli. 1991. Social problems at the Ujung Kulon National Park. *Voice of Nature* 93: 18-21.

Larson, J. S., P. R. Adamus, and E. J. Clairain Jr. 1989. *Functional assessment of freshwater wetlands: a manual and training outline.* Amherst: Environmental Institute Univ. of Massachusetts.

Latief, S. 1980. *Project statement studi konsumsi kayu bakar dan kayu pertukangan di beberapa desa sekitar cagar alam Gunung Honje dalam rangka pembuatan bufferzone di Taman Nasional Ujung Kulon.* Fakultas Kehutanan IPB, Bogor.

Lawton, J. H. 1989. Food webs. In *Ecological concepts: The contribution of ecology to an understanding of the natural world,* ed. J. M. Cherrett, 43-78. Oxford: Blackwell.

Lawton, J. H., and V. K. Brown. 1993. Redundancy in ecosystems. *Biodiversity and ecosystem function,* ed. E. D. Schulze, and H. A. Mooney, 255-270. Berlin: Springer.

LBN-LIPI. 1977a. *Ubi-ubian.* Bogor: Puslitbang Biologi - LIPI.

——. 1977b. *Berberapa jenis bambu.* Bogor: Puslitbang Biologi-LIPI.

——. 1979. *Binatang hama.* Bogor: Puslitbang Biologi-LIPI.

——. 1980. *Jenis rumput dataran rendah.* Bogor: Puslitbang Biologi - LIPI.

——. 1981. *Rumput pegunungan.* Bogor: Puslitbang Biologi - LIPI.

——. 1984a. *Polong-polongan perdu.* Bogor: Puslitbang Biologi - LIPI.

——. 1984b. *Kerabat beringin.* Bogor: Puslitbang Biologi - LIPI.

Lee, D. W., and J. B. Lowry. 1980. Solar ultraviolet radiation on tropical mountains: Can it affect plant speciation? *Am. Nat.* 115: 880-883.

Lee, V. H. 1983. Mosquitoes of Bali Island,

Indonesia: common species in the village environment. *SEA J. Trop. Med. Public Health* 14: 298-307.

Leefmans, S. 1914a. Naar het Bromobosch en den vulkaan Bromo. *Trop. Nat.* 14: 76-80.

——. 1914b. Van Tosari en het Tengergebergte. *Trop. Nat.* 3: 81-88.

——. 1916. Iets sover een javaanschee meikevier (*Leucopholis rorida* F.). *Trop. Nat.* 5: 17-22.

Leenhouts, P. W. 1956. Burseraceae. *Fl. Mal.* I 5: 209-296.

Lehmusluoto, P., and B. Mahbub. 1989. Three tropical crater lakes in Bali, Indonesia: A re-examination of some lakes visited by the German Sunda Expedition in 1929. *Archiv. Hydrobiol.* 114: 537-554.

Leinders, J. J. M., F. Aziz, P. Y. Sondaar and J. de Vos. 1985. The age of hominid-bearing deposits of Java: state of the art. *Geol. Mijnb.* 64: 167-173.

Lekagul, B., and J. McNeely. 1977. *Mammals of Thailand.* Bangkok: Association for the Conservation of Wildlife.

Lembaga Ekologi UNPAD, and Perum Otorita Jatiluhur. 1979. *Penelitian ekologi "Aquatic weeds" di Bendungan Curug: Pemanfaatan eceng gondok untuk biogas, 1978-1979.* Direktorat Pengairan Proyek Irigasi Jatiluhur, Bandung.

Lembaga Ekologi UNPAD. 1980. *Cagar Alam Gunung Tilu: Rencana pengelolaan 1980-1985.* Lembaga Ekologi Unpad, Bandung.

Lembaga Penelitian IPB. 1986. *Analisa dampak lingkungan pada Proyek Tulungagung.* Lembaga Penelitian IPB, Bogor.

Lembaga Penelitian UNDIP, and Bappeda Tingkat I Jawa Tengah. 1984. *Penelitian pengembangan wilayah Kepulauan Karimunjawa Kabupaten Daerah Tingkat II Jepara.* Lembaga Penelitian UNDIP, Semarang.

Lemmens, R. H. M. et al. 1989. *Plant resources of South-East Asia: Basic list of species and commodity groupings.* Wageningen: PROSEA.

Lesica, P., and F. W. Allendorf. 1992. Are small populations of plants worth preserving? *Conserv. Biol.* 6: 135-139.

Leuhery, D. 1989. *Bentuk-bentuk hutan kemasyarakatan dan interaksi penduduk dengan hutan di daerah Kaliurang lereng selatan gunung Merapi.* Fakultas Pasca Sarjana UGM, Yogyakarta.

Levington, J. S. 1982. *Marine ecology.* New Jersey: Prentice Hall.

Lewin, R. 1993. *Human evolution: An illustrated introduction.* Oxford: Blackwell.

Lie, G. L. 1958. Sedikit tentang daerah dis-

ekitar Gresik. *Penggemar Alam* 37: 67-77.

Lieftinck, M. A. 1934. An annotated list of the *Odonata* of Java, with notes on their distribution, habits and life-history. *Treubia* 14: 377-382.

——. 1937. Notes on a collection of Sphingidae collected by Messrs. M. and E. Bartels in Java (Lep.). *Treubia* 16: 37-44.

——. 1950. Further studies on Southeast Asiatic species of *Macromia*, with notes on their ecology, habits and life history and with descriptions of larvae and two new species (Odon., Epothalmiinae). *Treubia* 20: 657-716.

Liem, D. S. S. 1971. The frogs and toads of Tjibodas National Park, Mt. Gede, Java, Indonesia. *Phil. J. Sci.* 100: 131-161.

Lienhard, C. 1990. A new oriental species of Troctopsocidae (Insecta: Psocoptera). *Rev. Suisse Zool.* 97: 339-344.

Ligtvoet, W., R. H. Huges, and S. Wulffraat. 1996. Recreating a mangrove forest near Jakarta. *Land Water Int.* 84: 8-11.

Liley, N. R., and B. H. Seghers. 1975. Factors affecting the morphology and behaviour of guppies in Trinidad. In *Function and evolution in behaviour.* G. Baerends, C. Beer, and A. Manning, 92-118. Oxford: Clarendon.

Lilies, C. 1991. *Kunci determinasi serangga.* Yogyakarta: Kanisius.

Lim, B. H., and A. Sasekumar. 1979. A preliminary study on the feeding biology of mangrove forest primates, Kuala Selangor. *Malay. Nat. J.* 33: 105-112.

Lim, B. L. 1980. A study of small mammals in the Ciloto field station area, West Java with special reference to vectors of plague and scrub typhus. *SEA J. Trop. Med. Public Health* 11: 71-80.

Lim, F. L. K., and M. T. M. Lee. 1989. *Fascinating snakes of southeast Asia.* Kuala Lumpur: Tropical Press.

Lim, K. K. P., and P. K. L. Ng. 1991. Nepenthiphilous larvae and breeding habits of the sticky frog, *Kalophrynus pleurostigma* Tschudi (Amphibia: Microhylidae). *Raffles Bull. Zoology* 39: 209-214.

Lim, K. K. P., and F. L. K. Lim. 1992. *A guide to the amphibians and reptiles of Singapore.* Singapore: Singapore Science Centre.

Lim, M. T. 1978. Litterfall and mineral-nutrient content of litter in Pasoh Forest Reserve. *Malay. Nat. J.* 30: 375-380.

Lincoln, R. J., G. A. Boxshall, and P. F. Clark. 1982. *A dictionary of ecology, evolution and systematics.* Cambridge: Cambridge Univ. Press.

Ling, S. W. 1977. *Aquaculture in southeast Asia: a historical overview.* Washington, D.C.: Sea Grant Publications.

Litsinger, J. A., A. T. Barrion, and D. Soekarna. 1986. Upland rice insect pests: Their ecology, importance and control. *Progress in upland rice research.*

Liu, C.-D. 1950. Amphibians of western China. *Fieldiana Zool.* 2: 1-400.

Lobl, I. 1990. *Cyparium javanum* a new Scaphidiidae (Coleoptera) from Indonesia. *Elytron (Barc)* 4: 125-130.

Loiselle, B. A. 1990. Seeds in droppings of tropical fruit-eating birds: importance of considering seed composition. *Oecologia* 82: 494-500.

Lomer, C. J., A. B. Stride, R. Balfas, and L. I. Durwanti. 1992. The biology of cave-dwelling Hindola and some related Machaerotidae in Indonesia. *Bull. Entomol. Res.* (Accepted for publication):

Loogen, J. G. T. 1940. Herinneringen aan het Jang Hoogland. *Trop. Nat.* 29: 151-155.

—. 1941. Oekologie van eenige graswildernissen op den N. Diëng. *Trop. Nat.* 30: 65-70.

Loosjes, F. E. 1953. Monograph of the Indo-Australian Clausiliidae. *Beaufortia* 3: 1-226.

—. 1963. Supplement to a monograph of the Indo-Australian Clausiliidae. *Zool. Mededel.* 38: 153-169.

Lörzing, J. A. 1914a. Botanische verkenningen en tochten in het Wilis gerbergte. *Trop. Nat.* 14: 98-102, 120-123.

—. 1914b. De grot Goewa-Lawa bij Ngalean. *Trop. Nat.* 3: 172-177.

—. 1921. De Patoeha en zijn omgeving. *Trop. Nat.* 10: 97-105, 113-120, 134-141.

Lötschert, W., and G. Beese. 1981. *Collins guide to tropical plants.* London: Collins.

Lovelock, J. E. 1979. *Gaia: A new look at life on Earth.* Oxford: Oxford Univ. Press.

—. 1988. *Bumi yang hidup.* Jakarta: Yayasan Obor.

Lovric, B. A. 1990. Medical seminology and the semiotics of dance. *Rev. Indon. Malay. Aff.* 24: 136-194.

Lubis, M. 1969. Mysticism in Indonesia politics. In *Man, state and society in contemporary Southeast Asia,* ed. R. O. Tilman, 206-218. New York: Praegar.

—. 1975. *Harimau! Harimau!* Jakarta: Pustaka Jaya.

—. 1980. *Berkelana.* Pustaka Jaya: Jakarta.

—. 1988. Nurturing a love of nature. *Voice of Nature* 59: 1.

—. 1990. *Indonesia: Land under the rainbow.* Singapore: Oxford Univ. Press.

—. 1991. *Tiger.* Singapore: Select Books.

—. 1992. Budaya dan lingkungan hidup. In *Mochtar Lubis: Wartawan jihad,* ed. Atmakusumah, 404-415. Jakarta: Harian Kompas.

Lumbanbatu, D. T. F. 1979. *Aspek biologi reproduksi beberapa jenis ikan di Waduk Lahor, Jawa Timur.* Fakultas Perikanan IPB, Bogor.

Lundqvist, E. 1955. *In eastern forests.* London: Robert Hale.

Lütjeharms, W. J. 1937. Iets over de duinen van Parangtritis. *Trop. Nat.* 26: 85-92.

Luxmoore, R. 1989. *Indonesian trade in monitors.* Traffic, Cambridge.

MacArthur, R. H. 1955. Fluctuations of animal populations, and a measure of community stability. *Ecology* 36: 533-536.

—. 1972. *Geographical ecology: Patterns in the distribution of species.* New York: Harper & Row.

McArthur, R. H., and E. O. Wilson. 1967. *The theory of island biogeography.* Princeton, New Jersey: Princeton Univ. Press.

MacDonald, I. A. W., and G. W. Frame. 1988. The invasion of introduced species into nature reserves in tropical savannas and dry woodlands. *Biol. Conserv.* 44: 67-93.

Machbub, B. et al. 1992. *Major lakes and reservoirs in Indonesia - a report of the first field phase.* Expedition Indodanau, Bogor.

Machbub, B., H. F. Ludwig, and D. Gunaratnam. 1988. Environmental impact from agrochemicals in Bali, Indonesia. *Env. Monit. Assess.* 11: 1-24.

Mackie, C. 1988. *Tree cropping in upland farming systems: An agro-ecological approach.* USAID Indonesia, Jakarta.

Mackie, J. A. C. 1985. The changing political economy of an export crop: The case of Jember's tobacco industry. *Bull. Indo. Econ. Studies* 21: 112-139.

Mackie, J. A. C., and Z. Djumilah. 1989. East Java: Balanced growth and diversification. In *Unity and diversity: Regional economic development in Indonesia since 1970,* ed. H. Hill, 307-330. Singapore: Oxford Univ. Press.

MacKinnon, J. R. 1988. *Field guide to the birds of Java and Bali.* Yogyakarta: Gadjah Mada Univ. Press.

—. 1990. *Panduan lapangan pengenalan burung-burung di Jawa dan Bali.* Yogyakarta: Gadjah Mada Univ. Press.

MacKinnon, J. R., and M. A. Artha. 1982. A national conservation plan for Indonesia: I. Introduction. FAO, Bogor.

MacKinnon, J. R., and K. S. MacKinnon. 1986. *Review of the protected areas system in the Indo-Malayan realm.* Cambridge: IUCN.

MacKinnon, J. R., and K. Phillipps. 1993. *Field guide to the birds of the Greater Sundas.* Oxford: Oxford Univ. Press.

MacKinnon, J. R., and K. Phillipps. in press. *Panduan lapangan pengenalan burung-burung di Kalimantan, Sumatera, Jawa dan Bali.* Bogor: Birdlife.

MacKinnon, J. R., A. C. Smiet, and M. A. Artha. 1982. *A national conservation plan for Indonesia: III. Java and Bali.* FAO, Bogor.

MacKinnon, J. R. et al. 1986. *Managing protected areas in the tropics.* Gland: IUCN.

—. 1990. *Pengelolaan kawasan yang dilindungi di daerah tropika.* Yogyakarta: Gadjah Mada Univ. Press.

MacKinnon, K. S. 1983. Feasibility Study Phase II: Report of a WHO consultancy to Indonesia to determine population estimates of Cynomolgus or long-tailed *Macaca fascicularis* (and other primates) and the feasibility of semi wild breeding. *WHO Primate Resources Programme Report.* WWF, Bogor.

—. 1986a. *Alam asli Indonesia: Flora, fauna dan keserasian.* Jakarta: Gramedia.

—. 1986b. The conservation status of non-human primates in Indonesia. In *Primates: the road to self-sustaining populations.* K. Benirschke, 99-126. New York: Springer.

—. 1987. Conservation status of primates in Malesia with special reference to Indonesia. *Primate Conserv.* 8: 175-183.

—. 1991. Ujung Kulon, land of the Javan Rhino. *Voice of Nature* 93: 7-11.

—. 1992. *Wildlife of Indonesia.* Jakarta: Gramedia.

MacKinnon, K. S., C. Santiapillai, and R. Betts. 1990. *WWF Indonesia Programme statement on conservation priorities for Javan and Sumatran rhinos in Indonesia.* WWF, Jakarta.

MacKinnon, K. S., G. Hatta, H. Halim, and A. Mangalik. 1996a. *The ecology of Kalimantan.* Singapore: Periplus Editions.

—. 1994b. *Ekologi Kalimantan.* Jakarta: Yayasan Obor.

MacLeish, K. 1971. Java: Eden in transition. *Nat. Geog.* 139: 1-43.

MacNae, W. 1968. A general account of the fauna and flora of mangrove swamps and forests in the Indo-Pacific region. *Adv. Mar. Biol.* 6: 73-270.

Mahan, A., T. Suparno, and K. Carlander. 1978. Food habits of walking catfish and snakehead in Rawa Pening. *J. Satya Wacana Res.* 1: 374-380.

Maital, S. 1982. *Minds, markets amd money.* New York: Basic.

Majelis Permusyawaratan Rakyat. 1993. *Kete-*

tapan-ketetapan. Departemen Penerangan, Jakarta.

Maltby, E. 1986. *Waterlogged wealth: why waste the world's wet places.* London: Earthscan.

Manderson, L. 1974. *Overpopulation in Java: Problems and reactions.* Dept. Geography ANU, Canberra.

Mangoendihardjo, S. 1989. Some notes on the mass production and establishment of *Curinus coeruleus* in Central Java. *BIOTROP Spec. Publ.* 36: 313-316.

Mangoendihardjo, S., N. A. Sulistyo, and F. Wagiman. 1989. Bionomy of *Neochetina eichhornia*: A potential biocontrol agent for water hyacinth. *BIOTROP Spec. Publ.* 36: 235-241.

Mani, M. S. 1980. The animal life of highlands. *Ecology of Highlands*, ed. M. S. Mani, and L. E. Giddings, 149-159. The Hague: Junk.

Manik, R., and B. Tiensongrusmee. 1979. Integrated brackish water in farm system in Indonesia. *Bull. Brackishwater Aquacul. Dev. Cent.* 5: 369-376.

Mann, K. H. 1982. *Ecology of coastal waters: A systems approach.* Oxford: Blackwell.

Mannaerts, C., T. Rijnberg, and A. C. Smiet. 1990. Land use issues in the Cimandiri watershed, West Java. *SECM Special publications No. 3.* SECM, Bogor.

Mannetje, L. 't. and R. M. Jones. 1992. *Forages.* Bogor: PROSEA.

Manning, C. 1988. *The green revolution, employment, and economic change in rural Java: a reassessment of trends under the New Order.* Singapore: Institute of Southeast Asian Studies.

Manohara, D., D. Wahyuno, T. Kobayashi, and M. Oniki. 1993. Sooty leaf blotch of *Clausena excavata*, a new disease caused by *Mycovellosiella clausenae* sp. nov. *Trans. Mycol. Soc. Japan* 34: 423-427.

Mantra, I. B. The Balinese view of nature. In *National parks conservation and development: The role of protected areas in sustaining society,* ed. J. A. McNeely, and K. R. Miller, Washington, D.C.: Smithsonian Inst. Press.

—. 1984. Socio-economic problems of the Kampung Laut community in Central Java. *Indo. J. Geog.* 14: 103-113.

—. 1986. Population distribution and population growth in Yogyakarta Special Region. *Indo. J. Geog.* 16: 21-32.

Mantra, I. B., and Tukiran. 1976. Population projections by age and sex in special region of Yogyakarta, 1971-2001. *Indo. J. Geog.* 6: 1-19.

Manuaba, A. 1995. Enhancing the culture.

In *Bali: Balancing environment, economy and culture*, ed. S. Martopo and B. Mitchell. 437-464. Waterloo: Dept. Geography, University Waterloo.

Manuaba, A., N. Sunarta, and P. Prata. 1990. Tourism development in Bali and its impacts: a case study. In *International Seminar on Human Ecology, Tourism, and Sustainable Development*. 145-151. Denpasar: Human Ecology Study Group.

Manullang, B. O., J. Supriatna, and D. S. Hadi. 1982. Pengematan lutung (*Presbytis cristatus*) di hutan mangrove Tanjung Karawang, Jawa Barat. In Seminar II ecosistem mangrove, 238-242. Barurraden, 3 August 1982.

Maraviglia, N. 1990. *Indonesia: Family planning perspectives in the 1990s*. Washington, D.C.: World Bank.

Marchall, W. On some basic traditions in Nusantara cultures. In *Time past, time present, time future*, ed. D. S. Moyena, and H. J. M. Claessen, 69-77.

Margalef, R. 1969. Diversity and stability: a practical proposal and a model of interdependence. In *Diversity and stability in ecological systems*. eds G. M. Woodwell, and H. H. Smith,Upton, New York: Brookhaven.

Maria, L. 1988. Climbing Mt. Semeru. *Voice of Nature* 57: 39-45.

Maria, L., and N. Edwin. 1988. The Kesodo celebration at Mount Bromo. *Voice of Nature* 57: 33-36.

Marjatmo. 1987. Taman Wisata se-Indonesia. PHPA, Bogor.

Marks, P. 1953. Preliminary note on the discovery of a new jaw of *Meganthropus* von Koeningswald in the lower Middle Pleistocene of Sangiran, Central Java. *Indo. J. Nat. Sci.* 109: 26-33.

Marle, J. G., and K. H. Voous. 1988. *The birds of Sumatra: An annotated checklist*. London: British Ornithlolgists' Union.

Marr, C. 1993. *Digging deep: The hidden costs of mining in Indonesia*. London: International Campaign for Ecological Justice in Indonesia.

Marschall, W. 1988. On some basic traditions in Nusantara culture. In *Time past, time present, time future*, ed. D. S. Moyert, and H. J. M. Claessen, 69-77. The Hague: Nijhoff.

Marshall, A. G. 1971. The ecology of *Basilla hispida* (Diptera: Nycteribiidae) in Malaysia. *J. Anim. Ecol.* 40: 141-154.

Marshall, J. T. 1978. Systematics of smaller Asian night birds based on voice. *Ornithological monograph No. 25*.

Marshall, P. 1992. *Nature's web: an exploration of ecological thinking*. London: Simon and Schuster.

Marsono, D. 1989. Synecological considerations on rehabilitation of mangrove vegetation. *BIOTROP Spec. Publ.* 37: 171-180.

Marten, G. G. 1990. A nutritional calculus for home garden design: Case study from West Java. In *Tropical home gardens*, ed. K. Landauer, and M. Brazil, 169-185. Tokyo: United Nations Univ. Press.

Marten, G. G., and O. S. Abdoellah. 1988. Crop diversity and nutrition in West Java. *Ecol. Food Nutr.* 21: 17-43.

Martin, K. 1919. *Unsere palaeozoologische Kenntnis von Java*. Leiden: Brill.

Martin, P. S., and H. E. Wright. 1967. *Pleistocene extinctions: the search for a cause*. New Haven: Yale Univ. Press.

Martodigdo, S. et al. 1987. *Penelitian ekosistem hutan mangrove Pulau Nusa Sibelis, Segara Anakan Cilacap*. Fakultas Biologi UNSOED, Purwokerto.

Martodigdo, S., T. Sudibyaningsih, and T. P. Sinaga. 1986. *Studi pengembangan hutan bakau secara fungsional di daerah pantai Cilacap*. Fakultas Biologi UNSOED, Purwokerto.

Martodigdo, S., Suwarni, and P. Brahmana. 1986. *Struktur dan komposisi hutan mangrove di daerah pantai Tegal, Jawa Tengah*. Fakultas Biologi UNSOED, Purwokerto.

Martono. 1979. *Studi Anophelini (Colicidae, Diptera) di daerah tambak di Banyuwangi*. Sekolah Pasca Sarjana IPB, Bogor.

Martopo, S. 1984. Hydrological potential of the southern flank of Slamet Volcano, Central Java. *Indo. J. Geog.* 14: 55-66.

—. 1986. *Laporan penelitian pencemaran di daerah Kotamadya Yogyakarta*. PSL UGM, Yogyakarta.

Martopo, S., and B. Mitchell. 1995. *Bali: balancing environment, economy and culture*. Dept. Geography, University Waterloo.

Martosubroto, P. 1977. Hutan bakau dan peranannya dalam perikanan udang. In Prosiding Seminar II Perikanan Udang, 355-360. Jakarta, 15 March 1977.

Marwoto, R. M. 1988. Occurrence of a freshwater snail *Pomacea* sp. in Indonesia. *Treubia* 29: 275-276.

Marzali, A. 1992. The *urang sisi* of West Java: a study of peasants' responses to population pressure. Ph.D. diss., Boston Univ., Boston.

Marzuki, F. 1987. *Prinsip-prinsip budidaya pemeliharaan burung walet*. Biro Pusat Rehabilitasi Sarang Burung, Surabaya.

Mason, V. 1988. The Oriental pratincole

Glareola maldivarum in Bali. *Kukila* 3: 141.
—. 1990. Note on the large hawk-cuckoo, *Cuculus sparverioides*, in Bali. *Forktail* 5: 73.
—. 1992. *Bali bird walks.* Hong Kong: Apa.
—. 1993. A note on the occurence of zebra finch on Bali. *Kukila* 6: 132.
Mason, V., and F. Jarvis. 1989. *Birds of Bali.* Berkeley: Periplus Editions.
Matthews, A. 1983. *The night of Purnama.* Kuala Lumpur: Oxford Univ. Press.
Maull, H. W. 1992. Japan's global environmental policies. *The international politics of the environment,* ed. A. Hurrel, and B. Kingsbury, 354-372. Oxford: Oxford Univ. Press.
May, R. M. 1976. Island biogeography and the design of natural reserves. In *Theoretical evolution: principles and application,* ed. R. M. May, 163-186. Oxford: Blackwell.
—. October 1992. How many species inhabit the earth? *Scient. Amer.,* 42-48.
Mazak, V., C. P. Groves, and P. J. H. van Bree. 1978. On a skin and skull of the Bali tiger and a list of preserved specimens of *Panthera tigris balica* (Schwarz, 1912). *Zeits. Saügetierk.* 43: 108-113.
McArdle, B. H. 1990. When are rare species not there? *Oikos* 57: 276-277.
McCauley, D. S. 1982. *Soil erosion and land-use patterns among upland farmers in the Cimanuk watershed of West Java, Indonesia.* East-West Environmental and Policy Institute, Honolulu.
—. 1984. *Dryland gardens and soil conservation in the uplands of Java: A study of farmer's dryland use decisons in the Cimanuk watershed of West Java.* Center for Nat. Res. Mgmt. and Envt. Studies, Bogor Agricultural Univ., Bogor.
—. 1991. Watershed management in Indonesia: the case of Java's densely populated upper watersheds. in *Watershed resources management: studies from Asia and the Pacific.* K. W. Easter, J. A. Dixon, and M. M. Hufschmidt, 177-190. Singapore: Institute of Southeast Asian Studies.
McConnel, J., and I. Glover. 1988. A newly found bronze drum from Bali, Indonesia: some technical considerations. *Mod. Quat. Res. SE Asia* 11: 1-38.
McDonagh, S. 1986. *To care for the earth: A call to a new theology.* London: Geoffrey Chapman.
—. 1990. *The greening of the church.* London: Geoffrey Chapman and Orbis.
McDonald, P. F. 1976. *Response to population pressure: The case of the Special Region of Yogyakarta.* Gadjah Mada Univ. Press; Yogyakarta.

—. 1980. An historical perspective to population growth in Indonesia. In *Indonesia: Dualism, growth and poverty,* ed. R. G. Garnaut, and P. McCawley, 80-95. Canberra: Australian National Univ. Press.
McIntosh, J. L. 1986. Ecological issues in preproduction and production programs involving upland rice. In *Progress in upland rice research.* pp.463-474. Manila: IRRI.
McIntosh, J. L., S. Effendi, and A. Syarifuddin. 1977. Testing cropping patterns for upland conditions. *Cropping systems, research and development for the Asian rice farmer.* pp.201-221. Los Baños: IRRI.
McKean, P. F. 1989. Towards a theoretical analysis of tourism: Economic dualism and cultural involution in Bali. In *Hosts and guests: The anthropology of tourism,* ed. V. L. Smith, 119-138. Philadelphia: Univ. of Pennsylvania Press.
McKibben, B. 1992. *Berakhirnya alam.* Jakarta: Yayasan Obor.
McMillan, M. 1915. *A journey to Java.* London: Holden and Hardingham.
McNeely, J. A. 1978. Dynamics of extinction in southeast Asia. *BIOTROP Spec. Publ.* 8: 137-160.
—. 1980. Java's vanishing animals. *Int. Wildlife* 10 (March-April): 38-40.
—. 1988. *Economics and biological diversity.* Gland: IUCN.
—. 1992. *Ekonomi dan keanekaragaman hayati.* Jakarta: Obor.
McNeely, J. A., and P. S. Wachtel. 1988. *Soul of the tiger.* New York: Doubleday.
McNicoll, G. 1980. Institutional determinants of fertility change. *Pop. Dev. Rev.* 6: 441-463.
—. 1982. Recent demographic trends in Indonesia. *Pop. Dev. Rev.* 8: 811-820.
McTaggart, W. D. 1982. Land use in Sukabumi, West Java: Persistence and charge. *Bijdragen tot de Taal-, Land- en Volkenkunde* 138: 295-316.
—. 1983. Forest policy in Bali, Indonesia. *S'pore J. Trop. Geog.* 4: 147-162.
—. 1984. Some development problems in Bali. *Contemp. SE Asia* 6: 231-245.
—. 1988. Hydrologic management in Bali, Indonesia. *S'pore J. Trop. Geog.* 9: 96-111.
Meadows, D. H. et al. 1972. *The limits to growth: a report for the Club of Rome's project on the predicament of mankind.* New York: Universe.
—. 1982. *Batas-batas pertumbuhan.* Jakarta: Gramedia.
Medway, Lord. 1958. On the habits of *Arixenia esau* Jordan (Dermaptera). *Proc. R. ent. Soc. Lond. A.* 33: 191-195.

—. 1962a. The swiftlets *Collocalia* of Java and their relationships. *J. Bombay Nat. Hist. Soc.* 59: 146-153.

—. 1962b. The swiftlets (Collocalia) of Niah Cave, Sarawak. *Ibis* 104: 228-245.

—. 1967. A bat-eating bat, *Megaderma lyra* Geoffrey. *Malay. Nat. J.* 20: 107-110.

—. 1969. Studies on the biology of the edible nest swiftlets of southeast Asia. *Malay. nat. J.* 22: 57-63.

—. 1983. *Mammals of Malaya.* Kuala Lumpur: Oxford Univ. Press.

Mees, G. F. 1971. Systematic and faunistic remarks on birds from Borneo and Java, with new records. *Zool. Mededel.* 45: 225-242.

—. 1973. The status of two species of migrant swifts in Java and Sumatra (Aves: Apodidae). *Zool. Mededel.* 46: 197-207.

—. 1989. Remarks on the ornithological parts of Horsfield's zoological researches in Java. *Proc. K. Ned. Akad. Wet., Ser. C* 92: 367-378.

Meeth, P., and K. Meeth. 1989. Long billed plover on Bali in November 1973. *Dutch Birding* 11: 114-115.

Meier, R. L. 1976. A stable urban ecosystem. *Science* 192: 962-968.

Meijer, W. 1953. The study of hepatics in the Malaysian tropics. *Bryologist* 56: 95-98.

—. 1954. Plantensociologische waarnemingen in de top regionen van Gede en Pangrango. *Penggemar Alam* 34: 9-17.

—. 1954a. Bryologische brieven uit Indonesia III. Kijkjes in de levermosflora van de Noordhelling van de Pangerango boven Tugu, W. Java. *Buxbaumia* 8: 10-20.

—. 1954b. Bryologische brieven uit Indonesia IV. Een bezoek aan de toppen van Gede en Pangrango. *Buxbaumia* 8: 41-48.

—. 1955. Op zoek naar de boomvarens van Tjibodas. *Penggemar Alam* 35: 45-53.

—. 1958a. Opvallenda levermosvormen wit de flora van Tjibodas. *Buxbaumia* 12: 8-16.

—. 1958b. Notes on species of *Riccia* from the Malaysian region. *J. Hattori Bot. Lab* 20: 107-118.

—. 1959a. On some southeast Asiatic species of the genus *Plectocolea. J. Hattori Bot. Lab.* 21: 53-60.

—. 1959b. Notes on species of *Riccardia* from their type localities in western Java. *J. Hattori Bot. Lab.* 21: 61-78.

—. 1959c. Plant sociological analysis of montane rainforest near Tjibodas, West Java. *Acta Bot. Neerl.* 8: 277-291.

—. 1960. Notes on the species of *Bazzania* (Hepaticae) mainly of Java. *Blumea* 10: 367-

384.

—. 1976. Botanical explorations in Celebes and Bali. *Nat. Geog. Soc. Res. Rep. (Projects)* 60.

—. 1984. New species of *Rafflesia* (Rafflesiaceae). *Blumea* 30: 209-213.

Meijerink, A. M. J. 1977. A hydrological reconnaissance survey of the Serayu River basin, Central Java. *ITC J.* 4: 646-673.

Melisch, R. 1992. *Check-list of the land mammals of Java.* PHPA/AWB, Bogor.

Melisch, R., and I. W. A. Dirgayusa. in press. Some notes on the Javan surili (*Presbytis comata*) from two nature reserves in West Java, Indonesia. *Folia primatol.*

Melisch, R., Y. R. Noor, and W. Giesen. 1993. *Assessment of the importance of Rawa Danau for nature conservation and an evaluation of resource use.* Asian Wetland Bureau, Bogor.

Melisch, R., P. B. Asmoro, and L. Kusumawardhani. 1994. Major steps taken towards otter conservation in Indonesia. *IUCN Otter Specialist Group Bull.* 10: 21-24.

Melisch, R., L. Kusumawardhani, P. B. Asmoro, and I. R. Lubis. 1994. The role of otters (Mustelidae, Carnivora) in rice-fields and fisheries in West Java, Indonesia. Institute of Zoology, Univerity of Hohenheim.

Mence, A. J. 1975. Tigers in Java, Indonesia. *Tigerpaper* 2: 28.

Mendes, L. F., C. B. D. Roca, and M. G. Ricart. 1989. Description of *Cryptostylea* (Zygentoma: Ateluridae), a new termitophilous thysanuran genus from Java. *Garcia. de Orta. Ser. Zool.* 16: 205-210.

Mercer, D. E., and L. S. Hamilton. 1984. Mangrove ecosystems: some economic and natural benefits. *Ambio* 13: 14-19.

Mergner, H. 1971. Structure ecology and zonation of Red Sea reefs (in comparison with South Indian and Jamaican reefs). In *Regional variation in Indian coral reefs,* ed. D. R. Stodddart, and C. M. Yonge, 141-161. London: Zoological Society of London.

Mertens, R. 1930. Die Amphibien und Reptilien der Deutschen Limnologischen Sunda-Expedition. *Archiv. Hydrobiol. Suppl.* 8: 677-701.

—. 1936. Uber einige fur die Insel Bali neue Reptilien. *Zool. Anz.* 115: 126-129.

—. 1957. Amphibien und Reptilien aus dem aussersten westen Javas und von benachbarten Eilanden. *Treubia* 24: 83-105.

—. 1959. Amphibien und Reptilien von Karimun Jawa, Bawean und den Kangean Inseln. *Treubia* 25: 1-16.

Metzner, J., and N. Daldjoeni. 1987. *Ecofarming: Bertani selaras alam.* Jakarta: Yayasan

Obor.

Meyburg, B.-U. 1986. Threatened and near-threatened birds of prey of the world. *Birds of Prey Bull.* 3: 1-12.

Meyburg, B.- U., S. van Balen, J.- M. Thiollay, and R. D. Chancellor. 1989. Observations on the endangered Java hawk eagle *Spizaetus bartelsi*. In *Raptors in the modern world*, ed. B.-U. Meyburg, and R. D. Chancellor, Berlin: WWGEP.

Meyrick, E. 1969. *Exotic Microlepidoptera.* Hampton: Classey.

Michon, G., and J. M. Bompard. 1987. Indonesian agroforestry practices: A traditional contribution to the conservation of rain forest resources. *Rev. d'Ecol. la Terre et la Vie* 42: 3-38.

Michon, G., and F. Mary. 1990. Transforming traditional home gardens and related systems in West Java (Bogor) and West Sumatra (Maninjau). In *Tropical home gardens*, ed. K. Landauer, and M. Brazil, 169-185. Tokyo: United Nations Univ. Press.

Mikhola, H. 1986. Barn owl *Tyto alba* in Bali. *Kukila* 2: 95.

Miksic, J. 1991. *Borobudur*. Singapore: Periplus.

Milton, G. R. 1984a. *Market hunting of Javan coastal birds*. WWF, Bogor.

—. 1984b. Report on field trip to Ranca Danau, 3 - 7 April 1984. *Field report No. 2.* WWF, Bogor.

Milton, G. R., and A. Marhadi. 1984a. *Report on a field trip to Muara Angke, 11 March 1984.* Bogor, WWF.

—. 1984b. *Report on a field trip to Pulau Dua and Sawah Luhur, 25 February - 5 March 1984.* Bogor, WWF.

—. 1985. The bird life of the Nature Reserve Pulau Dua. *Kukila* 2: 32-41.

—. 1986. *A population census, nesting density, habitat utilization and management recommendation for birds nesting on the Nature Reserve Pulau Dua, West Java.* WWF - BKSDA VIII, Bogor.

Milward, R. C. 1915. *Note on the forests of Java and Madoera of the Dutch East Indies*. Calcutta: Superintendent Govt. Print. India.

Mitchell, B. 1995. Sustainable development strategy for Bali. In *Bali: Balancing environment, economy and culture*, ed. S. Martopo and B. Mitchell. 537-565. Waterloo: Dept. Geography, University of Waterloo.

Mitchell, D. S., and B. Gopal. 1991. Invasion of tropical freshwaters by alian aquatic plants. In *Ecology of biological invasion in the tropics*. P. S. Ramakrishnan, 139-154. New Dehli: International Scientific Publications.

Moerywati, B. R. A. 1991. *The legend of Nyai Roro Kidul*. Jakarta: Dir. Gen. Tourism.

Moffat, D. J. 1977. Development of the water resources of Bali: A master plan. Supporting report B: Soil and land classification. *Report of Indonesian Ministry of Public Works and Electric Power.* Ministry of Overseas Development. U.K.,

Moffett, M. W. 1986. Observations on *Lophomyrmex* ants from Kalimantan, Java and Malaysia. *Malay. Nat. J.* 39: 207-212.

Mohsin, A. K. M., and A. T. Law. 1980. Environmental studies of Kelang River. II Effects on fish. *Malay. Nat. J.* 33: 189-199.

Moiwen, M. 1990. Land degradation, soil conservation and trees. *FONC project communication No. 1990 - 2.* FONC - Fakultas Kehutanan UGM, Yogyakarta.

Molineaux, L. 1990. The epidemiology of mosquito-borne diseases. *Health and irrigation: Incorporation of disease control measures in irrigation, a multi-faceted task in design, construction and operation.* eds Oomen, J. M. V., J. de Wolf, and W. R. Jobin, 263-277. Wageningen: International Institute for Land Reclamation and Improvement.

Moncrief, L. W. 1978. The cultural basis for our environmental crisis. *Science* 170: 257-264.

Monk, K. A., and R. K. Butlin. 1990. A biogeographic account of the grasshoppers (Orthoptera: Acridoidea) of Sulawesi, Indonesia. *Tijd. Entomol.* 133: 31-38.

Moosa, M. K. 1980. Beberapa catatan mengenai rajungan di Teluk Jakarta dan Pulau-pulau Seribu. *Sumberdaya hayati bahari*, ed. Burhanuddin, M. K. Moosa, and H. Razak. LON - LIPI, Jakarta.

Morley, J. R., and J. R. Flenley. 1987. Late Cenozoic vegetational and environmental changes in the Malay Archipelago. In *Biogeographical evolution of the Malay Archipelago*. T. C. Whitmore, 50-59. Oxford: Clarendon.

Moro, D. S., Y. Irmawati, G. Reksodihardjo, and Setyowati. 1986. Pola sebaran moluska di mangrove Legon Lentah, Pulau Panaitan. Paper presented at Seminar III Ekosistem Mangrove, Denpasar, 5 August 1986.

Morooka, Y., and H. Mayrowani. 1990. *Upland economy in Java: A perspective of a soybean-based farming system.* Bogor: CGRPT.

Morrison, A. 1980. A note on Javanese aviculture. *Avicul. Mag.* 86: 108-110.

Morse, D. R., N. E. Stork, and J. H. Lawton. 1988. Species number, species abundance and body length relationships of arboreal beetles in Bornean lowland rain forest trees. *Ecol. Ent.* 13: 25-37.

Morton, J. 1990. *The shore ecology of the tropical Pacific.* Jakarta: UNESCO.

Morton, R. M. 1990. Community structure, density and standing crop of fishes in a sub-tropical Australian mangrtove area. *Mar. Biol.* 105: 385-394.

Moss, B. 1980. *Ecology of freshwaters.* Oxford: Blackwell.

Mouwen, M. 1990. Land degradation, soil conservation, and trees. Fakultas Kehutanan UGM, Yogyakarta.

MPA-UGM. 1992. *Pengukuran lapangan.* Mahasiswa Pencinta Alam UGM, Yogyakarta.

Mudjiman, A. 1986. *Budidaya ikan di sawah tambak.* Jakarta: Simplex.

Mukerji, A. K. 1989. Development of pine resin tapping industry. *Field document III - 2.* FAO Forestry Studies, Jakarta.

Mukhtar, A. S., and Pratiwi. 1986. Pola pemanfaatan lahan di daerah penyangga, serta pengaruhnya terhadap kelestarian Taman Nasional Gunung Gede Pangrango, Jawa Barat. *Bull. Pen. Hutan* 483: 1-16.

——. 1991. Diversity of tree species and its problem in Situgunung forest Gunung Gede Pangrango National Park West Java Indonesia. *Bull. Pen. Hutan* 533: 1-12.

Mulcahy, G. 1993. *Preliminary gibbon data. UEA Gunung Halimun Expedition 1992.* UEA, Norwich.

Muljadi, D., I. G. Ismail, and J. L. McIntosh. 1986. Farming systems reseach to conserve the environment of upper watersheds. In *Progress in upland rice research.* pp.171-176. Manila: IRRI.

Muller, K. 1992. *Underwater Indonesia.* Singapore: Periplus.

Muluk, C. 1979. A review of the fisheries on Lake Selorejo, East Java. Paper presented at the 5th International Symposium of Tropical Ecology, Kuala Lumpur, 1979.

——. 1985. Tambak systems in the Krawang coast of West Java: A case study of traditional coastal resource use. In *The traditional knowledge and management of coastal systems in Asia and the Pacific,* ed. K. Ruddle, and R. E. Johannes, 229-251. Jakarta:

Mulyadi. 1985. Fluktuasi komunitas fitoplankton di waduk Malakayu, Brebes, Jawa Tengah. *Berita Biologi* 3: 91-94.

——. 1986. *Fluktuasi kecepatan gugur daun Avicennia marina, dekomposisi dan kandungan kimianya dalam hutan mangrove Pulau Dua.* Puslitbang Biologi - LIPI, Bogor.

Munaan, A., A. Lolong, and B. Zelazny. 1989. Palm damage due to *Oryctes rhinoceros* and

virus incidence in trial plots in Central Java. *BIOTROP Spec. Publ.* 36: 95-98.

Munif, A. Pengaruh mangrove pada lagoon terhadap komposisi fauna nyamuk vektor malaria di daerah endemis Pameungpeuk, Jawa Barat. In Kumpulan Abstrak Kongres Nasional Biologi VIII, Purwokerto, 8 October 1987.

Munir, M. 1987. Gunung Halimun, West Java's languishing treasure house. *Voice of Nature* 47: 52-56.

Muphy, J. O., and P. H. Whetton. 1989. A re-analysis of a tree ring chronology from Java, Indonesia. *Proc. K. Ned. Akad. Wet., Ser. B* 92: 241-257.

Murphy, D. H. 1990. The natural history of insect herbivory on mangrove trees in and near Singapore. *Raffles Bull. Zoology* 38: 119-203.

Murray, D. V. 1982. Occurrence of the lesser adjutant stork *Leptophilos javanicus* in Bali, Indonesia. *Bull. Brit. Orn. Cl.* 102: 39-40.

Murray, M. G. 1990. Conservation of tropical rain forests: arguments, beliefs and convictions. *Biol. Conserv.* 52: 17-26.

Murtidjo, B. A. 1989. *Tambak air payau.* Yogyakarta: Kanisius.

Musser, G. G. 1981. A new genus of arboreal rat from West Java, Indonesia. *Zool. Verhand.* 189: 1-35.

——. 1982. The Trinil rats. *Mod. Quat. Res. S.E. Asia* 7: 65-85.

Musser, G. G., and C. Newcomb. 1983. Malaysian murids and the giant rat of Sumatra. *Bull. Am. Mus. nat. hist.* 206: 322-413.

Musser, G. G., J. T. Marshall, and Boeadi. 1979. Definition and contents of the Sundaic genus *Maxomys* (Rodentia: Muridae). *Am. Soc. Mammal.* 60: 592-606.

Muther, G., and H. Y. M. Ram. 1986. Floral biology and pollination of *Lantana camara. Phytomorph.* 31: 79-100.

Mydans, S. 1976. Rituals of Bali appease demons of an unseen world with food and wine, while masks scare away evil spirits. *Smithsonian* 7: 82-86.

Myers, J. P., S. L. Williams, and F. A. Pitelka. 1980. An experimental analysis of prey available for sanderlings (Aves: Scolopacidae) feeding on sandy beach crustaceans. *Can. J. Zool..* 58: 1564-1574.

Myers, N. 1979. *The sinking ark.* Oxford: Pergamon.

——. 1986. Tree-crop based agro ecosystems in Java. *For. Ecol. Manage.* 17: 1-12.

——. 1988. Threatened biotas: "hot spots" in tropical forests. *Environmentalist* 8: 187-208.

——. 1990. Mass extinctions: what can the past tell us about the present and future. *Palaeogeog. Palaeoclim. Palaeoecol.* 82: 175-185.

Myers, N., and T. J. Goreau. 1991. Tropical forests and the greenhouse effect: a management response. *Climatic Change* 18: 1-11.

Myers, R. F. 1991. *Micronesian reef fishes.* Guam: Coral Graphics.

Naamin, N. 1986. Impact of "tambak" aquaculture to the mangrove ecosystem and its adjacent areas with special reference to the North Coast of West Java. In *Mangroves of Asia and the Pacific: Status and management.* pp.355-365.

——. 1991. Penggunaan lahan mangrove untuk budidaya tambak, keuntungan dan kerugiannya. Prosidings Seminar IV Ekosistem Mangrove, S. Soemodihardjo et al., 49-57. Bandar Lampung, 7 August 1990. Jakarta: MAB-LIPI.

Naess, A. 1986. Intrinsinc value: Will the defenders to nature rise? In *Conservation biology,* ed. M. E. Soulé, 504-515. Sunderland, MA: Sinauer.

——. 1973. The shallow and the deep, longrange ecology movement. *Inquiry* 95-100.

——. 1985. Identification as a source of deep ecology attitudes. In *Deep ecology,* ed. M. Tobias. San Diego: Avant Books.

Nagendran, J. 1991. *Prokasih: A river cleaning program in Indonesia.* EMDI Project, Kantor Menteri KLH, Jakarta.

Nakata, S., and T. C. Maa. 1974. A review of the parasitic earwigs (Dermaptera, Arixenniina, Hemimerina). *Pacific Ins.* 16: 307-374.

NAS. 1977. *Leucaena: Promising forage and tree crop for the tropics.* Washington D.C.: National Academy of Sciences.

Nash, S. V. 1994. *Going for a song: the trade in SE Asian non-CITES birds.* Traffic, Cambridge.

Nasseef, A. O. 1986. The Muslim Declaration on Nature. In *The Assisi Declaration: Messages on Man and Nature from Buddhism, Christianity, Hinduism, Islam and Judaism.* pp.21-25. Gland: World Wide Fund for Nature.

Nässig, W. A., and Y. R. Suhardjono. 1989. A new species of the genus *Loepa* (Saturniidae) from Java. *Tinea* 12: 205-210.

Nasution, R. E. 1979. Kebun Raya cabang Purwodadi. *Bull. Kebun Raya* 4: 105-108.

Nasution, R. E., and S. Sastrapradja. 1976. Mengenal marga *Eria* dan jenis-jenisnya di pulau Jawa. *Bull. Kebun Raya* 2: 163-170.

Nasution, R. E., and Tarmudji. 1981.

Anggrek Eria Lamongan (*Eria lamonganensis* Rchb.f.) dari Gunung Lamongan. *Bull. Kebun Raya* 5: 5-6.

Nasution, S. H. 1990. Komunitas ikan pada anak sungai Cisadane. In *Biologi perairan sekitar Bogor.* F. Sabar, 15-20. Bogor: Balitbang Biologi Perairan - LIPI.

Nazif, M. 1988. Percobaan pengendalian *Acacia arabica* herbisida Indamin 720HC, Garlon 480EC dan trusi di Taman national Baluran, Jawa Timur. *Bull. Pen. Hutan* 499: 11-24.

Nee, S., and R. M. May. 1992. Dynamics of meta-populations: habitat destruction and competative coexistence. *J. Anim. Ecol.* 61: 37-40.

Neely, P. 1992. *Indo surf and lingo.* Noosa Heads: Indo Surf and Lingo.

Negroni, G., and L. Farina. 1993. L'élevage de grenouilles. *Cahiers Agric.* 2: 48-55.

Neill, W. T. 1973. *Twentieth century Indonesia.* New York: Columbia Univ. Press.

Nelson, A. J. 1989. Newly described material from Wadjak, Java, Indonesia. *Am. J. Phys. Anthro.* 78: 279-286.

Nelson, G. et al. 1991. *Toward sustainable development in the Segara Anakan region of Java, Indonesia.* Heritage Resources Centre - Univ. of Waterloo, Waterloo, Canada.

Nelson, J. G., E. LeDrew, C. Olive, and Dulbahri. 1995. Information for sustainable development. Coastal studies. In *Bali: Balancing environment, economy and culture,* ed. S. Martopo and B. Mitchell. 437-464. Waterloo: Dept. Geography, University of Waterloo.

Nelson, J. G., and R. Serafin. 1992. Assessing biodiversity: a human ecological approach. *Ambio* 21: 212-218.

Nelson, W. A. 1984. Effects of nutrition of animals on their ectoparasites. *J. Med. Entomol.* 21: 621-635.

New, T. R., M. B. Bush, and H. K. Sudarman. 1987. Butterflies from the Ujung Kulon National Park, Indonesia. *J. Lepidop. Soc.* 41: 29-40.

New, T. R., G. S. Farrell, N. W. Hives, and P. A. Horne. 1985. An early season migration of *Cotopsilia pomona,* (Lepidoptera: Pieridae) in Java, Indonesia. *J. Res. Lepidop.* 24: 84-85.

Ng, F. S. P., and C. M. Low. 1982. *Check list of endemic trees of the Malay Peninsula.* Forest Research Institute, Kepong.

Ng, P. K. L. 1988. *The freshwater crabs of Peninsular malaysia and Singapore.* Singapore: National Univ. Singapore.

——. 1989. *Terrathelphusa,* a new genus of semiterrestrial freshwater crabs from Borneo and Java (Crustacea: Decapoda: Brachyura: Sundathelphusidae). *Raffles Bull. Zoology* 37: 116-131.

—. 1991a. Conservation in Singapore. *Wallaceana* 64: 7-9.

—. 1991b. *A guide to freshwater life in Singapore.* Singapore: Singapore Science Centre.

Ng, P. K. L., and K. K. P. Lim. 1990. Snakeheads (Pisces: Channidae): Natural history, biology and economic importance. In *Essays in zoology.* L. M. Chou, and P. K. L. Ng, 127-152. Singapore: National Univ. of Singapore.

Ng, P. K. L., and A. J. Whitten. 1995. A new species of crab from Nusa Penida, Bali. *Trop. Biodiv.*

Nibbering, J. W. 1989. Forest degradation and reforestation in a highland area in Java. Proceedings of the IUFRO Conference, Canberra.

—. 1991. Crisis and resilience in upland land use in Java. In *Indonesia: resources, ecology and environment,* ed. J. Hardjono, 104-132. Singapore: Oxford Univ. Press.

Nielsen, G. L., and J. M. Widjaya. 1989. Modeling of ground-water recharge in Southern Bali, Indonesia. *Ground Water* 27: 473-481.

Nielsen, I. C., and H. C. F. Hopkins. 1992. Mimosaceae (Leguminosae - Mimosoideae). *Fl. Mal. I* 11: 1-226.

Nielsen, L. B. 1984. A critique of alternative tourism in Bali. In Proceedings of a Workshop on Alternative Tourism with a Focus on Asia, Ed. P. Holden, Chiang Mai, 26 April 1984. Bangkok: Ecumenical Coalition on Third World Tourism.

Nieuhof, I. L. 1682. *Voyages and travels to the East Indies 1653-1670.* Oxford: Oxford Univ. Press.

Nieuwenkamp, W. O. J. 1910. *Zwerftochten op Bali.* Amsterdam: Elsevier.

Nimpoeno, J. S. 12 February 1988. Masalah pengintegrasian buruh tani ke dalam sistem budaya akuatik di daerah genangan Saguling dan Cirata. Paper presented at Seminar Pemukiman Kembali Penduduk Saguling dan Cirata Melalui Pengembangan Perikanan, Jakarta.

Ninkovich, D., and L. H. Burckle. 1978. Absolute age of the base of the hominid-bearing beds in Eastern Java. *Nature* 275: 306-307.

Nisbet, I. C. T. 1968. The utilization of mangroves by Malayan birds. *Ibis* 110: 348-352.

Nisbet, L. J., and F. M. Fox. 1991. The importance of microbial biodiversity to biotechnology. In *The biodiversity of microorganisms*

and invertebrates: Its role in sustainable agriculture,* ed. D. L. Hawksworth, 229-244. Wallingford: CAB International.

Noerdjito, M., and A. Saim. 1979. Menyusur pinggang pulau Jawa. *Alam Kita* 4: 22-25.

Nofzsiger, J. 1978. Investigation of insects population at Rawa Pening Lake. *J. Satya Wacana Res.* 1: 381-391.

Nontji, A. 1984. *Biomassa dan produktivitas fitoplankton di perairan Teluk Jakarta serta kaitannya dengan faktor-faktor lingkungan.* Fakultas Pasca Sarjana IPB, Bogor.

—. 1987. *Laut nusantara.* Jakarta: Djambatan.

Nontji, A., and D. I. Hartoto. 1989a. *Ecology of a small tropical lake (Bogor, West Java).* MAB-UNESCO, Jakarta.

—. 1989b. *Limnologi Situ Bojongsari.* Bogor: PPPL-LIPI.

Noor, Y. R. 1991. *Buku panduan teknik pengelolaan lahan basah melalui kegiatan patroli lapangan.* Asian Wetland Bureau, bogor.

—. 1992. *Laporan penelitian koloni burung kokokan (Ardeidae) di Petulu, Bali.* Asian Wetland Bureau, Bogor.

Noor, Y. R., and T. Permana. 1990. *Pengamatan pendahuluan koloni burung air di Pulau Panjang, Kabupaten Jepara.* Dit. Jen. PHPA - AWB, Bogor.

Noordam, D. 1986. Aphids of Java. Part 2: *Sinomegoura,* (Homoptera: Aphididae) with a new species from *Coffea. Zool. Mededel.* 60: 39-61.

—. 1991. Hormaphidinae from Java (Homoptera: Aphididae). *Zool. Verhand.* 270: 3-52.

Noordam, D., and D. H. R. Lambers. 1985. Aphids of Java. Part 1: Introduction to five new species of *Taiwanaphis* with re-description of the genus (Homoptera, Aphididae). *Zool. Verhand.* 219: 3-46.

Nooteboom, H. P. 1962. Simaroubiaceae. *Fl. Mal. I* 6: 193-226.

—. 1975. Revision of the Symplocaceae of the Old World (New Caledonia excepted). Ph.D. diss., Leiden Univ., Leiden.

Norden, H. 1926. *Byways of the tropic seas: Wanderings among the Solomons and in the Malay Archipelago.* London: H.F. and G. Witherby.

Norton, B. G., and R. E. Ulanowicz. 1992. Scale and biodiversity policy: a hierarchical approach. *Ambio* 21: 244-249.

Noss, R. F. 1990. Indications for monitoring biodiversity: a hierarchical approach. *Conserv. Biol.* 4: 355-364.

Notodihardjo, M. 1987. Sistim pemantauan dan penilaian keberhasilan pengembangan wilayah sungai. Paper presented at Seminar Nasional II Himpunan Mahasiswa Pengairan, Malang, 28 November 1987. Fakultas Teknik UNIBRAW.

Notohadiprawiro, T., and A. A. Asmara. 1989. A geographical model of soil nutrient regimes. In Proceedings Nutrient Management for Food Crop Production in Tropical Farming Systems, Ed. J. van der Heide,, 63-72. Malang, 19 October 1987.

Notosoedarmo, S., and A. I. A. Kristyanto. 1978. Revised list of aquatic vegetation of the Rawa Pening Lake area, Central Java. *J. Satya Wacana Res.* 1: 329-344.

Notosoedarmo, and A. J. A. Kristyanto. 1989. *Tumbuhan-tumbuhan di Rawa Pening.* Fak. Sains, Univ. Kristen Satya Wacana, Salatiga.

Notowinarno. 1988. Bawean, forgotten outpost in the Java Sea. *Voice of Nature* 62: 6-10.

Notowinarto et al. 1986. *Inventarisasi: Biota goa-goa karst Gunung Kidul, Gombong Selatan, Pulau Nusa Kambangan.* Matalabiogama, UGM, Yogyakarta.

——. 1988. Memantau kelelawar di Goa Gombong Selatan. *Suara Alam* 55: 23-26.

NRC. 1983. *Little-known Asian animals with a promising economic future.* Washington D.C.: National Academy Press.

Nugroho, A. D., Djuwantoko, E. H. Haaften et al. 1991. *Baluran rapid appraisal.* Fakultas Kehutanan UGM, Yogyakarta.

Nugroho, T. 1986. *Studi karakteristik Goa Cipeureu di kawasan hutan Gunung Walat, Sukabumi, Jawa Barat.* Fakultas Kehutanan IPB, Bogor.

Nuitja, I. N. S. 1979. *Pengamatan habitat dan populasi penyu di Pantai Citirem.* Fakultas Perikanan IPB, Bogor.

——. 1983. *Studi ekologi peneluran penyu daging, Chelonia mydas L., di Pantai Sukamade, Kabupaten Banyuwangi.* Fakultas Perikanan IPB, Bogor.

Nuitja, I. N. S., and W. Ismail. 1984. Studi pendahuluan daerah peneluran penyu sisik (*Eretmochelys imbricata* L.) di Taman Nasional Bali Barat. *Lap. Pen. Perikanan Laut* 31: 49-54.

Nuitja, I. N. S., and J. D. Lazell. 1982. Marine turtle nesting in Indonesia. *Copeia* 3: 708-710.

Nuitja, I. N. S., and I. Uchida. 1983. Studies in the sea turtles. II. The nesting site characteristics of the hawksbill and green turtles. *Treubia* 29: 63-79.

Nuraini, S. 1986. Study on the growth of *Acropora aspera* (Dana) at Bandengan, Jepara, Central Java. *J. Pen. Perikanan Laut* 37: 101-105.

Nurdjana, M. L., B. Martosudarmo, and B.

Tiensongrusmee. 1977. Observations on diseases affecting cultured shrimp in Jepara, Indonesia. *Bull. Brackishwater Aquacul. Dev. Cent.* 3: 204-212.

Nurhadi, S. U., and S. Azul. 1987. The fall of the city of Banten: An environmental tragedy. *Voice of Nature* 53: 25-33.

Oberholser, H. C. 1917. The birds of Bawean Island, Java Sea. *Proc. US Nat. Mus.* 52: 183-198.

O'Brien, S. J., and J. F. Evermann. 1988. Interactive influence of infections of disease and genetic diversity in natural populations. *Trends Ecol. Evol.* 3: 254-259.

Ochse, J. J. 1977. *Vegetables of the Dutch East Indies (edible tubers, bulbs, rhizomes and spices included).* Canberra: Australian National Univ. Press.

O'Connor, C. T., and T. Sopa. 1981. *A checklist of the mosquitoes of Indonesia.* U.S. Naval Medical Research Unit, Jakarta.

Odum, E. P. 1969. The strategy of ecosystem development. *Science* 164: 262-270.

——. 1989. *Ecology and our endangered life support systems.* Sunderland: Sinauer.

Oetan, A. 1919. Naar den top van den Tjikorai. *Trop. Nat.* 8: 178-181.

Ohler, J. G. 1984. Coconut: tree of life. FAO, Rome.

Oka, I. N. 1979. Cultural control of the brown planthopper. *Brown planthopper: threat to rice production in Asia.* pp.357-369. Laguna: International Rice Research Institute.

——. 1980. Brown planthopper survey technique. *Rice improvement in China and other countries.* pp.287-291. Laguna: International Rice Research Institute.

——. 1983. The potential for the integration of plant resistance, agronomic, biological, physical/mechanical techniques, and pesticides for pest control in farming systems. *Chemistry and world food supplies: the new frontiers,* ed. L. W. Shemilt, 173-184. Oxford: Pergamon.

——. 1986. Plant resistance in rice pest management in farming systems in Indonesia. *IARD J.* 8: 20-25.

——. 1988. Role of cultural techniques in rice integrated pest management systems. *IARD J.* 10: 37-42.

——. 1990. Status and development of IPM policy in Indonesia: the case of the brown planthopper. Conference to introduce IPM concept in Thai rice cultivation to agricultural administrators and policy makers, Bangkok, 1990. Bangkok: FAO.

——. 1993. Pertanian berkelanjutan: pengalaman penerapan konsep PHT dan prospek pengembangannya dalam pendidikan tinggi pertanian. Paper at Lokakarya Nasional Pendidikan Tinggi Pertanian Masa Depan, Bogor, 8 December 1993.

Oka, I. N., and A. H. Bahagiawati. 1983. Perkembangan biotipe baru wereng coklat (*Nilaparvata lugens* Stal.) di propinsi Sumatera Utara, dan konsepsi penanggulangannya. Paper at Kongres Entomologi II, Jakarta, 24 January 1983.

—. 1988. Comprehensive program towards integrated control of *Leucaena* psyllid, a new insect pest of *Leucaena* trees in Indonesia. *IARD J.* 10: 23-30.

Oka, I. N., S. Partohardjono, and A. H. Bahagiawati. 1985. Agronomic techniques to conserve energy inputs for food production in developing countries. In *Alternative sources of energy for agriculture.* Taiwan: Food and Fertilizer Technology Centre.

Oka, I. N., and D. Pimentel. 1979. Ecological effects of 2,4-D herbicide: increased corn pest problems. *Contr. Centr. Res. Inst. Agric. Bogor* 49: 1-17.

Okajima, S. 1989. Five new species of the genus *Stephanothrips* (Thysanoptera: Phlaeothripidae) from Southeast Asia with a key to the East Asian species. *Jap. J. Entomol.* 57: 25-36.

Oldeman, L. R. 1977. An agro-climatic map of Java. *Contr. Centr. Res. Inst. Agric. Bogor* 17: 1-22.

Olembo, R. 1991. Importance of microorganisms and invertebrates as components of biodiversity. In *The biodiversity of microorganisms and invertebrates: Its role in sustainable agriculture,* ed. D. L. Hawksworth, 7-15. Wallingford: CAB International.

Olivier, J. 1928. De wilde zwijnen van Java. *Trop. Nat.* 17: 149-157.

Ong, J. E., W. K. Gong, and C. H. Wong. 1980. *Ecological survey of the Sungai Merbok estuarine mangrove ecosystem.* Penang: Univ. Sains Malaysia.

—. 1980. Contribution of aquatic productivity in a managed mangrove ecosystem in Malaysia. UNESCO Symposium on Mangrove Environment: Research and Management, Kuala Lumpur: Univ. Malaya Press.

—. 1985. Seven years of productivity studies in a Malaysian managed mangrove forest. Then what? *Coasts and tidal wetlands of the Australian Monsoon Region,* ed. K. N. Bardsley, J. D. S. Davie, and C. D. Woodroffe, 213-223. Darwin: ANU North Australia

Research Unit.

Ongkosongo, O. S. R. 1986. Some harmful stresses to the Seribu coral reefs, Indonesia. In Proceedings of the MAB-COMAR Regional Workshop on Coral Reef Ecosystem: Their Practices and Research/Training Needs, Bogor, 2 October 1986.

—. 1989. The evolution and distribution of mangrove forest in Java, Indonesia. *BIOTROP Spec. Publ.* 37: 285.

Ooi, J. B. 1986. The dimensions of the rural energy problem in Indonesia. *Appl. Geog.* 6: 123-147.

Oomen, J. M. V., J. de Wolf, and W. R. Jobin. 1990. *Health and irrigation: Incorporation of disease control measures in irrigation, a multifaceted task in design, construction and operation.* Wageningen: International Institute for Land Reclamation and Improvement.

Orr, D. W. 1991. The economics of conservation. *Conserv. Biol.* 5: 439-441.

Ostrofsky, M. L., and E. R. Zettler. 1986. Chemical defences in aquatic plants. *J. Ecol.* 74: 279-289.

Paine, R. T. 1995. A conversation on refining the concept of keystone species. *Cons. Biol.* 9: 962-964.

Pakpahan, A. M. 1992. Habitat and nest-site characteristics of water birds in Pulau Rambut Nature Reserve, Jakarta Bay, Indonesia. Ph.D. diss., Michigan State Univ., Ann Arbor.

Pakpahan, A. M., M. Thohari, and E. Sulistiani. 1991. Sarang, telur and perkembangan anakan kuntul kecil. Makalah pada Seminar Ilmiah dan Kongres Nasional Biologi X, Bogor, 24 September 1991.

Palmer, M. 1992. The Protestant tradition. *Christianity and ecology.* E. Breuilly, and M. Palmer, 89-96. London: Cassell.

Palowski, B., and H. Obro. 1976. Ground water study of a volcanic area near Bandung, Java, Indonesia. *J. Hydrol.* 28: 53-72.

Palte, J. G. L. 1980. *The development of Java's rural uplands in response to population growth.* Gadjah Mada Univ., Yogyakarta.

Pancho, J. V., and M. Soerjani. 1978. *Aquatic weeds of southeast Asia.* Quezon City: Nat'l Publ. Corp.

Pandu, P. 1986. *Penelitian tentang potensi jenis ikan hias pada habitat terumbu karang di Pulau Rambut, Kepulauan Seribu, Jakarta.* Fakultas Peternakan UNDIP, Semarang.

Pannekoek, A. J. 1940. Een merkwaadige waterval bij Soekaboemi. *Trop. Nat.* 29: 189-190.

—. 1941. De omgeving van het Telaga

Patengan (Zuidwest-Preanger). *Trop. Nat.* 30: 17-20.

—. 1949. Outline of the geomorphology of Java. *Tijds. K. Ned. Aard. Gen.* 66: 270-326.

Parrish, J. D. 1980. Effects of exploitation upon reef and lagoon communities. *Marine and coastal processes in the Pacific: Ecological aspects of coastal zone management.* Jakarta: UNESCO.

Parry, M. L., A. R. Magalhaes, and N. H. Ninh. 1992. *The potential socio economic effects of climate change.* Earthwatch GEMS UNEP, Nairobi.

Partomihardjo, T. 1986. Formasi vegetasi di Cagar Alam Pulau Dua, Banten, Jawa Barat. *Media Konservasi* 1: 10-16.

—. 1987. Dinamika musiman vegetasi savana Taman Nasional Baluran. In Kumpulan Abstrak Kongres Nasional Biologi VIII, Purwokerto, 8 October 1987.

Partomihardjo, T., and E. Mirmanto. 1986. Potensi dan permasalahan di Taman Nasional Baluran. *Duta Rimba* 13: 20-25.

Partomihardjo, T., and H. Tagawa. 1987. Vegetasi hutan daratan Pulau Panaitan, Taman Nasional Ujung Kulon, Jawa Barat. *Media Konservasi* 1: 49-57.

Pasang, H. 1989. *Kajian habitat owa abu-abu di Cagar Alam Gunung Halimun, propinsi Jawa Barat.* Fakultas Kehutanan IPB, Bogor.

Pattaratuna, A. 1976. *An ecological study on the green peafowl "Burung Merak" (Pavo muticus) in the game reserve Baluran, Banyuwangi, East Java, Indonesia.* BIOTROP, Bogor.

Payne, J., C. M. Francis, and K. Phillips. 1985. *A field guide to the mammals of Borneo.* Kuala Lumpur: WWF.

Pearce, D., and G. Atkinson. 1993. *Are national economies sustainable? Measuring sustainable development. Ecodecision* June: 64-66.

Pearce, D., and A. Markandya. 1990. Marginal opportunity cost as a planning concept in natural resource management. In *Environmental management and economic development,* ed. G. Scramm, and J. J. Warford, 39-55. Baltimore: John Hopkins University Press.

Pearce, D., and J. Warford. 1993. *World without end.* Oxford: Oxford Univ. Press.

Pearce, F. 1993a. How green is your golf? *New Scient.* 25 September: 30-35.

—. 1993b. Pesticide patent angers Indian Farmers. *New Scient.* 9 October 7.

Pearson, D. L., and F. Cassola. 1992. Worldwide species richness patterns of tiger beetles (Coleoptera: Cicindelidea): indicator taxon for biodiversity and conservation studies. *Conserv. Biol.* 6: 376-391.

Peluso, N. L. 1987. *Social forestry in Java: An evaluation.* Jakarta: Ford Foundation.

—. 1990. A history of state forest management. In *Keepers of the forest: Land management alternatives in Southeast Asia,* ed. M. Poffenberger, 27-55. West Harford, CT: Kumarian Press.

—. 1992a. *Rich forests, poor people: Resource control and resistance in Java.* Berkley: Univ. California Press.

—. 1992b. The ironwood problem: (mis)management and development of an extractive rain forest product. *Conserv. Biol.* 6: 210-219.

—. 1993. Coercing conservation? The politics of state resource control. *Global Env. Change* June: 199-217.

Peluso, N. L., M. Poffenberger, and F. Seymour. 1990. Reorienting forest management on Java. In *Keepers of the forest: Land management alternatives in Southeast Asia,* ed. M. Poffenberger, 220-236. West Hartford, CT: Kumarian Press.

PEMDA Bali. 1992. *Permanfaatan dan pengelolaan potensi sumber daya terumbu karang di Propinsi bali.* EMDI, Jakarta.

PEMDA Jateng. 1988. *Rencana induk Taman Nasional Laut Kepulauan Karimun Jawa.* Pemerintah Propinsi Dati I Jawa Tengah, Semarang.

Penny, D. H., and M. Ginting. 1984a. *Home gardens, farmers and poverty.* Yogyakarta: Gadjah Mada Univ. Press.

—. 1984b. *Pekarangan, petani dan kemiskinan: suatu studi tentang sifat dan hakekat masyarakat tani di Sriharjo pedesaan Jawa.* Yogyakarta: Gadjah Mada Univ. Press.

Perhutani. 1989. *Obyek rekreasi hutan: Wana wisata.* Jakarta: Perhutani Unit I Jawa Tengah.

—. 1990. Perum Perhutani 20 year plan: 1990-2009. Perum Perhutani, Jakarta.

—. 1992a. Silvofishery development in mangrove forests of West Java, Indonesia. *APAN News* 3: 8-9.

—. 1992b. *Sustained yield forest management in Indonesia (with a special emphasis to the island of Java).* Perum Perhutani, Jakarta.

Perhutani KPH Bantim. 1986. *Petunjuk singkat mendaki gunung Slamet.* Perum Perhutani KPH Banyumas Timur, Purwokerto.

Perrennou, C., P. Rose, and C. Pool. 1990. *Asian waterfowl 1990.* Slimbridge: IWRB.

Pesta, O. 1930. Zur kenntnis der Land- und Süsswasserkrabben von Sumatra and Java. *Archiv. Hydrobiol. Suppl.* 8: 92-108.

Petheram, R. J., and B. Lowry. 1983. *Mountains near Bogor: Notes on walks and climbs, peaks, plants and craters.* Bogor Expatriate Newcomers' Committee, Bogor.

Petheram, R. J., and A. Thahar. 1985. Land classification for livestock farming systems research and development in Java. *IARD J.* 7: 11-24.

Petr, T. 1987. Food fish as vector controls, and strategies for their use in agriculture. In *Effects of agricultural development on vector-borne diseases.* FAO, 87-92. Rome: FAO.

Peusens, M. C. I. 1989. *Recreation on the southern slopes of the volcano Merapi, Java, Indonesia.* Fakultas Kehutanan UGM, Yogyakarta.

Pezzey, J. 1989. *Economic analysis of sustainable growth and sustainable development.* World Bank, Washington D.C.

Pfeffer, P. 1965. Esquisse ecologique de la reserve de Baluran (Java Est). *Terre Vie* 112: 199-215.

PHPA. 1989. *Formulation of marine protected area management plan in the Segara Anakan, Cilacap.* PHPA, Bogor.

—. 1990. *Jenis kupu-kupu yang dilindungi Undang-Undang di Indonesia/Protected butterflies in Indonesia.* Jakarta: Dirjen PHPA.

—. 1991. *Laporan survai penilaian potensi sumberdaya alam laut di Nusa Barong dan sekitarnya, Jawa Timur.* PHPA, Bogor.

Phua, P. P., and R. T. Corlett. 1989. Seed dispersal by the lesser short-nosed fruit bat (*Cynopterus brachyotis*, Pteropodidae, Megachiroptera). *Malay. Nat. J.* 42: 251-256.

Picard, M. 1986. *Community participation in tourist activity on the island of Bali: Environment, ideologies and practices.* Paris: UNESCO.

—. 1990. Tourism: creating a new version of paradise. In *Bali: Island of the gods,* ed. Oey. E., 68-71. Singapore: Periplus.

Pienkowski, M. W. 1983. Surface activity of some intertidal invertebrates in relation to temperature and the foraging behaviour of their shorebird predators. *Mar. Ecol. Progr. Ser.* 11: 141-150.

Piepers, M. C., and P. C. T. Snellen. 1909. *The Rhopalocera of Java: Pieridae.* The Hague: Martinus Nijhoff.

—. 1910. *The Rhopalocera of Java: Hesperidae.* The Hague: Martinus Nijhoff.

—. 1913. *The Rhopalocera of Java: Damaidae, Satyridae, Ragadidae, Elymniadae.* The Hague: Martinus Nijhoff.

—. 1918. *The Rhopalocera of Java: Erycinidae, Lycaenidae.* The Hague: Martinus Nijhoff.

Pieters, D. 1959. Iets over de dieren van Pulau Panaitan en uit aangrenzende streken (Herten en hun belagers). *Penggemar Alam* 34: 25-35.

Pigeaud, T. G. T. 1962. *Java in the 14th century: a study in cultural history. The Nagara-Kertagama by Rakawi Prapanca of Majapahit, 1365 A.D. Vol. 4. Commentaries and recapitualtion.* The Hague: Nijhoff.

Piggott, A. 1979. *Common epiphytic ferns of Malaysia and Singapore.* Singapore: Heinnermann.

Piggott, A. G., and A. C. Piggott. 1988. *Ferns of Malaysia in colour.* Kuala Lumpur: Tropical Press.

Pingali, P. L., P. F. Moya, and L. E. Velasco. 1990. Prospects for rice yield improvement in the post green Revolution Philippines. *Phil. Rev. Econ. Bus.* 27: 85-106.

Plage, M. 1985. Return of Java's wildlife: In the shadow of Krakatau. *Nat. Geog.* 167 (June): 750.

Pluygers, L. A. 1952. *Natuurbescherming en wildbeheer: Speciaal met betrekking tot Indonesië.* Groningen: Wolters.

Po Milan, P., J. Margraf, and F. Göltenboth. in prep. *The ecology of the Philippines.*

Poffenberger, M. 1983. Changing dryland agriculture in eastern Bali. *Human Ecol.* 11: 123-144.

—. 1990. The evolution of forest management systems in Southeast Asia. In *Keepers of the forest: Land management alternatives in Southeast Asia,* ed. M. Poffenberger, 7-26. West Harford, CT: Kumarian Press.

Polak, B. 1949. De Rawa Lakbok, een entroof laagveen op Java. *Comm. Gen. Exp. Station, Buitenz.* 85: 1-60.

—. 1951. Construction and origin of floating islands in the Rawa Pening (Ambarawa). *Contrib. Gen. Agri. Res. Station, Bogor* 121: 1-11.

Polhaupessy, A. A. 1980. The palynological study of ancient Lake Bandung: A preliminary report. *Bull. Geol. Res. Dev. Cent.* 3: 19-23.

Polunin, I. 1987. *Plants and flowers of Singapore.* Singapore: Times Editions.

Polunin, N. V. C. 1983. The marine resources of Indonesia. *Oceanog. Mar. Biol. Annual Rev.* 21: 455-531.

Polunin, N. V. C., M. K. Halim, and K. Kvälvagnaes. 1983. Bali Barat: An Indonesian marine protected area and its resources. *Biol. Conserv.* 25: 171-191.

Polunin, N. V. C., A. H. Robinson, and M. H. Halim. 1980. Proposed Bali Barat National Park marine management plan. FAO, Bogor.

Polunin, N. V. C., and Sumertha N. 1984. Sea turtle populations in Thailand and Indonesa. In *Biology and conservation of sea turtles*, ed. K. Bjorndal, 353-362. Washington, D.C.: Smithsonian.

Ponder, H. W. 1990. *Javanese panorama: more impressions of the 1930's*. Singapore: Oxford Univ. Press.

Pons, T. L., J. H. H. Eussen, and I. H. Utomo. 1987. Ecology of weeds of rice. In *Weeds of rice in Indonesia*, ed. M. Soerjani, A. J. G. Kostermans, and G. Tjitrosoepomo, 15-23. Jakarta: Balai Pustaka.

Poore, M. E. D., and C. Fries. 1985. *The ecological effects of eucalyptus*. FAO, Rome.

Pope, G. G. 1983. Evidence of the age of the Asian hominids. *Proc. Nat. Acad. Sci* 80: 4988-4992.

——. 1985. Taxonomy, dating and paleoenvironment: the paleoecology of the early Far Eastern hominids. *Mod. Quat. Res. SE Asia* 9: 65-80.

——. 1988. Recent advances in Far Eastern palaeoanthropology. *Paleoanthro.* 17: 43-77.

——. October 1989. Bamboo and human evolution. *Nat. Hist.*, 49-57.

Porter, J. W., K. G. Porter, and Z. Batac-Catalan. 1977. Quantitative sampling of Indo-Pacific demersal reef plankton. Proc. 3rd Int. Coral Reef Symp., 105-112.

Posthumus, O. 1929. On palaeobotanical investigations in the Dutch East Indies and adjacent regions. *Bull. Jard. Bot. Buitenz.* 10: 374-384.

Postma, P. A. Ke Gunung Panandjakan (2770 m), Tengger (Djawa Timur). *Penggemar Alam* 39: 3-9.

Poulson, T. I., and W. B. White. 1969. The cave environment. *Science* 165: 971-981.

Power, M. E. 1983. Grazing responses of tropical freshwater fishes to different scales of variation in their food. *Env. Biol. Fishes* 9: 103-115.

PPA. 1972. *Suaka Margasatwa Tjikepuh*. PPA, Bogor.

Prahoro, P., and M. M. Wahyono. 1987. Kerusakan ekosistem perairan karang dan dampaknya terhadap sumberdaya ikan hias. In Kumpulan Abstrak Kongres Nasional Biologi VIII, Purwokerto, 8 October 1987.

Pramono, D., D. Sasangko, and B. Wirioatmodjo. 1989. Assesment of *Schizaphus rotundiventris* Signoret (Aphidae: Homoptera) as a promising biological control agent of the purple nutsedge. *BIOTROP Spec. Publ.* 36: 227-233.

Pranowo, H. A. 1985. *Manusia dan hutan: proses perubahan ekologi di lereng Gunung Merapi*. Yogyakarta: Gadjah Mada Univ. Press.

Praptohardiyo, K., and H. Muluk. 1982. Early development of the fish population and fisheries of Wonogiri reservoir, Central Java. Paper presented at the Seminar on Production and Exploitation of Open Waters (with special emphasis on reservoirs), Bogor, 15 June 1982.

Praseno, D. P., and Sukarno. 1977. Observation on beach erosion and coral destruction by remote sensing techniques. *Mar. Res. Indon.* 17: 59-68.

Pratiknyo, H. 1991. The diet of the Javan rhino. *Voice of Nature* 93: 12-13.

Pratiwi. 1987. Analisis komposisi jenis pohon di Taman Nasional Gunung Gede Pangrango Jawa Barat. *Bull. Pen. Hutan* 488: 28-34.

——. 1989. Vegetation analysis of the lava area from eruption product of Galunggung mountain, West Java, Indonesia. *Bull. Pen. Hutan* 512: 23-32.

Pratiwi, C. Anwar, and Y. Sumarna. 1986. Perkembangan regenerasi alam dan buatan hutan mangrove Cilacap. *Bull. Pen. Hutan* 482: 1-10.

Prawira, R. S. A. et al. 1972. Daftar nama pohon-pohon (List of tree species): Bali dan Lombok. *Laporan No. 145*. Lembaga Penelitian Hutan, Bogor.

Prawira, R. S. A., and Oetja. 1976. Daftar nama pohon-pohonan (List of tree species) Jawa-Madura. I: Jawa Barat. *Laporan No. 219*. Lembaga Penelitian Hutan, Bogor.

——. 1977a. *Daftar nama pohon-pohonan (List of tree species) Jawa-Madura. II: Jawa Tengah*. Lembaga Penelitian Hutan, Bogor.

——. 1977b. *Daftar nama pohon-pohonan (List of tree species) Jawa-Madura. III: Jawa Timur*. Lembaga Penelitian Hutan, Bogor.

Prawiradilaga, D. M. 1992. Feeding and dietary habits of the bar-winged prinia in ricefields. *Kukila* 6: 35-37.

Prawirakusuma, Y., and H. Alrasyid. 1981. Sistim pencegahan pencemaran industri dengan mengambil kasus pabrik pupuk urea P.T. Pupuk Kujang. Paper presented at Penataran dan Diskusi Masalah Pencemaran Industri Kimia, Bandung-Cikampek, Bandung-Cikampek, September 1910.

Prawiroatmodjo, S. W. 1985. Rusa Bawean, satwa khas Jawa Timur. *Suara Alam* 42: 41-42.

——. 1992. *Keindahan sebatang pohon*. Surabaya: Edumedia.

Prawirokusumo, S. W. 1987. Kalimas, tempat sampah terbesar. *Suara Alam* 47: 54-58.

Prayitno, S. A. Menjaga kelestarian perikanan di perairan Rawa Pening beserta budidaya perikanan. In Proceedings of the 2nd Seminar in Aquatic Biology and Aquatic Management of the Rawa Pening Lake, Salatiga, Central Java, 180-189. Salatiga, 29 October 1979. Salatiga: Satya Wacana Christian Univ.

Premo, D. 1985. The reproductive ecology of a ranid frog community in pond habitats of West Java, Indonesia. Ph.D. diss., Dept. of Zoology, Michigan State Univ., Michigan.

Premo, D. B., and A. H. S. Atmowidjojo. 1987. Dietary patterns of the crab-eating frog *Rana cancrivora* in West Java, Indonesia. *Herpetologica* 43: 1-6.

Preston, D. A. 1989. Too busy to farm: Under-utilisation of farm land in Central Java. *J. Dev. Studies* 26: 43-57.

Preston-Mafham, K. 1990. *Grasshoppers and mantids of the world.* London: Blandford.

Priatna, D. R., E. A. M. Zuhud, and H. Alikodra. 1989. Kajian ekologis *Rafflesia patma* Blume di Cagar Alam Leuweung Sancang Jawa Barat. *Media Konservasi* 2: 1-7.

Prihantono, H. 1981. *Laporan survey ke Guha Semar.* SPECAVINA, Malang.

Prime, R. 1992. *Hinduism and ecology.* London: Cassel.

Prins, H. H. T., and J. Wind. 1993. Research for nature conservation in south-east Asia. *Biol. Conserv.* 63: 43-46.

Priyantono, T., I. Effendi, and K. Budiono. 1980. The earthquakes of 2 November, 1979 and 16 April, 1980 in the Garut and Tasikmalaya areas, West Java. *Bull. Geol. Res. Dev. Cent.* 3: 13-17.

Proctor, J. et al. 1983a. Ecological studies in four contrasting lowland rain forests in Gunung Mulu National Park, Sarawak. II. Litterfall, litter standing crop and preliminary observations on herbivory. *J. Ecol.* 71: 261-283.

——. 1983b. Ecological studies in four contrasting lowland rain forests in Gunung Mulu National Park, Sarawak. I. Forest environment, structure and floristics. *J. Ecol.* 71: 237-260.

Program PHT. 1991. *Petunjuk percobaan lapangan: Musim PHT padi.* Program Nasional Pelatihan dan Pengembangan Pengendalian Hama Terpadu, BAPPENAS, Jakarta.

PSL UGM. 1986. *Studi pengendalian kerusakan lingkungan wilayah pesisir Bali bagian selatan.* PSL UGM, Yogyakarta.

PSL UNDIP, and Kantor Menteri Negara KLH. 1982. *Penelitian potensi dan pemanfaatan sumberdaya karang di Pulau Panjang*

dan Karimunjawa, Kabupaten Jepara, Jawa Tengah. PSL Undip, Semarang.

PSL UNIBRAW. 1982. Pergaruh kerusakan lingkungan hidup dan perkembangan dam terhadap pembangunan Selorejo. *Lingkungan dan Pembangunan* 2: 161-171.

——. 1983a. *Penelitian vegetasi, hidrologi dan sedimentasi proyek Bendungan Karangkates: Final report.* PSL UNIBRAW, Malang.

——. 1983b. *Penelitian vegetasi, hidrologi dan sedimentasi proyek bendungan Selorejo: Final report.* PSL UNIBRAW, Malang.

——. 1984a. *Penelitian vegetasi, hidrologi dan sedimentasi pada proyek pengembangan Kali Konto hulu: Laporan akhir.* PSL UNIBRAW, Malang.

——. 1984b. *Penelitian vegetasi, hidrologi dan sedimentasi pada proyek pengembangan Karangkates hulu: Draft final report.* PSL UNIBRAW, Malang.

——. 1986. *Penelitian limpasan air permukaan pengembangan Karangkates hulu: Final report.* PSL UNIBRAW, Malang.

Pudjoarianto, A. 1981. The invasion of newly formed land in the Segara Anakan area by mangroves species. Paper presented at the Seminar on Coastal Resources of Segara Anakan, Central Java, Yogyakarta, 18 August 1981.

Pujiati. 1987. *Studi populasi Jalak Bali* (Leucopsar rothschildi) *di Taman Nasional Bali Barat.* Fak. Hutan, IPB, Bogor.

Purba, M. 1976. *Suatu tinjauan tentang prospek dan masalah pembinaan margasatwa di Suaka Margasatwa Cikamurang.* Akademi Ilmu Kehutanan, Bandung.

Purbawiyatna, A. 1987. *Mempelajari kemungkingan distribusi macan tutul (Panthera pardus) di Resort Cibodas, Situgunung dan Bodogol, Taman Nasional Gunung Gede Pangrango.* Fakultas Kehutanan IPB, Bogor.

Purnomo, H. B. 1981. *Laporan Gua Seplawan.* HIKESPI, Bogor.

Purnomohadi, S. 1985. *Sistem interaksi sosial ekonomi dan pengelolaan sumberdaya alam, oleh masyarakat Badui, di desa Kanekes, Banten Selatan.* Fakultas Pasca Sarjana IPB, Bogor.

Purseglove, J. W. 1968. *Tropical crops: Dicotyledons.* London: Longman.

——. 1972. *Tropical crops: Monocotyledons.* London: Longman.

Purwaningsih, E. 1979. Urbanisasi dan hak atas tanah. *Urbanisasi: Masalah kota Jakarta,* ed. S. Muljanto. PSL UI, Jakarta.

Purwanto, A. 1984. *Studi pembentukan model daerah penyangga (bufferzone).* Fakultas Kehutanan IPB, Bogor.

——. 1985. *Studi beberapa jenis musang dan per-*

anannya dalam ekosistem hutan di resort hutan KPA Cibodas, Taman Nasional Gunung Gede-Pangrango, Jawa Barat. Fakultas Kehutanan IPB, Bogor.

Purwoko, A. 1991. *Status burung blekok (Ardeola speciosa) dan kuntul (Egretta garzetta) di Kodya Semarang.* Asian Wetland Bureau, Bogor.

Puthz, V. 1991. On Indo-Australian Steninae II (Insecta: Coleoptera: Staphylinidae). *Entomol. Abh. (Dres.)* 54: 1-46.

Putra, I. M. W. *Perilaku makan dan pengembaraan surili* Presbytis comata *di Cagar Alam Situ Patengan Jawa Barat.* Jurusan Biologi UNPAD, Bandung.

Putra, I. N. N. 1987. The realtive abundance of benthic macroinvertebrates in relation to eutrophication processes in Lake Rawa Pening, Central Java,Indonesia. *BIOTROP Spec. Publ.* 27: 25-42.

Putra, K. S. 1992. *The impact of coral mining in coral reef condition in the east and south coast of Bali, Indonesia.* Centre for Tropical Coastal Management, Univ. of Newcastle upon Tyne, Newcastle upon Tyne.

Putz, F. E., and N. M. Holbrook. 1988. Further observations on the dissolution of mutualism between *Cecropia* and its ants: the Malaysian case. *Oikos* 53: 121-125.

Quammen, M. L. 1982. Influence of subtle substrate differences on feeding by shorebirds on intertidal mudflats. *Mar. Biol.* 71: 339-343.

Quattro, J. M., and R. C. Vrijenhoek. 1989. Fitness differences among remnant populations of the endangered Sonoran topminnow. *Science* 245: 246-278.

Quinif, Y., and C. Dupuis. 1984. Morphologie souterraine du Gunung Sewu, Central Java, Indonésie. *Spelunca* 14: 18-24.

Quirk, K., A. I. A. Kristyanto, and K. D. Carlander. 1979. Observations on birds of Rawa Pening and vicinity. Salatiga, Satya Wacana Christian Univ.

Rabinowitz, A. R., and S. R. Walker. 1991. The carnivore community in a dry tropical forest mosaic in Huai Kha Khaeng Wildlife Sanctuary, Thailand. *J. Trop. Ecol.* 7: 37-47.

Rachman, F. A. 1990. *Laporan perjalanan penelusuran gua-gua pada kawasan karst Bali Selatan - Nusa Penida dan Nusa Lembongan.* HIKESPI, Bogor.

Rachmatika, I. 1987. Ekologi ikan kehkel *Glyptothorax platypogon* (Blgr) di Sungai Cisadane. *Zool. Indo.* 7: 1-6.

—. 1990. Tambra traditionally conserved. *Voice of Nature* 69: 56-57.

—. 1991. Tambra traditionally conserved.

Voice of Nature 89: 56-57.

Rackham, O. 1986. *The history of the countryside.* London: Dent.

—. 1990. *Trees and woodland in the British landscape.* London: Dent.

Raffles, T. S. 1817. *The history of Java.* Kuala Lumpur [Reprinted in 1988. Preface by J. Bastin]: Oxford Univ. Press.

Rahardjo, M. F. 1977. *Kebiasaan makanan, pemijahan hubungan panjang-berat dan faktor kondisi ikan hampal Hampala macrolepidota (C&V) di Waduk Jatiluhur, Jawa Barat.* Fakultas Perikanan IPB, Bogor.

—. 1987. Komunitas ikan di Waduk Bening, Jawa Timur: Suatu tinjauan ekobiologis. In Kumpulan Abstrak Kongres Nasional Biologi VIII, Purwokerto, 8 October 1987.

Raharjaningtrah, W. 1988. *Studi ekologi bluwok* (Mycteria cinerea) *di Kepetakan, Indramayu.* Jurusan Biologi UNPAD, Bandung.

Ramalingam, S. 1975. *A brief mosquito survey of Java.* Report WHO/VBC/74.504, Geneva: WHO.

Ramlan, A., and J. Iskandar. 1987. Pembangunan hutan kota ditinjau dari aspek lingkungan. Paper presented at Seminar Hutan Kota, Jakarta.

Ramseyer, U. 1977. *The art and culture of Bali.* Oxford: Oxford Univ. Press.

Rangkuti, N. 1987. Baduy, mandala yang makin terkikis. *Suara Alam* 51: 28-33.

Ranoemihardjo, B. S. 1986. Improved methods on finfish cultivation in brackish-water pond in Indonesia. *Bull. Brackishwater Aquacul. Dev. Cent.* 8: 1-11.

Rappard, F. W. 1948. De perperveider (*Ardea purpurea manillensis* Meyer) als bosvogel. *Tectona* 38: 306.

Rathbun, M. J. 1905. Les crabes d'eau douce. *Nouv. Arch. Mus. Hist. nat. Paris* 7: 159-323.

Ratnaningsih. 1987. Beberapa aspek permasalahan air baku untuk kebutuhan air minum di DKI Jakarta. *Keberadaan manusia dalam tata ruangnya: Kajian multidisiplin mengenai campur tangan manusia terhadap lingkungan hidup di perkotaan,* ed. M. A. Sumhudi. Lembaga Penelitian USAKTI, Jakarta.

Raunkiaer, C. 1934. *The life forms of plants and statistical plant geography.* Oxford: Oxford Univ. Press.

Raunkier, C. 1934. *The life forms of plants and statistical plant geography.* Oxford: Oxford Univ. Press.

Ravenholt, A. 1975. Bali: Microcosm for third world agriculture. *Common Ground* 1: 13-25.

Reader's Digest. 1971. *Concise encyclopedia of*

garden plants and flowers. London: Reader's Digest.

Real, L. A., and J. H. Brown. 1991. *Foundations of ecology.* Chicago: Univ. Chicago Press.

Recher, H. F., and W. Rohan-Jones. 1978. Wildlife conservation: a case for managing forests as ecosystems. Paper given at Proceeedings of the 8th World Forestry Congress, Jakarta,

Redtenbacher, J. 1908. *Die Insektenfamilie der Phasmiden, Vol. 3.* Leipzig:

Rees, W. E. 1989. Sustainable development: economic myths and ecological realities. *Trumpeter* 5: 133-138.

—-. 1993. Natural capital in relation to regional/global carrying capacity. In *Ecological economics: emergence of a new development paradigm.* pp.42-60. Ottawa: Univ. of Ottawa.

Reichart, C. V. 1985. A new species of *Anisops* from Java (Hemiptera: Notonectidae). *Int. J. Entomol.* 27: 235-238.

Reid, J. A. 1950. Some new records of anopheline mosquitoes from the Malay Peninsula with remarks on geographical distribution. *Bull. Raffles Mus.* 21: 48-58.

Reid, W. V. 1993. *Biodiversity prospecting: using genetic resources for sustainable development.* Washington D.C.: World Resources Institute.

Reid, W. V., and K. R. Miller. 1989. *Keeping options alive: the scientific basis for conserving biodiversity.* Washington D.C.: World Resources Institute.

Reksodihardjo, G., Y. Irmawati, and D. S. Moro. 1986. Pola sebaran moluska suku Potamididae di hutan mangrove Legon Lentah, Pulau Panaitan. Paper presented at Seminar III Ekosistem Mangrove, Denpasar, May 1908.

Reksowardojo, H. 1961. Penyu di Pantai Pangumbahan. *Penggemar Alam* 40: 16-24.

Rensch, B. 1930. Beitrage zur Kenntnis der Vogelwelt Bali. *Mitt. Zool. Mus. Berlin* 16: 530-542.

—-. 1931. Die molluskenfauna der Kleinen Sunda Inseln: Bali, Lombok, Sumbawa, Flores und Sumba. *Zool. Jahrb.* 61: 361-396.

—-. 1932. Die molluskenfauna der Kleinen Sunda Inseln: Bali, Lombok, Sumbawa, Flores und Sumba. *Zool. Jahrb.* 63: 1-130.

—-. 1933. Landmollussken der Deutschen Limnologischen Sunda-Expedition. *Archiv. Hydrobiol. (Suppl.)* 11: 739-758.

—-. 1934. Die molluskenfauna der Kleinen Sunda Inseln: Bali, Lombok, Sumbawa, Flores und Sumba. *Zool. Jahrb.* 65: 389-422.

—-. 1938. Neue Landschnechen aus der Insel Penida. *Zool. Anz.* 123: 302-306.

Repetto, R. 1986. Soil loss and population pressure on Java. *Ambio* 15: 14-18.

—-. 1989. Balance sheet erosion: How to account for the loss of natural resources. *Int. Env. Affairs* 1: 103-135.

—-. 1989. *Wasting assets (National resource accounting in the economy): II, the Indonesian resource accounts.* Washington D.C.: World Resources Institute.

RePPProT. 1989. *Regional physical planning programme for Indonesia, review of phase 1 results: Java and Bali.* Jakarta: Directorate General of Settlement Preparation, Ministry of Transmigration and London: Natural Resources Institute, Overseas Development Administration.

—-. 1990. *The land resources of Indonesia: A national overview.* Jakarta: Directorate General of Settlement Preparation, Ministry of Transmigration and London: Natural Resources Institute, Overseas Development Administration.

Republik Indonesia. 1993. *Garis besar haluan negara.* Jakarta: Sekretatiat Negara.

Research Institute UNPAD. 1986. *Environmental impact analysis of the Saguling Dam.* Padjadjaran Univ., Bandung.

Resosoedarmo, S., K. Kartawinata, and A. Soegiarto. 1989. *Pengantar ekologi.* Bandung: Remadja Karya.

Rhyther, J. H., and W. M. Dunstan. 1971. Nitrogen and phosphorus and eutrophication in the coastal marine environment. *Science* 171: 1008-1013.

Ricchiardi, E. 1992. On the Valginae, genus *Heterovalgus* with description of three new species (Coleoptera, Cetonidae). *Boll. Soc. Entomol. Ital.* 124: 115-120.

Ricklefs, R. 1990. *Ecology, 3th Ed.* New York: Chiron.

Rierink, A., and F. J. Appelman. 1940. Adjags. *Tectona* 33: 226-234.

Rifai, M. A. 1989. Mengintip taman zaman Syailendra dan Majapahit. *Asri* 69: 58-61.

Rifai, M. A., and E. A. Widjaja. 1979. An ethnobotanical observation on alang-alang in Bali. Sixth Asian Weed Science Society Conference.

Rijksen, H. D. 1984. Conservation: not by skill alone:the importance of a workable concept in the conservation of nature. *Environmentalist* 4: 52-59.

—-. 1989. *Project proposal for the conservation of lowland forest for integrated land use development of the Lemas project area.*

Rijnberg, T. F. R. 1992. *'s Lands Plantentuin*

Buitenzorg. Enschede: Rijnberg.

Rijsdijk. A., and L. A. Bruijnzeel. 1991. *Erosion, sediment yield and land use patterns in the upper Konto watershed, East Java, Indonesia. Part III Results of the 1989-1990 measuring campaign.* Directorat Jenderal Reboisasi dan Rehabilitasi Lahan, Jakarta.

Rindjin, I. G., and K. Sarna. 1990. The national park of western Bali as a tourist object and an object of study. In *International Seminar on Human Ecology, Tourism, and Sustainable Development.* 203-207. Denpasar: Human Ecology Study Group.

RINM. 1985. *Evaluation of forest land, Kali Konto upper watershed, East Java. Vol. I: Summary, conclusions and recommendation.* Research Institute for Nature Management, Leersum, Netherlands.

Rinpoche, L. N. 1986. The Buddhist Declaration on Nature. In *The Assisi Declaration: Messages on Man and Nature from Buddhism, Christianity, Hinduism, Islam and Judaism.* 5-7. Gland: World Wide Fund for Nature.

Risch, S. J., D. Andow, and M. A. Altieri. 1983. Agroecosystem diversity and pest control: data, tentative conclusions and new research directions. *Env. Entomol.* 12: 625-629.

Riswan, S. 1975. Vegetasi hutan di Cagar Alam Dungus Iwul, Jasinga, Bogor. In Prosiding II Seminar Biologi IV dan Kongres Biologi II, 127-132. Yogyakarta, 10 July 1975.

——. 1976. *Perjalanan ke Gunung Papandayan, 30 April - 2 Mei 1976.* Herbarium Bogoriense, Bogor.

Ritchie, L. D. 1991. Marine biology for Indonesian National Parks: Final report. *World Bank National Park Project.* World Bank, Jakarta.

RMI/PRC. 1986. *Land resource studies.* Jakarta: Directorate General of Representation and Land Rehabilitation, Department of Forestry.

Robbins, R. G., and J. Wyatt-Smith. 1964. Dry land forest formations and forest types in the Malay Peninsula. *Malay. For.* 27: 188-217.

Roberts, C. M., and N. V. C. Polunin. 1991. Are marine reserves effective in management of reef fisheries? *Rev. Fish Biol. Fisher.* 1: 65-91.

Roberts, S. J., S. J. Eden-Green, P. Jones, and D. J. Ambler. 1990. *Pseudomonas syzygii,* sp. nov., the cause of Sumatra disease of cloves. *Syst. Appl. Microbiol.* 13: 34-43.

Roberts, T. R. 1993. The freshwater fishes of Java, as observed by Kuhl and van Hasselt in 1820-23. *Zool. Verhard.* 285: 1-94.

Robertson, A. I., R. Giddin, and T. J. Smith. 1990. Seed predation by insects in tropical mangrove forests: Extent and effects on seed viability and the growth of seedlings. *Oecologia* 83: 213-219.

Robinson, A. H., R. C. Beudels, Anwar, and Kurnianto. *Nature reserves in Bali.* WWF, Bogor.

Robinson, A. H., D. Supriadi, and Anwar. 1982. Baluran National Park management plan: Revisions and recommendations. *Field report No. 45.* FAO, Bogor.

Robinson, H. C., and O. Thomas. 1917. A new mink-like *Mustela* from Java. *Ann. Mag. Nat. Hist.* 20: 261-262.

Robinson, J. G. 1990. Review of *Economics and biological diversity* by J. A. McNeely. *Ecology* 71: 410.

Roche, F. C. 1983. Cassava production systems on Java and Madura. Ph.D. diss., Stanford Univ., Palo Alto.

——. 1988. Java's critical uplands: Is sustainable development possible? *Food Res. Inst. Studies* 21: 1-43.

Roepke, W. 1935. *Rhopalocera Javanica: Geillistreed overzicht der dagulinder van Java (1935-1942).* Wageningen: H. Veeman en Zonen.

Roewer, C. F. 1928. Ein Javanischer Charontine. *Treubia* 10: 15-21.

Roger, P. A., K. L. Heong, and P. S. Teng. 1991. Biodiversity and sustainability of wetland rice production: role and potential of microoranisms and invertebrates. In *The biodiversity of microorganisms and invertebrates: Its role in sustainable agriculture,* ed. D. L. Hawksworth, 117-136. Wallingford: CAB International.

Rokhman, P. 1990. Anjing kintamani, ras asli Indonesia kini dikembangkan di Jakarta. *Trubus* 224: 139.

Rollet, B., and S. Sukardjo. 1976. *Preliminary analysis of a survey in the lower montane forest of Cibodas.* UNESCO, Jakarta.

Rollinson, D. H. L. 1984. Bali cattle. *Evolution of domesticated animals.* ed I. L. Mason, 28-34. London: Longman.

Romimohtarto, K., and M. K. Moosa. 1977. Fauna Crustacea dari P. Air, Pulau-pulau Seribu. *Teluk Jakarta: Sumber daya, sifat-sifat oseanologi, serta permasalahannya.* LON-LIPI, Jakarta.

Rondelli, M. R. et al. 1984. Stable isotope ratio as a tracer of mangrove carbon in Malaysian ecosystems. *Oecologia* 61: 326-333.

Rondo, M. 1982. Hewan bentos sebagai indikator ekologi di sungai Cikapundung, Bandung. ITB, Bandung.

Rondon, M. B., and J. P. Sumangil. 1991. Integrated pest management for golden snail. In *Environmental impact of the golden snail* (Pomacea *sp.) on rice farming systems in the Philippines.* B. O. Acosta, and R. S. V. Pullin. Manila: ICLARM.

Ronny, H. S. *Pembudidayaan ikan lele lokal dan dumbo.* Jakarta: Bhratara.

Room, P. M. 1990. Ecology of a simple plant-herbivore system: biological control of *Salvinia. Trends Ecol. Evol.* 5: 74-80.

Røpke, I. 1994. Trade development and sustainability - a critical assessment of the "free- trade dogma". *Ecol. Econ.* 9: 13-22.

Rosanto, R., and R. Priatna. 1982. The effect of rural inhabitants on G. Tilu Nature Reserve at West Java, Indonesia. Paper presented at the World National Parks Congress, Bali, 1982. Gland: IUCN.

Rosenweig, C., and D. Hillel. 1993. Agriculture in a greenhouse world. *Nat. Geog. Res. Ex.* 9: 208-221.

Ross, M. 1985. *A review of issues affecting the sustainable development of Indonesia's forest land, Vol. 2.* London: IEED.

Roth, L. M. 1985. The genus *Episymploce.* I: Species chiefly from Java, Sumatra and Borneo Kalimantan, Sabah Sarawak, Indonesia, (Dycioptera: Blattaria: Blattelidae). *Entomol. Soc. Can.* 16: 355-374.

—. 1991. The cockroach genera *Sigmella* Hebard and *Scalida* Hebard (Dictyoptera: Blattaria: Blattellidae). *Entomol. Scand.* 22: 1-30.

Rowe, W. E. 1983. A collection of Holothurians in the Leiden Museum from the East Indies and New Guinea, with the description of a new species of. *Zool. Mededel.* 57: 16.

Ruchiyat, Y. 1988. Development of a geothermal power plant and nature conservation in Kamojang, West Java. *Sustainable rural development in Asia*, ed. T. Charoenwatana, A. T. Rambo, A. Jintrawet, and P. Sorn-srivichai. Khon Kaen Univ., Bangkok.

Ruhiyat, Y. 1983. Socio-ecological study of *Presbytis aygula* in West Java. *Primates* 24: 344-359.

Rusli, I. 1991a. Kampung Naga: A village of taboos. *Voice of Nature* 91: 7-13.

—. 1991b. Farmers and the pesticide dillema. *Voice of Nature* 88: 18-23.

Rusmendro, H. 1980. *Perkembangan Cagar Alam Pulau Dua.* UNAS, Jakarta.

Ruttner, F. 1931. Hydrographische und hydrochemische Beobachtungen auf Java, Sumatra und Bali (German Sunda Limnological Expedition). *Archiv. Hydrobiol.* 8

(Supplement): 197-454.

—. 1932. Meerenonderzoek in Nederlandsch-Indie (Eenige resultaten der Duitsche Limnologische Soenda-Expeditie). *Trop. Nat.* 21: 151-157, 178-184.

—. 1938. Stabilitat und umschichtung in tropischen und temperierten seen. *Archiv. Hydrobiol.* 15 (Supplement): 178-186.

Ryadisoetrisno, B. 1992. *Konservasi dan masyarakat: diskusi dan rumusan Workshop keanekaragaman hayati Taman Nasional Gunung Halimun, Jawa Barat.* Jakarta: Biological Science Club UNAS.

Sabar, F., S. Wirjoatmodjo, D. I. Hartoto, and I. Rachmatika. 1984. *Laporan perjalanan ke hulu sungai Ciliwung, Kabupaten Bogor, Jawa Barat (4 - 11 September 1984).* Museum Zoologicum Bogoriense, Bogor.

Sachlan, M. 1978. Parasites, pests and diseases of fish fry. Paper presented at the Workshop on Diseases of Fish Cultured for Food in Southeast Asia, Cisarua, Bogor, 28 November 1978.

Sadjudin, H. R. 1992. Status and distribution of the Javan rhino in Ujung kulon National Park, West Java. *Trop. Biodiv.* 1: 1-10.

Saenger, P., E. J. Hegerl, and J. Davie. 1981. First report of the Global status of mangrove ecosystems. IUCN Commission of Ecology, Gland.

Saim, A. 1986. *Laporan perjalanan ke Suaka Margasatwa Banyuwangi Selatan, propinsi Jawa Timur (10 - 23 Maret 1986).* Museum Zoologicum Bogoriense, Bogor.

Sajudin, H. R. 1984. Monitoring populasi badak Jawa (*Rhinoceros sondaicus* Desm. 1822) di Semenanjung Ujung Kulon. *Laporan akhir penelitian periode 1982-1983.* Fakultas Biologi UNAS, Jakarta.

—. 1987. The Javan Rhino (*Rhinoceros sondaicus* Desm.) census in Ujung Kulon National Park. *Rimba Indonesia* 21: 16-25.

Sajudin, H. R., B. Djaja, and Y. K. Lo. 1981. *Sensus Badak Jawa (Rhinoceros sondaicus Desmarest 1822) di Semenanjung Ujung Kulon, Maret 1981.* UNAS/IUCN/WWF, Jakarta.

Saleh, C., and M. Munir. 1990. *Pembinaan masyarakat desa Ciusul di kawasan penyangga Cagar Alam Gunung Halimun Jawa Barat.* Biological Sciences Club, UNAS, Jakarta.

Salim, E. 1992. *Keynote address to the Report of the Indonesian country study on biological diveristy.* Kantor Menteri KLH, Jakarta.

Salm, R. V. 1987. Pulau Seribu: Paradise lost or paradise saved? *Conserv. Indo.* 5: 7-11.

Salm, R. V., and Halim. M. 1984. *Marine conservation data atlas.* WWF, Bogor.

Salvador, A. 1975. Un nuevo cecilido procedente de Java (Amphibia: Gymnophiona). *Bonn. Zool. Beitr.* 26: 366-369.

Sandy, I. M. 1964. *Pemakaian tanah di Bali sebelum letusan Gunung Agung Maret 1963.* Djakarta: Dinas Geografi Direktorat Topografi Angkatan Darat.

Sangat, H. M. 1982. Some ethnobotanical aspects of batik. *J. d'Agric. Trad. Bot. Appl.* 29: 41-56.

Sangat-Roemantyo, H., and S. Riswan. 1990. Javanese medicinal plants: the distribution and use. International Congress on Traditional Medicines and Medicinal Plants, Denpasar, 15 October 1990.

Sanguilla, W. M. 1976. The fauna of Pananjung Pangandaran Nature Reserve, with special reference on avifauna. In Report of the Training Course in the Management of Nature Reserve. Vol. IIIa, Bogor, 26 February 1910. Bogor: BIOTROP.

Santiapillai, C., and W. S. Ramono. 1992. Status of the leopard (*Panthera pardus*) in Java, Indonesia. *Tigerpaper* 36: 1-5.

Santiapillai, C., and H. Suprahman. 1984. On the proposed translocations of the Javan Rhinoceros (*Rhinoceros sondaicus* Desmarest 1822) from Ujung Kulon National Park to the Way Kambas Game Reserve in Southern Sumatra. *WWF/IUCN 3133 Report No. 6.* WWF - BKSDA II, Bogor.

—. 1986. The proposed translocation of the Javan rhinoceros *Rhinoceros sondaicus*. *Biol. Conserv.* 38: 11-19.

Santoso, E. 1985. *Ular berbisa dan pembelit raksasa di Jawa.* Bogor: Museum Zoologicum Bogoriense.

Santoso, H. B. 1989. *Budidaya bekicot.* Yogyakarta: Kanisius.

Santoso, N. 1985. *Studi populasi banteng* (Bos javanicus) *dan kerbau air* (Bubalus bubalis) *di Taman Nasional Baluran, Jawa Timur.* Fakultas Kehutanan IPB, Bogor.

Sarwita, S. 1983. Some aspects of ecology and fish population in man made lake Jatiluhur, West Java, Indonesia. *Int. J. Ecol. Env. Sci.* 9: 39-45.

Sarwono, E. 1988. Dunia flora yang tersembunyi di kaki Borobudur. *Suara Alam* 52: 14-19.

Sastrapradja, D. S. et al. 1989. *Keanekaragaman hayati untuk kelangsungan hidup bangsa.* Bogor: Puslitbang Biologi - LIPI.

—. 1990. The interactions between human life, tourism and biological diversity: the Balinese case. In *International Seminar on Human Ecology, Tourism, and Sustainable Development.* 53-66. Denpasar: Human Ecology Study Group.

Sastrapradja, S. 1975. Tropical fruit germplasm in South East Asia. In Proceedings of the Symposium on South East Asian Plant Genetic Resources, ed. J. T. Williams, C. H. Lamoureux, and N. W. Soetjipto, 33-46. Bogor, 20 March 1975. Bogor: LBN-LIPI.

Sastrapradja, S, ed.. 1977. *Sumber daya hayati Indonesia.* Bogor: LBN-LIPI.

Sastrapradja, S., and K. Kartawinata. 1975. Leafy vegetables in the Sundanese diet. In *Southeast Asian plant genetic resources,* ed. J. T. Williams, C. H. Lamordreux, and N. W. Soetjipto, 166-170. Bogor: BIOTROP - LBN LIPI.

Sastrapradja, S., and R. E. Nasution. 1977. Flora Bali: peranannya dalam kehidupan budaya dan kepariwisataan. Paper presented at Diskusi Panel Bali Pulau Taman, Bogor, 1920.

Sastrapradja, S., and M. A. Rifai. 1984. Kebun botani Serpong. *Warkat Warta Plasma Nutfah Indonesia* 3: 2-6.

Sastrapradja, S., E. A. Widjaya, S. Prawiroatmodjo, and S. Soenarko. 1977. *Berberapa jenis bambu.* Bogor: LBN - LIPI.

Sastrawibawa, S. 1979. *Some biological aspects of Mystus nemurus C.V. in Juanda Reservoir, West Java.* BIOTROP, Bogor.

Sastroamidjojo, K. 1993. The application of mangrove social forestry at Perum Perhutani Unit III West Java supporting the national food production. *Report of the regional expert consultation on participatory agroforestry and silvofishery systems in southeast Asia,* ed. H. Beukeboom, K. L. Chun, and M. Otsuka. APAN/FAO, Bogor.

Sastromulyono, B. 1993. *Indonesia: environment and heritage.* Jakarta: Dept. of Tourism, Post and Telecommunications.

Sastroutomo, S. 1992. *Pestisida: dasar-dasar dan dampak penggunaannya.* Jakarta: Gramedia.

Sastroutomo, S. S. 1990. *Ekologi gulma.* Jakarta: Gramedia.

Sastroutomo, S. S., and I. H. Utomo. 1985. Aquatic weed problems in Lake Rawa Danau, West Java. *BIOTROP Newsl.* 46: 8.

Sautereau de Chaffe, J. 1985. Jalons Javanais, ou Kali Suci, la riviere des Mille Montagnes. *Spelunca* 17: 27-33.

Savage, R. J. G., and M. R. Long. 1986. *Mammal evolution: An illustrated guide.* London: British Museum (Natural History).

Schaller, G. B. et al. 1990. Javan rhinoceros in Vietnam. *Oryx* 24: 77-80.

Scheffer, M., A. A. Achterberg, and B. Beltman. 1984. Distribution of macroinvertebrates in a ditch in relation to the vegetation. *Freshw. Biol.* 14: 367-370.

Scheibener, E. 1924. Over eenige bekende Javaansche zoetwater-en landslakken. *Trop. Nat.* 13: 90-95, 106-109.

Scheller, V. 1988. Two new species of Symphyla from the Krakatau Islands and the Ujung kulon peninsula. (Myriapoda: Symphylla: Scolopendrellidae, Scutigerellidae). *Phil. Trans. R. Soc. Lond B.* 322: 401-411.

Schenkel, R., L. S. Hulliger, and W. Sukohadi. 1978. Area management for the Javan Rhinoceros (*Rhinoceros sondaicus*): A pilot study. *Malay. Nat. J.* 31: 253-376.

Schenkel, R., and L. Schenkel-Hulliger. 1969. The Javan Rhinoceros (*Rhinoceros sondaicus* Desm.) in Udjung Kulon Nature reserve: Its ecology and behaviour. Field study 1967 and 1968. *Acta Tropica* 26: 97-135.

Schenkel, R. S. ,. 1982. Report on the possible causes of deaths of Javan rhinos in Ujung Kulon National Park in 1981. World Wildlife Fund, Bogor.

Schepman, M. M. 1932. Notes on some toadpoles, toads and frogs from Java. *Treubia* 14: 43-72.

Schereiber, A., R. Wirth, M. Riffel, and H. van Rompaey. 1989. *Weasels, civets, mongooses and their relatives: An action plan for the conservation of mustelids and viverids.* Gland: IUCN.

Schiffner, V. 1900. *Die Hepaticae der flora von Buitenzorg.* Leiden: Brill.

Schijfsma, K. 1932. Notes on some tadpoles, toads and frogs from Java. *Treubia* 14: 43-72.

Schiller, B. L. M. 1980. The green revolution in Java: ecological, socioeconomic and historical perspectives. *Prisma* 18: 17-93.

Schilling, T. 1952. *Tigermen of Anai.* London: George Allen and Unwin.

Schimper, A. F. W. 1903. *Plant geography upon a physiological basis.* Oxford: Clarendon.

Schmidt, F. H., and J. H. A. Ferguson. 1951. Rainfall types based on wet and dry period ratios for Indonesia and western New Guinea. *Verh. Djawatan Met. Geofisik, Jakarta* 42:

Schneider, S. H. 1993. Degrees of certainty. *Nat. Geog. Res. Ex.* 9: 173-190.

Schnepper, W. C. R. 1933. Vloedbosch culturen. *Tectona* 26: 907-919.

——. 1934. Uit een verslag van een reis naar Trouwers eiland, aan de Zuidkuist van Bantam. *Tectona* 27: 190-210.

Schoener, T. W. 1989. The ecological niche. In *Ecological concepts: The contribution of ecology to an understanding of the natural world,* ed. J. M. Cherrett, 79-113. Oxford: Blackwell.

Schreiber, A. et al. 1989. *Weasels, civets, mongooses, and their relatives: An action plan for the conservation of mustelids and viverids.* Gland: IUCN.

Schrenkenberg, K., M. Hadley, and M. I. Dyer. 1990. *Management and restoration of human-impacted resources.* Paris: UNESCO.

Schröter, C. 1929. Eine exkursion uis Tenggergebirge (Ost-Java). *Verhand. Nat. Ges. (Basel)* 40: 511-535.

Schuitemaker, J. P. 1950. *Bos en bosbeheer op Java.* Groningen: Wolters.

Schultz, G. A. 1985. Three terrestrial isopod crustaceans from Java, Indonesia (Oniscoidea: Philosciidae). *J. Nat. Hist.* 19: 215-224.

Schulze, E. D., and H. A. Mooney. 1993. Ecosystem function of biodiversity: a summary. *Biodiversity and ecosystem function,* ed. E. D. Schulze, and H. A. Mooney,Berlin: Springer.

Schupp, E. W. 1986. *Azteca* protection of *Cecropia*: ant occupation benefits juvenile trees. *Oecologia* 70: 379-385.

Schuster, J. 1911. Monographie der fossilen Flora der *Pithecanthropus*-Schichten. *Abhand. K. Bayer Akad. Wissen.* 25.

Schuster, W. H. 1950. Comments on the importation and transplantation of different species of fish into Indonesia. *Contr. Gen. Agric. Res. Sta. Bogor* 111: 1-31.

——. 1952. Fish culture in brackish water ponds of Java. *IPFC special publications No. I.*

Schütt, G. 1972. Fossil mammals of Java. IV: Zur kenntnis der Pleistozänen Hyän en Javas I. *Proc. K. Ned. Akad. Wet., Ser. B* 75: 261-287.

Schweizer, H. J. 1958. Die fossilen Dipterocarpaceen-Hölzer. *Palaeontographica* 105: 1-66.

Scidmore, E. R. 1899. *Java: The garden of the East.* Singapore [reprinted 1984]: Oxford Univ. Press.

Scott, D. A. 1989. *A directory of Asian Wetlands.* Gland: IUCN.

Scott, J. C. 1976. *The moral economy of the peasant.* New Haven: Yale Univ. Press.

Seal, U. S., and T. J. Foose. 1989. In Javan Rhinoceros *Rhinoceros sondaicus*: Population viability, analysis and recommendations, Bogor, 5 June 1989. Minnesota: Captive Breeding Specialist Group and Asian Rhinoceros Specialist Group.

Seidensticker, J. 1987. Bearing witness: Observations on the extinction of *Panthera tigris balica* and *Panthera tigris sondaica.* In *Tigers of the world: The biology, biopolitics, management, and conservation of an endangered species,* ed. R. L. Tilson, and U. S. Seal, 1-8. New Jersey: Noyes.

Seidensticker, J., and Suryono. 1980. *The Javan tiger and the Meru Beteri reserve: a plan for management.* WWF, Bogor.

Seifriz, W. 1923. The altitudinal distribution of plants on Mt Gedeh, Java. *Bull. Torrey Bot. Club* 50: 283-305.

Sémah, A. M. 1982. A preliminary report on a Sangiran pollen diagram. *Mod. Quat. Res. SE Asia* 7: 165-170.

——. 1984. Remarks on the pollen section of the Sambungmacan section (Central Java). *Mod. Quat. Res. SE Asia* 8: 29-34.

——. 1992. Study of a 4000 years pollen record of a core in the Ambarawa Basin, Central Java, Indonesia: Evidence of older grubbing periods. *C. R. Acad. Sci. ser. II, Univers. Sci. Terre* 315: 903-908.

Serrini, L. 1986. The Christian Declaration on Nature. In *The Assisi Declaration: Messages on Man and Nature from Buddhism, Christianity, Hinduism, Islam, and Judaism.* pp.9-14. Gland: WWF.

Setiadi, A. P., and I. Setiawan. 1992. *Studi Ekologi dan konservasi burung merak Jawa (Pavo muticus muticus) di Gunung Ringgit, Situbondo, Jawa Timur.* Jurusan Biologi, FMIPA, UNPAD, Bandung.

——. 1993. Ecology and conservation of green peafowl on Mount Ringgit, Pasir Putih, Java, Indonesia. *Bull. Orient. Bird Cl.* 17: 9-10.

Setiawan, A. 1985. *Interaksi antara masyarakat sekitar kawasan dengan hutan bambu dalam kawasan Tawan Nasional Meru Beteri, Jember, Jawa Timur.* Fakultas Kehutanan IPB, Bogor.

Setiawan, I. 1993. *Pola penggunaan tenggeran merak hijau (Pavo muticus muticus) terhadap bentuk sosialisasinya di Gunung Ringgit, Pasir Putih, Jawa Timur.* Jurusan Biologi, FMIPA UNPAD, Bandung.

Setiawati, T., and A. S. Mukhtar. 1990. Study on the behaviour of banteng *Bos javanicus* D'Alton in Cipalawah grazing ground, Leuweung Sancang Nature Reserve, Garut, West Java. *Bull. Pen. Hutan* 524: 27-36.

Settle, W. et al. in press. Managing rice through complexes of generalist natural enemies an alternative to the single-species approach. *Ecology.*

Setyawati, T., and A. S. Mukhtar. 1992. Productivity and carrying capacity on the grazing ground of herbivore animal in Baluran National Park, East Java. *Bull. Pen. Hutan* 549: 27-39.

Setyawati, Y. 1986. *Distribusi jenis-jenis kerang (Bivalvia) di pantai muara Sungai Cisengkeut, Desa Mekarsari, Kecamatan Cigeulis, Kabupaten Pandeglang, Jawa Barat.* Fakultas Perikanan IPB, Bogor.

Seymour, F.J., and D. Rutherford. 1993. Contractual agreements in the Java social forestry program. *Legal frameworks for forest management in Asia: case studies of community/state relations,* ed. J. Fox, 31-38. Honolulu: East -West Center.

Shepard, B. M., A. T. Barrion, and J. A. Litsinger. 1987. *Helpful insects, spiders and pathogens.* Laguna: IRRI.

——. 1991. *Mitra petani padi: serangga, laba-laba dan patogen yang membantu.* Jakarta: Program Nasional Pengendalian Hama Terpadu.

Sheppard, C., and S. M. Wells. 1988. *Coral reefs of the world. Volume 2: Indian Ocean, Red Sea and Gulf.* Nairobi/Gland: UNEP - IUCN.

Shodiq, N. 1987. *Kajian tentang interaksi beberapa jenis Muridae di Resort Hutan KPA Cibodas, Taman Nasional Gunung Gede-Pangrango.* Fakultas Kehutanan IPB, Bogor.

Shutler, R. J., and F. Braches. 1985. Problems in paradise: the pleistocene of Java revisited. *Mod. Quat. Res. SE Asia* 9: 87-97.

Sibley, C. G., and B. L. Monroe. 1990. *Distribution and taxonomy of birds of the world.* New Haven: Yale Univ. Press.

Sidiyasa, K. 1988. Beberapa aspek ekologi sawo kecik (*Manilkara kauki*) di Purwo Barat, Banyuwangi Selatan, Jawa Timur. *Bull. Pen. Hutan* 495: 1-19.

Sidiyasa, K., and Sukawi. 1984. *Jenis-jenis tumbuhan obat di hutan sekitar Rajegwesi dan Sukamade, Suaka Margasatwa Meru Betiri, Jawa Timur dan beberapa macam pengobatan tradisional oleh penduduk setempat.* Lembaga Penelitian Hutan, Bogor.

Sidiyasa, K., S. Sutomo, and R. S. A. Prawira. 1985. Struktur dan komposisi hutan Dipterocarpaceae tanah rendah di Cagar Alam Leuweung Sancang, Jawa Barat. *Bull. Pen. Hutan* 471: 37-48.

Sidiyasa, K., S. Sutomo, and Soewanda A. P. 1986. Permudaan alam jenis-jenis dari suku Dipterocarpaceae di Cagar Alam Leuweung Sancang, Jawa Barat. *Bull. Pen. Hutan* 475: 113-120.

Sieber, J. 1978. Freilandbeobachtungen und versuch einer bestandsaufnahme des Bali-Stars (*Leucopsar rothschildi*). *J. Ornithol.* 119: 102-106.

Sigit, A., and M. Baidiawi. 1987. Goa tanpa kelelawar. *Editor* 1: 34.

Sijatauw, W. E. 1973. *Rainfall atlas of Indonesia. Volume I: Java and Madura 1931-60.* Jakarta: Meteorology and Geophysics Institute.

Silitonga, H., Z. Hamzah, and S. Manan. 1987. The KPH Bondowoso: Agroforestry model in East Java; possible uses of coffee plants (*Coffea robusta* Lind.) as a source of income for local people. *Duta Rimba* 13: 79-80.

Silvius, M. J. and de Korte, J. 1990. Pelecaniformes in Indonesia: status, recent changes and management. *Paper presented at ICBP World Conference.* ICBP, Hamilton N.Z.

Silvius, M. J., and W. J. M. Verheught. 1989. The status of storks, ibises and spoonbills in Indonesia. *Kukila* 4: 119-132.

Silvius, M. et al. 1987. *Indonesian wetland inventory.* Bogor: PHPA/AWB.

Simanjuntak, L. 1992. Beras merah dan Indonesia hijau. In *Mochtar Lubis: Wartawan jihad,* ed. Atmakusumah,Jakarta: Harian Kompas.

Simberloff, D. 1986. Are we on the verge of a mass extinction in tropical rain forest? In *Dynamics of extinction.* D. K. Elliott, 165-180. New York: Wiley.

—. 1988. The contribution of population and community biology to conservation science. *Ann. Rev. Ecol. Syst.* 19: 473-511.

—. 1990. Review of *Economics and biological diversity* by J.A. McNeely. *Ecology* 71: 410-411.

—. 1992. Do species-area curves predict extinction in fragmented forest? In *Tropical deforestation and species extinction,* ed. T. C. Whitmore, and J. A. Sayer, 75-89. London: Chapman and Hall.

Simbolon, K. 1988. Social forestry in Indonesia: application and implementation. Faculty of Forestry, Kasetsart Univ., Bangkok.

Simkin, T. et al. 1981. *Volcanoes of the world: A regional directory, gazetter and chronology of volcanism during the last 10,000 years.* Stroudsburg, PA: Hutchinson Ross.

Simmonds, N. 1993. Terminal town. *Tjerutjuk* 2: 2-4.

Simon, H. T. 1992. *Profil desa PHT.* Program Nasional Pelatihan dan Pengembangan Pengendalian Hama Terpadu, Bappenas, Jakarta.

Simons, P. 17 July 1993. Could marigolds slay killer mosquitoes? *New Scient.,* 18.

Singagerda, M. M. 1991. *Kepariwisataan pantai di kawasan Parangtritis dan dampak lingkungannya.* Fakultas Pasca Sarjana, Universitas Gadjah Mada, Yogyakarta.

Singarimbun, M. 1976. *Penduduk dan kemiskinan: Kasus Sriharjo di pedesaan Jawa.* Jakarta: Bhratara Karya Aksara.

Singh, K. 1986. The Hindu Declaration on Nature. In *The Assisi Declaration: Messages on Man and Nature from Buddhism, Christianity, Hinduism, Islam and Judaism.* pp.15-19. Gland: World Wild Fund for Nature.

Sinukaban, N., O. Satjapradja, and S. S. Wastra. 1991. Effect of land use changing on hydrologic response in Manting sub watershed, Upper Konto watershed East Java. *Bull. Pen. Hutan* 544: 27-37.

Siregar, H., and O. Soemarwoto. 1976. Studies on *Panicum repens* L. in West Java. In *Aquatic weeds in South East Asia,* ed. C. K. Varshney, and J. Rzoska, 211-213. The Hague: Junk.

Situmorang, T., and A. Sudradjat. 1987. *Preparation of volcanic hazard zones of Mt. Semeru, East Java, using aerial photograph.* Volcanological Survey of Indonesia, Bandung.

Sjafrie, N. D. M. 1985. *Percobaan penanaman rumput laut (Gracilaria gigas Harv.) dalam tambak di Bali.* FMIPA UI, Jakarta.

Slametmuljana. 1976. *Majapahit.* Singapore: Singapore Univ. Press.

Sleumer, H. 1967. Ericaceae. *Fl. Mal. I* 6: 469-914.

Sloan, N. 1993. *Science and management review of tropical seagrass ecosystems in support of integrated coastal zone management in Indonesia.* EMDI-KLH, Jakarta.

Sloan, N. A., and A. Sugandhy. 1994. An overview of Indonesian coastal environmental management. *Coastal. Man.* 22: 215-233.

Sloan, N. A., A. Wicaksono, T. Tomascik, and H. Uktolseya. 1994. Pangumbahan sea turtle rookery, Java, Indonesia: towards protection in a complex regulatory regime. *Coastal Man.* 22: 251-264.

Smal, C. M. 1989. Barn owls (*Tyto alba*) for the control of rats in agricultural crops in the tropics. *BIOTROP Spec. Publ.* 36: 255-276.

Smalt, F. N. 1937. Periodike drooglegging van sawahs ter bestrijuing van malaria. *Mededel. Dienst. Volksg.* 26: 284.

Smiet, A. C. 1987. Tropical watershed forestry under attack. *Ambio* 16: 2-3, 156-158.

—. 1990a. Agro-forestry and fuel-wood in Java. *Env. Conserv.* 17: 235-238.

—. 1990b. Forest ecology on Java: conversion and usage in a historical perspective. *J. Trop. For. Sci.* 2: 286-302.

—. 1992. Forest ecology on Java: human impact and vegetation of montane forest. *J. Trop. Ecol.* 8: 129-252.

Smiet, A. C., H. G. J. Koop, and C. P. Thalen. 1989. *Human impact on mountain forests in the river Konto area: Vegetation and transect studies.* DHV Consultants, Jakarta.

Smiet, A. C., Suradji, and D. Yuwono. 1990. *Decrease in forest land in the Cimandiri watershed, West Java.* School of Environmental Conservation Management, Bogor.

Smith, B. J., and M. Djajasasmita. 1988. The land molluscs of the Krakatau Islands, Indonesia. *Phil. Trans. R. Soc. Lond. B.* 323: 379-400.

Smith, D. 1984. *Urban ecology.* London: Allen and Unwin.

Smith, J. M. B. 1970. Herbaceous plant communities in the summit zone of Mount Kinabalu. *Malay. Nat. J.* 24: 16-29.

—. 1986. Origins and history of the Malesian high mountain flora. *High altitude tropical biogeography,* ed. F. Vuilleumier, and M. Monasterio, 469-477. Oxford: Oxford Univ. Press.

Sody, H. J. V. 1925. Inlandsche (vogel-) namen. *Trop. Nat.* 14/15: 86-87, 107-109.

—. 1929. De adjag (*Cuon javanicus* Demarest). *Natuurk. Tijds. Ned.-Ind.* 89: 210-290.

—. 1931. Nogmaals: De adjag. *Natuurk. Tijds. Ned.-Ind.* 91: 50-56.

—. 1932. *Felis viverrina. Trop. Nat.* 21: 165.

—. 1933. On the mammals of Bali. *Natuurk. Tijds. Ned.-Ind.* 93: 56-95.

—. 1936. Enkele eerste aanteekeningen over de sporen der Javaansche zoogdieren. *Tectona* 29: 215-262.

—. 1937a. Notes on some mammals from Sumatra, Java, Bali, Buru and New Guinea. *Temminckia* 2: 211-220.

—. 1937b. Over de sporen der Javaanshe zoogdieren. *Tectona* 30: 724-730.

—. 1940. Voorplantingstijden der Javaansche zoogdieren. *Tectona* 33: 1-23.

—. 1941a. A new race of *Sus verrucosus* from Madoera island. *Treubia* 18: 393-394.

—. 1941b. *De Javaansche neushoorn Rhinoceros sondaicus historisch en biologisch.* Buitenzorg: 's Lands Plantentuin.

—. 1953. Vogels van het Javaanse djatibos. *Maj. Ilm. Al. Indo.* 109: 125-172.

—. 1955. Enkele opmerkingen over vogelcultuur of Java in het algemeen en over vogelplanten in het bijzonder. *Madj. Ilmu Alam Indo.* 111: 178-196.

—. 1956. De Javaanse bosvogels: Met enige in drukken over hun hoogteverspreding en over het verband tussen deze en de plantengroei. *Indo. J. Nat. Sci.* 1-18.

—. 1989a. Land mammals of Java. In *Henri Jacob Victor Sody (1892-1959): His life and work,* ed. J. H. Becking, 137-164. Leiden: Brill.

—. 1989b. Diets of Javanese birds. *Henri Jacob Victor Sody (1892-1959): His life and work.* ed J. H. Becking, 164-221. Leiden: Brill.

Soedarmi, S. 1965. *Report on the study of an aquatic insects at Singaparna, West Java, Indonesia.* Fakultas Perikanan IPB, Bogor.

Soedjiran, R., K. Kuswata, and S. Aprilani. 1989. *Pengantar ekologi.* Bandung: Remadja Karya.

Soedjito, H., S. Kadarsan, D. S. Sastrapradja, and Sumarsono. 1986. Air Ciliwung dan pembenahan jalur Bogor-Puncak. Paper presented at Kongres Ilmu Pengetahuan Nasional IV, Jakarta, 8 September 1986.

Soegiarto, A., and W. S. Atmadja. 1980. Potensi, pemanfaatan dan prospek pengembangan budidaya rumput laut di Indonesia. Makalah untuk Pekan Dagang Rumput Laut, Ambon, 8 September 1980.

Soegiarto, A., and N. V. C. Polunin. 1980. *The marine environment of Indonesia.* Bogor: World Wildlife Fund.

Soejono. 1992. Plant exploration to Meru Betiri National Park, East Java, Indonesia. Strategy of Indonesian Flora Conservation Conference, Bogor, 20 July 1992. Poster.

Soejono, F. X., and J. Hindarto. 1978. Beberapa sifat kimia dari air Danau Rawa Pening dan sungai-sungai di sekitar Rawa Pening. In Proceedings of the 1st Seminar on Aquatic Biology and Aquatic Management of the Rawa Pening Lake, Salatiga, Central Java, 20-32. Salatiga, 20 November 1978. Salatiga: Satya Wacana Christian Univ.

Soejono, R. P. 1984. Prehistoric Indonesia. In *Prehistoric Indonesia: A reader,* ed. P. van de Velde, 49-78. Dordrecht: Foris Publications.

Soekatno, T. 1980. Penemuan sepasang arca emas dari Purworejo. Pertemuan Ilmiah Arkeologi II, Jakarta, 25 February 1980. Jakarta: UI.

Soemandoyo, P. 1984. *Peranan vegetasi pakan badak di Ujung Kulon dalam kasus kematian badak Jawa (Rhinoceros sondaicus).* FKH IPB, Bogor.

—. 1985. Badak Jawa setelah tahun 1982. *Suara Alam* 33: 10-14.

—. 1988. The last Javanese tigers. *Voice of Nature* 58: 58.

Soemarwoto, O. 1973. *Ecological studies on aquatic nuisance at Jatiluhur Lake.* Institute of Ecology UNPAD, Bandung.

—. 1979. Exploitative city-rural relationship: a human ecological problem in economic development in Indonesia. *Oikos* 33: 190-195.

—. 1982. Mengapa lamtoro? Suatu analisis manfaat dan risiko lingkungan. Paper for Seminar Lamtoro Nasional I, Jakarta, 23 August 1982.

—. 1983. *Ekologi lingkungan hidup dan pembangunan.* Jakarta: Jambatan.

—. 1984a. The talun-kebun system: A modified shifting cultivation in West Java. *Environmentalist* 4: 96-103.

—. The talun-kebun: a man-made forest fitted to family needs. *Food Nutr. Bull.* 7: 48-51.

—. 1985b. The Javanese home garden as an integrated agro-ecosystem. *Food Nutr. Bull.* 7: 44-47.

—. 1985c. A quantitative model of population pressure and its potential use in development planning. *Demografi Indonesia* 12: 1-15.

—. 1985d. Constancy and change in agroecosystems. In *Cultural values and human ecology in Southeast Asia.* K. L. Hutterer, A. T. Rambo, and G. Lovelace, 205-218. Michigan: Ann Arbor.

—. 1987a. Homegardens: A traditional agroforestry system with a promising future. In *Agroforestry: A decade of development,* ed. H. A. Steppler, and P. K. R. Nair, 57-170. Nairobi: ICRAF.

—. 1987b. Traditional strategies of food security: threats and opportunities. Paper presented at the Nordic Conference on Environment and Development, Stocholm, 7 May 1987.

—. 1988a. Dams as agents of rural development. Sustainable rural development in Asia, ed. T. Charoenwatana, A. T. Rambo, A. Jintrawet, and P. Sorn-srivichai, 115-124. Khon Kaen Univ., Bangkok: Khon Kaen Univ.

—. 1988b. Pemukiman kembali penduduk Saguling-Cirata dengan pengembangan akuakultur dan perikanan. Paper presented at Seminar Pemukiman Kembali Penduduk Saguling dan Cirata Melalui Pengembangan Perikanan, Bandung, 12 February 1988.

—. 1990a. Sustainable tourist development. International Seminar on Human Ecology, Tourism and Sustainable Development, Denpasar, 20 March 1990. Denpasar: Bali Human Ecology Study Group.

—. 1990b. Reduction of forest loss in settlement projects and along transportation corridors. IPCC Workshop, Sao Paulo, 9 January 1990.

—. 1990c. Sustainable development in tourism. In *International Seminar on Human Ecology, Tourism, and Sustainable Development.* 40-52. Denpasar: Human Ecology Study Group.

—. 1991a. *Ekologi, lingkungan hidup dan pembangunan.* Jakarta: Djambatan.

—. 1991b. Human ecology in Indonesia: the search for sustainability in development. In *Indonesia: resources, ecology and environment,* ed. J. Hardjono, 212-235. Singapore: Oxford Univ. Press.

—. 1992. *Indonesia dalam kancah isu lingkungan global.* Jakarta: Gramedia.

Soemarwoto, O., and G. R. Conway. 1991. The Javanese homegarden. *J. Farm. Syst. Res. Ext.* 2: 95-118.

Soemarwoto, O., and B. A. Kurnani. 1984. Environmental impact assessment of Kamojang geothermal power plant, West Java. 3rd ASEAN Geothermal Energy Seminar, Bandung, 4 April 1930.

Soemarwoto, O., and I. Soemarwoto. 1983. A new approach in reforestation and erosion control in Indonesia. International Symposium on Strategies and Designs for Reforestation, Afforestation and Tree Planting, Wageningen, 19 September 1983.

—. 1984. The Javanese rural ecosystem. In *An introduction to human ecology research on agricultural systems in South East Asia.* A. T. Rambo, and P. E. Sajise, 254-287. Laguna: Univeristy of the Philippines.

Soemarwoto, O., Karyono, and G. Nugroho. 1976. Ecology and health in Indonesian villages. *Ekistics* 41: 235-239.

Soemarwoto, O. *et al.* 1990. Water quality suitability of Saguling and Cirata reservoirs for development of floating new cage aquaculture. In *Reservoir fisheries and aquaculture development for resettlement in Indonesia,* ed. B. Costa-Pierce, and O. Soemarwoto, 18-111. Manila: ICLARM.

Soepadmo. 1961. Perdjalanan ke Pulau Bali ditindjau dari segi ilmu tumbuh-tumbuhan. *Penggemar Alam* 40: 25-43.

Soepadmo, E. 1972. Fagaceae. *Fl. Mal. I* 7: 265-403.

Soepraptomo. 1957. Babi dalam hutan djati di daerah Tuban. *Penggemar Alam* 37: 41-44.

Soeriaatmadja, R. E. 1976. *Laporan feasibility study penyelamatan dan pengembangan* Cika-

murang. Dept. Biologi ITB, Bandung.
——. 1979. *Ilmu lingkungan.* Bandung: ITB Press.

Soerianegara, I. 1968. The causes of *Bruguiera* trees mortality in the mangrove forest near Cilacap, Central Java. *Rimba Indonesia* 13: 1-11.

——. 1971. Characteristics and classification of mangrove soils of Java. *Rimba Indonesia* 16: 141-150.

Soerjani, M. 1976. Aquatic weed problems in Indonesia with special reference to the construction of man-made lakes. In *Aquatic weeds in South East Asia.* C. K. Varshney, and J. Rzoska, 63-77. The Hague: Junk.

——. 1982. Jakarta and its urban environmental features (Paper no. 36). Paper presented at the ASAIHL Seminar on Human Ecology, Jakarta, 1982.

——. 1985. Principles of aquatic vegetation management. In Workshop on the Ecology and Management fo Aquatic Weeds, Jakarta, 26 March 1985.

——. 1987a. An introduction of the weeds of rice in Indonesia. In *Weeds of rice in Indonesia,* ed. M. Soerjani, A. J. G. Kostermans, and G. Tjitrosoepomo, 1-4. Jakarta: Balai Pustaka.

——. 1987b. Lingkungan hidup dan hutan kota. Paper presented at Seminar Hutan Kota, Jakarta, 15 December 1987.

——. 1987c. Water enrichment and the possible utilization of aquatic plants. *Adv. Limnol.* 28: 227-236.

——. 1993. *Gerakan Ciliwung bersih: Cetakan sejarah.* PPSML-UI, Jakarta.

Soerjani, M., A. J. G. Kostermans, and G. Tjitrosoepomo. 1987. *Weeds of rice in Indonesia.* Jakarta: Balai Pustaka.

Soerjowinata, H., and F. Göltenboth. 1977. River and lake water quality in the Rawa Pening area, Central Java. *J. Satya Wacana Res.* 1: 10-15.

Soeroto, K. S. 1985. *Konservasi lingkungan Gua Terawang, Gua Kijang, Gua Jero/Gua Kiskendo.* HIKESPI, Semarang.

Soeroyo. 1986. Struktur dan gugur serasah mangrove di Kembang Kuning, Cilacap. Paper presented at Seminar III Ekosistem Mangrove, Denpasar, May 1908.

Soeroyo, and S. Soemodihardjo. 1991. Tumbuhan gulma dan semai alami di hutan mangrove Segara Anakan, Cilacap. Prosidings Seminar IV Ekosistem Mangrove, S. Soemodihardjo, 161-168. Bandar Lampung, 7 August 1990. Jakarta: MAB-LIPI.

Soeroyo, and S. Sukardjo. 1991. Struktur dan komposisi hutan mangrove di Grajagan,

Banyuwangi. Prosidings Seminar IV Ekosistem Mangrove, S. Soemodihardjo et al., 129-136. Bandar Lampung, 7 August 1990. Jakarta: MAB-LIPI.

Soesilo, H. P. 1954. The problem of illegal deforestation on a large scale in Java. *Rimba Indonesia* 3: 109-119.

Soesilotomo, P. S. 1993. Perkembangan pengelolaan hutan payan KPH Madura. *Duta Rimba* 19: 16-21.

Soetardjo et al. 1985. *Series on seismology, Vol 5 - Indonesia.* Southeast Asia Association of Seismology and Earthquake Engineering, Jakarta.

Soetjipta et al. 1979. Kerapatan dan kelimpahan zooplankton (Cladocera, Copepoda, Ostracoda dan Rotifera) di dalam empat lingkungan Rawa Pening. In Proceedings of the 2nd Seminar in Aquatic Biology and Aquatic Management of the Rawa Pening Lake, Salatiga, Central Java, 24-51. Salatiga, 29 October 1979. Salatiga: Satya Wacana Christian Univ.

Soetjipto. 1977a. Penelitian limnologi di Bengawan Solo. Paper presented at Seminar Biologi V dan Kongres Biologi III, Malang, 7 July 1977.

——. 1977b. Tinjauan populasi ikan di Bengawan Solo hulu. Seminar Biologi V dan Kongres Biologi III, Malang, 7 July 1977.

Soetjipto, N. W., and G. G. Hambali. 1974. Tanaman-tanaman asing di cagar alam Telaga Warna. *Berita Biologi* 1: 12-26.

Soetranggono. 1991. Program for conservation and development of Javanese Fat-tail sheep in East Java. *Production aspects of Javanese fat-tail sheep in Indonesia,* ed. I. K. Sutama, and L. Iniguez, 1-8. Bogor: Indonesian Small Ruminant Network.

Soewadi, B., and D. Sastrapradja. 1980. Alang-alang and animal husbandry. *BIOTROP Spec. Publ.* 5: 157-178.

Soewito, T., and Z. Harahap. 1984. Cisadane: varietas padi tahan wereng coklat dengan rasa nasi enak. *Pemberitaan Penelitian Puslitbangtan* 6: 29-40.

Solbrig, O. T. 1991. *Biodiversity: scientific issues - collaborative research proposals.* Paris: UNESCO.

Sollart, K. 1986. The Javanese mixed home garden as a plant genetic resource. *FONC project communication No. 1986 - 10.* FONC - Fakultas Kehutanan UGM, Yogyakarta.

Somadikarta, S. 1986. *Collocalia linchi* Horsfield and Moore - a revision. *Bull. Brit. Orn. Cl.* 106: 32-40.

Sonck, C. E. 1987. A new *Taraxacum* species, *Taraxacum indonesicum,* new species from

Java. *Ann. Bot. Fennici* 24: 307-309.

Sondaar, P. Y. 1981. The *Geochelone* faunas of the Indonesian Archipelago and their paleogeographical and biostratigraphical significance. *Mod. Quaternary Res. SE Asia* 6: 111-120.

—. 1984. Faunal evolution and the mammalian biostratigraphy of Java. *Cour. Forsch. Inst. Senckenb.* 69: 219-235.

Sosromarsono, S. 1989. Biological control of agricultural pests in Indonesia. *BIOTROP Spec. Publ.* 36: 69-84.

Soulé, M. 1980. Thresholds for survival: maintaining fitness and evolutionary potential. In *Conservation: An evolutionary-ecological perspective*, ed. M. E. Soulé, and B. A. Wilcox, 151-170. Sunderland, Mass: Sinauer.

—. 1987. Where do we go from here? In *Viable populations for conservation*, ed. M. E. Soulé, 175-183. Cambridge: Cambridge Univ. Press.

Soulé, M. E., and B. A. Wilcox. 1980. Conservation biology: its scope and its challenge. In *Conservation biology: An evolutionary-ecological perspective*, ed. M. E. Soulé, and B. A. Wilcox, 1-8. Sunderland, Mass: Sinauer.

Sozer, R., and V. Nijman. 1995. Behavioural ecology, distribution and conservation of the Javan hawk-eagle *Spizaetus bartelsi* Stresemann, 1924. Institute of Systematics.

Sozer, R., and V. Nijman. in press a. Recent observations of the grizzled leaf monkey (*Presbytis comata*) and an extension of the range of the Javan gibbon (*Hylobates moloch*) in Central Java. *Trop. Biodiv.*

Sozer, R., and V. Nijman. in press b. The Javan hawk-eagle: new information on its distribution in Central Java and notes on its threats. *Trop. Biodiv.*

Spencer, J. W., and K. J. Kirby. 1992. An inventory of ancient woodland for England and Wales. *Biol. Conserv.* 62: 77-93.

Spennemann, A. W. 1923. Vogelleven in het oerbosch te Tjibodas. *Trop. Nat.* 12: 177-179.

—. 1924. Moeara Tjiasem 29 Mei - 2 Juni 1916. *Trop. Nat.* 13: 173-176.

—. 1925. Biologische notizen uber einige Javanische Vogel. *Treubia* 6: 12-19.

—. 1928. Jacht op larongs. *Trop. Nat.* 17: 4-7.

Srikosamatara, S. 1993. Density and biomass of large herbivores and other mammals in a dry tropical forest, western Thailand. *J. Trop. Ecol.* 9: 3-43.

SSC. 1987. *A guide to common garden animals.* Singapore: Singapore Science centre.

Stalker, P. 1984. Visions of poverty, visions of wealth. *New Scient.* 142: 7-29.

Start, A. N., and A. G. Marshall. 1975. Nectarivorous bats as pollinators of trees in West Malaysia. *Tropical trees: variation, breeding and conservation.* J. Burley, and B. T. Styles, 141-150. London: Academic.

Statistical Office of Bali Province. 1987. *Bali figures - 1987.* Statistical Office of Bali Province, Bali.

Statistical Office of East Java Province. 1987. *East Java figures - 1987.* Statistical Office of East Java Province, Surabaya.

Statistical Office of West Java Province. 1987. *West Java figures - 1987.* Statistical Office of West Java Province, Bandung.

Statistical Office of Central Java Province. 1988. *Central Java figures -1988.* Statistical Office of Central Java Province, Semarang.

Statistical Office of Jakarta Capital Area. 1988. *Jakarta Capital Area figures -1988.* Statistical Office of Jakarta Capital Area Province, Jakarta.

Statistical Ofice of Yogyakarta Special Area. 1988. *Yogyakarta Special Area figures - 1988.* Statistical Office of Yogyakarta Special Area Provice, Yogyakarta.

Steinmann, A. 1923. Microbiologische aantekeningen over het kratermeertje Telaga Warna. *Trop. Nat.* 12: 101-104.

Steinmann, A., and E. Scheibener. 1925. Zwerftochten over het schiereiland Penandjoeng. *Trop. Nat.* 14: 49-56.

Steinmann, H. 1989. *World catalogue of Dermaptera.* Dordrecht: Kluwer.

Stemmerik, J. F. 1964. Nyctaginaceae. *Fl. Mal. I* 6: 450-468.

Stewart, C. 1994. *An investigation into the ecology and behaviour of terrestrial crabs in Ujung Kulon National Park, West Java, and comparison with the Krakatau islands.* St. Anne's College, Oxford.

Stewart, W. D. P. 1991. The importance to sustainable agriculture of biodiversity among invertebrates and microorganisms. In *The biodiversity of microorganisms and invertebrates: Its role in sustainable agriculture*, ed. D. L. Hawksworth, 3-5. Wallingford: CAB International.

Stoddard, S. in press. Anglo-Australian speleological expedition to Java, 1986. *Cave Sci.*

—. 1985. Anglo-Australian speleological expedition to Java 1984. *Cave Sci.* 12: 49-60.

Stoller, A. 1975. *Garden use and household consumption patterns in a Javanese village.* Columbia Univ., New York.

Stoney, C., and M. Bratamihardja. 1990.

Identifying appropriate agroforestry technologies in Java. In *Keepers of the forest: Land management alternatives in Southeast Asia*, ed. M. Poffenberger, 145-160. West Hartford, CT: Kumarian Press.

Stork, N. E. 1988. Insect diversity: facts, fiction and speculation. *Biol. J. Linn. Soc.* 35: 321-337.

——. 1994. Inventories of biodiversity: more than a question of numbers. *Systematics and conservation evaluation.* P. Forey, C. J. Humphries, and R. I. Vane-Wright. Oxford: Oxford Univ. Press.

Storm, P. 1990a. *Mesolithic and Neolithic sites from Java: Human remains, artefacts and subrecent fauna.* Unpublished report,

——. 1990b. *Newly described Neolithic site from Java: The archaelogical site Hoekgrot; human remains, artifacts and the subrecent fauna of Java.* Unpublished report,

——. 1991. Microevolution of the human skull: implications of the Javanese Wadjak study for East Africa. In *Origins and development of agriculture in East Africa*, ed. R. E.Leakey and L. J. Slikkerveer, 25-37. Ames, Iowa: Iowa State Univ.

——. 1992. Two microliths from Javanese Wadjak Man. *J. Anthro. Soc. Nippon* 100: 191-203.

Stott, J. 1984. *Issues facing Christians today.* Basingstoke: Marshalls.

Strasters, B. 1914. De voornaamste kikvorschen Java. *Trop. Nat.* 3: 164-172.

Stresemann, E. 1912. Description of a new genus and new species of birds from the Dutch East Indies Islands. *Bull. Brit. Orn. Cl.* 31: 4-6.

——. 1913. Die vogels von Bali. *Nov. Zoologicae* 20: 325-387.

——. 1930. Eine vogelsammlung von Vulkan Papandajan (West-Jawa). *Treubia* 12: 425-430.

Struyvenberg, P. A. A., F. A. J. M. van Boxel, A. M. Polderman. 1986. Een bloedsuiger als ongewone oorzaak van epistaxis. *Ned. Tijds. Geneeskd.* 130: 791-792.

Stuart-Fox, D. J. 1982. *Once a century: Pura Besakih and the Eka Dasa Rudra festival.* Jakarta: Sinar Harapan.

Stuijts, I. 1984. Palynological study of Situ Bayongbong, West Java. *Mod. Quat. Res. SE Asia* 8: 17-27.

——. 1993. Late Pleistocene and Holocene vegetation of West Java, Indonesia. *Mod. Quat. Res. S.E. Asia* 12: 1-173.

Stuijts, I., J. C. Newsome, and J. R. Flenley. 1988. Evidence for late Quaternary vegetational change in the Sumatran and Javan

highlands. *Rev. Palaeobot. Palynol.* 55: 207-216.

Suadnya. 1990. *Mengenal subak.* Denpasar: Dinas PU.

Subagus, N. A. 1979. Obsidian industry in Leles, West Java - Preliminary report. *Mod. Quat. Res. SE Asia* 5: 35-41.

Subani, W. 1982. Penambangan karang serta dampaknya terhadap sumberdaya perikanan di perairan pantai selatan Bali. *Lap. Pen. Perikanan Laut* 23: 17-32.

Subani, W., and M. M. Wahyono. 1987. Kerusakan ekosistem perairan pantai dan dampaknya terhadap sumberdaya perikanan di pantai selatan Bali Barat, Timor, Lombok dan Teluk Jakarta. *Lap. Pen. Perikanan Laut* 42: 53-70.

Subroto. 1992. Varietas pisang di Daerah Istimewa Yogyakarta. Paper given at The strategy of Indonesian flora conservation, Bogor, 20 June 1992. Bogor: Kebun Raya Bogor.

Sudarmana. 1981. *Berburu belibis (Dendrocygna sp.) di daerah Karawang.* Museum Zoologicum Bogoriense, Bogor.

Sudjono, B. S. Utomo, and M. P. S. Tjondronegoro. 1976. Research note on the coastal villages of Kecamatan Kendal, Kabupaten Kendal (North Central Java). Paper read at the Seminar on Ecology of Coastal Villages, Semarang, 12 February 1976.

Sudomo, M., Baroji, and B. N. Sustriayu. 1985. Chemical control on malaria vector (*Anopheles aconitus*) in Central Java, Indonesia. *SEA J. Trop. Med. Public Health* 16: 153-162.

Sugardjito, J. 1984. *Laporan survey satwa mamalia besar di Taman Nasional Baluran.* Museum Zoologicum Bogoriense, Bogor.

Suhardjono, Y. R. 1989. Checklist of Indonesian Collembola. *Acta Zool. Asiae Orient.* 1: 1-22.

Suhardjono, Y. R., and L. Deharveng. 1992. *Siamura primadinae*, a new species of Neanurinae (Collembola: Neanuriidae) from East Java, Indonesia. *Raffles Bull. Zoology* 40: 61-64.

Suhardoyo. 1982. *Analisa vegetasi hutan lindung di Gunung Bukit Tunggul, RPH Bukanegara, Kabupaten Subang.* Akademi Ilmu Kehutanan, Bandung.

Suharsono. 1984. Coral death in the Java Sea, Indonesia. *Reef Encounter* 2: 6.

Suharsono, Sukarno, and Siswandono. 1986. Distribution diversity and species richness of Scleractinian corals around Kotok Kecil, Seribu Islands, Indonesia. *Oseanol. Indo.* 19: 1-16.

Suharti, M. 1990. *Acacia arabica*: Tanaman sekat bakar di Taman Nasional Baluran. *Duta Rimba* 16: 41-45.

Suharti, M., and E. Santoso. 1989. Analisa biaya pengendalian *Acacia arabica* di Taman Nasional Baluran Jawa Timur. *Bull. Pen. Hutan* 505: 1-8.

Suharti, M., E. Santoso, and M. Nazif. 1987. Sebaran anakan *Acacia arabica* di areal Taman Nasional Baluran, Jawa Timur. *Bull. Pen. Hutan* 490: 24-32.

Suharti, S., and H. Alrasyid. 1992. Shifting cultivation practice in West Java. *Bull. Pen. Hutan* 546: 35-45.

Suharto, A., and L. P. Irsa. 1988. Nature preserves in critical condition. *Voice of Nature* 59: 11-15.

Suharyanto, B. 1987. Tinjauan aspek manajemen dalam pelaksanaan reboisasi di KPH Malang. Paper presented at Lokakarya Pelaksanaan Rehabilitasi Lahan dan Konservasi Tanah Secara Terpadu di Sub DAS Konto, Malang, 11 March 1987.

Suhono, B. 1986. *Ular-ular berbisa di Jawa*. Jakarta: Antar Kota.

Sujatmaka. 1986. Mengendalikan serangan kutu loncat. *Trubus* 17: 36-38.

Sukadi, F., and D. Suseno. 1991. A review on the status of freshwater fish aquaculture genetics in Indonesia. Paper given at AGNA Meeting, Thailand, Bangkok, 3 October 1928.

Sukandi, T. 1990. Tumpangsari (taungya) in Indonesia: A bio-ecological review. *BIOTROP Spec. Publ.* 39: 99-106.

Sukardjo, S. 1977. Hutan di Cagar Alam Pananjung Pangandaran. *Biologica* 4: 30-34.

—. 1978. Mengenal hutan Gunung Tilu. *Bull. Kebun Raya* 3: 153-156.

—. 1979. Melihat keadaan beberapa cagar alam di Jawa Barat. *Kehutanan Indonesia* 6: 19-21, 31-38, 40-41.

—. 1982a. Ecosystem of mountain forests in Java. *Duta Rimba* 8: 3-12.

—. 1982b. Soil in the mangrove forest of the Cimanuk Delta, West Java (Indonesia). *BIOTROP Spec. Publ.* 17: 191-202.

—. 1983. The Pangandaran Nature Reserve and its endangered ecosystem. Paper presented at the Symposium of Wildlife Ecology in South-East Asia, Bogor, 7 June 1983.

—. 1984a. The present status of the mangrove forest ecosystem of Segara Anakan, Cilacap, Java. In *Tropical rain-forest: The Leeds symposium*, ed. A. C. Chadwick, and S. L. Sutton, 53-67. Leeds: Leeds Philosophical and Literary Society.

—. 1984b. Mangrove di Delta Cimanuk. *Suara Alam* 7: 40-41.

—. 1987a. Penelitian produktivitas hutan mangrove di cagar alam Cimanuk, Indramayu. I: Tanah-tanah mangrove di Kompleks Delta Cimanuk. *Ekologi tanah dan ekotoksikologi*, ed. S. A. Wirja, and Sunarto. Fakultas Biologi, Satya Wacana Christian Univ., Salatiga.

—. 1987b. Tanah dan status hara di hutan mangrove Tiris, Indramayu, Jawa Barat. *Rimba Indonesia* 21: 12-23.

—. 1987c. Produksi primer neto hutan mangrove Muara Angke-Kapuk, Jakarta. Makalah pada Seminar Laut Nasional II, Jakarta, 27 July 1987.

—. 1988. The relationship of litterfall to basal area and climatic variables of the *Rhizophora mucronata* Lmk forest plantation in Tritih, Segara Anakan, Cilacap, Central Java, Indonesia. Paper presented at JOA, Acapulco, 23 August 1988.

—. 1989a. Litterfall production and turn over in the mangrove forests in Muara Angke-Kapur, Jakarta. *BIOTROP Spec. Publ.* 37: 129-143.

—. 1989b. Tumpang sari pond as a multiple use concept to save the mangrove forest in Java. *BIOTROP Spec. Publ.* 37: 115-128.

—. 1990. Conservation of mangrove formations in Java. In *The plant diversity of Malesia*, ed. P. Baas, K. Klakman, and R. Geesink, 329-340. Dordrecht: Kluwer.

Sukardjo, S., and S. Akhmad. 1982. The mangrove forest of Java and Bali. *BIOTROP Spec. Publ.* 17: 113-126.

Sukardjo, S., and I. Santoso. 1982. Pulau Rambut, surga burung: Sampai kapan bisa bertahan. *Suara Alam* 16: 12-17, 29.

Sukardjo, S., and I. Yamada. 1992. Biomass and productivity of a *Rhizophora mucronata* Lamarck plantation in Tritih, Central Java Indonesia. *For. Ecol. Manage.* 49: 195-209.

Sukarme, K. T. 1986. Studi pendahuluan terhadap hutan bakau di Denpasar Selatan. Paper presented at Seminar III Ekosistem Mangrove, Denpasar, 5 August 1986.

Sukarno. ???**below**???. Fauna karang batu di terumbu karang Pulau Ayer dengan catatan tentang ekologinya. In Prosiding Teluk Jakarta: Sumberdaya, Sifat-sifat Oseanologis Serta Permasalahannya, ed. H. Malikusworo, K. Romimohtarto, and Burhanuddin, 292-309.

—. 1975. Penelitian pendahuluan tentang kepadatan karang batu di terumbu karang Pulau Ayer, Laut Jawa. In Prosiding II Seminar Biologi IV dan Kongres Biologi II, 175-

191. Yogyakarta, 10 July 1975.

Sukoraharjo, A. 1982. *Perbedaan keragaman dan kelimpahan rayap pada hutan alam dan hutan tanaman di Yanlappa (Jawa Barat).* Fakultas Kehutanan IPB, Bogor.

Sulistiadi, R. S. 1987. *Studi interaksi masyarakat dengan hutan mangrove Segoro Anak, Suaka Margasatwa Banyuwangi Selatan.* Fakultas Kehutanan IPB, Bogor.

Sulistijo. 1986. Percobaan berkebun rumput laut *Gracilaria* dalam tambak di Bali. Makalah pada Kongres Nasional Biologi Indonesia VII, Palembang, 29 July 1986.

Sulistijo, and S. W. S. Atmadja. 1980. Komunitas rumput laut di Tanjung Benoa, Bali. *Sumberdaya hayati bahari*, ed. Burhanuddin, M. K. Moosa, and H. Razak. LON - LIPI, Jakarta.

Sulito, A. S. B. 1985. *Studi habitat jalak putih (Leucopsar rothschildi) di Suaka Margasatwa Bali Barat.* Fakultas Kehutanan IPB, Bogor.

Sumandoyo, P. 1985. Harimau Bali yang tersisa. *Suara Alam* 7: 34, 46.

Sumardja, E. A. 1978. Feeding strategy of banteng. *BIOTROP Spec. Publ.* 8: 117-126.

—. 1989. Removing the buffalo population for improving banteng numbers in Baluran National Park, East Java, Indonesia. *Asian Wild Cattle Specialist Group Newsl.* 2: 5-6.

Sumardja, E., and K. Kartawinata. 1977. Vegetation analysis of the habitat of banteng (*Bos javanicus* d'Alton) at the Penanjung Pangandaran Nature Reserve, West Java. *BIOTROP Bull.* 13: 3-43.

Sumarjanto, and Priantoro. 1980. Keanekaragaman jenis ikan yang hidup di Sungai Kranji, Purwokerto. *Bull. Ilmiah UNSOED* 6: 10-15.

Sumarna, and Sudiono. 1973. *Konsumsi kayu bakar oleh rumah tangga, industri dan Perusahaan Jawatan Kereta Api di Jawa Timur.* Lembaga Penelitian Hasil Hutan, Bogor.

Sumertha, I. N. 1974. Perikanan penyu dan cara pengelolaan di Indonesia. *Dokumen. Kom. IPB* 8: 1-18.

—. 1976. Studfi habitat dan pembiakan penyu laut di pantai Pangumbahan, Kabupaten Sukabumi. Fakultas Perikanan IPB, Bogor.

Sunarno, B., and Rugayah. 1992. *Flora Taman Nasional Gede Pangrango.* Bogor: Herbarium Bogoriense.

Sundjoto, M. A. 1983. Pengaruh penggalian batu dan pasir terhadap kandungan sedimen tersuspensi aliran Sungai Ciapus daerah aliran sungai Cisadane. Fakultas Kehutanan IPB, Bogor.

Sungkawa, W., D. Natawiria, and A. P.

Suwanda. 1974. Pengamatan jalak putih (*Leucopsar rothschildi*) di Taman Perlindungan Alam Bali Barat. *Lap. Lembaga Penelitian Hutan* 195:

Sungkawa, W., and Wiriadinata. 1974. Inventarisasi rumput dan satwa liar di Cagar Alam Penanjung Pangandaran. *Laporan No. 193.* Lembaga Penelitian Hutan, Bogor.

Supratmo. 1945. *Animistic beliefs and religious practices of the Javanese.* Columbia Univ., New York.

Supriatna, J. 1981. *Ular berbisa Indonesia.* Jakarta: Bhratara Karya Aksara.

—. 1982. Jenis-jenis ular di hutan mangrove dan makanan ular tambak (*Cerberus rynchops*). In Prosiding Seminar II Ekosistem Mangrove, 172-174. Baturaden, 3 August 1982.

—. 1989. Chemical analysis of food plant parts of two sympatric monkeys (*Presbytis aurata* and *Macaca fascicularis*) in the mangrove forest of Muara Gembong, West Java. *BIOTROP Spec. Publ.* 37: 161-169.

—. 1992. Cagar Alam Gunung Halimun: kawasan konservasi, pendidikan dan penelitian pada masa datang. Workshop Cagar Alam Gunung Halimun, Jakarta, 23 January 1992.

Supriatna, J., Martarinza, and Sudirman. 1992. *Sebaran, kepadatan, dan habitat populasi lutung dan owa.* Dep. Biologi, UI, Depok.

Suprihayono. 1986. The effects of sedimentation on a fringing reef in north Central Java, Indonesia. Ph.D. diss., Univ. of Newcastle upon Tyne, Newcastle upon Tyne.

Supriyanto, E. 1989. Penyebaran *Rafflesia patma* Bl. di Cagar Alam Pananjung Pangandaran dan keberhasilan tumbuhnya. Akademi Kehutanan Bandung, Bandung.

Suratiningsih, S. 1986. *Tipe Cercaria yang terdapat pada beberapa siput (Gastropoda) di lima lokasi daerah Rawa pening Jawa Tengah.* Fakultas Biologi UGM, Yogyakarta.

Surjohudoyo, S. 1973. Life in a Javanese village. Seminar on peasant farming and their rural economy in village Indonesia, 13pp. Melbourne, 7 September 1973. Melbourne: Monash Univ..

Surya, I., and M. C. G. Clarke. 1985. The occurrence and mitigation of volcanic hazard in Indonesia as exemplified at the Mount Merapi, Mount Kelud and Mount Galunggung volcanoes. *Q. J. Erg. Geol. (London)* 18: 79-98.

Suryadarma, I. G. P. 1989. Analisis dan distribusi kenaneka-ragaman jenis tumbuhan sarana upacara dalam pola pekarangan masyarakat (studi kasus di Desa Marga, Tabanan, Bali). *J. Pendidikan* 19: 39-49.

—. 1991a. Efektivitas tata ruang tradisional Bali terhadap pengaturan siklus air alami. *Cakrawala Pendidikan* 10: 53-62.

—. 1991b. *Identifikasi macam pertanyaan-anjuran adat tradisi masyarakat Bali yang mendukung usaha konservasi keaneka-ragaman hayati.* FPMIPA IKIP, Yogyakarta.

—. 1992. *Efektivitas pola pekarangan adat Bali di desa agraris dan wisata terhadap usaha konservasi keaneka-ragaman hayati tumbuhan.* Pusat Penelitian IKIP, Yogyakarta.

Suryadi, H. 1990. Keong mas yang berpotensi menjadi hama. *Trubus* 248: 34-35.

—. 14 June 1992. Ada 2 jenis keong mas. *Kompas.*

Suryakusuma, H. 1980. *Guha Tuyo Ngrembes atau Imogiri.* SPECAVINA, Malang.

Susanto, H. 1989. *Budidaya kodok unggul.* Jakarta: Penebar Swadaya.

—. 1990. *Budidaya ikan di pekarangan.* Jakarta: Penebar Swadaya.

Susuki, T., and R. Ohtsuka, (Eds.). 1987. *Human ecology of health and survival in Asia and the south Pacific.* Tokyo: Univ. Tokyo Press.

Sutardjo, E. 1980. *Beberapa aspek biologi lalawak* Puntius bramoides *(C&V) di Waduk Jatiluhur, Jawa Barat.* Fakultas Perikanan, Institut Pertanian Bogor, Bogor.

Sutardjo, and Machfudz. 1971. Laporan survey sidat di Jawa Tengah bagian selatan. *Lap. Lembaga Penelitian Perikanan Darat* 52:

Sutarno. 1991. Dimensi etis-teologis dan etis-antropologis dalam pembangunan berwawasan lingkungan. Seminar lingkungan hidup, Satya Wacana Christian Univ., 26 November 1991. Salatiga: Yayasan Bina Darma.

Sutawan, N. et al. 1984. *Studi perbandingan subak dalam sistem irigasi PU.* UNUD, Denpasar.

Sutiyono. 1987. Silvicultural aspects of bamboo plantations at Kaliurang. *Bull. Pen. Hutan* 493: 14-20.

—. 1988. Silvicultur hutan bambu di hutan Soka, Banyuwangi. *Bull. Pen. Hutan* 497: 29-40.

—. 1989. A study of the potentials and socioeconomic significance of bamboo plantations in the sub-humid regency of Pati, Central Java, Indonesia. *Bull. Pen. Hutan* 512: 33-46.

Sutton, D. L. 1985. Aquatic vegetation for fish production. In Workshop on the Ecology and Management of Aquatic Weeds, Jakarta, 26 March 1985.

Sutton, S. L., and N. M. Collins. 1991. Insects and tropical forest conservation. In *The conservation of insects and their habitats.* N. M. Collins, and J. A. Thomas, 405-424. London: Academic.

Suwelo, I. S. 1973. *Buku Cagar Alam Pulau Rambut.* Dirjen. Kehutanan Dep. Pertanian, Jakarta.

Suyanto, A. 1984. Perlunya pelestarian kelelawar dan burung walet penghuni gua-gua Karangbolong-Buayan dan Nusakambangan. Mengapa seluruh kawasan batugamping Gombong Selatan perlu dilindungi, Karanganyar, 30 September 1984. Bogor: HIKESPI.

Suyanto, A. 1983. *Biospeliologi gua-gua di kabupaten Cilacap dan Kebumen.* HIKESPI, Bogor.

—. 1984. Upaya pelestarian sumber daya hayati gua kapur. Paper presented at Simposium Upaya Pengelolaan Sumber Daya Alam dan Pelestarian Kemampuan Daya Dukung Lingkungan Hidup di Daerah Gombong Selatan, Kabupaten Kebumen, Jawa Tengah, Semarang, 26 November 1984.

Suyenaga, J. 1991. Ratu Kidul, Goddess of the Southern Seas. In *Java,* ed. E. Oey, p.173. Singapore: Periplus Editions.

Suzuki, S. 1986. Three opilionid species *Arachnida* Opiliones Assamiidae and Gagrellidae from West Java, Indonesia. *Acta Arachnol.* 34: 41-48.

Swaine, M. D., and T. C. Whitmore. 1988. On the definition of ecological species groups in tropical forests. *Vegetatio* 75: 81-86.

Swaminathan, P. 1989. Allelopathy and *Acacia nilotica. J. Trop. For. Sci.* 2: 56-60.

Swarbrick, J. T., and B. L. Mercado. 1987. Weed science and weed control in Southeast Asia. *FAO plant production and protection paper 81.* FAO, Rome.

Swennen, C., and E. Marteijn. 1985. Feeding ecology studies in the Malay Peninsula. *Interwader East Asia/Pacific shorebird study programme annual report 1984,* ed. D. Parish, and D. Wells, 13-26. Kuala Lumpur: Interwader.

Swisher, C. et al. 1994. Age of the earliest-known hominids in Java, Indonesia. *Science* 263: 1118-1121.

Syafi'i, I., and S. Kasim. 1981. *Tingkat pendapatan dan keadaan sosial masyarakat sekitar hutan di daerah Mandilis Sanenrejo, Kecamatan Tempurejo, Kabupaten Jember.* Fakultas Pertanian IPB, Bogor.

Syahminudin, E. 1992. *Laporan hasil penelitian dan monitoring hutan mangrove di RPH*

Muara Gembong Tanjung Karawang Kabupaten DT II Bekasi, Jawa Barat. Biological Science Club UNAS, Jakarta.

Sya'rani, L. 1975. *Processing and production of Pecten.* Dirjen Perikanan, Jakarta.

——. 1983. Ecology of shallow water corral communities in the Java Sea, Indonesia. Ph.D. diss., Univ. of Newcastle upon Tyne, Newcastle upon Tyne.

Sya'rani, L., and N. G. Willoughby. 1985. The traditional management of marine resources in Indonesia, with particular reference to Central Java. In *The traditional knowledge and management of coastal systems in Asia and the Pacific,* ed. K. Ruddle, and R. E. Johannes, 253-264. Jakarta: UNESCO.

Syarifuddin, A. et al. 1986. Upland rice environments in Indonesia and the fitness of improved technology. In *Progress in upland rice research.* pp.7-13. Manila: IRRI.

Takenaka, O. 1985. Hemoglobin Bali Macaca, the first hemoglobin variant found in the crab-eating monkey *Macaca fascicularis* on Bali Island, Indonesia. *Primates* 26: 464-470.

Takken, W. 1990. Introduction. In *Environmental measures for malaria control in Indonesia - an historical review on species sanitation.* W. Takken, W. B. Snellen, J. P. Verhave, B. G. J. Knols, and S. Atmosoedjono, 1-3. Wageningen: Wageningen Agricultural Univ. Press.

Takken, W., and B. G. J. Knols. 1990. A taxonomic and bionomic review of the malaria vectors of Indonesia. In *Environmental measures for malaria control in Indonesia - an historical review on species sanitation.* W. Takken, W. B. Snellen, J. P. Verhave, B. G. J. Knols, and S. Atmosoedjono, 9-62. Wageningen: Wageningen Agricultural Univ. Press.

Takken, W., W. B. Snellen, and J. P. Verhave. 1990. Discussion: relevance of the Indonesian experience for modern-day malaria control. In *Environmental measures for malaria control in Indonesia - an historical review on species sanitation.* W. Takken, W. B. Snellen, J. P. Verhave, B. G. J. Knols, and S. Atmosoedjono, 155-158. Wageningen: Wageningen Agricultural Univ. Press.

Taman Nasional Bali Barat. 1989. *Studi penunjukan pohon inti sawo kecik* (Manilkara kauki) *di Taman Nasional Bali Barat.* Taman Nasional Bali Barat, Gilimanuk.

——. 1991. An action plan for the conservation of Bali starling and sawo kecik in Bali Barat National Park. Paper presented at the Workshop on Indonesian Biodiversity

Conservation and Management, Ciloto, West Java, 21 February 1991.

Tanner, E. V. J. 1983. Leaf demography and growth of the tree fern *Cyathea pubesecens. Bot. J. Linn. Soc.* 87: 213-227.

Tanner, E. V. J., and V. Kapos. 1982. Leaf structure of Jamaican upper montane rainforest trees. *Biotropica* 14: 16-24.

Tantra, I. G. M. 1982. Identifikasi flora Cagar Alam Sangeh, Bali. *Laporan No. 398.* Lembaga Penelitian Hutan, Bogor.

Tantra, I. G. M., and Anggana. 1977. Inventarisasi flora Cagar Alam Batukahu, Bali. *Lap. Lembaga Penelitian Hutan* 245: 1-54.

Tarrant, J. et al. 1987. *Natural resources and environmental management in Indonesia: an overview.* USAID, Jakarta.

Tarrant, J., and H. Poerbo. 1986. Integrated rural environment development programme: A case history of the application of action learning in upland development in Ciamis, West Java. Sussex: Univ. of Sussex.

Taufik, A. 1989. *Formulation of marine protected area management plan in the Segara Anakan, Cilacap.* PHPA, Jakarta.

Tay, K. S. 1989. *Mega-cities in the tropics: Towards an architectural agenda for the future.* Singapore: Institute of Southeast Asian Studies.

Taylor, E. H. 1960. On the caecilian species *Ichthyophis monochrous* and *I. Glutinosus. Univ. Kansas Sci. Bull.* 40: 37-120.

——. 1968. *The caecilians of the world.* Kansas: Lawrence.

Taylor, M. C., and F. B. Soetarto. 1993. *Integrated resource plan: Brantas river region, East Java.* EMDI, Jakarta.

Taylor, N. 1945. *Cinchona in Java.* New York: Greenberg.

Taylor, P. 1977. Lentibulariceae. *Fl. Mal. I* 8: 275-300.

Teguh, S. A. 1978. The food of *Osteochilus hasseltii* in the Rawa Pening Lake. *J. Satya Wacana Res.* 1: 24-32.

Tejwani, K. G., and C. K. Lai. 1992. *Asia-Pacific agroforestry profiles.* Asia-Pacific Agroforestry Network, FAO, Bogor.

Temminck, C. J. 1839. *Verhandingen over de natuurlijke geschiedenid der overzeesche bezittingen (11 Vols, 1839-1844).* Leiden:

Ten Houten, A., N. S. Aminah, T. Suroso, and I. G. Seregeg. 1980. *Effects of diflurobenzuron (DMS 1804) against malaria vectors breeding in lagoons in Bali, Indonesia.* WHO, Jakarta.

Teo, L. W., and Y. C. Wee. 1983. *Seaweeds of Singapore.* Singapore: Singapore Univ.

Press.

Terborgh, J. 1980. Keystone plant resources in the tropical forest. In *Conservation biology*, ed. M. Soulé,Sunderland Mass.: Sinauer.

Terborgh, J., and B. Winter. 1990. Some causes of extinction. In *Conservation biology: An evolutionary-ecological perspective*, ed. M. E. Soule, and Wilcox. B. A., 119-133. Sunderland Mass.: Sinauer.

Terra, G. J. A. 1953. The distribution of mixed gardening on Java. *Landbouw* 52: 163-223.

—. 1954a. Mixed garden horticulture in Java. *Malay. J. Trop. Geog.* 3: 33-43.

—. 1954b. Farm systems in South-East Asia. *Neth. J. Agric. Sci.* 6: 157.

Thalen, D. C. P., and A. C. Smiet. 1985. Quantified "Land-use policy options" in forest land evaluation for watershed management. *Neth. J. Agric. Sci.* 33: 89-103.

Thienemann, A. 1931. German Sunda limnological expedition. *Archiv. Hydrobiol. (Suppl.)* 8: 5-20.

—. 1951. Bilder aus der Binnenfischerei auf Java und Sumatra. *Archiv. Hydrobiol. (Suppl.)* 19: 529-620.

Thijsse, J. P. 1976. A plea for immediate activities to control the menacing deforestation and erosion in mountain areas on Java. *BioIndonesia* 2: 33-42.

Thiollay, J. M., and B. U. Meyburg. 1988. Forest fragmentation and the conservation of raptors: A survey on the island of Java. *Biol. Conserv.* 44: 229-250.

Thohari, M. 1988. Masih adakah harapan untuk menyelamatkan harimau Jawa *(Panthera tigris sondaica)*. *Media Konservasi* 2: 19-23.

Thomas, K. 1994. *Man and the changing world: Changing attitudes 1550-1800*. London: Penguin.

Thorhaug, A. 1985. Largescale seagrass restoration in a damaged estuary. *Mar. Poll. Bull.* 16: 55-62.

Thornton, I. W. B., T. R. New, and P. J. Vaughan. 1988. Colonization of the Krakatau Islands by Psocoptera (Insecta). *Phil. Trans. R. Soc. Lond. B.* 322: 427-443.

Thorton, I. W. B. 1984. *Psocoptera* and Wallace line collections from the island of Bali and Lombok. *Treubia* 29: 83-178.

Tiffen, M. 1991. *Guidelines for the incorporation of health safeguards into irrigation projects through intersectoral cooperation*. Geneva: WHO.

Tilsay-Rusoe, M. V., A. Sjarmidi, and S. Basmi. 1983. *Some aspects of behaviour and ecology of Macaca fascicularis in Pananjung*

Pangandaran and Pulau Peucang Reserve. BIOTROP, Bogor.

Tilson, R. L., and U. S. Seal. 1987. *Tigers of the world: The biology, biopolitics, management and conservation of an endangered species*. New Jersey: Noyes Publications.

Tim Penulis PS. 1991. *Budidaya dan prospek bisnis bekicot*. Jakarta: Penebar Swadaya.

Tinbergen, J., and R. Hueting. 1992. GNP and market prices: wrong signals for sustainable economic success that mask environmental destruction. In *Population, technoloy and lifestyle*. eds R. Goodland, H. E. Daly, and S. El-Serafy,Washington D.C.: Island Press.

Tirtakusumah, T. B. R. 1987. Peran pohon hutan bagi perkembangan perkotaan yang seimbang. *Keberadaan manusia dalam tata ruangnya: Kajian multidisplin mengenai campur tangan manusia terhadap lingkungan hidup di perkotaan*, ed. M. A. Sumhudi. Lembaga Penelitian Usakti, Jakarta.

Tjahjo, D. W. H. 1986. Luas peluang dan kompetisi pakan komunitas ikan di Waduk Bening, Jawa Timur. *Bull. Pen. Perikanan Darat* 5: 69-77.

Tjia, H. D. 1980. The Sunda shelf, southeast Asia. *Zeits. Geomorph.* 24: 405-427.

—. 1984. Holocene shorelines in the Indonesian tin island. *Mod. Quat. Res. S.E. Asia* 8: 103-117.

Tjinda, A. 23 November 1993. Pertamina galakkan pengeboran minyak di DKI dan Jabar. *Suara Pembaruan*, p. 8.

Tjitrosoedirdjo, S. S. 1989. The distribution and potential problems of *Mimosa pigra* in Indonesia. *Biotropica* 2: 18-24.

—. 1991. Notes on the vegetation and manual control of water hyacinth at Rawa Pening Lake, Central Java, 1989-90. *BIOTROP Spec. Publ.* 40: 165-175.

Tjitrosoedirdjo, S. S., and J. Wiraatmodjo. 1984. Water hyacinth management in Java, Indonesia. In Proceedings of the International Conference on Water Hyacinth, Ed. G. Thyacarajan, 176-192. Nairobi, Kenya: UNEP.

Tjitrosoedirdjo, S. S., and F. Widjaja. 1991. Aquatic weed management in Indonesia. *BIOTROP Spec. Publ.* 40: 25-38.

Tjitrosoep, G. 1984. Traditional classification of plants. *Environmentalist* 4 (supplement No. 7): 19-21.

Tjitrosoepomo, G. 1981. *Penelitian gulma air Waduk Sempor (Daerah Saluran Induk Sempor Timur)*. UGM, Yogyakarta.

Tjokrokusumo. 1981. Akibat-akibat pencemaran yang terjadi terhadap lingkungan.

Paper presented at Penataran dan Diskusi Masalah Pencemaran Industri Kimia, Bandung-Cikampek, 9 October 1981.

Tjolerowinoto, M. 1978. The human meaning of technological diffusion: The green revolution in eight Javanese villages. Ph.D. diss., Univ. of Pittsburgh, Pittsburgh.

Tjondronegoro, M. P. S. 1988. Pemukiman kembali penduduk Saguling dan Cirata melalui pembangunan perikanan. Paper presented at Seminar Pemukiman Kembali Penduduk Saguling dan Cirata Melalui Pengembangan Perikanan, Bandung, February 1988.

Tjondronegoro, S. M. P. 1984. *Social organisation and planned development in rural Java.* Singapore: Oxford Univ. Press.

Todd, N. J., and J. Todd. 1984. *Bioshelters, ocean arks, city farming: ecology as the basis of design.* San Francisco: Sierra Club.

Toha, M. 1985. Prinsip-prinsip dan strategi pemugaran hutan di Pulau Jawa. In Prosiding Lokakarya Pemugaran Kawasan Hutan di Pulau Jawa, 51-57. Semarang, 12 September 1985.

Tolboom, G. C. M. 1991. Patterns of agroforestry systems in various zones in the Gunung kidul area, Yogyakarta, Indonesia. Fakultas Kehutanan UGM, Yogyakarta.

Tomascik, T. et al. in prep. *Ecology of Indonesian Seas.* Singapore: Periplus.

Tootell, B. 1985. *"All four engines have failed": The true and triumphant story of Flight BA009 and the Jakarta incident!* London: Deutsch.

Toro, A. V. 1986. Beberapa catatan komposisi fauna Crustacea mangrove gugus Pulau Pari, Kepulauan Seribu, Jakarta. Paper presented at Seminar Ekosistem Hutan Mangrove. Jilid I, Jakarta, 1 February 1927.

—. 1987. Umur pertumbuhan dan catatan status populasi kepiting bakau niaga (*Scylla serrata*) Forskal di perairan mangrove Segara Anakan, Cilacap. In Kumpulan Abstrak Kongres Nasional Biologi VIII, Purwokerto, 8 October 1987.

Toro, A. V., and S. Sukardjo. 1987. Substrat udang windu, *Penaeus monodon* Fabricus, di perairan mangrove Segara Anakan, Cilacap. In *Ekologi tanah dan ekotoksikologi,* ed. S. A. Wirja, and Soenarto, 42-61. Salatiga: Satya Wacana Christian Univ.

—. 1989. Ecology of the large mangrove crab, *Scylla serrata* and its potential fishery in Segara Anakan mangrove waters. *BIOTROP Spec. Publ.* 37: 283.

Toro, V. 1991. Berberapa aspek ekologi udang windu *Penaeus monodon* Fabricius di perairan mangrove Segara Anakan, Cilacap,

Jawa Tengah. Prosidings Seminar IV Ekosistem Mangrove, S. Soemodihardjo, 117-128. Bandar Lampung, 7 August 1990. Jakarta: MAB-LIPI.

Torquebiau, E. 1992. Are tropical agroforestry home gardens sustainable? *Agric. Ecosyst. Environ.* 41: 189-207.

Townsend, C. R. 1980. *The ecology of streams.* London: Edward Arnold.

Treakle, K., and J. Sacko. 1989. Indonesia's pest management miracle. *Greenpeace* 14: 12-13.

Triarto, M. 1991. Sa Naga: Tradition and ceremony. *Voice of Nature* 91: 18-23.

—. 1991. Legon Pakis: A settlement within a national park. *Voice of Nature* 93: 22-23.

—. 1991. The environmental harmony. *Voice of Nature* 91: 14-17.

Trihadiningrum, Y. 1991. Pengelolaan kualitas air sistem Kali Blawi di daerah aliran sungai Bengawan Solo Hilir. Institut Teknologi Sepuluh Nopember, Semarang.

Triwibowo, R. 1979. Melongok lapangan penggembalaan satwa liar: Suaka Margasatwa Meru Betiri (Tempat hidup harimau Jawa terakhir?). *Kehutanan Indonesia* 6: 22-23,27.

Triyoga, L. S. 1991. *Manusia Jawa dan gunung Merapi: Persepsi dan sistem kepercayaan.* Yogyakarta: Gadjah Mada Univ. Press.

Tsukada, E. 1982a. *Butterflies of the Southeast Asian islands: III Satyridae, Libytheridae.* Tokyo: Plapac.

—. 1982b. *Butterflies of the Southeast Asian islands: I Papillionidae.* Tokyo: Plapac.

—. 1985a. *Butterflies of the Southeast Asian islands:IV Nymphalidae(I).* Tokyo: Plapac.

—. 1985b. *Butterflies of the Southeast Asian islands: II Pieridae, Danaidae.* Tokyo: Plapac.

—. 1991. *Butterflies of the Southeast Asian islands:V Nymphalidae (II).* Tokyo: Azumino B.R.I.

Tung, V. W. Y. 1983. *Common Malaysian beetles.* Kuala Lumpur: Longman.

Turnbull, C. January 1982. Bali's new Gods. *Nat. Hist.* 91: 26-29.

Turner, M. 1988. *In the wake of Sir Francis Drake.* Royal Geographic Society, London.

Tweedie, M. W. F. 1983. *The snakes of Malaya.* Singapore: Singapore National Printers.

Tyler, D. E. 1990. The taxonomic status of the Sangiran hominid mandibles of Java, Indonesia. *Am. J. Phys. Anthro.* 81: 309-310

Uecker, F. A., and D. A. Johnson. 1991. Morphology and taxonomy of *Phomopsis* on asparagus. *Mycologia* 83: 192-199.

Uji, T. 1992. Endemic and rare species of Rutaceae and the distribution in Indonesia.

Strategy of Indonesian Flora Conservation Conference, Bogor, 20 July 1992.

Ulmer, G. 1951. Köcherfliegen (Tricopteren) von den Sunda-Inseln. *Archiv. Hydrobiol. Suppl.* 19: 1-528.

Ulrich, H., and V. Pulpanky. 1976. *Ekologi und verhalten der Banteng (Bos javanicus) in Java.* Verlag Paul Parey, Hamburg.

Umbgrove, J. H. F. 1928. De invloed van den wind op de vorming der koraalriffen in de baai van batavia. *Trop. Nat.* 17: 123-130.

——. 1929. De koraalfiffen in de baai van Batavia. *Wet. Mededel. Dienst Mijnb. Ned.-Oost-Indië* 7: 1-69.

UNEP. 1992. Convention on biological diversity. United Nations Environment Programme, Nairobi.

Ungemach, H. 1969. Chemical rain studies in the Amazon region. *Simposio y foro de biologia tropical Amazonica.* pp.354-358. Association pro Biologia Tropical.

USDA. 1985. *Keys to soil taxonomy.* Soil Management Support Services, US Department of Agriculture, Washington D.C.

Usinger, R. L., and R. Matsuda. 1959. *Classification of the Aradidae.* London: British Museum (Natural History).

Utomo, B. S., and M. P. S. Tjondronegoro. 1986. Local resource development and stepwise transfer of technology among the Baduys. In Conference on Local Resource Management for Livelihood Security and Livelihood Enhancement, Bandung, 19 May 1986. Bandung: PSL ITB.

Utomo, T. 1983. Kondisi kualitas air perairan Waduk Wonogiri pada bulan Februari 1982 ditinjau dari aspek fisika, kimia dan biologi. Fakultas Peternakan dan Perikanan, Semarang.

Vaas, K. F. 1947. Biologische inventarisatie van de binnenvisserij in Indonesië. *Landbouw* 19 (522-543):

——. 1950. Iets over de zoetwatervisserij op Java, Borneo en Celebes. *Handel. Hydrobiol. Vereen. (Amsterdam)* 8: 1-20.

Vaas, K. F., and J. J. Schuurman. 1949. On the ecology and fisheries of some Javanese freshwaters. *Comm. Gen. Agricul. Res. Station, Buitenz.* 97: 1-60.

Valckenier-Suringer, J. 1916. Determinatietabel der boomsoorten welke in djatiwonden op Java voorkomen. *Tectona* 9: 879-887.

van Achterberg, C. 1992. Revisionary notes in the subfamily Homolobinae (Hymenoptera, Braconidae). *Zool. Mededel.* 66: 359-368.

van Assendelf, H. B. 1991. *Waterholes, mammals and human impact in the Alas Purwo National Park, East Java, Indonesia.* Fakultas Kehutanan UGM, Yogyakarta.

van Balen, S. 1984. Sight records of the black Baza *Aviceda leuphotes*; new record on Java. *Ardea* 72: 234.

——. 1986a. Short note on the occurrence of grey phase back-headed Bulbul, *Pycnonotus atriceps*, especially on Java. *Kukila* 2: 86-87.

——. 1986b. Some remarks on 'Milky storks and birds of the Java plain'. *Bull. Orient. Bird Cl.* 2: 20-21.

——. 1987a. Measures to increase wild bird populations in urban area in Java. 1: Nest site management. *Media Konservasi* 1: 17-20.

——. 1987b. Forest fragmentation and the survival of forest birds in Java. Proceedings of the Seminar Alumni Deutscher Akademischer Austauschdienst: Perlindungan, konservasi dan perbanyakan populasi satwa liar di Indonesia, 115-165. Bogor, 9 December 1987. Bogor: IPB.

——. 1987c. Birds of climax forest in western Java. *BIOTROP Spec. Publ.* 30: 177-187.

——. 1989a. Measures to increase wild bird populations in urban areas in Java. 2: Management of food supplies and bird plants. *Media Konservasi* 2: 87-95.

——. 1989b. *Report on a short visit to the Lebakharjo Protection Forest (Malang, E. Java), 25-28 October 1989.*

——. 1989c. The role of birds in the biological control of insect pests in Java. *BIOTROP Spec. Publ.* 36: 217-225.

——. 1989d. The terrestrial mangrove birds of Java. *BIOTROP Spec. Bull.* 37: 193-201.

——. 1991a. The Java hawk eagle *Spizaetus bartelsi. Birds of Prey Bull.* 4: 33-39.

——. 1991b. Faunistic notes from Bali with some new records. *Kukila* 5: 125-132.

——. 1992a. Bird watching areas: Gunung Gede Pangrango National Park, West Java, Indonesia. *OBC Bull.* 15: 27-29.

——. 1992b. The first breeding record of the Pacific black duck in Java. *Kukila* 6: 38-39.

——. 1992c. *Bali starling project activity report, Jan-Aug 1992.* ICBP, Bogor.

——. 1992d. Distribution, status and conservation of the forest partridges in the Greater Sundas (Indonesia) with special reference to the chestnut-bellied partridge (*Arborophila javanica*). *Gib. Faune Sauv.* 9: 561-569.

——. 1993a. The decline of the Brahminy kite *Haliastur indus* on Java. *Forktail* 8: 83-88.

——. 1993b. Special review of "A field guide to the birds of Borneo, Sumatra, Java and

Bali". *Bull. Orient. Bird Cl.* 18: 48-51.

—. 1993c. The identification of tit-babblers and red sunbirds on Java. *Bull. Orient. Bird Cl.* 18: 26-28.

—. 1993d. Studies of montane forests and their birds in east Java, Indonesia. *Bull. Orient. Bird Cl.* 18: 14-15.

van Balen, S., and Compost, A. 1988. Overlooked evidence of the short-toed eagle *Circaetus gallichus* on Java. *Kukila* 3: 44-46.

van Balen, S., and I. W. A. Dirgayusa. 1993. *Bali starling project activity report. September 1992-February 1993.* ICBP, Bogor.

van Balen, S., and V. H. Gepak. 1994. The captive-breeding and conservation programme of the Bali starling *Leucopsar rothschildi. Creative conservation: Interactive management of wild and captive animals,* ed. P. J. Olney, G. M. Mace, and A. T. C. Feistner, 420-430. London: Chapman and Hall.

van Balen, S., and D. A. Holmes. in press. Status and conservation of pheasants in the Greater and Lesser Sundas, Indonesia. 5th Pheasant Symposium, Lahore,

van Balen, S., and A. Lewis. 1991. Blue-crowned hanging parrot on Java. *Kukila* 5: 140-141.

van Balen, S., and Noske, R. 1991. Note on two sight records of the yellow-rumped flycatcher on Bali. *Kukila* 5: 142.

van Balen, S., E. T. Margawati, and Sudaryanti. 1988. A checklist of the birds of the Botainical Gardens of Bogor, West Java. *Kukila* 3: 82-91.

van Balen, S., D. M. Prawiradilaga, and M. Indrawan. in press. The distribution and status of green peafowl *Pavo muticus* in Java. *Biol Conserv.*

van Balgooy, M. 1989. Java. In *Floristic inventory of tropical countries,* ed. D. J. Campbell, and H. D. Hammond, 100-102. New York: New York Botanical Garden.

van Bemmel, A. C. V. 1944. The taxonomic position of *Cervus kuhlii* Müll. et Schl. *Dobuto Gaku-Iho (Treubia)* 1: 149-154.

—. 1953. One of the rarest deer in the world. *Beaufortia* 27: 1-5.

van Bemmelen, R. W. 1949. *The geology of Indonesia. Vol. IA.* The Hague: Martinus Nijhoff.

—. 1970. *The geology of Indonesia. Vol. 1A General geology of Indonesia.* The Hague: Nijhoff.

van Bemmel-Lieneman, N. A. 1940. De vogels van het Tenggergebergte. *Trop. Nat.* 29: 93-101.

van Benthem Jutting, T. 1929. On the occurrence of a *Cyclohelix* of Java. *Zool. Mededel.* 11: 1-3.

—. 1932. On prehistoric shells from Sampoeng cave, Central Java. *Treubia* 14: 103-108.

—. 1937. Non-marine Mollusca from fossil horizons in Java with special reference to the Trinil fauna. *Zool. Mededel.* 20: 83-180.

van Benthem Jutting, W. S. S. 1948. Systematic studies on the sub-marine Mollusca of the Indo-Australian archipelago. I: Critical revision of the Javanese operculate landshells of the families Hydrocenidae, Helicenidae, Cylophoridae. *Treubia* 19: 539-604.

—. 1950. Systematic studies on the non-marine Mollusca of the Indo-Australian archipelago. II: Critical revision of the Javanese pulmonate landshells of the families Helicariolidae, Pleurodontiadae, Fruticolidae. *Treubia* 20: 381-505.

—. 1951. Systematic studies on the non-marine Mollusca of the Indo-Australian archipelago. III: Critical revision of the Javanese Pulmonate landsnails of the families Ellobiidae to Limacidae with an appendix. *Treubia* 21: 291-435.

—. 1953. Systematic studies on the non-marine Mollusca of the Indo-Australian archipelago. IV: Critical revision of the freshwater bivalves of Java. *Treubia* 22: 19-73.

—. 1956. Systematic studies on the non-marine mollusca of the Indo-Australian archipelago. V: Critical revision of the Javanese freshwater gastropods. *Treubia* 23: 259-493.

van D. 1938. De bantengjacht op Penandjoeng. *Ned.-Ind. Jager* 8.

van Dam, A. A. 1990. Multiple regression analysis of accumulated data from aquaculture experiments: a rice-fish example. *Aquacul. Fish. Mgmt.* 21: 1-15.

van de Walle, E. 1973. Comments on Benjamin White's "Demand for labor and population growth in colonial Java". *Human Ecol.* 1: 241-242.

van den Bosch, A. C. 1938. Het verdwalen van dieren op bergtoppen. *Trop. Nat.* 27: 169-172.

van den Brink, L. M. 1982. On the mammal fauna of the Wajak Cave, Java (Indonesia). *Mod. Quat. Res. SE Asia* 7: 177-193.

—. 1983. *On the vertebrate fauna of the Goea Djimbe Cave, West Java (Indonesia).* Unpublished report.

van der Brink, J. 1980. The former distribution of the Bali tiger, *Panthera tigris balica* (Schwartz 1912). *Saugetierk. Mitt.* 28: 286-9.

van der Kroef, J. M. 1956. Population pressure and economic development in Indonesia. In *Demographic analysis: selected readings*, ed. J. J. Spengler, and O. D. Duncan, 739-754. Glencoe, Il.

van der Maarel, F. H. 1932. *Contribution to the knowledge of the fossils: Mammalian fauna of Java*. Utrecht: De Rijks-Universiteit te Utrecht.

van der Maessen, L. J. G., and S. Somaatmadja. 1989. *Plant resources of South-East Asia: Pulses*. Wageningen: Pudoc.

van der Mijn, F. E. 1988. *Natural forest land on the volcano Merapi, Java, Indonesia*. Fakultas Kehutanan UGM, Yogyakarta.

van der Paardt, T. 1929. Onbewoond Noord-West bali. *Tijds. K. Ned. Aard. Gen.* 46: 45-77.

van der Pijl, L. 1929. Tampomas. *Trop. Nat.* 18: 87-93.

——. 1933. De kalkflora van Padalarang. *Trop. Nat.* 22: 86-95.

——. 1937. Biological and physiological observations on the inflorescence of *Amorphophallus*. *Rec. Trav. Bot. Neer.* 34: 157-167.

——. 1939. The re-establishment of vegetation on Mt. Goentoer (Java). *Ann. Jard. Bot. Buitenz.* 48: 129-152.

——. 1957. Dispersal of plants by bats. *Acta bot. Neerl.* 5: 135-144.

van der Plas, F. 1971. Lemnaceae. *Fl. Mal. I* 7: 213-218.

van der Poel, P., and H. van Dijk. 1987. Household economy and tree growing in upland central Java. *Agrofor. Sys.* 5: 169-184.

van der Veen, R. 1936. Enkele merkwaardigheden van den Smeroekegel. *Trop. Nat.* 25: 191-193.

van der Zon, A. P. M. 1979. Mammals of Indonesia. FAO, Bogor.

van der Zon, A. P. M., and Supriadi. 1978. *Malang-Ijen-Raung feasibility study*. FAO, Bogor.

van Dijk, H. 1987. Social forestry and farming systems research: Two case studies on Java. *FONC project communication No. 1987 - 3*. FONC - Fakultas Kehutanan UGM, Yogyakarta.

van Dijk, J. W. 1949. *Different rate of erosion within two adjacent basins in Java*. Institute for Soil Research, Buitenzorg.

van Dillewijn, F. L. 1976. Some notes on the development of forestry on Java, with special reference to the Surakarta region. *FAO Upper Solo watershed management project*. FAO, Solo.

van Dreven, K. 1984. *Tourism in Bali: a closer look at the economic and environmental consequences*. Vakgroep Sociale Geografife van Ontwikkelingslanden, Geografisch Instituut, Utrecht.

van Helvoort, B. E. 1984. *A study on bird populations in the ecosystem of West Java, Indonesia: A semi quantitative approach*. Dept. Agriculture Univ. of Wageningen, Wageningen.

——. 1985a. A breeding record of great thick-knee *Esacus magnirostris* (Vieillot) on Bali. *Kukila* 2: 68-90.

——. 1985b. A leopard skin and skull *Panthera pardus* from Kangean Island, Indonesia. *Ziets. Saügetierk.* 50: 182-184.

——. 1986. *Laporan singkat peninjauan Taman Nasional Baluran*. Proyek Jalak Bali II, ——. 1987. Status and conservation needs of the Bali starling. *OBC Bull.* 5: 9-12.

van Helvoort, B. E., and Soetawijaya. 1987. First sight record of black-crested bulbul *Pycnonotus melanicterus* on Bali. *Kukila* 3: 52-53.

van Helvoort, B. E., M. N. Soetawijaya, and P. Hartojo. 1986. *The Bali Starling: a case for wild and captive breeding*. Proyek Jalak Bali II, Gilimanuk.

van Heurn, W. C. 1927. Iets over Buitenzorgsche vleermuizen. *Trop. Nat.* 16: 11-16.

——. 1931. Sawah-ratten-bestrijding in bergterrein. *Inst. Plantenziekten Alg. Proj. Landbouw Bull.* 23: 11-29.

van Hoesel, J. K. P. 1954. *Vipera russellii*, its zoological range and local distribution in Indonesia. *Trop. Nat.* 33: 133-9.

——. 1959. *Ophidia Javanica*. Bogor: Lembaga Pusat Penjelidikan Alam.

van Kampen, P. N. 1909. Das Vorkommon von *Rana hosii* Blgr auf Java. *Bull. Dept. Agri Ind.-Neer.* 25: 1-2.

——. 1923. *The amphibia of the Indo-Australian archipelago*. Leiden: Brill.

van Linschoten, J. H. 1598. *Discourse of voyages into the East and West Indies*. London: Wolfe.

van Lohuizen-de Leeuw, J. E. 1979. Contributions to the study of palaeolithic Patjitan culture, Java, Indonesia. *Stud. S. Asia Cult.* 6: 1-121.

van Marle, J. G. and K. H. Voous. 1988. *The birds of Sumatra*. Tring: British Ornithologists' Union.

van Meer Mohr, J. C. 1923. Beitrage zur Kenntnis de Pliozenfauna Java's (I). *Natuurk. Tijds. Ned.-Ind.* 83: 124-136.

——. 1924a. Beitrage zur Kenntnis de Pliozenfauna Java's (II). *Tijds. Ned.-Ind.* 84: 199-201.

——. 1924b. Uit het intieme leven der Javaansche veldrat (*Rattus brevicaudatus*) Horst et de Raadt. *Trop. Nat.* 13: 65-75.

——. 1926. Beitrage zur Kenntnis de Pliozenfauna Java's (III). *Natuurk. Tijds. Ned.-Ind.* 86: 1-4.

van Oye, P. 1921. Een wanderling naar Tjipanas bij Garut. *Trop. Nat.* 10: 58-62.

——. 1922. De op Java gekweekte karperachtige visschen. *Trop. Nat.* 11: 87-92, 102-106.

van Padang, M. N. 1931. De Merapi (Midden Java). *Trop. Nat.* 20: 99-103.

——. 1936. Het Dieng gebergte. *Trop. Nat.* Jubile Uitgave: 27-36.

——. 1938. De erupties van den Tjeremai in 1937. *Trop. Nat.* 27: 1-10.

van Peenen, P. F. D. 1974. Observations on *Rattus bartelsi. Treubia* 28: 83-117.

van Peneen, P. F. D., S. Atmosoedipuro, Iskandar, and J. Sarwo. 1974. Seasonal distribution of culicine mosquitoes near Jakarta. *J. Med. Entomol.* 11: 425-428.

van Setten van der Meer, N. C. 1979. *Sawah cultivation in ancient Java: Aspects of development during the Indo-Javanese period, 5th to 15th century.* Canberra: ANU Press.

van Steenis, C. G. G. J. 1928. Floristische indrukken van Tjibodas. *Trop. Nat.* 17: 199-207.

——. 1930. Eenige belangrijke plantengeographische vondsten op den Papandajan. *Trop. Nat.* 19: 73-91, 101-108.

——. 1931. Schets van de flora van den G. Tjibodas bij Tjiampea. *Trop. Nat.* 20: 188-191.

——. 1932a. Die Pteridophyten und Phanerogamen der Deutschen limnologischen Sunda Expedition. *Archiv. Hydrobiol. Suppl.* 11: 231-387.

——. 1932b. Fossiele bladafdrukken van den G. Papandajan. *Trop. Nat.* 21: 188-191.

——. 1933a. Report of a botanical trip to the Ranau region, S. Sumatra. *Bull. Jard. Bot. Buitenz.* 13: 1-56.

——. 1933b. Determinatietable voor de Indische zeegrassen. *Trop. Nat.* 22: 43-46.

——. 1934a. On the origin of the Malaysian mountain flora I. *Bull. Jard. Bot. Buitenz.* 13: 135-262.

——. 1934b. On the origin of the Malaysian mountain flora II. *Bull. Jard. Bot. Buitenz.* 13: 289-417.

——. 1934c. Het moerasboschje bij Tjitjadas, Res. Batavia. *Trop. Nat.* 23: 105-109.

——. 1934d. Enkele opmerkingen over het bergland aan de NW. Zijde van den Sanggaboewana (West Java). *Trop. Nat.* 23: 163-167.

——. 1935a. Open air hothouses in the tropics at 3100 m altitude? *Gdns' Bull. Straits Settl.* 9: 64-69.

——. 1935b. Eenige biologische waarnemingen op den Papandajan. *Trop. Nat.* 24: 141-147.

——. 1935c. Schets van de flora van het Eiland Dapoer (Duizend Eilanden). *Trop. Nat.* 24: 31-34.

——. 1936a. On the origin of the Malaysian mountain flora III. *Bull. Jard. Bot. Buitenz.* 14: 56-72.

——. 1936b. Landschap en flora in Indramajoe. *Trop. Nat.* 25: 111-123.

——. 1936c. Openlucht-broeikassen in de tropen op 3100 m boven zee. *Trop. Nat.* 25: 37-44.

——. 1936d. *Osbornia octodonta,* een wenig bekende mangrove-boom. *Trop. Nat.* 25: 194-196.

——. 1937a. De Papandajan met zijn krater en tegals. *Album van Natuurmonumenten in Nederlandsch-Indie.*

——. 1937b. Losse aanteekeningen over de Balische flora. *Trop. Nat.* 26: 69-78.

——. 1938a. Het Danoe-Meer, rapport van een dienstreis naar het Natuurmonument Danoe-Meer of Danoe-Moeras in Bantam. In *3 jaren indisch natuurleven, elfle verslag.* pp.214-222. Batavia: Ned. Ind. Vereen Natuurbescherming.

——. 1938b. Het gelambosch bij Angke-Kapuk (Batavia). *Tectona* 26: 889-901.

——. 1938c. Plantengeographie. *Atlas van tropisch Nederland.* van Steenis. C. G. G. J., Blad 7. Amsterdam: Koninklijk Nederlandsch Aardrijkskundig Genootschap.

——. 1939a. Klimplantsluiers en plantendekens als typen van inzakkende vegetatie. *Trop. Nat.* 28: 141-149.

——. 1939b. Erste naschrift. In *3 jaren indisch natuurleven.* pp.313-314. Jakarta: Nederlandsch-Indische Vereeniging bot Natuurbescherming.

——. 1940a. Vegetatieschetsen van het Idjen hoogland. *Trop. Nat.* 29: 157-161.

——. 1940b. Periodieke massabloei van *Strobilanthes. Trop. Nat.* 29: 88-91.

——. 1941a. Het veentje Rawa Gajonggang bij Tjibeureum boven Tjibodas (N. Gedeh). *Trop. Nat.* 30: 170-172.

——. 1941b. *Rafflesia rochussenii* in de Djampang. *Trop. Nat.* 30: 179-181.

——. 1942. Gregarious flowering of *Strobilanthes* in Malaysia. *150th Anniv. Vol. Ann. R. Bot. Gard. Calc.* : 91-98.

——. 1949a. Accounts of Javan plants collected by Hornstedt in 1783-1784. *Acta Hort. Bergiani* 15: 39-43.

——. 1949b. Podostemaceae. *Fl. Mal. I* 4: 65-68.

—. 1957. Outline of vegetation types in Indonesia and some adjacent regions. *Proc. Pacif. sci. Congr.* 8: 61-97.

—. 1972. *The mountain flora of Java.* Leiden: Brill.

—. 1981. *Flora untuk sekolah di Indonesia.* Jakarta: Pradnya Paramita.

van Steenis, C. G. G. J., and F. Ruttner. 1933. Die Pteridophyten und Phanerogrumen der Deutschen Limnologischen Sunda-Expedition. *Archiv. Hydrobiol. Suppl.* 11: 231-387.

van Steenis, C. G. G. J., and A. F. Schippers-Lammerste. 1965. Concise plant-geography of Java. In *Flora of Java: Vol. I*, ed. C. A. Backer, and R. C. Bakhuizen van den Brink,Groningen: Noordhoff.

van Steenis-Kruseman, M. J. 1950. Desiderata for future exploration. *Fl. Mal. I* 1: 107-116.

—. 1974. Desiderata for future exploration. *Fl. Mal. I* 8: III-IV.

van Strien, N. J. 1978. On the difference in the footprint of Javan and Sumatran Rhinoceros. *Tigerpaper* 5: 16-19.

—. 1981. *Birds of Pulau Dua and Pulau Rambut.* School of Environmental Conservation Management, Ciawi.

—. 1983. *A guide to the tracks of the mammals of Western Indonesia.* School of Environmental Conservation Management, Ciawi.

—. 1986. *Abbreviated checklist of the mammals of the Australian archipelago.* School of Environmental Conservation Management, Bogor.

van Tricht, B. 1929. *Levende antiquiteiten in West-Java.* Batavia: Kolff.

van Vreden, G., and A. L. Ahmadzabidi. 1986. *Pests of rice amd their natural enemies in Peninsular Malaysia.* Wageningen: Pudoc.

van Weers, D. J. 1979. Notes on Southeast Asian porcupines (Hystricidae: Rodentia) IV: On the taxonomy of the subgenus *Acanthion* F. Cuvier, 1823 with notes on the other taxa of the family. *Beaufortia* 29: 215-272.

—. 1985. *Hystrix gigantea*, a new fossil porcupine species from Java (Rodentia: Hystricidae). *Senckenb. Iethaca* 66: 111-119.

—. 1992. A new fossil porcupine *Hystrix vanbreei* n. sp. from the Pleistocene of Java, with notes on the extinct species of the Indonesian archipelago. *Senckenb. Iethaca* 72: 189-197.

van Zeist, W. 1984. The prospects of palynology for the study of prehistoric man in southeast Asia. *Mod. Quat. Res. SE Asia* 8: 1-15.

van Zeist, W., N. A. Polhaupessy, and I. M.

Stuijts. 1979. Two pollen diagrams from West Java, a preliminary report. *Mod. Quat. Res. SE Asia* 5: 43-56.

Vasudevan, P., and S. K. Jain. 1991. Utilization of exotic weeds: an approach to control. In *Ecology of biological invasion in the tropics.* P. S. Ramakrishnan, 157-175. New Dehli: International Scientific Publications.

Veldhuisen-Djajasoebrata, A. 1988. *Weavings of glory and might: The glory of Java.* Rotterdam: Museum voor Volkenkunde.

Vergara, N. T. 1982. *New directions in agroforestry: The potential of tropical legume trees.* Honolulu: East West Center.

Verheij, E. W. M., and R. E. Coronel. 1992. *Edible fruits and nuts.* Bogor: PROSEA.

Vermeulen, J. J. 1991. Notes on the nonmarine molluscs of the island of Borneo. 2: The genus *Ophistostoma* (Gastropoda: Prosobranchia: Diplommatinidae). *Basteria* 55: 139-163.

Vermeulen, J. and A. J. Whitten in prep. *Field guide to the land snails of Bali.* Leiden: Fauna Malesiana Stichtung.

Vermeulen, J. W. C., and A. L. Spaans. 1987. Feeding ecology of Javan Pond Heron *Ardeola speciosa* and Cattle Egret *Bubulus ibis* in North Sulawesi, Indonesia, with additional notes on the occurence of ardeids. *RIN Contrib. Res. Man. Nat. Res. 1987 - 2.*

Verstappen, H. T. 1953a. *Djakarta Bay, a geomorphological study on shoreline development.* The Hague: Tein.

—. 1953b. Oude en nieuwe onderzoekingen over de koraaleidelanden in de baai van Djakarta. *Tijds. K. Ned. Aard. Gen.* 70: 472-478.

—. 1954. Het kustgebied van noordelijk West-Java op de luchtfoto. *Tijds. K. Ned. Aard. Gen.* 71: 146-152.

—. 1956. Landscape development of the Ujung Kulon game reserve. *Penggemar Alam* 36: 37-51.

—. 1975. Landforms and inundations of the lowlands of south-central Java. *ITC J.* 4: 511-520.

—. 1988. Old and new observations on coastal changes of Jakarta Bay: An example of trends in urban stress on coastal environments. *J. Coastal Resources* 4: 573-587.

Verwey, J. 1930. Coral reef studies. I: The symbiosis between damselfishes and sea anenomes in Batavia Bay. *Treubia* 12: 305-366.

—. 1931a. Coral reef studies. II. The depth of coral reefs in relation to their oxygen consumption and their penetration of light in the water. *Treubia* 13: 169-198.

—. 1931b. Coral reef studies. III: Geomorphological notes on the coral reefs of Batavia Bay. *Treubia* 13: 199-216.

Verwij, J. 1987. *Consumption of fuelwood and agricultural wastes in rural household in Gunung Kidul.* FONC - Fakultas Kehutanan UGM, Yogyakarta.

Vickers, A. 1989. *Bali: A paradise created.* Singapore: Periplus.

Vink, G. J. *Dasar-dasar usaha tani di Indonesia.* Jakarta: Yayasan Obor.

Vink, W. 1957. Hamameliaceae. *Fl. Mal. I* 5: 363-379.

Vitousek, P. M., P. R. Ehrlich, A. H. Ehrlich, and P. A. Matson. 1986. Human appropriation of the products of photosynthesis. *BioScience* 36: 368-373.

Vogel, P. 1979. *Zur biologie des Bindenwaras (Varanus salvator) in West Javanischen Natuurschutsgebiet Ujung Kulon.* Basel: Aku-Foto-druck.

von Faber, F. C. 1927. *Die Kraterpflanzen Javas in physiologisch-ökologischer Beziehung.* Weltevreden: Lands-drukkerij.

von Koenigswald, G. H. R. 1976. Climatic changes in Java and Sumatra during the upper Pleistocene. *Kon. Ned. Akad. Wet. B* 79: 232-234.

Vorderman, A. G. 1893. Kangean-vogels. *Natuurk. Tijds. Ned.-Ind.* 59: 201-233.

Vorstman, A. G. 1928a. Some freshwater Bryozoa from West Java. *Treubia* 10: 1-13.

—. 1928b. Freshwater bryozoa from East Java. *Treubia* 10: 163-165.

Waagner, W., E. Koshier, and M. Meredith. 1983. *Java caves 1983: Report of a visit to Indonesia by Austrian and British cavers.* FINSPAC, Bogor.

Wagito. 1986. *Survey bufferzone/zone penyangga Taman Nasional Bromo-Tengger-Semeru tahap II.* UNEJ, Jember.

Wahono, R. I. 1972. *Structure of Metropolitan Jakarta: A study in ecology, population and social organization.* Jakarta: UI.

Wahyono, M. 1982. *Fungsi dan arti gua dalam masyarakat Jawa.* HIKESPI, Bogor.

—. 1987. Pencarian manusia purba di Jawa. *Suara Alam* 51: 15-19.

Wahyuni, E., and W. Ismail. 1987. Living condition of *Scylla serrata* Forskål around Tanjung Pasir waters, Tangerang, West Java. *J. Pen. Perikanan Laut* 38: 59-68.

Wahyuni, I. S., S. Nuraini, and W. A. Pralampita. 1987. Notes on the aquatic resources use in Segara Anakan (Cilacap). *J. Pen. Perikanan Laut* 39: 73-79.

Wainscoat, J. S. 1986. Evolutionary relationships of human populations from an analysis of nuclear DNA polymorphism. *Nature* 319: 491-493.

Walker, B. H. 1987. *Determinants of tropical savannas.* Oxford: IRL.

—. 1992. Biodiversity and ecological redundancy. *Conserv. Biol.* 6: 18-23.

Walker, S. H. et al. 1977. *Development of the water resources of Bali: A master plan.* Surbiton: Ministry of Overseas Development.

Walker, S. R., and A. R. Rabinowitz. 1992. The small mammal community of a dry-tropical forest in central Thailand. *J. Trop. Ecol.* 8: 57-71.

Wall, G. 1995. Forces for change: tourism. In *Bali: Balancing environment, economy and culture,* ed. S. Martopo and B. Mitchell. 57-74. Waterloo: Dept. Geography, University of Waterloo.

Wallace, A. R. 1860. On the zoological geography of the Malay archipelago. *J. Linn. Soc. Lond.* 14: 172-184.

—. 1869. *The Malay archipelago.* London: MacMillan.

Walter, H. 1971. *Ecology of tropical and subtropical vegetation.* Edinburgh: Oliver and Boyd.

Waltham, A. C. 1983. The caves of Gunung Sewu, Java. *Cave Sci.* 10: 55-96.

Waltner-Toews, D. 1990. An epidemic of canine rabies in central Java, Indonesia. *Prevent. Vet. Med.* 8: 295-304.

Waltner-Toews, D., A. Maryono, B. T. Akoso, S. Wisynu, D. H. A. Unruh. 1990. An epidemic of canine rabies in Central Java, Indonesia. *Prevent. Vet. Med.* 8: 295-304.

Wanner, H. 1973. Soil respiration in different types of Southeast Asian tropical rain forest. *Oecologia* 12: 289-302.

Ward, P. 1968. Origin of the avifauna of urban and suburban Singapore. *Ibis* 110: 239-255.

Wardoyo, S. T. H., R. T. M. Sutamihardja, and R. C. Tarumengkeng. 1979. A study on water pollution problem of the Surabaya River, East Java. In Proceedings of the 5th International Symposium of Tropical Ecology. Part I, Ed. J. I. Furtado, 629-633. Kuala Lumpur, 16 April 1979.

—. 1980. A case study on water pollution problem of the water pollution problem of the Surabaya river, East Java. In *Tropical ecology and development,* ed. J. I. Furtado, 629-633. Kuala Lumpur: Univ. of Malaya Press.

Waring, R. H. 1989. Ecosystems: fluxes of matter and energy. In *Ecological concepts: The contribution of ecology to an understanding of the natural world,* ed. J. M. Cherrett, 17-41. Oxford: Blackwell.

Wasser, H. J., and J. R. E. Harger. 1992. *Several environmental factors affecting the rainfall in Indonesia.* UNESCO/ROSTSEA, Jakarta.

Wastitodi, P. 1985. *Beberapa aspek pencemaran limbah industri-industri sekitar proyek irigasi Jatiluhur terhadap kualitas perairan Waduk Curug, Karawang, Jawa Barat.* UGM, Yogyakarta.

Waterloo, M. J. et al. 1993. The imp[act of converting grassland to pine forest on water yield in Viti Levu, Fiji. Hydrology of warm humid regions, ed J. S. Gladwell, 149. Yokohama, 13 July 1993. IAHS.

Watling, R. J. 1990. *Period report January - March 1990: National Parks management project.* Anzdec Consultants, Jakarta.

——. 1991a. *Resource information manual for Indonesian national parks.* World Bank National Park Project PHPA, Jakarta.

——. 1991b. *Fauna and vegetation for Indonesian national parks.* World Bank National Park Project PHPA, Jakarta.

Watson, J. G. 1928. Mangrove forests of the Malay Peninsula. *Malay. For. Rec.* 6.

Watson, R. E., and Kottelat, M. 1994. *Lentipes whittenorum* and *Sicyopus auxilimentus,* two new species of freshwater gobies from the western Pacific (Teleostei: Gobiidae: Sicydiinae). *Ichthyol. Explor. Freshwaters* 5: 351-364.

WCED. 1987. *Our common future.* Oxford: Oxford Univ. Press.

WCMC. 1990. *Red list of threatened animals.* Cambridge: World Conservation Monitoring Centre.

Webb, D. A. 1985. What are the criteria for presuming native status? *Watsonia* 15: 231-236.

Webb, L. J. 1959. A physiognomic classification of Australian rain forests. *J. Ecol.* 47: 551-570.

Weber, H. C., and Sunaryo. 1990. Balanophoraceae, extreme flowering plants with fungal characteristics. *Biol. Rundsch* 28: 83-86.

Weber, M. and L. F. de Beaufort. 1913. *The fishes of the Indo-Australian archipelago II. Malacopterygii, Myctophoidea, Ostariophysi: I Siluroidea.* Leiden: Brill.

——. 1916. *The fishes of the Indo-Australian archipelago III. Ostariophysi: II Cyprinoidea, Apodes, Synbranchi.* Leiden: Brill.

——. 1922. *The fishes of the Indo-Australian archipelago IV. Heteromi, Solenichthyes, Synentognathi, Percesoces, Labyrinthici, Microcyprini.* Leiden: Brill.

Wee, Y. C., and R. Corlett. 1986. *The city of the forest: Plant life in urban Singapore.* Singapore: Singapore Univ. Press.

Weesie, P. D. M. 1982. The fossil bird remains in the Dubois collection. *Mod. Quat. Res. SE Asia* 7: 87-90.

Weidenreich, F. 1978. *Giant early man from Java and South China.* New York: AMS Press.

Weitzel, V., and C. P. Groves. 1985. The nomenclature and taxonomy of the colobine monkeys of Java. *Int. J. Primatol.* 6: 399-409.

Weitzel, V., C. M. Yang, and C. P. Groves. 1988. A catalogue of primates in the Singapore Zoological Reference Collection. *Raffles Bull. Zoology* 36: 1-166.

Welcomme, R. L. 1988. International introductions of inland aquatic species. *FAO Fish. Tech. Pap.* 294: 1-318.

Wells, D. R. 1989. Eyebrowed thrushes *Turdus obscurus* in Bali. *Kukila* 4: 149.

Wells, M. 1992. Biodiversity conservation, affluence and poverty: mismatched costs and benefits and efforts to remedy them. *Ambio* 21: 237-243.

Wells, M., and K. Brandon. 1992. *People and parks: Linking protected area management with local communities.* Washington D.C.: World Bank.

Wells, S. M., R. M. Pyle, and N. M. Collins. 1983. *The IUCN invertebrate red data book.* Gland: IUCN.

Went, F. A. F. 1930. De watervallenplant van Java, *Cladopus nymani* Moller. *Trop. Nat.* 19: 53-60.

West, J. 1980. *Studi macan tutul dan burung gosong di Kepulauan Kangean, Jawa Timur.* Direktorat Perlindungan dan Pengawetan Alam, Bogor.

Weston, A. 1987. Forms of Gaian ethics. *Env. Ethics* 9: 217-230.

Westphal, E., and P. C. M. Jansen. 1989. *Plant resources of South-East Asia: A selection.* Wageningen: Pudoc.

Wetmore, A. 1940. Avian remains from the Pleistocene of Central Java. *J. Palaeontol.* 14: 447-450.

Wewalka, G., and O. Bistrom. 1989. *Hyphydrus jaechi,* new species and *Hyphydrus prinzi,* new species from Java and central Africa (Coleoptera: Dytiscidae). *Ann. Entomol. Fennica* 55: 129-132.

Wheatley, B. P., A. Fuentes, and D. K. H. Putra. 1993. The primates of Bali. *Asian Prim.* 3: 1-2.

Whendrato, I., and I. M. Madyana. 1989. *Budidaya bekicot.* Semarang: Eka.

White, A. 1985. *Nat. Res.* 11: 13-20.

White, A. T., L. M. Chou, M. W. R. N. de Silva, and F. Y. Guarin. 1990. *Artificial reefs for marine habitat enhancement in Southeast Asia.* ICLARM, Manila.

White, A. T., P. Martosubroto, and M. S. M. Sandorra. 1989. *The coastal environmental profile of Segara Anakan, Cilacap, South Java, Indonesia.* ICLARM, Manila.

White, L. 1967. Historical roots of our ecologic crisis. *Science* 155: 1203-7.

Whitmore, T. C. 1972a. Euphorbiaceae. In *Tree flora of Malaya, Vol. 1,* ed. T. C. Whitmore, 34-136. Kuala Lumpur: Longman.

——. 1972b. Aceraceae. In *Tree flora of Malaya, Vol. 2,* ed. T. C. Whitmore, 1-2. Kuala Lumpur: Longman.

——. 1972c. Hamamelidaceae. In *Tree flora of Malaya,* ed. T. C. Whitmore, 237-243. Kuala Lumpur: Longman.

——. 1977. *Palms of Malaya, 2nd ed.* Oxford: Oxford Univ. Press.

——. 1984. *Tropical rain forests of the Far East, 2nd ed.* Oxford: Clarendon.

——. 1989a. Phytogeography of the eastern end of Tethys. In *Gondwana and Tethys,* ed. M. G. Audley-Charles, and A. Hallam, 307-311. London: Geological Society.

——. 1989b. Forty years of rain forest ecology 1948-1988 in perspective. *GeoJournal* 19: 347-360.

——. 1989c. Guidelines to avoid remeasurement problems in permanent sample plots in tropical rain forests. *Biotropica* 21: 282-283.

——. 1989d. Canopy gaps and the two major groups of forest trees. *Ecology* 70: 536-538.

——. 1990. *An introduction to tropical rain forests.* Oxford: Clarendon.

——. 1991a. Tropical rain forest dynamics and its implications for management. In *Rain forest regeneration and management.* A. Gómez-Pompa, T. C. Whitmore, and M. Hadley, 67-89. Paris: UNESCO.

——. 1991b. Invasive woody plants in perhumid tropical climates. In *Ecology of biological invasion in the tropics.* P. S. Ramakrishnan, 35-40. New Delhi: International Scientific Publications.

Whitmore, T. C., and J. A. Sayer. 1992. Deforestation and species extinctions in tropical moist forests. In *Tropical deforestation and species extinction.* T. C. Whitmore, and J. A. Sayer, 1-14. London: Chapman and Hall.

Whitmore, T. C., I. G. M. Tantra, and U. Sutisna. 1989. *Tree flora of Indonesia: Checklist for Bali, Nusa Tenggara and Timor.* Forest Research and Development Centre, Bogor.

Whittaker, R.J., and S.H. Jones. 1994. The role of frugivorous bats and birds in the rebuilding of a tropical forest ecosystem,

Krakatau, Indonesia. *J. Biogeog.* 21: 245-258.

Whitten, A. J. 1982. Diet and feeding behaviour of Kloss Gibbons on Siberut Island, Indonesia. *Folia Primatol.* 37: 177-208.

——. 1989. Pelicans at Cengkareng. *Kukila* 4: 64.

Whitten, A. J. 1995. Natural areas and nature of inland Bali. In *Bali: Balancing environment, economy and culture,* ed. S. Martopo and B. Mitchell. 237-262. Waterloo: Dept. Geography, University of Waterloo.

Whitten, A. J., and S. J. Damanik. 1986. Mass defoliation of mangroves in Sumatra, Indonesia. *Biotropica* 18: 176.

Whitten, A. J., and C. McCarthy. 1993. List of the amphibians and reptiles of Java and Bali. *Trop. Biodiv.* 1: 169-177.

Whitten A. J., M. Mustafa, and G. S. Henderson. 1987a. *Ekologi Sulawesi.* Yogyakarta: Gadjah Mada Univ. Press.

Whitten, A. J., M. Mustafa, and G. S. Henderson. 1987b. *The ecology of Sulawesi.* Yogyakarta: Gadjah Mada Univ. Press.

Whitten, A. J., S. J. Damanik, J. Anwar, and N. Hisyam. 1987c. *The ecology of Sumatra, 2nd Ed.* Yogyakarta: Gadjah Mada Univ. Press.

Whitten, A. J., H. Haeruman, H. Alikodra, and M. Thohari. 1987d. *Transmigration in Indonesia.* Cambridge: IUCN.

Whitten, A. J., and J. E. J. Whitten. 1992. *Wild Indonesia.* London: New Holland.

——. 1993. A naturalist's guide to surf, sand and sea. *Bali.* E. Oey, 102-105. Singapore: Periplus.

Whyte, R. O. 1972. The Graminae, wild and cultivated, of monsoonal and equatorial Asia. *Asian Persp.* 15: 127-151.

——. 1974. Grasses and grasslands. In *Natural resources of humid tropical Asia.* pp.239-264. Paris: UNESCO.

Wibowo, C. 1985. *Studi beberapa perilaku rusa (Cervus timorensis) di Pulau Peucang, Taman nasional Ujung Kulon dan "game ranching" Hutan Tridharma Gunung Walat, Sukabumi.* Fakultas Kehutanan IPB, Bogor.

Wickham, T. W. 1992. *Indigenous environmental knowlege, institution and customary law: linking village sustainability and policy in Bali, Indonesia.* Bali Sustainable Development Project, Sanur.

——. 1993. Farmers ain't no fools: exploring the role of participatory rural appraisal to access indigenous knowledge and enhance sustainable development, research and planning. A case study of Dusun Pausan, Bali, Indonesia. Univ. of Waterloo, Waterloo, Ontario.

Wickramsinghe, E. D. 1990. Conservation of

endemic rain forest fishes of Sri Lanka: results of a translocation experiment. *Conserv. Biol.* 4: 32-37.

Widiarti, A., and R. Effendi. 1989. Socio-economic aspects of tambak forest in mangrove forest complex. *BIOTROP Spec. Publ.* 37: 275-279.

Widjaja, E. A., and Lester R. N. 1987. Morphological anatomical and chemical analysis of *Amorphophallus paeoniifolius* and related taxa. *Reinwardtia* 10: 271-280.

Widyanto, A. 19 June 1992. Musik alam di gua Tetabuhan. *Bisnis Indonesia*, p.8.

Widyanto, L. S. 1979. Prospect of some industrial and semi-industrial uses of water hyancinth (*Eichhornia crassipes* (Mart.) Solms). In Proceeding of the 2nd Seminar on Aquatic Biology and Aquatic Management of the Rawa Pening Lake, 336-349. Salatiga, 29 October 1979. Salatiga: Satya Wacana Christian Univ.

Wiegant, W., and B. E. van Helvoort. 1987. First sighting of *Tachybaptus novaehollandiae* on Bali. *Kukila* 3: 50-51.

Wiersum, K. F. 1975. *Introduction to the principles of forest hydrology and erosion with special reference to Indonesia.* Lembaga Ekologi, UNPAD, Bandung.

——. 1976. The fuelwood situation in the upper Bengawan Solo River Basin. *Upper Solo watershed management and upland development project report.* USAID Indonesia, Solo.

——. 1978. Bosbouw in Indonesie. *Ned. Bosbouw Tijds.* 50: 301-316.

——. 1980. Possibilities for use and development of indigenous agro-forestry systems for sustained land use on Java. In *Tropical ecology and development*, ed. J. Furtado, 515-521. Kuala Lumpur: Univ. of Malaya Press.

——. 1981. *Observations on agroforestry in Java, Indonesia: Report on an agroforestry course.* Fakultas Kehutanan, Gadjah Mada Univ., Yogyakarta.

——. 1982. Tree gardening and taungya on Java: Examples of agroforestry techniques in the humid tropics. *Agrofor. Sys.* 1: 53-70.

——. 1985. Effects of various vegetation layers in an *Acacia auriculiformis* forest plantation on surface erosion in Java, Indonesia. *Soil erosion and conservation*, ed. S. A. El-Swaify, W. C. Moldenhauer, and A. Lo. Soil Conservation Society of America, Ankeny, Iowa.

Wijaya, E. 1980. Indonesia. In *Bamboo research in Asia*. G. Lessard, and A. Chouinard, 63-68. Singapore: IDRC.

Wijaya, F., P. Suwignyo, and M. Taufan. 1991. Practical implementation of aquatic weed management in Situ Leutik Darmaga campus, Bogor, Indonesia. *BIOTROP Spec. Publ.* 40: 117-132.

Wijayakusuma, H. M. H. et al. 1992. *Tanaman berkhasiat obat Indonesia: Jilid I.* Jakarta: Pustaka Kartini.

——. 1993. *Tanaman berkhasiat obat Indonesia: Jilid II.* Jakarta: Pustaka Kartini.

Wilcox, B. A. 1980. Insular ecology and conservation. In *Conservation biology: An evolutionary-ecological perspective*, ed. M. E. Soulé, and B. A. Wilcox, 95-117. Sunderland, Mass.: Sinauer.

Wilde, C. J. M. 1938. Vogelparadijs op Java's Noordkust. *3 jaren indisch natuurleven*, ed. C. G. G. J. van Steenis, 299-315.

Wildt, D. E. et al. 1987. Reproductive and genetic consequences of founding isolated lion populations. *Nature* 329: 328-330.

Wilkinson, L. 1980. *Earthkeeping: Christian stewardship of natural resources.* Grand Rapids, MI: Eerdmans.

Willemse, C. 1955. Synopsis of the Acridoidea of the Indo-Malayan and adjacent regions (Insecta, Orthoptera). Part II. Family Acrididae, subfamily Catantopinae, part one. *Nat. Gen. Limburg* 8:

——. 1957. Synopsis of the Acridoidea of the Indo-Malayan and adjacent regions (Insecta, Orthoptera). Part II. Family Acrididae, subfamily Catantopinae, part two. *Nat. Gen. Limburg* 10:

Willemse, F. 1967. Additional data on some genera and species of Acrididea (Orthoptera, Acridoidea) from the Indo-Malayan region. *Tijd. Entomol.* 110: 381-399.

Willemsen, G. F. 1986. *Lutrogale paleoleptonyx* (Dubois 1908), a fossil otter from Java in the Dubois collection. *Proc. K. Ned. Akad. Wet., Ser. B* 89: 195-200.

Williams, E. 1957. *Hardella isoclina* Dubois redescribed. *Zool. Med.* 35: 235-240.

Willis, R. G., C. Boothroyd, and N. Briggs. 1984. The caves of Gunung Sewu, Java. *Cave Sci.* 11: 119-153.

Willoughby, N. G. 1986. Man-made litter on the shores of the Thousand Island Archipelago. *Mar. Poll. Bull.* 17: 224-228.

Wilson, D. E., and D. M. Reader. 1993. *Mammal species of the world.* Washington D.C.: Smithsonian Instution Press.

Wind, J. 1978. Banteng and deer problems in Indonesian and Baluran game reserves. Paper presented at the BIOTROP Training Course on Wildlife Population Dynamics, 15pp. Bogor, 6 March 1978.

——. 1991. *Buffer zone management for Indonesian national parks. Vol. 1. A concept for park protection and community participation.* World Bank National Parks Development Project/ DHV/RIN Consultancies, Bogor.

—. 1992. *Buffer zone management for Indonesian national parks. Vol. 2. tools for planning, monitoring and evaluation.* World Bank National Parks Development Project/ DHV/RIN Consultancies, Bogor.

Wind, J., and H. H. Amir. 1977. *Proposed Baluran National Park: Management plan 1978/79 - 1982/83.* FAO, Bogor.

Wind, J., and B. K. Soesilo. 1978. *Proposed Halimun Nature Reserve management plan 1979-1982.* FAO, Bogor.

Winemiller, K. O. 1990. Spatial and temporal variation in tropical fish trophic networks. *Ecol. Monogr.* 60: 331-367.

Winston, J. A., and B. F. Heimberg. 1986. Bryozoans from Bali, Lombok and Komodo, Indonesia. *Am. Mus. Nov.* 2847: 1-49.

Wiradi, G. 1986. Off-farm employment in the development of rural Asia. In ????, Ed. R. T. Shand, 309-326. Chiang Mai, 23 August 1983. Canberra: National Centre for Development Studies ANU.

Wiradinata, S., and E. Husaeni. 1987. Land rehabilitation with agroforestry systems in Java, Indonesia. In *Agroforestry in the humid tropics*, ed. N. T. Vergara, and N. D. Briones, 133-148. Honolulu: East-West Center.

Wirakusumah, S. 1987. *Suatu pemikiran program hutan kota untuk Jakarta.* UI, Jakarta.

Wiriadinata, H. 1977. Perjalanan ke Hutan Cisamporawangun, Lengkong dan Gunung Hanjuang Barat (Jawa Barat). *Bull. Kebun Raya* 3: 85-88.

Wiriosoepartho, A. S. 1984. Daya dukung lapangan perumputan satwa liar herbivora dan aspek konservasinya di Taman Nasional Baluran, Jawa Timur. *Bull. Pen. Hutan 428.*

—. 1985. Peranan daerah penyangga dalam pembinaan populasi margasatwa di Taman Nasional Ujung Kulon. *Bull. Pen. Hutan* 467: 45-59.

—. 1986a. Penggunaan habitat dan keragaman jenis burung merandai di Cagar Alam Pulau Rambut. *Bull. Pen. Hutan* 468: 51-60.

—. 1986b. Studi habitat dan populasi kalong (*Pteropus vampyrus* L.) di Pulau Rambut. *Bull. Pen. Hutan* 479: 11-27.

Wirjahardja, S. 1987. Rice cultivation in Indonesia. *Weeds of rice in Indonesia.* M. Soerjani, A. J. G. Kostermans, and G. Tjitrosoepomo, Jakarta: Balai Pustaka.

Wirjodarmodjo, H. 1978. Pengelolaan hutan bakau Cilacap. *Duta Rimba* 4: 11-23.

—. 1985. Pemugaran hutan di Pulau Jawa. In Prosiding Lokakarya Pemugaran Kawasan Hutan di Pulau Jawa, 31-50. Semarang, 12 September 1985.

Wirjodarmodjo, H., and M. Bratamihardja. 1984. Policies, strategies and designs of forest development on the island of Java. In Strategies and designs for aforestation, reforestation and tree planting, Ed. K. F. Wiersum, 363-374. Wageningen, 19 September 1983. Wageningen: Centre for Agricultural Publishing and Documentation.

Wisse, C. A. 1919. Trinil. *Trop. Nat.* 5: 7-9.

Wisseman, J. 1981. Review of N.C. van Setten van der Meer. Sawah cultivation in ancient Java: Aspects of development during the Indo-Javanese period, 5th to 15th century. *Indo. Circle* 24: 49-52.

Wisudo, and P. Bambang. 2 February 1992. Pengembangan wisata Karimunjawa: berpacu melawan kerusakan alam. *Kompas Minggu,* 8.

Witkamp, H. 1916. Een wandeling langs het bandjirkanaal van de Tjiliwong. *Trop. Nat.* 5: 7-9.

—. 1938. De Omgeving van Watoe Oeloh (Besoeki). *Trop. Nat.* 27: 185-192.

—. 1940. Nog iets over de vulkanruines bij Plered. *Trop. Nat.* 29: 56-58.

Wodzicka-Tomaszewska, M., S. Gardiner, A. Djajanegara, I. M. Mastika, and T. Rustanda. 1993a. *Small ruminant production in the humid tropics.* Solo: Univ. Sebelas Maret Press.

—. 1993b. *Produksi ruminant kecil di Indonesia.* Solo: Univ. Sebelas Maret Press.

Wood, J. J., and J. B. Comber. 1986. New orchids from Java and Bali. *Kew Bull* 41: 698-700.

Wood, J. J., and J. B. Comber. 1988. A new species of *Dendrobium* section Pedilonum from Java. *Orchid Rev.* 96: 51-53.

Wood, M. 1976. *Paphiopedilum victoria-regina. Orchid Rev.* 84: 133-143.

Wood, T. W. 1871a. On a new species of argus pheasant. *Ann. Mag. Nat. Hist.* 8: 67-68.

—. 1871b. A new species of argus pheasant. *Field* 8 April: 281.

World Bank. 1988. *Indonesia: The transmigration program in perspective.* Washington D.C.: World Bank.

—. 1990a. *Indonesia: Sustainable development of forests, land and water.* Washington D.C.: World Bank.

—. 1990b. *Vetiver grass: The hedge against ero-*

sion. Washington D.C.: World Bank.

——. 1993. *East Asia miracle.* Oxford: Oxford Univ. Press.

Wright, D. G. 1982. *A discussion paper on the effects of explosives on fish and marine mammals in the waters of the Northwest Territories.*

WRI/IUCN/UNEP. 1992. *Global biodiversity strategy.* Washington D.C.: World Resources Institute.

Wulijarni-Soetjipto, N. and J. S. Siemonsma. 1993. *Bibliography 2: Edible fruits and nuts.* Bogor: PROSEA.

Wustamidin. 1986. *Ekologi lahan tinggi, masalahnya serta usaha-usaha penanganannya di Kabupaten Jember.* Pusat Penelitian UNEJ, Jember.

——. 1988. *Manfaat daerah penyangga untuk pengembangan Taman Nasional Baluran.* Pusat Penelitian UNEJ, Jember.

Wuster, W., and R. S. Thorpe. 1989. Population affinities of the asiatic cobra species complex in SE Asia: Reliability and random sampling. *Biol. J. Linn. Soc.* 36: 391-409.

Wyeth, J., and C. Boothroyd. 1990. *The caves of Cipining.* Unpubl. ms.

Yafuso, M., and T. Okada. 1990. Complicated routes of the synhospitalic pairs of the genus *Colocasiomyia* in Java, Indonesia with descriptions of two new species (Diptera: Drosophilidae). *Esakia (Special Issue)* 1: 137-150.

Yalden, B. W., and P. A. Morris. 1975. *The lives of bats.* London: David and Charles.

Yamada, I. 1976a. Forest ecological studies of the montane forest of Mt. Pangrango, West Java I: Stratification and floristic composition of the montane rain forest near Cibodas. *Tonan Ajia Kenkyu* 13: 402-426.

——. 1976b. Forest ecological studies of the montane forest of Mt. Pangrango, West Java II: Stratification and floristic composition of the forest vegetation of the higher part of Mount Pangrango. *Tonan Ajia Kenkyu* 13: 513-534.

——. 1976c. Forest ecological studies of the montane forest of Mt. Pangrango, West Java III: Litterfall of the tropical montane forest near Cibodas. *SEA Studies* 14: 194-229.

——. 1977. Forest ecological studies of the montane forest of Mt. Pangrango. *SEA Studies* 15: 226-254.

——. 1990. The changing pattern of vertical stratification along an altitudinal gradient of the forests of Mt. Pangrango, West Java. In *The plant diversity of Malesia,* ed. P. Baas, K. Kalkman, and R. Geesink, 177-191. Dordrecht: Kluwer.

Yaman, A. R. 1991. *Sustainable development for forests and protected areas in Bali.* University of Waterloo, Waterloo.

Yamazaki, T. 1990. Three new species of *Torenia* from South-Eastern Asia. *J. Jap. Bot.* 65: 261-265.

Yates, B. F. 1993. The implementation of coastal zone management policy: Kepulauan Seruibu Marine Park, Indonesia. EMDI, Jakarta.

Yorke, C. D. 1984. Avian community structure in two modified Malaysian habitats. *Biol. Conserv.* 29: 345-362.

Yoshii, R., and Y. R. Suhardjono. 1989. Notes on the collembolan fauna of Indonesia and its vicinities. I. Miscellaneous notes with special references to Sairini and Lepidocyrtini. *Acta Zool. Asiae Orient.* 1: 23-90.

Yule, H. 1903. *The book of Sir Marco Polo the Venetian concerning the kingdoms and marvels of the East.* London: Murray.

Yusuf, E., and O. Rosjidian. 1983. *Laporan perjalanan ke Gunung Muria dan sekitarnya (tanggal 15 Februari s/d 1 Maret).* LBN - LIPI, Bogor.

Yusuf, R., M. Siregar, and Purwaningsih. 1988. *Struktur dan komposisi vegetasi hutan Gunung Kendeng dan Pasir Luhur Cagar Alam Gunung Halimun Jawa Barat.* Herbarium Bogoriense, Bogor.

Zaidi, I. H. 1981. On the ethics of man's interactions with the environment: an Islamic approach. *Env. Ethics* 3: 35-47.

Zanefeld, J. S., and D. G. Montague. 1958. "Boompjes-Eiland": Een koraleiland in de Java-zee. *Tijds. K. Ned. Aard. Gen.* 67: 715-745.

Zanefeld, J. S., and H. T. Verstappen. 1952a. A recent investigation about the geomorphology and the flora of some coral islands in the Bay of Djakarta (Part I). *J. Sci. Res.* 1: 38-43.

——. 1952b. A recent investigation about the geomorphology and the flora of some coral islands in the Bay of Djakarta (Part 2). *J. Sci. Res.* 1: 58-68.

Zerner, C. 1992. Development of the small-scale freshwater cage culture fishery in reservoirs in Java: legal, environmental and socioeconomic issues. In *Contributions to fishery development policy in Indonesia.* R. B. Pollnac, C. Bailey, and A. Poenomo, 78-86. Jakarta: Central Research Institute for Fisheries.

Zimmerman, H. C. O. 1938. Bali berindingen (herten tijgerafschot). *Ned.-Ind. Jager* 8: 49.

Zollinger, H. 1845a. Een uitstapje naar het

eiland Balie. *Tijds. Ned.-Ind.* 7 (4): 1-56.

——. 1845b. Eenige bijdragen tot de natuurlijke geschiedenis der *Rafflesia patma*. *Gen. Arch. Ned.-Ind.* 2: 553-554.

Zuhud, E. A. 1983. Studi beberapa aspek ekologi babi hutan *(Sus scrofa)* dan kerusakan yang disebabkannya di Desa Cinagara, calon daerah penyangga Taman Nasional Gunung Gede Pangrango. Fakultas Kehutanan IPB, Bogor.

——. 1988. Lingkungan hidup *Rafflesia zollingeriana* Kds. di Taman Nasional Meru Betiri. *Media Konservasi* 2: 25-30.

Zwerver. 1941. Naar het schiereiland 'Blambangan' of 'Poerwo'. *Ned.-Ind. Jager* 11: 17.

Index

948

For nine months of the year the residents of Sekartaji in southern Nusa Penida, Bali, must bring all their and their animals' water from a spring near the base of a 150-m sheer cliff. Plans are in place to provide piped water by hydraulic pumping from other springs. (By A.J. Whitten)

Inappropriate development of some tourist beaches has disturbed the natural dynamic process requiring the construction of unsightly groynes to slow erosion, but which lead to serious problems of their own; Sanur, Bali. (By R.E. Soeriaatmadja)

Severe pressures on land have led to the carving out of rice terraces on even the steepest slopes between Pacitan and Ponorogo, East Java. (By A.J. Whitten)

The ash ejected during the eruption of Mt. Galunggung, West Java, in 1982 devastated large areas. (By Alain Compost/Bruce Coleman Ltd.)

Java and Bali are dominated by volcanoes, the highest and one of the most active of which is Mt. Semeru, East Java. (By Gerald Cubitt/Bruce Coleman Ltd.)

In 1894 the water hyacinth was introduced from Brazil to Bogor Botanical Gardens from which it spread to become a major aquatic weed. (By C.B. and D.W. Frith/Bruce Coleman Ltd.)

The conversion of unstable slopes from forest gardens to terraced rice resulted in landslip which buried and killed 57 people. (By R.E. Soeriaatmadja)

Miocene tree trunks can be purchased at the roadside near Jasinga, West Java. (By A.J. Whitten)

Paphiopedilum glaucophyllum of low hills in West and East Java.

Dendrobium jacobsonii from mountains in eastern Java.

Endemic orchids of Java.

Bulbophyllum pahudii with a smell of rotten fish from West and Central Java.

Phalaenopsis javanica, a rare plant of West Java. (By J. Comber)

Banteng, here in Ujung Kulon National Park, West Java, are found in several of the conservation areas of Java and Bali. (By Gerald Cubitt/Bruce Coleman Ltd.)

The green peafowl is found across Java and is threatened in some areas by trapping. (By Dieter and Mary Plage/Bruce Coleman Ltd.)

The Javan lapwing is almost certainly extinct. Drawn from specimens in the National Museum of Natural History, Leiden. (By A.J. Whitten)

The kingfisher *Halcyon cyanoventris* is endemic to Java and Bali. (By Mary Plage/Bruce Coleman Ltd.)

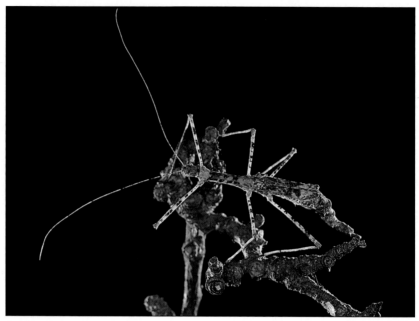

The Javan lichen stick insect *Orxines macklottii* is possibly endemic to Mt. Gede-Pangrango, West Java. (By A.J. Whitten)

The damselfly *Drepanosticta gazella*, photographed here in the forests of Mt. Papandayan, West Java, is endemic to western Java. (By J. van Tol)

The snail *Asperitas waandersiana,* endemic to East Java and Bali, is represented on Nusa Penida by this distinctive form. (By A.J. Whitten)

Bull racing is enormously important to many Madurese. (By Dieter & Mary Plage/Bruce Coleman Ltd.)

West of Jakarta lies the old town of Banten, centred on the now derelict Fort Speelwijk, which used to be the main port on Java until political and environmental problems caused its demise. (By Ingo Jezierski)

The culture of the Balinese is remarkably resistant against the forces of tourism. (By R.E. Soeriaatmadja)

The diversity of coral reefs is striking; here can be seen various species of hard and soft corals, crinoids, sponges, tunicates, and coralline algae. (by Jan Post)

Long-nose hawkfish, *Oxycirrhites typus*, are often found perched on black corals on drop-offs ready to pounce on passing prey. Black corals can be made into beautiful jewelry but the scale of the trade has resulted in their disappearance from many areas with consequent effects on the hawkfish. (by Jan Post)

The fish around the Tulamben wreck, eastern Bali, are tame and can be fed and photographed easily. (By T. Tonozuka)

The reefs around Nusa Penida are becoming increasingly popular dive sites. (By T. Tonozuka)

Purple anthias *Pseudanthias tuka* off Menjangan Island, Bali Barat National Park. (By T. Tonozuka)

Reef flats at low tide, such as here below Ulu Watu temple, Bali, provide gatherers with many forms of protein. (By A.J. Whitten)

Fishing using dynamite smashes coral and kills fishes indiscriminately. (By Mark Strickland)

An *Opheodesoma* sea cucumber in the *Enhalus* seagrass meadow of Sanur, Bali. Note that the large grains of sand are actually the tests of foraminifera. (By A.J. Whitten)

There are many mangrove replanting schemes in Java and Bali, and one of the earliest was near Cilacap, south Central Java. (By Alain Compost/Bruce Coleman Ltd.)

Rivers start as small forest streams of clean water; Mt. Tilu, West Java. (By A.J. Whitten)

Western Bali has extensive and little-visited forests. (By A.J. Whitten)

Moist deciduous forest in Meru Betiri National Park. (By Gerald Cubitt/Bruce Coleman Ltd.)

Dry deciduous forest dominated by flat-topped *Acacia leucophloea* trees in Baluran National Park. (By A.J. Whitten)

Casuarina trees and edelweiss *Anaphalis viscida* at 2600 m on the Yang Plateau, East Java. (By A.J. Whitten)

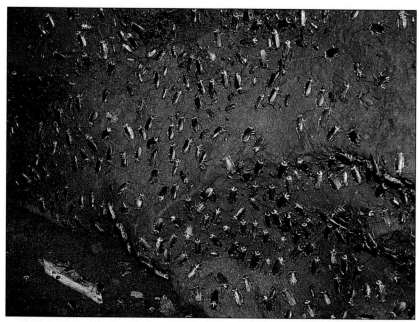

Hundreds of *Periplaneta* cockroaches on the walls of a cave below Mt. Selok, near Cilacap, where they feed on the faeces of the bats roosting above. (By A.J. Whitten)

Podocarpus-dominated forest at 1900 m below Mt. Argopura, East Java. (By A.J. Whitten)

Balanophora elongata is a root parasite found in the montane forests of western Java. (By Alain Compost/Bruce Coleman Ltd.)

A large chamber in Luweng Jaran Cave, Central Java. (By S. Stoddard)

The cultivation of irrigated rice on Bali has evolved into a highly productive, sophisticated and sustainable system. (By Sandro Prato/Bruce Coleman Ltd.)

Integrated pest management is advancing well in Java and Bali and this roadside sign near Banyuwangi, East Java, reminds farmers to protect spiders, an important group of predators on rice pests. (By A.J. Whitten)

Maize is a highly appropriate crop in the seasonal parts of Java and Bali. Much of the production goes into livestock feed. (By A.J. Whitten)

Maize growing in the shade of coconut trees, Ketapang, East Java. (By A.J. Whitten)

Forest gardens or *talun kebun* cover many of the hills in West Java. (By R.E. Soeriaatmadja)

Teak trees in the buffer zone west of Baluran National Park, East Java. (By A.J. Whitten)

Floating net culture of carp in Saguling reservoir, West Java. (By R.E. Soeriaatmadja)

A tiny teak seedling growing among young maize in an area of Perhutani tumpangsari near Cepu, East Java. (By A.J. Whitten)

A rubber tree being tapped in the early morning in an East Java plantation. (By Alain Compost/Bruce Coleman Ltd.)

Urban areas strike many contrasts between different income groups; Surabaya, East Java. (By A.J. Whitten)

The Java sparrow used to be very common but is now very scarce at least in part because they have been trapped to become cage birds like these. (By Hans Reinhard/Bruce Coleman Ltd.)

The Javan Gibbon is restricted to the remaining patches of lowland forest in West and Central Java, although many of these are too small for viable populations. (By Gerald Cubitt/Bruce Coleman Ltd.)

Eagles are now exceedingly rare on Java and Bali and most of them are dependent on forest, although this sea eagle is found primarily around the coasts. (By Alain Compost/Bruce Coleman Ltd.)

Lowland forest in or adjacent to Meru Betiri National Park being cleared and burned for smallholder agriculture. (By A.J. Whitten)

Tourists watching for endangered Bali starlings. The number of such visitors is severely limited to avoid undue disturbance. (By A.J. Whitten)

The endangered Javan rhino, symbol of Ujung Kulon National Park, is an elusive animal whose conservation has created heated debates. (By Dieter and Mary Plage/Bruce Coleman Ltd.)

Green turtles awaiting slaughter, Tanjung Benoa, Bali. The conservation of turtles would probably be served best by affording better protection to adults and nesting beaches. (By Ingo Jezierski)

The swamp forest in Rawa Danau reserve in western West Java is under many and varied threats. (By A.J. Whitten)

The dramatic sunrise view of Mt. Bromo, East Java, is all that is seen of the National Park by most of the 50,000 annual visitors, but the area has much more to offer. (By Gerald Cubitt/Bruce Coleman Ltd.)

The standing tree.

A 'bleeding' stump.

The timber tree *Dipterocarpus littoralis* is endemic to the prison island of Nusa Kambangan.

Large pieces of illegally-felled timber being ruined to prevent their sale.

Martoto, head of the island's prisons, standing over confiscated timber to be used as evidence against apprehended thieves. (By A.J. Whitten)

The lowland forests of Lebakharjo and Bantur urgently require to be gazetted as conservation areas. (By A.J. Whitten)

The inhabitants of remote parts of Nusa Penida probably have a better concept of the distinction between wants and needs than the rest of us. (By A.J. Whitten)

The consumption of Jakarta and its inhabitants is both inequitable and unsustainable, and dealing with this represents probably the greatest challenge in the future of Java and Bali. (By Ingo Jezierski)

Java

SUMATRA

Java Sea

Panjang

BAKAUHENI
Ferry (hourly)
Merak

Krakatau National Park
Cilegon

JAKARTA

Teluknaga
Bekasi
Kandanghaur
Indramayu

Serang
Tangerang
Cikampek
Jatibarang

Rangkasbitung
Cibinong
Botanical Gardens
D. Jatiluhur
Purwakarta
Jatiwangi
Cirebon
Pekalongan
Kenc

Labuan
Bogor
Plered
Cibodas Botanical Garden
Ciledug
Tegal
Pemalang
Batang
K

Badui Villages
WEST
Gn. Salak
2211m
Hot Springs & Volcano
Brebes
Slawi
D. Mejahayu
CENTRAL
Dien
Plat
JA

Gn. Halimun Reserve
Gn. Halimun
1929m
Cibadak
Cimahi
Bandung
D. Darma
Gn. Slamet
3432m
Wonosobo
Magel

Cianjur
S.Tanum
Bumiayu
Purbolinggo

Sukabumi
Hot Springs
Garut
Ciamis
Purwokerto
Sukaraja
Borob

Pelabuhan Ratu
D. Sipan-unjang
Gn.
Banjar
Wangon
Banyumas
Purworejo

Genteng National Park
Papandayan
2622m
Gn. Cikuray 2821m
Tasikmalaya
Kebumen

JAVA
Pangandaran National Park
Cilacap

Nusa Kambangan Reserve

Ujung Kulon National Park

Indian Ocean

Ujung Kulon
National Park
(Java)

EAST JAVA

Ketapang
Ferry (hourly)
Gilimanuk

Sunda Strait

Gn. Darahayu

Gn. Kadam
187m

Teluk Kasuaris

PULAU PANAITAN
Gn. Raksa
320m

Gn.Tajimalela
180m

Labuan

Panaitan Strait

Nyiur

Cikarang

PULAU PEUCANG

Gn. Telanea

Cigenter

Cipining

PULAU HANDEULEUM

Tamanjaya

Citadahan

Cikeusik

Cijungkulon

Gn. Guhabendeng
525m

Gn. Payung
480m

Karang Ranjang

Katejetan

Cisiih

Sumur

Cimanggu

Gn. Honje I

Cangkeuteuk

Gn. Honje II
582m

Citeluk

Gn. Cihadiyan
625m

Gn. Cibunua
Gn. Tilu
485m

Sompok

Tamanjaya Park Headquarters

Cikawung

Air Jeruk

Cegok

E

Indian
Ocean